LTE, WIMAX AND WLAN NETWORK DESIGN, OPTIMIZATION AND PERFORMANCE ANALYSIS

LTE, WIMAX AND WLAN NETWORK DESIGN, OPTIMIZATION AND PERFORMANCE ANALYSIS

Leonhard Korowajczuk
CelPlan Technologies, Inc., Reston, VA, USA

WILEY
A John Wiley & Sons, Ltd., Publication

This edition first published 2011

Registered office
John Wiley & Sons Ltd, The Atrium, Southern Gate, Chichester, West Sussex, PO19 8SQ, United Kingdom

For details of our global editorial offices, for customer services and for information about how to apply for permission to reuse the copyright material in this book please see our website at www.wiley.com.

Library of Congress Cataloging-in-Publication Data

Korowajczuk, Leonhard.
LTE, WIMAX, and WLAN network design, optimization, and performance analysis / Leonhard Korowajczuk.
p. cm.
Includes bibliographical references and index.
ISBN 978-0-470-74149-8 (cloth)
1. Wireless LANs. 2. IEEE 802.16 (Standard) 3. Long-Term Evolution (Telecommunications) I. Title.
TK5105.78.K67 2011
004.6 – dc22

2011007547

A catalogue record for this book is available from the British Library.

Print ISBN: 9780470741498
ePDF ISBN: 9781119970477
oBook ISBN: 9781119970460
ePub ISBN: 9781119971443
eMobi ISBN: 9781119971450

Typeset in 9/11pt Times by Laserwords Private Limited, Chennai, India
Printed in Malaysia by Ho Printing (M) Sdn Bhd

2 2012

I dedicate this book to my wife, Eliani, to my children, Cristine, Monica and Leonardo, my grandchildren, Julia, Paulo and Patrick and *in memoria* to my parents, Aleksander and Klara.

Contents

List of Figures

List of Tables

About the Author

Leonhard Korowajczuk has 40 plus years of experience in the telecommunication field working in R&D and Engineering areas. He graduated from UFRJ in 1969. His first assignments were in the Energy and FDM area at Standard Electrica S/A, followed by pioneer work on a PCM project at STC in England. He was part of the group that created the Telecom R&D Center (CPqD) in Brazil, where he did pioneer work on TDM switching. Next he joined Elebra S/A (later Alcatel do Brazil) where he was in charge of the Switching and Wireless Divisions. In 1992, he founded CelTec Tecnologia de Telecomunicações in Campinas, SP, Brazil, and in 1994 CelPlan Technologies in Reston, VA, USA, to provide design and optimization software for wireless operators. He was CTO of Comsat/Plexsys, where he was responsible for the development of advanced wireless equipment. Today he is CEO and CTO of CelPlan International, a company with subsidiaries in several countries, that provides design and optimization solutions for wireless operators. His team have done hundreds of designs of Cellular, PCS, WLAN, WiMAX and LTE networks worldwide. He is also the head of the Wi4Net division, which provides Citywide Video Surveillance Networks for Public Safety, using technologies like WLAN and WiMAX.

Preface

For nearly a hundred years telecommunications provided mainly voice services and very low speed data (telegraph and telex). With the advent of the Internet, several data services became mainstream in telecommunications; to the point that voice is becoming an accessory to IP-centric data networks. Today, high-speed data services are already part of our daily lives at work and at home (web surfing, e-mail, virtual private networks, VoIP, virtual meetings, chats...). The demand for high-speed data services will grow even more with the increasing number of people telecommuting.

Wireless circuit switched voice networks have experienced in the past two decades, an evolution towards mobility and today's users take for granted the universal availability of voice services. This demand is migrating to the data domain where 4G wireless networks have become essential.

Wireless networks became feasible with the advent of 1G networks (AMPS and ETACS) that provided analog voice services. With the increase in demand, more efficient technologies were required and 2G networks (TDMA, GSM, CDMA), designed for digital voice and higher spectral efficiency, were created.

The explosive demand of wireless services required even more spectrum efficient networks and the need for wireless data services started to emerge. 3G technologies (cdma2000, UMTS) developed to meet this demand were extensions of old voice switched networks and provided relatively low data speeds when compared to terrestrial networks. In these technologies, higher speeds were compromised in distance due to multipath effects.

4G technologies (WLAN, WiMAX, and LTE) are the first to break the high speed limitation for long distances by using OFDM technology. At the same time, these technologies have the advantage of being conceived as IP-based from the start.

Today's engineers have to be masters of multiple trades, as the different specialties converge. The design of a wireless network requires knowledge of business plans, networking, data applications, data protocols, data traffic, RF propagation, multiple wireless technologies, measurement techniques, optimization methodologies, among many other topics.

A question arises: what is the use of a book today if we can get all the information we need by browsing the World Wide Web?

The Internet provides, today, a wealth of information not equaled in the past by thousands of books. I used this resource constantly while writing this book, but it did not replace my collection of books for the following reasons:

- Internet information is presented in small topics and it is difficult to put everything together in a logical sequence. One of difficulties I found in writing this book was adopting the correct presentation sequence, so that one topic is based only on information previously provided and serves as a basis for subsequent topics.
- A significant part of the information available on the Internet is very superficial and may convey wrong interpretations.

- A significant amount of information is based on a single source and repeated in several sites.
- Information presented is generally not dated and may be obsolete.
- Browsers have mostly links to recent works and often important seminal works are not available.
- It takes time to browse and collect the necessary information on a specific topic.
- Internet information rarely goes deep enough in the majority of topics.

Even though I have tried to be thorough in the description of concepts and the theory behind wireless network design, it is impossible to cover all aspects and topics related to the subject. For readers who want more information on a specific subject, I suggest checking Wikipedia (www.wikipedia.com) as it is a good quick reference tool; for a more detailed view of certain topics, I recommend consulting the books listed in the Further Reading section.

One of the main questions that an author must ask himself when writing a book is how deeply to drill into a topic. A book has a limited number of pages, and it seems that they are never sufficient. My approach was to give a complete overview, including topics that may look beyond the scope of this work; however, I find them required in the day-to-day activities of the engineers who work with me. Topics related to 4G technologies were the ones I explored more deeply, as their comprehension is the basis of creating successful designs.

Design and optimization tasks are presented in detail and I found that the best way to illustrate them was to display the configuration screens of design and optimization tools. I am grateful to CelPlan Technologies for allowing me to use their tools to illustrate wireless design procedures.

As an engineer, I always endeavor to understand the physical meanings of mathematical equations and concepts, including the "whys" of the technology solution. I hope I was able to convey my understanding of these topics to the reader.

I have learned over the years that the best explanation is usually the simplest one. When I must resort to extensive mathematical equations or to a never-ending explanation, it is because I did not fully understand the subject or my explanation approach was incorrect.

I struggled in this book with whether an acronym should precede the name or vice versa. After swaying from one approach to the other I decided to let it flow naturally, so whatever comes first when I am writing, I maintain. I apologize if this may create some confusion to the reader, but I found it more natural to write in this manner.

This book is intended to work as a tutorial and as a reference guide. It can be used for the training of engineers, academic classes or as reference for consultants, vendors, and operators.

Acknowledgements

I would like to express my thanks to CelPlan Technologies, Inc. for allowing me to describe its planning methodology and to use CelPlanner Suite tool dialogs and screen shots to illustrate concepts and procedures.

I would like to express my gratitude to my daughter Cristine Korowajczuk, for revising my text and providing valuable comments and corrections throughout the book.

The methodology explained in this book was developed over the past twenty years by a group of CelPlan partners, mainly me, Aluisio Ribeiro, Leila Ribeiro, Paulo Leite and Wagner Mello.

I would like to express my gratitude to Paola Durant and Mary Rizzo, two excellent English teachers who spent time revising parts of my text.

Leonhard Korowajczuk

List of Abbreviations

Acronyms and abbreviations have become a must in technical literature to replace extensive names, and are equivalent to a nickname. This list covers many areas, as some acronyms can have different meanings. In the text we have repeated the full form of the abbreviations many times, as we always feel that the reader may have difficulties remembering their meaning.

3GPP	3rd Generation Partnership Project
3GPP2	3rd Generation Partnership Project 2
64QAM	64 Quadrature Amplitude Modulation
8QAM	8 Quadrature Amplitude Modulation
16QAM	16 Quadrature Amplitude Modulation
32QAM	32 Quadrature Amplitude Modulation
AAS	Advanced Antenna System
AAS	Adaptive Antenna Steering
AAS	Adaptive Antenna System
ACK	Acknowledgement
ACL	Access Control List
ACLR	Adjacent Channel Leakage Rejection
ADC	Analog to Digital Converter
ADS	Air Data Speed
ADT	Air Data Tonnage
AES	Advanced Encryption Standard
AES-CCMP	AES- CTR CBC MAC Protocol
AGC	Automatic Gain Control
AIFS	Arbitration Inter Frame Space
AIP	All IP
AMC	Adaptive Modulation and Coding
AMC	Adjacent Mapping of Sub-Carriers
AMPS	Advanced Mobile Phone Service
AM-RLC	Acknowledged Mode Radio link Control
AMS	Adaptive MIMO Switching
ANR	Automatic Neighbor Relation
AoI	Area of Interest

AP	Access Point
AP	Aggregation Point
ARIB	Association of Radio Industries and Business (Japan)
ARP	Address Resolution Protocol
ARPANET	Advanced Research Projects Agency
ARQ	Automatic Repeat reQuest
AS	Access Stratum
ASCA	Adjacent Subcarrier Allocation
ASCII	American Standard Code for Information Interchange
ASN	Access Service Network
ASN-GW	ASN- Gateway
ASP	Application Service Provider
ATIS	Alliance for Telecommunications and Industry Solutions (USA)
ATM	Asynchronous Transfer Mode
AuC	Authentication Center
AWGN	Additive White Gaussian Noise
B	Block
BA	Block ACK
BBS	BroadBand Services
BCCH	Broadcast Channel
BE	Best Effort
BER	Bit Error Rate
BFN	Beam Forming Network
BG	Block Group
BGP	Border Gateway Protocol
BLAST	Bell Labs Layered Space Time
BLER	Block Error Rate
BOSS	Back Office Support System
BPSK	Binary Phase-Shift Keying
BR	Bandwidth Request
BS	Base Station
BSC	Base Station Controller
BSS	Business Support System
BSS	Basic Service Set
BSSID	Basic Service Set ID
BTC	Block Turbo Code
BTS	Base Terminal Station
CA	Collision Avoidance
CA	Coordination Function
CAC	Channel Access and Control
CAMEL	Customized Application for Mobile network Enhanced Logic
CAPEX	Capital Expenditure
CAZAC	Constant Amplitude Zero Auto Correlation
CB	Coding Block
CBC	Cipher Block Chaining
CC	Convolutional Code
CC	Chase Combining
CCA	Clear Channel Assessment
CCCH	Common Control Channel

CCD	Cyclic Delay Diversity
CCE	Control Channel Element
CCK	Complementary Code Keying
CCM	CTR with CBC-MAC
CCSA	China Communications Standards Association
CD	Collision Detection
CDD	Cyclic Delay Diversity
cdf	cumulative distribution function
CDM	Code Division Multiplex
cdma2000	code division multiple access for beyond year 2000
CEMS	Configuration Element Management System
CEPT	Conférence des administrations Européenes des Postes et Télécommunications
CERN	Centre Européenne pour la Recherche Nucléaire
CFI	Control Format Indicator
CFP	Contention Free Period
CH	Chase Combining
CHAP	Challenge Handshake Authentication Protocol
CI	CRC Indicator
CID	Connection Identifier
CIDR	Classes Inter-Domain Routing
CINR	Carrier to Interference and Noise Ratio
CL	CLient
CN	Core Network
COMPASS	named used by the Chinese GNSS
CORE	Core Network
CP	Cyclic Prefix
CPE	Customer Premises Equipment
CPS	Common Part Sublayer
CPU	Central Processing Unit
CQI	Channel Quality Indicator
CRC	Cyclic Redundancy Code
C-RNTI	Cell Radio Network Temporary Identity
CS	Customer Station
CS	Carrier Sense
CS	Convergence Sublayer
CS	Circuit Switched
CSD	Cyclic Shift Delay
CSI	Channel State Information
CSMA	Carrier Sense Multiple Access
CSN	Connectivity Service Network
CTC	Convolutional Turbo Code
CTR	Counter Mode
CTS	Clear To Send
CW	Contention Window
CW	Continuous wave
DA	Destination Address
DAC	Digital to Analog Converter
DARPA	Defense Advanced Research Projects Agency
dBd	dB in relation to dipole antenna

dBi	dB in relation to isotropic antenna
DCCH	Dedicated Control Channel
DCD	Downlink Channel Descriptor
DCF	Distributed Coordination Function
DCI	Downlink Control Information
DCS	Dynamic Channel Selection
DDC	Digital Down Converter
DF	Decision Feedback
DFFT	Discrete Fast Fourier Transform
DFS	Dynamic Frequency Selection
DFT	Discrete Fourier Transform
DFT-S-OFDM	DFT- Spread- OFDM
DHCP	Dynamic Host Control Protocol
DiffServ	Differentiated Services
DIFS	Distributed InterFrame Space
DIUC	Downlink IUC
DL	Downlink
DL	Downlink
DL-MAP	Downlink Map
DL-SCH	Downlink Shared Channel
DMB	Digital mobile Broadcast
DNS	Domain Name System
DoA	Direction of Arrival
DRB	Data Radio Bearer
DRS	Downlink Reference Signal
DRS	Demodulation Reference Signal
DRX	Discontinuous Reception
DS	Downstream
DS	Distribution System
DSA	Distributed System Architecture
DSB	Dual Side Band
DSC	Diversity Selection Combining
DSCA	Distributed Subcarrier Allocation
DSL	Digital Subscriber Line
DSP	Digital Signal Processor
DSSS	Direct Sequence Spread Spectrum
DTCH	Dedicated Traffic Channel
DTIM	Delivery Traffic Indication Message
DTP	Data Transfer Protocol
DTR	Data Tonnage Rate
DTS	Data Transfer Speed
DVB-H	Digital Video Broadcasting- Handheld
DVB-T	Digital Video Broadcasting- Terrestrial
DVRP	Distance Vector Routing Protocol
DwPTS	Downlink Pilot Time Slot
E	Erlang
EAP	Extensible Authentication Protocol
EAP-TTLS	EAP- Tunneled Transport Layer Security
EARFCN	Absolute Radio Frequency Channel Number

EC	Encryption Control
EDCA	Enhanced Distributed Channel Access
EDGE	Enhanced Data rates for GSM Evolution
EDRR	Enhanced Deficit Round Robin
E-field	Electrical field
EGC	Equal Gain Combining
EGP	External Gateway Protocol
EIFS	Extended Inter Frame Space
EIR	Equipment Identity Register
EKS	Encryption Key Sequence
EMC	Electro Magnetic Compatibility
EML	Element Management Layer
eNB	Evolved Node Base Station
eNode B	Evolved Node Base Station
eNodeB	Evolved Node Base Station
EPC	Evolved Packet Core
EPS	Evolved Packet System
EQM	Equal Modulation Scheme
ertPS	enhanced real time Polling Service
ESF	Extended Sub header Field
ESS	Extended Service Set
ETSI	European Telecommunications Standard Institute
EUI	Extended Unique Identifier
E-UTRA	Evolved- Universal Terrestrial Radio Access
E-UTRAN	Evolved- Universal Terrestrial Radio Access Network
EVDO	Evolution Data Optimized
EVM	Error Vector Magnitude
FA	Foreign Agent
FCH	Frame Control Header
FDD	Frequency Division Duplex
FDM	Frequency Division Multiplex
FEC	Forward Error Correction
FER	Frame Error Rate
FFSE	Fairly Shared Spectrum Efficiency
FFT	Fast Fourier Transform
FIFO	First In First Out
FIN	Finish
FIR	Full Incremental Redundancy
FSK	Frequency Shift Keying
FSTD	Frequency Switched Transmit Diversity
FT	Fourier Transform
FTP	File Transfer Protocol
FUSC	Full Use of Sub-Carriers
GALILEO	named used by the European GNSS
GAN	Generic Access Network
GBR	Guaranteed Bit Rate
GCB	Geographic Census Bureau
GERAN	GSM/Edge Radio Access Network
GGSN	GPRS Gateway Support Node

GIS	Geographic Information System
GLONASS	GLObal NAvigation Sputnik System
GMSC	Gateway MSC
GMSK	Gaussian Minimum Shift Keying
GNSS	Global Navigation Satellite System
GP	Guard Period
GPRS	General Packet Radio Service
GPS	Global Positioning System
GRE	Generic Routing Encapsulation
GSM	Global System for Mobile Communications
GSM	Groupe Spécial Mobile
GTP	GPRS Tunneling Protocol
GUTI	Global Unique Terminal Identity
HA	Home Agent
HARQ	Hybrid ARQ
HARQ	Hybrid Automatic Repeat reQuest
HCCA	HCF Controlled Channel Access
HCF	Hybrid Coordination Function
HCS	Header Check Sum
HeNB	Home eNB
H-FDD	Half-Frequency Division Duplex
H-field	Magnetic field
HLR	Home Location Register
HPA	High Power Amplifier
HSCSD	High-Speed Circuit Switched Data
HSDPA	High Speed Downlink Packet Access
HSPA	High Speed Packet Data
HSPA+	High Speed Packet Data Plus
HSS	Home Subscriber Server
HSUPA	High Speed Uplink Packet Data
HT	High Throughput
HT GF	HT Green Field
HTM	HT Mixed
HTML	Hyper Text Markup Language
HTTP	Hyper Text Transfer Protocol
I	In-phase
IAB	Internet Architecture Board
IANA	Internet Assigned Number Authority
IBSS	Independent Service Set
ICCB	Internet Configuration Control Board
ICMP	Internet Control Message Protocol
ICS	Implementation Conformance Statement
IdCell	Identification of the Cell
iDFFT	inverse of Discrete Fast Fourier Transform
iDFT	inverse Discrete Fourier Transform
IEEE	Institute of Electrical and Electronics Engineers
IETF	Internet Engineering Task Force
IFDMA	Interleaved Frequency Division Multiple Access
IFFT	Inverse Fast Fourier Transform

IFS	Inter Frame Space
IGMP	Internet Group Message Protocol
IGP	Internal Gateway Protocol
IGRP	Interior Gateway Routing Protocol
IM	Interference Matrix
IMAP	Internet Message Access Protocol
IMEI	International Mobile Equipment Identity
IMS	IP Multimedia Subsystem
IMSI	International Mobile Subscriber Identity
IMT2000	International Mobile Telecommunications for beyond year 2000
InARP	Inverse Address Resolution Protocol
InfraBSS	Infrastructure Basic Service Set
IntServ	Integrated Services
IP	Internet Protocol
IPDS	IP Data Speed
IPDT	IP Data Tonnage
IPv4	IP version 4
IPv6	IP version 6
IR	Incremental Redundancy
IRC	Internet Ready Chat
IRNSS	Indian Regional Navigation System
IS-2000 1XRTT	Information System-2000 Single Carrier Radio Transmission Technology
ISI	Inter Symbol Interference
IS-IS	Intermediate System to Intermediate System
ISM	Industrial, Scientific and Medical Equipment
ISO	International Standards Organization
ISP	Internet Service Provider
ITS	Intelligent Transportation System
ITU	International Telecommunication Union
IUC	Interval Usage Code
IV	Initialization Vector
K2D	Korowajczuk 3D propagation model
K3D	Korowajczuk 2D propagation model
KPI	Key Performance Indicator
KPI	Key Parameter Indicator
L2TP	Layer 2 Tunneling Protocol
LAC	Layer 2 Tunneling Protocol Access Concentrator
LB	Long Block
LD	Linear Detector
LDAP	Lightweight Directory Access Protocol
LDP	Label Distribution Protocol
LDP	Linear Diversity Pre-coding
LDPC	Low-Density Parity Check
LEN	Length field
LLC	Logical Link Control
LNA	Low Noise Amplifier
LNS	Layer2 Tunneling Protocol Network Server
LPCP	Linear Pre-Coding and post -Coding
LPP	LTE Positioning Protocol

LPPD	Linear Programming Detector
LRG	Likelihood Receiver Generator
LSP	Label Switch path
LSRP	Link-State Routing Protocol
LTE	Long Term Evolution
LTE-A	LTE Advanced
LTS	Long Training Sequence
MA	Multiple Access
MAC	Message Authentication Code
MAN	Metropolitan Area Network
MBGP	Multi Protocol BGP
MBMS	Multimedia Broadcast Multicast Service
MCC	Mobile Country Code
MCCH	Multicast Control Channel
MCH	Multicast Channel
MCS	Modulation and Coding Scheme
MDI	Medium Dependent Interface
MDIX	Medium Dependent Interface crossed
MFLOPS	Million of Floating-point operations Per Second
MIB	Master Information block
MIME	Multipurpose Internet Mail Extension
MIMO	Multiple Input to Multiple Output
MIPS	Million of Integer operations Per Second
MISO	Multiple In to Single Out
MLD	Maximal Likelihood Detector
MLSD	Maximum Likelihood Sequence Detection
MM	Market Modeling
MME	Mobility Management Entity
MMF	Max-Min Fairness
MNC	Mobile Network Code
MPDU	MAC Protocol Data Unit
MPLS	Multi Protocol Label Switching
MRC	Maximal Ratio Combining
MS	Mobile Station
MSC	Mobile Switching Center
MSDP	Multicast Source Discovery Protocol
MSDU	MAC Service Data Unit
MSSE	Minimum Mean Square Error
MSTR	Maximum Sustained Traffic Rate
MTCH	Multicast Traffic Channel
MUD	Multiple User detection
N/A	Not Available
NACK	Not Acknowledged
NAS	Non Access Stratum
NAT	Network Address Translators
NAV	Network Allocation Vector
NB	Node with Base station
NCP	Network Control Program
NCSA	National Center for Supercomputing

NDP	Neighbor Discovery Protocol
NF	Noise Figure
NHT	Non HT
NIC	Network Interface Card
NM	Neighborhood Matrix
NML	Network Management Layer
NMS	Network Management System
NNTP	Network News Transfer Protocol
Node B	Node Base Station
Non-GBR	Not Guaranteed Bit Rate
NRT	Non-Real Time
nrtPS	non-real time Polling Service
NRZ	Non Return to Zero
NTP	Network Time Protocol
OFDM	Orthogonal Frequency Division Multiplex
OFDMA	Orthogonal Frequency Division Multiple Access
OFDSA	Orthogonal Frequency Division Single Access
OFUSC	Optional FUSC
OoB	Out of Band
OPEX	Operational Expenditure
OPUSC	Optional PUSC
OSA	Open Service Access
OSI	Open System Interconnection
OSPF	Open Shortest Path First
OSR	Over-Subscription Ratio
OSS	Operation Support System
OTUSC	Optional Tiled Usage of Subchannels
PA	Percentage of Area
PAP	Peak to Average Power Ratio
PAPR	Peak to Average Power Ratio
PAR	Peak to Average Ratio
PAT	Port Address Translation
PBCH	Physical Broadcast Channel
PC	Point Coordinator
PC	Personal Computer
PCCH	Paging Control Channel
PCF	Point Coordination Function
PCFICH	Physical Control Format Indicator Channel
PCI	Physical Cell ID
PCRF	Policy Control and Charging Rules Function
PDAS	Pilot and Data Allocation Scheme
PDCCH	Physical Downlink Control Channel
PDCP	Packet Data Convergence Protocol
pdf	probability density function
PDN	Packet Data network
PDN-GW	PDN Gateway
PDSCH	Physical Downlink Shared Channel
PDU	Protocol Data Unit
PER	Packet Error Rate

PermBase	Permutation Base
PGW	Packet data network GateWay
PFICH	Physical Format Indicator Channel
P-GW	PDN Gateway
PHICH	Physical Hybrid ARQ Channel
PHY	Physical Layer
PIFS	Point Inter Frame Space
PIM-DM	Protocol independent Multicast Dense Mode
PIM-SM	Protocol independent Multicast Sparse Mode
PING	Echo Request
PIR	Partial Incremental Redundancy
PLCP	Physical Layer Convergence Procedure
PLF	Polarization Loss Factor
PLL	Phase Locked Loop
PLMN	Public land Mobile Network
PMCH	Physical Multicast Channel
pmf	probability mass function
PMK	Pair-wise Master key
PMP	Portable Multimedia Player
POP	Post office Protocol
PoP	Point of Presence
PP	Percentage of Population
PPDU	PLCP Protocol Data Unit
PPP	Point to Point Protocol
PRACH	Physical Random Access Channel
PRBS	Pseudo Random Binary Sequence
PRS	Primary Reference Source
PS	Power Save
PS	Packing Sub header
PS	Packet Switching
PSC	Prediction Service Class
PSDU	PLCP Service Data Unit
PSS	Primary Synchronization Signal
PSTN	Public Switched Telecommunications Network
PTP	Precision Timing Protocol
PUCCH	Physical Uplink Control Channel
PUSC	Partial Usage of Subchannels
PUSC-DL	Partial Usage of Subchannels Downlink
PUSCH	Physical Uplink Shared Channel
PUSC-UL	Partial Usage of Subchannels Uplink
Q	In-quadrature
QoS	Quality of Service
QPSK	Quadrature Phase-Shift Keying
QZSS	Quasi Zenith Satellite System
RA	Receiving STA Address
RACH	Random Access Channel
RADIUS	Remote Authentication Dial in User Service
RAP	Random Access Preamble
RAT	Radio Access Technology

RB	Resource Block
RBT	Random Back-off Time
RC	Raised Cosine
RCPC	Rate Compatible Convolutional Code
RCTP	Rate Compatible Punctured Turbo Code
RED	Random Early Detection
REG	Resource Element Group
Rel	Release
RF	Radio Frequency
RFC	Request for Comments
RFH	RF Head
RFP	Request for Proposal
RFQ	Request for Quote
RIP	Routing Information Protocol
RIR	Regional Internet registries
RL	Return Loss
RL	Reflection Loss
RLC	Radio Link protocol
RMS	Root Mean Square value
RNC	Radio Network Controller
RoHC	Robust Header Compression
ROI	Return of Investment
RPC	Remote Procedure Call
RPF	Reflected Power Factor
RRC	Radio Resource Control
RRM	Radio Resource Management
RS	Reference Signal
RS-CC	Reed-Solomon Convolutional Code
RSCP	Received Signal Code Power
RSL	Received Signal Level
RSRP	Reference Signal Received Power
RSSI	Receive Signal Strength Information
RSV	ReSerVed bit
RSVP	Resource Reservation Protocol
RT	Real Time
RTG	Receive Transition Gap
RTP	Real Time Protocol
rtPS	real time Polling Service
RTS	Request To Send
RTSP	Real -time Streaming Protocol
RX	Receive
RXLEV	Receive Level
RZ	Return to Zero
S1	S1 interface
S1AP	S1 Application Protocol
SA	Source Address
SAE	System Architecture Evolution
SAG	Service Activation Gateway
SAP	Service Access Point

SB	SuBscriber
SB	Short Block
SBC	Single Board Computer
SC	Service Class
SC	Selection Combining
SCH	Synchronization Channel
SCM	Spatial Channel Model
SCME	Spatial Channel Model Extended
SC-OFDM	Single Carrier -OFDM
SC-OFDMA	Single Carrier -OFDMA
SCTP	Stream Control Transmission protocol
SD	Sphere Detector
SD	Sphere Decoding
SDP	Session Description protocol
SDU	Service Data Unit
SEMS	Service Element Management System
SFBC	Space Frequency Block Code
SFD	Start Frame Delimiter
SFID	Service Flow Identifier
SFN	Single Frequency Network
SG	Smart Grid
SGSN	Serving GPRS Support Node
S-GW	Serving Gateway
SIC	Successive Interference Cancellation
SIFS	Short Inter Frame Space
SIMO	Single In to Multiple Out
SINC	Sine Cardinal or Sinus Cardinalis
SIP	Session Initiation Protocol
SISO	Single In to Single Out
SKA	Shared Key Authentication
SLA	Service Level Agreement
SLB	Server Load Balancing
SM	Smart Meter
SM	Spatial Multiplexing
SME	Small and Medium Enterprise
SML	Service Management Layer
SMTP	Simple Mail Transfer Protocol
SN	Sequence Number
SNA	System Network Architecture
SNIR	Signal to Noise and Interference Ratio
SNMP	Simple Network Management Protocol
SNR	Signal to Noise Ratio
SOAP	Simple Object Access Protocol
SOFDMA	Scalable OFDMA
SON	Self-Organizing Network
SP	Service Plan
SR	Scheduling Request
SRB	Signalling Radio Bearer
SRRC	Square Root Raised Cosine

SRS	Sounding Reference Signal
SS	Subscriber Station
SS	Service Set
SS	Security Sublayer
SS7	Signalling System 7
SSB	Single Side Band
SSH	Secure Shell
SSID	Service Set Identifier
SSM	Source Specific Multicast
SSS	Secondary Synchronization Signal
ST	Slot Time
STA	Service Target Area
STA	Station
STBC	Space Time Block Code
STC	Space Time Coding
STS	Short Training Sequence
SUI	Stanford University Interim
SVD	Singular Value Decomposition
SW	Software
SYN	Synchronized sequence Number
SYNC	Synchronization
T	Tract
TA	Transmitting STA Address
TA	Tracking Area
TAC	Tracking Area Code
TACS	Total Access Communication System
TAI	Tracking Area Identity
TB	Transport Block
TCL	Transit Control List
TCP	Transmission Control Protocol
TDD	Time Division Duplex
TDG	Traffic Distribution Grid
TDL	Tapped Delay Line
TDM	Time Division Multiplex
TEK	Traffic Encryption Key
Telnet	Telecommunications Network
TEMS	Traffic Element Management System
TG	Target Area
TIA	Telecommunications Industry Association
TIM	Traffic Indication Map
TKIP	Temporal Key Integrity Protocol
TLS	Transport Layer Security
TMN	Telecommunications Management Network
TPC	Transmit Power Control
TRD	Transmit and Receive Diversity
TS	Terminal Station
TSD	Transmit Selection Diversity
TSF	Timing Synchronization Function
TSL	Transmitted Signal level

TTA	Telecommunications Technology Association (Korea)
TTC	Telecommunications Technology Committee (Japan)
TTG	Transmit Transition Gap
TTI	Transmit Time Interval
TUSC	Tiled Usage of Subchannels
Tx	Transmit
U	User
UCD	Uplink Channel Descriptor
UCM	Uplink Collaborative MIMO
UDP	User Datagram Protocol
UE	User Equipment
UEQM	Unequal Modulation Scheme
UGS	Unsolicited Grant of Service
UIUC	Uplink IUC
UL	Uplink
UL_PermBase	Uplink Permutation Base
UL-MAP	Uplink Map
UL-MIMO	Uplink MIMO
UL-SCH	Uplink Shared Channel
UM-RLC	Unacknowledged Mode Radio Link Control
UMTS	Universal Mobile Telecommunication System
U-NII	Unlicensed National Information Infrastructure
UP	Upstream
UP	Uplink
UpPTS	Uplink Pilot Time Slot
URS	Uplink Reference Signal
US	UpStream
USA	United States of America
USB	Universal Serial Bus
USC	User Service Class
UTRA	UMTS Terrestrial Radio Access
UTRAN	UMTS Terrestrial Radio Access Network
VANC	VoLGA Access Network Controller
VoIP	Voice over IP
VoLGA	Voice over LTE via Generic Access
VPN	Virtual Private Network
VRRP	Virtual Router Redundancy Protocol
VSWR	Voltage Standing Wave Ratio
WAVE	Wireless Access for Vehicular Environment
WCDMA	Wideband Code Division Multiple Access
WDPT	Wireless Design and Planning Tool
WDS	Wireless Distribution System
WEP	Wired Equivalent Privacy
WFQ	Weighted Fair Queuing
WiMAX	Worldwide Interoperability for Microwave Access
WINNER	Wireless world Initiative New Radio
WLAN	Wireless Local Area Network
WM	Wireless Medium
WO	Wireless Overhead

WPA	Wi-Fi Protected Access
WRED	Weighted RED
WWW	World Wide Web
X2	X2 interface
XML	eXtensible Markup Language
ZCC	Zero Cross Correlation
ZF	Zero Forcing

Introduction

Wireless communications are becoming a major factor in our daily lives. The unbelievable processing power achieved by personal computers, the rise of the World Wide Web, the miraculous search engines and a myriad of applications, is changing the way we conduct our lives and increasing the demand for seamless wideband communications.

A series of technological innovations happened in the past twenty years that have significantly impacted our lives, but they occurred so smoothly that we adapted to them very easily:

- Personal computers came first and became the cornerstone that allowed the other innovations to follow. Word editors, Spreadsheets and Presentation software increased efficiency tenfold.
- Today, we count on the availability of wireless networks wherever we go. The offering became ubiquitous and wireless is displacing landlines, soon to be the dominant method of communications. Public pay phones are being virtually removed.
- Internet connectivity is a must and e-mails are overwhelmingly the main form of written communication. Soon hand-written letters will be a thing of the past also.
- GPS location is not yet fully explored, but paper maps are disappearing and many new and exciting applications are being developed.

All these fields are still evolving at a fast pace, one prompting the development of the other. Particularly in the wireless field the demand for ubiquitous broadband wireless communications is increasing and it will replace, over time, the existing wireless infrastructure.

New technologies had to be conceived to provide the throughput required by this new demand. Radio frequency spectrums had to be re-assigned for these technologies and new networks had to be designed. These networks, different from the existing ones, are data-centric, more precisely IP (Internet Protocol) centric, as IP became the *de facto* standard for wireless communications, be it voice or data.

The design of these networks requires a new understanding of the basic assumptions that define user demand and, physical network constraints and an in-depth understanding of the technologies available. The design task became ten times more complex than the one used in the design of existing voice networks.

The new network designer has to revise many of the old concepts and extend them to cover much broader bases. It is imperative to understand the "why" and "how" of each physical effect, each proposed solution and each process. Only then, a proper design of a wireless broadband network can be achieved.

LTE, WiMAX and WLAN Network Design, Optimization and Performance Analysis, First Edition.
Leonhard Korowajczuk.

Chapters 1 to 11 and Chapter 22 revise the basic concepts involved in the design of new networks. Chapters 12 to 15 describe the main wireless broadband technologies: WLAN, WiMAX and LTE (in the order they appeared). Chapters 16 to 21 cover the design process step by step, which is applicable to all technologies.

Traditional technologies, such as CDMA and GSM, evolved in the late 1990s to provide data support, respectively through EVDO and HSPA. Although both solutions were IP-centric, their shortcomings soon became evident: not enough throughput and limited range due to multipath effects.

In parallel, a new technology based on orthogonal multiplexing (OFDM) made possible a whole new wireless broadband market: WLAN (trademarked as Wi-Fi). Specified by the IEEE (Institute of Electrical and Electronic Engineers), it is present in the majority of homes and businesses as the indoor extension of the wired Internet. This technology was originally conceived to provide wireless broadband at short distances (up to 20 meters) and for few users, although it is been used in some cases at distances of a few kilometers. Wi-Fi uses a non hierarchical per packet contention-based access, which limits its throughput significantly. One of the main reasons for WLAN's success was its low cost and ease of deployment.

Based on the WLAN (Wi-Fi) success, the IEEE started developing a new standard for outdoor coverage, which became known as WiMAX. This standard became commercial and has addressed the main shortcomings of WLAN, by establishing a hierarchical structure.

Later, the ITU (International Telecommunication Union) created its own OFDM version, known as UMTS (Universal Mobile Telecommunication System) version 8 or LTE (Long Term Evolution), which addressed some of the perceived deficiencies in the WiMAX specification.

Political and patent issues limited the development of both technologies. Everyone would benefit if the two entities joined forces and created a single standard. This has not yet happened, but that hope is still in sight. Patent laws that protected the technological development in the past are hindering it today. A revision of patent laws is required, so they can be adapted to the fast pace of today's technological evolution.

In this book, we present the basic concepts that are used in WLAN, WiMAX and LTE, as they are very similar and, then describe how those concepts were implemented in each technology. WLAN was included in the book because it has an important role as the last distribution link of the broadband wireless network mainly in indoor environments.

There is no consensus in the literature on how to classify the different technologies. We adopted the following classification in this book:

- CDMA (IS-95) and GSM: 2G
- cdma2000 and GSM/GPRS/EDGE: 2.5 G
- EVDO and HSPA: 3G
- WLAN, WiMAX and LTE: 4G

Because a WBN (Wireless Broadband Network) has several orders of magnitude in more parameters to be defined than a traditional network, a good designer has to be familiar with all the basic telecommunications concepts and their physical and practical implementations. Only after understanding these concepts it is possible for an engineer to approach the design of a WBN.

When first designing a WBN, major effort and time are required to model the market, the services, the environment and the infrastructure, before even starting the design effort. The lack of results in this phase may be frustrating and expensive for some, but it is an essential part of the design process and taking short-cuts usually results in a poor design.

The design of these networks is a multi-disciplinary issue and requires a deep understanding of all aspects involved. In this book I bring this issues together and show how they inter-relate, so a proper design of broadband wireless networks can be achieved.

Traditional wireless designers, sometimes frustrated with the poor results of prediction tools, abandoned the design in lieu of continuous measurements and network adjustments. This might have worked for voice networks at huge expense, but the market growth was so vast that speed of deployment replaced cost and quality. Now, everyone has access to wireless and the revenues have stabilized, but the traffic continues to increase with the adoption of new data centric applications. It is expected that network usage will increase by 100 times, while revenues will approximately double.

Although new spectrum bands are being made available, the new demand requires much greater spectrum efficiency, and only a proper design can accommodate this requirement. Yet, how can we rely on prediction tools if they failed in predicting much simpler networks? We must understand why they did not produce the expected results, and focus on fixing these issues.

Prediction tools were victims of their own initial success, when they predicted well the first cellular deployments, using very simple models. Networks density's increase, however, required a much bigger modeling effort, better databases and more advanced prediction and simulation algorithms. The leading tools in the market did not provide these advances; neither did the busy designers spend the time required to properly model the networks. Brute force was the preferred approach.

Imagine if someone decided to construct a building without a blueprint or floorplans, figuring out how to add more space at each step of the way, instead. That is how most of our existing networks are being designed today. Just as a building is constructed, a wireless design can only be properly done if time is spent modeling its constraints and requirements. Only then a design can be done, but before it is deployed, it is essential to predict the outcome. It is our intention in this book to address all the steps, from conception to implementation, of an economical and efficient broadband wireless network. I will relate here some of our (mine and of my team at CelPlan) experiences in designing these networks and describe a methodology that avoids the most common pitfalls.

Terrain and traffic databases have to be revised and in many cases re-done to provide the information required by a design.

Once all bases are covered, the design process can start, which, by itself, also requires many new skills. The use of specialized tools is essential due to the complexity of the task (versions used for voice networks fail miserably in the design of a WBN). We would like to thank CelPlan Technologies, Inc. for allowing us to use their CelPlanner Suite set of tools to exemplify design procedures and illustrate a planning tool configuration and its outputs.

A properly done design is an essential element of a WBN operation. It can provide significant savings in CAPEX (Capital Expenditure) and OPEX (Operational Expenditure). The investment in the initial design can be 12–15% of the initial investment, but it can bring savings of more than 25% on CAPEX and OPEX and can be the difference between a failed or successful network. Furthermore, a well-designed network can prove to be a big advantage in relation to competitors that did not take the same care in their design.

1

The Business Plan

1.1 Introduction

Wireless broadband networks are very different from the traditional voice networks, hence should not be deployed as an extension of those. A greenfield operator should start the conception of a new network by building a business case. An existing network benefits also from a proper business case, even if it is done during its operational life.

A properly designed business plan requires a small investment upfront, but substantiates the investment and can be used to leverage capital. Thus investors are not surprised by unexpected cash flow requirements or by unforeseen technical or operational issues. Figure 1.1 illustrates the main components of a business plan.

A business plan has three main components, described in detail in the next sections:

- the market plan;
- the engineering plan;
- the financial plan.

1.2 Market Plan

Understanding the market is essential to define the product offering and its acceptance by the market. This should be done through market research, which could be exploratory or confirmatory.

- In the exploratory case, options are left wide open and the results from the research will define the outcome.
- In the confirmatory case, a set of assumptions is made and are confirmed or not by the research.

A market research is divided into three areas:

- market information: where information is collected;
- market segmentation: where demographic, psychographic, ethnographic and lifestyle information is gathered;
- market trends: where market evolution over time is predicted.

LTE, WiMAX and WLAN Network Design, Optimization and Performance Analysis, First Edition.
Leonhard Korowajczuk.

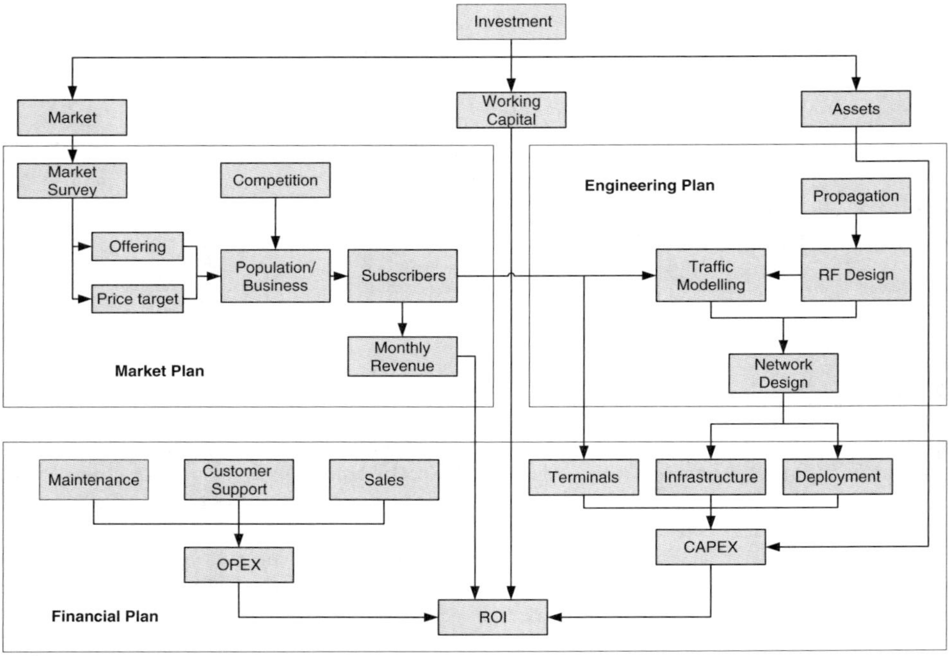

Figure 1.1 Business plan.

Market research can be done in four phases:

- *Market scan*: collection and analysis of available data that can contribute to the subject. Optionally customer visits can be done at the location where they use the service (businesses or residences), to ask broad questions about their satisfaction with existing services and their willingness to accept alternative offerings.
- *Options generation*: unconstrained options should be formulated to define all possible offerings.
- *Option selection*: each option should be evaluated based on the previously collected data and the best ones selected. The proper technique for this selection is choice modeling, which categorizes the data for each choice.
- *Selected options evaluation*: a customer survey should be done, with questions specific to each option.

Market research should be done by a specialized professional or company, as many of the network assumptions are based on it. It should be done periodically for existing deployments as well, so the service can be adapted to customer expectations and expansions can be properly planned.

The outcome of market research is the market plan, which should aid network designers by specifying the following items:

- *Service target area (STA)*: area in which service should be provided. It can constitute a single continuous area or several separate areas. These areas should be then divided in sub-areas classified by characteristics such as type of service expected and demand.

- *Product*: product to be offered, its features and restrictions. This includes service plans and its SLA (service level agreement).
- *Service coverage*: coverage area.
- *Client demographics* for the STA.
- *Client evolution* over the years.

1.3 The Engineering Plan

The engineering plan defines the design that fulfills market plan requirements. A complete design should be done, even if the equipment vendor is not yet defined.

Many vendors want to do an initial estimate of the number of cells required for a deployment, for budgetary reasons. The most common question asked to the network designer is: What cell size should be considered for the budgetary quote?

There are many factors that affect cell size:

- RF signal propagation, which depends on the environment and is mistakenly used as the sole criterion.
- Location where service will be provided (rooftop, outdoor, indoor).
- Spectrum availability and, consequently, expected interference.
- Equipment to be used.
- Amount of traffic to be carried in each location and its distribution.

These items interact with each other and cannot be treated separately. As an example, if the traffic to be carried is high, we need to resort to higher modulation schemes that require stronger signals and are more prone to interference. We generally give a range that can be applied. A common mistake is to consider a uniform traffic distribution, which leads to significant under-estimation of the infrastructure required.

Table 1.1 gives an idea of the variability of number of sites required in different scenarios. We strongly suggest that an initial design be done, so more precise numbers are used.

Ideally, a drive test should be conducted to collect measurements and calibrate RF propagation models for the area. Default propagation parameters can be used, but this will cause some imprecision.

The design step requires the designer to become familiar with the operator's intentions and with all facilities and restrictions of the area and of the license. A questionnaire should be sent to the operator, followed by an interview to gather the required information. This information guides the design effort. The following is a list of the main questions that should be answered:

- What is the spectrum available, its regulations and restrictions?
- What geographical data bases are available and is their quality good enough?
- What are the deployment plans?
- Are there any preferred vendors?
- What are the arrangements for wireline, Internet and backhaul connections?
- What are the site deployment restrictions?

The traffic-carrying capacity of the initial design must first be verified by using a noise rise figure to account for interference, as at this stage the network optimization has not yet been carried out. A traffic simulation can pinpoint traffic flow issues which should be corrected by redesign.

The cells' footprint should be enhanced and network resources (neighbors, frequencies, codes and parameters) should be optimized.

Table 1.1 Number of sites for an initial design

	Cell radius (km)		Effective area (km^2)		Cells/100 km^2	
Scenario	min	max	min	max	max	min
Rooftop	1	5	2.20	54.98	46	2
Outdoor ground	0.5	1	0.55	2.20	182	46
In-vehicle	0.3	0.7	0.20	1.08	506	93
Indoor window	0.2	0.4	0.09	0.35	1137	285
Indoor	0.1	0.25	0.02	0.14	4548	728

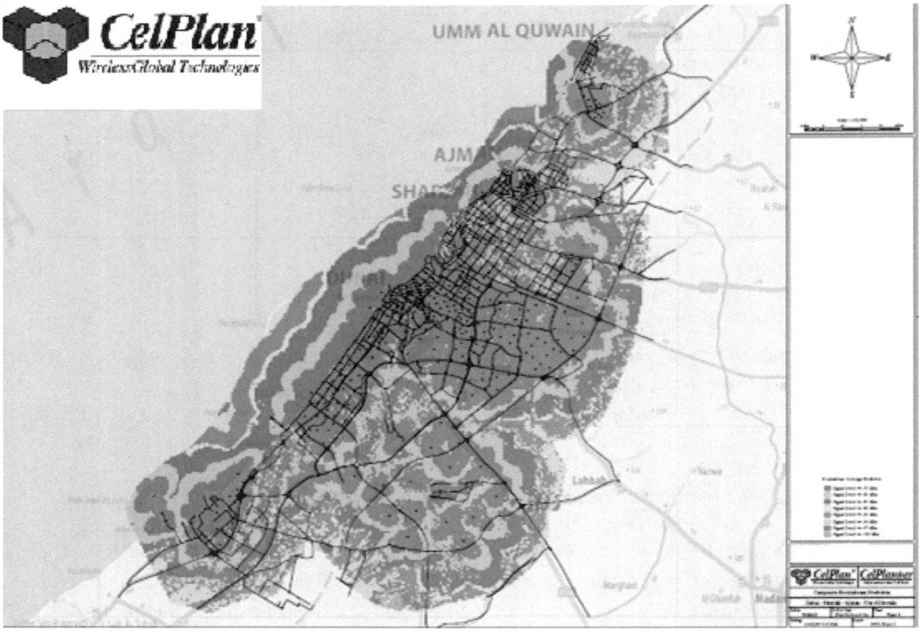

Figure 1.2 Planning tool prediction.

Finally, a performance analysis should be done and KPIs (Key Performance Indicators) should be compared with SLA (Subscriber Level Agreement) requirements.

The engineering plan must be updated during the life of the equipment, as it will play an important role in SON (self-organizing network) features to be introduced in most networks in the near future.

The design for the engineering plan should be done using a professional planning tool and experienced engineers. Broadband wireless designs require expertise and cannot be done in the same way as narrowband designs. A screenshot of such a planning tool is shown in Figure 1.2.

1.4 The Financial Plan

The financial plan analyzes the venture's financial feasibility and requirements. There are many specialized software packages that generate a financial plan according to the technology. These software packages are very good for initial ballpark estimates and can be updated as the project matures. Since they rely on many estimates, such as spectrum efficiency and penetration rates, which are very subjective, their inputs must be based on solid market and engineering plans, otherwise they can lead to any type of conclusion. It is strongly recommended that these software packages are used after or in parallel with the market and engineering plans. Screenshots from a financial planning tool are shown in Figure 1.3.

1.4.1 Capital Expenditure (CAPEX)

CAPEX summarizes capital investments per year, based on the market plan and engineering plan. Non-operational capital investments, such as office furniture, cars and vehicles, should also be considered. The main items that constitute the CAPEX are:

- Spectrum purchase (if any)
- Site construction and development
- Site infrastructure (power, batteries, air conditioning)
- Base station equipment
- Core equipment
- Backhaul equipment

1.4.2 Operational Expenditure (OPEX)

OPEX summarizes operational expenses, including leases, rents, operation and maintenance personnel.

- Site rental costs
- Site and backhaul maintenance costs
- Backhaul fees (fiber lease)
- Internet access costs
- Wireline interconnection costs
- VoIP termination costs
- CPE installation costs
- CPE subsidies
- Billing costs
- Customer care costs
- Engineering team costs
- Marketing costs
- Sales commission costs
- Promotion costs
- Bad debt
- Financial costs
- Administration staff
- Indirect costs

1.4.3 Return of Investment (ROI)

The required investment and its return can then be calculated on a yearly basis. Several other financial indicators can be calculated, such as the income statement and balance sheet.

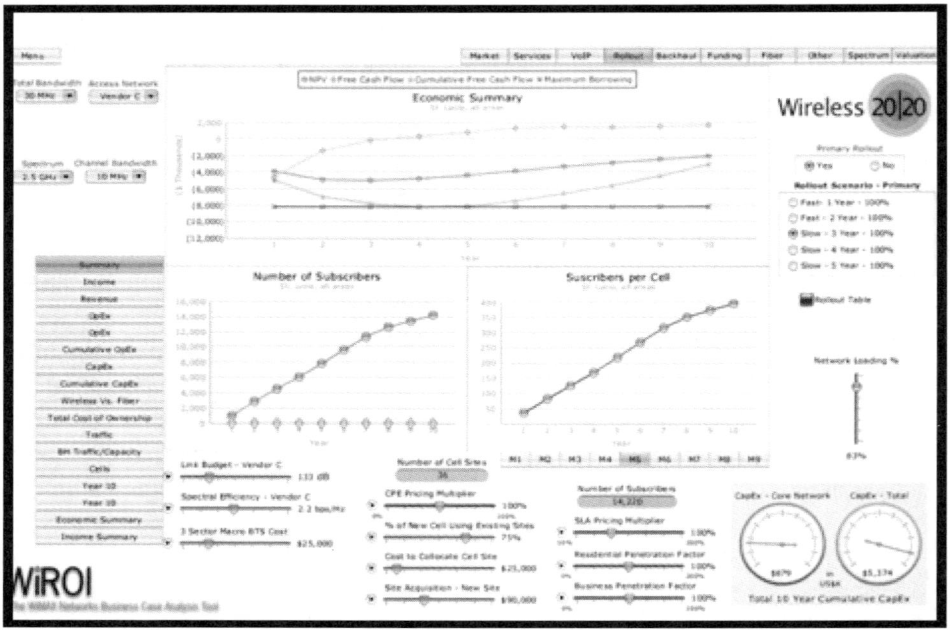

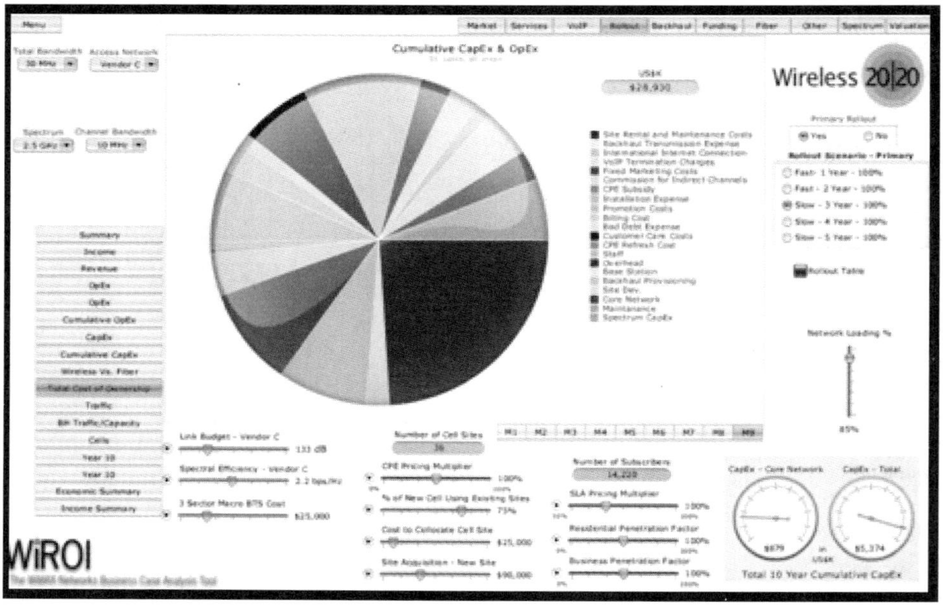

Figure 1.3 Financial planning tool screenshots.

1.5 Business Case Questionnaire

Unfortunately, in many cases market research is not done and the engineering information is not available either. In this case the designer has to obtain the information himself. This can be done by researching public information, local government agencies and by interviewing network entrepreneurs and operators. The conclusions and assumptions should be listed and approved by the client.

Typical questions to be asked are:

1. Define geographically your areas of interest.
   - Use a polygon to mark them on a map
2. Where do you intend to provide service?
   - Outdoor rooftop
   - Outdoor ground
   - Indoor window
   - Indoor
3. What is your investment potential?
   - Feasibility study
   - Pre-launch
   - Year 1, 2, 3
4. Do you intend to deploy the network at once or in phases?
   - How many phases?
5. Who are your target clients?
   - Residential
   - Stores
   - Small businesses
   - Medium businesses
   - Large businesses
   - Hotels
   - ISPs
6. List specific application that may use your services, for example, meter reading.
7. Does the area have video rental services?
8. Do you intend to offer services to tourists?
9. Do you plan to deploy additional technologies (WIMAX and LTE) in the future?
10. How does the area population fluctuate during in-season and off-season periods?
11. Do you expect to have nomadic clients?
12. Do you expect to have mobile clients?
13. List specific localities where you intend to provide service, for example, airport, coffee shops.
14. Do you have specific locations of large potential clients?
15. Do you intend to provide maritime service? Marinas? Near the coast?
16. Does someone else use the same spectrum as you? In your area? In nearby areas?
17. How do you intend to provide backhaul?
18. Where will your main equipment be installed?
19. Have you defined a marketing strategy?
20. Have you defined your sales channels?
21. Where will your sales stores be located?
22. Who will provide maintenance?
23. Do you have preferred vendors?
24. Do you have terrain and demographic databases available?
25. Do you know if this data is available from agencies in the area?

26. Do you have access to microwave frequencies for backhaul?
27. How much spectrum do you have access to and at what frequency ranges?
28. Is your spectrum owned or leased?
29. Are there restrictions for the use of the spectrum?
30. Do you have deployment commitments?
31. Where is the PoP (Point of Presence of optical fiber) available?
32. How is the connection to the Internet made? How is it charged?
33. What is the price per minute/kb for the Internet connection?
34. Do you have any agreement for site locations? Are you planning to negotiate one?
35. Any preferred sites?
36. What restrictions exist to deploy new sites? New towers?
37. What prevailing materials are used in are dwellings?
38. What kind of terminals do you plan to support?
    - rooftop
    - window
    - desktop
    - standalone
    - USB
    - PC card
    - embedded
    - phones
39. Do you plan to commercialize user terminals?
40. Do you plan to subsidize user terminals?
41. Describe the process to get licenses to build in the area? New towers? New poles?
42. Do you have to follow special construction codes? Proof against hurricanes, earthquakes?
43. How do you plan to process your billing?
44. How are you going to interconnect to the landline carrier? Are there fees?
45. Do you plan to provide Wi-Fi extensions?
46. What policy do you plan to implement to control network usage (downloads)?
47. What service plans do you envisage?
48. Do you plan to limit or charge for tonnage?
49. How many subscribers do you expect to have at signing? After 1 year? After 2 years?
50. Do you have a list of tower facilities in the area?
51. Do you have a list of high rise buildings in the area?
52. Do you plan to provide video backhaul services, that is, public surveillance? If yes, under which conditions?
53. How many ISPS are in the area?
54. Do you intend to provide service to ships? Where?
55. Do you plan to provide service to nearby areas?
56. Are you planning to rent phones to tourists?

1.6 Implementing the Business Plan

After a business case is prepared, the engineering plan can be used as a base for an RFQ (request for quote) if only budgetary numbers are required or an RFP (request for proposal) if firm number are desired.

Consultants and professional companies can tailor those documents to each operator's needs. A well-prepared RFP allows the selection of the most appropriate vendor for each deployment. One of the merits of an RFP is to make possible the comparison between different solutions.

After the proposals have been received, they should be analyzed, which is a major task because the understanding of the proposals requires a lot of experience with technology options and deficiencies. It is wise to select a short list of vendors and then arrange meetings with them.

Contract negotiations are also complex due to the many technicalities involved in wireless broadband deployments. The network deployment has to be followed up closely, as it will be hard to change things after the deployment has been made. It is strongly recommended that an expert closely follows the deployment.

System acceptance test is an extremely complex task, because the system is lightly loaded at acceptance time and many of load-related issues cannot be detected easily. This is where a planning tool is essential, by comparing the results for a lightly loaded system and extrapolating them for a loaded system.

Meanwhile all sales, marketing and administrative structure should be put in place. Then you are ready to proceed.

2

Data Transmission

Data transmission plays an important role in wireless broadband networks, and understanding this process is crucial to correctly dimension the network. The bulk of the data traffic handled by the networks will be the Internet, using TCP/IP protocol. Besides, all interconnections to the wireless system will come from wired networks, where the Ethernet prevails.

2.1 History of the Internet

In the late 1960s, it became obvious that there was a need to interconnect computers. In 1970, the ISO (International Standards Organization) developed a reference model called OSI (Open System Interconnection), which defined a seven layers model. This model became the reference for comparing different protocols, but its full implementation was extensive and was not practical for the majority of the applications. This model was defined by a committee and lacked practical implementations. OSI-defined protocols were then developed by several entities such as the ITU-R (CCITT) X.25 for packet switching and EISA/TIA-232 and 422. Large computer manufacturers implemented proprietary OSI-based protocols such as SNA (System Network Architecture) from IBM or DSA (Distributed System Architecture) from Honeywell Bull and others.

In the USA, the first attempts to interconnect different computer platforms at different locations were sponsored by DARPA (the Defense Advanced Research Projects Agency) through the implementation of the ARPANET (Advanced Research Projects Agency Network) in 1969. This network connected four universities, using Interface Message Processors (precursors of today's routers) at each location to store and forward packets of data. The hardware was implemented by BBN Technologies (Bolt, Beranek and Newman), a Massachusetts company. The design of the network was set so that it should only provide routing and transmission capabilities and that the remaining functionalities should stay on the periphery. The basic functionality was provided by the (NCP) Network Control Program, developed by Vinton Cerf, and which could run on several hosts.

ARPANET was an open field for testing and implementing new ideas and solutions. Not imposing a pre-defined architecture led to the development of many protocols, which had to prove their benefits through peer acceptance. This was the case in 1972 of Telnet (Telecommunications Network), developed by NCSA (National Center for Supercomputing Applications). In 1973, the FTP (File Transfer Protocol) was standardized to transfer files between computers. Robert Khan and Vinton Cerf developed in 1974 the basic specification of a new protocol called TCP (Transmission Control Protocol).

LTE, WiMAX and WLAN Network Design, Optimization and Performance Analysis, First Edition.
Leonhard Korowajczuk.

In 1974, DARPA contracted BBN Technologies, Stanford University and the University College of London to develop operational versions of this protocol and after four rounds the TCP/IP v4 (Transmission Control Protocol/Internet Protocol was released in 1978. ARPANET fully migrated to this protocol by 1983.

The DARPA network grew to 230 IMPs (Interface Message Processor or, as defined by Bob Kennedy, Interfaith Message Processor) by 1980, interconnecting mainly universities and government agencies. In March 1982, the US Department of Defense declared TCP/IP the standard for all military computer networking. The astounding success of this network led to the creation of the acronym Internet, by abbreviating Interconnected Networks. With the increase in the number of participants, in 1979, DARPA created a body to oversee the technical and engineering development of the Internet, the Internet Configuration Control Board (ICCB), which in 1992 became the Internet Architecture Board (IAB).

The expansion of the network to commercial interests began in 1989, with the interconnection of MCI mail, UUNET, PSINet, Compuserve, Sprintnet and many others.

The commercial availability of standardized routers, the ability of TCP/IP to work over any network and its rigorous implementation on all Operating Systems (OS) were responsible for the popularization of the Internet.

Internet provided the interconnectivity between computers, but did little to standardize the content to be exchanged. The World Wide Web addressed this issue and was invented by British scientist Tim Berners-Lee, in 1989, as a web of hypertext documents to be viewed by browsers using client–server architecture. Berners-Lee was working at CERN (Centre Européenne pour la Recherche Nucléaire) in Geneva at the time and the first web site went on-line on 1991. In 1993, CERN announced that the World Wide Web (WWW) access would be free to everyone.

WWW popularity increased with the introduction of the Mosaic web browser in 1993, designed by Marc Andreessen and Eric Bina at NCSA (National Center for Supercomputing Applications). It also supported other protocols, like FTP (File Transfer Protocol), Usenet and Gopher, and was licensed free for non-commercial applications.

Andreessen with others started Mosaic Communications that became later Netscape. Spyglass, Inc. licensed the technology from NCSA and produced its own web browser called Spyglass Mosaic. It was later bought by Microsoft and renamed Internet Explorer.

Initially to find Internet content NCSA had a session called "What's New" in its server. The first search engine was developed in 1990 by Alan Emtage and was called Archie (or archive without a v, as it indexed FTP archives). The program downloaded directory listings of all files located on public anonymous FTP sites, creating a searchable database of file names, but without indexing them. Next, Gopher, a TCP/IP application, was made available in 1991 by the University of Minnesota, and was designed to distribute, search and retrieve documents over the Internet.

Only in 1994 did the first WWW search engines become available. They would let users search for any word in any web page. WebCrawler and Lycos were the first ones, followed by AltaVista, InfoSeek, Excite, Magellan and Yahoo.

In 1997, Larry Page and Sergey Brin developed a search engine called Google that became popular in the academic arena, due to its non-commercial look. In this engine, web pages are ranked according to the weighted sum of pages that link to them.

Today the Internet, the World Wide Web and search engines are an essential part of our lives and have permeated all levels and ages of our society.

2.2 Network Modeling

The most known network reference model was proposed by ISO (International Organization for Standardization) and is called the OSI (Open System Interconnection) reference model. It has a very

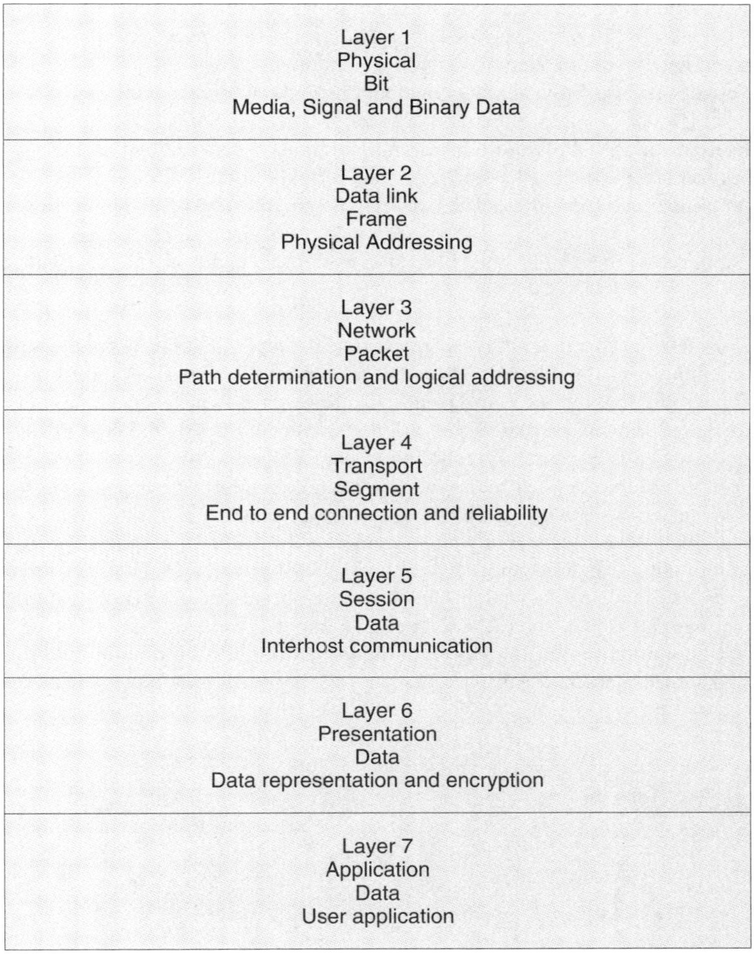

Figure 2.1 OSI network modeling reference layers.

didactical approach but this does not mean that it is the best implementation. It separates network functionalities in seven layers. Each layer receives services from the layer below and provides services to the layer above. The layers are illustrated in Figure 2.1.

- *Layer 1*: Physical layer (PHY) – Defines electrical and physical specifications of the device providing the service: its interconnection to the physical medium, the signal that will travel in the medium and the medium itself. In a wireless medium, this layer specifies the characteristics of the signal to be transmitted (modulation schemes, processing, power) and the antennas. It includes auxiliary signals that may be sent to help retrieve the original data.
- *Layer 2*: Data link layer – Defines logical procedures and functionality to transfer data between physical network entities and detect and correct errors incurred in the first layer. It can be connection oriented, when an end-to-end physical or logical connection is established prior to the data exchange

(e.g. circuit switching) or connectionless when data can be delivered independently of a previous connection negotiation. In a wireless medium, this layer creates packet envelopes that carry layer 3 data between network entities and uses entities' addresses.

- *Layer 3*: Network layer – Provides functional and procedural means to transfer data over multiple networks.
- *Layer 4*: Transport layer – Provides transparent data transfer between end users.
- *Layer 5*: Session layer – Establishes, manages and terminates the connections between computers.
- *Layer 6*: Presentation layer – Translates data representation between application and network format.
- *Layer 7*: Application layer – In this layer the end user interacts directly with the software application.

OSI layers define network functionalities separately (one per layer), but this is not the most economical way of implementing real-life solutions. Practical network implementations mixed functionalities of different layers in a single layer and consequently had fewer layers. Layer mappings vary from author to author, thus the allocations presented here are not universal. Such is the case of the Internet, and its mapping is presented in Figure 2.2, according to our interpretation.

Internet specifications are defined by protocols, which define formats, fields, addresses and procedures. Such specifications started as RFC (Request for Comments) at the beginning of ARPANET in 1969. The original document was a simple memorandum to be commented on by peers. Today anyone can contribute to an RFC and they are published by the IETF (Internet Engineering Task Force). RFCs are classified as informational, experimental, best current practice, historic or unknown. There are more than 5,000 published RFCs. The IETF adopted some of these RFCs as Internet standards. RFCs are submitted in ASCII (American Standard Code for Information Interchange) format. This simple and informal methodology has proven very powerful and is one of the reasons for the Internet's success.

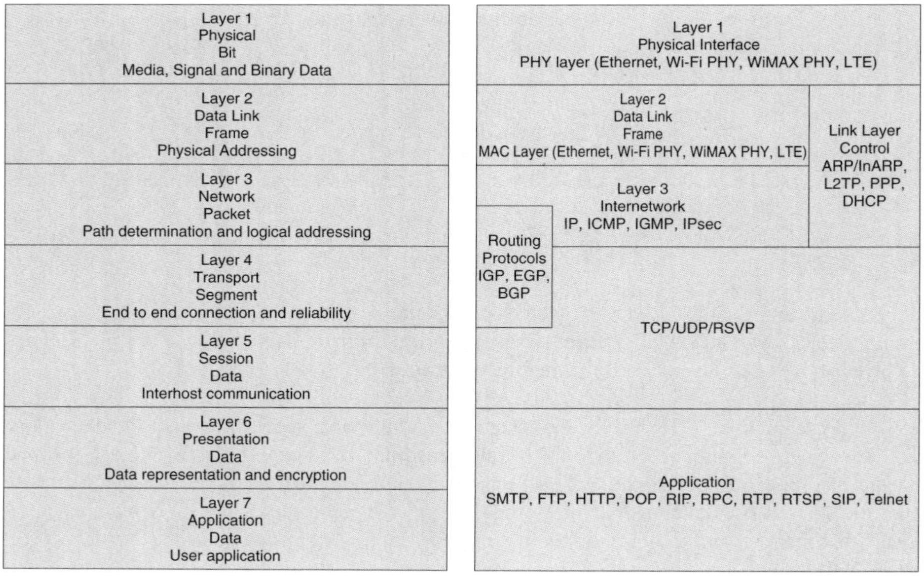

Figure 2.2 OSI and Internet network modeling reference layers.

2.3 Internet Network Architecture

The proposed Internet network architecture, illustrated in Figure 2.3, is extremely simple. Its main virtue is that it is a democratic network that does not have a central control and can grow autonomously like a living being. Computers are connected in a star or bus configurations to routers, which in turn provide the interface between those computers and the network, connect to other routers and direct data packets according to routing tables. The interconnection between routers uses high capacity links and may be done over very long distances, as across the world. These long connections can traverse several routers and take only 200 ms in travel time.

2.3.1 *Router*

Routers are electronic networking devices that route and forward information between computers throughout a network. They are layer 3 devices and work in two planes:

- *Control plane*: routers learn which outgoing interface is most appropriate to forward a specific information to a certain destination.
- *Forwarding plane*: routers process received information and send to the route established by the control plane.

Routers implement different routing protocols and have to be programmed with routing tables.

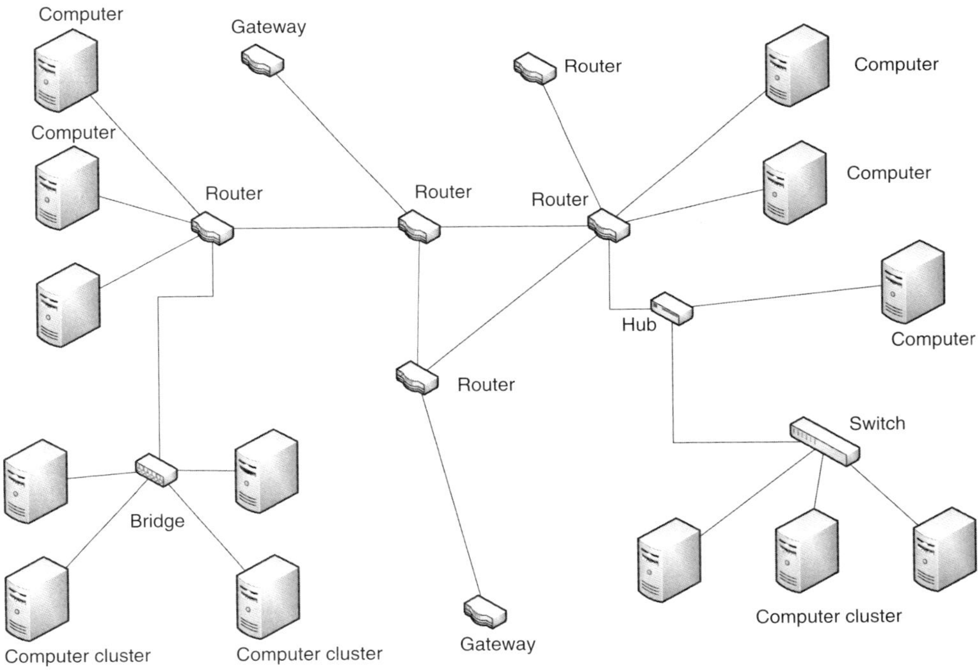

Figure 2.3 Internet network architecture.

2.3.2 *Hub*

Hubs are networking electronic devices that interconnect ports (twisted pairs or optical fiber) from multiple devices, making them act as a cluster. Hubs operate as layer 1 and execute the function of a repeater. They retransmit to all ports the activity at each of its ports, including collisions. Hubs are used as wiring extension devices and today are being replaced by switches.

2.3.3 *Bridge*

A bridge connects multiple network segments at layer 2. Differently from a hub, a bridge analyzes the data address and content and only sends good packets to a destination port. Unlike routers, bridges do not make assumptions about the destination address, as they are only concerned with neighbor connections. Bridges use device MAC addresses to assemble a routing table. When a packet is received with a layer 2 address, the bridge performs a flooding, by sending a packet to all of its ports and once an acknowledgement is received from one of the ports, it maps it as a destination for this address, so the next packets are sent only to this port.

2.3.4 *Switch*

Switches have the same functionality as a bridge, but to be more efficient they do not check the integrity of the whole packet, but only of its header. This was very important in the first days of the technology when processing power was at a premium. Today switches are affordably priced and are replacing hubs and bridges.

There are vendor-specific multi-layer switches that operate above layer 2, but they are only used in specialized deployments, customized to specific applications.

2.3.5 *Gateway*

Gateways are in principle protocol converters that interconnect dissimilar networks. This term has been loosely used and today is also applied to devices that connect LANs (Local Area Networks) to WANs (Wide Area Networks) and WANs to WANs. Gateways may work on some or all seven OSI layers. A PC or a router can perform the function of a gateway.

2.4 The Physical Layer

The most important physical layer in use today is the Ethernet and is described here. The other wireless physical layers are described in the technology sections of this book.

2.4.1 *Ethernet PHY*

The amazing growth of the Internet required a common way to interconnect routers and computers and Ethernet became the de facto standard for wired LANs. It was developed by XEROX in 1975, as a multipoint data communication system with collision detection. It inventors are Robert Metcalfe, Chuck Thacker and Butler Lampson. Metcalfe later founded 3Com and, after joining forces with DEC and Intel, proposed it as a standard to the IEEE in 1980. Support was also given by other standard bodies, like the ISO, and it was published as IEEE 802.3 "Carrier Sense Multiple Access with Collision Detection for LAN" (CSMA-CD LAN).

Table 2.1 Ethernet physical layer interfaces

Name	Standard	Category	Medium	Mbit/s	distance (m)
1BASE	802.3(11)	Legacy Ethernet	copper twisted pair	1	25
10BASE-T	802.3(14)	Regular Ethernet	copper twisted pair	10	100
100BASE-T	802.3(21)	Fast Ethernet	CAT5 twisted pair	100	100
1000BASE-T	802.3(40)	Gigabit Ethernet	CAT6 twisted pair	1000	100
1000BASE-SX	802.3	Gigabit Ethernet	multi-mode fiber	1000	550
1000BASE-LSX	802.3	Gigabit Ethernet	single-mode fiber	1000	2000

Ethernet specifications cover the physical layer and part of the data link layer. The IEEE specification is divided into sections, each covering a specific implementation. The most popular implementations are listed in Table 2.1. "BASE" stands for baseband signal, "T" for twisted pair cable, "SX" for multi-mode fiber and "LSX" for single-mode fiber.

The 10BASE-T interface uses a +2.5 V (Volt) and −2.5 V signal, 100BASE-T interface uses a +1 V, 0 V, −1 V signal and the 1000BASE-T uses +2 V, +1 V, 0 V, −1 V, −2 V signal in their interface.

Cables carry the signal from one device to another, including the wireless elements. A designer should have an understanding about cables, their features and limitations. Regular twisted cables use gauge 24 and are specified for frequencies up to 10 MHz, CAT5 cables are specified for frequencies up to 100 MHz, and CAT6 cables for frequencies up to 500 MHz.

Devices that terminate the Ethernet connections are called the NIC (Network Interface Card) and its interface is called MDI (Medium Dependent Interface) or MDIX, where X stands for crossed. A straight interface connects a TX to a TX port, while a crossed interface connects a TX to an RX port. Network devices implement the straight interface, whereas hubs, switches and routers implement the cross interface, in such a way that a transmit pin is connected to a receive pin, so a straight cable can be used, while direct interconnections between devices require a cross cable. Tables 2.2 to 2.5 show the pins and pairs defined for the straight and cross wirings between MDI and MDIX combinations.

This multiple possibility of interfaces and cables was prone to create confusion, so recent specifications recommend that network devices use auto MDI/MDIX sensing and adapt to the interface, permitting any cable to be used. As some interfaces do not use all pairs, there are cables with only two pairs.

Table 2.2 Ethernet MDI straight wiring

TIA/EIA-568B T568A MDI straight wiring						
Pin	Pair	Polarity	Color	10BASE-T	100BASE-T	1000BASE-T
1	3	A+	white/green	x	x	x
2	3	A−	green	x	x	x
3	2	B+	white/orange	x	x	x
4	1	C+	blue			x
5	1	C−	white/blue			x
6	2	B−	orange	x	x	x
7	4	D+	white/brown			x
8	4	D−	brown			x

Table 2.3 Ethernet MDIX straight wiring

TIA/EIA-568B T568B MDIX straight wiring						
Pin	Pair	Polarity	Color	10BASE-T	100BASE-T	1000BASE-T
1	2	B+	white/orange	x	x	x
2	2	B−	orange	x	x	x
3	3	A+	white/green	x	x	x
4	1	D+	white/brown			x
5	1	D−	brown			x
6	3	A−	green	x	x	x
7	4	C+	blue			x
8	4	C−	white/blue			x

Table 2.4 Ethernet MDI wiring crossed

TIA/EIA-568B T568A MDI wiring crossed						
Pin	Pair	Polarity	Color	10BASE-T	100BASE-T	1000BASE-T
1	2	B+	white/orange	x	x	x
2	2	B−	orange	x	x	x
3	3	A+	white/green	x	x	x
4	4	D+	white/brown			x
5	4	D−	brown			x
6	3	A−	green	x	x	x
7	1	C+	blue			x
8	1	C−	white/blue			x

Table 2.5 Ethernet MDIX wiring crossed

TIA/EIA-568B T568B MDIX wiring crossed						
Pin	Pair	Polarity	Color	10BASE-T	100BASE-T	1000BASE-T
1	3	B+	white/green	x	x	x
2	3	B−	green	x	x	x
3	2	A+	white/orange	x	x	x
4	4	D+	white/brown			x
5	4	D−	brown			x
6	2	A−	orange	x	x	x
7	1	C+	blue			x
8	1	C−	white/blue			x

2.5 The Data Link Layer

The Data Link Layer is divided by OSI in MAC (Medium Access and Control) and LLC (Logical link Control) sub-layers. Both sub-layers may be combined inside one MAC extended layer.

Preamble	Start of Frame Delimiter	Destination MAC address	Source MAC address	802.1Q header (optional)	Ethernet Type/ Length	Payload Data And padding	CRC32	Interframe gap
7 octets 10101010	1 octets 10101011	6 octets	6 octets	4 octets	2 octets	46 to 1500 octets	4 octets	12 octets

Figure 2.4 Ethernet packet format.

2.5.1 Ethernet MAC

Ethernet MAC defines how information will be packed before being sent over the Ethernet.

2.5.1.1 Ethernet Packet Format

The Ethernet packet format is defined in IEEE 802.3 and is shown in Figure 2.4. The overhead per packet is 48 octets. The specification of this packet should be considered as part of the Data Link layer together with other MAC (Medium Access Control) layers and it is presented here to keep all Ethernet specifications together. The smallest frame has 64 bytes and the longest 1518 bytes.

2.5.1.2 Transmission Algorithm

A simple algorithm is used to transmit data by a user. An Ethernet NIC can simultaneously transmit and monitor the transmitted signal.

- Main transmission procedure:
  - Frame is ready for transmission.
  - If medium is idle, start transmission.
  - If medium is not idle, wait until it becomes idle plus the inter-frame gap of 9.6 μs.
  - If a collision occurs go, to collision procedure.
  - Reset transmission counters and end frame transmission.
- Collision procedure:
  - Continue transmission until minimum packet time is reached (jam signal) to ensure that all receivers detect collision.
  - Increment retransmission counter.
  - If the maximum number of retransmissions is reached, abort transmission.
  - Otherwise wait a random back-off period based on the number of previous collisions.
  - Restart main transmission procedure.

Wireless connections use different protocols that perform a similar functionality as the Ethernet protocol and they will be described in the following chapters. As the Ethernet data transmission rate is much higher than the wireless data rate, we do not need to be concerned with the overhead of this protocol.

2.5.1.3 MAC Address

Layer 2 protocols take care of transmitting messages from one machine to the next and they do not need to know the final destination of the message as they are only responsible for transferring data over

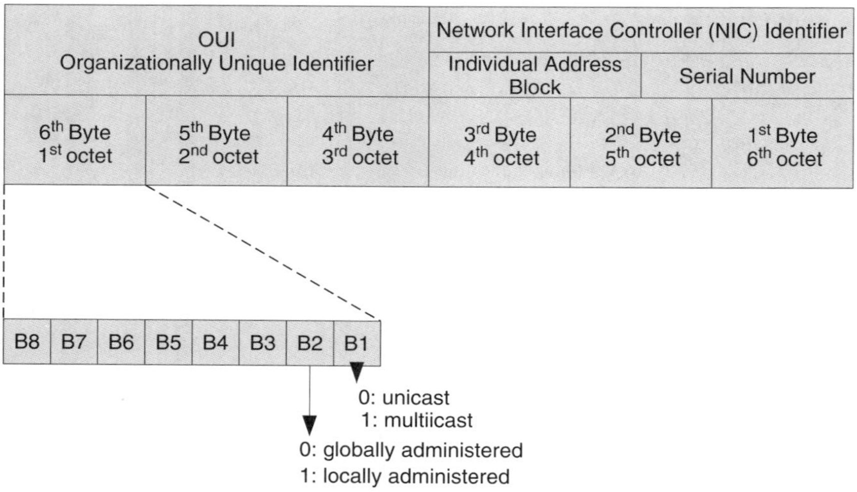

Figure 2.5 Ethernet MAC address.

network neighboring segments. Each data transfer is done between two NICs. This implies that NICs should have embedded a unique hardware address, called Medium Access Control (MAC) address.

The MAC address is 48 bits (6 octets) long and its format is shown in Figure 2.5. It is represented by six groups of two hexadecimal digits separated by colons or hyphens. An example is 32:41:36:AB:16:08. The first three octets identify the hardware vendor organization and are assigned and managed by the IEEE (Institute of Electrical and Electronics Engineers). The last three octets identify the specific hardware and are managed by the organization that manufactures the hardware. Each unit produced should have a unique number; so many organizations use it as a manufacturing serial number.

Although this is the general procedure, MAC addresses can be assigned by the local organization if they are not going to interconnect with external networks and in this case they are considered as locally administered and this is indicated by the second bit of the first octet.

MAC addresses can be changed in some network applications, as is the case of wireless stations that start with their MAC address to establish a communication and then clone (or spoof) the MAC address of the device connected to them.

MAC addresses are also known as MAC-48 or EUI-48 (Extended Unique Identifier) and can provide potentially 2^{48} addresses, but the practical number is much smaller as they are allocated in blocks and many are not used. To solve this problem, IEEE created EUI-64, which is used in the new IPv6 protocol and provides additional addresses.

2.6 Network Layer

Interworking protocols can be considered layer 3 protocols as they identify data source and destination. They add an envelope to the data received from the transport layer, which identifies source and destination using the Internet Protocol (IP) addressing.

2.6.1 Internet Protocol (IP)

This is the protocol most used today. Data from the upper layer protocol is encapsulated in datagrams. Datagrams are packets that are sent through networks without assurance of transmission reliability (unreliable networks).

IP is a connectionless protocol, as it sends datagrams without establishing a physical or logical connection. The protocol design assumes that the network infrastructure is inherently unreliable and has a dynamic availability of links and nodes. There is no central entity that tracks the state of the network. The Internet Protocol provides a "best effort" delivery and a transmission using datagrams is subject to data corruption, lost datagrams, duplicate arrivals, or out-of-order packet delivery.

The IPv4 protocol checks the message header check sum and discards defective datagrams. IPv6 does not check the header to improve forwarding speed.

2.6.1.1 IP Addresses

Addresses used in Internet Protocol (IP) were created to identify computers on a network. The Internet Assigned Number Authority (IANA) manages the IP address space allocations globally and cooperates with five Regional Internet Registries (RIRs) to allocate IP address blocks to Local Internet Registries (ISPs).

The IP address was originally established as a 32-bit number and is used in version 4 of the IP protocol (IPv4). Due to the popularity of the Internet, it was clear that this addressing space would be insufficient so a newer version was released IPv6and it uses a 128-bit IP address. The transition from IPv4 to IPv6 is expected to happen gradually over the years, so new equipment is supposed to support both.

Several measures were taken to delay address exhaustion such as temporary (dynamic) address assignment and use of private address spaces inside private networks. This created the need for Network Address Translators (NAT).

Some of the bits on the IP address identify sub-networks and the number of bits used for this is indicated in dot-decimal or CIDR (Classes Inter-Domain Routing) notation, appended to the IP address. This dot-decimal notation indicates how many addresses are reserved for the network at each octet. The CIDR states how many bits are reserved for the network. The following examples illustrate the different representations.

iPv4 address: 192.168.100.1;

iPv4 address with dot-decimal notation: 192.168.0.0/255.255.255.0;

IPv4 address with CIDR notation: 192.168.0.0/24;

IPv6 address: 2001: DB8:0:0: 0:0:0:0;

IPv6 with CIDR notation: 2001: DB8::/48.

The number of bits assigned to the host depends on the type of corporation. Table 2.6 gives address allocation ranges according to their use.

A computer can have a static IP address assigned to it, so each time the computer boots up, it will use the same address. This is not efficient as a significant number of computers are generally off network. A more efficient method is assigning the IPs dynamically on an as-needed basis from a block of reserved IPs. Those assignments are done by a local server using the DHCP protocol (Dynamic Host Configuration Protocol).

Table 2.6 IP address ranges per use

CIDR	Host bits	Subnet mask	Hosts in the subnet	Typical usage
/8	24	255.0.0.0	16,777,216	Few very large corporations
/17 to /19	15 to 13	255.255.128.0 to 255.255.224.0	32,768 to 8192	ISPs/Large businesses
/20 to /21	12 to 11	255.255.240.0 to 255.255.248.0	4096 to 2048	Small ISPs/Large businesses
/24 to /25	8 to 7	255.255.255.0 to 255.255.255.128	256 to 2128	Large LANs
/26 to /28	6 to 4	255.255.255.192 to 255.255.255.240	64 to 16	Small LANs
/29	3	255.255.255.248	8	Smallest multi host
/30	2	255.255.255.252	2	Point to point

2.6.1.2 IP Network Address Translation

Network Address Translation (NAT) was developed to enable multiple hosts in a private network to use a single public IP address. When traffic is received by a router from the private network, it tracks the source address and its port and maps it onto the public IP address and another port. When a reply is received, it re-maps the address and sends the reply to the private network. The router has to change not only the address but also has to recalculate the check sum of the packet and this requires processing power, which is limited in low end routers.

A basic NAT translates only the IP address while a more elaborate NAT performs also Port Address Translation (PAT). NAT can be dynamic or static or a mixture of both. NAT is also used to protect private networks from unauthorized external access, as only accessed IPs have internal access for a short time (timer expiration).

There are several drawbacks in using NAT, as it may not work with all protocols (like FTP), although it should perform well with TCP and UDP.

2.6.1.3 Firewalls

Firewalls are designed to block unauthorized access to computers and can be implemented in hardware or software. All messages that pass though the firewall are examined and the ones that do not comply with pre-established criteria are blocked. The main techniques used today are:

- *Packet filter*: each packet is analyzed and is accepted or rejected according to specific user-defined rules.
- *Application gateway*: certain applications are filtered, so the user cannot access or be accessed by them.
- *Circuit-level gateway*: security mechanisms are applied when a TCP or UDP session is being established, but once established no further checking is done.
- *Proxy server*: intercepts all messages entering and leaving the network and hides the true network addresses using NAT procedures.

2.6.2 Internet Control Message Protocol (ICMP)

This protocol is used by network computers to send error and test messages related to datagrams. Typical error messages are:

- Echo reply
- Destination unreachable

- Source quench
- Redirect message
- Echo request
- Time exceeded
- Trace route

2.6.2.1 PING (Echo Request)

This message sends a certain amount of data to a destination, which, after being received, is retransmitted back. This is the most popular ICMP administration utility used to test whether a specific host is reachable across the IP network and to calculate the round-trip time.

2.6.3 Multicast and Internet Group Message Protocol (IGMP)

This protocol is used by network computers to send error and test messages related to multicast groups. It is equivalent to ICMP but for multicast. A multicast requires an IP multicast group address. A receiver informs the address that it wants to join the group using the IGMP (Internet Group Management Protocol). The source will then send the datagrams to the group addresses. Routers around the receiver build a tree from join group requests, so they can appropriately route datagrams.

2.6.4 Link Layer Control (LLC)

LLC includes a series of protocols that resolve addressing issues.

2.6.4.1 Address Resolution Protocol (ARP)

This protocol translates IP addresses into MAC addresses. It is an IPv4 link layer protocol.

2.6.4.2 Inverse Address Resolution Protocol (InARP)

This protocol translates MAC addresses into IP addresses. It is an IPv4 link layer protocol.

2.6.4.3 Neighbor Discovery Protocol (NDP)

This is a protocol used in IPv6 to replace the ARP protocol used in IPv4.

2.6.4.4 IP Assignment and Dynamic Host Configuration Protocol (DHCP)

Static IP assignment consumes many addresses that are seldom used. A cleverer IP approach is the dynamic IP assignment procedure. In this procedure a block of IPs is assigned to a server, which performs temporary IP address allocation, when requested by computers. Dynamic Host Configuration Protocol (DHCP) is used to provide this functionality. When becoming active on a network, clients broadcast a message within the physical subnet to discover available DHCP servers. The server then sends an IP offer to the client. A client may receive IP offers from different servers, and, after selecting one, it sends an acceptance to one of the servers. The servers who are not selected will time out and release the IP they reserved. The selected server then sends a message confirming the allocation and its duration.

2.6.4.5 Tunneling, Virtual Private Networks and Layer 2 Tunneling Protocol (L2TP)

Tunneling means that one network protocol encapsulates a different payload protocol. It is equivalent to someone placing a letter inside another envelope and sending it to the same address. Tunneling protocols may use data encryption to transport unsecure payload protocols over a public network, providing VPN (Virtual Private Network) functionality.

L2TP is such a tunneling protocol, and its entire payload is sent inside an UDP datagram. The VPN connection has to be authenticated before data can be exchanged. A typical application is a laptop that wants to connect to an enterprise network remotely. The laptop uses regular Ethernet packages, which will be encapsulated by L2TP and addressed to the enterprise network, where they will be removed from the capsule and presented and connected to the enterprise Ethernet network.

2.6.4.6 Point to Point Protocol (PPP)

This protocol establishes a direct connection between two nodes.

2.6.4.7 GPRS Tunneling Protocol (GTP)

GTP is used to carry General Packet Radio Service (GPRS) within GSM (Global System for Mobile Communications, previously Groupe Spécial Mobile) and UMTS (Universal Mobile Telecommunication Systems) networks. It is in reality a very simple IP-based tunneling protocol.

2.7 Transport Protocols

These protocols are concerned with the end-to-end connection (application to application).

2.7.1 User Datagram Protocol (UDP)

UDP is a very simple protocol for data exchange as it does not guarantee reliability, ordering or data integrity. It just packs the data in datagrams and sends them to the destination. It is a very efficient protocol, with little overhead and processing and is ideal for applications that are not affected by receiving some frames in error or by missing frames. This protocol identifies the sender and destination ports.

2.7.2 Transmission Control Protocol (TCP)

TCP concerns only with the end-to-end connection and provides a reliable and orderly stream of bits. TCP controls segment size, flow control, rate and network traffic congestion. TCP relies on IP to carry its messages, segmenting the data and numbering sequentially these segments. The TCP header format of these segments is shown in Figure 2.6.

A TCP data transfer has several phases:

- Connection establishment: server and client synchronize to a random sequence number SYN (Synchronized Sequence Number).
- Data transfer is made according to following directives:
  - Retransmission of lost packets.
  - Discard of duplicate packets.

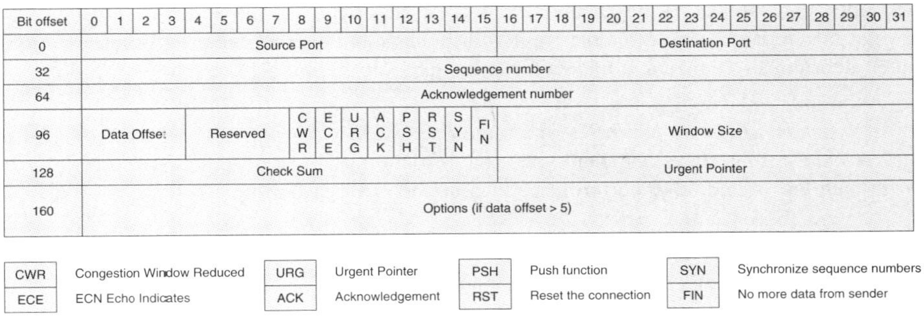

Figure 2.6 Transmission control protocol header.

- Error-free data transfer.
- Ordered data transfer.
- Control flows to avoid host overflow.
- Congestion control through sliding window.

- Connection termination: A four-way handshake is done with a FIN (Finish) and
- ACK (Acknowledgement) package from each side.

2.7.2.1 Port and Sockets

A port is a logical process-specific software construct serving as a communications endpoint used by UDP and TCP. A port is identified by its number. The Internet Assigned Numbers Authority (IANA) is responsible for assigning and registering port numbers. The following are the port ranges:

- Well-known ports: 0 through 1023.
  - Port 23: Telnet
  - Port 53: DNS (Domain Name System)
  - Port 80: WWW
- Registered ports: 1024 to 49151.
- Dynamic or private ports: 49152 to 65535.

An Internet socket is an end point of a bidirectional communications flow. A socket address is a combination of an IP address and a port number and is represented by a 32-bit number.

2.8 Routing Protocols

When a device wants to communicate with another device outside its subnet over an IP network, it must pass its address, the destination address and the data to a router. The IP router must know how to transfer this data to the destination or at least forward it towards the destination. The router will use a routing algorithm to find the best route to the destination.

2.8.1 Basic IP Routing

Devices are either configured with IP addresses of their default gateways or they look for routers connected to its subnet. When a device wants to send data, it compares the destination address with its own address and subnet. If they match, the data is sent directly to the destination (using ARP), if they do not match, it is because the data is in a different subnet, so it is sent to a router.

Router algorithms initialize and maintain routing tables, containing:

- *network identifier*: network addresses of a remote network;
- *interface number*: interface that should be used to route traffic to a specific network;
- *metric*: path length and other performance measurements of each route.

In static routing, a network administrator pre establishes the routes. In dynamic routing, routers communicate with each other and update their tables frequently to accommodate network changes. This is done, for example, using the Routing Information Protocol (RIP).

The following metrics can be used to find the best route:

- path length/hop count
- reliability
- delay
- bandwidth
- load
- communication cost

Routers are organized in a hierarchy within one administrative authority forming an Autonomous System, where Interior Gateway Protocols (IGPs) are used. The connection between the Autonomous System is done using External Gateway Protocols (EGP). Two types of algorithms are used by IGPs:

- Distance Vector Routing Protocol (DVRP).
- Link-State Routing Protocol (LSRP).

2.8.2 Routing Algorithms

The main routing protocols are listed next.

2.8.2.1 Remote Information Protocol (RIP)

This is an interior dynamic routing protocol used in local and wide area networks and uses a Distance-Vector (DV) algorithm to calculate the best route using Distance Vector Routing Protocol (DVRP). It is still used but considered obsolete s it was superseded by the Open Shortest Path First (OSPF) protocol and the OIS (IS-IS) protocol.

2.8.2.2 Interior Gateway Routing Protocol (IGRP)

IGRP is a CISCO proprietary protocol that uses DVRP algorithm. It is mostly used by large enterprise networks.

2.8.2.3 Intermediate System to Intermediate System (IS-IS)

IS-IS is an internal IGP OSI protocol adopted by IETF. It is more commonly used between large ISPs (Internet Service Providers).

2.8.2.4 Border Gateway Protocol (BGP)

BGP is the core Internet external routing protocol.

2.8.2.5 Open Shortest Path First (OSPF)

OSPF is an advanced IGP that uses LSRP (Link State Routing Protocol) algorithms to find the best route. It uses a path vector protocol, based on paths, network policies and rule sets.

2.9 Application Protocols

The definition of application is vague and classifications vary. Both presentation and application layers can be included into this category. In principle, these protocols are related to the data to be transmitted and implement peculiarities related to this data.

2.9.1 Applications

Typical applications are text, voice, real time events, messages, streams, e-mails, and so on.

2.9.2 Data Transfer Protocols

The main Data Transfer Protocols (DTPs) are described next.

2.9.2.1 TELNET (TELecommunications NETwork)

This protocol provides a bidirectional interactive communications facility. It is used to interact with software utilities, verify logs and even to chat.

2.9.2.2 File Transfer Protocol (FTP)

This protocol is used to exchange and manipulate files over a TCP/IP network. It is a client–server protocol and uses separate data and control connections. FTP supports user authentication (password-based) or anonymous access.

2.9.2.3 Trivial File Transfer Protocol (TFTP)

This is a simplified FTP protocol ideal to transfer small amount s of data. It is built over UDP.

2.9.2.4 Hypertext Transfer Protocol (HTTP)

This protocol is designed for distributed and collaborative hypermedia information systems. It is a client–server protocol, where a web browser acts as a client while a web site acts as a server.

- *Hypertext Markup Language (HTML).* This is a markup language for web pages. Web pages can be written in HTML, and is built by HTML elements in the form of <tag> content to be rendered as </tag>.

- *Extensible Markup Language (XML).* While HTML has a fixed structure to specify its elements based on pre-defined tags, XML allows users to define their own tags.
- *Java*. This is a programming language developed by Sun Microsystems (Oracle Corporation today), based on C and C++ and highly portable to different platforms. It is used in many web-based applications.

2.9.2.5 Simple Mail Transfer Protocol (SMTP)

E-mails can be sent by an e-mail client to an e-mail server using SMTP.

2.9.2.6 Multipurpose Internet Mail Extension (MIME)

This is an extension to the SMTP protocol that as well as text, includes, non-text attachments, message bodies with multiple parts and non ASCII characters.

2.9.2.7 Post Office Protocol (POP)

E-mails are sent using to an e-mail server and are stored in the recipient's e-mail box. An e-mail client retrieves the messages using the POP protocol. This protocol is being replaced by the more powerful IMAP.

2.9.2.8 Internet Message Access Protocol (IMAP)

E-mails are sent to an e-mail server and are stored in the recipient's e-mail box. An e-mail client retrieves the messages using the IMAP protocol.

2.9.2.9 Internet Relay Chat (IRC)

This application allows real-time exchange of text messages between individuals or a group of individuals.

2.9.2.10 Network News Transfer Protocol (NNTP)

This protocol is used to transport Usenet new and article between servers and to provide user access to read this news. Usenet is an Internet discussion system, distributed between a constantly changing set of servers, to which clients can post and read news.

Usenet is a world-wide distributed discussion system over the Internet. It is similar to a Bulletin Board System (BBS) but does not have a central server or administrator. It was developed in 1979 by Duke University, NC, is organized in news groups and can be used to distribute text and binary files. Its popularity is winding down. There are many Usenet providers, one being Google Groups.

2.9.2.11 Gopher

This application was designed for distributing, searching and retrieving documents, being a predecessor to WWW.

2.9.3 *Real Time Protocols*

Several applications require real time action and special protocols and specific applications were developed for it.

2.9.3.1 Real Time Transport Protocol (RTP)

This is a standardized protocol delivering audio and video over the Internet. It is the technical foundation for VoIP.

2.9.3.2 Real Time Streaming Protocol (RTSP)

It is a network control protocol designed for use in entertainment and communication systems that control streaming media servers.

2.9.3.3 Network Time Protocol (NTP)

This is protocol used to synchronize computer clocks over packet switched variable latency data networks.

2.9.3.4 Voice Over Internet Protocol (VoIP)

This is a general term for a family of transmission technologies for delivery of voice communications over IP networks. VoIP systems employ audio codecs to digitize the audio and session control protocols to set up, control and tear down calls. The most popular vocoders are listed in Table 2.7.

- *H.323*: this is an ITU recommendation that defines the protocols to provide audio visual communications over packet-based networks. It uses a mix of TCP and UDP as transport mechanism.
- *Quality of Service (QoS)*: certain services, such as audio and video require a minimum performance in the delivery of packages. This performance is called the required Quality of Service. It is

Table 2.7 Most popular vocoders

ITU spec.	Rate (kHz)	Bit rate (kbit/s)	Latency (ms)
G.711	8	64	0.125
G.722	16	64	4
G.722.1	16	24,32	40
G.722.1C	32	24,32,48	40
G.723	8	24	
G723.1	8	5.3, 6.3	37.5
G726	8	16, 24, 32, 40	
G.729	8	6.4	
GSM FR	8	13	30
GSM HR	8	5.6	25
GSM AMR	8	4.75, 5.15, 5.9, 6.7, 7.4, 7.95, 10.2, 12.2	25
Speex	8,16,32,48	2.15 to 44.2	30

Note: Speex is an open source codec that is not restricted by patents.

minimally implemented in the Internet and a QOS study group concluded that increasing bandwidth is probably more practical then implementing a fully blown QoS system. The RSVP protocol was developed to support QoS.

- *Resource Reservation Protocol (RSVP)*: this protocol was designed to reserve resources across a network, to facilitate the compliance with required QoS parameters. It is rarely used today.

2.9.4 Network Management Protocols

The main Network Management Protocols are described next.

2.9.4.1 Domain Name Server (DNS)

The Domain Name System is used to assign domain names to groups of Internet users and maps those names to IP addresses by designating distributed authoritative name servers for each domain. It is the equivalent of an Internet phone book.

A domain name consists of one or more parts, called labels that are concatenated and delimited by dots. The hierarchy of domains descends from left to right, being each label to the left a sub-domain of the one on the right. A label can have 63 characters and the full domain can have 253 characters.

2.9.4.2 Simple Network Management Protocol (SNMP)

This is an UDP-based protocol used for network management and monitoring.

2.9.4.3 Remote Procedure Call (RPC)

This protocol executes a computer program subroutine in an address space in another computer.

2.9.4.4 Secure Shell (SSH)

This is a protocol that allows data to be exchanged using a secure channel between two networks.

2.9.4.5 Transport Layer Security (TLS)

This protocol provides security to Internet communications.

2.9.4.6 Session Description Protocol (SDP)

This is a protocol used to initialize streaming media parameters.

2.9.4.7 Session Initiation Protocol (SIP)

This is a protocol used to initialize multimedia communication sessions with voice and video.

2.10 The World Wide Web (WWW)

A WWW site is identified by a name and an IP address, which together are referred as the uniform resource identifier (URI). This identifier consists of two parts: the uniform resource name (URN) and the uniform resource locator (URL). The URL consists of a scheme name (protocol used), followed by : //, a host name (or its IP address), a port number, the path of the resource to be fetched or the program to be run, and optionally a ?query strip or an #anchor (place from where a page should be displayed. An example is showed below

URL sintax: scheme://
username:password@domain:port/path?query#anchor

As WWW is one of the most popular applications, it is important to consider what is a satisfactory performance for users utilizing this service. The following performance parameters are considered as guidelines to express user satisfaction waiting to access a web page:

- ideal response time: 0.1 s
- highest acceptable response time: 1 s
- unacceptable response time: 10 s

3

Market Modeling

3.1 Introduction

Detailed market modeling (MM) is essential for wireless designs as it provides information about network users, their location and traffic demand. It also details the operators' offering and matches it to the subscribers' demand. This data will then be used in dimensioning the network and evaluating its performance. A Wireless Design Planning Tool (WDPT) is essential to create and store this data.

In this book we use the following definitions for the relationship between the population and the wireless operating company:

- *Clients or customers*: defines people or businesses that can be potentially served by a wireless operator.
- *Subscribers*: defines people or companies that have subscribed to the service.
- *Users*: persons who use the service through a subscription, as there may be one or more persons using the same wireless subscription.

The business plan gives general guidelines about the target population and types of services to be provided. Market modeling has to detail this information so it can be used in the design process. Data for MM can be obtained from many different sources, such as statistical institutes, geographical services, tax departments, transit departments, commercial credit companies. This data is valuable, but comes in different formats and needs to be converted to the WDPT format.

Collected data does not provide all the information and the designer has to estimate many values. This can be done based on judgment alone, or be substantiated by small experiments that validate or guide the estimates. Because the analysis is statistical and a huge amount of data is considered, small imprecisions tend to even out and do not affect the final results.

The market should be modeled in terms of services, subscribers, applications, user equipment, voice and data traffic and wireless infrastructure. The main items are:

- service plans and their characteristics;
- potential clients (subscribers) and how they are distributed;
- applications that will be used by customers and its traffic;
- user equipments (terminals) used in the network;
- peak, hourly and daily traffic expectations.

LTE, WiMAX and WLAN Network Design, Optimization and Performance Analysis, First Edition.
Leonhard Korowajczuk.

Before MM can be done, we need to establish data traffic definitions and its representation in a clear way.

3.2 Data Traffic Characterization

Data traffic is complex and requires novel methodologies to be specified. The industry is familiar with the classical voice traffic definition, and it is only natural that the industry wanted to extend this concept to data traffic. The following sections explain why this extension cannot be done and which new procedures are required to express data traffic. Packet data is defined within the scope of adaptive modulation networks, the concept of tonnage is introduced and the confusion between different data rates is clarified.

3.2.1 Circuit-Switched Traffic Characterization

Circuit-switched networks establish fixed connections between a source and a destination that last for the duration of a call. Consequently the circuit is considered busy for the call duration.

Traditionally circuit-switched traffic was defined in Erlang (E) and represented the circuit's occupancy rate. This unit is named after Agner Krarup Erlang, a Danish engineer, pioneer in traffic engineering and queueing theory, with key publications in 1909 and 1917. Being a rate, the Erlang unit is a dimensionless quantity.

When a certain number of sources offer traffic to a network that has infinite resources, the average number of simultaneous busy sources over a period of time represents the offered traffic (demand) and is expressed in Erlang. For a single user, the occupancy rate of a single circuit represents its offered traffic in Erlang. A typical residential voice user generates traffic around 90 s per hour that is 90/3600 of the time or 0.025 E (25 mE).

In circuit switching, the main application, actually the only one to be dimensioned, was voice, which has always a similar behavior. User-offered traffic can be represented by a single occupancy parameter, expressed in Erlang. Because circuits had a constant offering, the only issue was how to accommodate user traffic in time.

A network limits the offered traffic to the carried traffic, which represents the average number of circuits that are simultaneously busy. Network traffic limitation is caused by congestion, and there are three ways congested calls can be modeled.

- Rejected calls go away and are not retried (Erlang B model or Erlang loss model).
- Rejected calls are retried within a short time span (Extended Erlang B model).
- Rejected calls are queued (Erlang C model).

Erlang developed mathematical equations (Erlang B and C) to model the above scenarios and proposed the Erlang distribution to model them. This distribution is a particular case of the Gamma distribution, which is used to model, among other things, the size of insurance claims and rainfall, and is a two parameters distribution, that has a scale parameter θ and a shape parameter k. Integer values of k result in the Erlang distribution, which is usually solved by recursion. For this reason, extensive tables were generated for ranges of sources and circuits. Those tables were published in books and are essential for circuit-switched traffic dimensioning.

3.2.2 Packet-Switched Traffic Characterization

Circuit-switched traffic is expressed in Erlang by the ratio of the occupancy of the circuit. In wireless systems, the equivalent to the circuit is the radio channel. Wireless 2G and 3G generations use a single modulation scheme, resulting in a constant throughput as long as the RF communication can be established. The information carried is voice and the network can be modeled as a regular circuit-switched network, with traffic expressed in Erlang. Even packet data can be similarly modeled as there are few data services available for this technologies and only one modulation scheme.

In 4G wireless broadband networks, however, the radio channel uses multiple coding schemes and is adaptive, so a single throughput number cannot be associated with a radio channel. This has serious implications in expressing user traffic, as it should be now expressed independently of the network capacity.

3.2.2.1 Traffic Volume

Traffic has to be expressed in volume, known as traffic tonnage. Tonnage is expressed in kB (thousand of Bytes) per time interval, or can be averaged in a constant stream of kbit/s (thousand of bit per second), defining the Data Tonnage Rate (DTR).

3.2.2.2 Traffic Latency and Jitter

Latency is the overall data transmission delay, measured from the moment data is offered to the wireless network to the moment it is delivered to the user. Applications can be divided into two groups in relation to latency:

- *Non-Real Time (NRT)*: includes applications that can tolerate larger latencies, such as web browsing, file transfer and e-mails. The only latency constraint here is user satisfaction. An acceptable value for this type of latency in a wireless network is 100 ms.
- *Real Time (RT)*: includes applications that require latency to be lower than a specified value. This is the case of VoIP (Voice over IP), movies, audio and gaming. The majority of RT applications accept a latency value of 75 ms, which includes the wireline IP network delay. An acceptable value only for the wireless network is 30 ms.

In addition to latency, jitter (data phase variation) is also important, but it only plays a role for latency values that are close to the tolerable limit, so keeping the latency within reasonable limits allows the designer to ignore jitter. The goal is to keep the latency at less than half of its limit value, including the need to accommodate retransmissions.

3.2.2.3 Traffic Data Error Rate

The provision of error-free transmission requires very large S/N (Signal to Noise Ratio) margins to compensate for fast fading deeps. This margin is required during very short periods, and it is more efficient to allow the occurrence of errors as long as they can be corrected on the receive side.

Data Error Rate (DER) is defined by the ratio between the number of errors and the total amount of data transmitted. It can be expressed in BER (Bit Error Rate), FER (Frame Error Rate), PER (Packet Error Rate) or BLER (BLock Error Rate) depending on which scale the analysis is performed.

Error correction can only be performed by the transmission of redundant information, which is then used on the received side to recover the corrupted data. The most common error correction technique is Forward Error Correction (FEC), in which redundant bits are added to each packet or data block, before transmitting the data. FEC efficiency depends on its size relative to the data sent, and this will define up to what data rate it can correct errors. Typical redundancy rates are one to two times the amount of data sent. These high redundancy rates are still more efficient than the increase in S/N ratios required to achieve the same overall throughput.

When the error rate is very low, FEC becomes inefficient, the residual errors can be corrected by resending the data, a process called Automatic Repeat Request (ARQ). In this case, it is essential to detect the presence of errors and this is done using a Cyclic Redundancy Check (CRC), also known as a polynomial check sum. CRC was proposed by W. Peterson in 1961, and it is done by a polynomial long division operation in which the remainder is used as CRC. An "n" bit CRC can correct single errors with a maximum length of n bits.

RF BER level is selected to optimize network throughput and it was found that a better throughput is generally obtained with higher BER levels. NRT applications generally target a BER of 10^{-2} while an RT of 10^{-3}, due to the more stringent latency requirements of RT. The final goal is to deliver both applications with a PER of 10^{-6}, which corresponds to a BER of 10^{-8} for 100 bit packets. This is achieved using FEC, CRC and ARQ techniques.

Some RT applications can live with low error rates, as is the case of voice and video; hence they usually do not apply further error correction (UDP/IP), to avoid increasing latency. NRT applications that carry numeric or alphanumeric data cannot have errors, and the final correction is left to the data stack, which use protocols that check the integrity of data, such as TCP/IP.

Wireless broadband networks are expected to carry hundreds of data applications, including IP voice. Each of them has different requirements and should be modeled separately. These applications can then be associated with users and an average tonnage calculated per user. The adaptive characteristic of data traffic allows for some simplification and similar users can be represented by their average behavior.

In data applications, the concept of a call is replaced by a session, which may last long periods and have large discontinuities, in which no data is sent. A large number of applications are traffic adaptive, so the demand (traffic offered) adjusts to the network capacity. This inherent adaptation provides continuity to a session, but may cause user dissatisfaction if it slows down the application use. Traffic is characterized statistically by distributions that represent the arrival rate and duration of an event.

Circuit-switched voice traffic is represented by a Poisson distribution for call arrival and a Rayleigh distribution for call duration. Packet Switched data sessions, however, have to be specified at session, burst and packet level. Sessions are defined by the length of time the user is connected to a destination, bursts represent periods of user activity and packets represent the actual data sent by users. Sessions and bursts usually have a Poisson arrival rate, but follow a long-tailed Pareto distribution for its duration. Packets generally follow a Poisson arrival with a Constant or Rayleigh distribution duration.

3.2.3 *Data Speed and Data Tonnage*

There is significant confusion in the industry about data transmission performance in wireless networks. This performance can be expressed according to different parameters that can be measured using specialized tools, which generate data packets at specific rates and detect the arrival rate. The confusion arises as some tools measure speed (instant data rate), whereas others measure tonnage (average data rate), and both parameters may be expressed in kbit/s. This confusion can be partially avoided if speed values are expressed in different units, like speed values in kbit/s and tonnage values in MB/hour. Both parameters have different values for incoming (downlink) and outgoing (uplink) data.

These parameters are illustrated in Figure 3.1 and listed below:

- *Data Transfer Speed (DTS)*: this represents the instantaneous rate with which data is transferred when it is scheduled to be transmitted.
- *Air Data Speed (ADS)*: a wireless network always transfers data at the maximum instantaneous data rate (speed) allowed by the RF channel, so it can maximize the channel throughput.

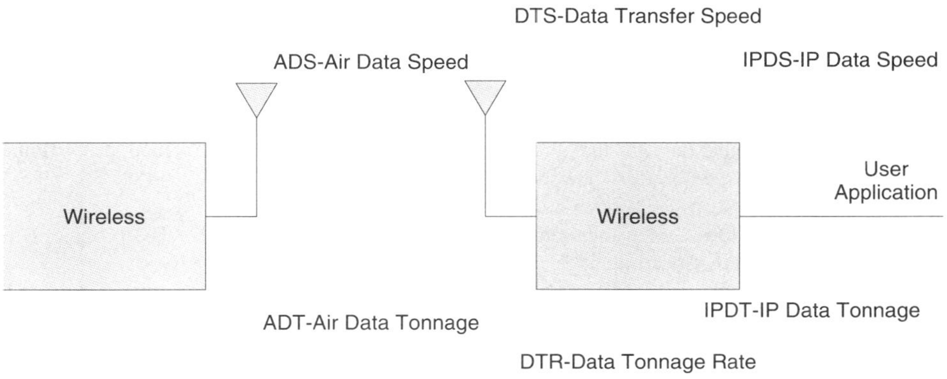

Figure 3.1 Data speed and tonnage parameters.

- *IP Data Speed (IPDS)*: the speed at which data is delivered to the user is defined by the IP network to which it is connected, even if it is the IP circuitry inside the radio. This speed is still measured at the air interface, but the wireless protocol overhead is discounted. Technology marketers publicize this figure for the best possible scenario, and many people believe that this rate reflects the quality of service provided by the network.
- *Data Tonnage Rate (DTR)*: this represents the amount of data transferred during a relatively long period of time, which can be specified as a quarter of an hour, half hour or one hour.
- *Air Data Tonnage (ADT)*: ADT is the amount of data actually transferred on the air interface, including the wireless protocol overhead, FEC and ARQ procedures. It is significantly larger than the IPDT, described below, as it includes the wireless protocol overhead. This overhead will be calculated in each technology and should be considered by the network designer when dimensioning the network.
- *IP Data Tonnage (IPDT)*: IPDT is the amount of IP data transferred by the user application. It can be evaluated by specialized applications that verify the amount of data that transferred over a period of time.

The parameter that best reflects user satisfaction is the IPDT, which is significantly lower than the ADS, although ADS is generally the figure published by operators.

The ratio between the IPDT and ADS is called the Wireless Overhead (WO) and should be considered in the design process. Typical OW values vary between 0.25 and 0.4, depending on the technology implementation and the interference level expected.

3.3 Service Plan (SP) and Service Level Agreement (SLA)

A service plan specifies the performance of the service offered to customers and is backed up by a more detailed service level agreement. The SP must be able to express the offering in a simple way that the public can understand. Generally this is done through plans, using different titles, as exemplified below:

- *Platinum Plan*: oriented towards small and medium businesses with up to 8 users. Recommended terminal types are rooftop or window mounted.

- *Gold Plan*: oriented towards small business and residences with more than one user. Recommended terminals are rooftop, window mounted and desktop.
- *Silver Plan*: oriented towards residential use, with desktops and laptops.
- *Silk Plan*: oriented towards portable use, with laptops, multimedia players, palmtops and phones.

An SLA provides technical data for the above plans. Transfer Speed (IPDS), Average Daily Tonnage (IPDT) and Quality of Service (QoS) are the parameters used to define service.

A suggestion for an SLA is shown in Table 3.1.

Many operators like to use as a marketing number the IPDS (IP Data Speed). The problem is that this number varies with the location, so it can only be expressed statistically.

A broadband spectrum of 10 MHz supports one 10 MHz TDD channel or two 5 MHz FDD channels. Considering today's technologies, the IPDT will be about 12 Mbit/s for the spectrum. For TDD considering a 1/3 UL/DL ratio, we get for the downlink 9 Mbit/s and for the uplink 3 Mbit/s. For FDD, we will have 6 Mbit/s for both. This gives a clear spectrum use advantage to TDD.

Table 3.2 gives the maximum IPDT for different Over-Subscription Ratios (OSR), which correspond to different service plans.

A wireless service has a different IPDT in different geographical areas and this should be expressed statistically as exemplified in Figure 3.2.

The SLA tonnage for the previous example is calculated in Section 0. Tonnage can be expressed in MB (10^6 bytes) or MiB (1.024×10^6 bytes).

$$1\,\frac{\text{MB}}{\text{hour}} = 2.22\,\frac{\text{kbit}}{\text{s}}$$
$$1\,\frac{\text{MiB}}{\text{hour}} = 2.27\,\frac{\text{kbit}}{\text{s}}$$

Table 3.1 Example of a Service Level Agreement

	IPDS		IPDT		QoS- Maximum application latency (ms)		RF BER		User PER
	Target transfer speed (kbit/s)		Average target daily tonnage (MB)				Bit error rate		Packet error rate
Service plan	Incoming	Outgoing	Incoming	Outgoing	RT	NRT	RT	NRT	RT & NRT
Platinum	900	300	800	182	30	100	10^{-3}	10^{-2}	10^{-6}
Gold	450	150	200	46	30	100	10^{-3}	10^{-2}	10^{-6}
Silver	225	75	100	23	30	100	10^{-3}	10^{-2}	10^{-6}
Silk	125	37.5	50	11	30	100	10^{-3}	10^{-2}	10^{-6}

Table 3.2 IPDT per user exemplified for different service plans

	10 MHz TDD 75% Downlink			10 MHz TDD 25% Uplink		
Plan	MSTR (Mbit/s)	OSR	IPDT (Mbit/s)	MSTR (Mbit/s)	OSR	IPDT (Mbit/s)
Platinum	9	10	0.90	3	10	0.30
Gold	9	20	0.45	3	20	0.15
Silver	9	40	0.23	3	40	0.08
Silk	9	80	0.11	3	80	0.04

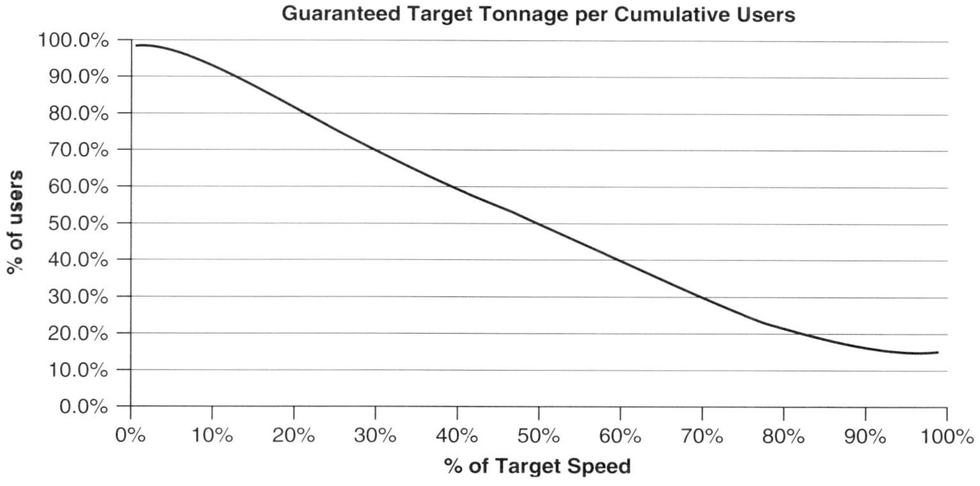

Figure 3.2 Guaranteed target tonnage (IPDT) per cumulative users.

Even if a formal SLA does not exist, the designer must determine performance parameters to guide the design process.

3.4 User Service Classes

A wireless network can have thousands of users and it is not possible to represent the traffic of each one of them. Fortunately, users can be divided into groups that present a similar behavior and can be analyzed together. These groups should be identified and characterized in User Service Classes (USC).

The factors that characterize a group of users are:

- *Similar service characteristics*: defined by applications used, their traffic demand and QoS (Quality of Service) required. Traffic demand is influenced by the type of user equipment.
- *Similar user equipment*: defined by the radio type and antenna location (indoor, outdoor, in car, hand-held, desktop, ground level, building floor...).
- *Similar RF environment*: defined by environmental attenuations (penetration, nearby obstruction, rain ...) and fading characteristics.
- *Similar traffic characteristics*: defined by type of user (business, residential, professional, youth, adult ...) and service plan.

A service class is then defined by a Service, User Equipment (User Terminal), RF Environment and Traffic Distribution Grid (TDG). The number of SCs configured for a given network should be kept as low as possible, but still enough to characterize the multiple types of users.

User traffic cannot be expressed by a simple parameter and the best way to express traffic in a wireless broadband network is to calculate the number of users of each SC according to each service type.

3.5 Applications

Applications software is essential to run data-based solutions. They are based on text or multi-media data and each one generates different amounts of traffic (tonnage).

3.5.1 Application Types

There are hundreds of different data applications that can be used, and the following list presents some of the most common ones. Voice, in VoIP format, also becomes a data application:

- Web browsing (NRT)
- E-mail (NRT)
- Instant Messaging/Skype (NRT)
- Micro blogging (social networking) (NRT)
- Infrastructure (NRT)
- Tunneling (VPN) (NRT)
- Online gaming (RT)
- Peer to peer (NRT)
- Audio download (RT)
- Video download (RT)
- Video streaming (RT)
- Remote meeting (NRT)
- File sharing (NRT)
- VoIP (RT)

Some of these applications are Real Time (RT) and have stringent latency requirements, while others are Non-Real Time (NRT) and do not have such stringent requirements. Applications do not carry QoS information, nor do the IP packets. It is the IP data protocol used to carry the data that is considered by the wireless network as defining the QoS requirements of each application.

3.5.2 Applications Field Data Collection

Each application should be modeled statistically, by collecting data about it. An application is defined by periods of activity, called sessions. Inside each session there are several bursts of activity, each composed of packets. Sessions, burst and packets are defined by a distribution, inter-arrival time and event length. The possible statistical distributions are: constant, Poisson, Rayleigh, exponential and Pareto. These statistics can be obtained by monitoring actual user activity using a link monitor that records usage for different network protocols, as illustrated in Figure 3.3 for a single user and Figure 3.4 for a small enterprise.

Each of these applications has different traffic patterns, for example, web browsing is one of the most common applications today and is heavy in incoming traffic but has little outgoing traffic.

E-mails have a similar traffic pattern, which will vary according to the terminal type, as some terminals do not allow attachments (or restrict them), mainly because they cannot deal with them. All these different patterns can be represented through one of the statistical distribution algorithms mentioned previously.

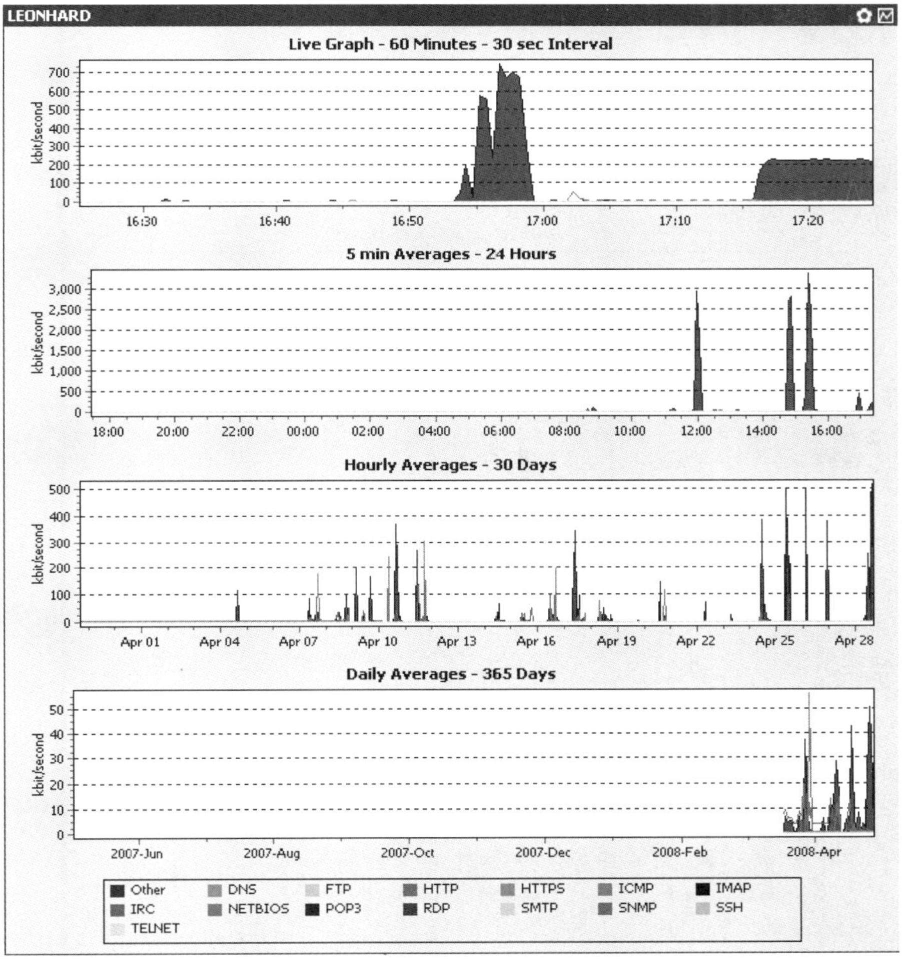

Figure 3.3 Single user traffic statistics.

3.5.3 Application Characterization

To organize and guide explanations, this section illustrates the characterization of an application using actual screens from a design tool. It is important to stress that these are unconstrained traffic specifications, limited only by network throughput, which should be dimensioned properly so as not to affect user needs.

An example of a web browsing application is shown in the planning tool dialog in Figure 3.5 for the session level, Figure 3.6 for the burst level and Figure 3.7 for the packet level. In this example, the service is identified as Web Surfing Unconstrained, with NRTPS (non-real time Polling Service)

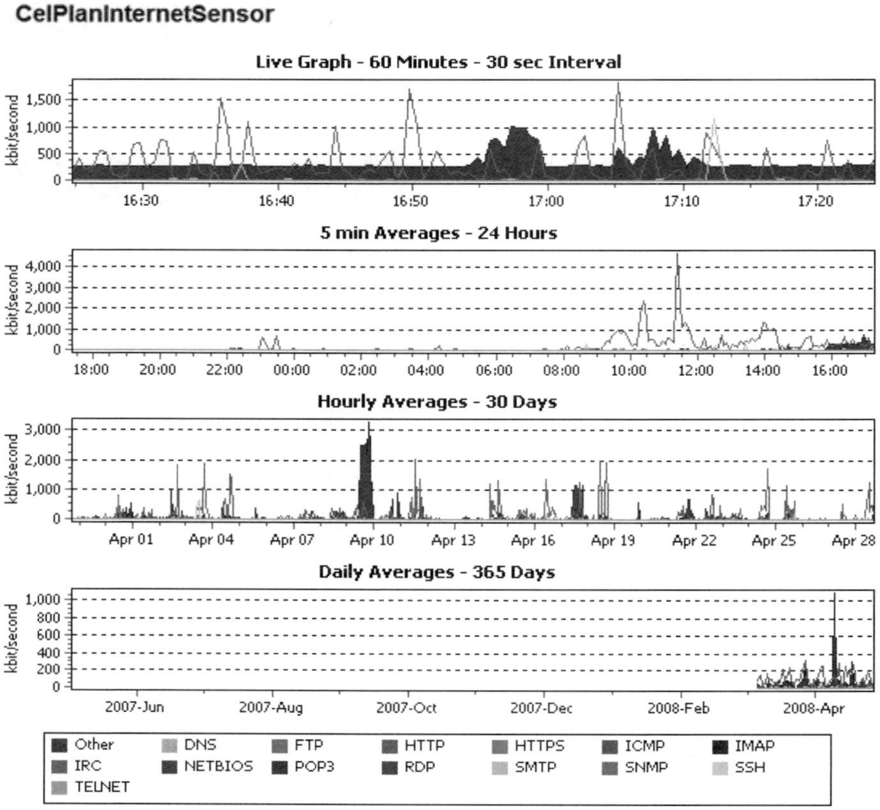

Figure 3.4 Small enterprise traffic statistics.

as QoS (Service Type in the figures). The QoS defines the frame resource allocation methodology and is described in detail later for each technology.

Because multiple types of services must be considered in the traffic simulation of the network, the Service Priority must somehow be defined; in this example, the Traffic Weight field is used. This priority determines which services are allocated first by the network, according to scheduling procedures.

This example defines web surfing in three levels: session, burst, and traffic. A session is defined by the whole period in which a user is engaged with a terminal; during this time, the user may be active, idle with allocated resources, or dormant without allocated resources. A burst is the part of the session in which the user has resources allocated, regardless of being active or idle. A packet is the part of the burst where the user is actively transmitting or receiving data.

Session, burst and packet statistics are specified in the traffic part of the dialog box. Input parameters are entered by the designer and dependent parameters are calculated using formulas specified in Table 3.3.

The traffic simulation also needs to consider how long it takes for the network to release resources after a user becomes inactive. This is usually defined by release timer and set-up delay values, which are considered in relation to burst establishment.

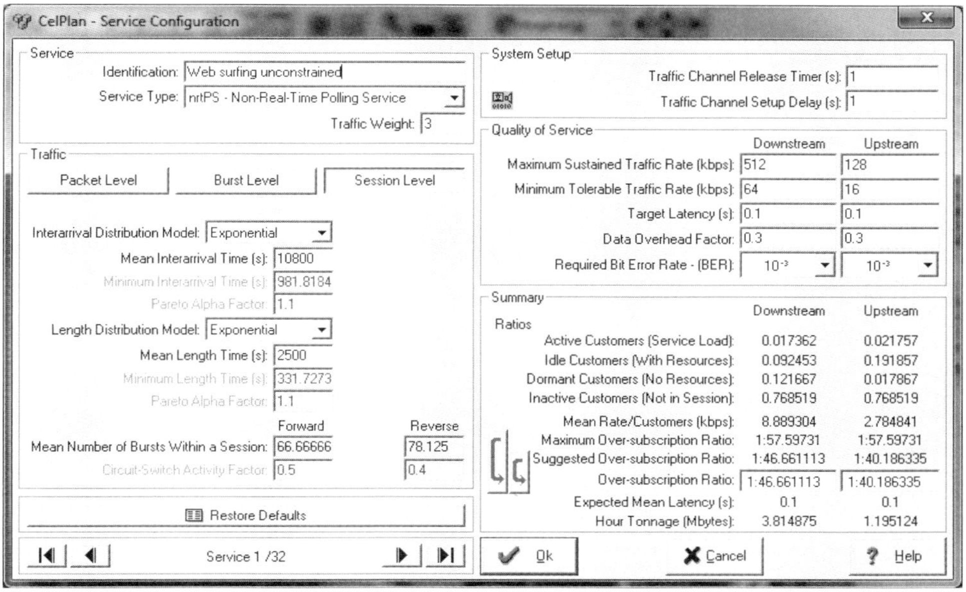

Figure 3.5 Web browsing application characterization – session level.

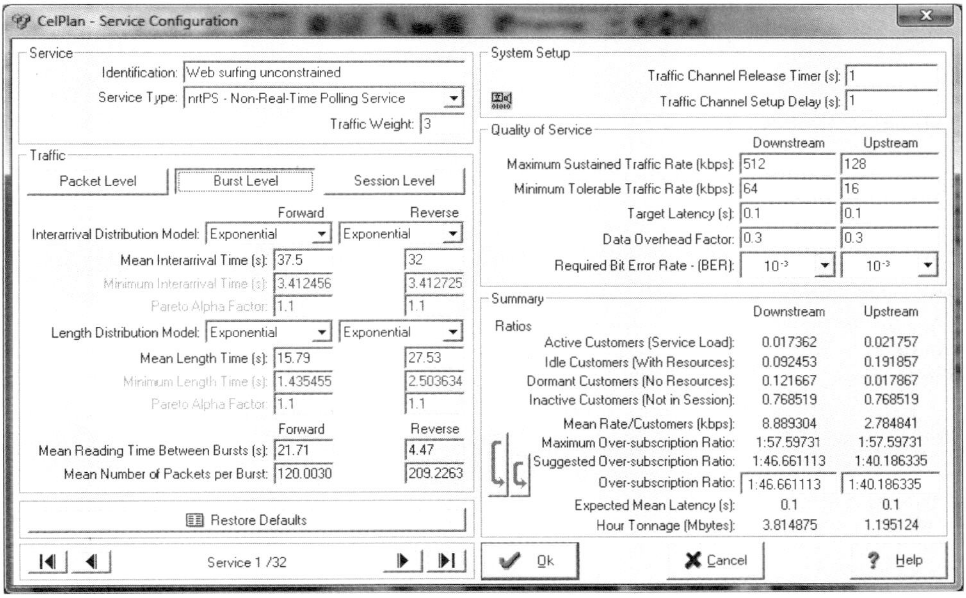

Figure 3.6 Web browsing application characterization – burst level.

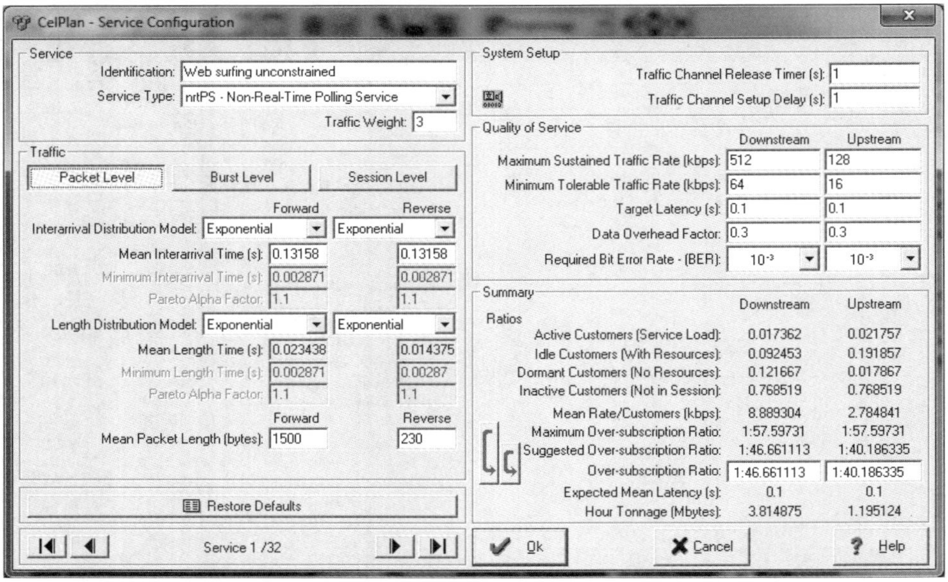

Figure 3.7 Web browsing application characterization – packet level.

Services can have multiple QoS requirements and the following is a list of the main parameters that should be considered:

- *Maximum sustainable traffic rate (MSTR)* specifies the maximum data rate that the service can sustain (provide data) and is limited by the application, the wireline or the wireless network, whichever is smaller. When the limitation is caused by the wireless portion, it corresponds to IPDT and should be confirmed when KPI (key parameter indicator) values are calculated, after performing network traffic simulation.
- *Minimum tolerable traffic rate (MTTR)* specifies the minimum tonnage required by an application to provide service and if the data rate falls below this level, the session is dropped. For applications that require a constant rate, this value is equal to the MSTR.
- *Maximum latency (target latency)* specifies the maximum acceptable latency for the service.
- *Data overhead factor* specifies the overhead imposed by the wireless protocol. The overhead should consider FEC, MAC and ARQ and varies per technology. These calculations are done when describing the technologies in Chapters 12 to 14.
- *Target Bit Error Rate (Required Bit Error Rate)* is the error rate targeted by the FEC correction. Any remaining errors are corrected by the ARQ process and by the TCP/IP protocol.

These parameters determine whether the system can provide service at each location, thus they should be carefully considered by network designers. The Mean Rate per Customer can be calculated based on these service statistics; and, with this rate, the ratio of active, idle (with allocated resources), dormant (without allocated resources), and inactive users is calculated.

The over-subscription ratio (OSR), described next, can also be estimated based on the above parameters. The usage of the maximum ratio possible leads to extremely long delays, thus, designers must calculate a ratio that satisfies the desired latency using queueing theory. This value is mainly

Table 3.3 Service configuration parameters

Service configuration	Acronym	Unit	Equation	Downlink	Uplink
Packet Level					
Mean Inter-arrival Time	PMIT	s	input	0.1316	0.1316
Mean Length Time	PMLT	s	MPL*8/MSTR/1000	0.0234	0.0146
Mean Packet Length	MPL	bytes	input	1500	230
Packet Delivery Rate	*PDR*	*packet/h*	*BMNP*BDR*	*2667*	*5449*
Burst Level					
Mean Inter-arrival Time	BMIT	s	input	37.5000	32.0000
Mean Length Time	BMLT	s	input	15.7900	27.530
Mean Reading Time Between Bursts	MRTBB	s	BMIT-BMLT	21.7100	4.47000
Mean Number of Packets per Burst	BMNP		BMLT/PMIT	120.003	209.226
Burst Delivery Rate	*BDR*	*burst/h*	*SDR*SMNB*	*22.2222*	*26.0417*
Session Level					
Mean Inter-arrival Time	SMIT	s	input	10,800	10,800
Mean Length Time	SMLT	s	input	2500	2500
Mean Number of Bursts per session	SMNB		SMLT/BMIT	66.6667	78.1250
Session Delivery Rate	*SDR*	*sessions/h*	*3600/SMIT*	*0.3333*	*0.3333*
System Set-up					
Traffic Channel Release Time	TCRT	s	input	1	1
Traffic Channel Set-up Time	TCST	s	input	1	1
Quality of Service					
Maximum Sustained Traffic Rate	MSTR	kbps	input	512	128
Minimum Tolerable Traffic Rate	MTTR	kbps	input	64	16
Target Latency	TL	s	input	0.1000	0.1000
Data Overhead Factor	DOF		input	0.3000	0.3000
Required Bit Error Rate	BER		input	0.0010	0.0010
Summary					
Active Customers (Service Load)		AC	MRPC/MSTR	0.0174	0.0218
Idle Customers (with resources)		IdC	(((BMLT+TCRT+TCST)* SMNB)/SMIT)-AC	0.0925	0.1919
Dormant Customers (No resources)		DC	1-(InC+idC+AC)	0.1217	0.0179
Inactive Customers (not in session)		InC	1-SMLT/SMIT	0.7685	0.7685
Mean Rate/customer	MRPC	kbps	SMNB*BMNP*MLP*8/ SMIT/1000	8.8891	2.7848
Maximum Oversubscription ratio			MSTR/MRPC	57.599	45.963
Suggested Oversubscription ratio				46.66	40.18
Expected Mean Latency				0.1	0.1
Hourly Tonnage		MiB	MRPC*3600/8/1024	3.9063	1.2238

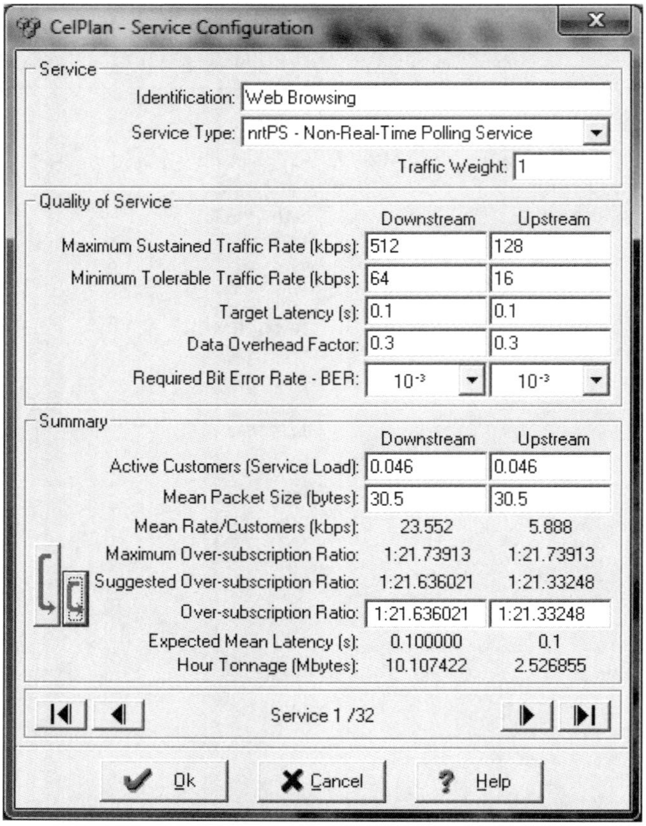

Figure 3.8 Application or service group characterization – simplified dialog.

informative, as a reliable design tool allocates users dynamically, mainly because the MSTR is a target value and is calculated by the tool during the traffic simulation process.

Instead of performing a detailed traffic analysis, where one must choose distribution models and define statistical parameters (Table 3.3) for session, burst, and packet levels, designers might choose to simplify the analysis and determine only a mean packet size for each service type, along with the active customers' ratio (Figure 3.8).

To define the throughput required by an application, it is necessary to consider the terminal type being used. Data throughput varies not only with the application but also with the terminal used, for example, a web service has a larger throughput on a desktop PC than on a palmtop, because of processor speed, and ease of use (i.e. users take longer to input data and do not browse for long periods when using smaller terminals). This relationship will be described later in this chapter.

3.6 Over-Subscription Ratio (OSR)

In traditional circuit-switched voice networks, the ratio between the total number of users and the active users is called the over-subscription ratio (OSR). Circuit-switched voice has a relatively constant traffic per user, typically 0.025 E (Erlang) per residential user. Considering that one circuit can carry one Erlang, mathematically 40 users would fit in a circuit and the OSR would be 1. In practice, users have to be accommodated in time and this leads to large blockages if such a high OSR is used. Data can be queued before being sent and we can use Erlang C formula to calculate the OSR value for a certain queuing time.

In wireless broadband, both factors of the OSR ratio are variable, as each user group has different traffic characteristics and each network radio has different and time variable capacity. OSR calculation cannot be done in the traditional way, but the industry is demanding that somehow this factor should be expressed. One of the difficulties is the traffic dissimilarity between users and this can be solved by expressing OSR within uniform groups of users, referred to as service classes (SC) in this book.

The traffic capacity (tonnage) of a radio depends on the modulation schemes that can be allocated to users. Each service class will be allocated a maximum sustained traffic rate (MSTR), which will correspond to the instantaneous IPDT allocated for each user.

User tonnage can be calculated based on service usage statistics, thus allowing calculation of the OSR for different latency times.

Increasing IPDT or its equivalent MSTR becomes one of the designer's goals.

3.7 Services Summary

The large number of applications makes it impractical to consider all applications separately, so it is recommended to group them according to their QoS, and similar QoS groups can be grouped together. This leaves us mainly with two types of services: real time (RT) and non-real time (NRT). When real time services represent a small fraction of the total, they can also be grouped together with non-real time ones, as by having a larger priority they will be scheduled first and the latency issue will not be a concern.

When grouping multiple services into just a few categories, it becomes very difficult to generate detailed statistics for a mix of applications, so a simpler approach as in the example of Figure 3.8 is usually the best choice. In this case, the tightest QoS parameters should be extracted from each individual application, to guarantee that all of them can be served appropriately.

3.8 RF Environment

The RF path loss provides the average loss value from the transmitter to a given location. The actual received signal, however, is influenced by several other environmental factors that define the RF channel to this location:

- *Human body attenuation*: the human body can block RF energy directed to the radio, depending on the type of user terminal and its position in relation to the user himself.
- *Penetration attenuation*: users can be at different locations indoors and the RF signal is impacted by walls and furniture in the propagation path.
- *Rain precipitation*: mainly impacts frequencies around 10 GHz and above.
- *Shadow fading*: path loss is calculated as an average to a pixel (square or rectangular area to be predicted), but there are signal variations even within each pixel, which characterize the shadow fading. These variations can be obtained from measurements. The pixel resolution is defined by its latitudinal dimension; typical resolutions are 3 m, 10 m, and 30 m. The longitudinal dimension varies with the geographical latitude.
- *Multipath fading*: this is the signal variation effect modeled by several channel models that predict fading for specific conditions. In real life, these conditions change constantly and should be modeled

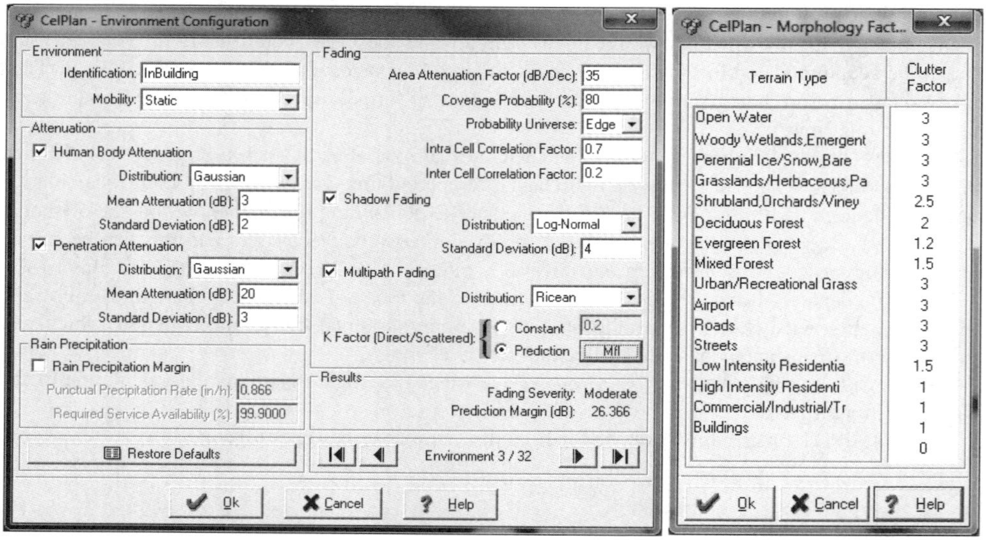

Figure 3.9 Sample dialog box for user environment configuration.

on a per pixel basis. The planning tool used as an example in this book uses the k factor prediction to estimate the ratio of the direct signal to scattered signals according to nearby surroundings. In this prediction, the channel is modeled with a Ricean distribution with its respective k factor, which can make it behave like a Rayleigh channel (totally non-line of sight) on one extreme, to Gaussian (full line of sight).

All these factors are statistically defined and are used together in combination. The tool used in the example displays the average prediction margin obtained from these parameters as a sanity check for users. Figure 3.9 shows the environmental characteristics configured in a planning tool.

3.9 Terminals

Terminal is a generic name for the equipment used by the end user, composed of an IP modem, a radio, and an antenna (which can be packaged together or separately).

3.9.1 Terminal Types

Typical terminal offerings are listed below:

- *Rooftop terminal*: the deployment of the antennas is done on buildings' rooftops. The RF part is, in most cases, integrated to the antenna or just below it. The modem can be on the rooftop or indoors. Antennas use narrow beams and can be pointed precisely at the transmitter. Typical antenna beamwidth is 15°.
- *Window terminal*: the antenna is mounted on a window, preferably one facing the nearest base station. Typical antenna beamwidth is 30°.

- *Desktop terminal*: the antenna is placed on a desktop and it still needs to be adjusted to the best azimuth in respect to the transmitter. Antenna beamwidth can be as large as 45°.
- *PC card/USB terminal*: omni antenna integrated to a PC (personal computer), a PC card, or USB (Universal Serial Bus) device that can be placed anywhere.
- *Portable Multimedia Player (PMP)*: omni antenna is integrated into the terminal.
- *Palmtop or hand-held*: similar to a PC, also with integrated omni antenna, but the terminal itself has a much less efficient processor and user interface.
- *Phone*: also with omni antenna, but designed for voice, usually with a very rudimentary data (text) interface.

A common misconception is to think that all terminals are used at ground level, when, quite often, users are located on different building floors. RF propagation varies significantly with height, and, to properly represent the network, representative heights of the market should be chosen and modeled in different SCs.

3.9.2 Terminal Specification

The customer terminal must be configured in a way that the main installation characteristics are defined, such as transmit and receive losses/gains and antenna parameters. The antenna height above ground in Figure 3.10 should be determined and considered in the terminal specification.

The customer terminal dialogue defines installation characteristics such as transmit and receive losses/gains and antenna parameters. The antenna height above ground is included in this dialog. It also defines the radio model used by the terminal. The terminal configuration dialog is shown in Figure 3.11.

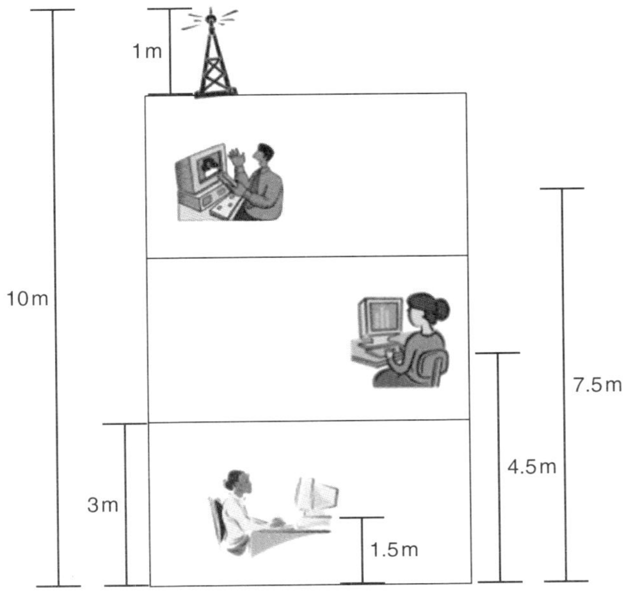

Figure 3.10 User terminal height above ground.

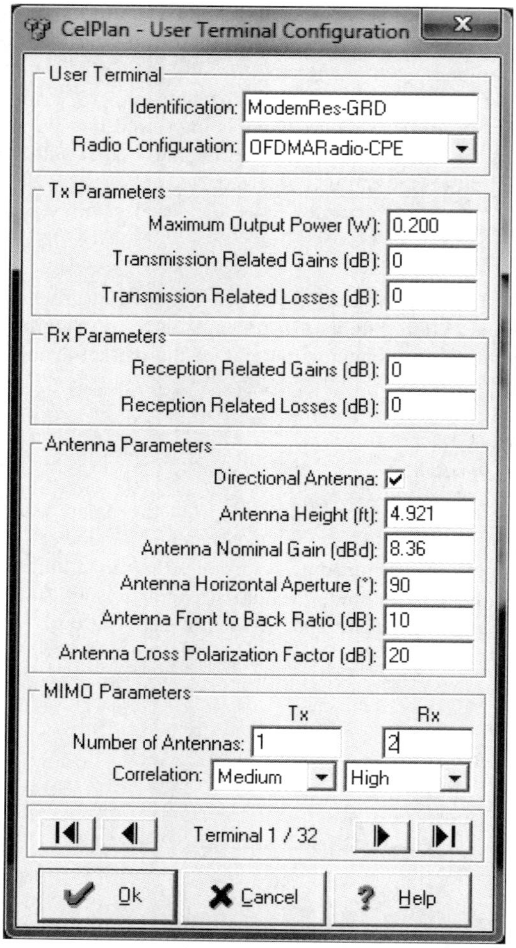

Figure 3.11 Sample dialog box for user terminal configuration.

3.9.2.1 Radio Configuration

The radio model has many parameters that must be defined for proper simulation by the planning tool. The following list shows the main parameters that must be considered for a proper simulation. Figure 3.12 illustrates the configuration of these parameters in a screen from the planning tool used as an example.

- technology standard
- main standard characteristics definition
- modulation schemes supported
- permutations supported
- frame structure

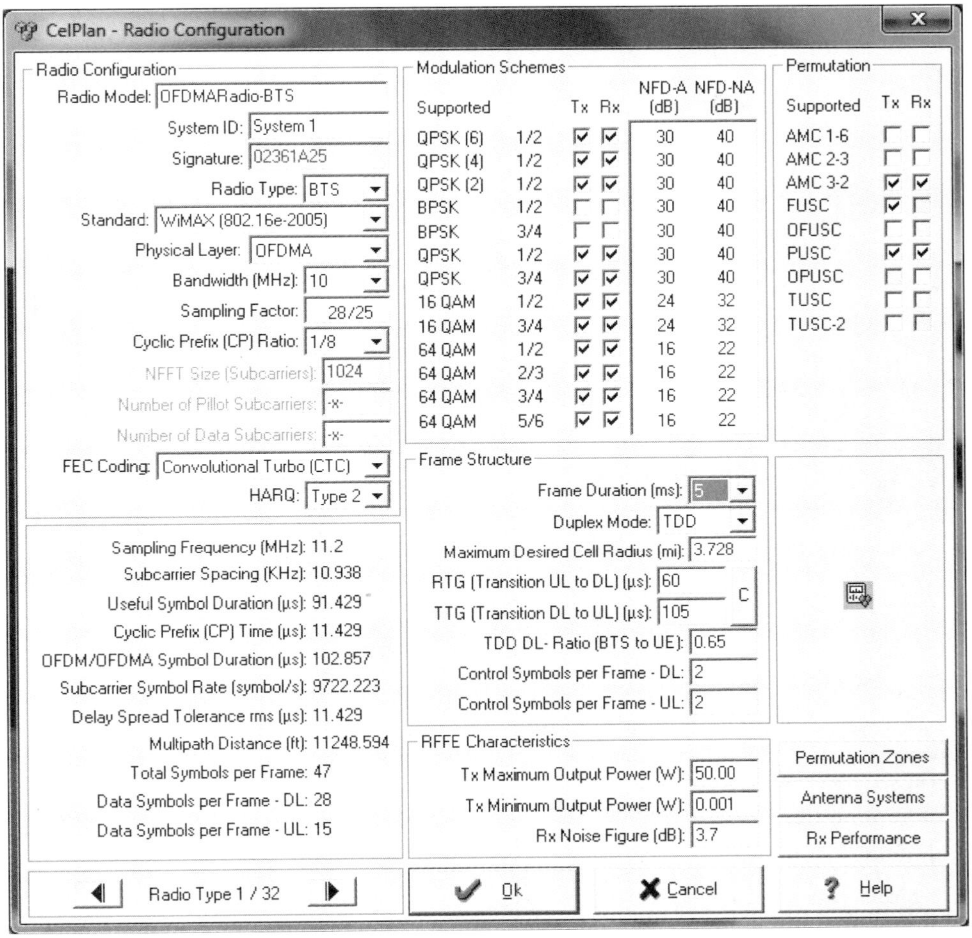

Figure 3.12 Sample dialog box for user terminal radio configuration.

- RFFE (RF front end) characteristics
- supported antennas systems
- RX performance

3.9.2.2 Permutation Zones Configuration

Certain technologies support zones configuration to allow greater reuse of channels closer to the center of the site coverage area, and lower reuse on the periphery (Figure 3.13). WiMAX (802.16e and above) supports zones in its specification; LTE does not refer directly to zones but gives vendors freedom to implement similar procedures to control resource coordination; as a formal implementation strategy has not been defined by the standard, planning tools can model this by applying the zones concepts also to LTE.

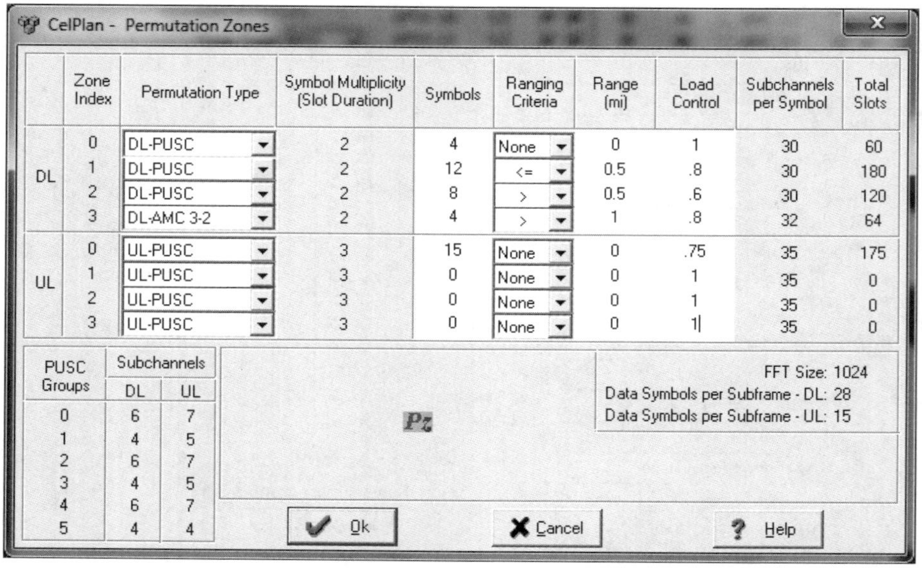

Figure 3.13 Permutation and zones configuration.

3.9.2.3 Antenna System Configuration

Antenna systems are the different MIMO options supported by a given radio. The following is a list of the main techniques used today. Figure 3.14 shows a typical screen for the configuration of supported antenna systems, breaking down the main categories in the most commonly used types:

- RX Diversity
- TX Diversity
- DL Spatial Multiplexing
- Adaptive MIMO Switching
- UL Collaborative Spatial Multiplexing
- Advanced Antenna Steering – Beamforming

3.9.2.4 Performance Specification

All these configuration options involve the issue of radio link performance, that is, what throughput can be achieved at each given CINR (Carrier to Interference and Noise Ratio). Receive sensitivity can also be used to represent performance by adding the CINR to the noise floor of the radio. A performance dialog example is shown in Figure 3.15.

Performance should be defined for each of the different modulation schemes (13 are supported in this example), for different channel fading models (4 ranges in this example), and different BER (Bit Error Rate) requirements (5 in this example).

Performance figures should be calculated for different FEC (Forward Error Correction Codes) gain or loss, mobility, HARQ (Hybrid ARQ), and permutations. The effect of the antenna system on throughput and CINR, for different antenna correlations (4 ranges in this example), has also to be considered.

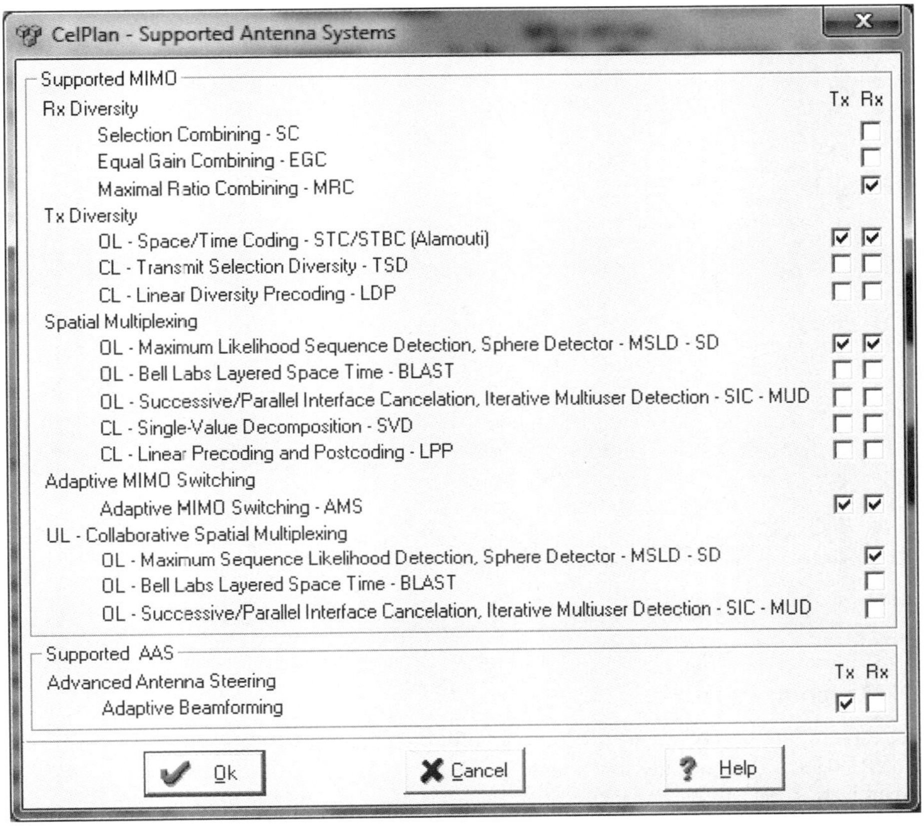

Figure 3.14 MIMO and antenna steering techniques.

This gives thousands of combinations which should be sampled by designers to perform a sanity check of the tool assumptions. In the tool used as an example, CINR gains and losses are given by default tables provided by the software, which can also be edited by the designer. The following is a list of the base tables used in this example. The tables themselves might be presented in different formats or combinations depending on the planning tool being used; the parameters they represent, however, must be somehow included in radio performance calculations:

- base CNIR
- FEC
- mobility
- permutation
- HARQ
- MIMO
- RX diversity
- TX diversity
- DL spatial multiplexing
- UL collaborative MIMO

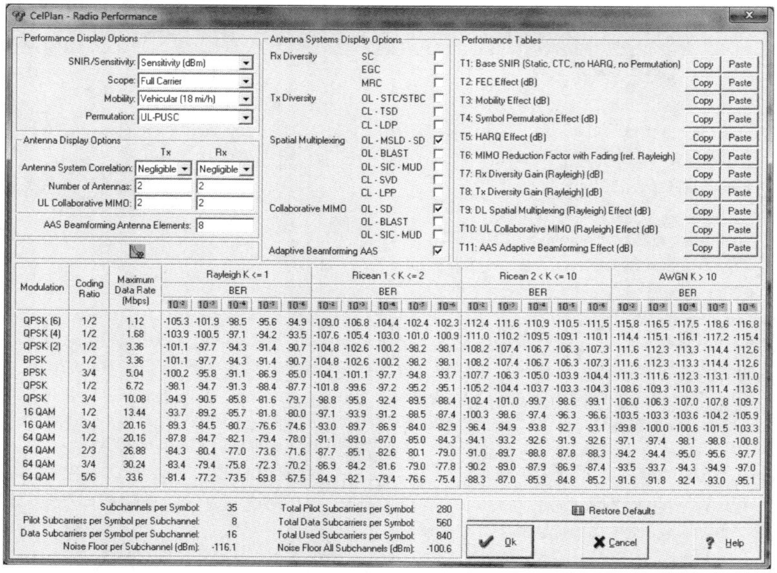

Figure 3.15 Sample table for RX performance analysis.

3.10 Antenna Height

The location of the subscriber antenna is essential to determine its RF received signal strength information (RSSI) and consequently its throughput capability. Antenna height, specifically, is one of the important factors, because the RF path may change substantially with height, by clearing obstructions, and, thus, affecting signal level and interference. RF predictions should consider every height by calculating the propagation path at each height. Simple, height-related, power correction factors are deceiving and do not provide a realistic outcome, as signal improvement is not linear with height.

It is impossible to predict points for every single height, thus users are grouped into representative heights, which are then predicted for all points in the area. Generally, between two and four heights are chosen to represent an area, with ground level as one of them. Other heights are selected according to the number of floors for buildings in the area. Three heights are usually enough, and users are bundled within these heights.

3.11 Geographic User Distribution

Users are distributed non-uniformly over the entire target area. Fixed terminals can be assigned to a specific location but portable terminals can move from one location to another during different hours of the day. User location has to be statistically represented and some aspects of this representation are discussed next.

3.11.1 Geographic Customer Distribution

Once a subscriber's traffic statistics are known, this traffic must be geographically located according to the hour of the day. A design done for a single peak hour does not fully exercise the network and

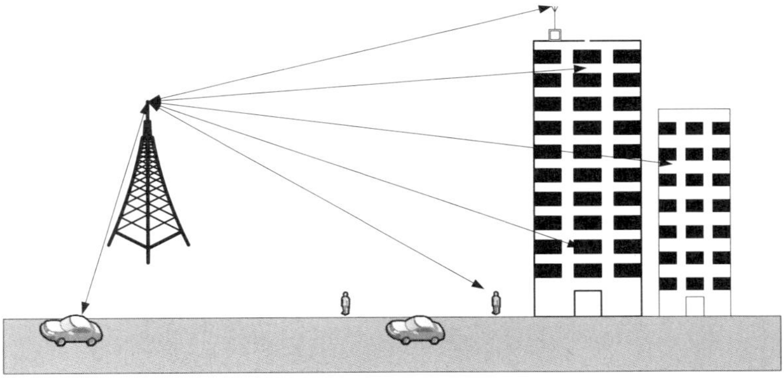

Figure 3.16 Customer distribution in different environments.

may leave many areas unanalyzed. It is recommended to analyze network performance for at least two peak hours. Traffic grids are used for this purpose and are organized in layers, each representing a uniform set of users.

3.11.1.1 Distribution of Customers

Geographically, traffic can be distributed indoors, outdoors, and inside vehicles. Indoor traffic can also be vertically distributed between building floors.

Although the exact customer distribution is unknown, the network has to be designed to accommodate them throughout the whole area of interest (AoI) or target area (TG). The user base should be defined not only in numbers of users but also by their geographical distribution (horizontal, vertical, and encapsulated), as illustrated in Figure 3.16.

3.11.1.2 Customer Horizontal Distribution

Quantitative horizontal customer distribution is defined by regions, generally obtained from the local Geographic Census Bureau (GCB). GCBs specify geographical polygons (regions) with a given set of attributes, such as population, households, and SMEs. Marketing plan assumptions can then be used to estimate the number of customers within each region.

As previously mentioned, users are not uniformly spread within a region and have to be further distributed for a more precise location. Morphology data specifies the type of clutter existing in an area, such as vegetation, buildings, and flat areas. This data can be used to distribute customers within each region, for example, business customers are in built-up areas, whereas users in vehicles are on streets. This additional distribution is very important as it concentrates traffic in certain areas, impacting cells' load.

Census Bureau regions are shown in Figure 3.17 with their respective attributes. Figure 3.18 shows horizontal user distribution after converting region attributes into users and spreading them according to morphological proportions.

3.11.1.3 Customer Vertical Location

Indoor customers can be above ground level and this has to be represented when modeling the network, as this distribution has a large impact on the network design. Generally, two to four height levels are

Figure 3.17 Horizontal distribution of customers (regions).

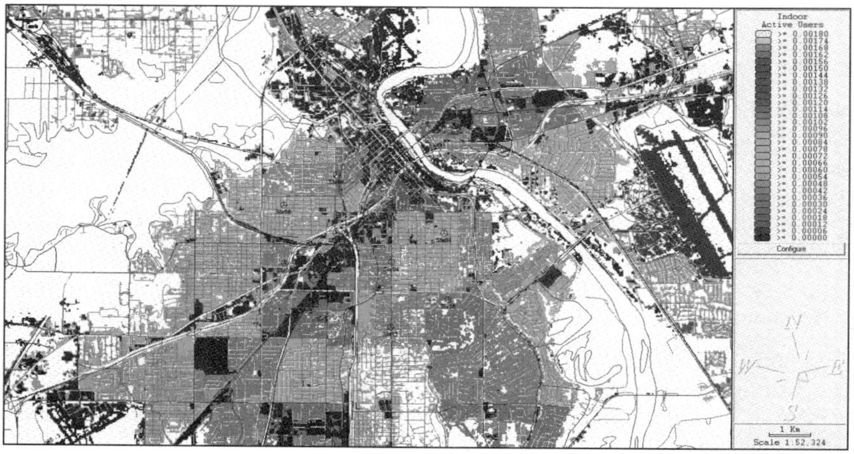

Figure 3.18 Horizontal distribution of users after spreading by morphology.

sufficient. The following list gives a set of representative heights that could be selected to model a target area containing tall buildings:

- ground floor (1.5 m)
- intermediate low floor (10 m)
- intermediate high floor (30 m)
- top floor (60 m)

Customers in vehicles can be also above ground, and this has to be represented mainly when there are long elevated highways or bridges.

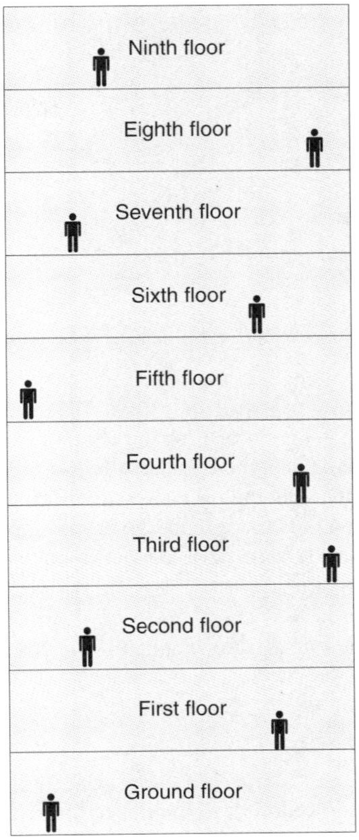

Figure 3.19 Vertical distribution of customers.

Vertical location of customers is a sub-set of the horizontal distribution, usually done by intersecting building height layers with the horizontal distribution of customers, that is, there should be no customers located on the ninth floor in a rural area. Distribution factors must then be applied to each height so as not to count the same customer many times, for example, if using the four heights given in the prior example, designers can define that 10% of the total number of users are located in the top floor, 20% on high floors, 30% on low floors, and 40% at ground level.

The vertical distribution of customers is depicted in Figure 3.19. The RF signal has to be predicted for each location as a ground level prediction does not apply to users at higher elevations. It is expected that the majority of users of a 4G network will be in elevations above ground.

3.11.1.4 Customer Encapsulation

Customers can also be classified according to the encapsulation of their equipment's antennas; the most common antenna encapsulation types are listed below:

- *rooftop*: customer's antenna is on the rooftop of a house or a building;
- *outdoor*: customer's antenna is outside any construction;

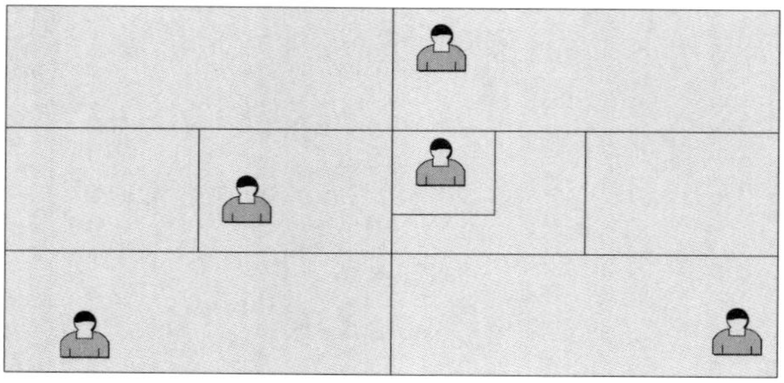

Figure 3.20 Customer encapsulation.

- *shallow indoor*: customer's antenna is on or near the window;
- *deep indoor*: customer's antenna is anywhere inside a house or building;
- *enclosed indoor*: customer's antenna is indoor enclosed by RF obstructions, like an elevator;
- *underground*: customer's antenna is below ground level, as in a garage;
- *in-car*: customer's antenna is inside a car. An antenna mounted on the outside of a car body is considered an outdoor antenna.

Encapsulation is usually represented as a factor applied to horizontal and vertical distributions. The encapsulation is illustrated in Figure 3.20.

3.11.1.5 Customer Movement

Customers should also be classified according to their speed in relation to the environment:

- *fixed*: a customer at a permanent fixed location, which may allow use of directional antennas;
- *nomadic*: customer at a fixed location while using the service, but can move between each network access;
- *mobile*: customer moves while using the service and the speed impacts fading characteristics (Doppler Effect and fading time). Several speeds may be considered.

3.11.1.6 Customer Terminal

Customers can use different terminal sets, such as portable phones, laptops, outdoor CPEs, and indoor CPEs, each with a different radio and antennas that have to be characterized to properly model the network.

3.11.2 Customer's Distribution Layers

The distribution of customers does not express traffic directly; instead, it generates traffic grids, which are then used to calculate traffic values for each service. Traffic grid layers have to be created according to their location and height above ground. Some examples of traffic layers are listed on the next page:

- outdoor pedestrian;
- indoor ground level (1.5 m);
- indoor third floor 10 m above ground;
- indoor tenth floor 30 m above ground;
- in vehicle;
- fill-in covers all other areas where customer presence is rare, such as water, forests, fields and deserts.

These layers could be sufficient to characterize traffic for a whole city. The remaining distribution of customers can be done by applying multipliers to these layers, that is, percentages can be assigned for different terminal types, movements, and encapsulations.

3.12 Network Traffic Modeling

This is a complex task that should be done in several steps. The steps presented in the following list are only a guideline and should be adapted to local particularities. Each of the steps is described in detail next using a fictional network:

- unconstrained busy hour data user traffic;
  - application type: non-real time and real time traffic;
  - user type: personal and business traffic;
- traffic constraint factor per terminal type;
- expected number of users per terminal type;
- busy hour traffic per subscription;
- daily traffic per subscription;
- service plan tonnage ranges;
- number of subscriptions per service plan;
- total number of users;
- mapping of portable terminal users;
- users' area mapping;
- hourly traffic variation;
- prediction service classes;
- mapping service classes traffic layers;
- network traffic per layer.

3.12.1 Unconstrained Busy Hour Data User Traffic

An unconstrained data demand is not limited by the communications network, being only limited by the user interaction with applications. In practice, the data demand obtained over regular landline broadband Internet links can today be considered as unconstrained. This allows us to map average user traffic per application, and although user usage patterns vary with age and culture, this average is good enough to analyze the network statistically. Even so, we need to distinguish between a personal user (home user) and a business user, as the application use differs between them. An estimate for both types of users is shown in Tables 3.4 and 3.5. This unconstrained use has to be adjusted by the designer for each network, according to local particularities. All non-real time applications are accumulated as they represent similar traffic and the same applies to real time traffic.

Table 3.4 Unconstrained BH personal user traffic

Applications	Unconstrained BH personal user incoming traffic (MiB)	Unconstrained BH personal user outgoing traffic (MiB)	QoS
Web browsing	1.7	0.36	NRT
E-mail without attachments	0.26	0.05	NRT
E-mail with attachments	0.58	0.12	NRT
Instant Messaging/Skype	0.21	0.04	NRT
Micro blogging (social networking)	0.21	0.04	NRT
Infrastructure (automatic SW updates)	0.68	0.14	NRT
Tunneling (VPN)	0.07	0.01	NRT
Online gaming	0.08	0.02	NRT
Peer to peer	0.27	0	NRT
Total NRT (MiB)	4.06	0.78	NRT
Audio download	0.29	0.06	RT
Video streaming	1.34	0.28	RT
Video download	1.11	0.24	RT
Remote Meeting	0.19	0.04	RT
File sharing	0.28	0.06	RT
VoIP	0.51	0.51	RT
Total RT (MiB)	3.72	1.19	RT

Table 3.5 Unconstrained BH business user traffic

Applications	Unconstrained Business BH user incoming traffic (MiB)	Unconstrained Business BH user outgoing traffic (MiB)	QoS
Web browsing	2.7	0.75	NRT
E-mail without attachments	0.8	0.22	NRT
E-mail with attachments	1.4	0.39	NRT
Instant Messaging/Skype	0.7	0.19	NRT
Micro blogging (social networking)	0.05	0.01	NRT
Infrastructure (automatic SW updates)	0.68	0.19	NRT
Tunneling (VPN)	1.4	0.39	NRT
Online gaming	0	0.00	NRT
Peer to peer	0	0.00	NRT
Total NRT MiB	7.73	2.15	NRT
Audio download	0.3	0.10	RT
Video streaming	0.1	0.03	RT
Video download	0	0.00	RT
Remote Meeting	1.22	0.39	RT
File sharing	0	0.00	RT
VoIP	1.5	0.48	RT
Total RT MiB	3.12	1.01	RT

Table 3.6 Traffic constraint factor by terminal type

Terminal	Mobility	Efficiency (%)
Rooftop (R)	Fixed	100
Desktop (D)	Fixed	80
Laptop (L)	Portable	60
Palmtop, phone, PMP (P)	Portable	30

Table 3.7 Expected number of users per terminal type

Terminal	Personal Users per Subscription (PUS)	Business Users per Subscription (BUS)
Rooftop (R)	2	6
Desktop (D)	1.5	2
Laptop (L)	1	1
Palmtop (P)	1	1

3.12.2 Traffic Constraint Factor per Terminal Type

Unconstrained user traffic requires a high throughput connection and a generous man–machine interface, such as desktops. From the options in the wireless arena, rooftop connections usually provide the best throughput, most likely using desktops to interface with the network.

Desktop units (window-mounted and desktop antennas) still have a generous user interface, but may suffer in terms of throughput due to the antenna location. Laptops offer a slightly more restrictive user interface and have a smaller gain antenna. Palmtops, phones, and portable multimedia players have a poor user interface, combined with an even lower antenna gain. Table 3.6 estimates an average tonnage constraint factor for the different type of terminals. This constraint factor limits user tonnage per terminal type. Surfing the web from a desktop with a rooftop connection, for example, should allow users to generate about three times more tonnage than doing the same from a palmtop.

3.12.3 Expected Number of Users per Terminal Type

A terminal represents one subscription, but it can be accessed by several users. This access can be simultaneous through a switch or at different times. An estimated number of users per terminal is given in Table 3.7 and varies for personal (home) and business applications.

3.12.4 Busy Hour Traffic per Subscription

To establish the tonnage that different terminals require, the single user traffic must be multiplied by the expected number of users per terminal. As each terminal corresponds to one subscription, the total traffic per subscription is as shown in Table 3.8.

Table 3.8 Busy hour traffic per subscription (or terminal)

Terminal	Traffic type	BH personal incoming traffic per subscription	BH personal outgoing traffic per subscription	BH business incoming user per subscription	BH business outgoing traffic per subscription
Rooftop (R)	NRT	8.1	1.6	46.4	12.9
Desktop (D)	NRT	4.9	0.9	12.4	3.4
Laptop (L)	NRT	2.4	0.5	4.6	1.3
Palmtop (P)	NRT	1.2	0.2	2.3	0.6
Rooftop (R)	RT	7.4	2.4	18.7	6.0
Desktop (D)	RT	4.5	1.4	5.0	1.6
Laptop (L)	RT	2.2	0.7	1.9	0.6
Palmtop (P)	RT	1.1	0.4	0.9	0.3
Rooftop (R)	Total	15.6	11.8	65.1	18.9
Desktop (D)	Total	9.3	3.2	17.4	5.0
Laptop (L)	Total	4.7	1.2	6.5	1.9
Palmtop (P)	Total	2.3	0.6	3.3	0.9

Table 3.9 Daily traffic per subscription (or terminal)

Terminal	Traffic type	Daily personal incoming traffic per subscription	Daily personal outgoing traffic per subscription	Daily business incoming traffic per subscription	Daily business outgoing traffic per subscription
Rooftop (R)	NRT	81	16	464	129
Desktop (D)	NRT	49	9	124	34
Laptop (L)	NRT	24	5	46	13
Palmtop (P)	NRT	12	2	23	6
Rooftop (R)	RT	74	24	187	60
Desktop (D)	RT	45	14	50	16
Laptop (L)	RT	22	7	19	6
Palmtop (P)	RT	11	4	9	3
Rooftop (R)	Total	156	118	651	189
Desktop (D)	Total	93	32	174	50
Laptop (L)	Total	47	12	65	19
Palmtop (P)	Total	23	6	33	9

3.12.5 Daily Traffic per Subscription

There is a relationship between the busy hour traffic and the daily traffic. This ratio can be obtained from measurements in existing networks. This ratio varies between 9% and 12%. This example assumes 10% as the ratio. The daily traffic per terminal is then given in Table 3.9.

3.12.6 Service Plan Tonnage Ranges

There are many different tonnages in, and they can be grouped in a few similar ranges. This is shown in Table 3.10, where four ranges were established and associated with four service plans. These values were used in Table 3.1.

Table 3.10 Service plans and tonnage ranges

Service plan	Daily incoming tonnage (MiB)	BH incoming tonnage (kbit/s)	Daily outgoing tonnage (MiB)	BH outgoing tonnage (kbit/s)
Platinum	800	182	200	46
Gold	200	46	60	14
Silver	100	23	32	7
Silk	50	11	12	3

3.12.7 Number of Subscriptions per Service Plan

Once the service plans have been defined, the number of subscribers per service plan can be estimated. In live networks, this number is easily obtained from operator records.

Subscribers choose a plan according to their tonnage requirements, and it is wise for the operator to give general guidelines for non-technical subscribers. An example of number of subscribers per service plan is shown in Table 3.11.

3.12.8 Total Number of Users

Most terminals present two different traffic levels, a higher one when in business use and a lower one when in personal use. The exceptions are terminals like phones, which, here, are assigned the proportion of 25% for business and 75% for personal use. Based on this assumption and using the number of terminal per type from Table 3.7, it is possible to calculate the number of users in the network per type of user and of terminal, as shown in Table 3.12.

3.12.9 Mapping of Portable Terminal Users (MPU)

Terminals can be divided into fixed and portable. Fixed terminals are used indoors, but portable terminals can be used in a variety of locations. This is exemplified in Table 3.13.

Table 3.11 Number of subscriptions per service plan

Service plan	Rooftop (R)	Desktop (D)	Laptop (L)	Phone (P)
Platinum	500	0	0	0
Gold	100	1000	0	0
Silver	0	2000	500	0
Silk	0	0	1000	10,000

Table 3.12 Total number of users in a network (TNU)

	Total number of users (TNU)			
	Rooftop (R)	Desktop (D)	Laptop (L)	Phone (P)
Business (B)	3000	2000	500	2500
Personal (P)	200	3000	1000	7500

Table 3.13 Mapping of portable users (MPU) to different location types

Mapping of portable users (MPU)				
	Laptop personal (LP)	Laptop business (LB)	Phone personal (PP)	Phone business (PB)
Indoor (I)	0.7	0.8	0.4	0.5
Outdoor (O)	0	0	0.2	0.1
Vehicle (V)	0.3	0.2	0.3	0.3
Commercial (C)	0	0	0.1	0.1

Table 3.14 Area mapping (AM)

Area mapping (AM)				
Layers	Layer area	Multiplier (floors)	Total area	%
Indoor ground P (IGP)	10000	1	10000	43
Indoor 3rd floor P (I3P)	2000	4	8000	35
Indoor 10th floor P (I10P)	500	10	5000	22
Indoor ground B (IGB)	2500	1	2500	26
Indoor 3rd floor B (I3B)	1000	4	4000	42
Indoor 10th floor B (I10B)	300	10	3000	32

3.12.10 Users' Area Mapping

Indoor users are located inside buildings that can be multi-floor; a uniform distribution can be assumed throughout the floors. Not all floors can be represented in the RF analysis for processing reasons, thus a few anchor floors are considered to represent the area of several floors. Table 3.14 estimates the indoor area of each anchor floor for personal (residential areas) and business areas.

The area calculation can be simplified, by defining areas in which buildings of different height exist. For this example, areas should be defined for single and two-floor buildings, three to nine floors, and ten floors or higher. This area definition can be done by creating polygons (regions) that encompass the buildings. The area of each of the three layers should be multiplied by the number of floors it represents, to get the total area of each layer, as shown in Table 3.14. This process has to be done for residential areas and business areas. The final result is the percentage of the indoor traffic to be allocated to each anchor layer. This height grouping concept is illustrated in Figure 3.21 for four different height groups.

3.12.11 Hourly Traffic Variation

Traffic can be grouped according to the locations it is carried and can be classified into the following categories:

- *Personal*: represents traffic connected to individuals, and includes personal and residential traffic.
- *Business*: represents traffic conducted for business purposes while at work. It is common to limit the size of targeted businesses to small and medium enterprises (SME).
- *Commercial*: represents traffic in stores, restaurants and entertainment venues.

- *Vehicular*: represents traffic carried in vehicles (cars, buses, trucks).
- *Outdoor*: represents traffic carried outdoors (streets, parks, wilderness, water).

Traffic is not constant over the day and even assuming it were constant during the busy hour is an approximation. Possible traffic variations per type of area are shown in Table 3.15 and Figure 3.22. Table values express the percentages of the busy hour traffic of each category that is active during different hours of the day.

3.12.12 Prediction Service Classes (PSC)

Prediction service classes should be established to represent network behavior in terms of traffic and RF performance. In terms of traffic, the two subscriber types are represented by personal and business layers, and the two traffic QoSs are represented by non-real time (NRT) and real time (RT) traffic. In terms of RF, the representation of multi-level indoor, outdoor and vehicle completes the picture.

A prediction service class is defined by a service, a terminal, an environment and a traffic layer. In this example we have two types of services differentiated by their QoS requirements: non-real time services which do not require stringent latencies and real time service which have latency requirements.

Terminals are grouped into three basic types: rooftop, desktop, and portables. Rooftop terminals have their antennas at rooftop height, but as only three height levels are considered, all personal rooftop installations are placed at 10 m high and all business installations at 30 m high. Desktop terminals are considered at three heights: 1.5 m, 10 m, and 30 m. Portable terminals group laptops, portable multi-media devices, and phones. These types of terminals are considered at ground level, unless their traffic includes desktops, in which case they are considered at the respective desktop height.

Table 3.16 exemplifies 22 prediction service classes (PSCs). Each PSC has a traffic layer associated with it. Figure 3.23 shows PSCs and their relationship to different terminals, environment and traffic layers.

Figure 3.21 Height grouping illustration.

Table 3.15 Hourly busy hour multiplier (HM)

Hour	Personal (P) (%)	Business (B) (%)	Commercial (C) (%)	Vehicular (V) (%)	Outdoor (O) (%)
0	50	10	5	5	0
1	30	10	5	5	0
2	20	10	5	5	0
3	20	10	5	5	0
4	20	10	5	5	0
5	20	10	5	10	0
6	20	10	5	50	0
7	25	20	5	80	0
8	25	40	10	100	10
9	25	60	30	90	30
10	25	70	30	30	40
11	25	80	40	40	50
12	40	40	60	60	100
13	40	50	60	50	50
14	30	80	40	40	30
15	35	90	20	30	30
16	35	100	20	20	50
17	40	70	70	80	70
18	60	40	100	100	80
19	70	30	90	80	60
20	80	20	50	60	20
21	100	20	30	50	10
22	90	10	10	30	0
23	70	10	5	10	0

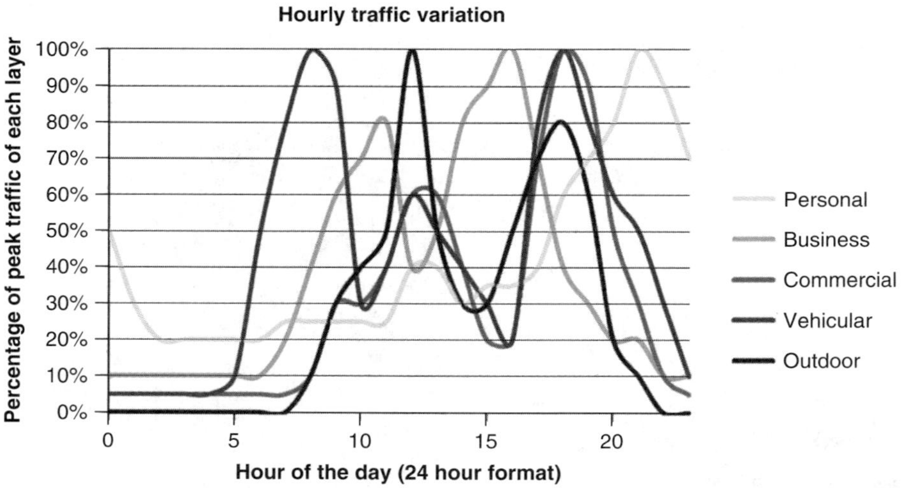

Figure 3.22 Hourly traffic variation.

Table 3.16 Traffic layers composition

	Layers	Users' mapping
1	Indoor ground P NRT	AM(IGP)*HM(IGP)*(TNU(P,D)+MPU(I,LP)*TNU(P,L) +MPU(I,PP)*TNU(P,P))
2	Indoor 3rd floor P NRT	AM(I3P)*HM(I3P)*(TNU(P,D)+MPU(I,LP)*TNU(P,L) +MPU(I,PP)*TNU(P,P))
3	Indoor 10th floor P NRT	AM(I10P)*HM(I10P)*(TNU(P,D)+MPU(I,LP)*TNU(P,L) +MPU(I,PP)*TNU(P,P))
4	Indoor ground B NRT	AM(IGB)*HM(IPB)*(TNU(B,D)+MPU(I,LB)*TNU(B,L) +MPU(I,PB)*TNU(B,P))
5	Indoor 3rd floor B NRT	AM(I3B)*HM(I3B)*(TNU(B,D)+MPU(I,LB)*TNU(B,L) +MPU(I,PB)*TNU(B,P))
6	Indoor 10th floor B NRT	AM(I10B)*HM(I10B)*(TNU(B,D)+MPU(I,LB)*TNU(B,L) +MPU(I,PB)*TNU(B,P))
7	Outdoor ground NRT	HM(O)*(MPU(O,PP)*(TNU(P,P)+MPU(O,PB)*TNU(B,P))
8	Outdoor rooftop P NRT	HM(OR)*TNU(P,R)
9	Outdoor rooftop B NRT	HM(OR)*TNU(B,R)
10	Vehicle NRT	HM(V)*(MPU(V,LP)*TNU(P,L)+MPU(V,LB)*TNU(B,L) +MPU(V,PP)*TNU(P,P)+MPU(V,PB)*TNU(B,P))
11	Commercial NRT	HM(C)*(MPU(C,PP)*TNU(P,P)+MPU(C,PB)*TNU(B,P))
12	Indoor ground P RT	AM(IGP)*HM(IGP)*(TNU(P,D)+MPU(I,LP)*TNU(P,L) +MPU(I,PP)*TNU(P,P))
13	Indoor 3rd floor P RT	AM(I3P)*HM(I3P)*(TNU(P,D)+MPU(I,LP)*TNU(P,L) +MPU(I,PP)*TNU(P,P))
14	Indoor 10th floor P RT	AM(I10P)*HM(I10P)*(TNU(P,D)+MPU(I,LP)*TNU(P,L) +MPU(I,PP)*TNU(P,P))
15	Indoor ground B RT	AM(IGB)*HM(IGPB)*(TNU(B,D)+MPU(I,LB)*TNU(B,L) +MPU(I,PB)*TNU(B,P))
16	Indoor 3rd floor B RT	AM(I3B)*HM(I3B)*(TNU(B,D)+MPU(I,LB)*TNU(B,L) +MPU(I,PB)*TNU(B,P))
17	Indoor 10th floor B RT	AM(I10B)*HM(I10B)*(TNU(B,D)+MPU(I,LB)*TNU(B,L) +MPU(I,PB)*TNU(B,P))
18	Outdoor ground RT	HM(O)*(MPU(O,PP)*(TNU(P,P)+MPU(O,PB)*TNU(B,P))
19	Outdoor rooftop P RT	HM(OR)*TNU(P,R)
20	Outdoor rooftop B RT	HM(OR)*TNU(B,R)
21	Vehicle RT	HM(V)*(MPU(V,LP)*TNU(P,L)+MPU(V,LB)*TNU(B,L) +MPU(V,PP)*TNU(P,P)+MPU(V,PB)*TNU(B,P))
22	Commercial	HM(C)*(MPU(C,PP)*TNU(P,P)+MPU(C,PB)*TNU(B,P))

3.12.13 Traffic Layers Composition

The composition of traffic layers may be complex as user traffic has to be divided between PSCs. Table 3.16 shows how each of the 22 traffic layers could be assembled in this example. The general format is table (row parameter, column parameter). Tables, rows and columns are identified by abbreviations in Table 3.12 (TNU), Table 3.13 (MPU), Table 3.14 (AM) and Table 3.15 (HM).

Terminals can be divided into fixed and portable. Fixed terminals are used indoors, while portable terminals can be used in a variety of locations. Indoor users are located inside buildings that can be multi-floor and a uniform distribution is assumed throughout the floors. Not all floors can be represented in the RF analysis due to the amount of processing required, thus a few anchor floors are

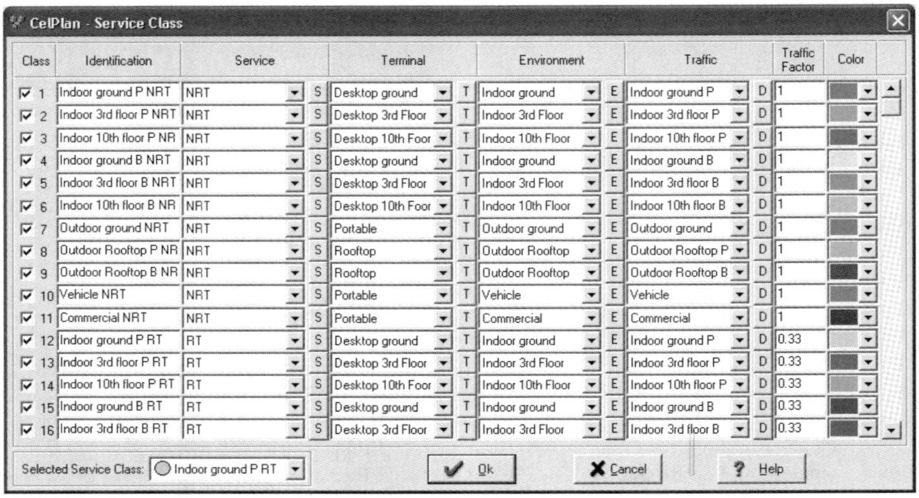

Figure 3.23 Service class representation in prediction tool dialog box.

considered to represent the area of several floors. Estimates are made for the indoor area of each floor for personal (residential) and business areas.

The area calculation can be simplified, by defining the areas in which buildings of different heights exist. For example, areas should be defined for single and two floor buildings, three to nine floors and ten floors or higher. This area definition can be done by creating polygons (regions) that encompass the buildings. The area of each of the three layers should be multiplied by the number of floors it represents, to get the total area of each layer, as shown in Table 3.16. This process has to be done for residential and business areas. The final result is the percentage of indoor traffic to be allocated to each anchor layer.

3.12.14 Network Traffic per Layer

Using Table 3.4, Table 3.5 and Table 3.16, it is possible to calculate the number of users (personal and business) and the total BH traffic (incoming-downlink and outgoing-uplink). The last two columns of Table 3.17 give an idea of the average tonnage per user of each layer expressed in kbit/s. Each user is represented twice in this table, once for the NRT traffic and again for the RT traffic. The traffic numbers were calculated for 16:00 (4 p.m.). The network in the example has 15,100 subscriptions, which results in 19,700 users. At 16:00, there are only 10,846 active users.

3.13 KPI (Key Performance Indicator) Establishment

The performance of the network can be verified against the SLA (service level agreement), by calculating key performance indicators (KPI). The SLA parameters are the following:

- Target speed (IPDS), dependent on RF coverage quality.
- Target tonnage (IPDT) within specified QoS (latency and BER), dependent on network capability to cope with offered traffic.

Table 3.17 Network traffic per layer

	Prediction service class	Traffic layer	Hour multiplier	Hour multiplier (HM) (%)	Number of users personal	Number of users business	BH traffic incoming (MiB)	BH traffic outgoing (MiB)	BH traffic Incoming per user (kbit/s)	BH traffic Outgoing per user (kbit/s)
1	Indoor ground P NRT	Indoor ground P	Personal	35	1020	0	4139	795	9	2
2	Indoor 3rd floor P NRT	Indoor 3rd floor P	Personal	35	816	0	3312	636	9	2
3	Indoor 10th floor P NRT	Indoor 10th floor P	Personal	35	510	0	2070	398	9	2
4	Indoor ground B NRT	Indoor ground B	Business	100	0	987	7628	2119	18	5
5	Indoor 3rd floor B NRT	Indoor 3rd floor B	Business	100	0	1537	11,880	3300	18	5
6	Indoor 10th floor B NRT	Indoor 10th floor B	Business	100	0	1153	8910	2475	18	5
7	Outdoor ground NRT	Outdoor ground	Outdoor	50	750	125	4011	853	10	2
8	Outdoor rooftop P NRT	Outdoor rooftop P	Personal	35	70	0	284	55	9	2
9	Outdoor rooftop B NRT	Outdoor rooftop B	Business	100	0	3000	23,190	6442	18	5
10	Vehicle NRT	Vehicle	Vehicular	20	510	170	3385	763	11	3
11	Commercial NRT	Commercial	Commercial	20	150	50	996	224	11	3
12	Indoor ground P RT	Indoor ground P	Personal	35	1020	0	4139	795	9	2
13	Indoor 3rd floor P RT	Indoor 3rd floor P	Personal	35	816	0	3312	636	9	2
14	Indoor 10th floor P RT	Indoor 10th floor P	Personal	10	510	0	2070	398	9	2
15	Indoor ground B RT	Indoor ground B	Business	5	0	987	7628	2119	18	5
16	Indoor 3rd floor B RT	Indoor 3rd floor B	Business	5	0	1537	11,880	3300	18	5
17	Indoor 10th floor B RT	Indoor 10th floor B	Business	5	0	1153	8910	2475	18	5
18	Outdoor ground RT	Outdoor ground	Outdoor	0	750	125	4011	853	10	2
19	Outdoor rooftop P RT	Outdoor rooftop P	Personal	10	70	0	284	55	9	2
20	Outdoor rooftop B RT	Outdoor rooftop B	Business	5	0	3000	23,190	6442	18	5
21	Vehicle RT	Vehicle	Vehicular	0	510	170	3385	763	11	3
22	Commercial RT	Commercial	Commercial	0	150	50	996	224	11	3

An operator may additionally specify the following requirements:

- Percentage of area (PA) with service, typically, a 90% value is specified.
- Percentage of population (PP) with service, typically, a 90% value is specified.

Network KPIs are calculated by the design software performance evaluation features. Speed and tonnage are evaluated through dynamic simulation, described later in Chapter 21. Percentages of area and population are evaluated by comparing signal levels footprints to the target area and population footprint polygon.

3.14 Wireless Infrastructure

Wireless service requires implementation of a specific type of infrastructure. The infrastructure nomenclature used in this book is described here.

There are two types of wireless service:

- *Point-to-point (PP)*: In this case, both ends of the wireless connection are known, and, usually, at fixed locations. The ends are known as radio nodes (RN). An RN may house one or more radios within. The connection between RNs is called a radio link, and is identified by transmit and receive nodes. It is illustrated in Figure 3.24.
- *Point-to-multipoint (PMP)*: In this case, one end of the wireless connection is the central point, while the many other ends are customer ends. The central end is called a base station (BS or BTS); and the customer ends are called CS (customer stations), SS (subscriber stations), MS (mobile stations), UE (user equipment), or TS (terminal stations). It is illustrated in Figure 3.25.

PMP connections can be to fixed, nomadic, or mobile customers. The connection from the base station (TX-transmit) to the customer (RX-receive) is called downstream (DS) or downlink (DL); the connection from the customer (TX) to the base station (RX) is called upstream (US) or uplink (UL). The downlink (DL) and uplink (UL) terms are more applicable to voice circuits.

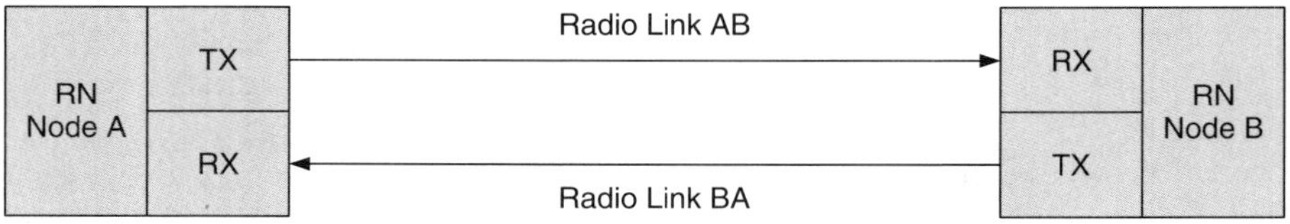

Figure 3.24 Point-to-point infrastructure.

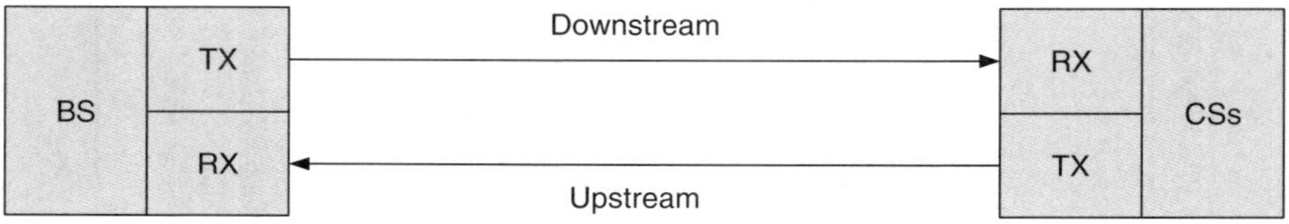

Figure 3.25 Point-to-multi-point infrastructure.

Wireless infrastructure has an overlapping nomenclature that varies from one technology to another. The main components are specified below:

- *Sites*: locations that have one or more Base Stations.
- *Base Stations (BS)*: locations that have one or more sectors; sometimes called a cell, although, lately, the term cell is more used to designate a single sector. Other names for a Base Station are Node B, BTS, eNode B, and eNB.
- *Sector*: a set of radios and antennas that have the same coverage area, can also be called a cell.
- *Radio*: a transceiver connected to the sector-antenna system, transmitting on a specific central frequency and receiving on the same or on another frequency. A radio can be also called a carrier.

The wireless infrastructure is specified in more detail in Chapter 17.

4

Signal Processing Fundamentals

There are a few important principles that are essential to the comprehension of signal processing in wireless systems. This chapter describes each of these concepts in detail. It is important for network designers to have a good understanding of these principles to be able to properly dimension network resources.

4.1 Digitizing Analog Signals

Analog signals carry a lot of redundancy and are hard to retrieve from noise whereas digital data conveys information as a stream of ones and zeros, which can be more easily recovered. Digital information can represent numbers, codes, images, text or even analog signals. To do so, analog signals have to be digitized and the fundaments to do it are established by the sampling theorem.

Digital data is generally grouped into packets; each packet carrying data and a source and destination address, IP (Internet Protocol) being a typical example of this.

To digitize and analog a signal two questions have to be answered: can a continuous analog signal be fully represented by discrete samples? And, if so, how many of them are required?

Many authors have contributed to this topic, but two papers are considered fundamental. In 1928, Harry Nyquist published a paper entitled "Certain topics in telegraph transmission theory", where he demonstrated that $2*B$ independent pulse samples can be sent through a system with a bandwidth B. Then, in 1949, Claude Shannon in his paper, "Communications on presence of noise" demonstrated the double of Nyquist's paper, stating that a signal of bandwidth B can be defined by $2B$ samples. Many authors call the sampling theorem the Nyquist–Shannon theorem.

The sampling theorem states that if a bandwidth-limited continuous signal is sampled at a rate twice its bandwidth B, it is possible to reconstruct the original signal from these samples. These samples, being discrete, can have their value digitized, by transforming the analog signal into a digital stream of data. Figure 4.1 shows a signal being sampled with an interval T, Equation (4.1) gives the Nyquist sampling frequency and Equation (4.2) gives the Nyquist sampling period.

$$f_s = 2B \qquad \text{(4.1) Nyquist sampling frequency}$$

$$T_s = \frac{1}{f_s} \qquad \text{(4.2) Nyquist sampling period}$$

LTE, WiMAX and WLAN Network Design, Optimization and Performance Analysis, First Edition.
Leonhard Korowajczuk.

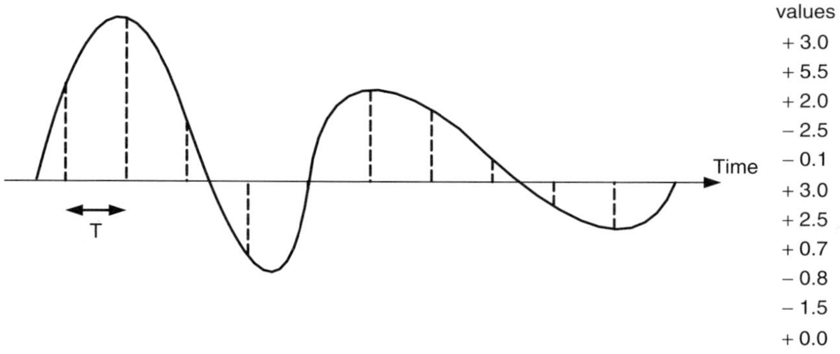

Figure 4.1 Sampled waveform.

Where:

$B =$ Maximum bandwidth frequency.
$f_s =$ Nyquist sampling frequency.
$T_s =$ Nyquist sampling period.

Sampling a function x(t) creates a spectrum with a periodic function X(f) as illustrated in Figure 4.2. This spectrum has a base spectrum and images of it are spaced by f_s (the sampling frequency). Those images are alias of the base spectrum. Any of the aliases can be filtered as shown in Figure 4.2 and still convey all the required information.

The base spectrum can be recovered by using a band pass filter, but if the original signal extends beyond $\frac{f_s}{2}$ the images will interfere between themselves, distorting the original signal. An anti-alias (or anti-aliasing) filter is used to limit the signal, but like any other filter, it is not perfect and will require some additional bandwidth to filter the signal to acceptable levels. For this reason the sampling frequency should be increased to accommodate the bandwidth required by the filter. There is no harm in over-sampling a signal. This amount of over-sampling is defined by the type of filter implemented.

The reconstruction of the analog signal can be done by integrating between samples, as illustrated in Figure 4.3. Usually a simple RC circuit is used.

The sampling frequency applies to the signal bandwidth B, even if this bandwidth does not start at zero-frequency. When a signal is moved to a carrier f_c, the sampling frequency f_s range is defined by:

$$\frac{2f_c - B}{m} \geq f_s \geq \frac{2f_c + B}{m+1} \qquad \text{(4.3) Sampling frequency range}$$

Where m is any positive integer that results in $f_s \geq 2B$.

This under-sampling still returns the base band waveform, but the information about the carrier frequency is lost as a consequence of violating the Nyquist–Shannon theorem. This means that signals with a bandwidth of 100 KHz on a 10 MHz carrier, can be sampled by a frequency in any of the 99 frequency range shown in Table 4.1.

In real life, an analog signal is digitized by an ADC (analog to digital converter) and the signal is recovered from the digital samples by a DAC (digital to analog converter). ADC/DACs are mainly specified according to:

- sampling frequency (e.g. 48 kHz);
- resolution (in bits) used to express the amplitude level of each sample (e.g. 8 bits);
- conversion speed (e.g. 100,000 samples per second).

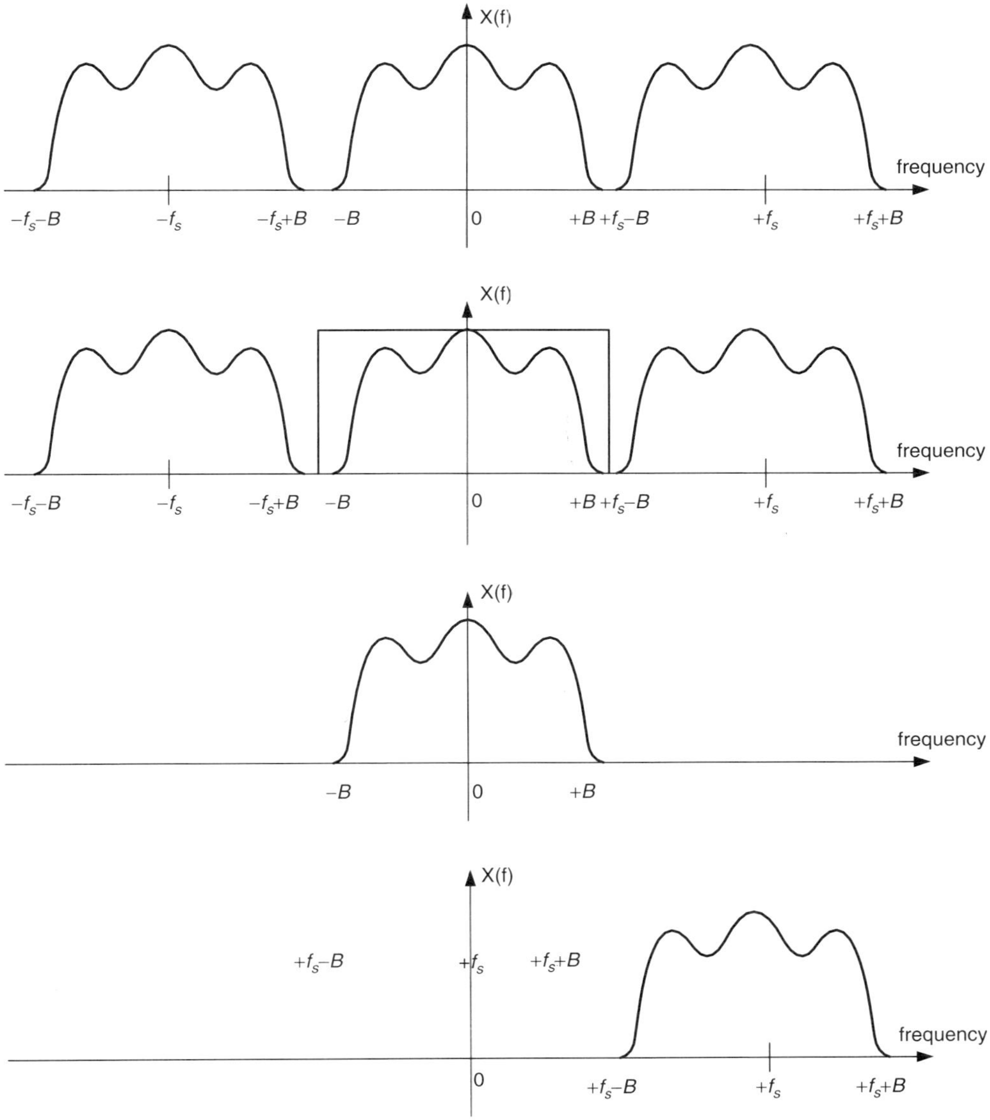

Figure 4.2 Spectrum of a sampled waveform.

Let's assume that we have a signal whose bandwidth is limited from 100 kHz to 110 kHz and the anti-aliasing filter needs to attenuate the signal in 3 kHz by a 20 dB (1%). The total bandwidth is then 16 kHz and the sampling frequency should be between 32,333 Hz and 32,286 Hz (for an m of 6).

When the input waveform is known, such as a sine wave, the samples can be mathematically calculated directly in digital form. This is usually done by DSPs (Digital Signal Processors). This shortcut is essential for the feasibility of the new wireless technologies.

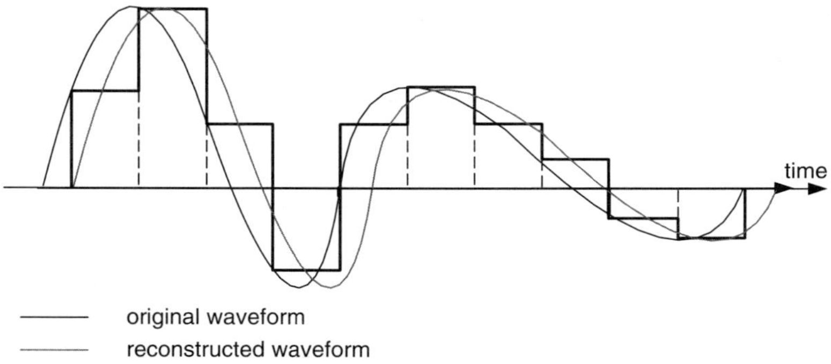

Figure 4.3 Reconstructed waveform.

Table 4.1 Sampling table

m	$f_{s\max}$ (KHz)	$f_{s\min}$ (kHz)
1	19,900.00	10,050.00
2	9,950.00	6,700.00
3	6,633.33	5,025.00
4	4,975.00	4,020.00
5	3,980.00	3,350.00
6	3,316.67	2,871.43
7	2,842.86	2,512.50
8	2,487.50	2,233.33
9	2,211.11	2,010.00
10	1,990.00	1,827.27
50	398.00	394.12
60	331.67	329.51
70	284.29	283.10
80	248.75	248.15
90	221.11	220.88
91	218.68	218.48
92	216.30	216.13
93	213.98	213.83
94	211.70	211.58
95	209.47	209.38
96	207.29	207.22
97	205.15	205.10
98	203.06	203.03
99	201.01	201.00

4.2 Digital Data Representation in the Frequency Domain (Spectrum)

To transmit digital data we need to understand its main properties and digital data is a sequence of ones and zeros that have a constant data rate. The time domain representation of a digital signal is easy to visualize, but it is also important to visualize its frequency domain representation, or, in other words, its spectrum.

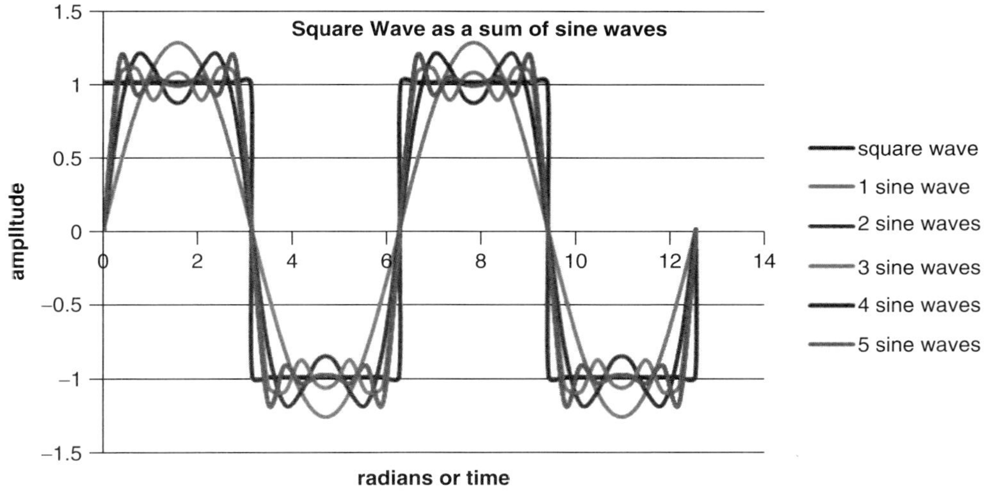

Figure 4.4 Square wave as a sum of sine waves.

Joseph Fourier demonstrated in 1807 (in his *Mémoire sur la propagation de la chaleur dans les corps solides*) that a periodic signal can be decomposed into a sum of simple oscillating functions (sine and cosine), as shown in Figure 4.4. This series is called the Fourier series and allows us to derive the spectrum (frequency domain) from a signal (time domain). Figure 4.4 shows a square wave approximated by the sum of one, two, three, four and five odd multiples of a base sine wave. The Fourier series does this approximation.

As explained in Chapter 22, sine waves can be represented by complex exponentials. It is, in fact, possible to express any continuous functions in the time domain as a sum of discrete complex exponentials (sine waves) in the frequency domain. A Fourier Transform (FT) is the operation that is used to do this operation. The inverse of the FT (iFT) generates the time domain signal from the frequency domain spectrum.

The analysis of a discrete signal (time-limited) is done by a Discrete Fourier Transform (DFT), which only considers the components required to generate one segment of what would be an infinite periodic function. The signal to be analyzed should have non-zero values and have a limited duration (period).

The way a DFT is expressed mathematically requires a large number of calculations, which can take an excessive time to complete. Algorithms were proposed that significantly reduce the number of operations to calculate a DFT, known as Fast Fourier Transform (FFT). Basically they avoid repetitive calculations done on sine waves that are multiples of the base one.

The processing of digital signals requires significant mathematical manipulation, which can be done using a regular CPU (Central Processing Unit). CPU architecture is not optimized to perform huge amounts of mathematical calculations, so Digital Signal Processors (DSPs) are used instead. They were conceived to perform this task and can perform millions of operations per second. There are two types of DSPs, the ones that only do fixed-point operations and the ones that do floating-point operations. The throughput of the first ones is expressed in MIPS (Million of Integer Operations Per Second), while the second ones are expressed in MFLOPS (Million of FLoating-point Operations Per Second). Typical numbers are in the range from 50 to 500 million operations per second.

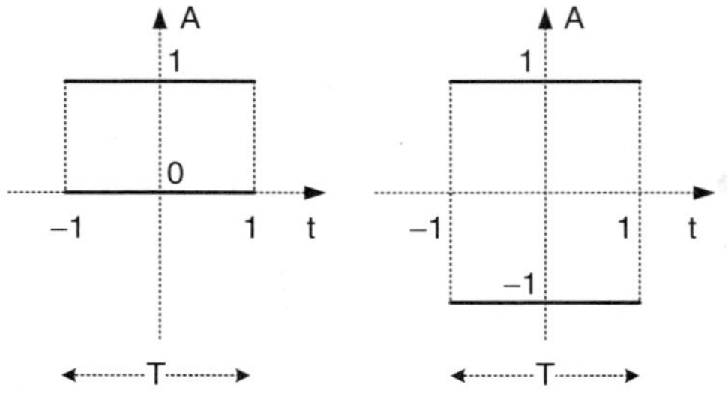

Figure 4.5 RZ and NRZ representation.

We will start analyzing a single unit of information that can represent a value of 1 or zero and has duration defined by T (bit). First, we will convert the bit to a Non-Return to Zero (NRZ) format to eliminate the DC component as represented below. The bit is centered at the origin. Both representations are shown in Figure 4.5.

The Discrete Fourier Transform of this signal results in the Sinc (Sinus Cardinalis) function defined in Equation (4.4).

$$\sin c(Tf) = \frac{\sin(T\pi f)}{T\pi f} \qquad \text{(4.4) Sinc function}$$

The Sinc function is equivalent to the sin $(x)/x$ function, but the value for $x = 0$ is predefined as 1. This function has zero value for integer values of Tf and decays with $1/(Tf\pi)$ as shown in Figure 4.6. The first peak carries the relevant information, whereas the other peaks are aliases and can be filtered. Actually the peak value is enough to retrieve the information sent.

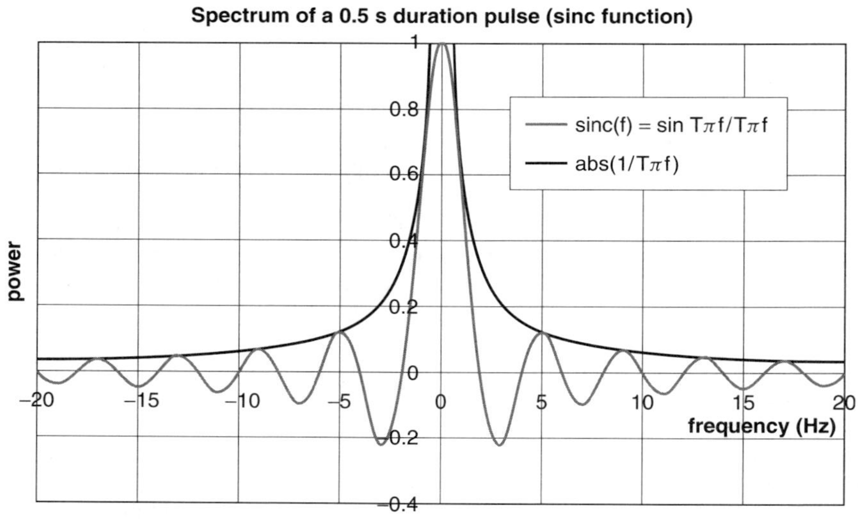

Figure 4.6 Spectrum of a 0.5 s duration pulse (sinc function).

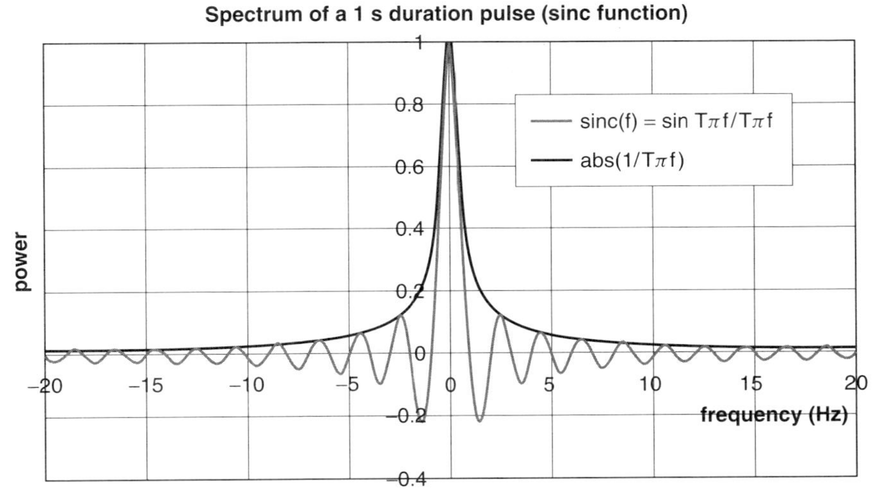

Figure 4.7 Spectrum of a 1 s duration pulse (sinc function).

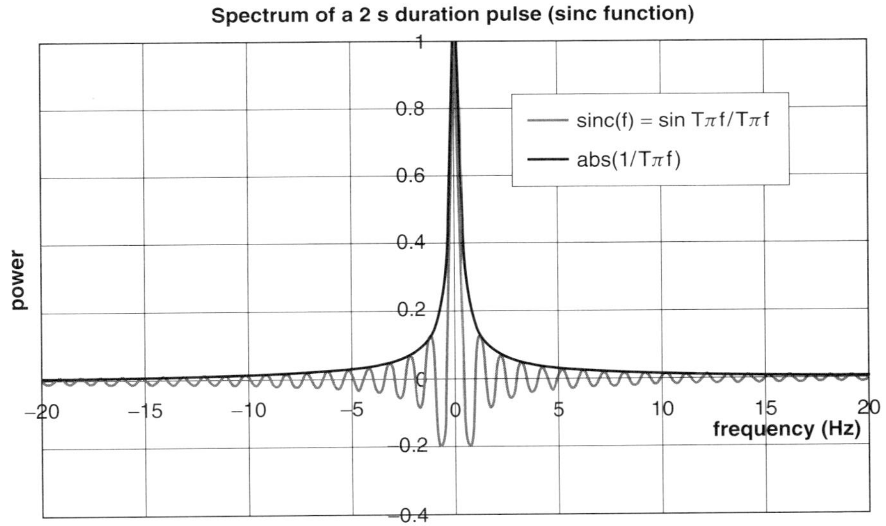

Figure 4.8 Spectrum of a 2 s duration pulse (sinc function).

In Figure 4.6 the pulse duration is 0.5 s and the Sinc function nulls occur every 2 Hz. Figure 4.7 considers a pulse duration of 1 s and nulls occur every 1 Hz. Figure 4.8 considers 2 s pulses and nulls occur every 0.5 Hz. The larger the pulse duration, the shorter is the bandwidth B of the relevant spectrum and vice versa. Equation (4.5) shows the pulse bandwidth. An envelope curve has been

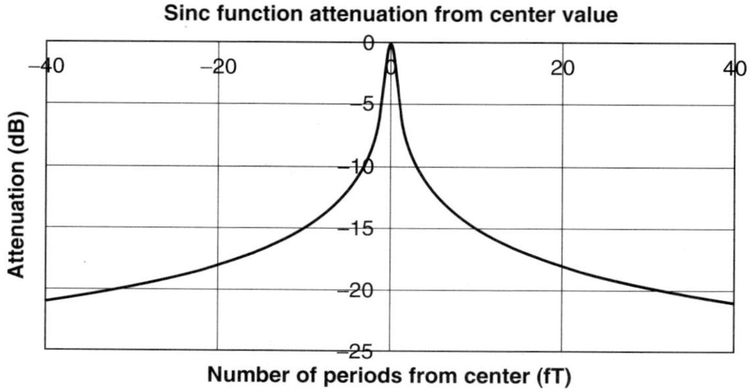

Figure 4.9 Sinc function attenuation from center expressed in number of subcarriers.

added in each graph to show the decay of the power with frequency.

$$B = \frac{1}{T} \qquad \text{(4.5) Pulse bandwidth}$$

Figure 4.9 gives the spectrum envelope attenuation with frequency in dB and is expressed by Equation (4.6).

$$\text{att(d B)} = 10(\log(\text{abs}(1/T\pi f))) \qquad \text{(4.6) Sinc function attenuation}$$

The Sinc function drops to 1% (20 dB) after 30 periods. Higher attenuations will require the use of a band pass filter.

4.3 Orthogonal Signals

An important property of signals is their orthogonality, meaning that they can be detected independently of each other. Two signals are considered orthogonal if their product over an entire period (dot product) is null. A dot product is the result of the integration of the regular product of two functions or its samples, taken over an integer number of periods.

4.3.1 Sine and Cosine Orthogonality

A sine and cosine are orthogonal to each other, as demonstrated in Equations (4.7) and (4.8). We first multiply both functions and then we integrate the resulting curve, obtaining a sum of zero. This can be done by multiplying the sine wave samples and adding the result for an integer number of periods.

$$\sin x . \cos x = \sin 2x \qquad \text{(4.7) Product of a sine by a cosine}$$

$$\int_0^{2\pi} \sin 2x = 0 \qquad \text{(4.8) Integral of the product of a sine by a cosine}$$

In this case, the orthogonality is only preserved if both signals have the same phase.

4.3.2 *Harmonically Related Signals' Orthogonality*

Another important set of orthogonal functions comprises any harmonically (multiple) related signals. This is expressed in Equation (4.9).

$$\int_0^{2\pi} \sin x . \sin nx = 0 \qquad \text{(4.9) Harmonically related signal orthogonality}$$

This orthogonality is preserved independently of the phase relationship between the signals. Orthogonality also holds when signals are represented by its samples, and this property is used by the DSPs that process digital signals.

Orthogonal signals (or their samples) can be added and the combined signal can be verified for correlation with known signals:

- An auto-correlation is achieved when the combined signal is multiplied and integrated against a known signal and the result is a value different from zero.
- A low cross-correlation is achieved when a known signal is not present in the combined signal, resulting in zero integration.

Table 4.2 lists four harmonically related signals that are multiple of 1 Hz: 1 Hz, 2 Hz, 3 Hz and 5 Hz. The frequency of 4 Hz has been excluded on purpose. Each frequency needs to be represented by at least 10 samples (2* 5 Hz) per period to digitally represent all the signals. Next, the samples are added in the sum column. The sum is then multiplied by each frequency and the products totalized. The only frequency that results in a zero sum is f4, which was not present in the sum composition.

When the investigated frequency is available in the sum (auto-correlation), the value of its samples will be squared, resulting into a large positive signal. The other frequencies samples will provide positive and negative values that will cancel each other. Figure 4.10 shows the sum of the four sine waves used in this example.

Orthogonality properties are the base of wireless communications and of the OFDM concept.

Table 4.2 Sum of sine waves

time (s)	alpha	radians	sin f1	sin f2	sin f3	sin f5	sum	sum*sin f1	sum*sin f2	sum*sin f3	sum*sin f5	sum*sin f4
0	0	0	0	0	0	0	0	0	0	0	0	0
0.083	30	0.524	0.500	0.866	1.000	0.500	2.866	1.433	2.482	2.866	1.433	0.000
0.167	60	1.047	0.866	0.866	0.000	−0.866	0.866	0.750	0.750	0.000	−0.750	0.000
0.250	90	1.571	1.000	0.000	−1.000	1.000	1.000	1.000	0.000	−1.000	1.000	0.000
0.333	120	2.094	0.866	−0.866	0.000	−0.866	−0.866	−0.750	0.750	0.000	0.750	0.000
0.417	150	2.618	0.500	−0.866	1.000	0.500	1.134	0.567	−0.982	1.134	0.567	0.000
0.500	180	3.142	0.000	0.000	0.000	0.000	0.000	0.000	0.000	0.000	0.000	0.000
0.583	210	3.665	−0.500	0.866	−1.000	−0.500	−1.134	0.567	−0.982	1.134	0.567	0.000
0.667	240	4.189	−0.866	0.866	0.000	0.866	0.866	−0.750	0.750	0.000	0.750	0.000
0.750	270	4.712	−1.000	0.000	1.000	−1.000	−1.000	1.000	0.000	−1.000	1.000	0.000
0.833	300	5.236	−0.866	−0.866	0.000	0.866	−0.866	0.750	0.750	0.000	−0.750	0.000
0.917	330	5.760	−0.500	−0.866	−1.000	−0.500	−2.866	1.433	2.482	2.866	1.433	0.000
							sum	6	6	6	6	0

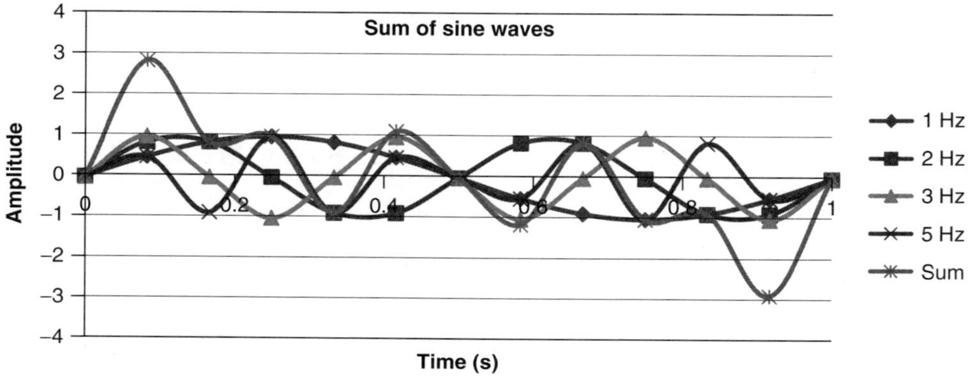

Figure 4.10 Sum of sine waves.

4.4 Combining Shifted Copies of a Sine Wave

Combined non-orthogonal signals, such as phase-shifted copies of a sinusoidal waveform result in a sinusoidal waveform that is phase shifted itself. The final phase shift is the average of the individual components phase shifts. This is illustrated in Figure 4.11 and Figure 4.12.

In Figure 4.11, seven shifted versions of a sine wave are combined and the resulting waveform is also a sine wave, but with a phase of 135°. In Figure 4.12, the same shifted versions are differently attenuated and the combined signal, though different from the first, is also a sine wave, but with a phase of 45.5°. This property is very important because it reflects what happens when the multipath is received.

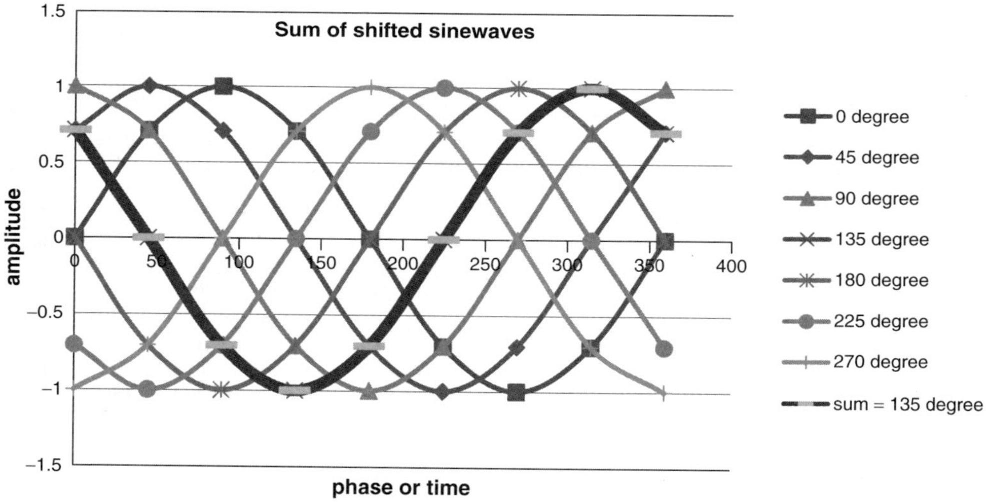

Figure 4.11 Shifted sine waves and combined sine wave.

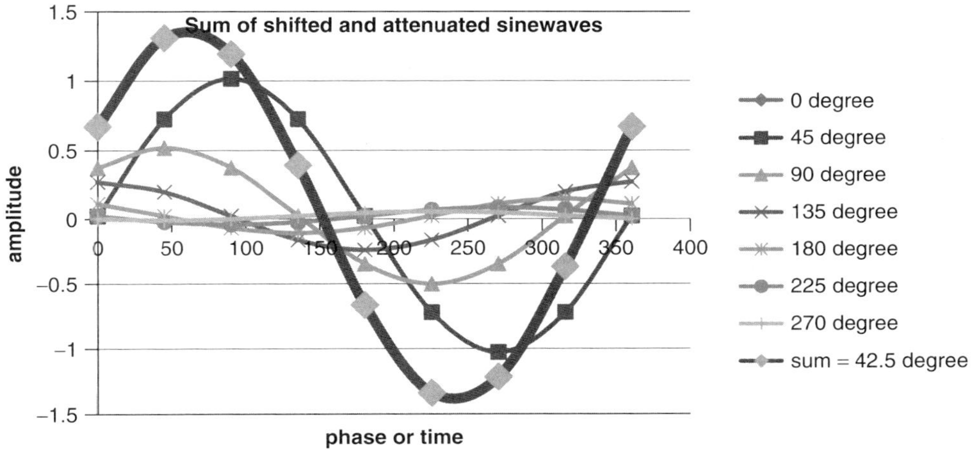

Figure 4.12 Shifted and attenuated sine waves and combined sine wave.

4.5 Carrier Modulation

The process of loading digital information on a carrier is called modulation. In the modulation process, sets of bits are combined into symbols and assigned to carrier states (phase x energy) forming a constellation. These constellations can be represented in polar form, showing the phase and magnitude in the same diagram, as illustrated in Figure 4.13. The distance between constellation points represents how much noise the modulation can accommodate.

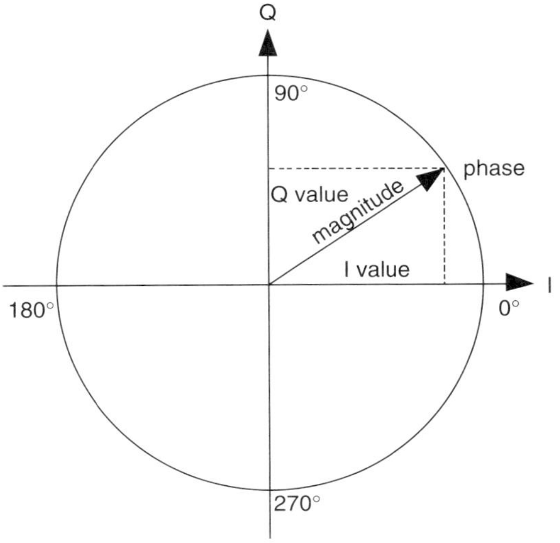

Figure 4.13 Polar and rectangular constellation.

The most common modulation methods are PSK (Phase Shift Keying) and QAM (Quadrature Amplitude Modulation). PSK can be considered a particular case of QAM, thus we will focus on QAM.

In QAM, information bits are associated with carrier amplitudes and phases, and each technology specifies its own map for this association. This map is called the modulation constellation because it resembles a formation of stars. A modulation can have many states, so more than one bit can be assigned to the same state. Bits assigned to a state are called baud or symbol.

The most common modulation types are:

- BPSK Binary Phase-Shift Keying.
- QPSK Quadrature Phase-Shift Keying.
- 16QAM 16 Quadrature Amplitude Modulation.
- 64QAM 64 Quadrature Amplitude Modulation.

Other modulations are possible, such as 8QAM and 32 QAM, but they offer similar S/N performance as respectively 16QAM and 64QAM, but at lower throughputs. Table 4.3 gives the number of bits that can be mapped for each modulation type.

Each state in the constellation is defined by amplitude and phase, and can be represented by Equation (4.10).

$$\text{state} = A\cos(2\pi f_m t + \varphi) \qquad \text{(4.10) Constellation states}$$

Generating and detecting a phase component is very difficult, so the amplitude and phase information are recorded in two orthogonal functions, as shown in Figure 4.14.

The implementation of this amplitude and phase relationship is done by combining a sine and a cosine, according to Equation (4.11).

$$S(t) = I(t)\cos(2\pi f_m t) - Q(t)\sin(2\pi f_m t) \qquad \text{(4.11) Constellation states using I and Q signals}$$

Where:

$I(t)$ In-phase signal, represents the constellation value for the I axis.

$Q(t)$ Quadrature signal, represents the constellation value for the Q axis.

f_m Modulation frequency.

An example of this function is shown in Figure 4.14 for QPSK and bit sequence 00 (I = 6, Q = 6) and in Figure 4.15 for 16QAM and bit sequence 0100 (I = −2, Q = −6). In both figures the combined waveforms of I and Q axis have different phases and amplitudes.

The constellations for the main modulations are shown in Figure 4.16 with the bit assignment used in WiMAX. Wi-Fi assignments have 1 and 0 reversed.

Table 4.3 Number of bits per modulation scheme

Modulation	Number of bits per symbol
BPSK	1
QPSK	2
16QAM	4
64QAM	6

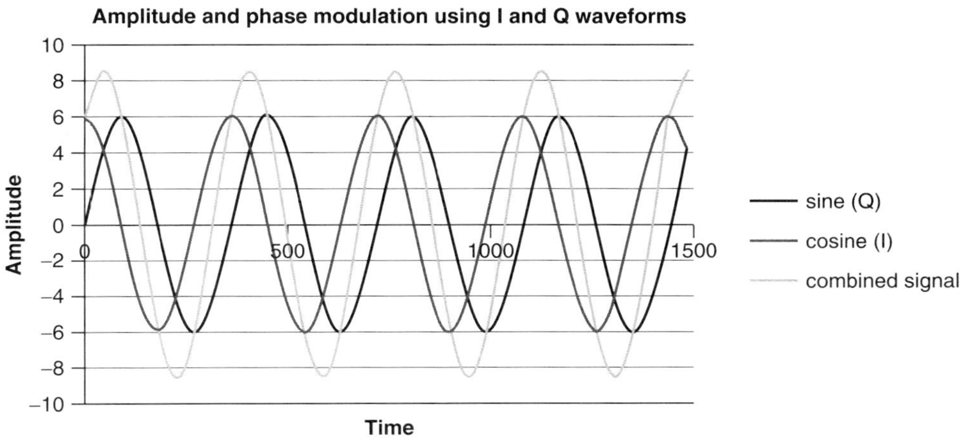

Figure 4.14 Amplitude and phase modulation using I and Q waveforms for QPSK.

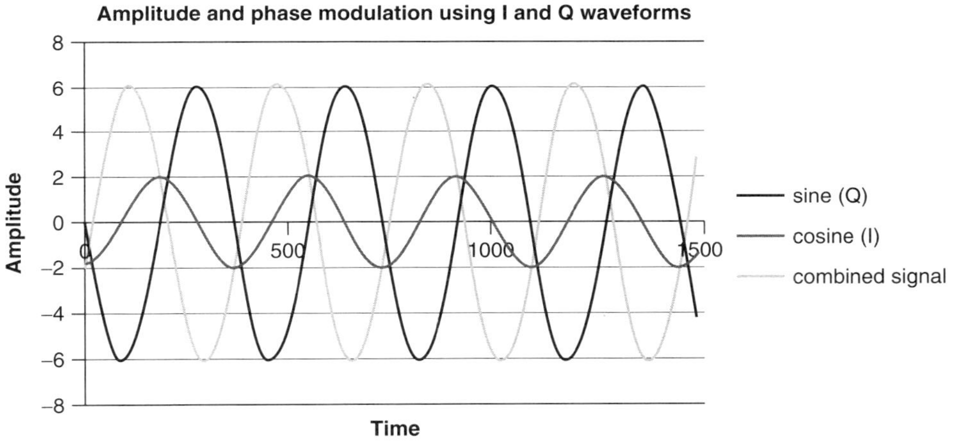

Figure 4.15 Amplitude and phase modulation using I and Q waveforms for 16QAM.

A Gray code (devised by Frank Gray of Bell Labs, in 1953) is used to assign bit sequences to states. This code has the property of changing only 1 bit between adjacent states and this improves the chances of correctly detecting the state in the receive side. Additionally, the relative amplitude between modulation types can be adjusted for the same peak power or for the same average power. The examples in Figure 4.16 all use approximately the same peak power. Bits are mapped from left to right, so if we have a "10" as a sequence of bits, it will be mapped to I = −6 and Q = 6 in the QPSK constellation. The "0" is the LSB (Least Significant Bit).

Constellations with same average power are obtained by multiplying each state amplitude by a factor: BPSK factor = 1, QPSK factor = 1/sqrt(2), 16QAM factor = 1/sqrt(2) and 64QAM factor = 1/sqrt(42).

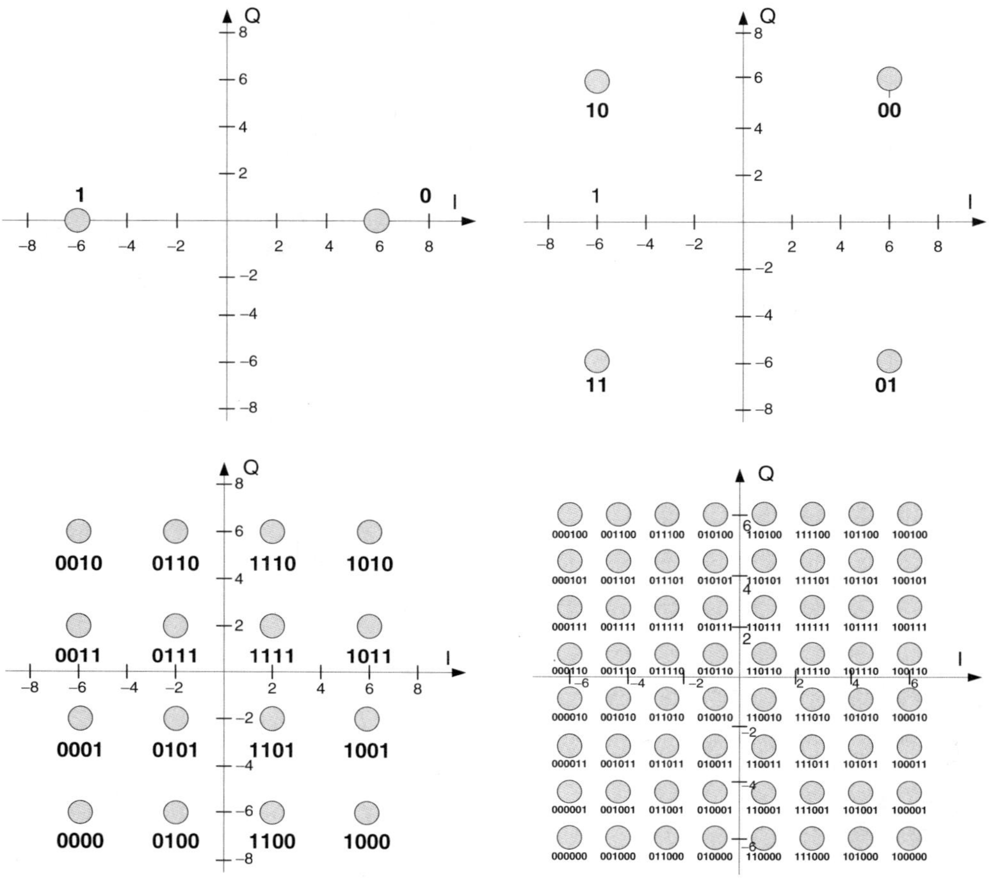

Figure 4.16 Modulation constellations for BPSK, QPSK, 16QAM and 64QAM.

Each constellation point is defined by the amplitude and phase of the carrier, but as mentioned earlier, detecting phase components is very difficult, thus the phase information is also sent as a frequency component, so the amplitude and phase are sent as different signals.

The amplitude modulates the carrier, starting from a specific phase (cosine in the formula), whereas the phase information modulates a 90° shifted carrier (sine in the formula) called in quadrature. As the signals are orthogonal to each other, it is possible to extract amplitude and phase information by just detecting the presence of the sine or cosine in the I and Q waveforms.

Amplitude or phase imbalance between the I and Q waveforms creates a side band, whereas a DC offset (due to distortion) results in carrier leakage. Figures 4.17 to 4.20 show the resultant waveforms for different modulations and bit sequences.

This modulation is done at base band and an additional up-conversion will be done before the RF signal is combined, so the I and Q streams are kept separate, until the final stage. There is a trend to use zero-IF architecture as today's DACS can provide samples up to 5 MHz. The DACs oversample internally I and Q signals (typically by 4 times), using interpolation, this spreads the aliases and gives more room for the filters to act. This is illustrated in Figure 4.21.

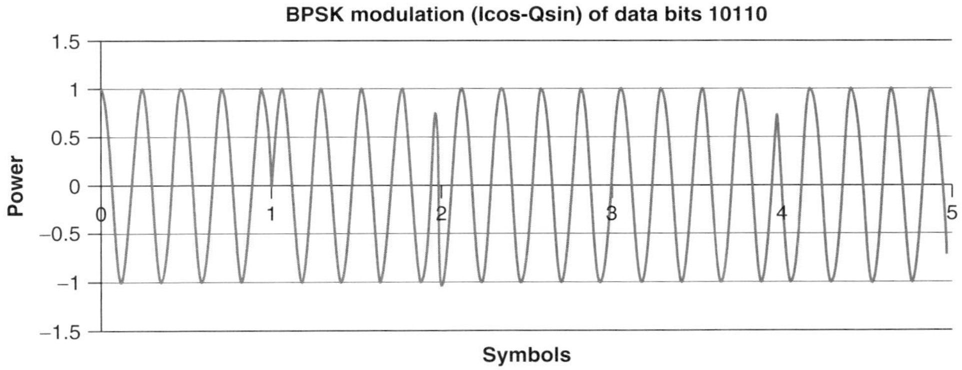

Figure 4.17 BPSK modulation of data bits 10110.

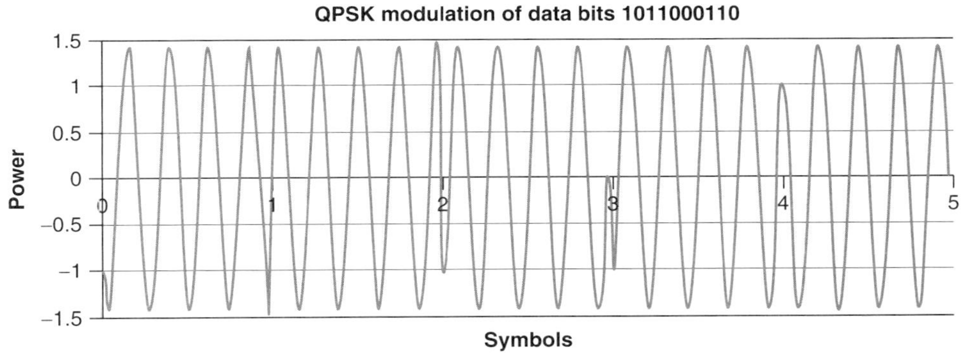

Figure 4.18 QPSK modulation of data bits 1011000110.

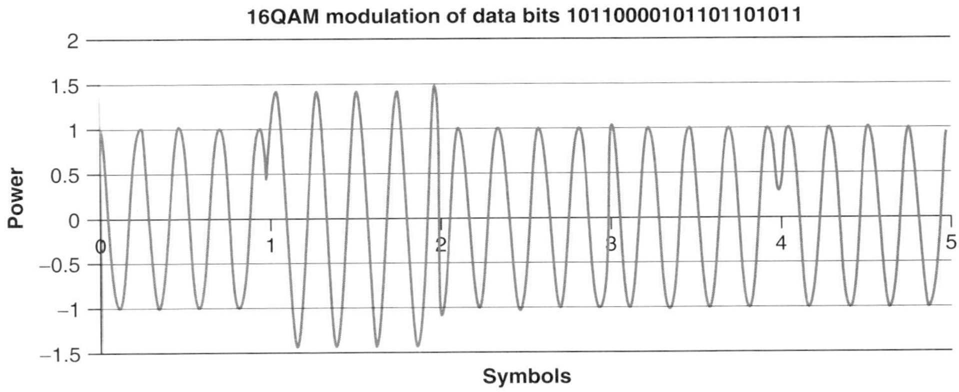

Figure 4.19 16QAM modulation of 10110000101101101011.

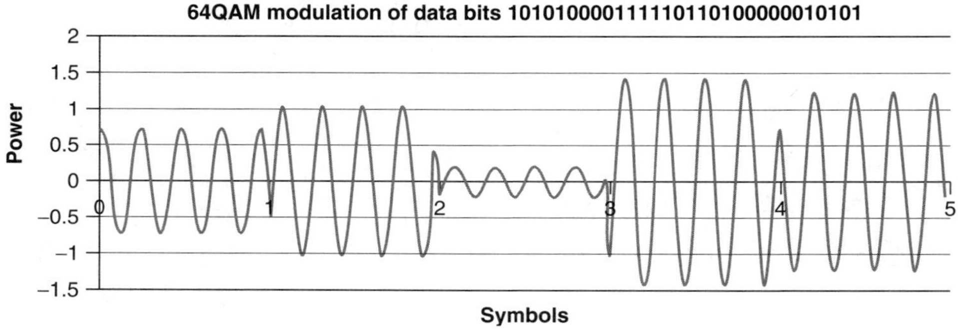

Figure 4.20 64QAM modulation of 101010000111110110100000010101.

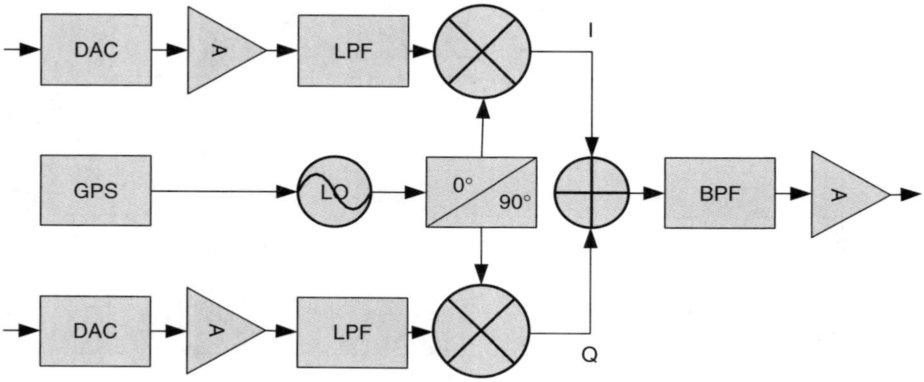

Figure 4.21 I and Q modulation.

Equations (4.12), (4.13), and (4.14) show the modulation process and how only one side band is left in the process without the need for additional filters.

$$\sin(\omega_c)\cos(\omega_m) = \frac{1}{2}(\sin(\omega_c + \omega_m) + \sin(\omega_c - \omega_m)) \quad \text{(4.12) I modulated carrier}$$

$$\cos(\omega_c)\sin(\omega_m) = \frac{1}{2}(\sin(\omega_c + \omega_m) - \sin(\omega_c - \omega_m)) \quad \text{(4.13) Q modulated carrier}$$

$$\sin(\omega_c)\cos(\omega_m) + \cos(\omega_c)\sin(\omega_m) = \sin(\omega_c + \omega_m) \quad \text{(4.14) I + Q modulated carrier}$$

Figure 4.22 illustrates the I and Q base band signals in the frequency domain (0 to 10 MHz) modulating a carrier (244 MHz). I and Q branches use the same carrier, but the Q branch carrier is shifted by 90°. The result is a DSB (Dual Side Band) signal, with the suppression of the carrier. Next I and Q signals are combined and one of the lower side bands is cancelled, resulting in an SSB (Single Side Band) signal.

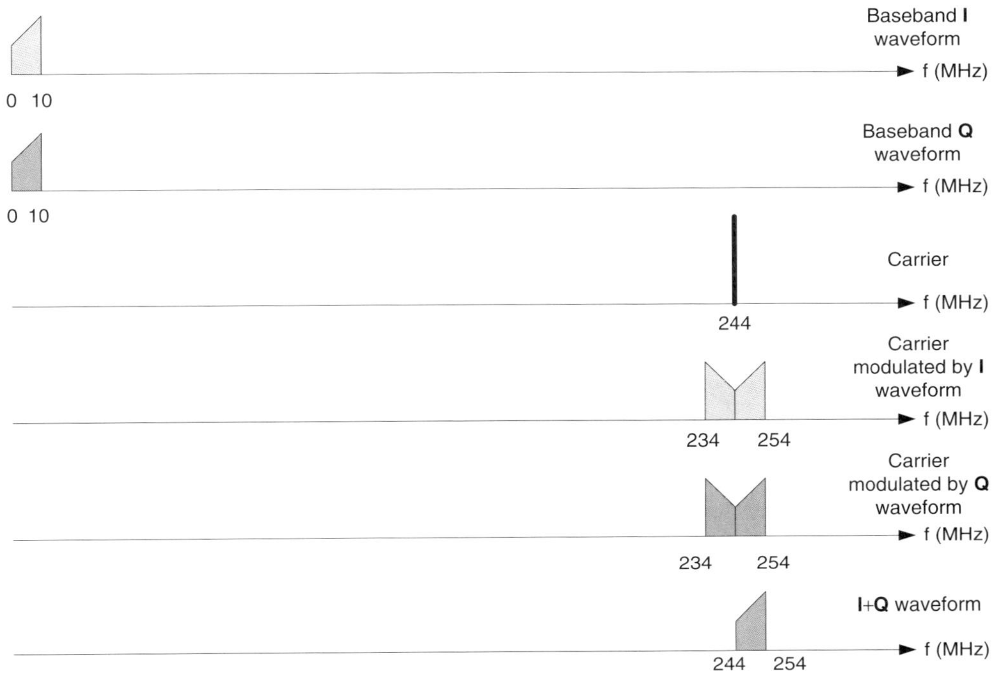

Figure 4.22 IF modulation of I and Q signals.

The representation above is ideal, but for it to happen, the circuits must be very precise. An amplitude imbalance between I and Q results in the lower sideband reappearing, and the same applies to phase variations of the quadrature carrier. A DC offset (due to waveform distortion) results in carrier leakage.

5

RF Channel Analysis

This chapter describes the signal to be transmitted and its interaction with the RF channel where it propagates. The material contained here presents aspects relevant to an RF network designer to allow him to properly perform his tasks. This book assumes that the reader is familiar with the subject; otherwise we suggest reading Chapter 9 of my book, *Designing cdma2000 Systems* published by Wiley in 2004.

5.1 The Signal

Digital wireless communications transmit digital information (sets of 1 and 0s) by changing the amplitude and phase of a carrier (phase modulation). This phase shift is done by combining two sinusoids shifted by 90°, as explained in Chapter 4. Amplitude variation of those sinusoids allows for the generation of sinusoids with different amplitude and phase shifts. This is called quadrature phase modulation or quadrature amplitude modulation (which also includes phase). Quadrature stands for the two sinusoids shifted by 90°.

To calculate the spectrum of each symbol transition in phase modulation, we can consider a continuous sine wave multiplied by a rectangular pulse that has one symbol width, as shown in Figures 5.1 and 5.2.

The spectrum of a phase-modulated signal results from the convolution of a sine wave with a rectangular function and is represented by the sinc function (sine cardinalis) defined in Equation (5.1) for which the bandwidth is given in Equation (5.2).

$$\text{sinc}(\pi t B) = \frac{\sin(\pi t B)}{\pi t B} \qquad \text{(5.1) Sinc function}$$

$$B = 1/T \qquad \text{(5.2) Bandwidth}$$

where:

B = bandwidth.
T = symbol duration.

Figure 5.3 plots the spectrum of a phase-modulated carrier (sinc function), normalized to the carrier frequency. This spectrum has the property of having zero energy at regular intervals (equal to the inverse of the symbol duration).

LTE, WiMAX and WLAN Network Design, Optimization and Performance Analysis, First Edition.
Leonhard Korowajczuk.

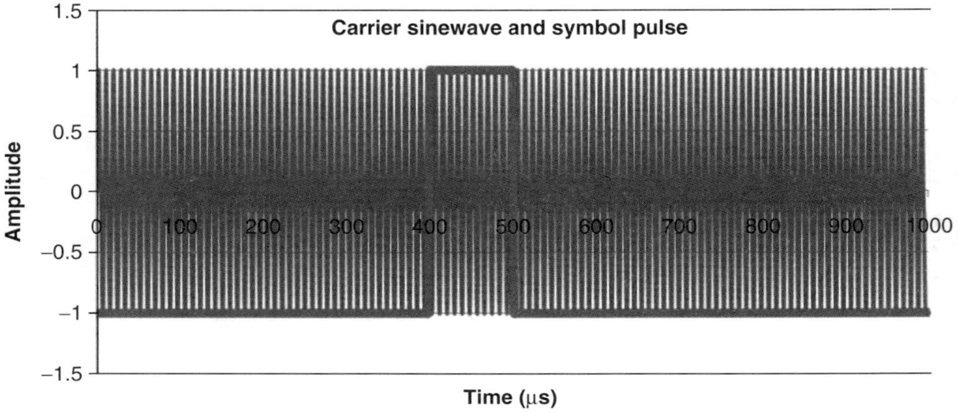

Figure 5.1 Carrier sine wave and symbol pulse.

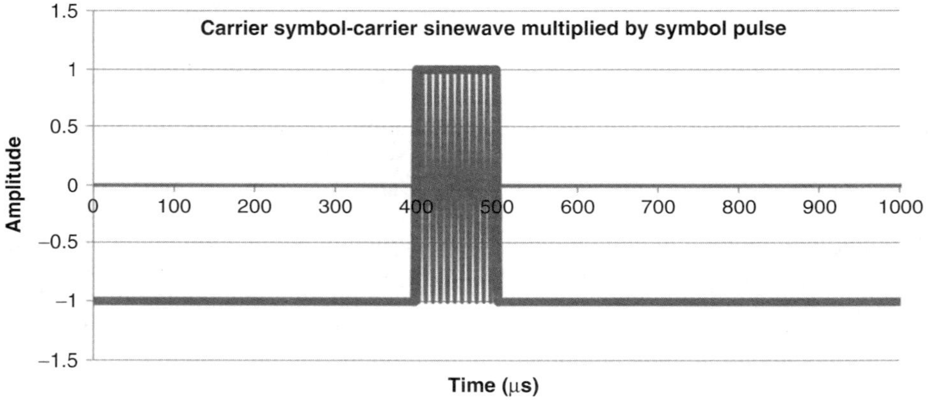

Figure 5.2 Carrier symbol-carrier sine wave multiplied by symbol pulse.

When additional symbols are added, sharp transitions occur at each symbol boundary as illustrated in Figure 5.4, but the signal spectrum can still be represented by the sinc function.

The transitions that generated the high frequency components can be smoothed by a low pass filter with a linear phase, not to distort the signal. This is illustrated in Figure 5.5 for the same transition as above.

Low pass filters can be used for this purpose. The most common being the Raised Cosine (RC) filter, defined by Equations (5.3) and (5.4). The factor α (filter roll-off factor) is a measure of the excess bandwidth required by the filter expressed in fraction of the bandwidth.

$$h(t) = \text{sinc}\left(\frac{t}{T}\right) \frac{\cos\left(\frac{\pi\alpha t}{T}\right)}{1-(2\alpha t/T)^2} \qquad \text{(5.3) RC filter (time domain)}$$

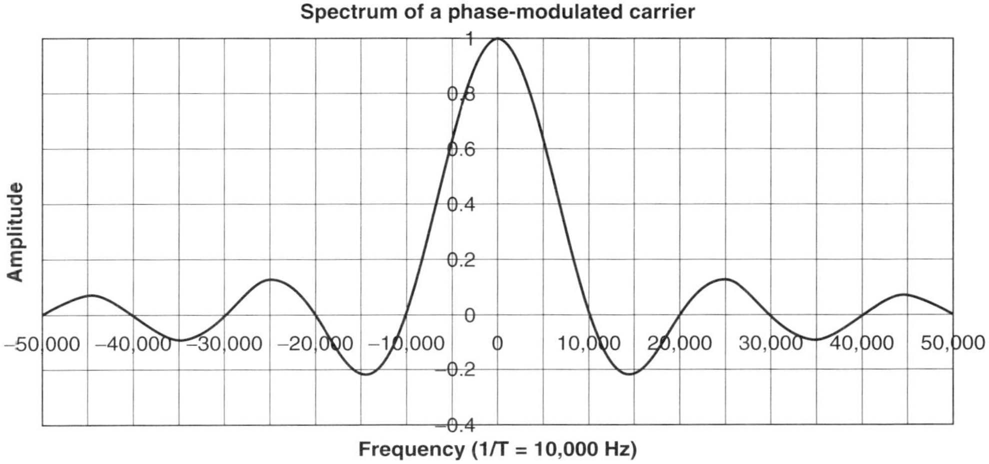

Figure 5.3 Spectrum of a phase-modulated carrier.

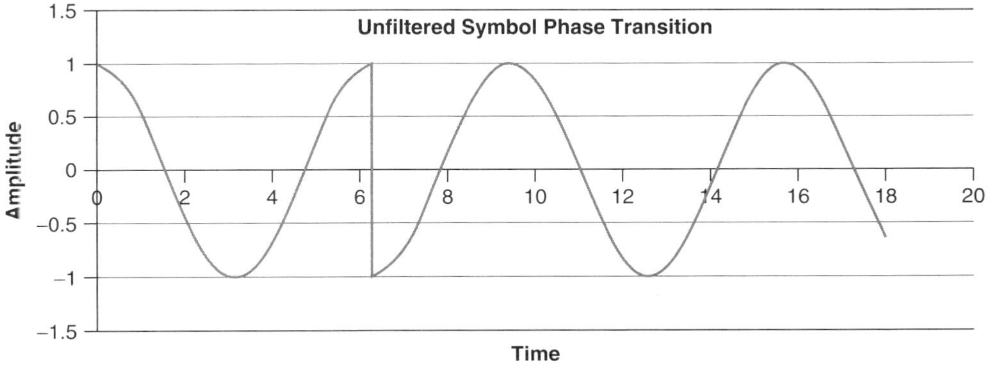

Figure 5.4 Unfiltered between symbols phase transition.

$$H(f) = \frac{T}{2}\left(1 + \cos\left(\frac{\pi T}{\alpha}|f|\frac{1-\alpha}{2T}\right)\right) \qquad \text{(5.4) RC filter (frequency domain)}$$

where:

$T =$ symbol duration.
$\alpha =$ filter roll-off factor.

Figures 5.6 and 5.7 show the frequency and time response of the raised cosine filter. A roll-off factor of zero reduces the high frequency components and narrows the main bandwidth. The use of this roll-off helps to meet stringent emission masks.

The application of this filter helps the reduction of emissions, but on the receive side, the noise is still received equally over the whole bandwidth. To minimize noise, the raised cosine filter was

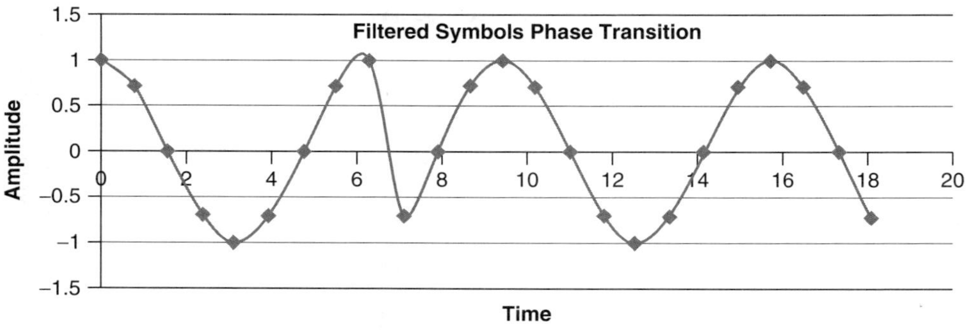

Figure 5.5 Filtered between symbols phase transition.

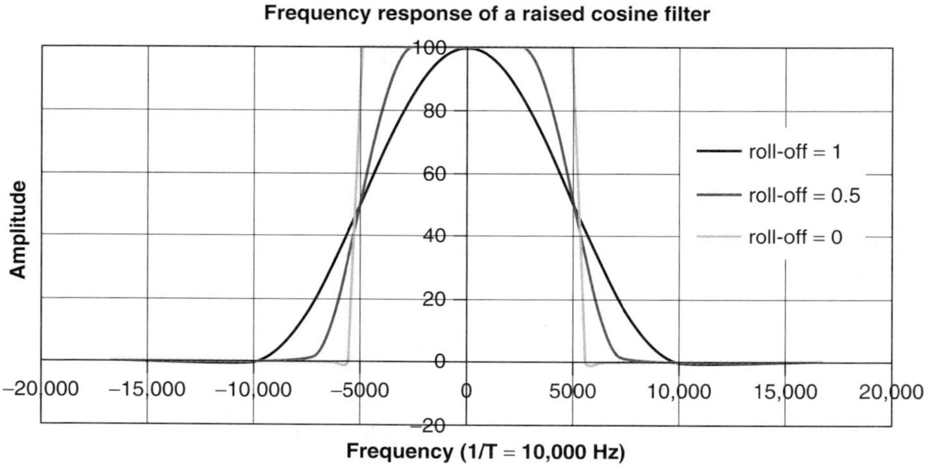

Figure 5.6 Frequency response of a raised cosine filter.

replaced by a square root raised cosine filter (SRRC) at the transmitter and a matched filter at the receiver. This results in a total response equivalent to the raised cosine filter with the addition of noise filtering in the receiver.

The equation for the square root cosine raised filter is the same as for the cosine raised filter with the application of a square root. The frequency response of such filter is shown in Figure 5.8.

Each of these carriers represents an OFDM sub-carrier. Sine waves that are multiples of each other are orthogonal and do not interfere with each other, although this orthogonality is only valid when the analysis is done over one or multiple full cycles.

A base frequency is applied to the first sub-carrier and by spacing the other sub-carriers by the inverse of symbol duration, will result in harmonically related sub-carriers. The interference caused by the transitions can be controlled and does not interfere with adjacent sub-carriers. This allows us

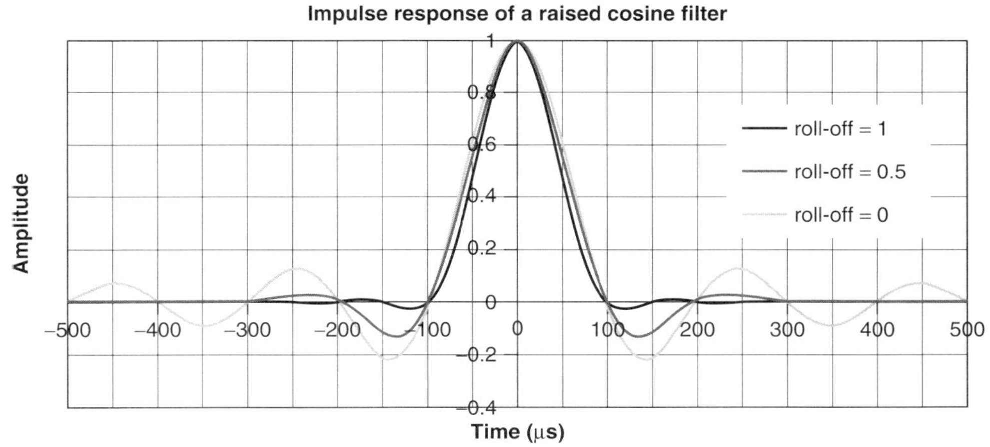

Figure 5.7 Impulse response of a raised cosine filter.

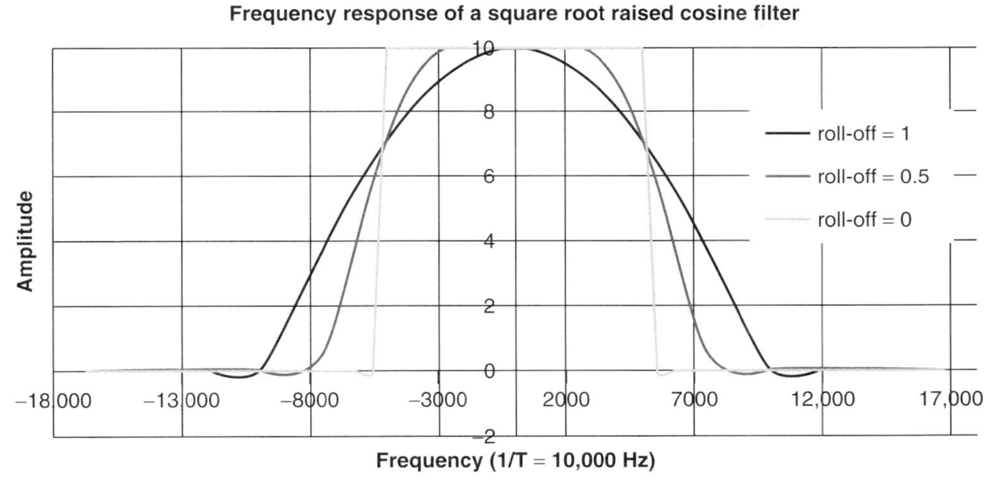

Figure 5.8 Frequency response of a square root raised cosine filter.

to construct an OFDM carrier with many sub-carriers spaced by a frequency interval equal to the inverse of the symbol duration, as illustrated in Figure 5.9.

A stream of symbols can then be represented by the superposition of these individual spectrums, each centered in a different frequency (sub-carrier). Making the frequencies coincide with the spectrum nulls avoids interference and allows for a very tight packing without the need for filters.

The time domain representation of the OFDM carrier is shown in Figure 5.10.

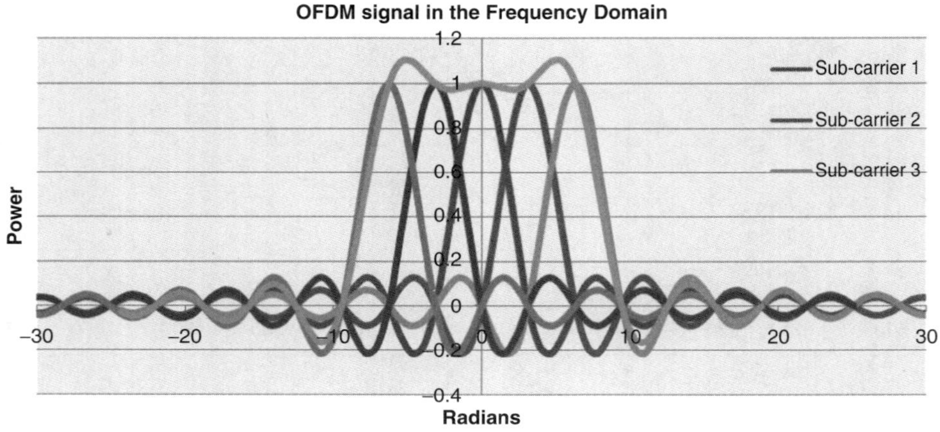

Figure 5.9 OFDM signal in the frequency domain.

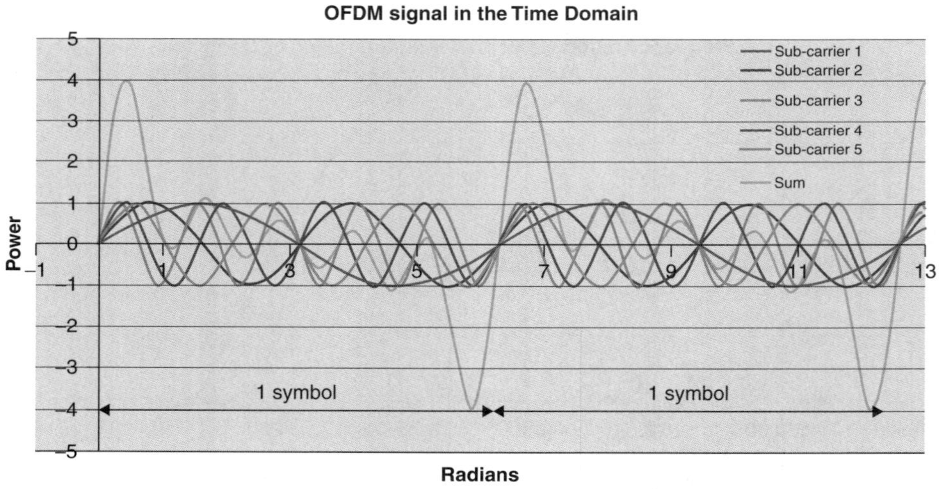

Figure 5.10 OFDM signal in the time domain.

Knowing the transmitted waveform helps enormously to detect the received signal. Summarizing, it is known that:

- Sub-carriers are orthogonal to each other as they are harmonically related. The sub-carriers can be detected by multiplying the received signal by the sub-carrier frequency and integrating the result. The orthogonality prevails even if the phase changes between sub-carriers.
- Each symbol has a duration that encompasses a multiple of full cycles of each sub-carrier frequency. This assures that the integration above is valid.
- Each sub-carrier phase is defined by orthogonal sine and cosine signals. The original phase can be detected by multiplying the received signal by the sine and cosine of the sub-carrier frequency.

Table 5.1 Bandwidth and noise floor of wireless technologies

	WiMAX	LTE	TDMA	GSM	CDMA	UMTS
Nominal bandwidth (kHz)	10	15	30	200	1500	5000
Actual bandwidth (kHz)	10	15	12	160	1250	4300
Noise floor (dBm)	−133.98	−132.21	−133.18	−121.93	−113.01	−107.64

The detection process uses the energy of the signal around the sub-carrier frequency to detect the signals. The received signal is significantly attenuated during the propagation from the transmitter and receiver, but as long as it is above the noise level in its bandwidth, it can be amplified and detected. Lower bandwidths have less noise and can propagate further. Table 5.1 gives the noise floor for different technologies.

The received signal detection is further complicated by distortions and interferences caused by the RF channel and we will analyze those effects next.

5.2 The RF Channel

When an RF signal is applied to an antenna, energy is radiated into free space. This energy propagates outward in all directions (for an isotropic antenna) and is subject to reflections, diffractions and refractions until it reaches the receiver. The receiver antenna then captures part of this energy as the received RF signal.

A transmitted signal can be broken in its sinusoidal components, so it suffices to analyze only one of its component frequencies. This frequency can be expressed as a vector characterized by its amplitude and phase, which can be represented in its complex trigonometric or exponential forms, as shown in Equation (5.5).

$$\boldsymbol{s} = s\cos\varphi + is\sin\varphi = se^{i\varphi} \quad \text{(5.5) Transmitted sinusoid}$$

The received signal can be represented in the same form as shown in Equation (5.6).

$$\boldsymbol{r} = r\cos\varphi + ir\sin\varphi = re^{i\varphi} \quad \text{(5.6) Received sinusoid}$$

The relationship between the transmitted and received signal represents the RF channel response, which can be represented by a complex multiplicative distortion, as indicated in Equation (5.7).

$$\boldsymbol{h} = \alpha\, e^{i\theta} \quad \text{(5.7) RF channel response}$$

The received signal can then also be represented by Equation (5.8).

$$\boldsymbol{r} = \boldsymbol{hs} \quad \text{(5.8) Received signal}$$

The RF channel response is not constant, due to variations in frequency and time, as illustrated in Figure 5.11. Those variations impair the signal detection and have to be deal with.

Predicting the variations and equalizing the channel is one way to deal with the issue, but this is easier said than done. Several techniques have been developed for this, but none is perfect and all of them require lots of processing time and power. It is easier to adjust channel parameters from one symbol to the next, but it is hard to compensate variations within one symbol. Besides, the number of iterations grows exponentially with the number of overlapped symbols.

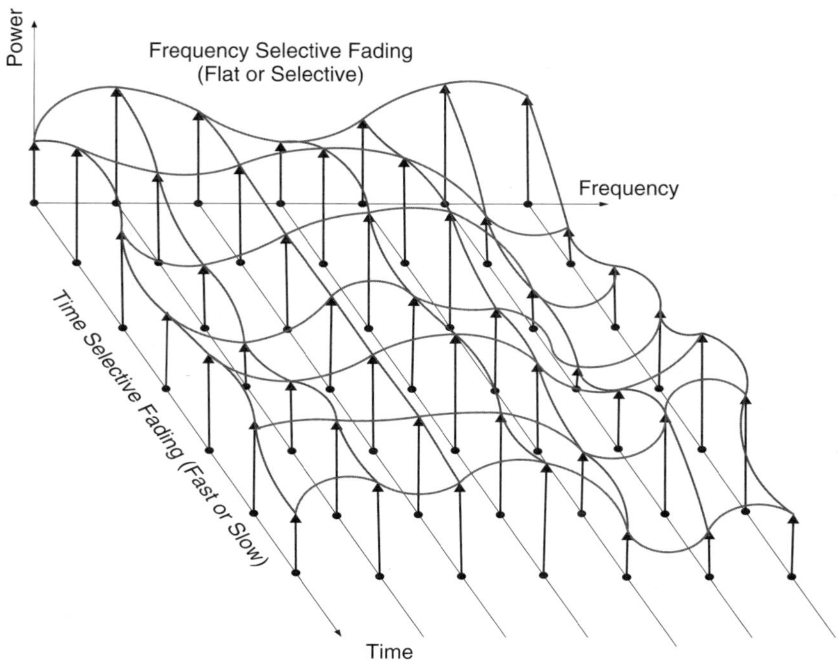

Figure 5.11 RF channel representation in frequency, time and power domains.

Another option is to restrict the symbol length in frequency and time, so it is present during bandwidths and times in which channel variations are smaller and the channel can be considered flat. This requires understanding the causes of channel variation and the extent to which the channel can be considered flat in frequency and time domains.

For the RF designer, the understanding of these impairments is key when making decisions between scenarios, analyzing prediction results or deciding the best location for deployment and orientation of antennas.

5.3 RF Signal Propagation

This section assumes previous knowledge of basic propagation mechanisms.

5.3.1 Free Space Loss

An RF signal has a free space loss given by Equation (5.9) and is shown in Figure 5.12 on a logarithmic scale.

$$L_{dB} = 32.444 + 20\log_{10} f_{MHz} + 20\log_{10} d_{km} \qquad \text{(5.9) Free space loss}$$

The loss curve has a constant slope of 20 dB per decade. In real life the slope will be higher than this.

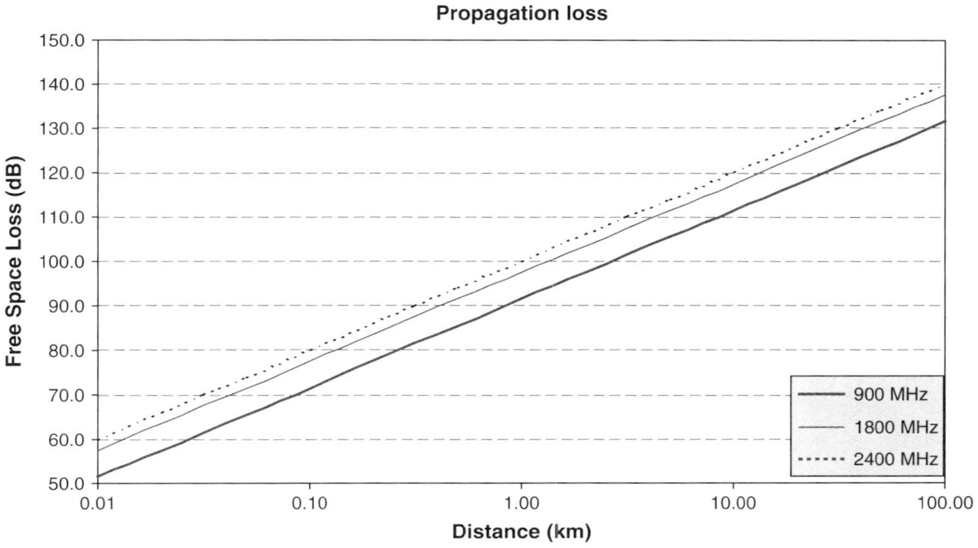

Figure 5.12 Free space propagation loss for different frequencies.

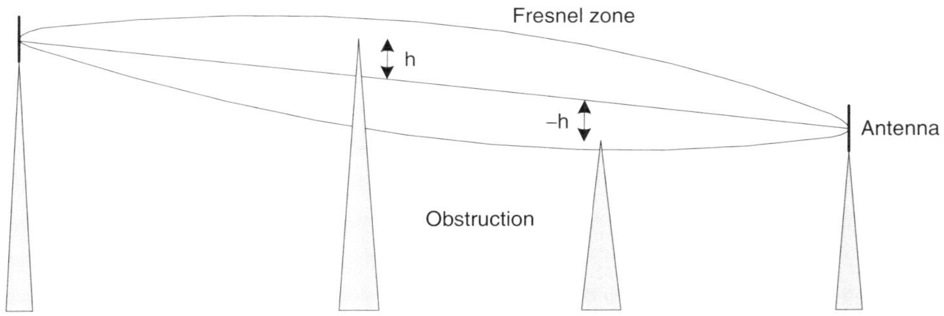

Figure 5.13 Fresnel zone depiction.

5.3.2 Diffraction Loss

RF waves go around obstructions, providing a signal behind them, but with some loss. This is called diffraction loss and has to be calculated for every position behind the obstruction.

Diffraction loss depends how much of the Fresnel zone is penetrated by the obstruction. The Fresnel zone is an ellipsoid between the transmitter and receiver radiation centers, defined by the carrier wavelength and the distance between both. Figure 5.13 shows the Fresnel zone and the relative height of the obstructions in relation to the LOS line. In Figure 5.13, h is the height of the obstruction, measured above the line connecting the transmitter and receiver radiation centers. Obstructions that are below the LOS line have a negative height.

A Fresnel zone defines a volume in which reflected or diffracted rays arrive within a specific out of phase range in relation to the LOS signal. Zone 1 corresponds to a range of 0 to 90°, zone 2 between 90° and 270° and range 3 between 270° to 450°. Zone 1 signals are constructive, zone 2 signals are destructive and zone 3 signals are constructive again.

The radius of the Fresnel zones varies with the carrier wavelength, which is given by Equation (5.10).

$$\lambda = c/f \qquad \text{(5.10) Carrier wavelength}$$

$\lambda =$ Carrier wavelength.
$c =$ Speed of light $= 3 \times 10^8$ m/s.
$f =$ Carrier frequency.

Fresnel zone radii are calculated by Equation (5.11).

$$r_F = \sqrt{\frac{n\lambda d_1 d_2}{d}} \qquad \text{(5.11) Fresnel zone radius}$$

where:
$r_F =$ n^{th} Fresnel zone radius.
$n =$ number of the Fresnel zone.
$d_1 =$ distance from transmitter to obstacle.
$d_2 =$ distance from obstacle to receiver.
$d =$ distance from transmitter to receiver.

Table 5.2 gives the first Fresnel zone radius at the mid-point between transmitter and receiver.

A factor (ν) is used to normalize the diffraction loss equations, so general formulas can be obtained. This factor is shown in Equation (5.12).

$$\nu = -h\sqrt{\frac{2d}{\lambda d_1 d_2}} \qquad \text{(5.12) Normalization factor}$$

where:
$h =$ height of the obstruction above the line connecting transmitter and receiver radiation centers. Obstructions that are below the LOS line have a negative height.
$d =$ total distance.
$d_1 =$ distance to obstruction.
$d_2 =$ distance from obstruction.

Table 5.2 Fresnel zone radius at 50% distance (m)

		Distance (m)			
f (GHz)	λ (m)	10	100	1000	10,000
0.7	0.429	1.04	3.27	10.35	32.73
1	0.300	0.87	2.74	8.66	27.39
2.5	0.120	0.55	1.73	5.48	17.32
3.5	0.086	0.46	1.46	4.63	14.64
5	0.060	0.39	1.22	3.87	12.25

The diffraction loss for different ranges of the normalization factor (ν) is given by Equations (5.13), (5.14) and (5.15).

$\nu <= 0$

$$L(\nu) = 0 \quad (5.13)\ \nu <= 0$$

$0 < \nu <= 4$

$$L(\nu) = 0.0056\nu^6 + 0.0906\nu^5 + 0.5406\nu^4 - 1.339\nu^3 + 0.1008\nu^2 + 8.5679\nu + 6.1154 \quad (5.14)\ 0 < \nu <= 4$$

$\nu > 4$

$$L(\nu) = 20\log\left(\frac{0.225}{\nu}\right) \quad (5.15)\ \nu > 4$$

Tables 5.3 and 5.4 show the diffraction loss at 1 GHz for links of 100 m and 1000 m respectively. The loss is shown for several obstruction heights and locations along the link.

Table 5.3 Diffraction loss for 1 GHz at 100 m for different distance ratios

		Obstruction height (m)					
Obstruction height (m)		−1.64	0.00	2.74	5.00	10.00	20.00
d_1 (m)	d_2 (m)	Diffraction loss (dB)					
1	99	0	6	33	38	44	50
2	98	0	6	30	35	41	47
5	95	0	6	26	31	37	44
10	90	0	6	23	29	35	41
20	80	0	6	21	26	32	38
30	70	0	6	20	25	31	37
40	60	0	6	19	24	30	36
50	50	0	6	19	24	30	36

Table 5.4 Diffraction loss for 1 GHz at 1 km for different distance ratios

		Obstruction height (m)					
		−5.20	0.00	8.66	5.00	10.00	20.00
d_1 (m)	d_2 (m)	Diffraction loss (dB)					
10	990	0	6	33	28	34	40
20	980	0	6	30	25	31	37
50	950	0	6	26	21	27	34
100	900	0	6	23	19	25	31
200	800	0	6	21	17	22	28
300	700	0	6	20	15	21	27
400	600	0	6	19	15	20	26
500	500	0	6	19	15	20	26

5.3.3 Reflection and Refraction

When an RF wave hits an obstruction, part of its energy propagates through it (is refracted) and part is reflected. The amount of energy of each component is influenced by the type of material and the incidence angle, which is defined as the angle to the perpendicular to the surface. The reflection and refraction behavior is different when the electrical field is parallel or perpendicular to the scattering plane.

A vertical antenna's electrical field is perpendicular to the ground (vertical polarization), whereas a horizontal antenna's is parallel to ground (horizontal polarization). This is illustrated in Figure 5.14.

The ratio of the reflected to the incident power is expressed by the reflected power factor and is presented, respectively, in Figures 5.15 and 5.16 for parallel and perpendicular incidence and for different materials.

A vertical antenna signal significantly reflects from the ground, but not much from walls. A horizontal antenna signal reflects little from ground, but significantly from walls. Signals change polarities when they reflect and reach ground and walls at different angles and different polarities. An estimated average loss of 3 dB for ground reflections and 6 dB for wall reflections can be considered.

The general guidelines for signal propagation are the following:

- Conductive and dielectric material reflect incident RF waves, varying with the wave's incidence angle.
- Plane surfaces larger than ten wavelengths are good reflectors, whereas surfaces with irregularities less than one wavelength will scatter the signal.
- Flat surfaces reflect better than irregular ones and a surface can be considered reflective if it has flat planes larger than the wave wavelength.
- Signals refract around buildings also. An isolated, very high building may cause little obstruction.
- At long range, signals propagate over the morphology and not through it.

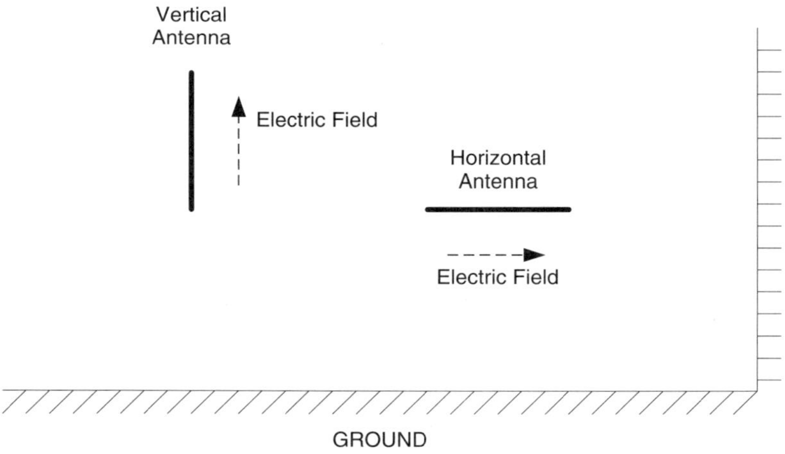

Figure 5.14 Electrical field direction in relation to antenna polarization.

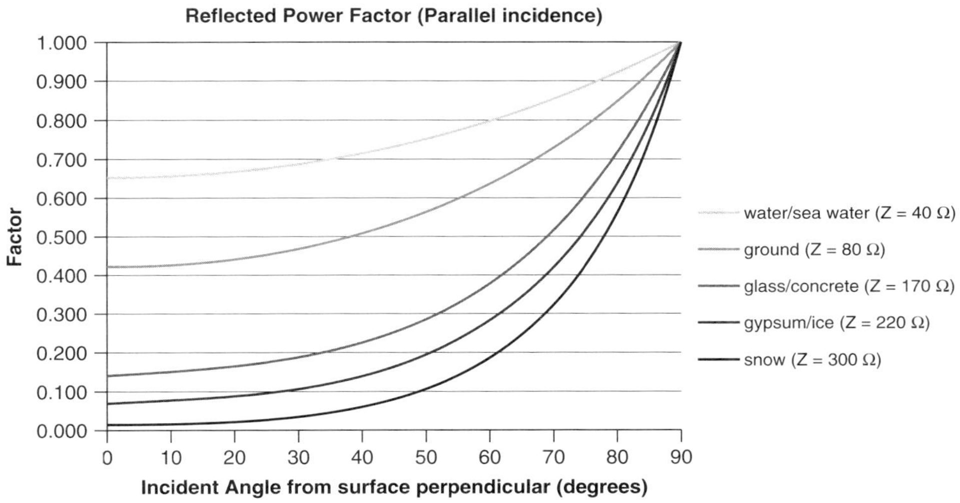

Figure 5.15 Reflected power factor for parallel incidence.

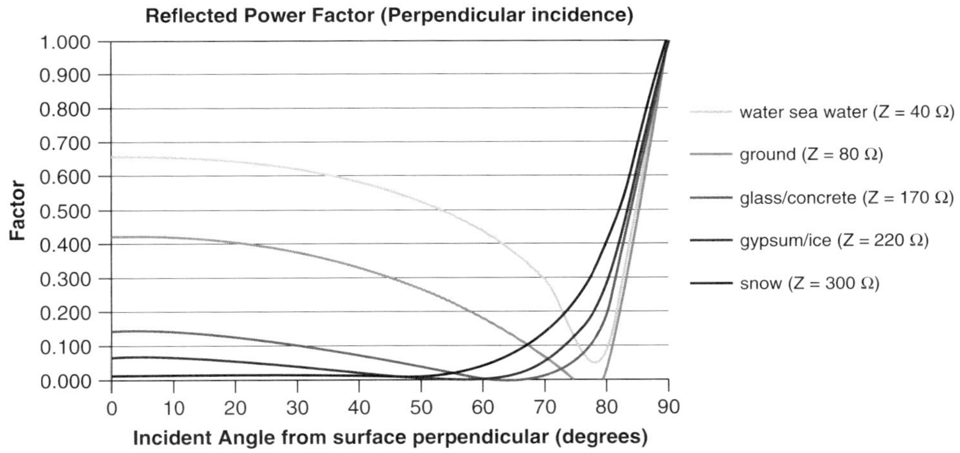

Figure 5.16 Reflected power factor for perpendicular incidence.

5.4 RF Channel in the Frequency Domain

5.4.1 Multipath Fading

A major impairment of the received signal is caused by multipath, which can be constructive or destructive, as signals arrive out of phase and may enhance or cancel each other. The received signal is the result of the sum of many paths, in which the signal was diffracted and reflected several times.

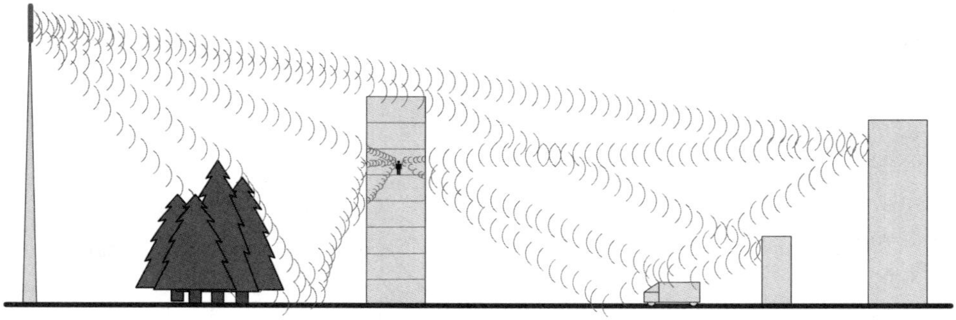

Figure 5.17 Multipath depiction.

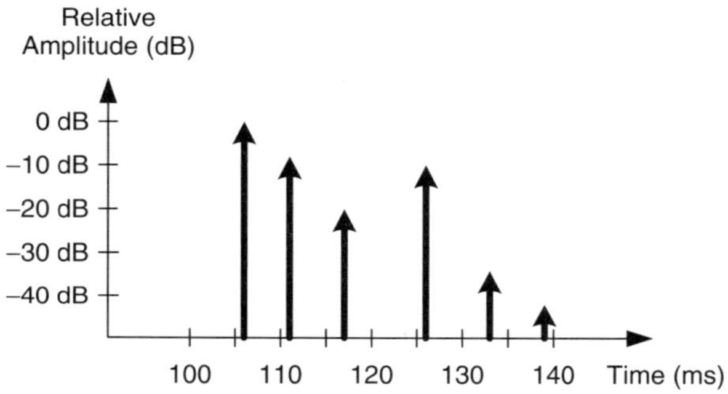

Figure 5.18 Multipath components arrival times.

Figure 5.17 depicts several signal multipaths arriving at a single location. These multipaths are caused by single or multiple refractions, reflections and diffractions. The designer should evaluate the possible multipaths in each deployment as an understanding on how signals are attenuated and reflected is important to properly characterize the RF channel.

Figure 5.18 shows multipath components' arrival times. Several copies of the same signal are received. Each of the received copies causes a phase shift on the signal, which is negligible if the signal is small compared to the main one. Signals 10 dB below the main signal should be discarded.

Figure 5.19, Figure 5.20 and Figure 5.21 show different phase combinations for two paths of the same transmitted signal. Short phase delays reinforce the signal, but delays with a phase shift of 180 degrees will null the signal. A difficulty that arises is how to express a single multipath delay at a certain point, due to the different amplitudes of each path. As explained in Chapter 4, the phase of the resultant sinewave can be calculated by power weighting the phase contribution of each sine wave. We can apply the same reasoning for the time delay. The average value is then used to calculate the RMS value and this is considered as the RMS delay spread.

Figure 5.22 shows the resultant RMS power of a main sinewave and a shifted copy of it, for a complete cycle of phase shifts, expressed in dB. Shifts up to 120° do not have much impact on the resultant signal, thus the channels can be considered flat within this range (about 1/3 of the cycle), while shifts between 120° and 240° are destructive (deep fading area). Larger shifts between 240° and 360° result in flat channels again. The deep is very pronounced when both signals have the same amplitude. Figure 5.23 shows the deep, when one signal has 50% of the amplitude of the main signal.

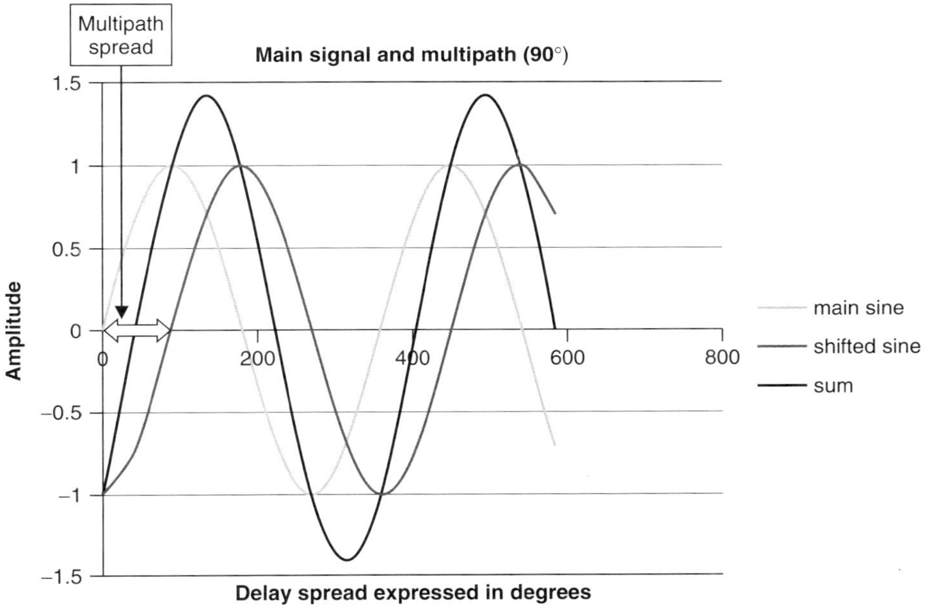

Figure 5.19 Main signal and a 90° multipath combination.

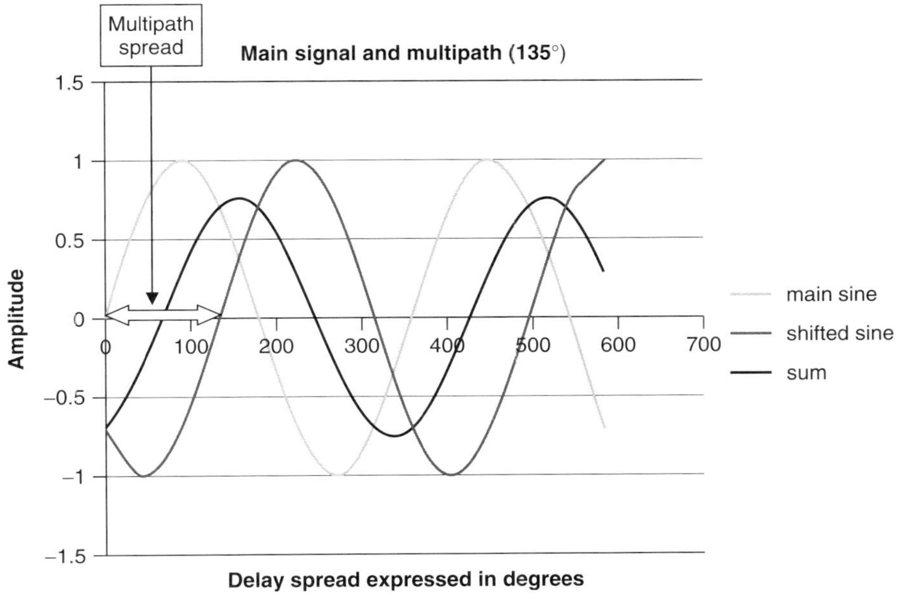

Figure 5.20 Main signal and a 135° multipath combination.

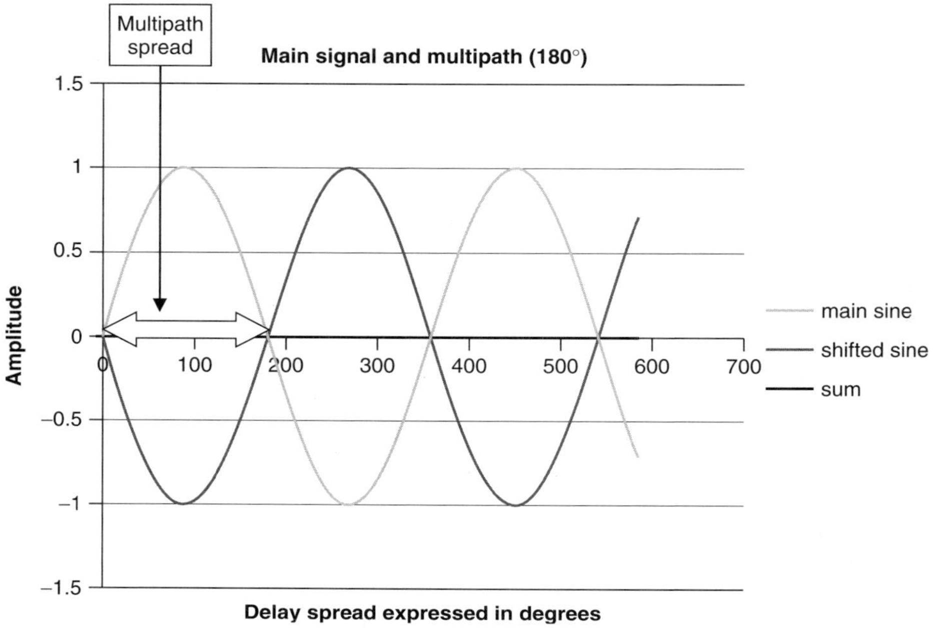

Figure 5.21 Main signal and a 180° multipath combination.

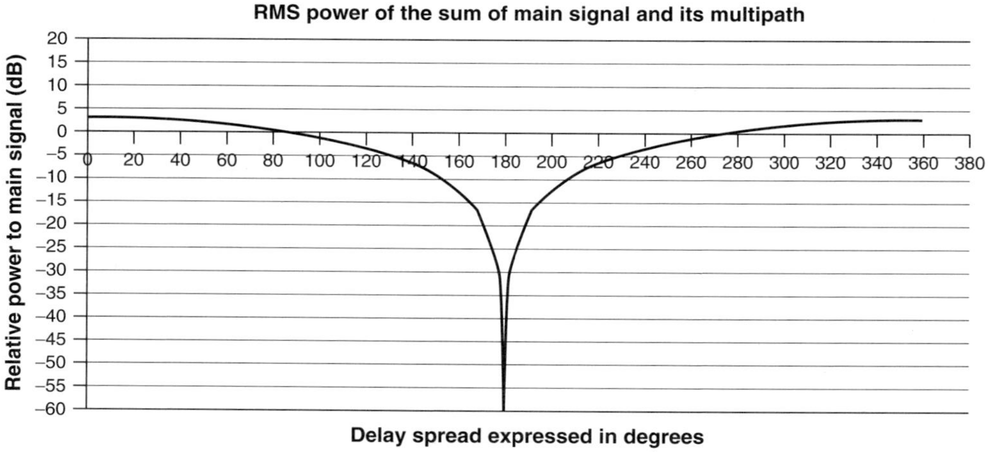

Figure 5.22 RMS power of the sum of same amplitude main signal and its multipath.

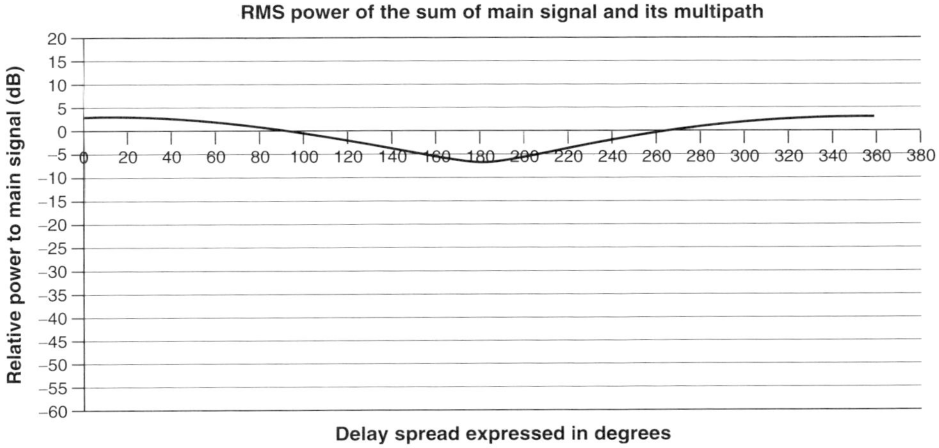

Figure 5.23 RMS power of the sum of main signal and its 50% amplitude multipath.

Table 5.5 Multipath fading distance for different frequencies

Bandwidth (kHz)	1 cycle (μs)	1/3 cycle (μs)	1/3 cycle (m)
1.0	1000	333.3	100,000
5.0	200	66.7	20,000
7.5	133	44.4	13,333
10.0	100	33.3	10,000
15.0	67	22.2	6667
200	5	1.7	500
1000	1.00	0.33	100
5000	0.20	0.07	20
10,000	0.10	0.03	10
100,000	0.0100	0.0033	1.00
1,000,000	0.0010	0.0003	0.10
2,500,000	0.00040	0.00013	0.04
3,500,000	0.00029	0.00010	0.03
5,000,000	0.00020	0.00007	0.02

There is little we can do to control the multipath in an environment, but we can restrict the channel bandwidth, so it will not be seriously affected by the multipath.

Table 5.5 shows the extent of the fading distance (1/3 of the cycle) for different bandwidths.

As shown in Table 5.5, for OFDM, the 10 kHz subcarrier bandwidth tolerates about 10,000 m of multipath; an UMTS system, however, with its 5000 kHz channel, only tolerates about 20 m. UMTS has, however, much more complex equalization capabilities to try to compensate for this (it can equalize about 5 to 6 symbols).

For the carrier frequencies, it is impossible to avoid fading, because the multipath distance is very small, but a small position adjustment will reposition the receiver outside the fading zone. This is not true, however, for the lower modulating frequencies in which distances involved are large. Multipath fading effect affects the whole band and its impact should be considered on the same frequency and on different frequencies inside the band.

- Same frequency signals are affected by the multipath length spread (difference between the signal paths).
- Different frequency signals are affected by the whole path length, as each frequency arrives with a different phase due to the whole path propagation. This phase variation effect is deterministic and can be compensated for by a band equalizer.

It is possible to build channel equalizers, to counteract the effect of multipath, but because it has a dynamic effect (varies with time), but it is difficult and expensive to implement. Ideally, the channel bandwidth should be such that multipath can be avoided altogether, and this means that the path propagation should be smaller than the multipath distance (1/3 of the bandwidth cycle).

As an example, assume a 3.5 GHz carrier with a channel bandwidth of 10 kHz. In this case, the carrier has a fading distance of 30 cm, and its fading can be adjusted by moving the receiver antenna a few centimeters. The modulating frequency of 10 kHz allows providing service up to 10 km, without significant multipath effect. If a band frequency equalizer is used, a 10 km multipath spread can be supported. At the other extreme, a 1 MHz bandwidth can be used, but this reduces our multipath to 100 m, which is too short a distance even for the multipath spread.

Figure 5.24 and Figure 5.25 show the maximum distance that still avoids multipath issues for different frequencies. Note that this distance applies to the whole path if frequency bandwidth equalization is not used, and only up to the distance spread if it is used.

The duration of the path or the multipath spread, whichever is more restrictive, is called the coherence bandwidth and is defined by the inverse of this time. In the literature, the coherence bandwidth is expressed as the bandwidth for which a full cycle shift occurs and is given by Equation (5.16).

$$B_c = 1/\sigma_\tau \qquad \text{(5.16) Coherence bandwidth}$$

where:

σ_τ = RMS delay spread.

τ = delay spread for signals within a 10 dB window.

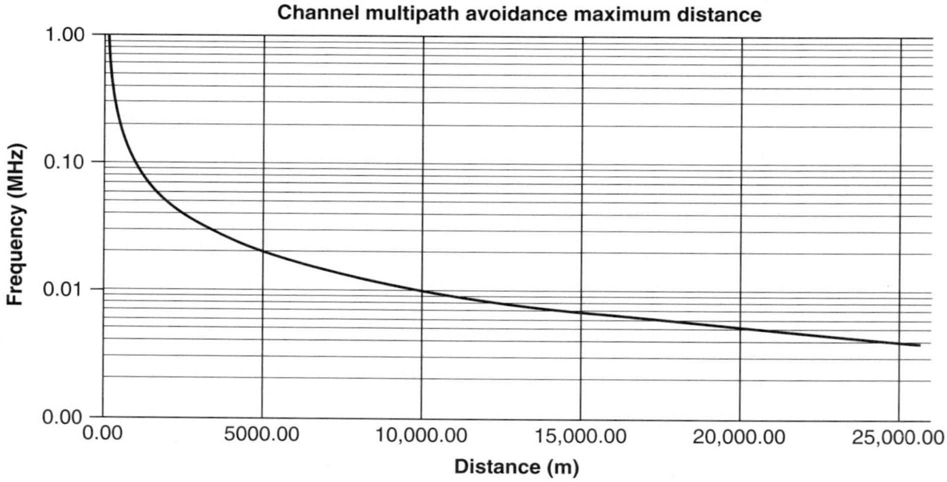

Figure 5.24 Channel multipath avoidance maximum distance.

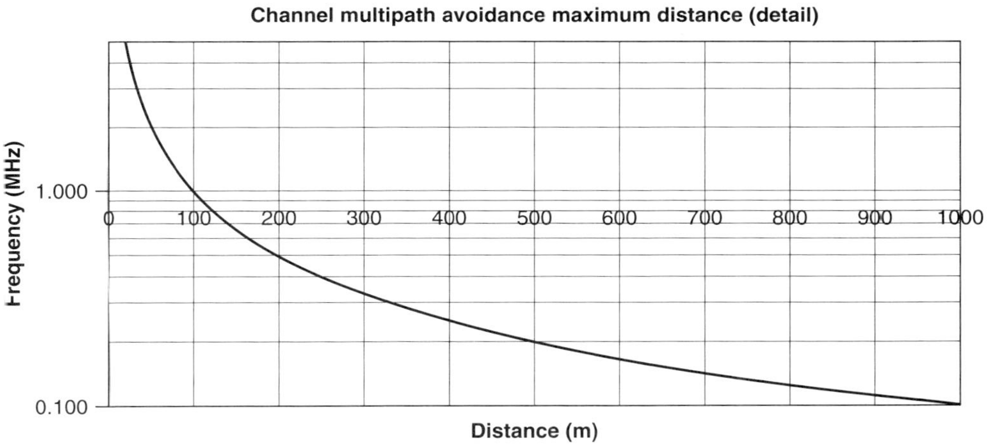

Figure 5.25 Channel multipath avoidance maximum distance (detail).

The RMS delay spread is given by Equation (5.17):

$$\sigma_\tau = \sqrt{\overline{\tau^2} - \overline{\tau}^2} \qquad \text{(5.17) RMS delay spread}$$

The coherence bandwidth can then be expressed for 50% correlation between end frequencies (the channel varies by 3 dB), or for 90% correlation (0.5 dB channel variation) and is defined respectively by Equations (5.18) and (5.19):

Coherence bandwidth for 50% correlation (3 dB variation).

$$B_{c,50\%} = 1/5\sigma_\tau \qquad \text{(5.18) Coherence bandwidth 50% correlation}$$

Coherence bandwidth for 90% correlation (0.5 dB variation).

$$B_{c,90\%} = 1/50\sigma_\tau \qquad \text{(5.19) Coherence bandwidth 90% correlation}$$

The 1/3 cycle criteria can be used to define a flat channel, which corresponds to a variation of about 6 dB and is defined by Equation (5.20):

Coherence bandwidth for 25% correlation (6 dB variation).

$$B_{c,25\%} = 1/3\sigma_\tau \qquad \text{(5.20) Coherence bandwidth 25% correlation}$$

Table 5.6 gives the coherence bandwidth with 25% correlation for several multipath distances.

Multipath spread varies significantly from one area to another and has to be evaluated by the designer on a case by case basis. Typical design values are presented in Table 5.7 for different scenarios.

The ideal bandwidth is between 33 kHz and 2 kHz, based on Table 5.8.

Small bandwidths are prone to frequency tolerance issues. Commercial crystals can be obtained with a stability of 10^{-7}, which represents a deviation of 500 Hz at 5 GHz. The minimum bandwidth should be at least 20 times this limit (10 kHz).

Table 5.8 presents the maximum multipath distance for the main wireless technologies for the rural scenario of Table 5.7. Some technologies extend this distance by using channel equalization techniques.

Table 5.6 Coherence bandwidth for several multipath distances

Multipath (km)	Coherence bandwidth (kHz)
1	100
2	50
4	25
6	17
8	13
10	10
12	8
14	7
16	6
18	6
20	5

Table 5.7 Typical multipath used for design

Environment	Multipath RMS delay spread (ns)	RMS spread distance (m)
Indoor	≤ 270	≤ 81
Urban	≤ 2100	≤ 630
Suburban	≤ 3500	≤ 1050
Rural	$\leq 10,000$	≤ 3000

Table 5.8 Coherence bandwidth for different technologies

Technology	Nominal bandwidth (kHz)	Used bandwidth (KHz)	Maximum multipath distance(m)
WIMAX	10.5	10	10,000
LTE	15.5	15	6667
TDMA	30	12	8333
GSM	200	160	625
CDMA	1500	1250	80
UMTS	5000	4500	22

5.4.2 Shadow Fading

Shadow fading is also known as long-term fading and is caused by morphology obstructions. When the first prediction models were developed, shadow fading was a very important factor because the predictions were done per distance (over a circle) and this factor gave the variation over the circle. Today predictions are done on a much smaller pixel basis, and shadow fading was reduced to express the variations inside a pixel. Shadow fading distribution is considered log-normal, meaning that its value in dB follows a Gaussian distribution. Shadow fading can be measured in the field, during the propagation model calibration process and depends on pixel size.

5.5 RF Channel in Time Domain

In a static environment, the multipath effect would be constant, and in a well-designed system the modulating frequency does not fade, which means that a carrier fading free spot can be easily found. In real life, the environment is never static and the fading pattern will be affected by changes in the environment.

It is possible to adjust the system gain for changes between symbols, but changes during one symbol period disrupt its correct detection. Channel variations cannot be reduced in time, but an adequate symbol size shorter than those variations can be chosen.

The variations in time domain are caused by the movement of the receiver and transmitter, or by changes in the environment. These factors cause a change in the multipath components and, consequently, a different fading pattern. Changes of fading patterns over time are known as time selective fading.

To characterize changes in time, the speed of the elements (receiver, transmitter, and environment) causing the changes has to be determined. This section uses the term "relative speed" to describe this concept. Cars, trees, and receiver movement are the major contributors to these changes. This section explains how the channel is affected by fading due to the relative speed of these elements.

5.5.1 Wind Effect

The wind causes trees to move, impacting the relative speed of the system, and, thus causing fading. This movement is relatively slow and varies with the speed of the wind. Assuming a range of movement for the tree of 1 meter, the fading time can be calculated by Equation (5.21). Table 5.9 correlates the expected fading durations to different wind speeds.

$$F_t = \frac{d}{v} \qquad \text{(5.21) Fading time due to trees}$$

where:

$D =$ range of tree movement (m).
$v =$ speed of the wind (m/s).
$F_t =$ fading time (s).

5.5.2 Vehicles Effect

Vehicles reflect the signal on their sides and their speed causes these reflections to impact fading differently. This impact can be calculated by Equation (5.22).

$$F_t = \frac{l}{v} \qquad \text{(5.22) Fading time due to vehicles}$$

Table 5.9 Trees effect on fading duration

Wind speed (km/h)	Wind speed (m/s)	Fading time (ms)
10	2.778	360
5	1.389	720
1	0.278	3600
0.1	0.028	36,000
0.01	0.003	360,000

Table 5.10 Vehicle movement effect on fading duration

Vehicle speed (km/h)	Vehicle speed (m/s)	Fading time (ms)	
		1 m long (car)	4 m long (truck)
120	33.3	30	120
100	27.8	36	144
80	22.2	45	180
60	16.7	60	240
40	11.1	90	360
20	5.6	180	720
10	2.8	360	1440
5	1.4	720	2880
1	0.3	3600	14,400

where:

l = length of the vehicle (m).
v = speed of vehicle (m/s).
F_t = fading time (s).

Table 5.10 summarizes the expected fading effect caused by the movement of cars and trucks.

5.5.3 *Doppler Effect*

Movement of the receiver or transmitter causes a change in the perceived frequency (Doppler effect), resulting in a Doppler frequency shift that causes a phase slip, which, over a certain number of cycles, builds up to a full cycle. Halfway through this time, the system reaches an anti-phase (fading) and the signal is then built up again until reaching the full cycle. The time it takes for the initial slip to reach a full cycle slip is known as Coherence Time (T_c).

The variation in the perceived frequency due to the Doppler effect is expressed by Equation (5.23) Doppler frequency change.

$$\Delta f = \frac{v}{\lambda} = \frac{fv}{c} \qquad \text{(5.23) Doppler frequency change}$$

where:

Δf = frequency variation (Hz).
v = vehicle speed (m/s).
λ = wavelength given by c/f (m).
f = frequency (Hz).
c = speed of light (m/s).

Table 5.11 shows the Doppler shift at various speeds for a frequency of 1 GHz, for a receiver moving in the direction of the transmitter. The frequency shift is reduced if the movement is not in the direction of the source.

The frequency with which these variations happen can be calculated for each of the different effects described in this section. This allows the determination of the fastest changing cycle. Table 5.12 calculates this fading frequency for a 1 GHz carrier based on different relative speeds of the system. In Table 5.12, N represents the number of cycles passed until the accumulated phase slip reaches 1 full

Table 5.11 Doppler shift

Relative speed (km/h)	Relative speed (m/s)	Δf (Hz)
120	33.3	111.1
100	27.8	92.6
80	22.2	74.1
60	16.7	55.6
40	11.1	37.0
20	5.6	18.5
10	2.8	9.3
5	1.4	4.6
1	0.3	0.9
0.1	0.0	0.1
0.01	0.0	0.01

Table 5.12 Coherence time of a 1 GHz carrier for different relative speeds of the system

Relative speed (km/h)	Δf (Hz)	N (cycles)	Coherence time T_c (ms)
120	111.1	9.0E+06	9
100	92.6	1.1E+07	11
80	74.1	1.4E+07	14
60	55.6	1.8E+07	18
40	37.0	2.7E+07	27
20	18.5	5.4E+07	54
10	9.3	1.1E+08	108
5	4.6	2.2E+08	216
1	0.9	1.1E+09	1080
0.1	0.1	1.1E+10	10,800
0.01	0.01	1.1E+11	108,000

cycle. T_c, or coherence time, is the period during which the phase shifts one period (360 degrees/full cycle) due to the frequency shift caused by the Doppler effect. Because the Doppler shift varies with the wavelength, the coherence time also varies with the frequency.

Table 5.13 calculates the Doppler effect (coherence time) for different frequencies, showing that the fading caused by it can vary from 3 ms to approximately 1 second. The Doppler effect inflicts the most constraint on channel variation with time, which is 2 ms (for 5 GHz).The coherence time is then defined by Equation (5.24).

$$T_c = \frac{1}{\Delta f} \qquad \text{(5.24) Coherence time}$$

In the literature, we find variations between 0.179 and 0.423 as acceptable limits of the coherence time. Based on the findings presented previously for the coherence bandwidth, a similar relationship can be assumed.

The coherence time can be expressed for 12.5% correlation between the end frequencies (the channel varies by 3 dB), or for 25% correlation (0.5 dB variation) and is defined respectively by Equations (5.25) and (5.26):

Table 5.13 Summary of Doppler effect

Relative speed (km/h)	Coherence time (ms) 1 GHz	3 GHz	5 GHz
120	9	3	2
100	11	3	2
80	14	4	3
60	18	5	4
40	27	8	5
20	54	15	11
10	108	31	22
5	216	62	43
1	1080	309	216

Coherence time for 12.5% correlation (12 dB variation) is defined by:

$$T_c = \frac{1}{2.5\Delta f} \qquad \text{(5.25) Coherence time for 12.5\% correlation}$$

Coherence time for 25% correlation (6 dB variation) is defined by:

$$T_c = \frac{1}{3\Delta f} \qquad \text{(5.26) Coherence time for 25\% correlation}$$

Use of 75% correlation is recommended for a 5 GHz carrier frequency, and it results in a coherence time of 300 μs as maximum symbol duration. The symbol duration is inversely proportional to the bandwidth, resulting in a minimum bandwidth of 3.3 kHz is obtained, well within the 10–30 kHz range previously established.

Coherence time for 75% correlation (2 dB variation) is defined by Equation (5.27).

$$T_c = \frac{1}{6\Delta f} \qquad \text{(5.27) Coherence time for 75\% correlation}$$

5.5.4 *Fading Types*

Fading can be classified according to the relation of the coherence time to the symbol duration and the coherence bandwidth to the channel bandwidth. This is shown in Figure 5.26.

Fading rapidity can be measured by examining how often fading crosses a given threshold (level), in what is known as level crossing rate. The approximate rate for different fade severities at different speeds in a Rayleigh channel is defined by Equation (5.28) and typical values are shown in Table 5.14.

$$N_c = \sqrt{2\pi}\Delta f \rho^2 \qquad \text{(5.28) Number of fading crossings}$$

where:

N_c = number of crossings below fade intensity threshold, defined by ρ.
Δf = Doppler frequency variation (Hz).
ρ = fade severity (dB).

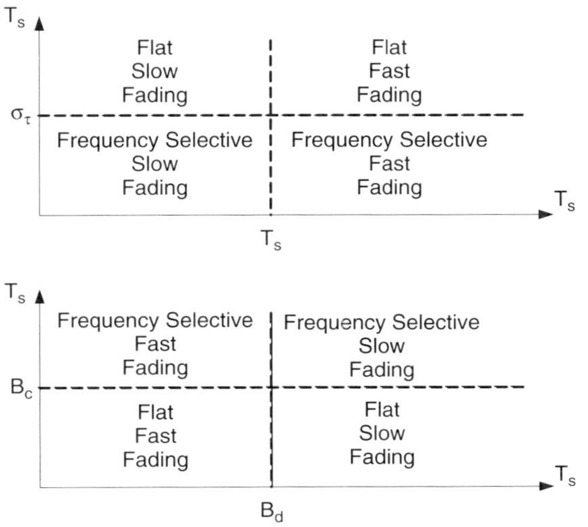

Figure 5.26 Fading classification.

Table 5.14 Level crossing rate according to receiver speed

Level crossings/second				
	Fade intensity (dB)			
speed (km/h)	−3	−6	−10	−20
0.1	1	0	0	0
1	5	2	1	0
10	52	22	8	1
100	522	217	82	8
200	1044	434	164	16
240	1253	520	196	19

The approximate fade duration for different severities at different speeds in a Rayleigh channel is defined by Equation (5.29) and typical values are shown in Table 5.15.

$$\overline{\tau} = \frac{(1 - e^{-\rho^2})}{N_c} \qquad (5.29)\ \text{Average fade duration}$$

$\overline{\tau}$ = average fade duration (s).
N_c = number of crossings below fade intensity threshold, defined by ρ.
ρ = fade severity (dB).

Table 5.16 shows that individual fade duration decreases with increase in speed; according to Table 5.15, however, the number of fades increases with speed. This brings the sum of the fades to approximately the same value for all speeds, varying only according to the fade severity, as illustrated in Table 5.16.

Table 5.15 Fade duration according to receiver speed

Fade duration (μs)				
	Fade intensity (dB)			
Speed (km/h)	−3	−6	−10	−20
0.1	703,237	320,029	124,048	12,343
1	70,324	32,003	12,405	1,234
10	7032	3200	1240	123
100	703	320	124	12
200	352	160	62	6
240	293	133	52	5

Table 5.16 Total fade duration (cumulative per second)

	Fade intensity (dB)			
	−3	−6	−10	−20
Total fade duration (s)	0.3671	0.0694	0.0102	0.0001

Fading at low speeds and at high speeds is illustrated in Figure 5.27 and Figure 5.28 respectively.

5.5.5 Multipath Mitigation Procedures

Multipath can be minimized by using some mitigation techniques:

- Use of directional antennas to reduce the amplitude of multipath signals.
- Analysis of possible reflectors to avoid pointing the antenna to them.
- Use of diversity antennas and techniques.

5.5.6 Comparing Multipath Resilience in Different Technologies

Table 5.17 compares the throughput of different technologies and the respective multipath distance spread for one symbol duration. Table 5.17 shows that WiMAX and LTE support high speed and large delay spread whereas other technologies have to trade speed by range. The cell radius can be estimated from the distance spread by multiplying it by 3. The table values are approximate, just to give an idea of the trade-offs.

UMTS, HSPA, cdma and EVDO require band equalizers that can equalize up to three symbols and consequently improve the multipath range threefold.

5.6 RF Channel in the Power Domain

Power level and power controls are two variables that can fundamentally impact the efficiency of a WiMAX network. Transmit power limits are regulated by local or national agencies, such as FCC in the United States; however, it is important that it is adapted to the maximum power of the SS or MS devices to provide a balanced link.

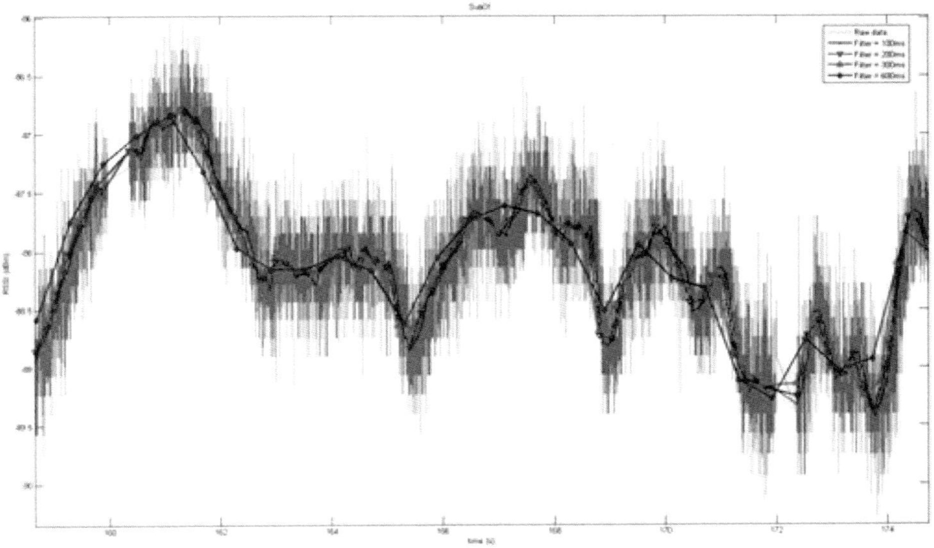

Figure 5.27 Fading at low speed.

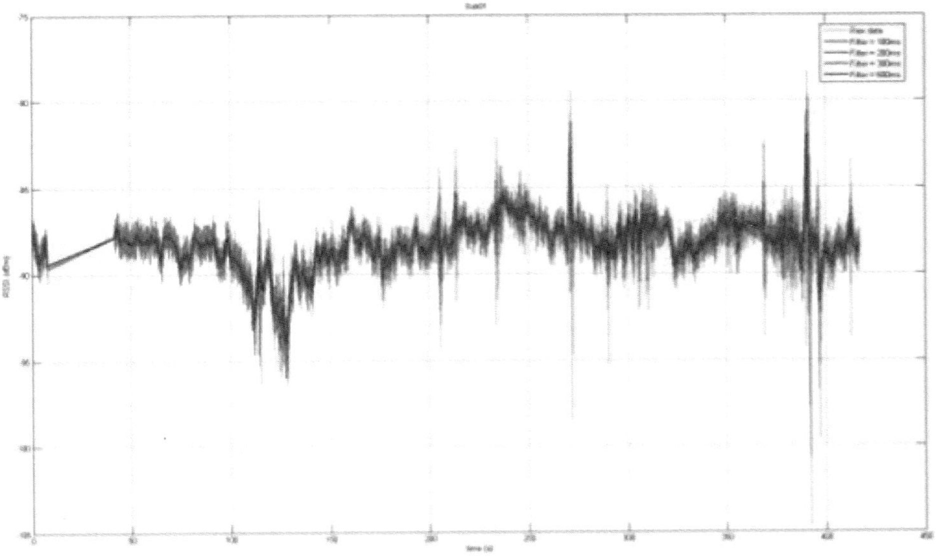

Figure 5.28 Fading at high speed.

Table 5.17 Technology comparison table

Technology	Bandwidth (MHz)	Symbol duration (us)	Highest modulation	Maximum theoretical data rate per sub-carrier (Mbps)	Theoretical absolute maximum spectral efficiency (bit/Hz)	Maximum multipath distance spread for 75% correlation (m)	Coherence bandwidth (KHz)	Coherence time for 5 GHz 120 km/h 75% correlation (μs)
GSM/GPRS/EDGE	0.16	3.7	QPSK	0.54	3.38	370	270	666
UMTS	3.844	0.3	QPSK	7.69	2.00	26	3846	666
HSDPA	3.844	0.3	16QAM	15.38	4.00	26	3846	666
CDMA	1.288	0.8	QPSK	2.56	1.99	78	1282	666
EVDO	1.288	0.8	16QAM	5.13	3.98	78	1282	666
WLAN (Wi-Fi)	0.3125	3.2	64QAM	1.25	4.00	320	156	666
WiMAX (sub-carrier)	0.0105	95.2	64QAM	0.04	4.00	9524	10	666
LTE (sub-carrier)	0.0155	64.5	64QAM	0.06	4.00	6452	10	666

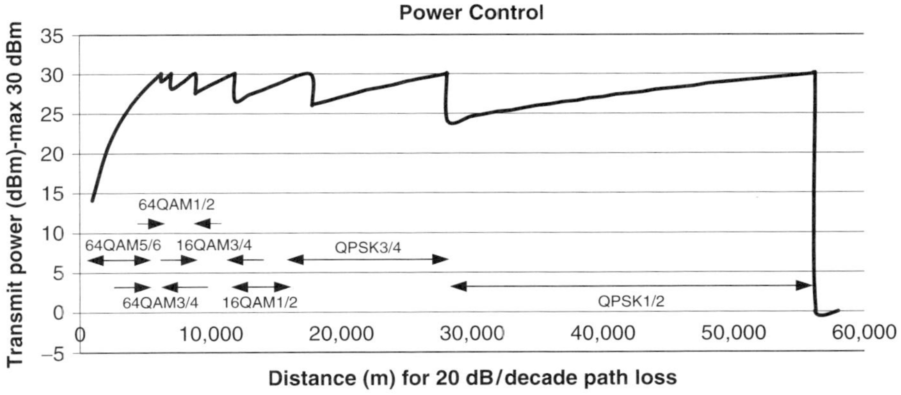

Figure 5.29 Variation of transmitted power with distance and modulation schemes for free space.

Pilot and other control signals are usually BPSK modulated and are boosted few dB (typically 2.5 dB) above the average value specified for other modulations.

SSs/MSs are classified by the WiMAX Forum in four distinct power classes: 20 dBm, 23 dBm, 27 dBm and 30 dBm. Similar values apply to LTE and WLAN.

Power control is used in both the downstream and upstream with a dynamic range of up to 50 dB. Figure 5.29 shows the variation of transmitted power with distance, for a 20 dB/decade path loss. Distances displayed in the graph change significantly with the path loss slope.

SS/MS devices can basically operate in three different modes:

- *Normal mode*: the device is constantly monitoring downlink messages and updating its ranging values.
- *Sleep mode*: the device goes into sleep mode and periodically wakes and updates its ranging values.
- *Idle mode*: the device unregisters and goes into sleep mode. It must be paged for incoming data and re-register again for outgoing data. This mode is useful in situations where the device is moving (e.g. highways) to avoid excessive registrations.

Chapters 12, 13 and 14, describe each technology and provide additional specifications for them.

5.7 Standardized Channel Models

Pre-defined channel models are required to test equipment and algorithms in similar conditions. They are simplified versions of expected real-life situations and focus on some channel aspects required for testing specific equipment features. They are not suitable to be used in network planning. The commonest models are described here and additional models are presented in the chapters describing WiMAX and LTE technologies.

5.7.1 3GPP Empirical Channel Model

In this model, the path loss is a function of distance, frequency, BS antenna height, MS antenna height and a factor for used for two different environments. Between 1 and 20 delayed paths are considered following exponential power decay. Each multipath corresponds to a cluster of M sub-paths corresponding to local scatterers.

The AoD (Angle of Departure) considered is narrow and the AoA (Angle of Arrival) is uniformly distributed. The final channel is created by summing all sub-path components.

5.7.2 3GPP2 Semi-Empirical Channel Model

The ITU has defined several RF channels to be used in evaluation comparisons of different solutions:

- *Pedestrian A*: flat fading model corresponding to a single Rayleigh fading at 3 km/h.
- *Pedestrian B*: comprises four delay paths (0, 0.11, 0.19, 0.41 μs and relative power of 1, 0.107, 0.012, 0.0052) at 3 km/h.
- *Vehicular A*: comprises four delay paths (0, 0.11, 0.19, 0.41 μs and relative power of 1, 0.107, 0.012, 0.0052) at 30 km/h.
- *Vehicular B*: comprises six delay paths (0, 0.2, 0.8, 1.2, 2.3, 3.7 μs and relative power of 1, 0.813, 0.324, 0.158, 0.166, 0.004) at 30 km/h.

5.7.3 Stanford University Interim (SUI) Semi-Empirical Channel Model

Six typical channels are specified for the USA with three multipath fading delays.

5.7.4 Network-Wide Channel Modeling

All these models represent particular situations but, in a real network, there are many more variations. A standard channel model can be used for many applications such as testing equalizers, but they are not suitable for network design.

When there is direct Line of Sight (LOS), Gaussian fading is expected; non-LOS scenarios tend to follow Rayleigh fading. There are many other situations in between, however, so the best way to represent the channel variation is through statistical distributions, the most appropriate in this case being Ricean.

In Ricean distributions, the k factor allows modeling of different distributions from Rayleigh all the way to Gaussian. This is illustrated in Figure 5.30.

The Ricean distribution allows modeling all the different scenarios by adjusting the k factor. A Rayleigh distribution is represented by a factor smaller than 1, whereas a log-normal distribution is represented by a factor larger than 10.

The k factor is used to express the environment at each location by analyzing its surroundings. The k factor should be estimated based on the following parameters:

- LOS
- distance
- antenna height
- antenna type
- clutter factor

The Ricean Power Distribution Function (PDF) is given by Equation (5.30).

$$p(r) = \frac{r}{\sigma^2} e^{\frac{(r^2 + A^2) I_o \left(\frac{rA}{\sigma^2}\right)}{2\sigma^2}} \qquad \text{(5.30) Ricean PDF}$$

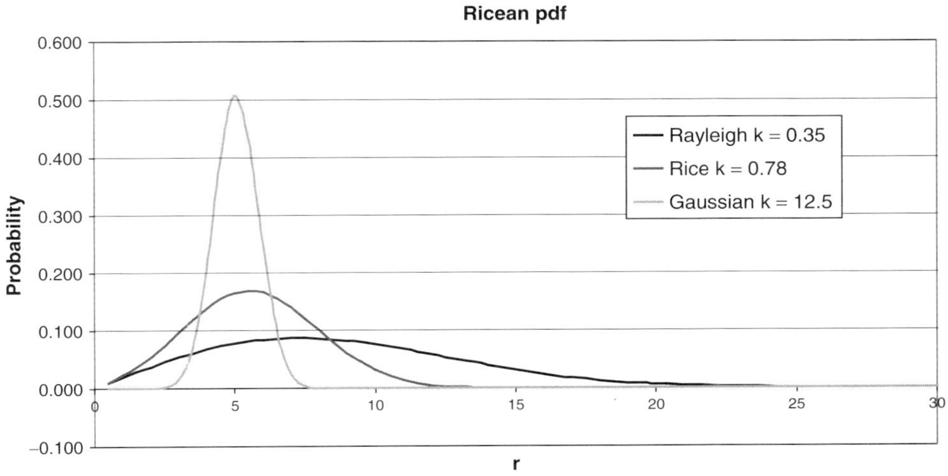

Figure 5.30 Ricean distribution.

where:

$p(r)$ = SNR(Signal to Noise Ratio) probability.
r = Signal to Noise Ratio.
σ = Standard Deviation.
A = Peak amplitude of the dominant signal.
K = Distribution factor.

The Ricean distribution k factor is defined in Equation (5.31).

$$k = \frac{A^2}{2\sigma^2} \quad \text{(5.31) Ricean distribution } k \text{ factor}$$

In practice, k has to be estimated following guidelines presented in Equations (5.32) for LOS and (5.35) for NLOS. Those equations were empirically obtained.

For LOS:

$$k = 10 F_s F_h F_b d^{-0.5} \quad \text{(5.32) } k \text{ factor for LOS}$$

where:

F_s = Morphology factor (1 to 5), lower values apply to multipath rich environments.
F_h = Receive antenna height factor.
F_b = Beam width factor.
d = Distance.
h_r = Receive antenna height.
b = Beam width in degrees.

$$F_h = (h_r/3)^{0.46} \quad \text{(5.33) Receive antenna height factor}$$

$$F_b = \left(\frac{b}{17}\right)^{-0.62} \quad \text{(5.34) Beam width factor}$$

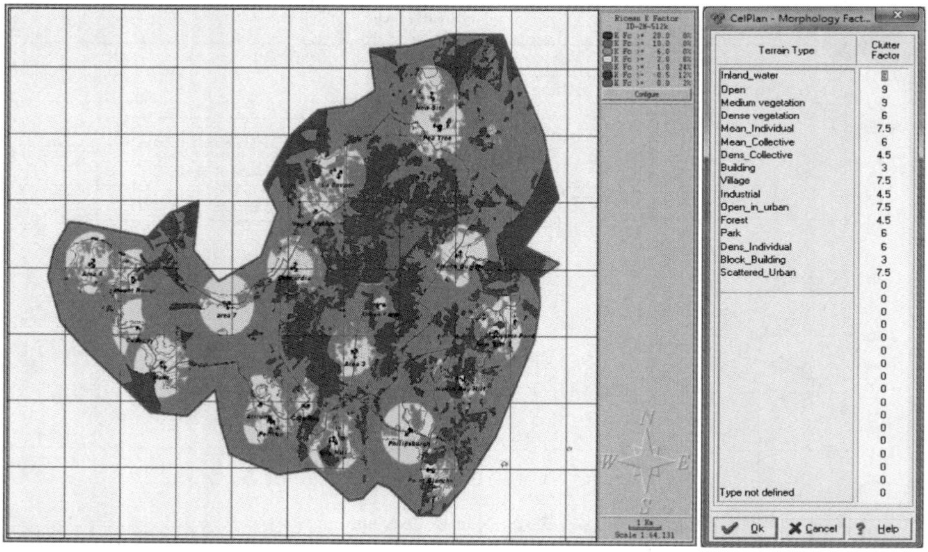

Figure 5.31 Ricean k factor (Ricean distribution) plot.

For NLOS:

$$k = 0 \quad (5.35)\ k \text{ factor for NLOS}$$

Based on the above, the k factor can be predicted on a pixel by pixel basis and be used to calculate the fading at each pixel. An example of this k factor prediction is shown in Figure 5.31, as well as the clutter factors for different morphologies.

5.8 RF Environment

The propagation prediction assumes an outdoor transmitter and a receiver in certain morphology. Propagation parameters are derived from measurements done without immediate obstructions and filtering fading. This means that the prediction values are representative of the average signal level in a certain location. Advanced prediction models, like Korowajczuk 3D (see Chapter 6) can predict the signal level outdoor and indoor at different heights.

When doing final predictions, environment losses at each location should be added to the predictions. Environmental losses should be represented by a distribution and its standard deviation.

The main components of the environment are:

- human body attenuation
- penetration attenuation
- rain precipitation attenuation
- fading attenuation

Figure 5.32 presents a sample dialogue box used to configure these parameters.

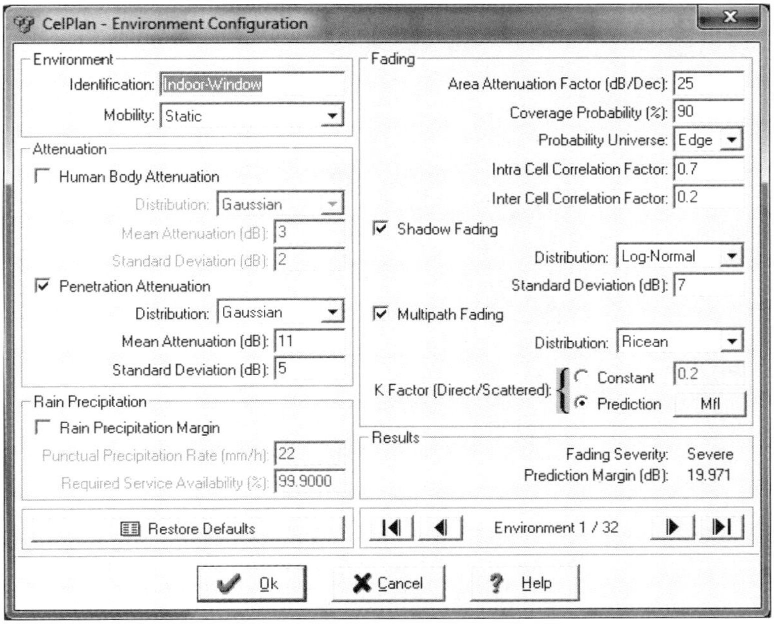

Figure 5.32 Environment configuration dialogue.

5.8.1 *Human Body Attenuation*

The human body attenuation depends on the terminal type; for phones, it has a typical mean attenuation of 3 dB with 2 dB standard deviation.

5.8.2 *Environment Penetration Attenuation*

Environmental penetration attenuation is any additional loss caused by the environment in addition to the morphology final factor and the penetration loss considered in the Korowajczuk 3D propagation model (see Chapter 6). For other models that do not have penetration and morphology final factors, the environment penetration attenuation can be used to replace them.

5.8.3 *Rain Precipitation*

Rain precipitation only affects frequencies above 10 GHz, so it is important to be considered in microwave point to point links. Figure 5.33 shows the rain precipitation in various areas of the world.

5.8.4 *Environment Fading*

Fading expresses RF signal variations and should be defined for each environment. Fading calculations are complex and are described in the next section.

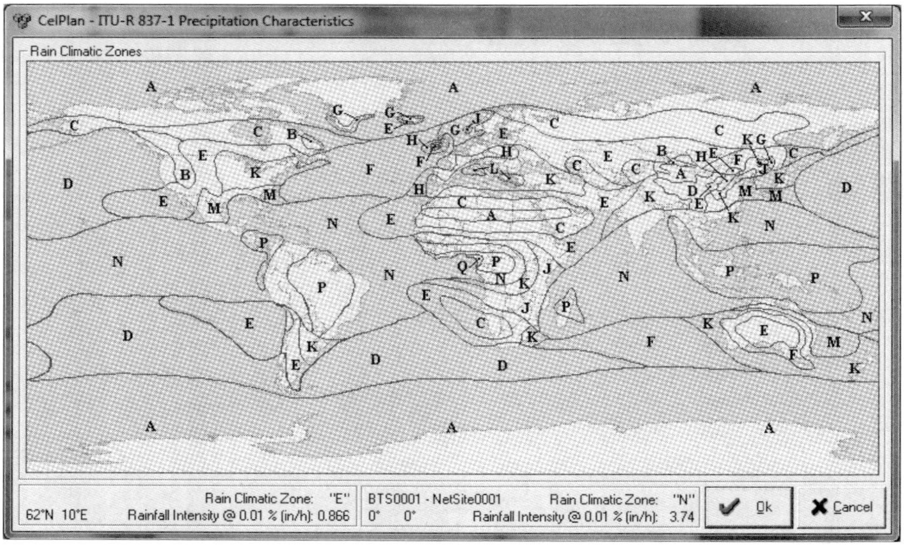

Figure 5.33 Rain precipitation map.

5.9 Fading

The two most common sets of modeling approaches for terrestrial wireless communications (excludes satellite and spatial communication links) can be classified as:

- *Point-to-point*: fixed line-of-sight (LOS) wireless links. Point-to-Point category includes most microwave links, which typically are designed with LOS, and the fading effects observed in those cases are mostly caused by multipath effects such as: (1) interference between direct rays in a varying refractive index gradient atmospheric medium; (2) ground-reflected components; and (3) partial reflections from atmospheric elevated layers. In addition to multipath, other effects that cause fading in signals in this category may include antenna decoupling, earth surface intrusion (even in LOS condition), and precipitation (rain) in the propagation path. Performance models that deal with fading margins in this context include the ITU-R 530-8 and Barnett-Vigants multipath outage probability models.
- *Point-to-multipoint*: mobile and fixed wireless links that may or may not have LOS. Point-to-multipoint category consists of mobile and fixed point-to-multipoint communication links, where the propagation path may vary with time, and the LOS condition may not be secured at all times. The typical received signal is a combination of multiple rays that have traveled different paths, most of them indirect, some reflected off buildings and other surrounding objects, and others diffracted from rounded objects and knife-edges, such as building corners.

Multiple samples of the same signal arrive at the receiver with different delays and amplitudes, depending on the path they have traveled, and combine constructively or destructively at different occasions (multipath). This set of elements that cause such reflection and diffraction effects varies with time as the mobile moves or suffers micro motion around its position, while the surrounding environment may also change its geometry. Performance models that deal with fading margins in this context typically include different ways to statistically model fading. Four different statistical distributions are required to model the different fading types:

1. short-Term fading (Rayleigh distribution);
2. short-term fading (Ricean distribution);
3. long-term fading (Log-Normal distribution);
4. combined short-term and long-term fading (Suzuki distribution).

With respect to the multipath delay spread described above, the impact of multipath fading depends on the nature of the transmitted signal when compared to the fading characteristics. If the propagation channel has a flat response over the frequency domain compared to the bandwidth of the transmitted signal, the spectral characteristics of the transmitted signal are preserved at reception, and the channel is referred to as a flat fading channel. In the opposite scenario, when the channel response is narrower (in the frequency domain) than the transmitted signal bandwidth, besides the variation in amplitude there is also a distortion induced by inter-symbol interference (ISI), and this effect is referred to as frequency selective fading.

Most systems are assumed to suffer flat fading for the purpose of performance analysis, otherwise bandwidth equalization is required. Techniques such as OFDM are used to allow the transmission over large bandwidths while mitigating the effect of frequency selective fading, by the use of multiple smaller carriers, which can be modeled as individually suffering flat fading. The models described here refer to flat fading channels.

With respect to the coherence time of the fading, a channel is referred to as fast fading when the received signal varies rapidly compared to the symbol duration (i.e. the coherence time of the channel is smaller than the symbol period of the transmitted signal). Conversely, a channel is referred to as slow fading when the channel impulse response changes at a rate much slower than the transmitted baseband signal, and may be considered static over one or several reciprocal bandwidth intervals. Most channels at large bandwidths are considered slow fading, because the user mobility (and the mobility of objects in the channel) present small velocities compared to the large transmitted baseband signal bandwidth.

5.9.1 Fading Types

5.9.1.1 Short-Term Fading

The expression "short-term" describes the signal behavior with respect to its time variations (for the same receiver location):

- *Gaussian, or ideal channel*: when there is only one stationary dominant signal, that is, the only noise present at the receiver is the Additive Gaussian White Noise (AWGN) developed at the receiver. This situation is almost impossible to achieve in the mobile environment.
- *Rayleigh channel*: where all components are indirect, that is, the received signal is a combination of the indirect rays (this is the worst case scenario). The power distribution function (p.d.f.) of the received signal envelope, when the multipath components are independent, follows a Rayleigh distribution.
- *Ricean channel*: this is the intermediary situation between the Gaussian channel and the Rayleigh channel. The received signal includes a dominant stationary component (typically the LOS direct path) plus additional indirect paths (multipath components).

5.9.1.2 Long-Term Fading

In addition to the signal variations over time at a given location, another way to describe signal strength distribution is to observe how the received signal varies with location, even for points that are equidistant from the transmitter. These variations are typically caused by signal shadowing (obstructions) on the different propagation paths that surround the transmitter.

These shadow-related variations with location in the coverage area of a transmitter are described by the log-normal distribution (which means a Gaussian distribution around the mean received signal value, when described in dB).

It is important to notice that, even though the shadowing log-normal distribution is usually referred to, in the literature, as the distribution associated with "long-term fading", it consists, in fact, of two different effects. While the terms "short-term fading" or "long-term fading" refer to variations over time (for a given location), the log-normal distribution refers to a variation over a set of locations (which may be defined by a circumference, the whole area within a circle, or the resolution area of a grid bin).

The term "long-term fading" is used interchangeably with shadowing because this effect is noticed and measurable only if there is no short-term fading present, or if short-term fading is filtered during measurement post-processing. Considering that the term "fading" refers specifically to time variations of the signal, a better term to define this effect would be "shadowing dispersion", to avoid the misleading connotation of the term fading. Nevertheless, to keep consistency with the current literature, in this document, the term long-term fading will be used to refer to this effect.

The standard deviation of the log-normal distribution associated with shadowing depends on the area of scope over which it has been calculated and has to be consistently applied to that scope. In other words, if predicting signal strength as a function of distance, statistics must be processed for multiple distances, to find the average received signal and standard deviation at each distance.

On the other hand, if running a "per-pixel" type of prediction, which already considers clutter and shadowing effects at each different propagation path, the correct standard deviation corresponds to the signal variations that are expected within the analysis resolution (e.g. 30 × 30 meter pixels). One common mistake found in propagation studies is the application of distance-based standard deviations (calculated over the entire circumference, and most commonly referred to in books) on top of a pixel-based prediction, which leads to exaggerated shadowing fading margins.

5.9.1.3 Combined Short-Term and Long-Term Fading

Short-term and long-term (shadowing dispersion) fading are not exclusive effects. If we filter (low pass) over time the signal strength variations at each location, the effects of short-term fading are removed, and the local variations in signal strength may be used to obtain the standard deviation of the shadowing dispersion in the area.

There are multiple approaches that propose performance (outage) models to deal with both short-term and long-term fading simultaneously, such as the Suzuki distribution.

5.9.2 Fading Probability

The calculation of fading margin uses the fading probability density distribution (pdf) (log-normal, Rayleigh, Rician, or Suzuki), which are temporal distributions of the signal envelope at one given location, and integrate this probability over multiple locations. Because these distributions refer to point-to-multipoint applications, where multiple receivers surround a transmitter and vice versa, two approaches are typically used to calculate these probabilities: border (or edge) and area.

Considering a variable W representing the power received at a given location, and a threshold W_0 (both in dBm) and assuming that both the probability density function $p(W)$ of W, and its mean signal strength M_W are known (this mean signal strength can be determined by the use of prediction models or measurements); the two most common approaches to obtain the probability of outage are the following:

5.9.2.1 Edge

The probability that the received signal power W is above a threshold W_0 is defined by Equation (5.36). This proportion is referred to as β and is defined as:

$$\beta = \text{prob}(W \geq W_0) = \int_W^{\infty} p(W)\,dW \qquad \text{(5.36) Fading probability above threshold}$$

Or, as a function of the fading margin which is expressed in Equation (5.37):

$$\beta = \text{prob}(W) \geq M_W - M_g \qquad \text{(5.37) Fading as function of fading margin}$$

where:

W = actual received signal level (dBm).

M_w = mean signal strength of points located at the pixel (dBm). In CelPlanner predictions, this is the level that is calculated by the propagation models as the mean value expected at a given location. Even though the median is not the same as the mean, it is usually said that this is the 50% confidence level value, that is, roughly 50% of real measurements at a location are expected to be below and 50% above the calculated M_w value.

W_0 = threshold for the received signal level (dBm). This is the level that is displayed as the "predicted dBm" value at a given location, with the fading margin already discounted, compared to the M_w value calculated by the propagation model, that is, W_0 is typically a value smaller than M_w, and the difference between them is called the "fading margin".

M_g = Fading margin, expressed in dB in Equation (5.38).

$$M_g(dB) = M_W - W_0 = 10\log\left(\frac{m_W}{W_0}\right) \qquad \text{(5.38) Fading margin}$$

where:

W_0 = threshold for the received signal level (mW).

m_W = signal strength at that location (mW).

In other words, the confidence level b showed above expresses the probability of the actual signal (W) to be larger than the predicted signal ($W_0 = M_w - M_g$) at any given location, given that the mean of the signal at the location is M_w. That is, it expresses the percentage of time where the signal suffers fading effects that have amplitudes smaller than M_g. In that context, it is usual to describe the confidence level distributions as a function of the margin M_g.

This probability calculation can be solved for any point or set of points for which the distribution is known. In practice, typical applications of this approach include: (1) a set of points on a circumference of a given radius (in the simplistic assumption of propagation models based on distance), or (2) a set of points within an analysis pixel, in more elaborate analysis that take into account different clutter and diffraction effects that vary per pixel.

5.9.2.2 Area

In this approach, the proportion of locations within the circular area defined by the radius L where the received signal W is above the threshold W_0 is obtained. The calculation assumes that mobiles are uniformly distributed within the cell area, that is, the proportion of locations is the same as the proportion of mobiles. This proportion is referred here as μ and is defined in Equation (5.39).

$$\mu = \frac{1}{A}\int_A \text{prob}(W \geq W_0)\,dA \qquad \text{(5.39) Area probability 1}$$

where $A = \pi L^2$ and *dA* is an infinitesimal area ($dA = l dl d\theta$). Therefore, the desired probability is the average probability of W exceeding W_0 over the entire circular area, as expressed by Equation (5.40).

$$\mu = \frac{1}{\pi L^2} \int_0^L \int_0^{2\pi} \text{prob}(W \geq W_0) l dl d\theta \qquad (5.40) \text{ Area probability 2}$$

This results in the area probability expressed in Equation (5.41).

$$\mu = \frac{2}{L^2} \int_0^L \text{prob}(W \geq W_0) \qquad (5.41) \text{ Area probability 3}$$

The area approach was developed at a time when predictions were typically circular, and signal strength predicted values were associated with a whole circumference contour. In those circumstances, it was desirable to evaluate the probability of having signal strength above a required threshold inside the circular area, that is, all points contained in the circle were taken into account for the "averaging area calculation", including those close to the transmitter at the center of the circle (with very strong signal) as well as those close to the edge of the circle. Therefore, it is easy to see that fading margins had to be smaller in the area approach (as probabilities of being above a threshold were higher for the same local mean).

5.9.3 Fading Distributions

The four fading distributions are described next, including their outage calculations.

5.9.3.1 Log Normal Distribution (for Long-Term Fading)

The log normal distribution is used to describe signal variations that are location related, typically caused by signal shadowing (obstructions). If the excess path loss is defined as the ratio between the actual received signal and that which would be received in free space, this variable will have a Gaussian distribution with parameters that depend on the type of environment and obstructions (buildings, tunnels, hills, trees, etc.) found in the propagation path. In the logarithmic scale (dB), this distribution is known as the log normal distribution. Therefore the envelope of the received signal W, measured in dB, has a log normal probability density function given by Equation (5.42).

$$p(W) = \frac{1}{\sqrt{2\pi}\sigma_W} \exp\left[-\frac{1}{2}\left(\frac{W - M_W}{\sigma_W}\right)^2\right] \qquad (5.42) \text{ Log normal distribution}$$

where M_W and $(\sigma_W)^2$ are, respectively, the mean and variance of W given in decibels. Typical values of σ_W vary with the scope of the distribution. For grid analyses of 100 m × 100 m (about 3 sec resolution), typical values of σ_W are found to be between 2.5 and 3 dB, while at 1 arc sec resolution (30 m × 30 m grids), they smaller than 2 dB.

- Outage probability calculation – edge approach.

The formula below expresses the outage probability of the received signal W exceeding a given threshold W_0 at any point on the border (edge) of a circle around the transmitter, given that the average signal level at that border is m_w, as shown in Equation (5.43).

$$\beta = \text{prob}(W \geq W_0) = \frac{1}{2}\left[1 - \text{erf}\left(\frac{W_0 - M_W}{\sqrt{2}\sigma_W}\right)\right] \qquad (5.43) \text{ Edge outage for log normal distribution}$$

The value of the error function erf( ) is given by Equation (5.44).

$$\mathrm{erf}(x_0) = 2\int_0^{x_0} \frac{1}{\sqrt{\pi}} \exp(-x^2)\,dx \qquad \text{(5.44) Error function}$$

The parameters used in Equations (5.43) and (5.44) are explained below:

$\beta =$ confidence level of a given threshold on the edge approach.
$W =$ actual received signal level (dBm).
$\sigma_W =$ standard deviation of W (dispersion of variations in dB) around the local mea.
$M_w =$ mean signal strength of points located at the pixel (dBm).
$W_0 =$ threshold for the received signal level (dBm).

Considering the definition of fading margin in dB expressed in Equation (5.45) the confidence level for the log normal distribution can be rewritten as expressed in Equation (5.46).

$$M_g(dB) = M_W - W_0 = 10\log\left(\frac{m_W}{W_0}\right) \qquad \text{(5.45) Fading margin in dB}$$

$$\beta = \mathrm{prob}(W \geq M_W - M_g) = \frac{1}{2}\left[1 - \mathrm{erf}\left(\frac{-M_g}{\sqrt{2}\sigma_W}\right)\right]$$

(5.46) Log normal distribution confidence level

- Outage probability calculation – area approach.

The formula in Equation (5.47) expresses the outage probability of the received signal W exceeding a given threshold W_0 at any point within the area of a circle around the transmitter, given that the average signal level at the border of that circle is m_w, is:

$$\mu = \frac{1}{2}\left\{1 + \mathrm{erf}\left(\frac{M_W - W_0}{\sqrt{2}\sigma_W}\right) + \exp\left[\frac{2(M_W - W)10\alpha\log e + 2\sigma^2}{100\alpha^2\log^2 e}\right] * \left[1 - \mathrm{erf}\left(\frac{(M_W - W)10\alpha\log e + 2\sigma^2}{(10\alpha\log e)\sqrt{2}\sigma_W}\right)\right]\right\}$$

(5.47) Area outage for Log normal distribution

The parameters of (5.47) are described next:

$m =$ confidence level of the threshold W_0.
$W_0 =$ threshold for the received signal level (dBm).
$M_w =$ mean signal strength of points located at the edge (dBm).
$a =$ local attenuation factor (typically 3 to 8) Obs. This parameter gives the standard loss per decade in the area (e.g. if a = 4, the path loss grows with distance at a rate of 40 dB per decade).

Γ(a,x) is the incomplete gamma function, that can be numerically approximated with good accuracy.

5.9.3.2 The Rayleigh Distribution (for Short-Term Fading)

For Rayleigh channels, it is assumed that there is no dominant signal (direct path) at the receiver location. All components are indirect, that is, the received signal is a combination of the indirect rays.

The power distribution function (pdf) of the received signal envelope follows a Rayleigh distribution described by Equation (5.48).

$$p(W) = \frac{1}{m_W} \exp\left(-\frac{W}{m_W}\right) \qquad \text{(5.48) Rayleigh distribution}$$

where W is the power received at a location (in watts or mW), and m_w is the mean signal strength at that location.

- Outage probability calculation – edge approach.

The formula in Equation (5.49) expresses the outage probability of the received signal w exceeding a given threshold W_0 at any point on the border of a circle around the transmitter, given that the average signal level at that border is m_w, is:

$$\beta = \text{prob}(W \geq W_0) = \exp\left(-\frac{W_0}{m_W}\right) \cdot \qquad \text{(5.49) Edge outage for Rayleigh distribution}$$

where:

$\beta =$ confidence level of a given threshold on the edge approach.
$W_0 =$ threshold for the received signal level (mW).
$W =$ actual received signal level.
$m_w =$ mean signal strength of points located at that distance (mW).

Since the previous equation is invertible, the fading margin may be expressed as an equation as well, which is given by Equation (5.50) and in dB by equation (5.51).

$$\frac{W_0}{m_W} = -\ln(\beta) \qquad \text{(5.50) Rayleigh distribution edge margin}$$

$$\text{Margin}(dB) = M_W - M_0 = 10\log\left(\frac{m_W}{W_0}\right) = -10\log[-\ln(\beta)] \qquad \text{(5.51) Rayleigh edge margin in dB}$$

where

$\beta =$ confidence level of a given threshold on the edge approach.
$w_0 =$ threshold for the received signal level (mW).
$W_0 =$ threshold for the received signal level (dBm).
$m_w =$ mean signal strength at that location (mW).
$M_w =$ mean signal strength of points located at the border (dBm).

- Outage probability calculation – area approach.

The formula in Equation (5.52) expresses the outage probability of the received signal W exceeding a given threshold W_0 at any point within the area of a circle around the transmitter, given that the average signal level at the border of that circle is m_w.

$$\mu = \frac{2}{\alpha}\left(\frac{m_W}{W_0}\right)^{2/\alpha} \Gamma\left(\frac{2}{\alpha}, \frac{W_0}{m_W}\right) \qquad \text{(5.52) Rayleigh distribution area outage}$$

where:

$m =$ confidence level of the threshold w_0.
$w_0 =$ threshold for the received signal level (mW).

$m_w =$ mean signal strength at that location (mW).
$\alpha =$ environment standard attenuation (typically varies from 2 to 8) Obs. This parameter gives the standard loss per decade in the area (e.g. if a = 4, the path loss grows with distance 40 dB per decade).

Γ(a,x) is the Incomplete Gamma Function, given by Equation (5.53).

$$\Gamma(x, y) = \int_0^y t^{x-1} \exp(-t)\, dt \qquad (5.53) \text{ Gamma function}$$

5.9.4 The Rician Distribution (for Short-Term Fading with Combined LOS and NLOS)

The outage probability y and fading margin for Ricean Distribution are generally calculated using the Edge approach. Area calculations are better done using the Rayleigh distribution.

When using Rician distribution, designers have the choice of calculating the *k* factor based on a predicted or constant value. When working with predicted *k* factor, designers must set up a clutter factor for each morphological type.

When the option of *k* factor Constant is selected, the Prediction Margin shown in the bottom of the dialog corresponds to the sum of the all the attenuation (human, penetration) and fading (shadow and multipath). When the option From Prediction is selected, the output of Prediction Margin shows only what could be called a minimum prediction margin, because that value includes all other factors (human, penetration, shadow fading) but not multipath fading because that is calculated through the *k* factor prediction and varies per morphology.

5.9.4.1 The *K* Factor Prediction

The *k* factor is calculated in two circumstances:

- As a prediction to display as on raster output (available from the Prediction menu, called "k factor (linear)".
- As an "auxiliary prediction", calculated internally for use as an input in the calculation of the fading margin.

Designers should note that the *k* factor is service class dependent because it is associated with the environment configuration. For every pixel, the *k* factor (linear) is calculated by Equation (5.54) for LOS and Equation (5.55) for NLOS.

$$k_{\text{mean}} = F_s * F_h * F_b * k_0 * d^{\gamma} \qquad (5.54)\ k \text{ factor LOS}$$

$$k_{\text{mean}} = 0 \qquad (5.55)\ k \text{ factor NLOS}$$

where:

$F_s =$ the morphology factor.
$F_h =$ a receive antenna height factor.
$F_b =$ the antenna beamwidth factor.
$K_0 =$ a regression coefficient (constant = 10).
$\gamma =$ a regression coefficient (constant = −0.5).

d is the distance between the CPE location (pixel) and the sector that provides the strongest individual signal, which is calculated as the individual power dBm coming from each sector (even in the DVB-H systems, it does not look at C/(N+I), or aggregate power)

For the morphology factor (F_s), most authors suggest a value of 1 in the summer (trees full of leaves) and 2.5 in the winter (no leaves).

- open areas: $F_s = 2.5$ to 3.0.
- dense foliage: $F_s = 1.0$.
- intermediate density: F_s between 1 and 2.5.

The receive antenna height factor (F_h) is considered by most authors as expressed by Equation (5.56).

$$F_h = \left(\frac{h_{rx}}{3}\right)^{0.46}$$ (5.56) Antenna height factor

where h_{rx} = CPE receive antenna height in meters.

The F_h factor is the same when receiving (downlink) and transmitting (uplink), for the same path loss prediction and fading margin in both directions (symmetry).

The antenna beamwidth factor (F_b) is proposed by most authors as expressed by Equation (5.57).

$$F_b = \left(\frac{b}{17}\right)^{-0.62}$$ (5.57) Antenna beamwidth factor

where b = antenna beamwidth in degrees.

The beamwidth (b) is automatically taken from the parameter Antenna Horizontal Aperture of each CPE.

5.9.5 The Suzuki Distribution (for Combined Long- and Short-Term Fading)

Most channels experience a local signal variation (short-term fading) that can be assumed, in the worst case, to have a Rayleigh distribution (typically expected in outdoor scenarios). Considering that the mean value of the signal W received at a location is given by m_w, the probability dense function of this local mean, over a given area, follows a log normal distribution, which has an area mean value given by M_{WA} and standard deviation σ_{WA}. The pdf of W is therefore given by the density of W conditional to the local mean m_w (Rayleigh) averaged over all possible values of mw (i.e. averaged over the lognormal distribution). This is expressed by Equation (5.58).

$$p(W) = \sqrt{\frac{\pi}{8\sigma_W^2}} \int_{-\infty}^{\infty} \frac{W}{m/10} \exp\left(\frac{\pi w^2}{4 \times 10^{m/10}}\right) \exp\left[-\frac{1}{2}\left(\frac{m_W - M_{WA}}{\sigma_W}\right)^2\right] dm_W$$

(5.58) Suzuki distribution

5.9.6 Traffic Simulation with Fading

Snapshot initialization: for each session, the tool creates a vector with the prediction from each sector in the system whose prediction radius extends to the call (session) location.

To consider shadow fading margins during the simulation, the following procedure is performed for each snapshot:

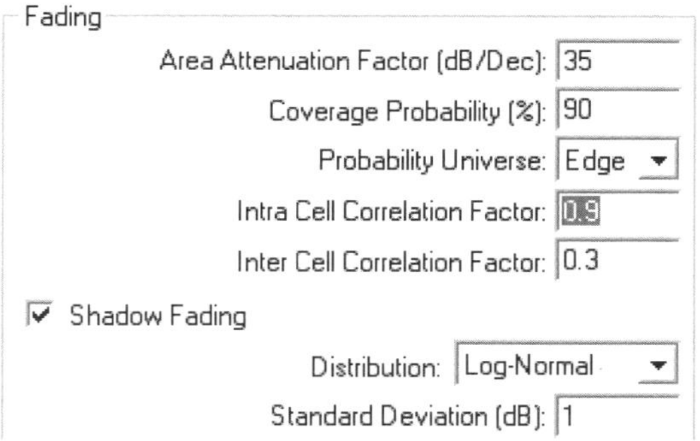

Figure 5.34 Fading configuration.

- Draw penetration loss and human body attenuation.

These values are selected (random draw) once for each session within a snapshot and do not vary with server selection. Because neither parameter can assume negative numbers (gains), they are limited to a minimum value of zero.

- Draw shadow fading.

Shadow fading values are generated as a different random variable for each sector. However, sectors of the same site are correlated, and therefore a correlation factor between sectors of same site is required. The intra-cell correlation factor represents the correlation between sectors of the same site, while the inter-cell correlation factor represents the correlation between sectors of different sites. These factors vary between 0 and 1, and the inter-cell factor has to be smaller than or equal to the intra-cell factor. An example of a dialogue used to configure theses parameters is shown in Figure 5.34.

To give more insight to the meaning of the correlation factors, it is important to highlight some interesting points:

- when the intra-cell correlation factor is equal to 1, fading values are identical among sectors of same site;
- when the inter-cell correlation factor is equal to 1 (which implies the intra-cell factor is also 1), fading values are for all sites and sectors;
- when the intra-cell correlation factor is equal to 0 (which implies the inter cell factor is also 0), all components are independently random.

In the context of one session, one value of shadow fading is generated for each sector, and is applied to all predictions coming from that sector. To generate fading values from each sector in the system, CelPlanner proceeds as follows, for each session.

- Generate a random number (Fcomm), from 0 to 1, that is the common part of the random fading value for a session. Therefore, it is applied to all fading values of that session (independent of which sector/site it comes from). This can be abstracted as related to the physical shadowing fading causes located close to the mobile, such as a valley, or trees, that is, this fading "belongs to the session" and does not vary with the direction the signal is coming from.
- Generate a random variable (Fsite) for each different site in the list of servers of a call. This factor is related to the shadow fading experienced on the path from the site location to the session (mobile

location). Therefore, the same value is used for signals from all sectors that come from the same location (same site). This factor has no effect when the inter-cell correlation factor is set to 1, and, in this case, Fcomm is applied to all elements.
- For the same session, fading coming from different sectors may have some components that are uncorrelated with each other, for example, the antennas may not be exactly in the same position for each sector, or diversity configurations may exist. This variable is related to the individual sector component on the shadow fading, and is generated independently as a random variable (Fsector) for each sector serving the session. This factor has no effect when the intra-cell correlation factor is set to 1.

The signal coming from each individual sector is then affected by shadow fading by combining these three components (Fcomm, Fsite, Fsector).

- Draw multipath fading.

The multipath fading calculation follows a very similar procedure to the shadow fading calculation.

- Calculate delta for path loss.

The final delta to be used for path loss calculation is a combination of the values described in the previous steps:

- Penetration loss per session.
- Body loss per session.
- Shadow fading (per sector and session).
- Multipath fading (per sector and session).

6

RF Channel Performance Prediction

6.1 Advanced RF Propagation Models

Propagation models have evolved over the years as wireless deployments and the density increased, but not enough to be used for 4G deployments. The complexity of the wireless markets requires a new breed of propagation models that will provide precise and consistent results at a pixel basis. These models not only have to work across multiple quality terrain databases, from low, canopy resolutions to high, building-level resolutions but also must be able to simultaneously predict signal levels outdoors and indoors. This new breed of RF propagation models addresses many of the weak points in empirical and physical models. These new models can be considered as hybrids and their main characteristics are described in the following sections.

6.1.1 Terrain Databases

Terrain databases are defined by topography and morphology layers. Topography represents the terrain altitude, whereas morphology, or clutter, represents anything that is above ground (e.g. vegetation, trees, and buildings). These databases store raster data defined by pixels of a certain resolution.

Terrain databases resolution is expressed in arc seconds or meters and defines the size of the smaller pixel (rectangle) that carries information. Arc seconds representation uses variable size pixels, which get narrower at higher latitudes, and are well suited to represent a spherical surface. Meter representation uses square pixels, but has a limited range, due to the distortion in representing a spherical surface. The arc sec representation is preferable.

Figure 6.1 shows a geographical grid with 15" resolution over a topography representation. Figure 6.2 shows a 1" resolution grid, which is the same resolution as the topography database. Figure 6.3 shows the same grid, but applying interpolation between bins.

Topography is typically represented with horizontal resolutions of 30 m to 5 m and the vertical resolution is generally 1 m. Morphology can be represented in different ways, described below, which

LTE, WiMAX and WLAN Network Design, Optimization and Performance Analysis, First Edition.
Leonhard Korowajczuk.

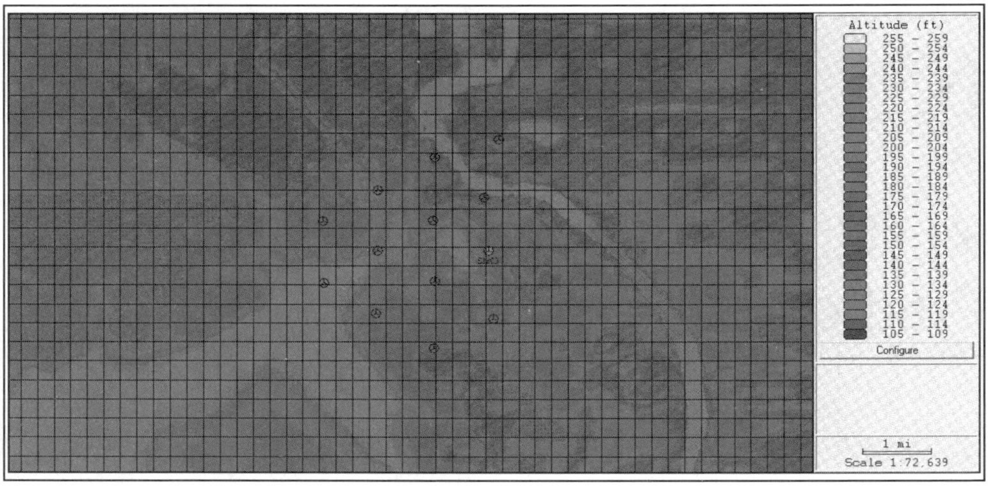

Figure 6.1 Geographical grid with 15 arc second resolution.

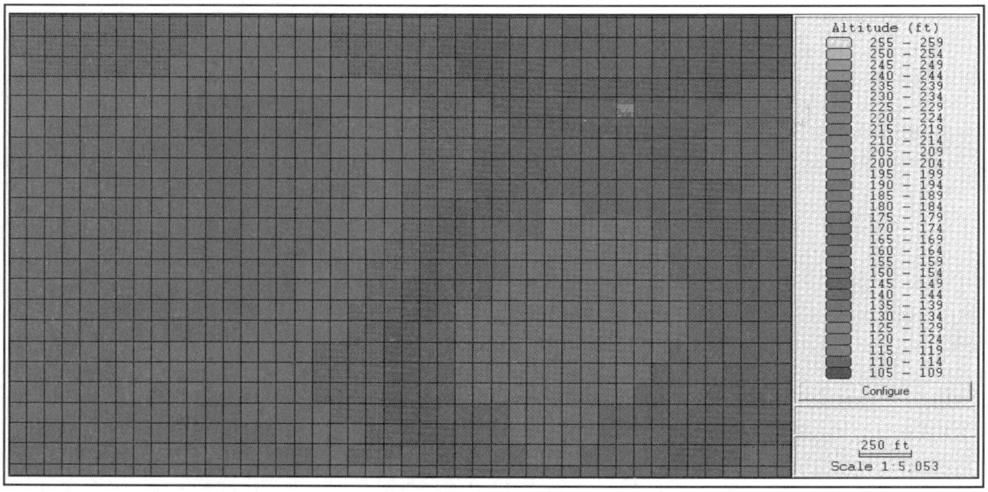

Figure 6.2 Geographical grid with 1 arc second resolution.

define the database quality. Horizontal morphology resolution defines how well terrain features (clutter) are represented. This representation can be divided into four categories:

- *Canopy morphology*: this is equivalent to throwing a sheet over the terrain and capturing macro features, such as residential areas, building areas, forests, but missing streets and spacing between constructions. It is characterized by a low resolution and may or may not have morphology heights associated with it. Typical resolutions are 250 m, 90 m or 30 m. The advantage is that it is cheap to produce.

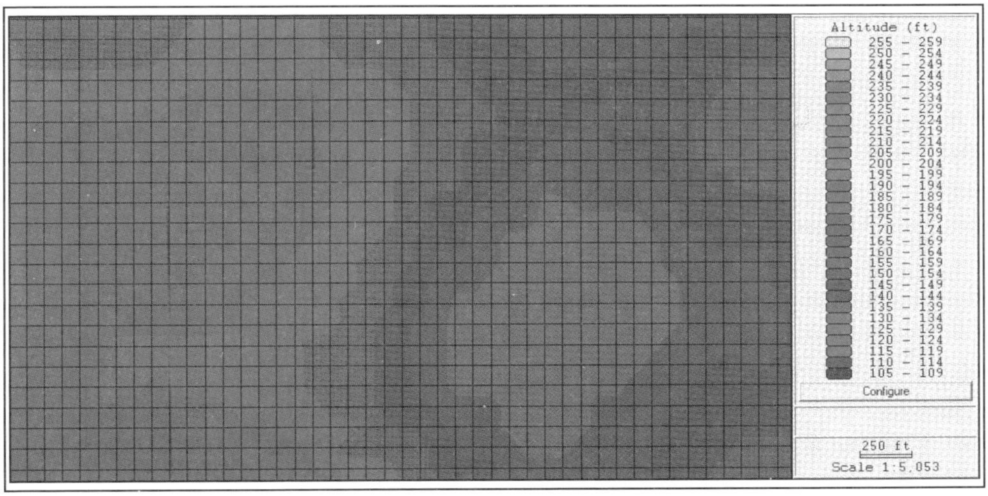

Figure 6.3 Geographical grid with 1 arc second resolution and interpolation between bins.

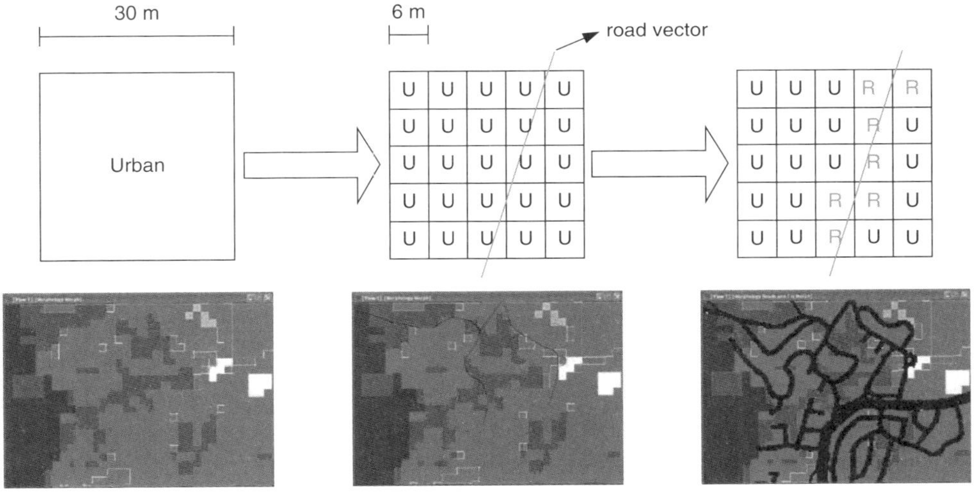

Figure 6.4 Morphology carving process.

- *Carved morphology*: this is based on canopy morphology with average heights over-sampled to a higher resolution and with streets and roads carved in it. Typical horizontal resolutions are 5 m and 3 m. Vertical resolutions are 5 m, but it is difficult to be precise as it represents average heights. It is slightly more expensive than the canopy morphology. This process is illustrated in Figure 6.4, in which a morphology database with 30 meter resolution is over-sampled to a 6 meter resolution and street vectors are applied to determine pixels that represent streets and their morphology type is changed to street.

- *Building level morphology*: this is based on the digitization of buildings in an area. The horizontal resolution is usually between 3 m and 1 m, with a vertical resolution of 1 m. It is very expensive and time-consuming to produce, thus only small areas are generated.
- *Mixed morphology*: the previous three types can be simultaneously used over a large area. Ideally, the different morphologies should not be combined into a single database; instead, they should be used as separate layers with the prediction software deciding, on a case-by-case basis, which one to use.

It is important to highlight that databases should be based on recent data (e.g. current maps). Different databases used in the same design must use the same coordinate system and datum; mixing databases of unknown origin or projection leads to poor results.

The representation of morphology heights in geographical databases is not very precise, and variations of several meters can be observed, depending on the reference used to estimate the building height (popular methods include triangulation from aerial photographs, and shadow length calculation).

Because of this inaccuracy, rooftop-installed antennas should always be stated in terms of height above ground AGL, and not above morphology (AML). Figure 6.5 shows the difference reference points for defining the antenna height. A common mistake in network design is to use AML to define the transmitter antenna height and then find base station antennas buried inside buildings due to differences in the actual building height. Figure 6.5 illustrates antenna height references, in relation to sea level (AMSL, Above Mean Sea Level), in relation to ground level (AGL, Above Ground Level) and in relation to building top level (AML, Above Morphology Level).

6.1.2 Antenna Orientation

A common mistake is to specify antenna orientation using a compass, because it indicates the magnetic north and not the true north required in a design. Also compass indications vary according to the location on the Earth.

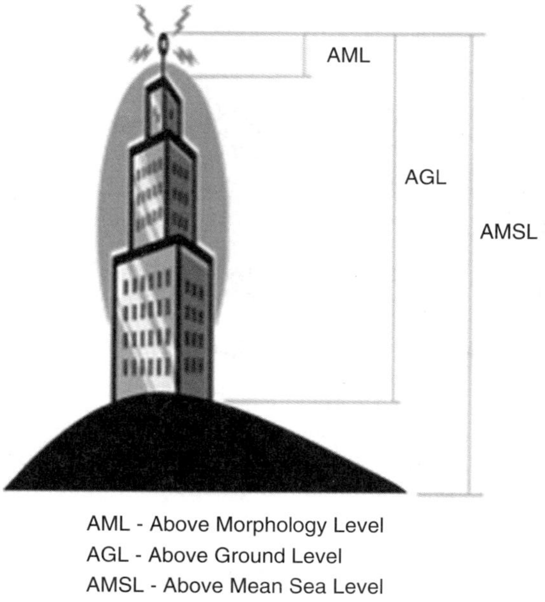

Figure 6.5 Antenna height references.

The reason for the magnetic north variation is that between 28,000 and 5000 km below the surface of the Earth lies a molten metallic region, much of which is iron. Due to the magnetic properties of iron and the complex fluid motion of this layer, the magnetic north of the Earth is not constant in time; instead, it follows the slow movement of the magnetic field. The angle formed between the directions of the true north (represented in maps and globes) and the magnetic north is known as the magnetic declination.

To aid in navigation and any other activity that requires some kind of geographical orientation (e.g. antenna installation azimuth in wireless networks), an IGRF (international geomagnetic reference field) model is compiled every five years or so as a collaboration of several international institutes that work with geomagnetic measurements. Each IGRF model has a validity of approximately five years; in late 2009, IGRF-11 was released, with values valid from 2010–2015.

Figure 6.6 shows a magnetic declination map for the year 2005. On the map, the numbers represent the difference, in degrees, between magnetic north and true north.

A good planning tool should offer designers some means of comparing true and magnetic north during the design process. A common mistake is to use, in the design, antenna orientation angles provided by a field team (collected using a compass). Depending on the part of the globe where the network is located, the difference between the angle measured using a compass and the true north used by the tool can be off more than 20 degrees.

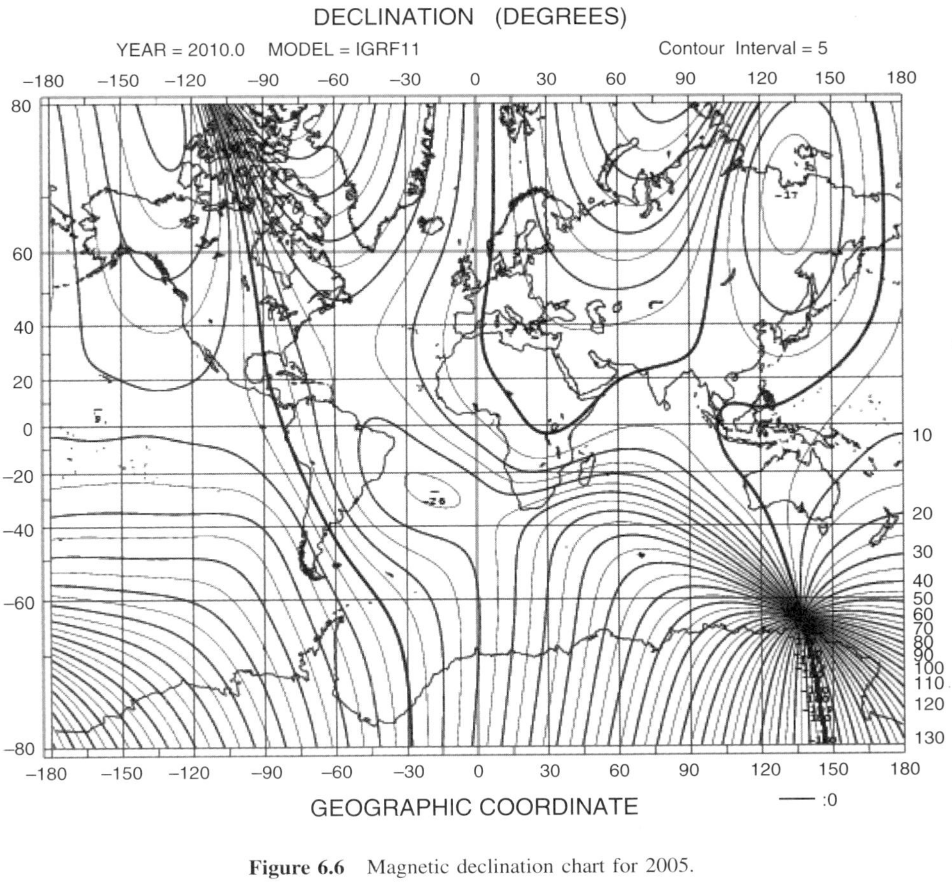

Figure 6.6 Magnetic declination chart for 2005.

6.1.3 Propagation Models

A propagation model is a set of algorithms that uses different parameters to estimate the path loss between two points. Some design tools call every set of propagation parameters a propagation model, but this is confusing, as it does not individualize the algorithms used by different models.

Certain propagation models perform better with low resolution databases, whereas others require good resolution ones. Design tools usually define propagation models per site, thus the best model can be chosen for each one, based on the databases used.

6.1.3.1 Propagation Model Prediction Parameters

RF prediction models rely on equations whose coefficients have to be adjusted for different environments. This is achieved through a calibration process that adjusts these parameters, so results are consistent with field measurements.

RF propagation models perform their predictions using information from terrain databases, so model propagation parameters are database dependent, and should be derived for each database. A significant update in a terrain database requires a parameter re-calibration.

The calibration of prediction parameters using mixed morphology databases allows for two prediction possibilities.

- Keep morphology layers separate and generate prediction parameters for each morphology layer. Predictions use the best resolution layer for each pixel.
- Mix morphology layers and derive parameters for the combined database.

An important factor to be considered when choosing a prediction model is the reusability of prediction parameters. Some models require measurement collection for each site in the system; some even require measurements on a sector basis to provide acceptable predictions. Advanced prediction models tend to require just a sampling of sites, to calculate prediction parameters that can then be applied to all sites in the area. These are the models described in this section. The calibration procedure is presented in details later in the chapter.

6.1.4 Prediction Layers

RF predictions give the average signal level calculated at every pixel. Multipath fading is averaged over time and pixel shadow fading is averaged over the pixel area. This allows for a consistent path loss value for every pixel, which corresponds to an average value available 50% of the time and on 50% of the area.

Additionally, each pixel can have multiple path losses associated with it, when we consider the receiver antenna location and height in relation to the morphology. Those losses can be divided into the following:

- *Over-morphology propagation losses*: represent the additional loss (above free space), when the signal has its Fresnel zone touching the morphology and should be considered by the propagation model.
- *Morphology penetration losses*: represent the loss by penetrating the morphology enclosure (walls, trees ...) and should be specified in the model, but if not supported by it can be emulated in the environment.
- *Morphology propagation losses*: represent the loss when propagating inside morphologies (treed area, rooms ...) and should be considered in the propagation model for single type morphologies and in the environment for multi-type morphologies.

Morphology losses are considered in the propagation model and in the environment specification. The designer has to specify for each service class where he considers the receiver to be, and how he will split the losses between the propagation model and the environment, as morphology-related losses have to be added to the path loss.

We can divide morphologies in four groups, in terms of loss analysis:

- *Low height morphologies*: this includes morphologies, such as water, grass, bare rock ... The receiver is always above the morphology. Additional losses should be specified in the environment.
- *Mixed height morphologies*: this includes morphologies as streets (with and without trees), parks, forests ... The morphology height should be expressed as an average, and the receiver may be located immersed or above the morphology. Morphology losses can be specified either in the model or environment.
- *Mixed type morphologies*: this includes morphologies such as residential, commercial areas. In this case there are open areas (grass, parking), tree area and building areas. There is usually a single height associated with this morphology, but multiple heights can be used. Morphology losses can be specified either in the model or the environment. It is up to the designer to decide where in the morphology he wants to represent the receiver (outdoor or indoor).
- *Multiple height morphologies*: this includes buildings and some other constructions, where the user can be located at different levels, but always indoor. Predictions have to be done for multiple building heights and morphology losses ideally should be considered in the model. Building heights are usually grouped in ranges as explained in Chapter 3.

Due to the above considerations, a prediction file should have several prediction layers:

- one layer for each transmitter (sector) covering 360° at ground level;
- one additional layer for each transmitter for each receiver height above ground.

A three-sectored site prediction for a four-height range area will have 12 prediction layers, covering each 360°.

6.1.5 Fractional Morphology

Classical models consider the role that morphology plays in RF propagation by changing some of the model parameters, but a single category is used for the whole area. Extensions of these models, such as the general model, consider a single morphology type per path or, in more recent implementations, a single morphology per receiver location.

This implies that a parameter used to characterize a propagation path applies to a mix of morphologies and, because this mix varies significantly between paths, it cannot properly represent all mixes. Only single morphology sites can be properly represented by the traditional methods. Every site has to be measured to adjust the predictions and, even so, only average values can be adjusted, presenting large variations for individual points.

In a single path between a transmitter and a receiver, many different morphology types are usually found, as illustrated in the geographical profile in Figure 6.7.

Figure 6.8 shows signal level predictions at different heights over a geographical profile using the Lee model and Figure 6.9 provides the legend for the levels used in the profile. Color saturation variations are used within each signal range to illustrate level variations. Only terrain obstructions are considered in this model, observe that the morphology height does not impact the signal level. This model, and other similar ones, such as Okumura-Hata, and Walfish-Ikegami, only predict well at ground level and do not benefit from high resolution databases.

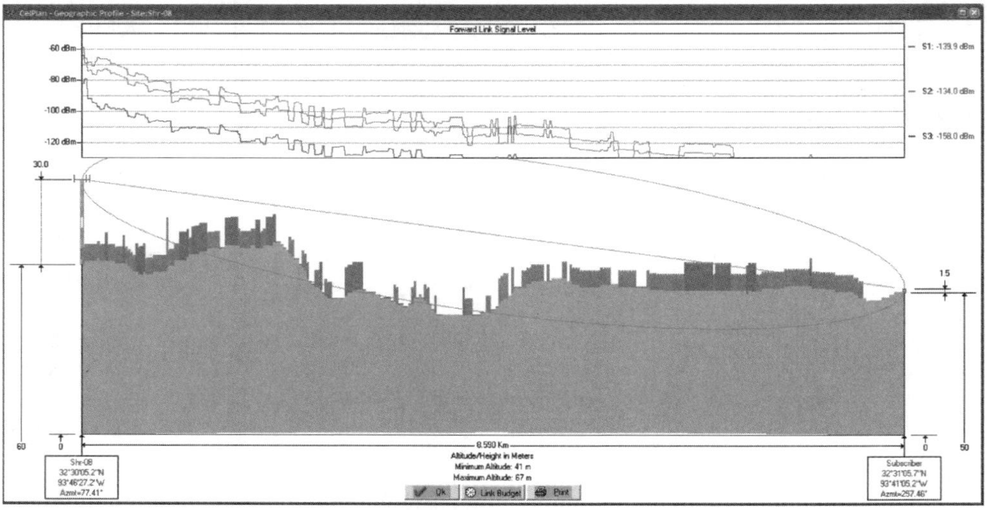

Figure 6.7 Terrain geographical profile showing the Fresnel zone.

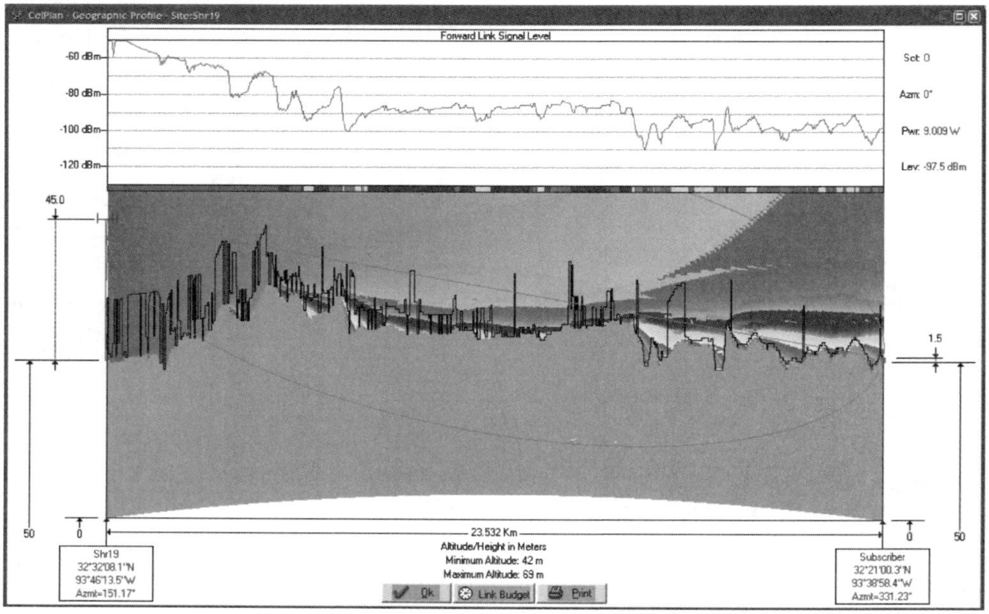

Figure 6.8 Terrain geographical profile for Lee's model.

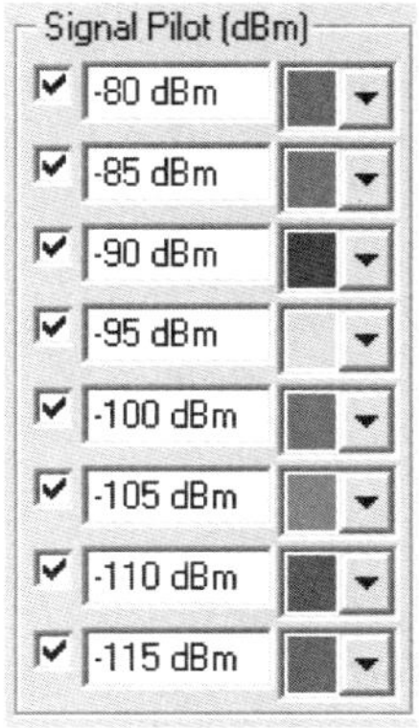

Figure 6.9 Legend for propagation loss profile.

Aware of this issue, CelPlan Technologies, in 1993, developed a fractional morphology method that considered the effect of different morphologies in the propagation path. In this method, propagation parameters are assigned to morphologies (instead of sectors) and can be reused for new sites and new areas. This method can be applied to any of the classical propagation models using single or multiple slopes. Figure 6.10 illustrates the concept of fractional morphology.

Table 6.1 shows fractional morphology parameters, assuming an area characterized by m different morphology types (m_a, m_b, ... m_m). Each type has two path slopes that characterize loss due to morphology according to distance.

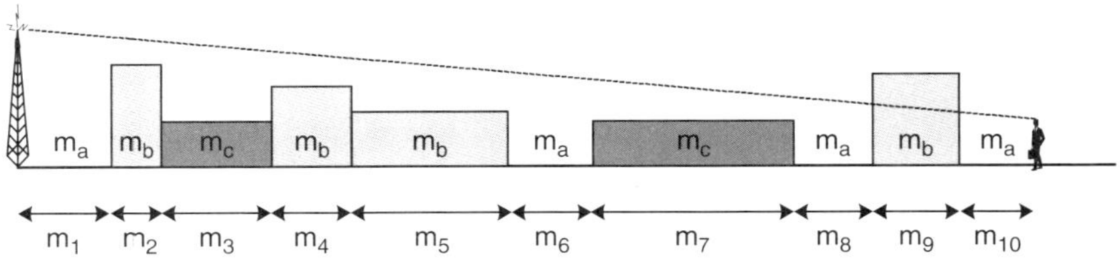

Figure 6.10 Fractional morphology concept.

Table 6.1 Fractional morphology parameters

Morphology	Slope 1	Slope 2
ma	S1ma	S2ma
mb	S1mb	S2mb
mc	S1mc	S2mc
–	–	–
mm	S1mm	S2mm

In the example of Figure 6.10, $m_1, m_2, \ldots. m_{10}$ represent the morphology types. A start (d_s) and an end distance (d_e) are assigned to each of these types. The slope break distance (d_b) varies depending on the propagation model being used. This distance represents the point before which slope 1 is used, and after which slope 2 is used. The overall loss is given by Equations (6.1) and (6.2).

For $d <= d_b$, slope 1 (S_1) is used:

$$Loss_1 = \sum_{m_i=1}^{n} S_{1mi} \log\left(\frac{d_{smi}}{d_{emi}}\right) \qquad (6.1) \text{ Path loss for slope 1}$$

For $d > d_b$, slope 2 (S_2) is used:

$$Loss_2 = \sum_{m_i=1}^{n} S_{2mi} \log\left(\frac{d_{smi}}{d_{emi}}\right) \qquad (6.2) \text{ Path loss for slope 2}$$

CelPlan applied the fractional morphology method to Lee's model with great success. In this model, the first slope is defined by the signal strength at 1 mile and the second slope, by the loss in dB/per decade of distance as shown in Figure 6.11. These parameters can be reused between sites and even between cities.

6.1.6 *Korowajczuk 2D Model for Outdoor and Indoor Propagation*

The use of fractional morphologies has significantly improved predictions but there were still many situations in which the outcome was not satisfactory:

- antenna heights much higher or smaller than the morphology did not result in good predictions;
- the effect of canyons (streets) was not well represented;
- the dual slope dichotomy caused situations in which the loss was not the same when the transmitter and receiver were reversed;
- small cells were not properly predicted;
- signals inside buildings were not predicted;
- signals at various building floors were not predicted.

A new approach was required to cope with these issues. Physical models address some of these issues but do not have the capability to predict the variety of situations encountered in real life. Korowajczuk introduced the Korowajczuk 2D model to overcome the shortcomings of traditional models when predicting real life networks.

One of the problems diagnosed with the propagation models available is how the morphology is considered to affect the signal. RF propagation happens through RF waves. These waves are longitudinal waves, as illustrated in Figure 6.12.

Sound waves are also longitudinal as illustrated in Figure 6.13. RF waves vary the density of magnetic and electrical fields similarly to the way sound waves propagate varying air pressure.

In two dimensions, this propagation can be represented by concentric circles or ellipses as shown in Figure 6.14. Figure 6.14 shows that the propagation in a real environment is obstructed and distorted by the morphology, indicating that the morphology height should be considered an obstruction. The issue is that morphologies are not continuous and sometimes are not even compact (trees), so the knife-edge treatment for obstructions does not fully apply. Therefore, the Korowajczuk model proposes a morphing factor (m_m) that adapts the knife-edge loss to a morphology edge loss.

The Huygens-Kirchhoff theory says that, as long as 0.6 of the first Fresnel zone is not obstructed, free space propagation can be considered between transmitter and receiver. This chapter refers to this zone as the inner part of the Fresnel zone as shown in Figure 6.15.

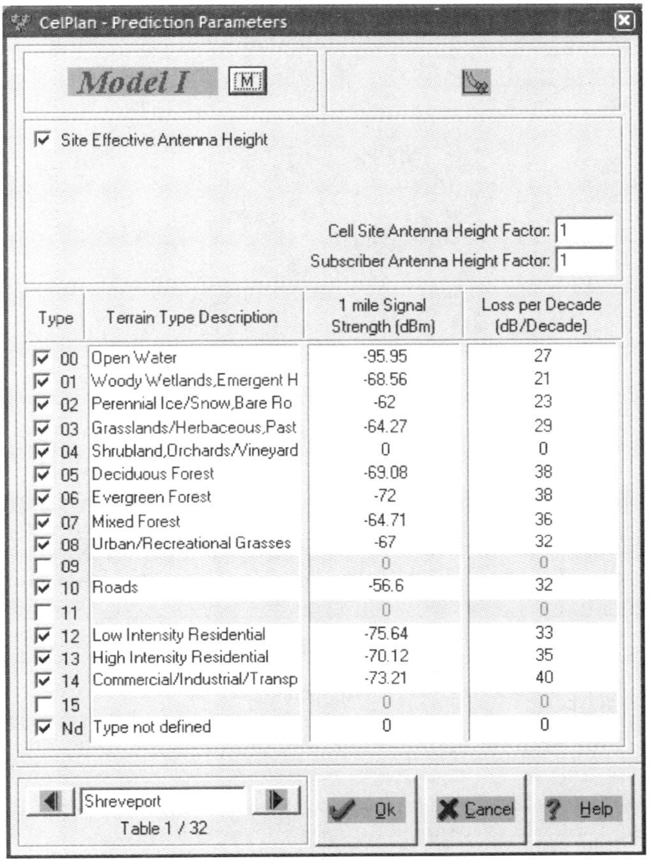

Figure 6.11 Fractional morphology parameters for Lee's model.

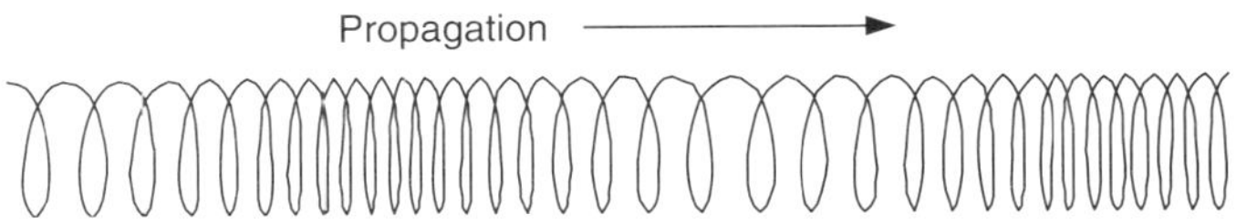

Figure 6.12 Longitudinal wave.

All methods (empirical and physical) used to analyze the effect of multiple obstructions have limitations and should be used only within their limits. The best compromise to find diffraction losses is to use the Deygout or Korowajczuk methods limited to three peaks. The peaks should be determined using the morphology, but the losses should be calculated for the topography and morphology at each of the points. The three peaks are determined by choosing the three points (topography + morphology) that intrude more into the Fresnel zone, that is, that cause the most obstruction.

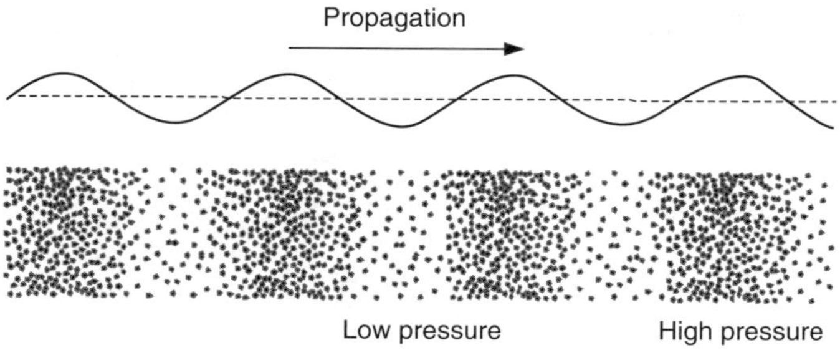

Figure 6.13 Sound motion through air molecules.

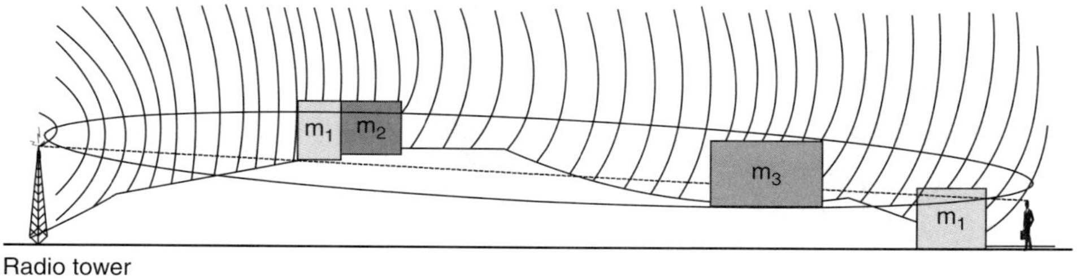

Figure 6.14 Wave propagation over morphology.

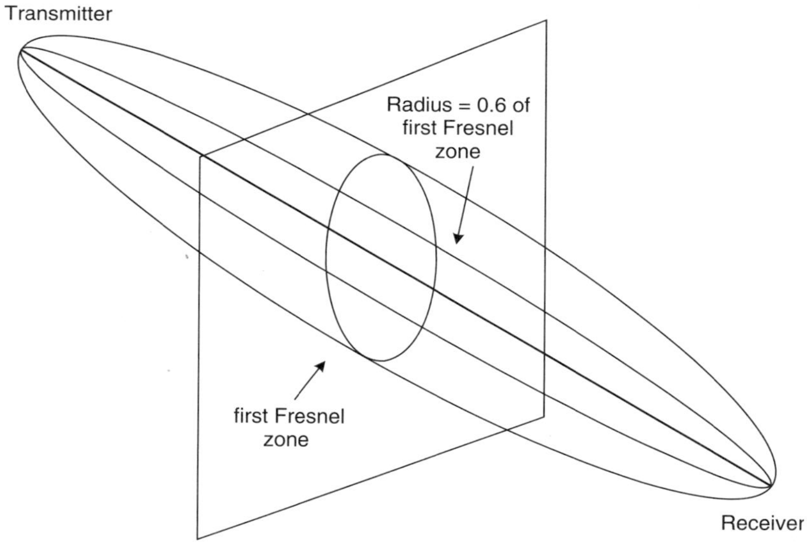

Figure 6.15 Fresnel zone representation.

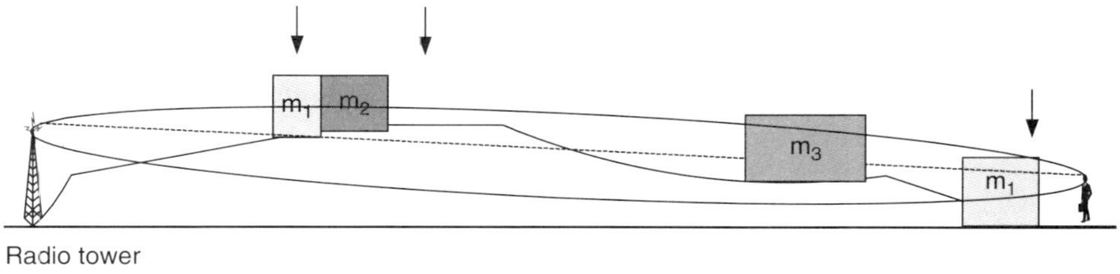

Figure 6.16 Diffraction considering terrain and morphology.

To apply these diffraction methods when considering morphology, the transmitter is automatically assigned a clearance equivalent to the antenna height in all directions around it. If the morphology where the receiver is located is higher than the receiver height, the diffraction is calculated to the top of that morphology as shown in Figure 6.16.

The selected diffraction model, Deygout or Korowajczuk, is then applied to calculate the loss at each point for topography and morphology heights separately. The difference between both models is that, in the latter, the loss attributed to the morphology should be multiplied by the morphing factor of the corresponding morphology type as in Equation (6.3).

$$LD = D_{1t} + m_m D_{1m} + m_m D_{2m} + m_m D_{3m} \quad \text{(6.3) Diffraction loss}$$

where:

D_{1t}, D_{2t}, D_{3t} = the three terrain obstructions.
D_{1m}, D_{2m}, D_{3m} = the three morphology obstructions.
m_m = the roundness factor.

The morphology diffraction adds a loss to the terrain diffraction, but as the morphology is not a continuous knife edge, a morphing factor (or roundness factor) was added to adjust the contribution of the morphology.

After the diffraction loss is determined, the Korowajczuk propagation model requires the division of the propagation loss into four parts (Figure 6.17):

- initial distance
- almost free space zone
- obstructed zone
- penetration zone

Initially, a path based on three diffractions is established and will be analyzed for further losses analysis.

The initial distance represents the path segment immediately close to the antenna, where no obstructions should be expected. It is usually set to the height above ground of the transmit antenna. The loss for this distance is always considered as free space as shown in Equation (6.4), initial distance loss.

$$L_d = 32.44 - 20\log(f) - 20\log(d_i) \quad \text{(6.4) Initial distance loss}$$

For the remaining path, distances for which the ellipsoid of 0.6 of the Fresnel zone does not touch the morphology (almost free space zone), a slope equal or higher than free space is assigned. This

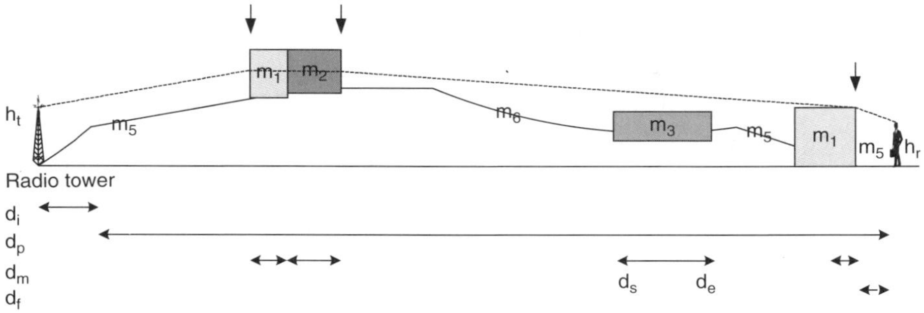

Figure 6.17 Propagation loss according to Korowajczuk model.

slope varies for different areas and depends on local conditions. A slope S_p is assigned for each path segment as in Equation (6.5).

$$L_p = \sum_{m_i=1}^{n} S_p \log\left(\frac{d_{smi}}{d_{emi}}\right) \quad \text{(6.5) Loss where Fresnel zone does not touch morphology}$$

For the distances where the inner Fresnel ellipsoid is obstructed by the morphology (obstructed zone) a different slope is assigned for each morphology type as in Equation (6.6).

$$L_m = \sum_{m_i=1}^{n} S_{mi} \log\left(\frac{d_{smi}}{d_{emi}}\right) \quad \text{(6.6) Loss where Fresnel zone touches morphology}$$

Finally, a penetration loss is considered. This loss represents the combination of all losses/gains in the environment surrounding the receiver. If the receiver is higher than the morphology, the penetration loss represents all signals that are added to the main signal, for example, signals reflected from other morphologies or from the ground. If the receiver is inside the morphology, the loss is applied proportionally to the height difference between receiver and top of morphology as in Equation (6.7).

$$L_f = p_m \log(h_{morph} - h_r) \quad \text{(6.7) Penetration loss}$$

The final path loss is the sum of the diffraction loss and the loss calculated for each of the four parts in the propagation path as indicated in Equation (6.8).

$$L = L_D + L_d + L_p + L_m + L_f \quad \text{(6.8) Total path loss}$$

Even though this procedure is computationally intensive, it can be executed by today's computers in a short time. All slopes are empirically calculated from measurements. Typical values depend on the area and the database used.

A typical parameters table for the Korowajczuk model is shown in Figure 6.18.

The following list maps the terms used in the dialogue of Figure 6.18 with the formulas presented in this section:

- Diffraction roundness factor is m_m.
- Propagation loss (dB/decade) is L_p.

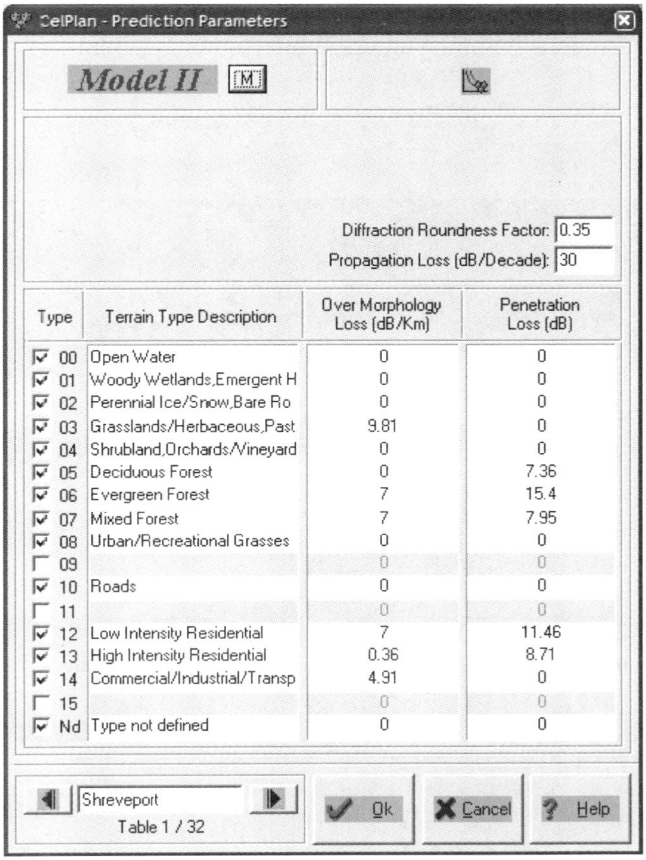

Figure 6.18 Korowajczuk model propagation parameters.

- Over morphology loss (dB/km) is S_m.
- Penetration loss (dB) is p_{mi}.

Because it considers the morphology height, this model allows the calculation of path loss for users not only at the street level but also on different floors of a building. Figures 6.19 and 6.20 show the signal levels predicted for different user heights according to the legend of Figure 6.21. Color saturation variations are used inside each range to illustrate level variations inside it. The figures clearly show the diffraction loss caused by morphologies and the signal loss when penetrating morphologies. The model predicts the signal inside buildings and between building floors with good precision, as can be seen in Figure 6.19 and Figure 6.20.

The Korowajczuk 2D model requires the availability of good terrain databases with at least carved streets and average morphology heights. For databases that are coarser than this, other models might get better results.

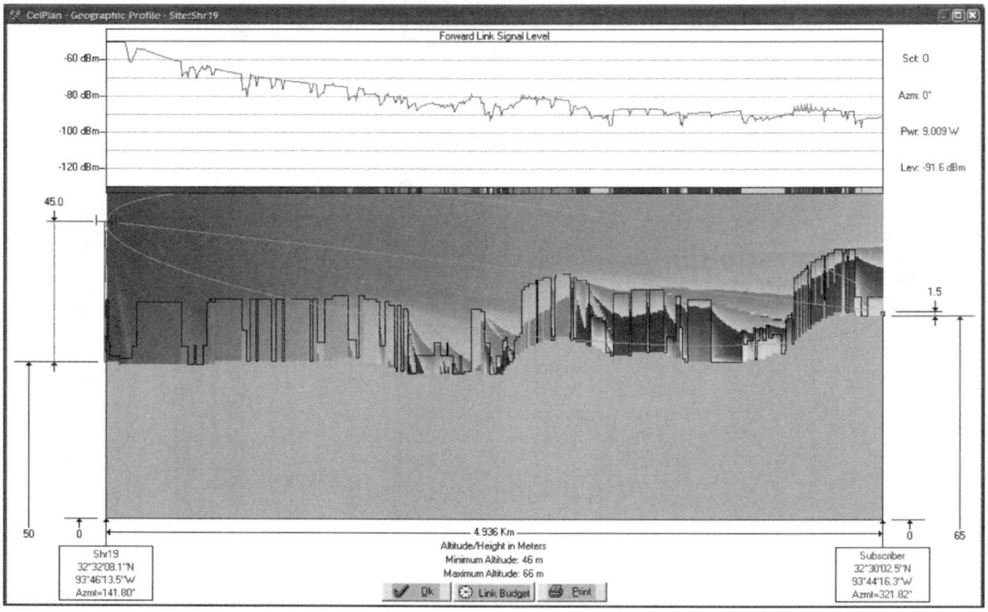

Figure 6.19 Korowajczuk model propagation loss profile (short distance).

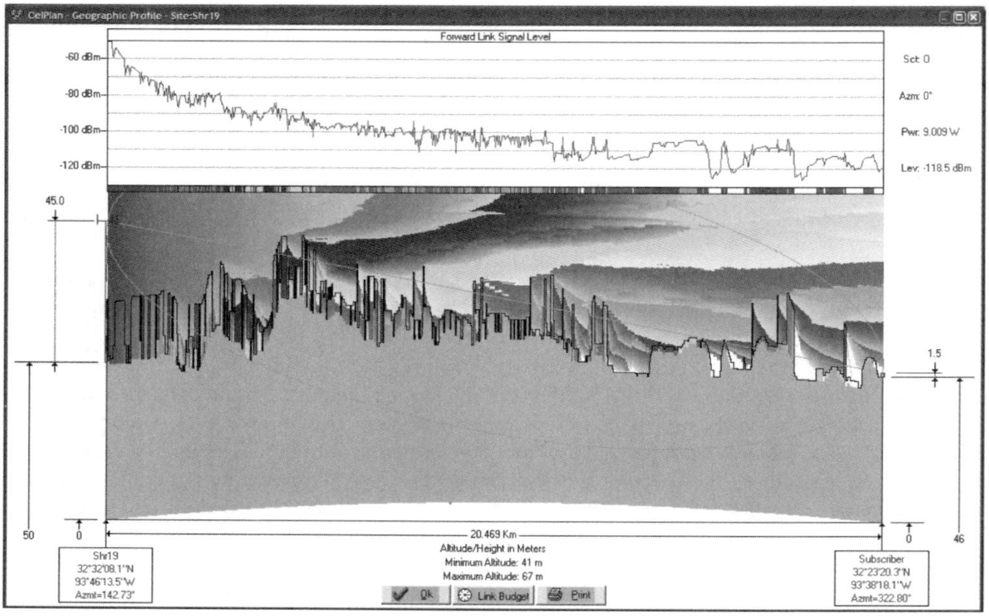

Figure 6.20 Korowajczuk model propagation loss profile (large distance).

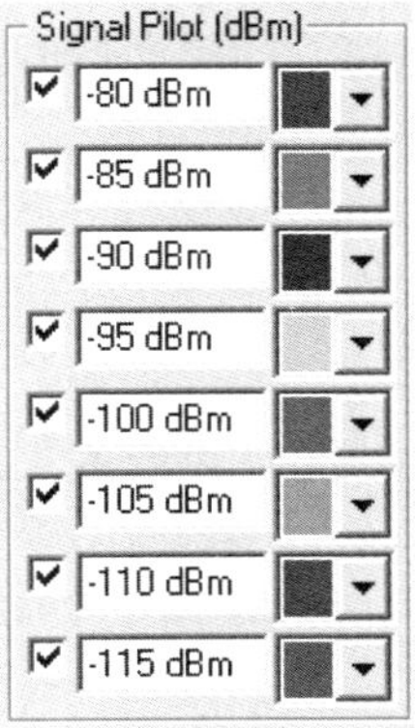

Figure 6.21 Legend for propagation loss profile.

6.1.7 Korowajczuk 3D Model

The Korowajczuk 2D model gives, in general, better results than the other models described here, but, in some situations in which very detailed building databases are used, it does not perform well when predicting signals behind high, isolated buildings. This happens because the RF wave goes around the building (instead of going over it) and the RF profile considered in the 2D model only looks into the vertical plane, as represented in Figure 6.22. Additionally, certain morphologies, such as residential areas and single tree lines, do not add diffraction losses and should not be considered in diffraction calculations.

This model improves on the 2D model by adding another dimension, the inclined horizontal planes between transmitter and receiver. Thus three possible propagation paths are considered and the sum of them is used as the result (Figures 6.23 and 6.24). Each of the paths is calculated as in model K2D.

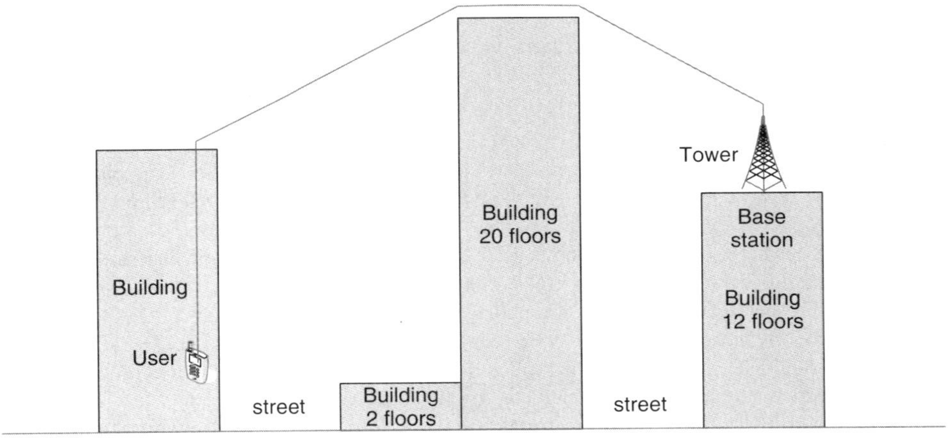

Figure 6.22 Korowajczuk 2D model RF path calculation.

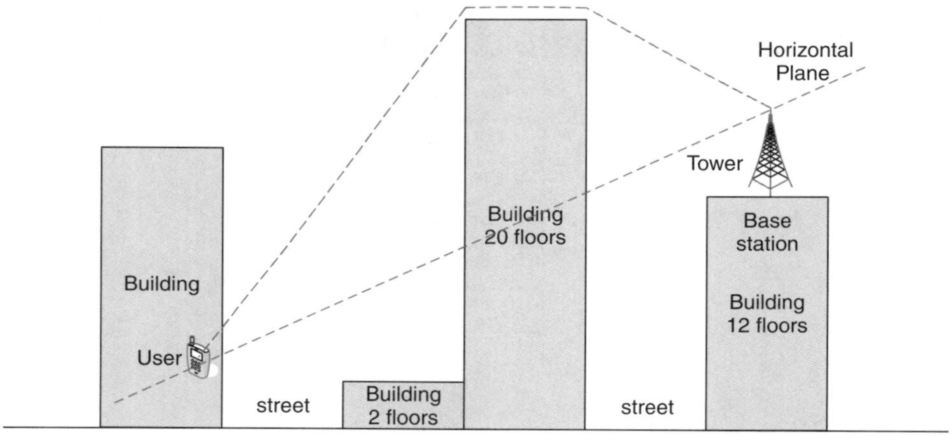

Figure 6.23 Korowajczuk 3D model RF path calculation on the vertical plane.

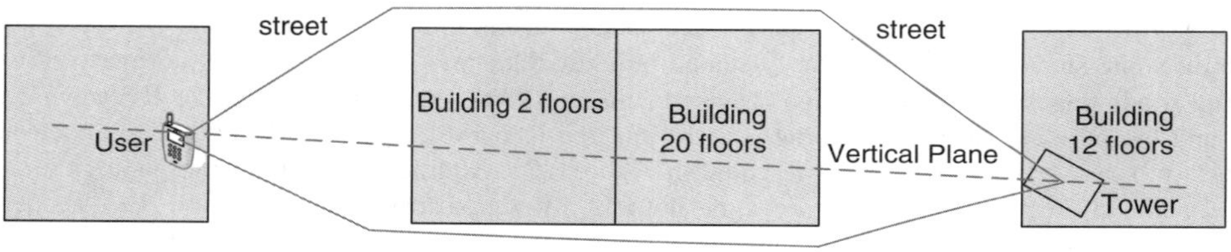

Figure 6.24 Korowajczuk 3D model RF path calculation on the horizontal plane.

The morphology-based diffraction is applied only to morphologies specified by the designer and each of these morphologies now has its own diffraction factor multiplier.

The K3D model is more complex than other models because it has to analyze three paths instead of one, leading to an average increase in prediction time of about 30%.

6.1.7.1 Three Break Points

The model supports up to three break points, which allows for a better adjustment to changing environments. Those break points are defined by the user, which will define them based on measurement slopes in the area or in similar areas.

Each of the areas between the break points has a different average propagation loss slope (expressed in dB/decade), and over-morphology-loss (Mfl loss). This loss is applied to the length in distance in which the first Fresnel zone touches the morphology. This is illustrated in Figure 6.25.

The diffraction switch (check boxes in the configuration screen in Figure 6.27) indicates which morphologies should be considered when calculating diffraction. The diffraction factor multiplier (Diffr Factor) adjusts the loss initially calculated by the knife edge method.

The penetration loss factor (PenL) represents the signal loss inside the last morphology type, that is, where the user terminal is located; it is only considered when the receiver is embedded in the

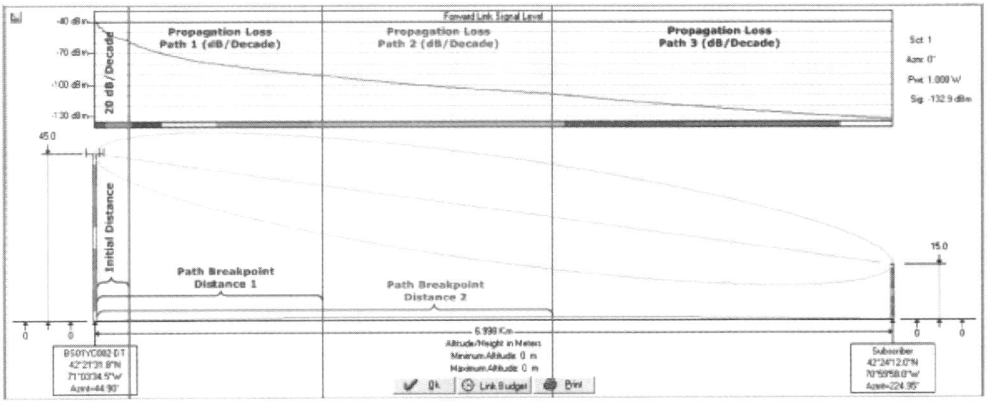

Figure 6.25 Model 3D three slopes.

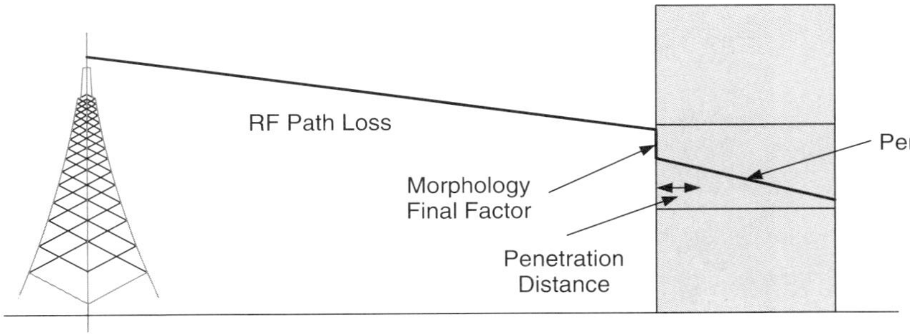

Figure 6.26 Model 3D penetration loss and morphology final factor loss.

morphology, that is, the terminal is lower than the surrounding clutter. It is given by Equation (6.9) and illustrated in Figure 6.26.

$$\text{Penetration loss} = \text{PenL} * \log(\text{morphology penetration distance} + 1) \quad (6.9)$$ Penetration loss factor

The morphology final factor loss (FnFac) represents the loss from penetrating the morphology encapsulation, usually between the exterior of a building and its interior. Typical values for different materials are given in Table 6.2.

An example of the propagation parameters for this model is shown in Figure 6.27.

A cross-section of the Korowajczuk 3D model is presented in Figure 6.28. RF signals follow the antenna pattern and are attenuated by building walls and then further attenuated inside the building. Building tops diffract the signal, but this diffraction is attenuated by signals that come around the building. Color thresholds are shown in Figure 6.29.

A downstream site prediction for the Korowajczuk 3D model is shown in Figures 6.29 and 6.30. The prediction shows the RF penetrating street canyons and one side of a street with a stronger signal than the other side, as expected in real life.

Table 6.2 Final factor loss for different construction materials

	Attenuation in first room (dB)				Std
Construction material	1 GHz	2 GHz	3 GHz	5 GHz	dB
Glass	10	12	14	16	6
Wood and plaster	12	15	18	20	7
Steel and plaster	15	18	20	22	8
Masonry	18	20	22	25	9
Concrete	20	22	23	30	10

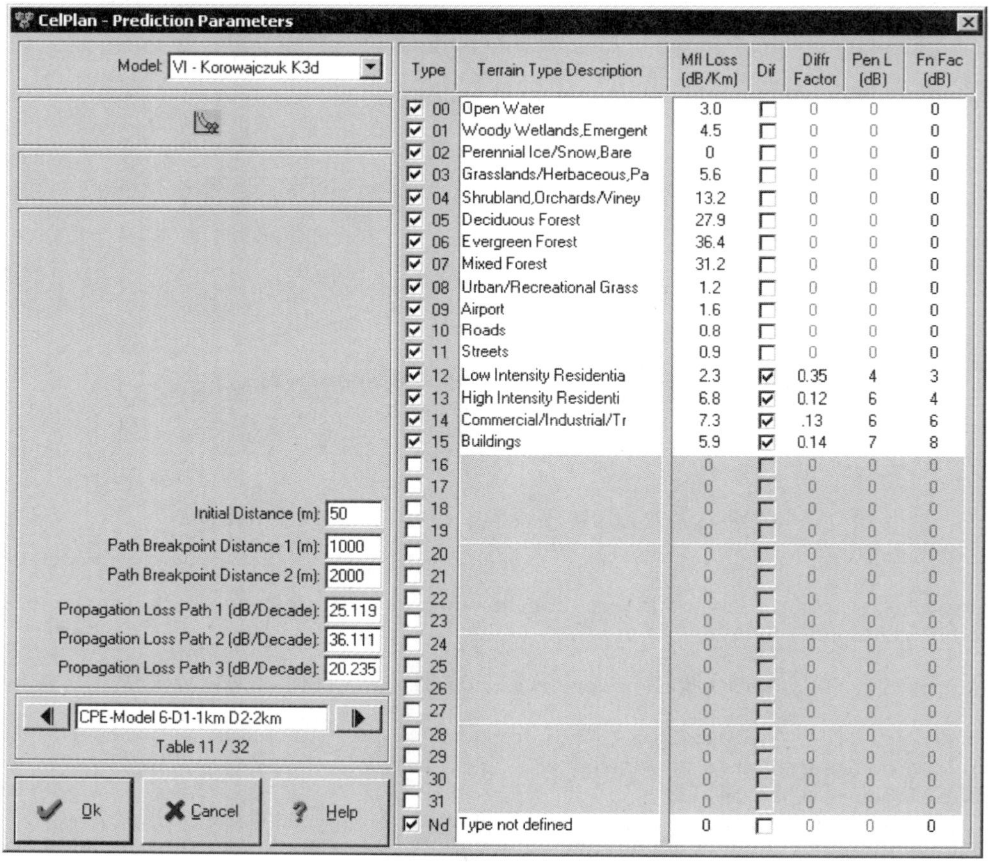

Figure 6.27 Korowajczuk 3D propagation model parameters.

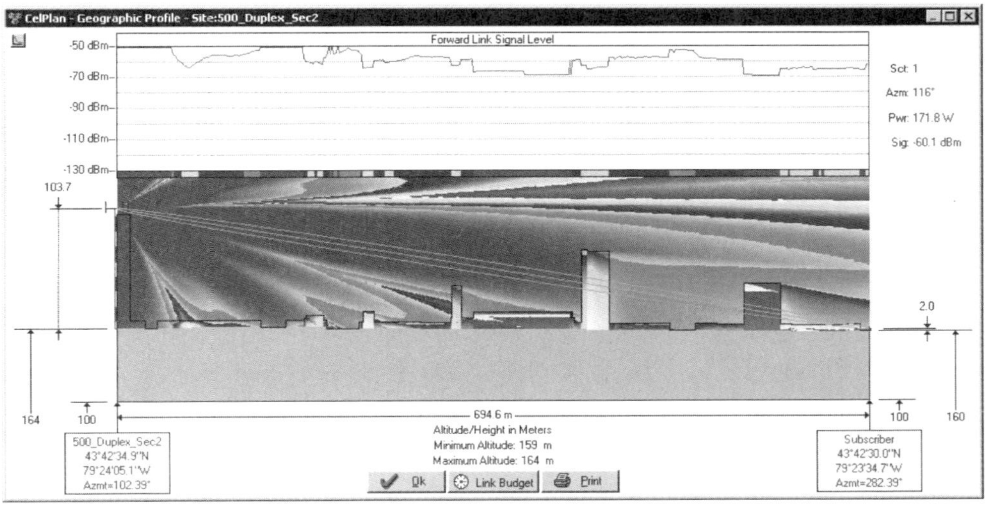

Figure 6.28 Korowajczuk 3D profile.

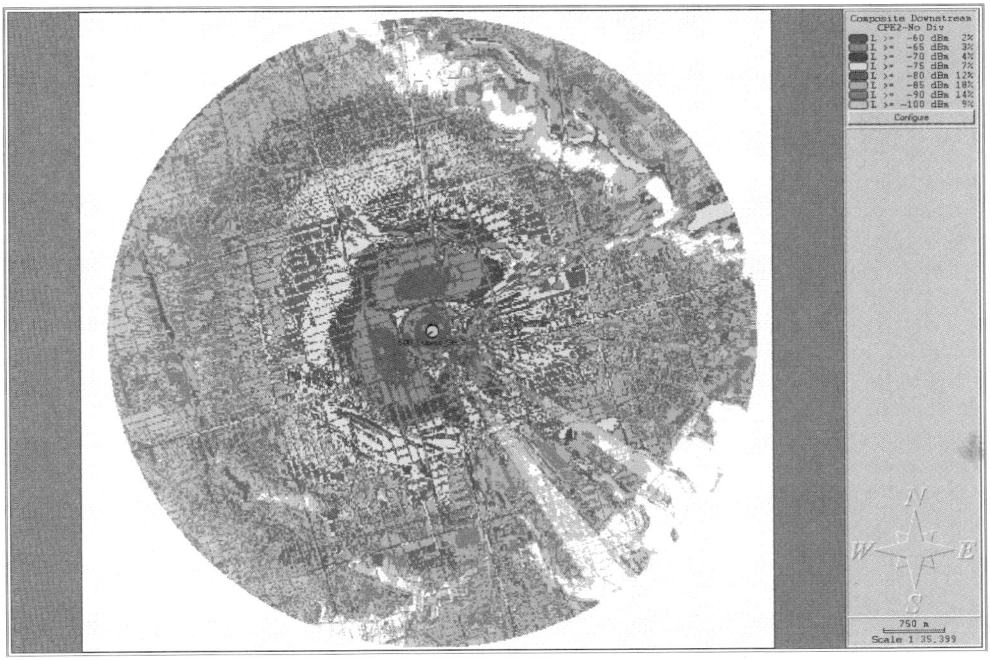

Figure 6.29 Korowajczuk 3D signal level prediction.

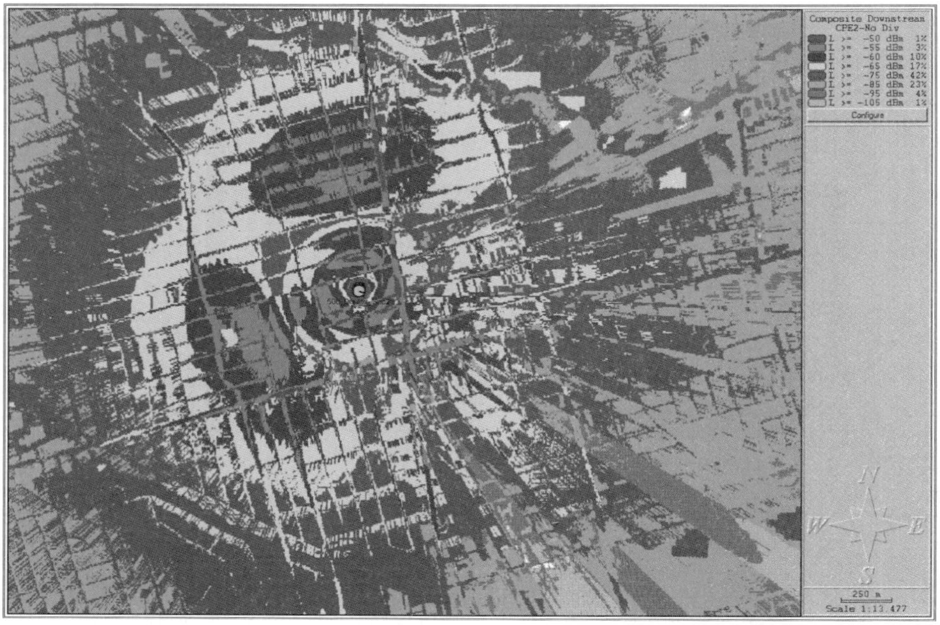

Figure 6.30 Korowajczuk 3D signal level prediction detail.

6.1.8 CelPlan Microcell Model

This model was developed for microcell propagation predictions. Microcell antennas are usually placed below the morphology height surrounding it and aim to cover specific areas. Ray tracing techniques can be used in this case, but the prediction time is very large. This model uses regular predictions and applies factors to the amount of morphology crossed by the direct path between transmitter and receiver.

Microcells are used in high traffic areas and should have small footprints. The usual technique to achieve this is by using sites with reduced power and with antennas placed at low heights. The morphology surrounding microcells is much higher than the cells themselves and the canyons in which microcells are located are usually narrow. This means that the diffraction loss over the morphology is high and the transmission is primarily done through the canyons.

At the beginning of the path, propagation happens mainly through and around morphology obstructions. However, after a certain distance, the propagation mechanism becomes mixed: part of the signal comes over the morphology and part comes through and around the obstructions. At even further distances, the propagation over morphology becomes predominant. This means that several propagation mechanisms play a role in a microcell analysis; therefore multiple slopes are required to represent this.

The distances involved are small, which means that the Fresnel zone is also small, therefore the analysis of direct rays is a reasonable assumption. The propagation inside canyons (streets) and open spaces can be considered as free space propagation. For propagation outside the canyons, where the signal comes through and around obstructions, one option for the analysis is to use ray-tracing techniques. However, these techniques require accurate databases that are rarely available and never represent all morphology nuances, mainly when vegetation is involved.

CelPlan found a good correlation between the morphology composition of a direct ray connecting transmitter and receiver and the signal received at the antenna, as illustrated in Figure 6.31. A certain

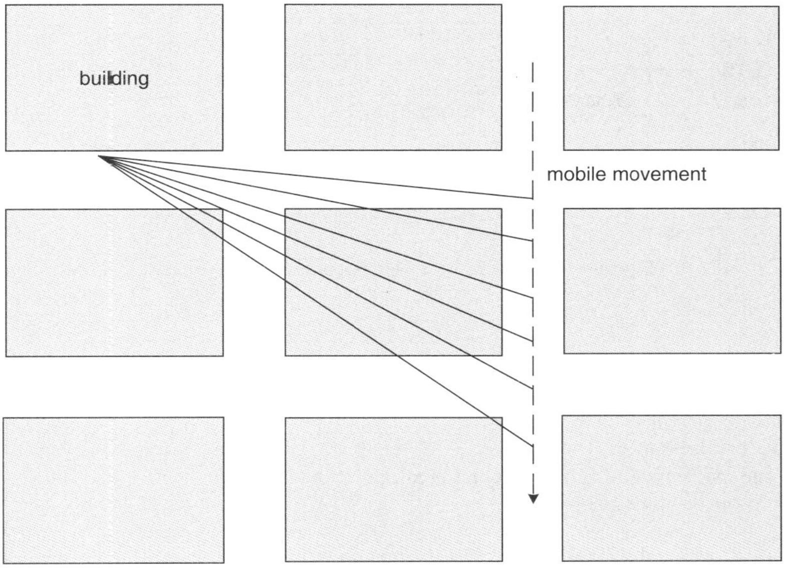

Figure 6.31 Microcell model diagram (top view).

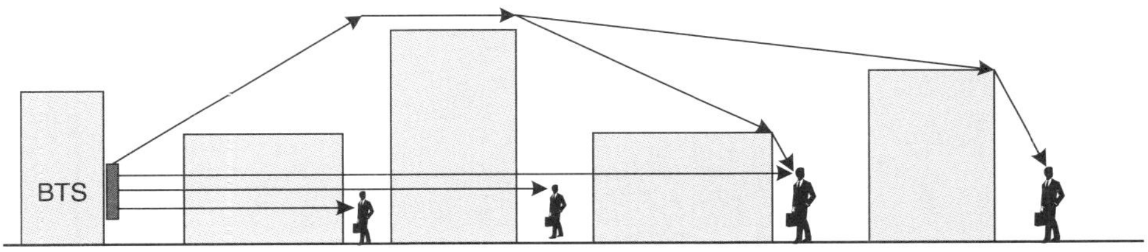

Figure 6.32 Microcell model diagram (profile view).

slope is assigned to each of the morphologies, representing the obstruction to the passage of the signal through or around it.

Figure 6.32 shows the different paths that the strongest signal can take. Direct paths are routed around street corners as illustrated in Figure 6.33.

To calculate the microcell propagation loss, the topography-based diffraction loss is calculated using Deygout with up to three diffraction points. The microcell model developed at CelPlan uses a fractional morphology multiple slope approach, with the break distances for each slope defined by Equations (6.10), (6.11) and (6.12).

$$d_2 = 4h_t h_r(d - d_p)/(\lambda^2 \sqrt{(h_t - h_r)^2 + (d - d_p)^2}) \quad (6.10)\ \text{Plane Earth break distance (second break point distance)}$$

$$d_1 = d_2/2 \quad (6.11)\ \text{First break point distance}$$

$$d_3 = 3d_2 \quad (6.12)\ \text{Third break point distance}$$

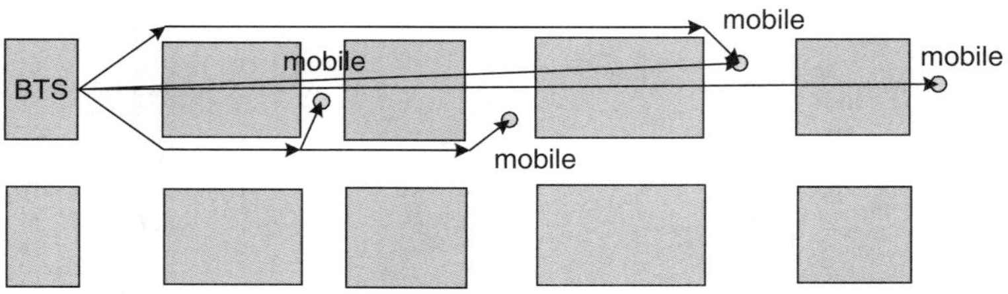

Figure 6.33 Microcell model diagram (bird's-eye view).

where:

d = total path distance.
d_p = distance to last diffraction point.
d_i = antenna height distance.
d_1 = first break point distance.
d_2 = plane earth break distance (second break point distance).
d_3 = third break point distance.

The distances d_1, d_2, and d_3 begin at the last diffraction point or at the transmitter, whereas d always begins at the transmitter. The significance of these distances can be understood in Figures 6.32 and 6.33 as the main propagation component changes from a path below rooftops to a mix and then to a path above rooftops. Each segment has a slope attributed to it. These slopes are empirically calculated from measurements. The distances are calculated separately for each diffraction segment.

Let's assume the following path losses:

$$0 < d \leq d_i : S_i = 20$$

$$d_i < d \leq d_1 : S_1$$

$$d_1 < d \leq d_2 : S_2$$

$$d_2 < d \leq d_3 : S_3$$

Equation (6.13) shows the diffraction calculation. The overall loss for CelPlan's microcell model is given by Equation (6.14).

$$D = D_1(T, R) + D_2(T, 1) + (32, R) \qquad \text{(6.13) Diffraction loss}$$

$$L = 32.44 + 20\log(f) + D + L_d + L_{11} + L_{21} + L_{31} + L_{12} + L_{22} + L_{32} + L_{13} + L_{23} + L_{33} \qquad \text{(6.14) Overall loss}$$

Equations (6.15) to (6.18) show the calculation of each of the losses involved in Equation (6.14). In these equations, n represents a diffraction segment.

$$L_i = 20\log(d_i) \qquad \text{(6.15) Initial loss}$$

$$L_{1n} = S_1 \log(d_1) \qquad \text{(6.16) First loss}$$

$$L_{2n} = S_2 \log(d_2) \qquad \text{(6.17) Second loss}$$

$$L_{3n} = S_3 \log(d_3) \qquad \text{(6.18) Third loss}$$

Figure 6.34 shows the parameter configuration table for CelPlan's microcell model.

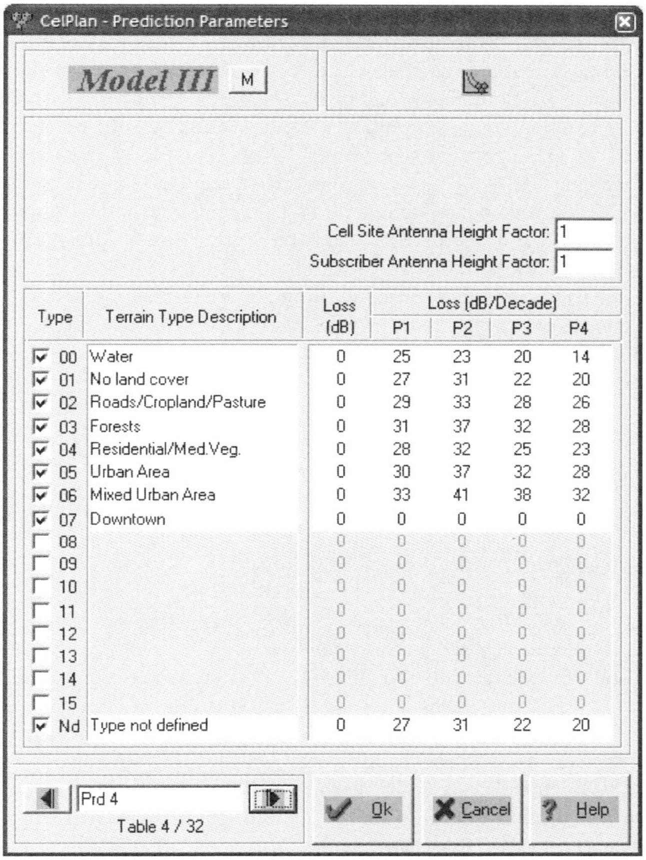

Figure 6.34 CelPlan microcell model propagation parameters.

6.2 RF Measurements and Propagation Model Calibration

Propagation models are empirical models that require particular parameters for each area. Traditionally, propagation models require different parameters for each sector, but this was overcome by the use of fractional morphology. This solution allows the same sets of propagation parameters to be used over vast areas, as long as they have similar characteristics.

Even with fractional morphology, though, different site configurations might require different sets of parameters. The factors to be considered when grouping sites that can use the same propagation parameters are:

- similar height over average terrain;
- similar antenna surrounding obstructions.

Propagation parameters can be derived from the literature, based on previous experience, or be extracted from actual field measurements; the latter being the best option to obtain more accurate results. This method is called propagation parameter calibration, in which predictions are done for each measured point and propagation parameters are adjusted to reproduce the collected path loss values.

Predictions use terrain databases to calculate the received signal, so calibrated propagation parameters are tied to a specific terrain database and cannot be used with different databases.

Morphologies play an important role in model calibration, and must be properly identified. Ideally the databases should be prepared with this in mind. It is common to have several morphological types identified in the database, which have the same RF behavior and could be joined into a single type, increasing the overall number of samples for it. A typical example is low grass, open area, bare rock, and so on. Too many morphology types reduce the number of samples per morphology and consequently reduce the calibration quality. In general, eight types should be enough for a single morphology layer and 16 types should suffice when multiple layers are used.

The same morphology digitized with different resolutions may have different propagation parameters for each resolution, thus their types should be identified differently.

6.2.1 RF Measurements

Advanced propagation models require the definition of many propagation parameters, which have to be derived from measurements. RF channel measurements have to reflect the average path attenuation at short and long distances, so interferer signals can be predicted. This requires measurements to be done with a narrowband channel (low noise floor) and be filtered for fast fading. A CW (continuous wave) transmitter should be used with a bandwidth of few Hz, so the noise floor at the receiver is kept very low, increasing the measurement distance. Care should be taken that no other transmission is happening at the chosen frequency and that transmission at neighboring frequencies will not overload the receiver. An initial drive test without a transmitted signal should be done first to satisfy these conditions, followed by the actual transmission drive test. It is always important to analyze drive test data for inconsistencies.

Alternatively, a scanner for a specific technology can be used, as it will identify the channel source. The advantage of this alternative is that only valid measurements will be collected, but the disadvantage is that low level signals will not be detected and this will make it difficult to properly calibrate the propagation model. The scanner will also identify areas prone to interference and this will filter CW measurements in this areas.

Ideally a mix of CW and scanner collected measurements should be used. It is important that all morphology types be present in significant quantities in the paths between transmitter and receiver. Paths should be chosen to represent the terrain well and streets should be driven preferably in both directions.

Outdoor drive tests limit the data to this environment, so indoor measurements should be collected and compared to nearby outdoor measurements.

Measurements should be collected with an associated time and location stamp, so they can be processed to eliminate redundant measurements, and be filtered for fading, distance, antenna nulls, noise floor and by locations not well represented in the database.

Measurement positioning should be adjusted (due to GPS error), so they fall into morphologies that were actually measured. In urban areas the GPS location should be assisted by a dead reckoning system.

Collected measurements should be filtered for fast fading, according to the multipath limits shown in Chapter 5. A common mistake is to filter fast fading by applying the usual value of 40 λ. This value is, however, an integration formulated by W. Lee for cellular urban applications in the 800 MHz band only, and does not apply to other frequencies. Lee calculated the 40 λ parameter by integrating measurements over a distance of 12 meters, which he found representative for the area that he was analyzing.

Figure 6.35 shows the procedure that should be followed to collect measurements.

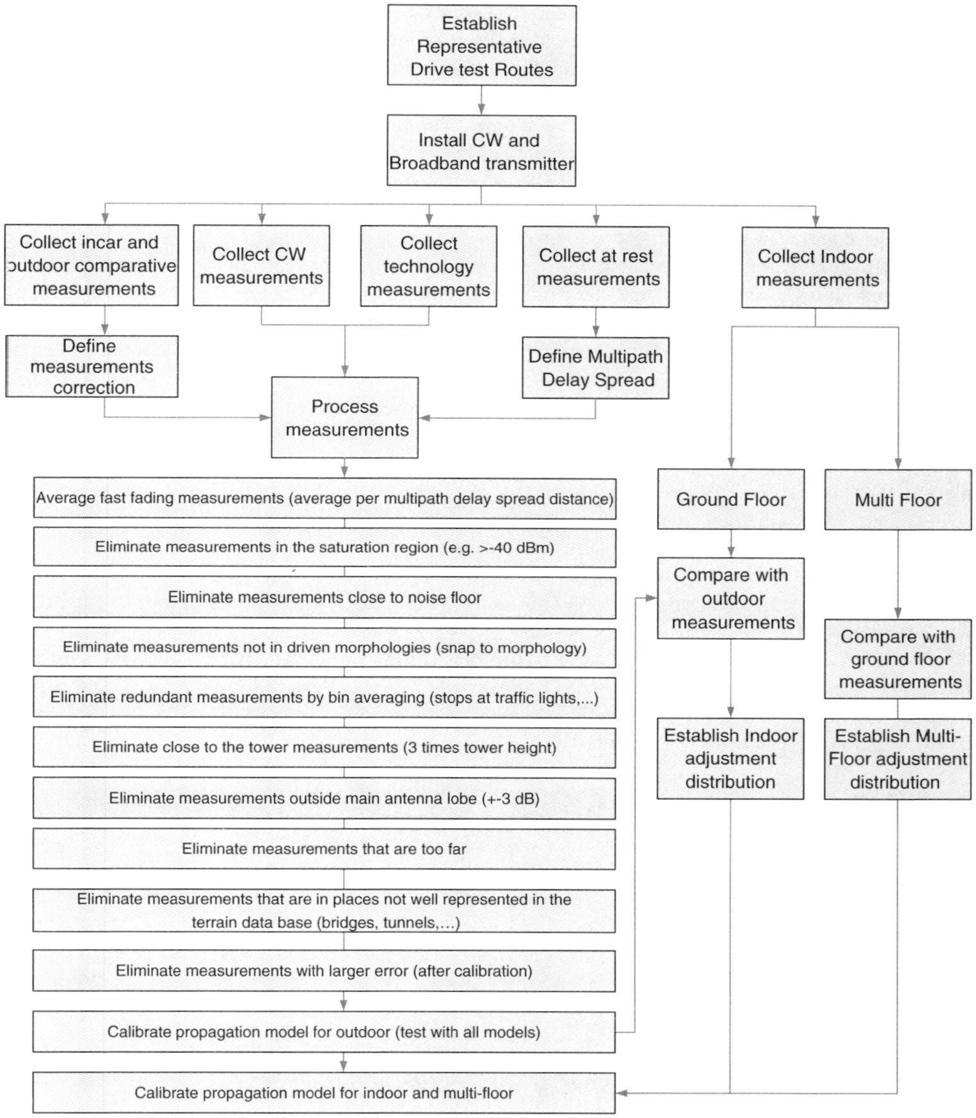

Figure 6.35 RF measurement drive test collection procedure.

Test sites should be chosen to represent the different environments in the area. When fractional morphology propagation models are used, as few as five sites can be sufficient. With other models this number can be as high as 50 or more.

Test routes should be determined to sample all morphologies in the area and measurements should be done in both directions of streets. Other parameters such as target speed and measurement device setting should be established and documented. Proximity to other transmitters should be avoided as they can overload the receivers. Filters should be used in the receiver to limit out-of-band signals.

The transmitter should be properly installed and the transmit power at the antenna should be measured and documented, preferably with a picture, as wrong settings at the transmitter are the reason for many disastrous measurement campaigns. The drive test vehicle should also be properly installed, with adequate power and antenna installation over a ground plane. Cables should be checked for losses before starting any measurements. The drive test vehicle environment should be characterized and measurements performed inside and outside the vehicle, so a correction factor can be established to adjust the measurement values.

Once the actual drive test campaign starts, it is strongly recommended that the measurement files be examined daily, so issues can be detected sooner rather than later. Some of the items that should be verified are saturation, low signals, and lack of coordinates.

In parallel, static measurements should be collected at several locations to characterize multipath delay spread. This is done by collecting samples at several frequencies at a high rate. Fading duration can then be determined. This measurement is important not just for calibration but also for network parameters adjustment.

Indoor measurements should be collected at ground level and at other floors; allowing for the calculation of adjustments in the propagation models and environment for indoor multi-level predictions. Ideally, these measurements should be divided into three categories: at windows, close to windows, and deep indoor (e.g. more than one wall between transmitter and receiver).

Figures 6.36 and 6.37 show sample filter configuration dialogues for signal and morphology; additionally filtering can also be done based on coordinates. Figure 6.37 shows a set of measurements that have been adjusted to the morphologies.

An additional filter should be applied after the first calibration in which measurement with large prediction errors are scrapped, as this will eliminate atypical samples, and the model should then be re-calibrated. Figure 6.38 shows measurement analyses screens.

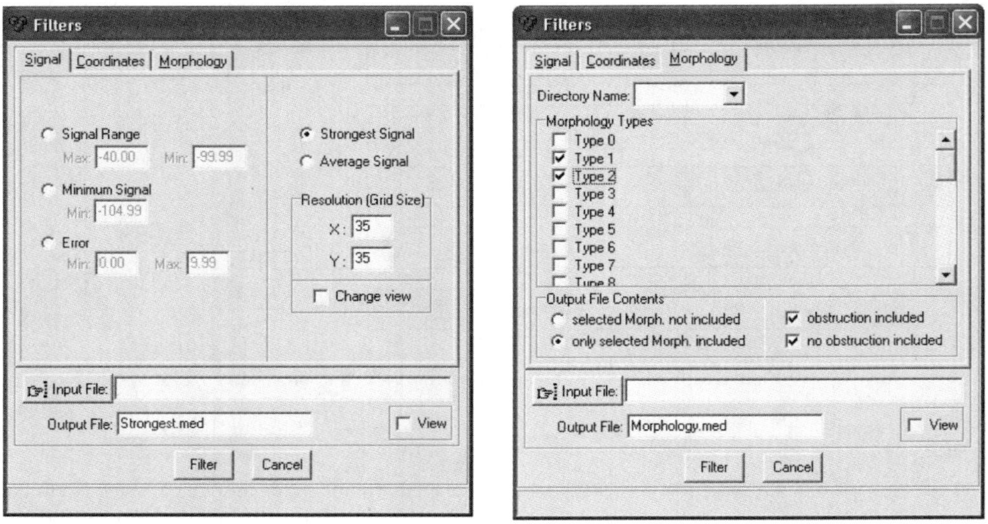

Figure 6.36 Measurement filters dialogue box.

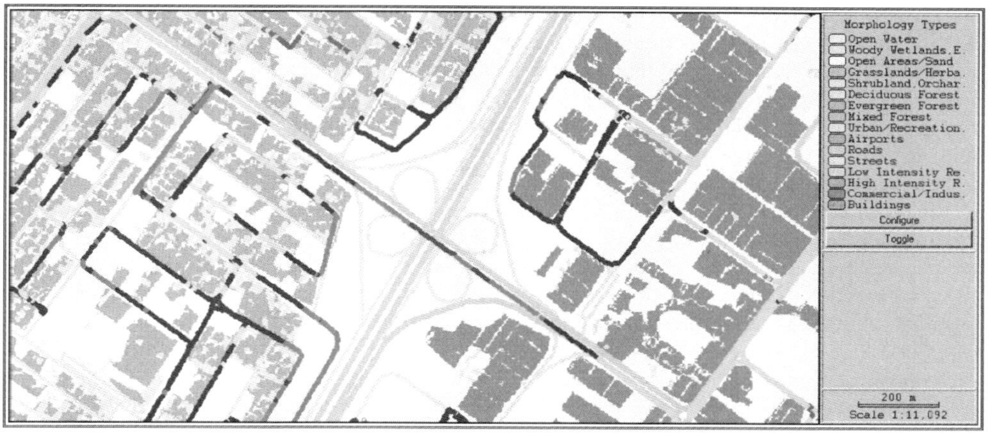

Figure 6.37 Drive test collection (snap to morphology).

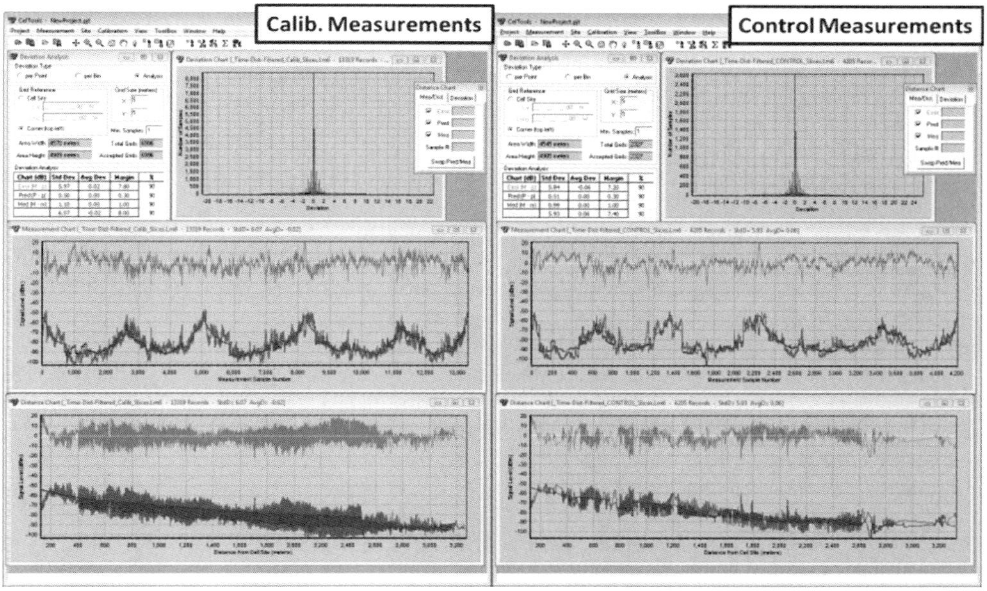

Figure 6.38 Measurement analysis.

6.2.2 *RF Propagation Parameters Calibration*

Once the measurements have been validated, the calibration process can begin. It is very important that the transmitter and receiver (power, antenna, ...) site data be well characterized at this point, as it will be used during the calibration process.

Each measured value can be mapped to a mathematical equation that represents the path loss from the base station to the subscriber unit. These mathematical equations are defined by the propagation model and initially have all propagation parameters as unknowns, while the outcome of the equation is the value of the measurement at the point. For a morphology database with 16 types, equations representing the Korowajczuk 2D model have 34 unknowns, for the Korowajczuk 3D model, 67 unknowns.

Each sample point creates a new equation, thus forming a matrix with the equations representing a given set of collected data. This matrix can be solved interactively targeting zero average deviation between measurements and predictions, while, at the same time, minimizing the standard deviation.

Because the solution of this matrix is a mathematical process, the best result can give values outside physical expectations, thus the calculation should offer, as an option, the use of user-configurable constraint ranges for each parameter. An example of a propagation model calibration dialogue box is shown in Figure 6.39.

The calibration process is based on measurements done at ground level, so only ground level parameters can be automatically calibrated. The calibration engine should provide the age of participation of each parameter in the total calculation, so the confidence level of each parameter can be evaluated. It is up to the designer to then adjust the parameters to reflect indoor and multi-floor situations.

Alternatively, the designer can generate measurements indoor and at multiple height levels, based on the outdoor measurements by applying the correction adjustment factors previously estimated. These factors should be applied using the statistical distribution considered for each parameter. This will allow the automatic calibration of parameters for all scenarios.

Because the solution to this matrix is a mathematical process, the best result can give values outside physical expectations, thus the calculation should offer, as an option, the use of user-configurable constraint ranges for each parameter.

Figures 6.40 and 6.41 show the measured and predicted signals, as well as result of the calibration process, respectively for the measurements used for calibration (calibration set) and measurements used to verify calibration quality (control set).

Figure 6.42 shows measurements at specific points compared to the average measured value in a prediction bin. In Figure 6.42, capital letters stand for the average value inside a specific bin: M for measurement. Small letters represent the punctual value, that is, the exact location where the sample was collected. "M-m" represents the difference, in dB, of the average of the samples within a bin of the grid, compared to each actual sample within that same bin.

Figure 6.43 shows the comparison between measurements and predictions for bins in the area. The capital letters M (measurement) and P (prediction) stand for the average value inside the grid (pixel or bin) specified in the dialogue of Figure 6.43. The small letters represent the punctual (the exact location were the measurement was done) value of each measurement. The screen shows the average and standard deviation for four categories (expressed in dB):

- Med (M-m): This category shows the variation of the measurements within the specified grid size (7 m in the example). It represents signal variation inside the average grid bin. For a larger grid, this value will increase.
- Pred (P-p): This category shows the average predicted value per grid compared to the punctual value of each measurement. The predicted value has a smaller deviation than the measured value as the database used in the predictions is not as rich as the real life.
- Dev (p-m): This category represents the actual difference between the predicted and measured value for each location in which a measurement was done. This is the value most representative of the deviation between measured values and predicted values.
- Error (M-p): This category gives the average value of the measurements of the grid compared to the predicted values inside the same grids.

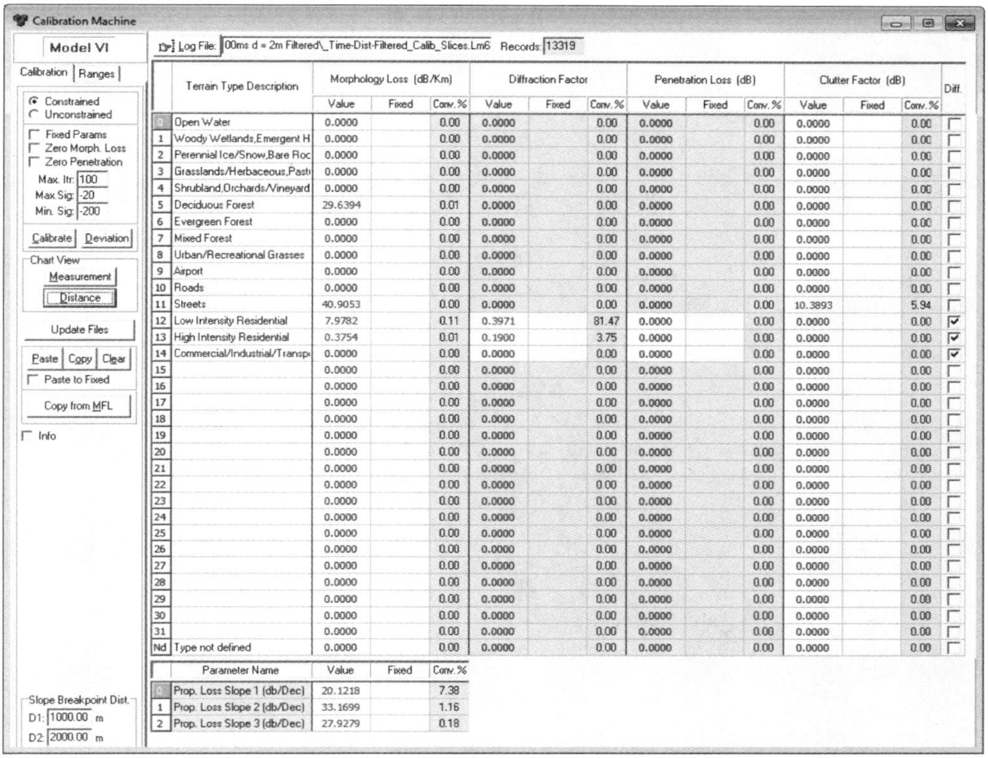

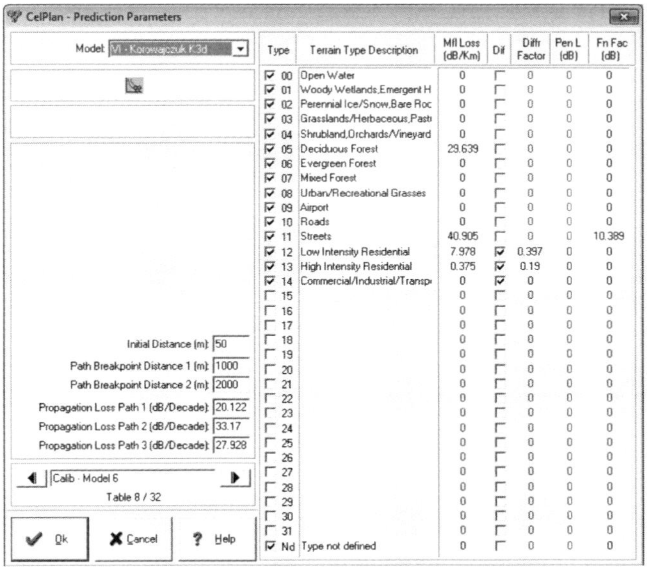

Figure 6.39 Propagation model calibration dialogue box.

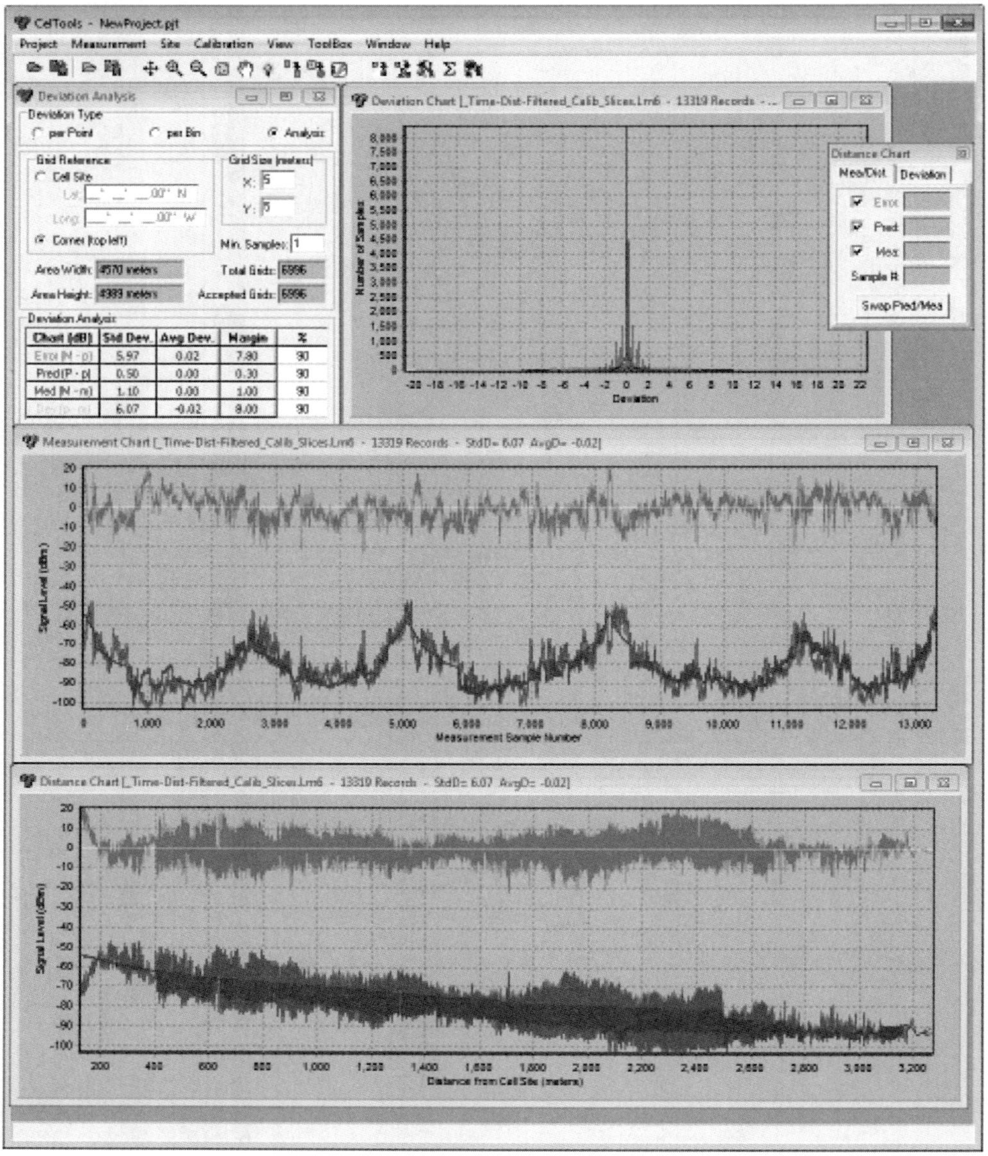

Figure 6.40 Measured × predicted signal comparison calibration set.

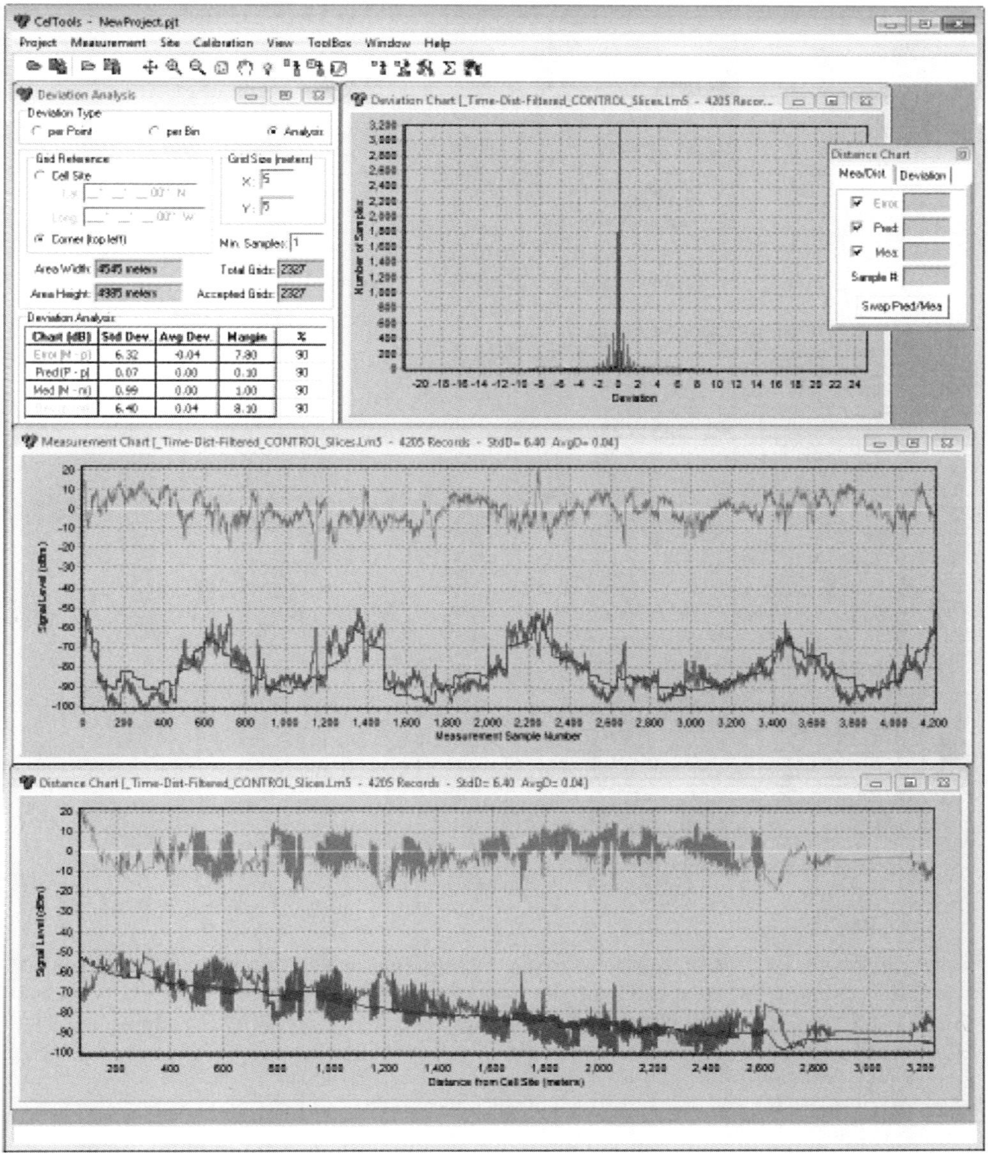

Figure 6.41 Measured × predicted signal comparison control set.

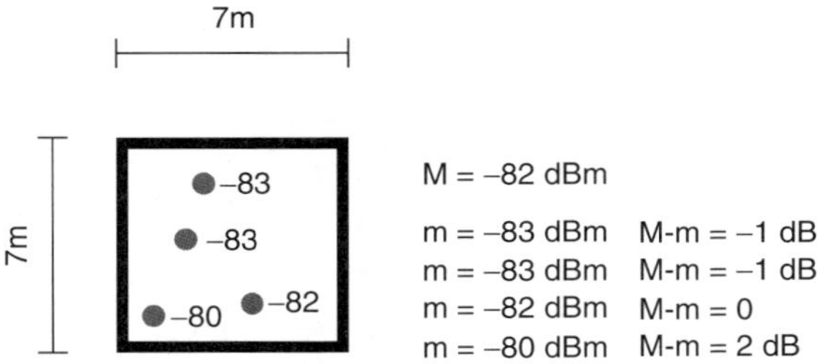

Figure 6.42 Average bin value (M) to measured location value (m) relationship.

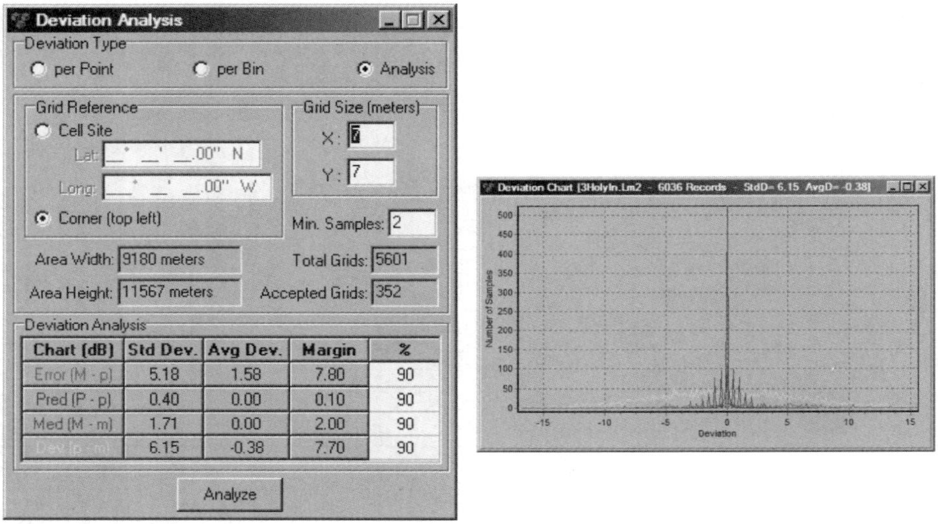

Figure 6.43 Prediction deviation analysis.

It is very important that the average value be close to zero, as this assures that in average the predictions are good. The standard deviation represents the difference between each measurement and its predicted value. The margin column indicates that in this example 90% of the measurements are within 7.7 dB of the averaged value.

Acceptable deviation values for the samples in the calibration group are 0.5 dB for the average deviation and 8 dB for the standard deviation. For the samples in the control group, this can be increased to 2 dB for the average deviation and 10 dB for the standard deviation.

This is where the fractional morphology reuse method excels as it provides similar values to both groups of measurements, while non-fractional calibrations produce very bad results for the control group.

6.3 RF Interference Issues

Any optimization methodology requires, as input, how cells interfere with each other. A cell is defined for this purpose as a base station and its potential users are distributed geographically in the cell according to a specific traffic pattern. Expressing this interference correctly is essential but extremely complex to do. Many short cuts were proposed by the industry, such as simple C/I relationships, cost functions, and others, but none of them represents the interference statistical variation well. This results in poor optimizations that do not perform well when deployed.

Many operators resort to continuously drive testing the network to optimize it. This is expensive and also ineffective, as only part of the network can be tested. Other operators rely on performance data collected by the switch, which is cost effective, but incapable of locating the issues geographically and giving few indications on how to fix issues. A precise interference modeling methodology was required to allow the optimization to be done by predictive tools.

The method described in the next sections of this chapter was developed by CelPlan Technology engineers and provides a novel methodology to deal with interference issues. This is a sophisticated and complete method, but still within the capabilities of modern processors.

6.3.1 Signal Level Variation and Signal to Interference Ratio

The signal level received at a location is constantly varying over time due to fading effects. This signal variation can be expressed by its mean value and its statistical distribution. This applies to the desired signal and also to the interfering signals, as shown in Figure 6.44. This figure shows the desired signal and three interferers. The last interferer shows the interference expected from the two others if sub-carrier permutation is used, as the effect of this permutation is the averaging of the received interference.

Signal reception requires a Signal to Noise and Interference Ratio (SNIR) but it is not possible to establish a single SNIR in Figure 6.44. The use of average values may lead to erroneous conclusions, as they do not express the moments in which the SNR is bad. This can be remedied by associating each SNIR value to a time percentage in which it is not reached, that is, there is an outage of that value. As an example, we can say that a received SNIR is below 17 dB 30% of the time and below 12 dB 5% of the time. Outage values can be calculated based on signal and interference distributions.

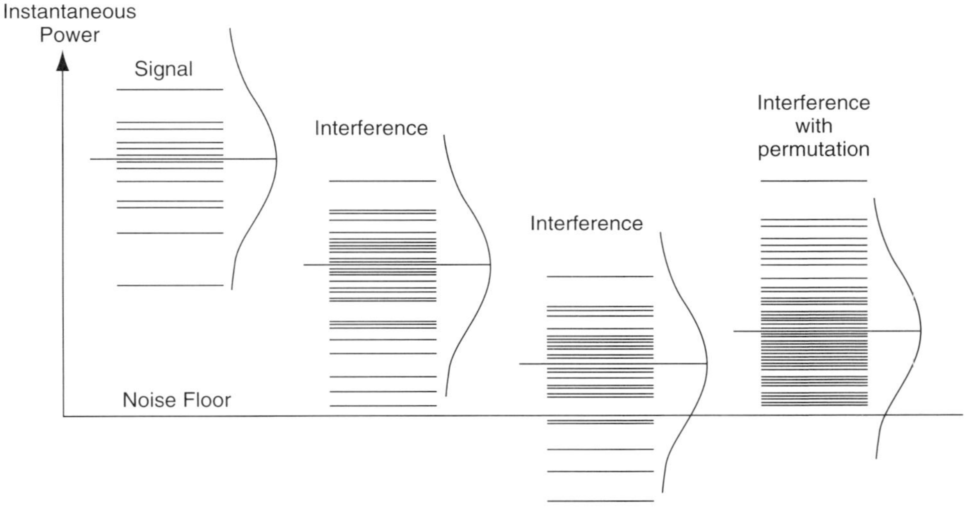

Figure 6.44 Desired signal and three interferers.

Signal level distributions can be obtained by collecting measurements over certain periods of time at different locations. These distributions are relatively constant for a network and can be approximated by Gaussian distributions. Figure 6.45 illustrates signal and interference distributions. These curves are used to determine the outage, which is expressed as the likelihood of the SNIR being smaller than the desired threshold.

This results in a new distribution for the SNIR from which the outage can be calculated. This is shown in Figure 6.46. The table shown in this dialog box provides outages for different values based

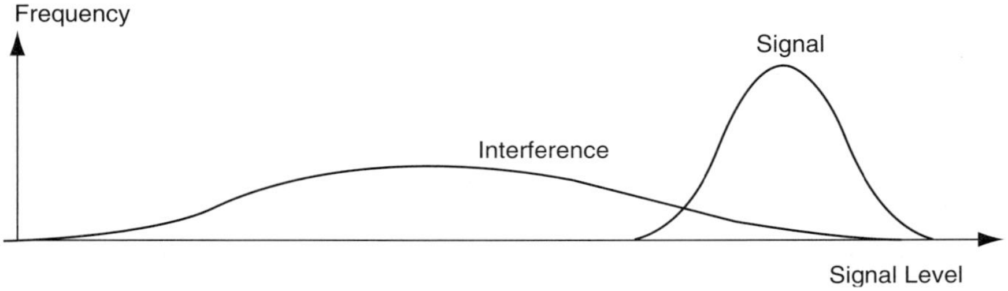

Figure 6.45 Signal and interference distribution curves.

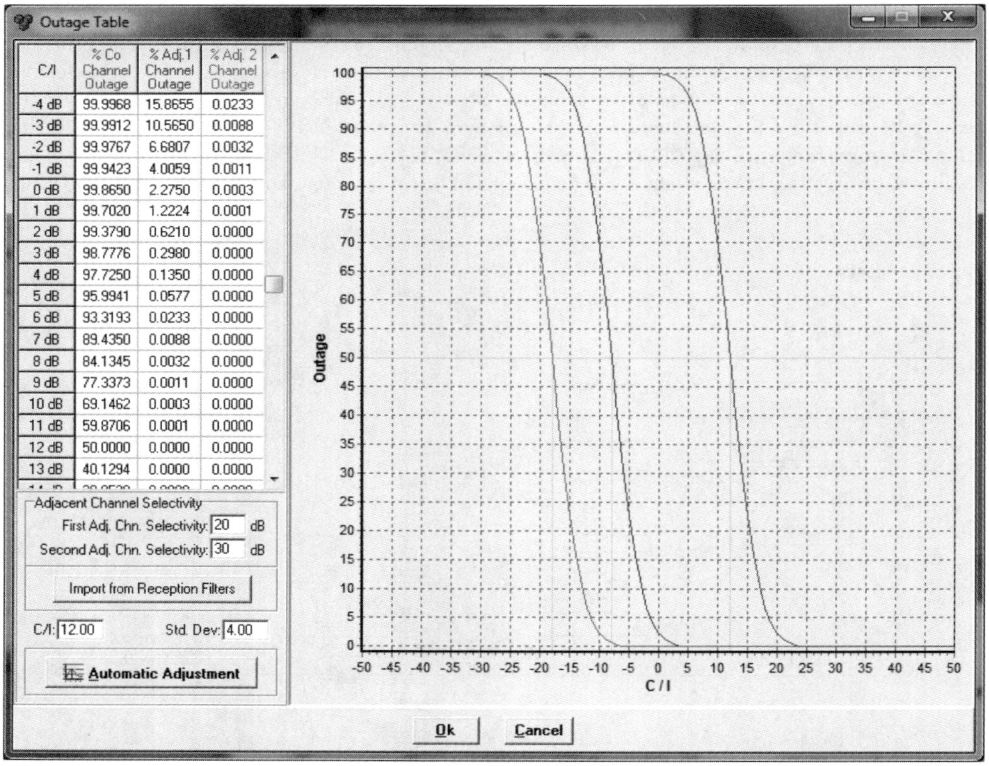

Figure 6.46 SNIR distribution curve and outage table configuration.

on the specified standard deviation for the SNIR distribution. Typical values of SNIR deviation are between 6 and 12 dB.

Figure 6.46 shows the outage table for a SNIR (C/I) requirement. Although SNIR values vary for each location, a value has to be chosen by the designer as representative of the network configuration, to which the network is optimized. In this example 12 dB was chosen. The received signal is subject to fading and will have a variation around its average value, which in this example was set to 4 dB.

Reading the curve, we conclude that a received co-channel signal with an average SNIR (C/I) of 15 dB results in an outage of 20%, a 12 dB SNIR gives an outage of 50% (as expected) and a 5 dB SNIR gives an outage of 95%. The same can be applied to the first and second adjacent channels, but in their case the channel selectivity should be added to the SNIR.

The traditional way of adding a margin to the prediction is misleading and cannot be used when we are analyzing the interaction between two or more signals. Margins can be used to calculate which signal levels result in specific outages, but they should not be used to calculate SNIR.

Simple ratios of average values may say that two cells do not interfere and this may be true for 50% of the time, but they may interfere 40% of the time and this interference should be considered.

Outage is the best way to express interference. Outage can be related to traffic, by multiplying the affected traffic by the time outage, resulting in a traffic outage. Time outages can be statistically added, while traffic outages can be simply accumulated.

6.3.2 *Computing Interference*

In a wireless network interference does vary over time also due to the location of the interferers. This is even more applicable in WiMAX where transmitters change at every frame.

In the downlink scenario illustrated in Figure 6.47 we define for analysis purposes one cell as Interfered (Ed) and another as Interferer (Er). Customers of both cells receive their respective signals, but one interferes with the other. Interference is show in only one direction, to simplify the analysis. The interfering signal varies with the position of the customer due to power control. As customers change locations, the level of interference also changes. This change is not as large in WiMAX due to the adaptive modulation scheme used and the restricted range of the power control.

In the uplink scenario the same happens, but now the level change is much larger as the path loss between the interfering customers to the interfered cell also plays a role. This is illustrated in Figure 6.48.

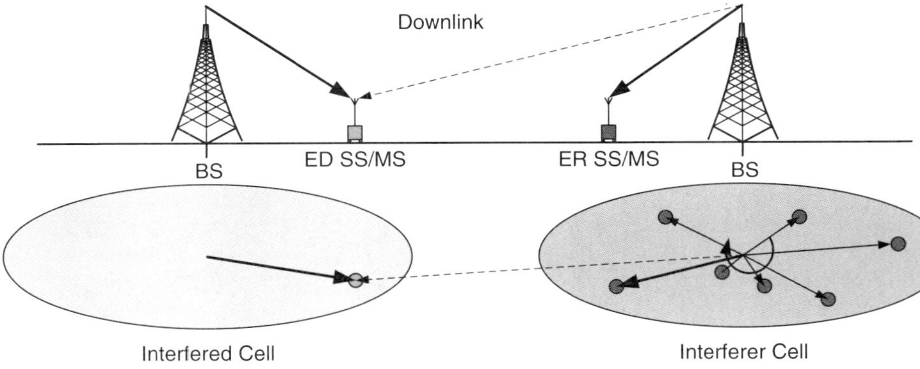

Figure 6.47 Downlink interference.

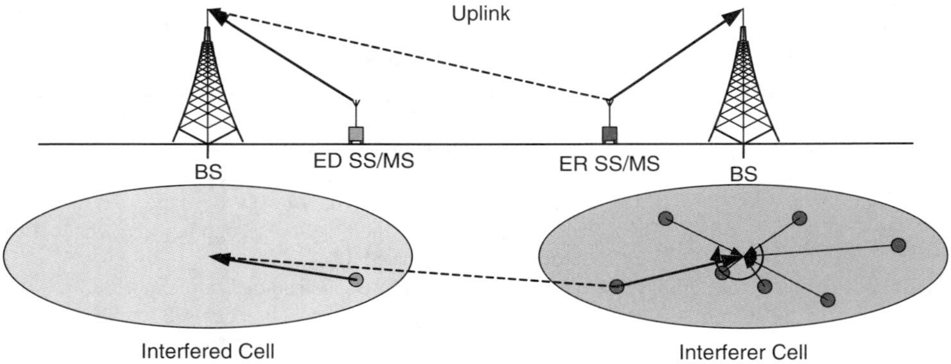

Figure 6.48 Uplink interference.

When the sub-carrier permutation is added to the mix, the interference can change from symbol to symbol.

Additionally, traffic load has to be considered when performing the analysis, as there may be moments without transmission and, consequently, without interference. It is the sub-carrier permutation that allows us to benefit from silent moments, by averaging the interference.

Another important observation is that the interference in the downstream and the upstream is not symmetrical, as illustrated in Figure 6.49. The paths for downlink and uplink interferences may suffer different losses, as illustrated. In Figure 6.49 the downlink interference is high, whereas the uplink is low.

6.3.3 Cell Interference Statistical Characterization

There is a need to assess the average value of the average Transmitted Signal Level (TSL) from a cell. This is done as illustrated in Figure 6.50 by calculating the power to each pixel in the cell service area and doing a traffic weighted sum. Dark represents the area where the cell is the preferred server (i.e. carries nearly all the traffic) and the light area is where the cell is a secondary server (i.e. carries only a small portion of the traffic). In other words, the light area represents the handover area.

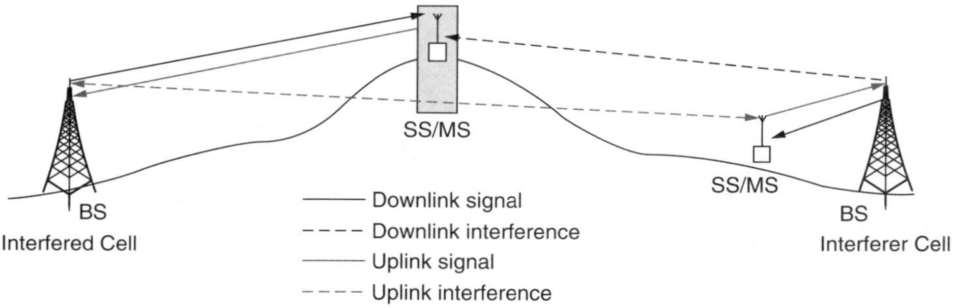

Figure 6.49 Downlink and uplink interference comparison.

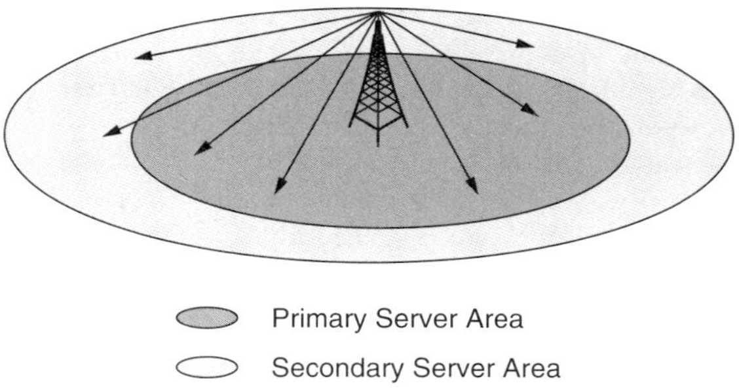

Figure 6.50 Primary and secondary service areas of a site.

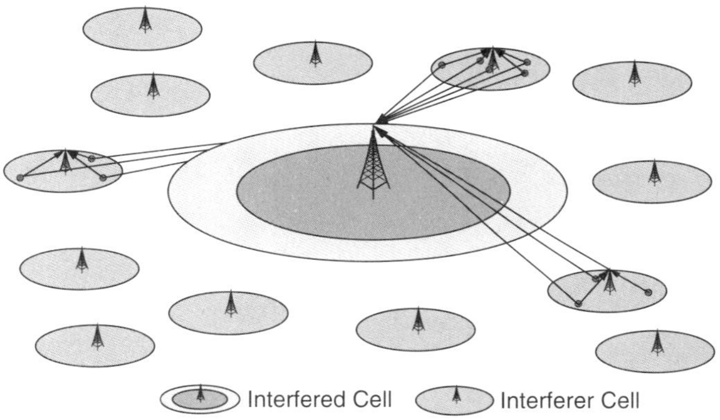

Figure 6.51 Average received signal level assessment.

It is also necessary to assess the average Received Signal Level (RSL) by a cell. This is done as illustrated in Figure 6.51 by calculating the power received from each customer location in another cell and performing a traffic weighted sum.

Both calculations (TSL and RSL) should assume a noise rise, defined by the designer, at each cell to define the required transmit power.

6.3.3.1 Computing Outage

Once the interference can be well represented by traffic outages, the outage between each cell pair can be calculated. This is done by calculating the outages at every pixel and populating outage matrixes, as outages can be added. We will describe here the implementation used by CelPlan Technologies to exemplify how outage can computed.

6.3.3.2 Pixel Outage

To better understand the creation of an outage matrix, it is important to first comprehend the analysis that must be done at each and every pixel. Initially, a reference SNIR is established. Then, received signals at each pixel should be listed in order of strength and classified as preferred server, potential server, interferer or don't care. The preferred server is the strongest server at the pixel. Potential servers are the ones with signals above the handover threshold. Interferers are the ones that due to its lower signal level are below their handover threshold and could not be servers at the pixel. Don't care signals are the ones that are well below the noise floor.

Servers are then associated with a fraction of the pixel traffic. The preferred server gets the largest fraction based on a mobility factor previously defined by the designer; the remaining traffic is split between the potential servers in proportion to their relative signal strength. The reference SNIR outage for each pair server–interferer (all servers are also interferers at this step) is calculated, traffic weighted and recorded to an outage matrix.

The downlink interference should be calculated using the TSL as the average signal level transmitted from the interfering cell. The uplink interference is calculated using the RSL as the average received level by the interfered cell.

6.3.4 Interference Outage Matrix

The process described in the previous section has to be repeated for all pixels in the network and the following matrixes should be built:

- downlink matrix
- co-channel interference
- first adjacent channel interference
- second adjacent channel interference
- cross-polarization channel interference
- uplink matrix
- co-channel interference
- first adjacent channel interference
- second adjacent channel interference
- cross-polarization channel interference
- overall matrix

The overall matrix combines all the above matrixes, weighted by traffic.

The outage matrix expresses the potential interference that the signals from (downstream) and to (upstream) one sector controller have to other sector controllers. This interference is represented by an outage in relation to a pre-specified QoS (SNIR threshold), multiplied by the traffic affected. When complete, the matrix provides a good indication of interference between any pair of sectors. This information is then used to optimize the distribution of resources for each sector.

CelOptima is one of the tools used to calculate interference matrixes and the dialogue to configure its calculation is shown in Figure 6.52.

The outage matrix table is displayed in Figure 6.53, in table form and in graphical form. This type of display is unique to CelOptima.

The interference relationship can be also visualized graphically, as shown in Figure 6.54. It can be seen that there are many interferers to a single cell, besides just the neighboring cells. The interference is displayed in colors defined in the outage table. Not always the closest cell is the strongest interferer. This stresses the importance of an automatic tool, such as CelOptima, to calculate the interference patterns.

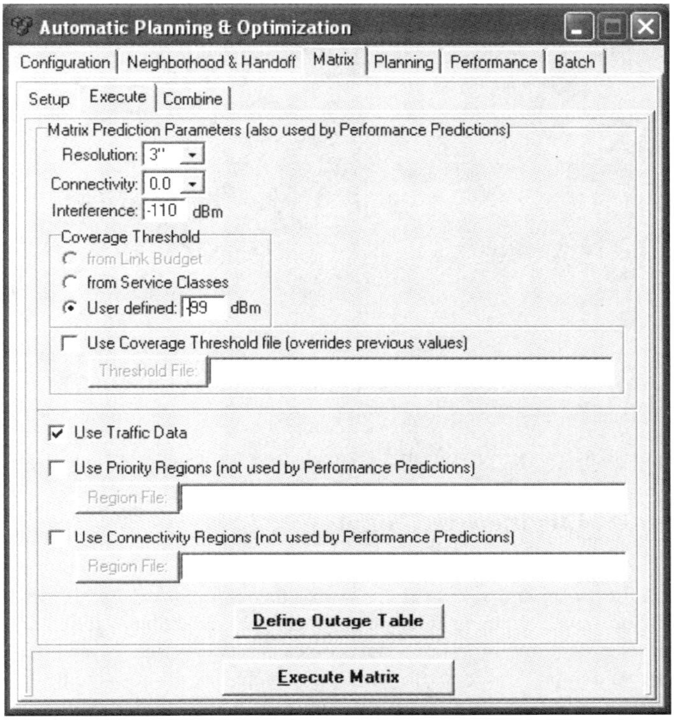

Figure 6.52 CelOptima matrix configuration screenshot.

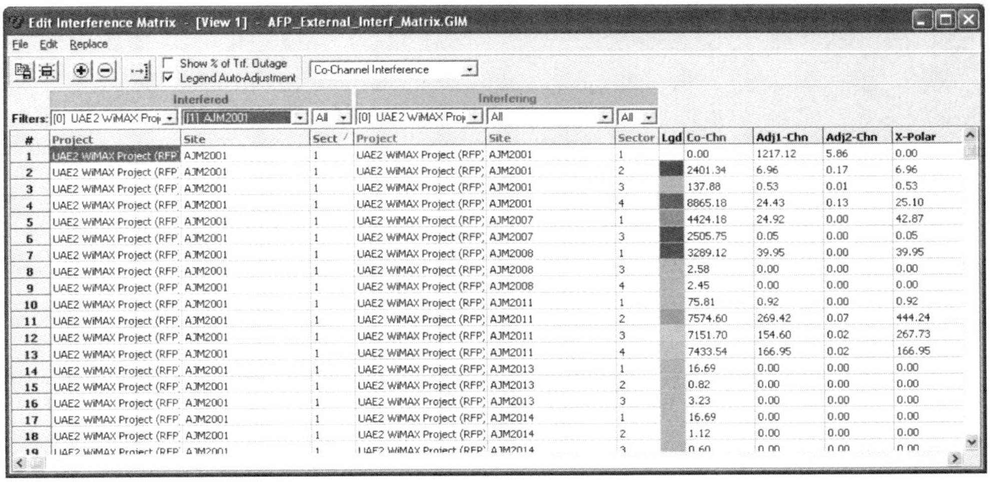

#	Project	Site	Sect	Project	Site	Sector	Lgd	Co-Chn	Adj1-Chn	Adj2-Chn	X-Polar
1	UAE2 WiMAX Project (RFP	AJM2001	1	UAE2 WiMAX Project (RFP)	AJM2001	1		0.00	1217.12	5.86	0.00
2	UAE2 WiMAX Project (RFP	AJM2001	1	UAE2 WiMAX Project (RFP)	AJM2001	2		2401.34	6.96	0.17	6.96
3	UAE2 WiMAX Project (RFP	AJM2001	1	UAE2 WiMAX Project (RFP)	AJM2001	3		137.88	0.53	0.01	0.53
4	UAE2 WiMAX Project (RFP	AJM2001	1	UAE2 WiMAX Project (RFP)	AJM2001	4		8865.18	24.43	0.13	25.10
5	UAE2 WiMAX Project (RFP	AJM2001	1	UAE2 WiMAX Project (RFP)	AJM2007	1		4424.18	24.92	0.00	42.87
6	UAE2 WiMAX Project (RFP	AJM2001	1	UAE2 WiMAX Project (RFP)	AJM2007	3		2505.75	0.05	0.00	0.05
7	UAE2 WiMAX Project (RFP	AJM2001	1	UAE2 WiMAX Project (RFP)	AJM2008	1		3289.12	39.95	0.00	39.95
8	UAE2 WiMAX Project (RFP	AJM2001	1	UAE2 WiMAX Project (RFP)	AJM2008	3		2.58	0.00	0.00	0.00
9	UAE2 WiMAX Project (RFP	AJM2001	1	UAE2 WiMAX Project (RFP)	AJM2008	4		2.45	0.00	0.00	0.00
10	UAE2 WiMAX Project (RFP	AJM2001	1	UAE2 WiMAX Project (RFP)	AJM2011	1		75.81	0.92	0.00	0.92
11	UAE2 WiMAX Project (RFP	AJM2001	1	UAE2 WiMAX Project (RFP)	AJM2011	2		7574.60	269.42	0.07	444.24
12	UAE2 WiMAX Project (RFP	AJM2001	1	UAE2 WiMAX Project (RFP)	AJM2011	3		7151.70	154.60	0.02	267.73
13	UAE2 WiMAX Project (RFP	AJM2001	1	UAE2 WiMAX Project (RFP)	AJM2011	4		7433.54	166.95	0.02	166.95
14	UAE2 WiMAX Project (RFP	AJM2001	1	UAE2 WiMAX Project (RFP)	AJM2013	1		16.69	0.00	0.00	0.00
15	UAE2 WiMAX Project (RFP	AJM2001	1	UAE2 WiMAX Project (RFP)	AJM2013	2		0.82	0.00	0.00	0.00
16	UAE2 WiMAX Project (RFP	AJM2001	1	UAE2 WiMAX Project (RFP)	AJM2013	3		3.23	0.00	0.00	0.00
17	UAE2 WiMAX Project (RFP	AJM2001	1	UAE2 WiMAX Project (RFP)	AJM2014	1		16.69	0.00	0.00	0.00
18	UAE2 WiMAX Project (RFP	AJM2001	1	UAE2 WiMAX Project (RFP)	AJM2014	2		1.12	0.00	0.00	0.00

Figure 6.53 Interference matrix table.

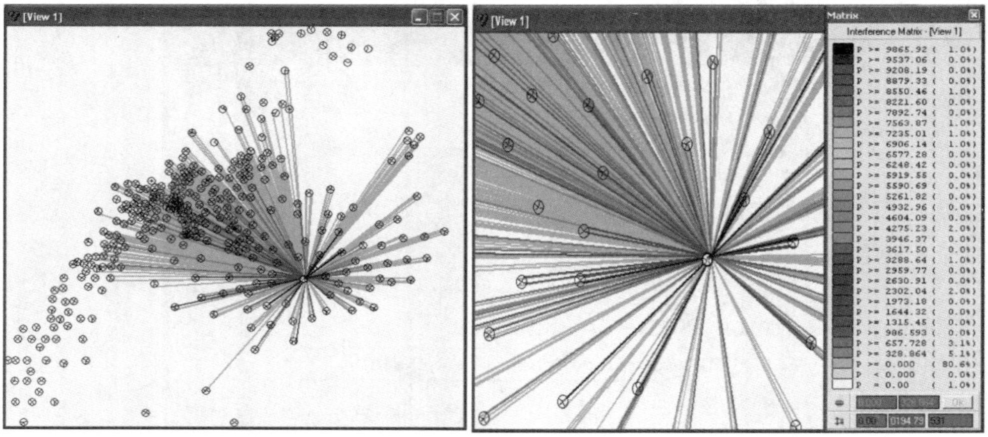

Figure 6.54 Interference matrix representation for a single site and detail.

6.4 Interference Mitigation Techniques

The use of large bandwidth increases spectral efficiency in terms of data throughput. The previous statement is correct for broadband systems only when considering an isolated cell; in a network with multiple cells, however, the large bandwidth of a WIMAX carrier reduces the total number of carrier available (limited spectrum) and interference becomes a major issue, affecting the data throughput. The following paragraphs describe the two main techniques for reducing this interference in broadband networks.

6.4.1 Interference Avoidance

In the interference avoidance technique, resources are split between users, so they do not interfere with each other. This can be achieved by segmenting a carrier, a concept known as segmentation.

The WiMAX Standard IEEE Std. 802.16e identifies up to three segments for each carrier. This leads to a practical increase in resources, because the carrier is now multiplied by three; however, it also implies a threefold reduction of throughput capacity (not considering the reduction by interference). Segmentation also improves carrier adjacency interference. To help receivers to tune to a specific subset of subcarriers, additional pilots are added within each segment. LTE does not support segmentation per se, but it allows for coordination between cells in terms of resource block allocation.

Interference avoidance should be planned by optimization tools considering the segments as a resource multiplier and accounting for carrier adjacency interference reduction.

6.4.2 Interference Averaging

Interference is not evenly distributed along the network or along the subcarriers. When a fixed allocation of subcarriers is used, some connections suffer severe interference while others suffer very little. This is even more noticeable at light traffic loads.

Interference mitigation can be done by averaging through a pseudo-random use of resources, so interferers and interfered rotate and the overall interference is averaged between all users. With less interference, error correction codes (e.g. FEC) are more effective and more capable of restoring system capacity.

Interference averaging should be used with care and be assessed by strong design tools that can verify network performance beforehand, because a network can go from an acceptable to a disastrous performance just by a slight increase in interference. Interference averaging can only be properly evaluated through dynamic traffic simulation, when the varying network interference is taken into consideration.

The pseudorandom distribution of subcarriers can be achieved in many ways and, to give users more flexibility, several subcarrier permutation schemes were included in the standard and are known as pilot and data allocation schemes. These schemes define the set of pilots and data subcarriers that carry data and how they are mapped to sub-channels. The pilot content is known and is used to estimate the RF channel; this knowledge is then used to extract the data. Pilot and data allocation schemes are described in the next sections.

Interference averaging is available in WiMAX, but is not used explicitly in LTE.

6.5 RF Spectrum Usage and Resource Planning

Network resources are items that are available in limited numbers or denominations and thus have to be carefully distributed to network elements. Resource examples are varied and include: OFDM carriers, antenna tilts, handover thresholds and power thresholds. This section analyzes the optimization and planning of these resources. A summary of each optimization step is presented; more details are given in Chapter 20 of this book on network optimization.

6.5.1 Network Footprint Enhancement

When designing a network, cells are located to provide traffic and area coverage, but after the initial results have been obtained, the footprint of each cell should be optimized. This means that the overlap between cells should be minimized, keeping only enough overlap between adjacent (neighbor) cells to complete handovers.

Each cell footprint can be adjusted through the following parameters:

- antenna height
- antenna type
- antenna azimuth
- antenna tilt
- transmit power (EIRP)

An initial enhancement should not consider resource allocations (e.g. frequency plan), so more freedom is given for footprint optimization, thus allowing a better frequency and code planning later.

After the initial enhancement, the resources allocation can be performed. A final enhancement should then be done considering these resources. Enhancement optimization tools should be able to support traffic and service classes.

6.5.2 Neighborhood Planning

Neighborhood planning is important because it determines the operational overlap between cells and the extent of the interference generated by mobiles. Usually neighborhood is defined geographically into three dimensions and this is called the topological neighborhood. This neighborhood relationship is important as it is the one that will occur when handover happens. Non-topologically neighbor cells can serve the same areas and be major interferers to each other, so they should be considered as interference neighbors.

Ideally, neighborhood planning should be based on a specific number of server cells at each pixel, as this offers handover options. However, neighbors can, and should, be trimmed in quantity to avoid too many handover options, which could significantly increase network interference. The best approach to eliminate neighbors is to rank them by common traffic.

When eliminating neighbors, however, one should take care not to cut topological neighbors, as all of them should be considered at each sector, as there are geographical situations in which the overlap between cells is small but they still have to hand off to each other.

6.5.3 Handover Planning

Ideal handover thresholds are hard to visualize and only an automatic tool can calculate them properly. The calculation of these thresholds should consider signal levels at each cell border, compare it to neighboring cells and, based on this, calculate the ideal threshold per sector or per pair of neighbors, depending on the selected implementation.

6.5.4 Paging Zone Planning

When a network receives data for a user, it has to locate it and this is done through paging, which is a broadcast signal. Sending this message to all base stations represents a large overhead, so the paging procedure is divided into phases; being done first only in the zone the user was contacted last and then extending it to neighboring zones and finally paging it over the whole network. There is no rule to define the size of a paging zone.

Paging zones should be determined by the designer, based on overall traffic patterns. The design tool should be used to choose the cells that should belong to each zone with the provision of an appropriate overlap. The overlap is required to cope with users that are at the border of two zones.

6.5.5 Carrier Planning

OFDM carriers, when used for resource planning, are called carriers, frequencies, or channels. This book uses the term frequency planning as the other options may cause confusion. Frequencies and codes used for interference avoidance should be planned together, whereas codes used for interference averaging should be planned first.

6.5.5.1 Clusters

Traditionally, planning was done according to specific patterns, called reuse patterns. Cells were assigned to a regular grid of hexagons and those were sub-divided into sectors (usually three). Frequencies were then distributed according to specific patterns, defined by the number of BSs in each cluster, the number of sectors in each BS, and the number of frequencies available. Ideally, there should be one frequency for each sector in a cluster. Regular hexagon clusters can be built with one, three, and seven hexagons, which, in turn, can be split into three sectors each. The different configurations then require three, nine, and 21 resources respectively.

A configuration with one three-sectored hexagon, using three frequencies requires interference averaging techniques and has to use load control to work. This configuration was extended further by using a single frequency in all sectors (reuse of one), but, for this to work, sectors must be very lightly loaded traffic-wise.

A configuration with three three-sectored hexagons with nine frequencies can be implemented, although it will be subject to some amount of interference and may still require some type of load control.

A configuration with seven three-sectored hexagons with 21 frequencies provides the best interference avoidance and can be implemented without any load control.

Designs using standard clusters are not optimized in terms of spectrum usage. Much more effective spectrum usage is obtained by placing sites at areas with high traffic concentration, which generates an irregular site distribution, making a fixed cluster unfeasible. Planning of networks like this requires specialized tools, but the spectrum usage can be significantly increased.

6.5.5.2 Segments

Wireless broadband uses wide channels, so operators usually have only a few channels to work with. One way to increase the number of resources available is to segment the sub-carriers in the main carrier, so each segment can be allocated separately, resulting in interference avoidance. Although LTE does not explicitly use segments, they should be conceived for use by the allocation algorithm.

6.5.5.3 Zones

Some wireless broadband technologies require subscribers to adjust their timing to the BS timing, thus allowing calculation of the user distance from the BS. This knowledge allows resource reuse to be further increased by limiting the distance from the BS in which it is assigned. This distance delimitation is known as a zone. Although LTE does not specifically support zones, they can and should be used by the frequency allocation algorithm.

6.5.5.4 Carrier and Code Reuse Patterns

The Frequency Reuse Scheme (FRS) is defined by the number of BSs per cluster, number of sectors per BS, number of frequency channels per sector, and, optionally, number of segments. When the number of segments is not included, no segmentation should be assumed. Some possible configurations are listed below.

- FRS = 3,3,9 This is a low interference configuration.
- FRS = 1,3,3 This is a high interference configuration in which some load sharing should be made.
- FRS = 1,3,1 This configuration should only be used as a last resource and implies heavy load sharing. A sector cannot carry more than one-third of its traffic capacity, so, in reality, it performs as a 1,3,3 scheme.
- FRS = 1,3,1,3 This is the same configuration as the previous one but using different segments at each sector.

Figures 6.55 and 6.56 illustrate a basic block for the 3,3,9 reuse and a combination of several of these blocks.

Regular reuse patterns such as this are used in this chapter for illustration only, as in real life, BSs are not uniformly distributed and carrier planning should be done with an automatic planning tool that considers all factors, including traffic.

Figure 6.57 shows a 1,3,1 reuse block. This configuration only works for very light loads using interference averaging; otherwise, segmentation has to be used as shown on the right side of the figure.

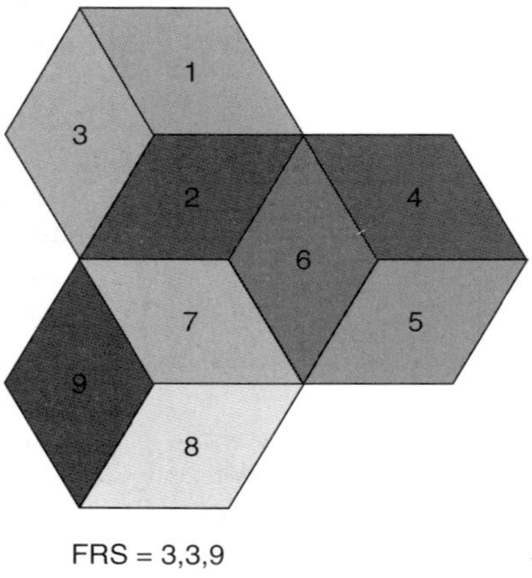

Figure 6.55 Basic 3,3,9 reuse block.

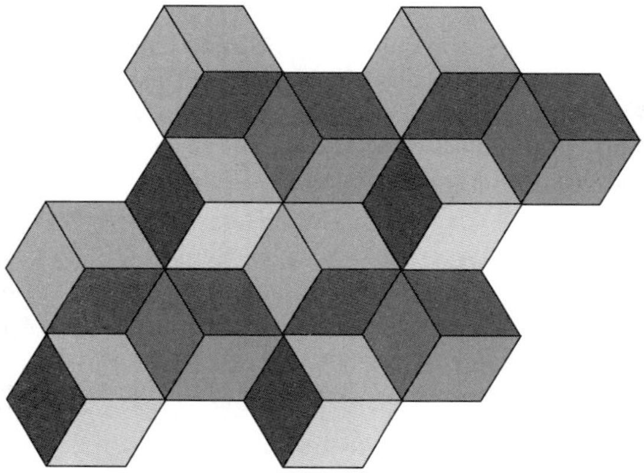

Figure 6.56 Combination of 3,3,9 reuse blocks.

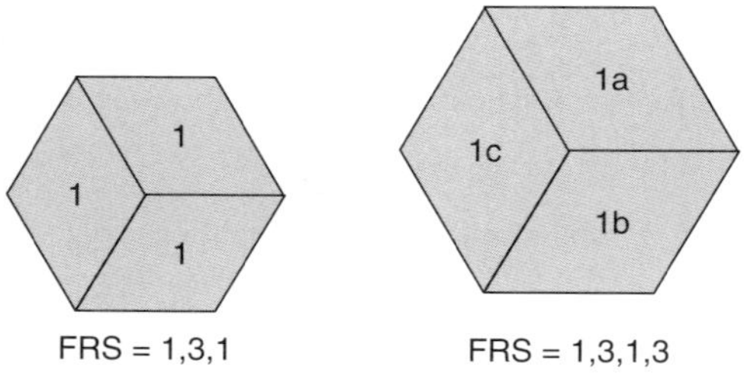

Figure 6.57 Example of 1,3,1 reuse block without segmentation (left) and with segmentation (right).

Several strategies can be used when performing a frequency plan:

- Frequencies can be reserved for point to point connections.
- Frequencies can be reserved for the cell core coverage.
- Frequencies can be partially used (segmented).
- Frequencies can be allocated by zones.
- Frequencies can be partially loaded when using interference averaging.

This approach is exemplified in Figure 6.58. In this example, three carriers are available. Carrier three is used for point-to-point rooftop connections. Carrier two is used for coverage close to the cell, whereas carrier one is used in the outskirts of the cell to avoid interference between adjacent cells.

A carrier plan may look simple enough to do manually, but its complexity is always underestimated and only an automatic design is able to take all aspects into consideration. For the automatic design to be successful, it should consider the traffic of the different service classes on a pixel basis, use neighborhood and handover thresholds, consider statistical interference and all the schemes and solutions previously presented. This certainly is not an easy task.

Network optimization is a long and interactive process and the process may take several days for a network with 1000 BS, mainly because the result of a plan can only be evaluated after the KPI analysis is done.

The decision about using segmentation, zoning, and load control should be taken by the designer based on the resources available and the density of the network. In many cases it is not a trivial decision and we suggest that different scenarios be tested over a regular grid that approximates the

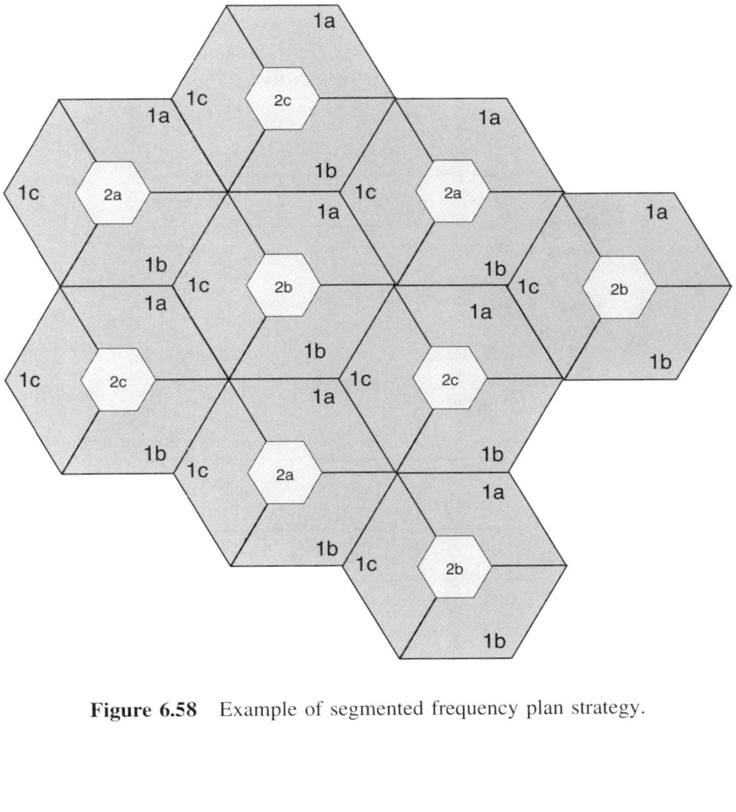

Figure 6.58 Example of segmented frequency plan strategy.

most congested areas of the actual deployment. These tests can be done quickly over flat terrain and using free space propagation, just to assert the capabilities of each scenario, to help in the decision of selecting one of the possibilities.

Carrier optimization with segmentation can be done without considering segmentation initially, and then considering the planning segmentation by splitting the carrier segments between the sectors. In this case segments are planned as codes.

Another approach is to split the carrier in three and do the planning accordingly. Both approaches should be tried to see which gives better results.

6.5.6 Code Planning

There are several identifications that have to be assigned to sectors, which are available in restricted numbers. These identifications are broadly called codes and should be planned similarly to the carrier planning.

Each technology has different code planning requirements, such as IDCell, and PermBase; each of these codes is discussed in more detail in their respective technology sections.

6.5.7 Spectrum Efficiency

In Chapter 5, carrier resources usage was analyzed for each domain (frequency, time and power), determining what percentage of the resources is allocated for the actual data and to assure data integrity, that is, the carrier overhead. A simplified calculation is listed in Table 6.3 for WiMAX. Calculations for LTE are presented in Chapter 14 with the description of that technology.

There is another overhead directly related to the data retrieval and we call it data overhead. It is listed in Table 6.4 for the minimum and maximum coding overhead.

As seen in these tables, only 8% to 20% of the carrier capacity is available for the actual data to be transmitted and, even though this may look like a low amount, this is more than most other

Table 6.3 Carrier overhead

Carrier overhead	Percentage
Guard bands	18
Pilot DL and UL	23
Cyclic prefix	13
TDD partition	5
TDD gap	3
OFDMA preamble and mapping	10
Total for support	72
Available for data	28

Table 6.4 Data overhead

Data overhead	Minimum (%)	Maximum (%)
Coding	17	50
MAC overhead	3	5
HARQ	10	15
Total	30	70
Available for data	20	8

technologies can claim. There is plenty of room for new ideas to optimize spectrum usage, so one can rest assured that there is plenty of room for new technology generations to appear in the future.

6.6 Availability

Availability defines a percentage of area or time during which a specific service is supported, with the desired quality, by a wireless network. Area and time availability apply mainly to mobile services, while fixed services consider only time availability, as the user is in a known location and the connection will be only installed if service is available.

The availability concept is usually considered for high frequency microwave (>10 GHz) point-to-point links, in which fog, rain and snow seriously disrupt the communication link. Multipath fading does also have some effect, although much smaller than the one occurring in point to multipoint links.

Point-to-point links' typical availabilities are specified between 99.9% (3 nines) and 99.999% (5 nines) of time. This gives, respectively, an outage of 8.76 hours per year and 5 minutes and 15 seconds a year. This implies relatively long service outages, during which the service is not available, as the whole outage can happen in one single event.

Fixed location point-to-multipoint links are usually operated at lower frequencies (<10 GHz) and are not subject to weather-related attenuation, but instead have to cope with more severe fading, caused by multipath. This fading is distributed over time and results in receive errors. Some services (like voice and video) can live with a certain amount of errors; others require the total elimination of errors.

When we browse the Internet or transfer files, we do not expect to get the information with errors, but they exist in any wireless system. These errors are corrected by the wireless and higher levels protocols, before being sent to its destination.

Error correction implies retransmissions and consequently creates delays (latencies) in the delivery of the information. Data availability in this network is expressed as a time percentage in which a specified latency value is not exceeded. This availability is calculated for the wireless link, and the eventual remaining errors will be corrected by higher level protocols, like TCP/IP or the application protocols themselves.

Service availability calculations are becoming important due to the deployment of mission critical applications, like power plant Smart Grids.

Utility plants are deploying wireless networks to gather information and control their plants remotely. These applications can be classified into three categories:

- *Smart metering*: In this application metering devices at the customer's premises are read with a certain periodicity. Assuming an hourly periodicity, high latency values are acceptable. A typical packet has a length of 15 bytes and acceptable latencies range in minutes. Availability can be as low as 90%.
- *Smart monitoring*: In this application high resolution, high sensitivity, low frame video will be used to monitor transmission lines. A one frame per second will be consider, which requires a latency of less than 1 second. The size of the packet will limited by the IP packet size of 1400 bytes. Acceptable availability is in the order of 99%.
- *Smart grid*: In this application, metering and control devices are actuated and this requires a fast response. A maximum latency of 100 ms (five 50 Hz cycles) is considered. A typical packet has a length of 15 bytes. Required availability has to be in the order of 99.9% to 99.999%, depending on the event criticality.

The availability calculation has to undergo the following steps:

1. Radio sensitivity should be characterized for different error rates, as shown in Figure 6.59 and Figure 6.60, respectively for an AWGN and a Rayleigh channel.

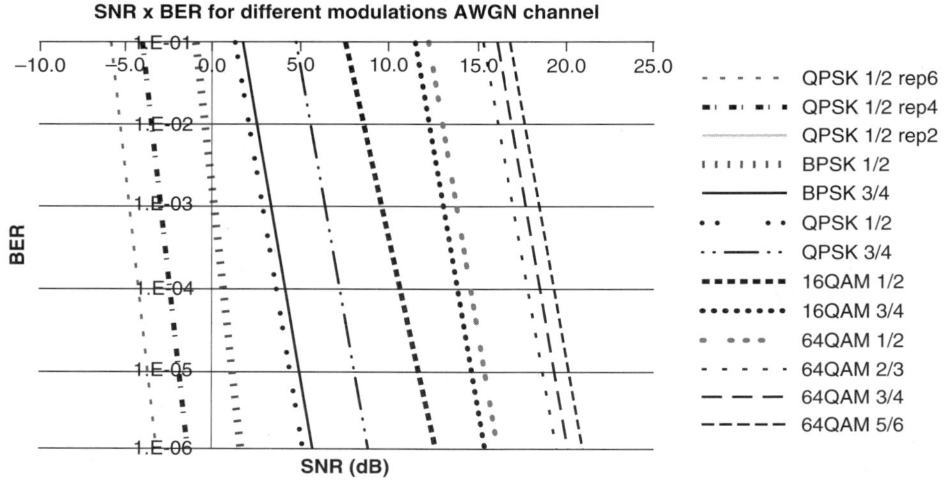

Figure 6.59 SNR required for different BER on an AWGN channel.

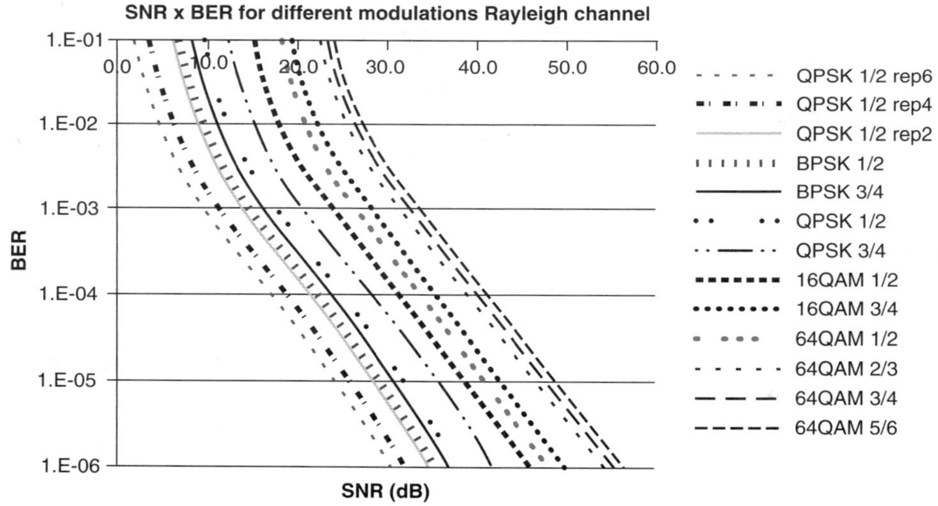

Figure 6.60 SNR required for different BER on a Rayleigh channel.

2. The wireless message length should be established from the application message length. In this calculation, the protocol's overheads, like CRC and FEC, should be considered. As an example, a 15 bytes application message becomes a 295 bytes (2360 bits) wireless message by adding TCP/IP and wireless MAC protocols, as indicated in Figure 6.61.
3. The wireless fading should be characterized in terms of signal level distribution, by its standard deviation. The number of standard deviations required for each of the availabilities can then be established, as a margin to be applied over the receiver sensitivity. Figure 6.62 shows signal variation with fading, while Figure 6.63 gives the signal amplitude distribution.

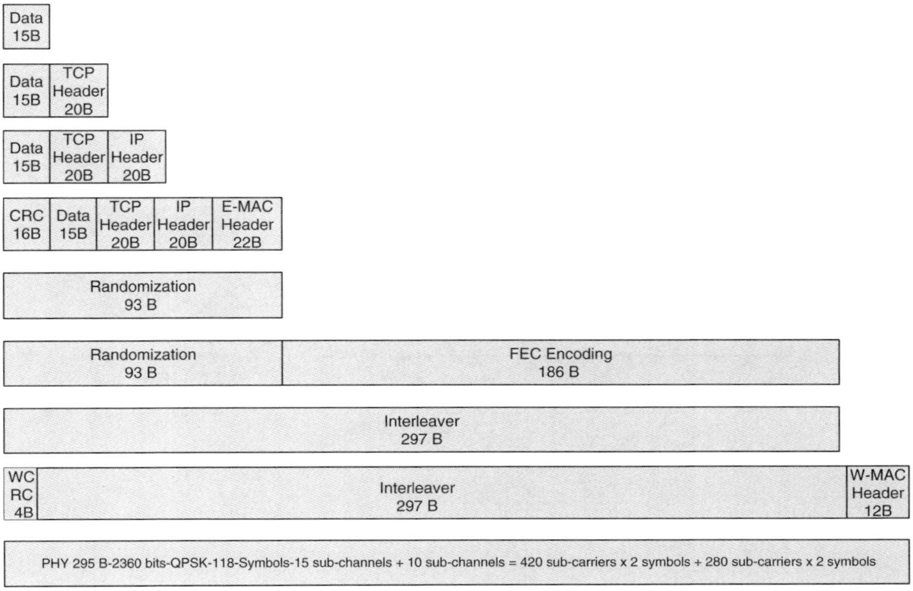

Figure 6.61 Message overhead.

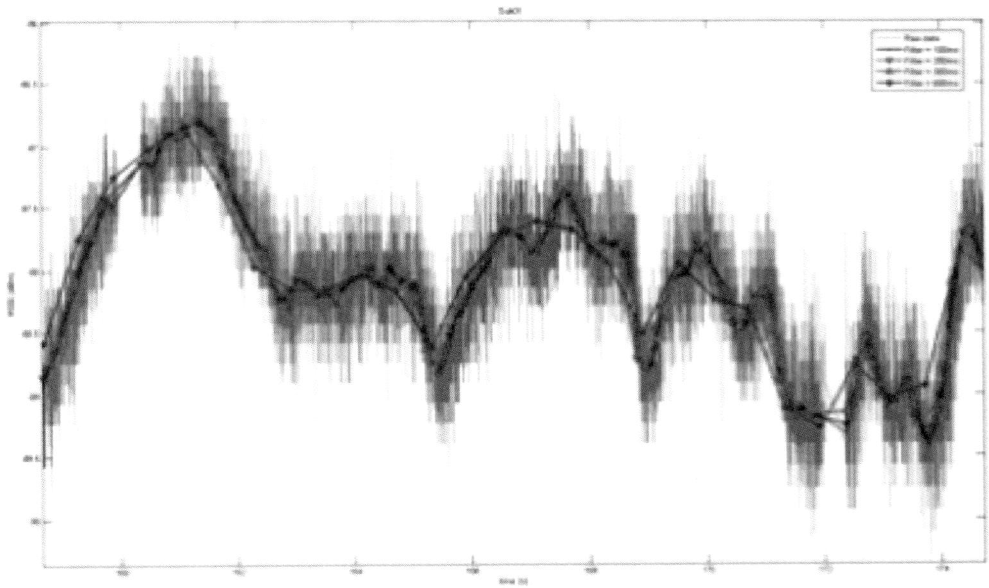

Figure 6.62 Signal variation due to fading.

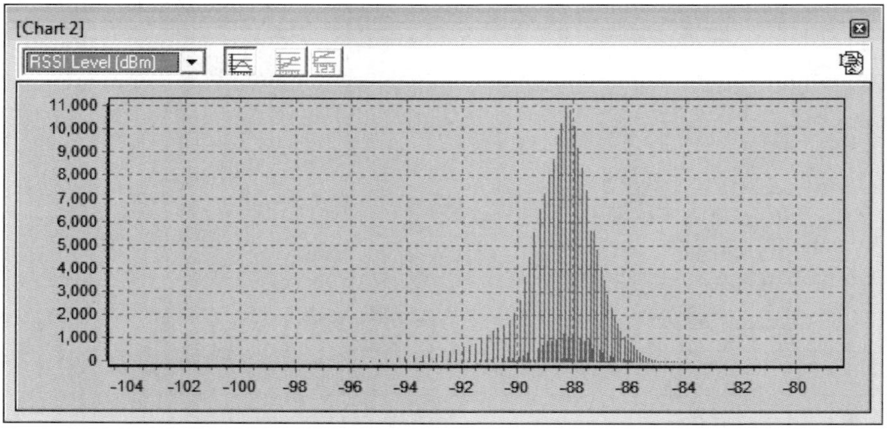

Figure 6.63 Fading distribution.

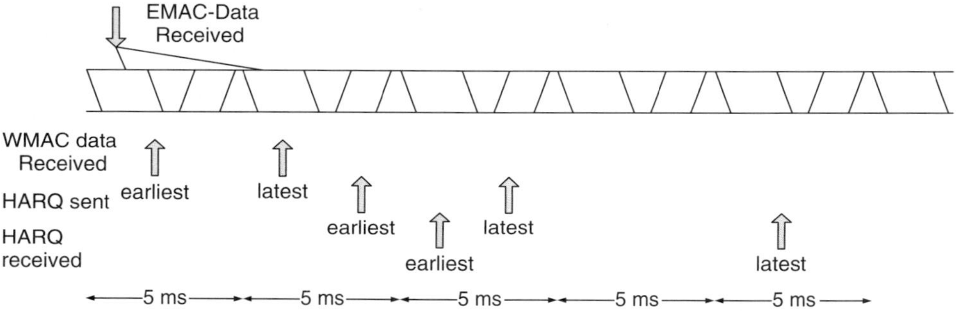

Figure 6.64 HARQ processing delay example for 5 MHz WiMAX.

4. This margin is related to the error correction technique used, which for WiMAX and LTE is based on HARQ (Hybrid Automatic Repeat Request), and implies in a certain latency as illustrated in Figure 6.64.
5. The algorithm that allocates the message the OFDM sub-channels plays an important role in the definition of the time required to correct residual errors in the wireless segment.
6. The delay caused by the HARQ can be calculated for one or more cycles, as illustrated in Figure 6.64, for a WiMAX channel. The latency for a one HARQ cycle, considering a 5 ms frame is 15 ms and for two HARQ cycles is 25 ms.
7. The margin required for each of the availability can then be established as illustrated in Figure 6.65.

Table 6.5 and Table 6.6 give the receiver sensitivity for one HARQ (15 ms) and two HARQ (2 ms) latency, for different availabilities.

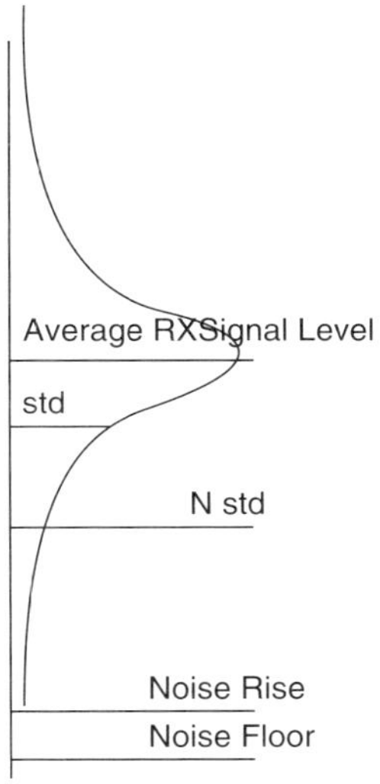

Figure 6.65 Margin calculation for certain availability.

Table 6.5 Receiver sensitivity (signal threshold) for various availabilities and 1 HARQ latency

Availability (%)	Error rate	SNR for QPSK1/2 rep 2 (dB)	Fading std (dB)	1 HARQ cycle Number of std	Margin (dB)	Noise floor +noise rise (dBm)	Average signal threshold (dBm)
99	10^{-2}	5	1.44	2.33	8.35	−99	−90.65
99.9	10^{-3}	12	1.44	3.09	16.45	−99	−82.55
99.999	10^{-5}	25	1.44	4.27	31.14	−99	−67.86

Table 6.6 Receiver sensitivity (signal threshold) for various availabilities and 2 HARQ latency

Availability (%)	Error rate	SNR for QPSK1/2 rep 2 (dB)	Fading std (dB)	2 HARQ cycle Number of std	Margin	Noise floor +noise rise (dBm)	Average signal threshold (dBm)
99	10^{-2}	5	1.44	1.28	6.85	−99	−92.15
99.9	10^{-3}	12	1.44	1.86	14.67	−99	−84.33
99.999	10^{-5}	25	1.44	2.73	28.93	−99	−70.07

The above procedure relates sensitivity and latency to information availability for point to multipoint networks, essential for the deployment of Smart Grids and other similar applications.

7

OFDM

Wideband wireless technologies, such as HSPA and EVDO, have difficulty in providing high data throughput over large distances due to multipath interference. There was clearly a need for another technology that could live with large multipath. Orthogonal Frequency Division Multiplex (OFDM) was the natural candidate, as although nominally broadband, it is subdivided into narrowband sub-carriers. Simultaneously the time was right for OFDM introduction as the hardware performance required to implement it became available, through powerful DSPs (Digital Signal Processor) and fast ADCs (Analog to Digital Converter). This chapter describes the basic principles of OFDM, which apply to all technology implementations described later in this book.

7.1 Multiplexing

Multiplexing is a technique that allows a medium to carry different streams of information at the same time. Multiplexing can be done in frequency (FDM, Frequency Division Multiplex), time (TDM, Time Domain Multiplex) or code (CDM, Code Division Multiplex) domains.

In FDM, different frequencies, called carriers, are used to carry information, and each frequency is assigned to a different user. It is necessary that the interference between carriers, the Inter Symbol Interference (ISI), be kept within values smaller than the signal to noise levels required to recover the information (between few dB to 30+ dB). To carry the maximum possible amount of information per Hz, each carrier spectrum has to be limited by filtering, so carriers can be placed close together. This approach requires the separation between carriers of several times the useful bandwidth (two to three times being typical), even when using sharp filters. Since 1957, several proposals have been put forward based on multiplexing orthogonal carriers, which by definition do not interfere with each other.

The idea is to multiplex carriers separated by the period interval, so the peak value of one carrier coincides with the nulls of all others, as shown in Figure 7.1. In this case, the carriers are said orthogonal, as they do not interfere with each other.

This multiplexing scheme was then called OFDM (Orthogonal Frequency Division Multiplex) and allowed the multiplexing of several carriers without the use of filters. This was a big advantage in terms of spectrum usage, but the digital signal processing required (explained later) was beyond the available technology at the time. Only in late 1990s CPUs (Central Processing Units) and DSPs (Digital Signal Processors) reach the required processing capabilities.

LTE, WiMAX and WLAN Network Design, Optimization and Performance Analysis, First Edition.
Leonhard Korowajczuk.

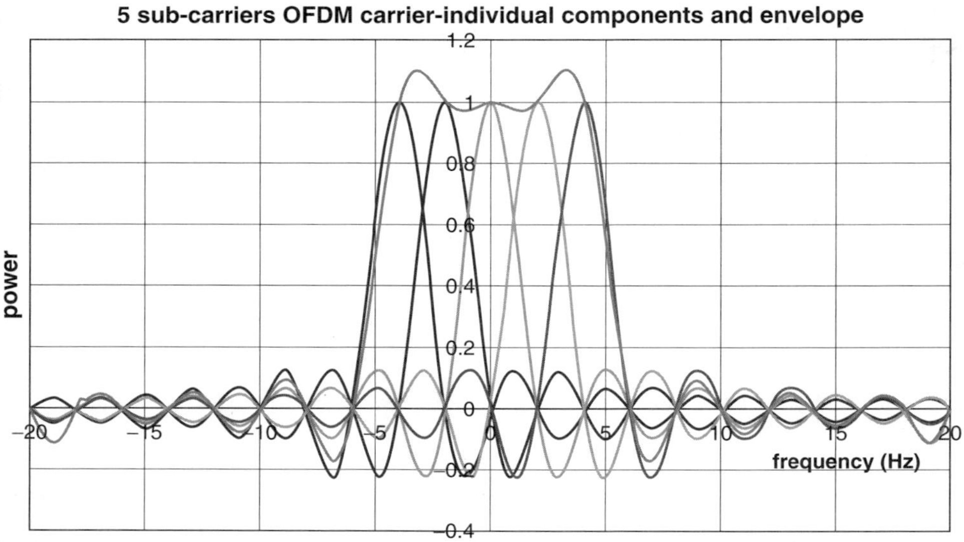

Figure 7.1 Five subcarriers forming an OFDM carrier.

Individual carriers are called sub-carriers and the combination of all sub-carriers is called an OFDM carrier. The main parameters of such a carrier are:

$$T_S = \frac{1}{\Delta f} \qquad \text{(7.1) OFDM carrier parameter relationship}$$

where :

T_s = Symbol duration.

Δf = Frequency step between sub-carriers.

The OFDM multiplexing process is summarized in the block diagram in Figure 7.2.

A data bit stream is mapped to symbols, which are then divided into two groups of values, I and Q, according to the constellation that is being used. I and Q values are then assigned to the sub-carriers that make up the sub-channel being used. These sub-carriers, in turn, undergo an iDFFT (inverse Discrete Fast Fourier Transform).

7.1.1 Implementation of an Inverse Discrete Fast Fourier Transform (iDFFT)

The intention of the iDFFT is to create a time domain waveform that represents the sum of several sub-carriers modulated by a cosine (I waveform) or sine (Q waveform). This waveform is represented by its samples, so the first thing we need to find out is the minimum number of samples according to the Nyquist–Shannon theorem (twice the difference between the lowest and highest frequencies). At each sample the values of all frequencies are calculated and added together. The result is the samples of the waveform in the time domain that combine all sub-carriers. The Fast algorithm (in iDFFT) uses the property that the samples are the same for consecutive periods, so the cosine and sine have to be calculated only for the first period of each frequency. This drastically decreases the processing time, from an [N^2] amount to an [N] amount of calculations.

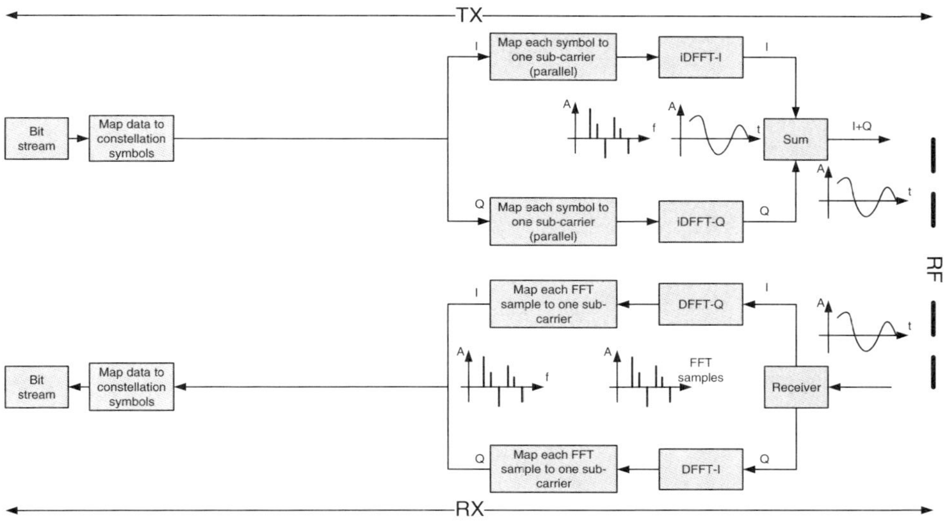

Figure 7.2 Multiplexing and de-multiplexing I and Q streams.

I and Q waveforms are then combined (as they are orthogonal) and sent to the antenna. No considerations are made about moving the signal to higher bands and vice versa as it does not affect the basic operations described here.

On the receive side the reverse is done and the time domain waveform is analyzed for content, using a DFFT (Discrete Fast Fourier Transform). Although the mathematics is very complex, the actual implementation is simple.

7.1.2 Implementation of a Discrete Fast Fourier Transform

The aim of the DFFT is to find out the original components that were used to generate the waveform. The input waveform is brought to its baseband and sampled, so we need to process only its samples.

The orthogonality property is used and the waveform is multiplied by the cosine of each sub-carrier frequency in the I branch and the sine of each sub-carrier frequency in the Q branch. Results for each frequency are added up over a full period and the sum gives the relative amplitude between the sub-carriers. The Fast algorithm in (DFFT) uses the property that only one period has to be calculated for each frequency, and this reduces the number of calculations, from an [N^2] amount to an [N] amount of calculations. The FFT used should be a power of 2 immediately larger than the number of sub-carrier used and the extra carriers should be nulled.

I and Q waveforms are then combined (as they are orthogonal) and each carrier is mapped to its constellation symbol, so the original bit stream can be recovered.

In the time domain, I and Q signals are represented by their samples. For a 1 bit/s data rate, the carrier separation is 1 Hz. For five carriers, the highest frequency is 10 Hz and the required sampling rate is 20 sample/s (ignoring the anti-aliasing margin). In Figures 7.3 to Figure 7.7, the I and Q components are combined for four carriers, at 1 Hz, 2 Hz, 3 Hz and 5 Hz. The fourth carrier (4 Hz) is not used. All carriers have the same amplitude and the larger amplitude waveform is the combined signal. The single digit Hz frequencies are used as an example; any harmonic frequency can be used.

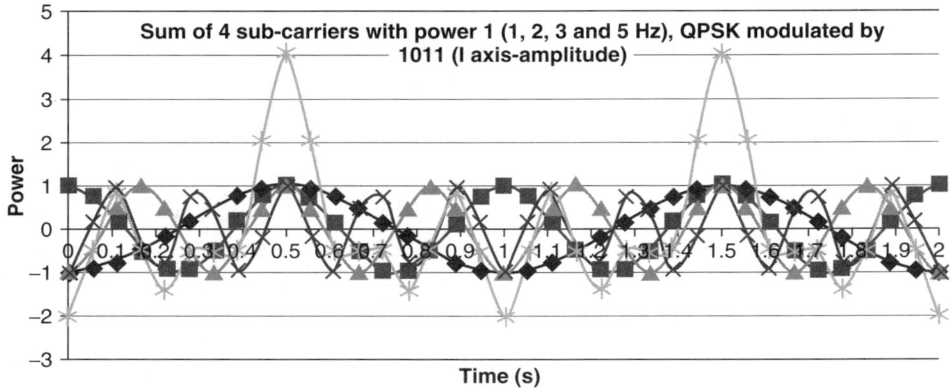

Figure 7.3 Four subcarriers forming I signal of an OFDM carrier.

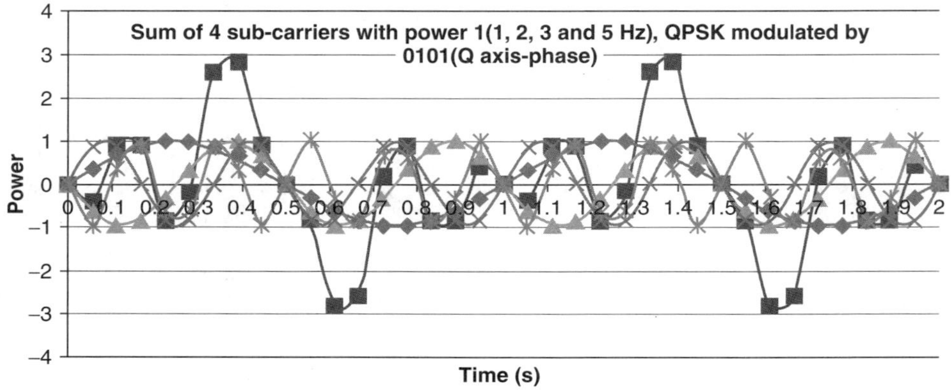

Figure 7.4 Four sub-carriers forming Q signal of an OFDM carrier.

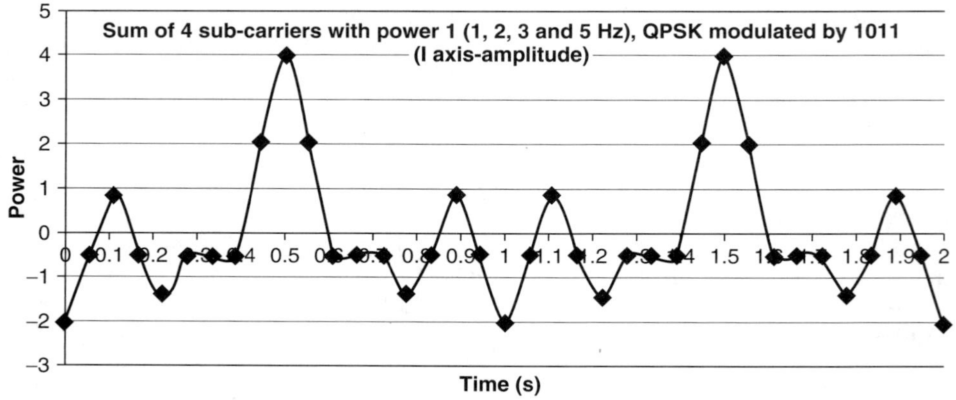

Figure 7.5 I signal of an OFDM carrier.

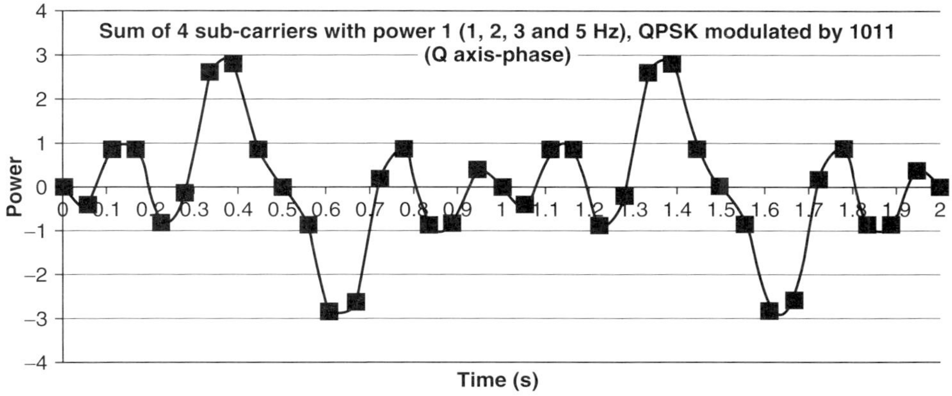

Figure 7.6 Q signal of an OFDM carrier.

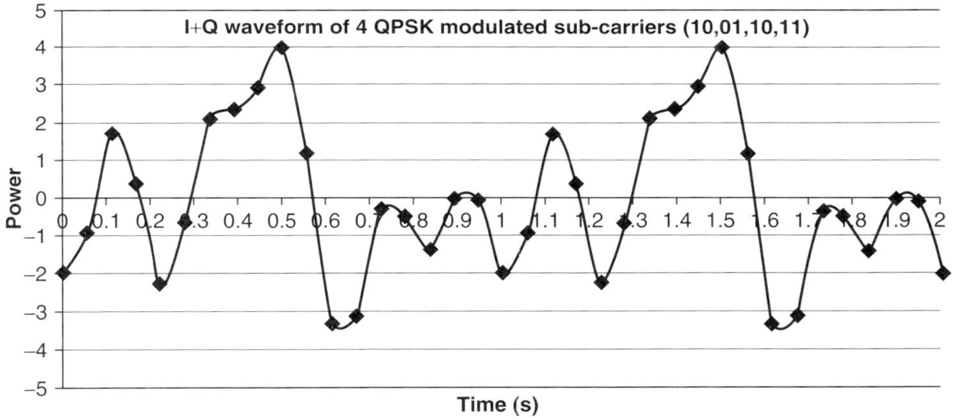

Figure 7.7 I+Q signal of an OFDM carrier.

The time domain waveforms are periodical and their sum results also in a periodical waveform. This property will be useful later on.

7.1.3 Peak to Average Power Ratio (PAPR)

Examining the combined waveform in Figure 7.7, it can be seen that it peaks several times at the average power, and consequently the output waveform can have a very large Peak to Average Power Ratio (PAP, PAPR or PAR). Table 7.1 estimates the PAPR value for different numbers of sub-carriers.

Maximum PAPR occurs when all the components have the same phase, and because phase is irrelevant to orthogonality, it is possible to use a random starting phase for each component and thus reduce the PAPR value.

The probability of reaching the maximum PAPR is negligible and can be corrected by the FEC (Forward Error Correction) code. In Table 7.1, the column "Practical PAPR" is an estimate of the

Table 7.1 Peak to average power ratio

Sub-carriers	Maximum PAPR (dB)	Practical PAPR (dB)
4	6	3.0
8	9	4.5
16	12	6.0
32	15	7.5
64	18	9.0
128	21	10.5
256	24	12.0
512	27	13.5
1024	30	15.1
2048	33	16.6

required back-off in the HPA (High Power Amplifier) for 99.9% of the OFDM symbols. PAPR is more critical in the subscriber unit, and is one of the reasons for the low nominal power used in WiMAX. LTE has adopted a different scheme for the subscriber unit that minimizes this issue, the SC-ODFM, described in the next section.

7.1.4 Single Carrier OFDM (SC-OFDM)

A solution has been proposed to minimize the PAPR issue, called Single Carrier OFDM (SC-OFDM). The name SC-OFDM is misleading as DFT-S-OFDM (Discrete Fourier Transform-Spread-OFDM) better represents the process, but SC-OFDM is used as a marketing name.

In this solution, blocks of symbols to be transmitted are transformed into a time domain waveform using an FFT (Fast Fourier Transform) process. The samples of this time domain waveform are then mapped to a block of sub-carriers in an iFFT (inverse Fast Fourier Transform) process. The FFT processed samples are still added, but statistically to a slightly lower number due to the spreading caused by the DFT. The PAPR improvement is around 2.5 dB over the regular OFDM process. The DFT-S-OFDM brings several disadvantages, like the minimum block size and additional overhead caused by having to place pilots (reference signals) outside the block. LTE had chosen this solution for the uplink. WiMAX considered it for the 802.16m version, but the disadvantages outweighed the advantages and the solution was dropped.

The DFT-S-OFDM multiplexing block diagram is shown in Figure 7.8.

The bit stream is mapped to the desired constellation symbols and I and Q sequences are generated. I and Q values form the samples of a time domain waveform, by serializing the information of each sub-channel sub-carrier. This waveform is represented by its samples, so, first, we have to find out the minimum number of samples according to the Nyquist–Shannon theorem (twice the difference between the lowest and highest frequencies).

The waveform has to be modified from an NRZ (Non Return to Zero) format to a RZ (Return to Zero) format, so sequences of 1s and 0s can be represented without a DC component. A DFFT (described previously) is performed to obtain the frequency domain representation of this waveform, resulting in assignments for each sub-carrier. Next, the process is identical to the one used for the regular OFDM. An extra stage is added on the receive side to convert the sub-carrier values to the time domain, obtaining the samples for the serialized data. Those samples are then mapped to the constellation and to the resultant bit stream.

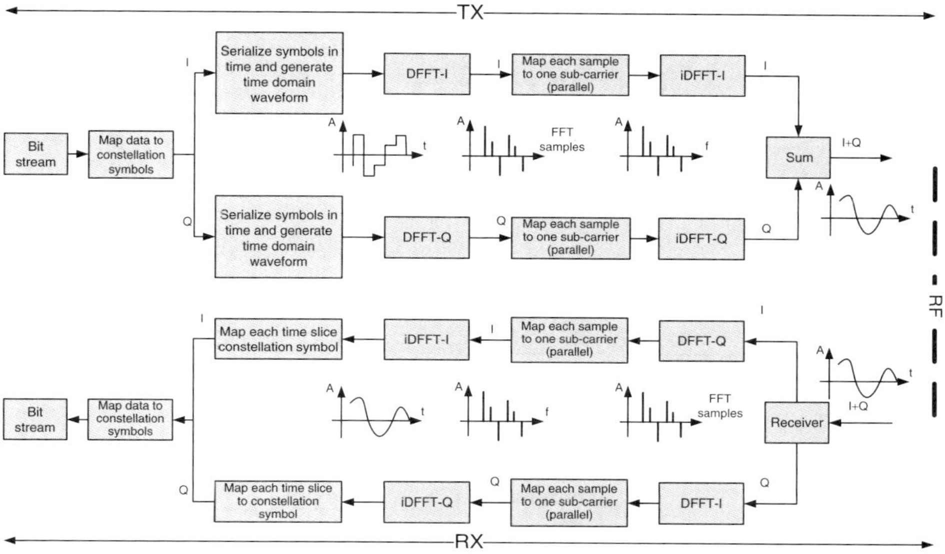

Figure 7.8 DFT-S-OFDM block diagram.

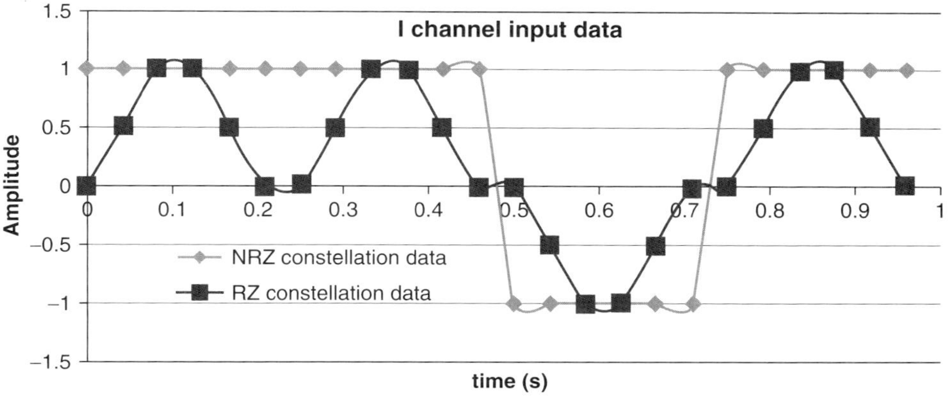

Figure 7.9 I channel input data example.

This is illustrated in Figure 7.9 to Figure 7.14; the first three figures showing the transmitted I signal, and the next three, the received one.

The PAPR improvement achieved by using SC-OFDM is about 30% of the regular OFDM back-off requirement. The practical PAPR varies from 1 to 5 dB depending on the number of sub-carriers. Figure 7.15 shows the PAPR for regular and SC-OFDM for 64 sub-carriers. It can be seen that SC-OFDM requires about 2.5 dB less back-off than regular OFDM.

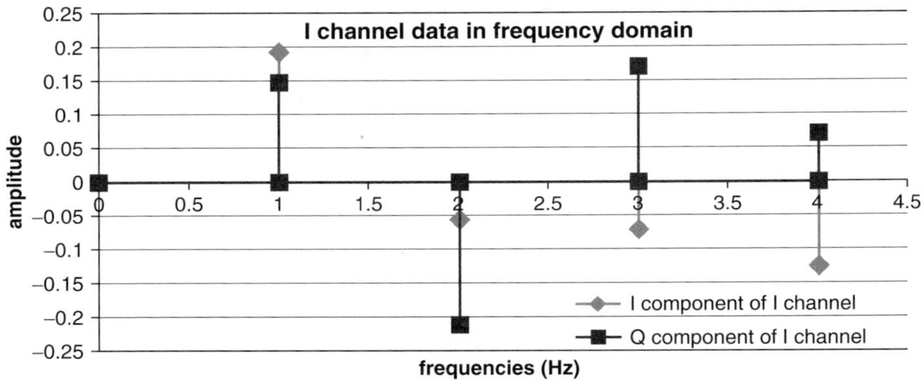

Figure 7.10 I channel data in frequency domain.

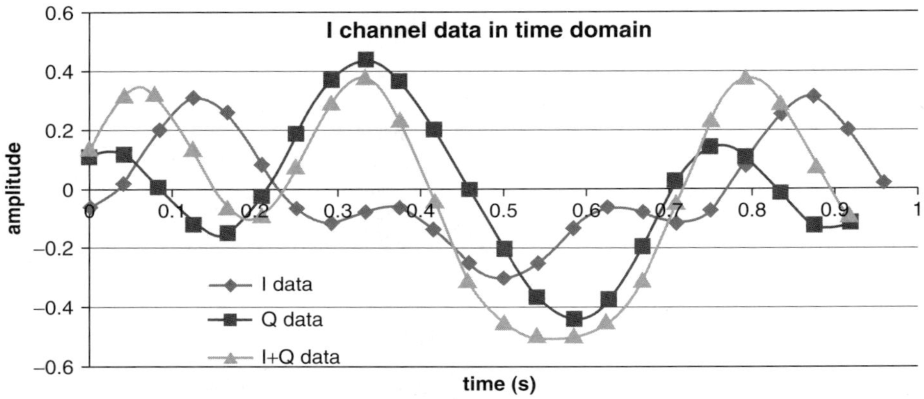

Figure 7.11 I channel data in time domain.

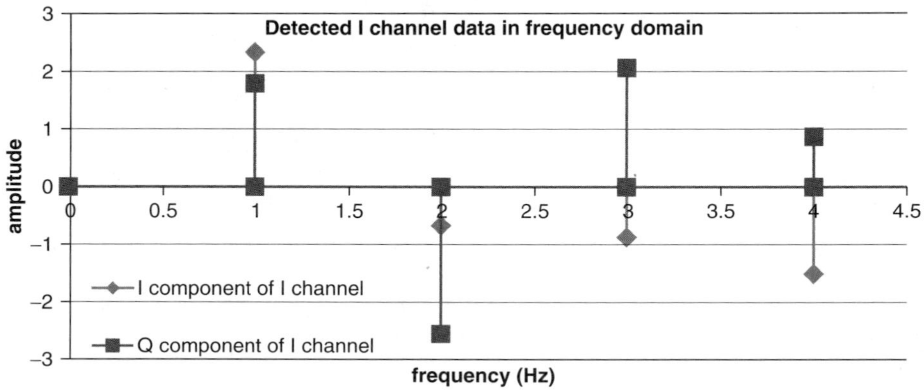

Figure 7.12 Detected I channel data in frequency domain.

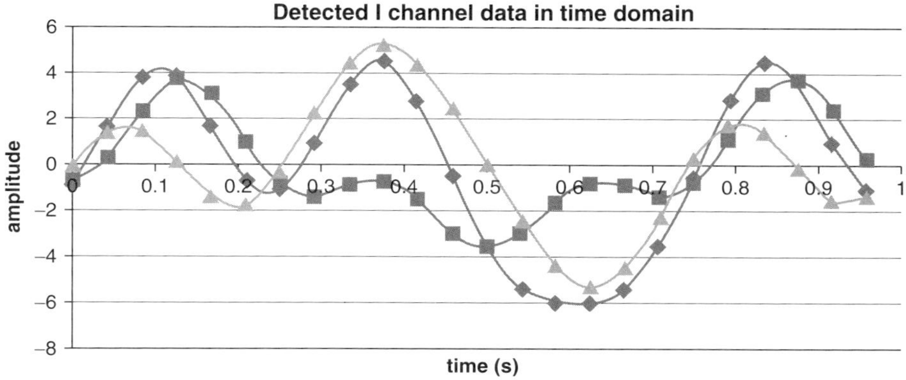

Figure 7.13 Detected I channel data in time domain.

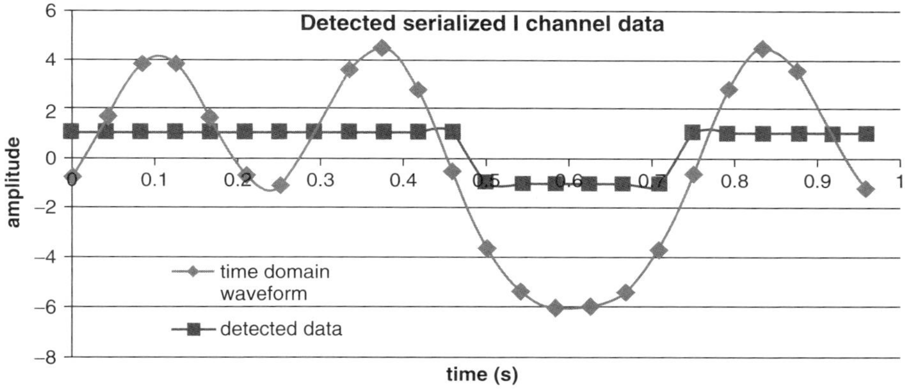

Figure 7.14 Detected serialized I channel data.

7.2 Other PAPR Reduction Methods

A possibility that is worth investigating is to reduce PAPR by using different delays for each subcarrier, so the peaks' coincidence is avoided. This requires that different delay patterns be tested for each data pattern to be transmitted. An average PAPR reduction of 3.5 dB can be obtained by examining a few delay patterns.

7.3 De-Multiplexing

Extracting the components of the above waveforms is done by using their orthogonality property. Because the modulation used is known, the possible components are also known. The waveform samples can then be multiplied by the samples of each component and the results accumulated; if the component is present, the result is different from zero; otherwise it is zero.

To extract phase and amplitude information, a known reference signal should be sent together with the other signals, so its extraction can be used to correct RF channel distortions. The phase distortion

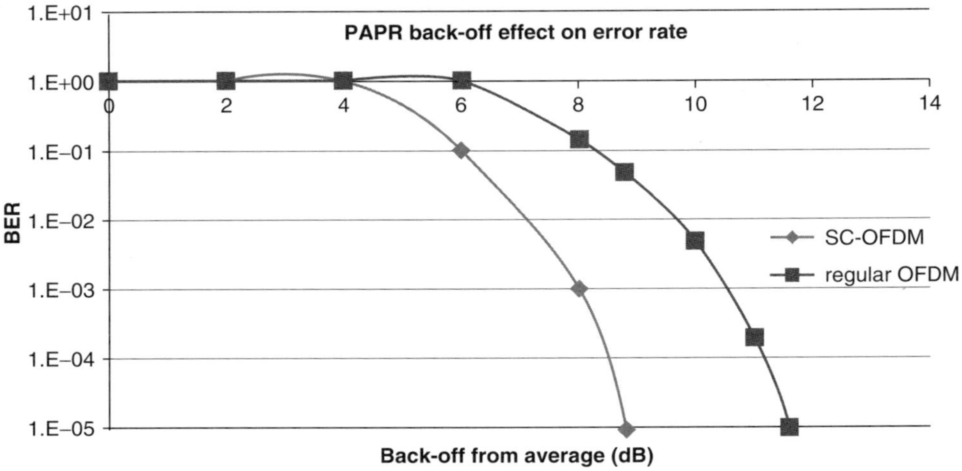

Figure 7.15 PAPR back-off effect on error rate.

should be corrected first, so the sine- and cosine-modulated I and Q signals can be extracted. Then, their amplitudes should be adjusted.

7.4 Cyclic Prefix

Chapter 5 showed that several copies of the same signal are mixed in the receiver due to multipath, and delayed copies will interfere with the adjoining symbols resulting in ISI (Inter Symbol Interference). The same chapter gives guidelines on how to estimate this delay.

The delay signals cause inter-symbol interference and an intra-symbol inference as illustrated in Figure 7.16. The inter-symbol interference is caused by the delayed paths of the previous symbol. A guard period would be enough to eliminate it, as seen in Figure 7.16. The intra-symbol interference is caused by the delayed samples of the symbol itself.

In Chapter 4 we saw that delayed sinewaves, when combined, will result in a shifted sinewave and considering that the signal can be decomposed in sine waves, the same would be applied to the signal. The only condition is that the waveforms should exist over a whole period (one symbol duration).

The guard interval does not satisfy this condition as there is not a single interval in which all multipaths are available for an entire period. This requirement can be achieved if the transmitted signal is repeated for the period of the multipath. Fortunately the iDFFT calculation results in a cyclical signal and it is enough to leave it running longer (for the duration of the multipath). The transmitted symbol becomes then longer than the required symbol duration, but this is compensated at the receiver, by eliminating the extension.

This process is called Cyclic Prefix (CP). In reality, as can be seen in Figure 7.17, the combined waveform repeats itself at the end of the symbol, so it is enough to extend the sampling period by the length of the estimated compromised symbols. This is illustrated in Table 7.2. Inter-symbol and intra-symbol interference and cyclic prefix are shown in tabular form and in Figure 7.17 in graphical form.

The cyclic prefix increases the total symbol duration during the transmission process, but it does not affect the orthogonality between the sub-carriers, as the possibly compromised samples are removed before submitting the signal to the receive FFT. It is important to note that the signal at this point is just a series of numbers as shown in Table 7.2.

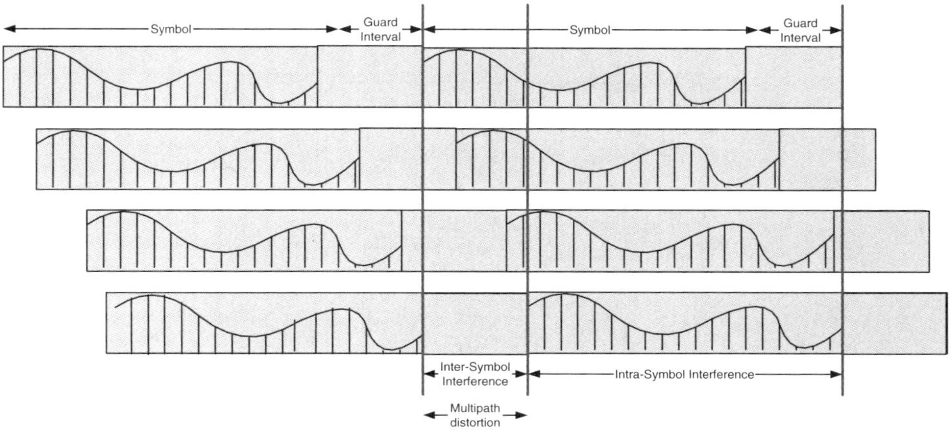

Figure 7.16 Multipaths using a guard interval.

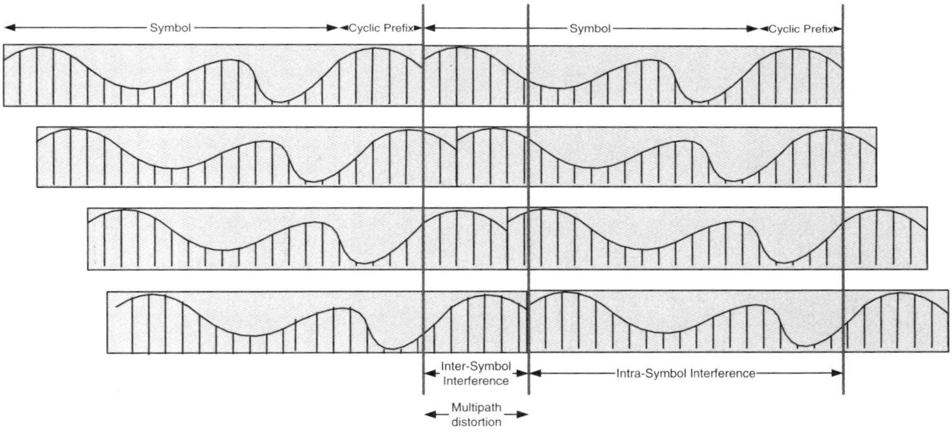

Figure 7.17 Multipaths using the cyclic prefix as a guard interval.

7.5 OFDMA

The information to be transmitted has to be directed to the appropriate destination and this can be done in different ways:

- *Circuit switching*: in this type of circuit, the switching is done for the whole duration of the call or session, so the destination address can be informed during the call/session establishment. This method is not very efficient, because circuits are dedicated to the call/session even if there is no information to send. In this type of switching it is very hard to accommodate variable and discontinuous throughputs, making it inefficient for digital communications.
- *Packet switching*: in this type of circuit, the switching is done per packet or multiple packets, and the destination of each packet is included in it. This method has an additional overhead due to the packet protocol and assignment (or access) mapping, but it can share the bandwidth better,

Table 7.2 Inter-symbol and intra-symbol interference and cyclic prefix

<table>
<tr><th colspan="3">Sample value</th></tr>
<tr><td rowspan="12">Symbol period</td><td>128</td><td rowspan="5">Inter-symbol interference</td></tr>
<tr><td>35</td></tr>
<tr><td>47</td></tr>
<tr><td>32</td></tr>
<tr><td>−56</td></tr>
<tr><td>−98</td><td rowspan="12">Intra-symbol interference</td></tr>
<tr><td>−73</td></tr>
<tr><td>...</td></tr>
<tr><td>...</td></tr>
<tr><td>...</td></tr>
<tr><td>64</td></tr>
<tr><td>66</td></tr>
<tr><td rowspan="5">Cyclic prefix</td><td>128</td></tr>
<tr><td>35</td></tr>
<tr><td>47</td></tr>
<tr><td>32</td></tr>
<tr><td>−56</td></tr>
</table>

mainly in case of digital data. To minimize the assignment overhead, time frames are established and channels are allocated in them. Two options are available:

- *OFDSA* (Orthogonal Frequency Division Single Access): In this access method, the channels in a frame are allocated for the whole frame duration. This term is not commonly used, and OFDM (Orthogonal Frequency Division Multiplex) is used instead. This raises confusion as OFDMA, explained next, is also an OFDM circuit.
- *OFDMA* (Orthogonal Frequency Division Multiple Access): In this access method, a frame can be further divided and multiple channel allocations can be done during the duration of a single frame.

7.6 Duplexing

There are two directions of transmission that have to be accommodated in a bi-directional wireless communication system, one from the base stations to the customers (downstream) and another from the customers to the base stations (upstream). Downstream can also be called Downlink, although the latter is more appropriate for circuit switching and the same applies to Upstream and Uplink.

This direction multiplexing is called duplexing and there are three basic methods to implement it: FDD, TDD and H-FDD.

7.6.1 FDD (Frequency Division Duplexing)

In this method, the multiplexing of the two flows is done using two different frequency bands, one for Downstream and another for Upstream. This allows for a continuous transmission on both bands, with the precaution explained next.

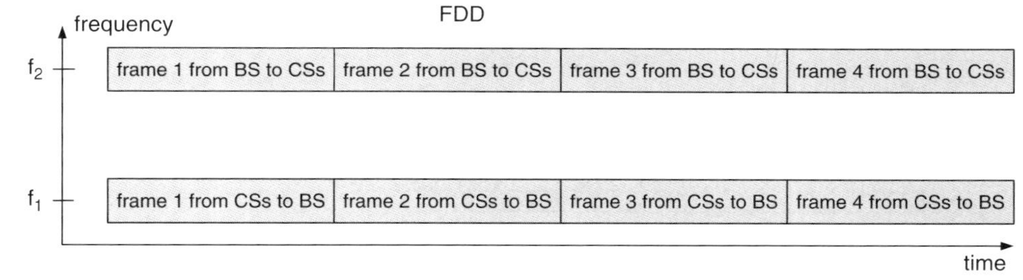

Figure 7.18 Frequency division duplex.

FDD duplexing is illustrated in Figure 7.18, where the two streams are represented. Each side of the stream has a transmitter and a receiver. The signal strength differences between both streams is huge (can be more than 100 dB), and even though they are in different bands, it is very difficult to protect the receiver from the signal of its own transmitter. This can only be achieved if significant frequency spacing exists between both, so filtering/blocking can be done. The usual minimum spacing is in the order of 45 MHz.

Many FDD bands were allocated in the twentieth century, using the same bandwidth for both directions. This creates inefficiency for digital data, as it tends to be asymmetrical, and one band ends up congested whereas the other still has availability.

7.6.2 TDD (Time Division Duplexing)

In this method the multiplexing is done in time and only one band is used. This method is favorable for digital communications as it allows balancing both communications streams. TDD is illustrated in Figure 7.19.

In TDD, separate transmission times are allocated for downlink and uplink transparently to users. This is illustrated in Figure 7.20. The allocation cycle is defined by a frame period, which is divided into a downlink sub-frame and an uplink sub-frame. The length of these sub-frames, in turn, is a multiple of the symbol duration. The two sub-frames can have different durations to accommodate asymmetrical traffic in the downstream (DS) and the upstream (US).

The ratio between the upstream sub-frame duration and the downstream sub-frame duration is the TDD ratio and is unique for a network, that is, all BSs use the same ratio; this avoids interference between downlink and uplink signals. A typical ratio is 60%, but this value should be adjusted by network designers based on the desired service configurations.

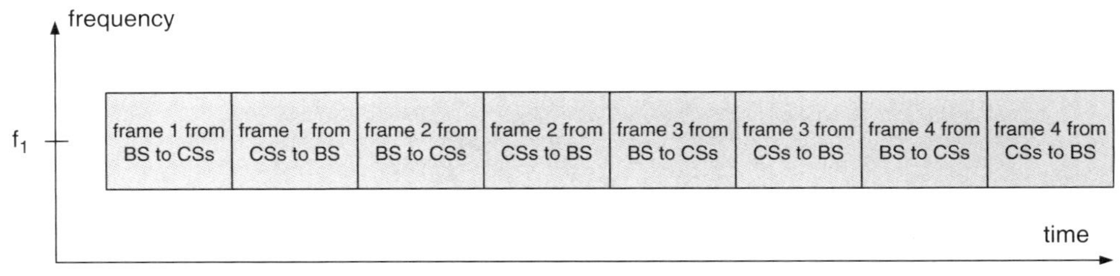

Figure 7.19 Time division duplex.

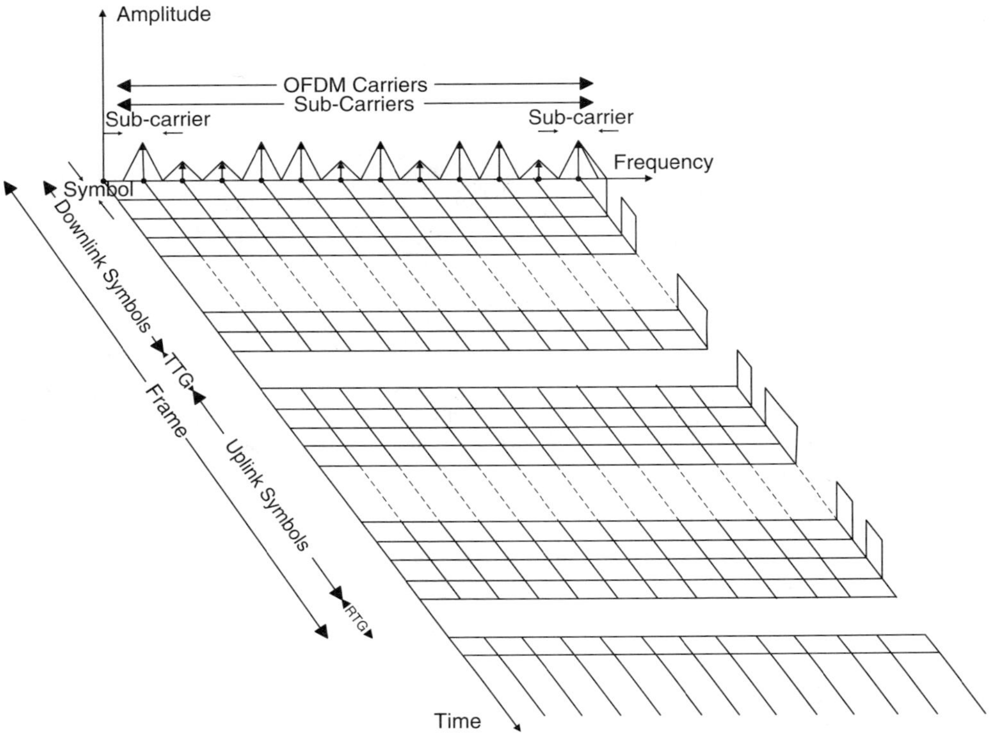

Figure 7.20 TDD Transmission in OFDM.

The split in DS and US adds some inefficiency to TDD transmission as the TDD ratio represents an average value of network throughput demand; this may cause the system to be under-utilized at moments in which the demand differs from this average. This loss is estimated to be 5%.

In TDD systems, both upstream and downstream directions share the same RF conditions of the channel (unlike FDD systems, because of the use of distinct frequencies for each direction). This allows measurement data to be collected in the upstream, to be used for RF-tuning of the downstream.

In H-FDD, the BS and SS also use different frequencies, but they do not transmit at the same time. This overcomes the requirement of a large frequency separation but uses frequencies only 50% of the time, allocating the remaining 50% to another cell. This time allocation is illustrated in Figure 7.21 for two cells A and B.

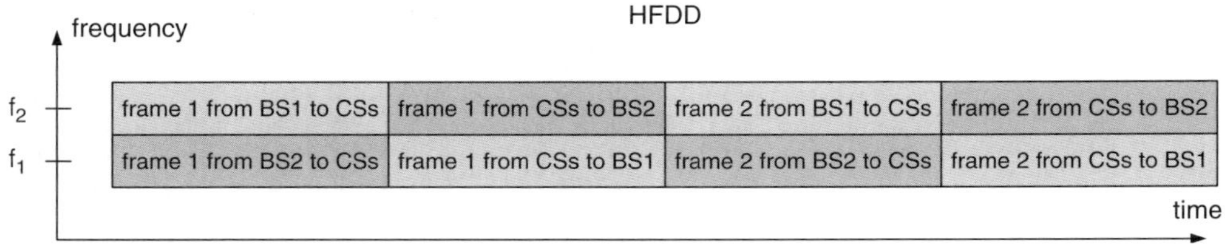

Figure 7.21 H-FDD time allocation of a frequency channel.

Both FDD and its H-FDD variation use a fixed frame duration for downstream and upstream transmission. This works well for symmetrical services that require the same data rate for both directions, however, it is not suitable for applications such as the Internet, in which MSs and SSs offer an asymmetrical demand to the network, requiring different throughput for the up and downstream. Even though asymmetrical frequency bands could be considered in FDD, this is not easily done, as frequency allocations were already done in the past using symmetrical bands. For such applications, TDD mode is more efficient because it offers adaptive distribution of frame duration for up and downstream transmissions. One advantage of FDD, however, is the fact that users do not need to wait for their assigned sub-frame, thus reducing system latency.

7.7 Synchronization

There are two ways to manage the air access in wireless systems, unframed and framed. The unframed solution is used in WLAN (Wireless Local Access Network) and the framed solutions are used by WiMAX (Wireless Microwave Access) and LTE (Long-Term Evolution).

7.7.1 Unframed Solution

This solution relies on contention, by collision detection and collision avoidance techniques. This reduces system throughput, mainly with traffic increase, but its advantage it that it does not require a central controller. The synchronization is achieved for each transmitted packet by a preamble that precedes the data.

7.7.2 Framed Solution

In this solution, time is divided in continuous frames. A small area of the frame is still dedicated to contention access. The frame is generated at the base stations and subscriber stations have to synchronize to it, at symbol and frame levels.

Subscriber stations' upstream synchronization adjusts the stations' transmit time, so their transmission arrives at the correct frame time at the base station.

In TDD, frames are divided into two sub-frames: downstream and upstream. Base stations should synchronize their frame start to avoid conflict between downstream and upstream transmissions. A detailed description is provided for WiMAX in Chapter 13. LTE does not specify a methodology for this synchronization, leaving it to the vendor's discretion.

Frames can be further divided into zones, which delimit usage of specific characteristics, such as interference mitigation and use of resources. Zones can be used to limit the use of resources to specific cell areas. WiMAX specifications consider the zone concept and have provisions for it. LTE does not specify zones, but considers that vendors will prevent interference by scheduling resources accordingly between CPEs. Future LTE specifications may include zones in the SON (Self-Organizing Networks) concept.

7.7.2.1 Subscriber Station Synchronization

There are two types of synchronization functions: timing synchronization and frequency synchronization.

- Timing synchronization is less critical in OFDM than in other technologies because the symbol time is larger and the equalization is easier (the multipath spread is a fraction of the symbol time).

- Frequency synchronization, however, is more critical than in other technologies because the subcarriers are very close to each other.
- Downstream synchronization is done through pre-defined sequences sent by the base stations in preambles or in defined locations in the frame. This procedure is further described in Chapters 12, 13, and 14 for each technology.
- The upstream synchronization of subscriber stations uses the contention access area of the frame to get synchronization feedback from the base station and adjust its timing. This process is called ranging. Minor misalignments are absorbed by the cyclic prefix.

7.7.2.2 Inter-Base Station Synchronization

The easiest and cheapest way to perform this synchronization is to use GPS (Global Positioning System), which provides a Stratum 1 (derived from the main national Reference Clock- stratum 0) Primary Reference Source (PRS) with a precision of 10^{-12}. As an alternative, the IEEE 1588 Precision Timing Protocol is currently under trial.

GPS uses a NMEA (National Marine Electronics Association) 0183 standardized format, which consists of an ASCII (text) string transmitted at 4800 baud. One message contains the current time with one second resolution, receiver status, latitude, longitude, speed over ground, heading (track), date, magnetic variation in degrees and check sum. Another message contains the receiver's mode, number of satellites and quality of the reading (DOP dilution of precision). Table 7.3 shows synchronization requirements for different technologies.

7.8 RF Channel Information Detection

The transmitted RF signal is sent to the antenna and then irradiated to the air. This signal propagates and sends its energy in all directions according to the antenna pattern. This irradiated energy reaches the receiver at different moments because some energy may reach the receiver in a direct line and some may be diffracted, reflected, and refracted through different paths. Each path renders a different amplitude and phase, which may vary with frequency and time (frequency and time selective). The received energy is a combination of all these paths. Noise and interference from other sources also add to the signal.

The first step is to recover the base-band signal and for this the exact frequency has to be recovered using a PLL (phase locked loop). To recover the data it is necessary to recover the original waveform, and the only way is to equalize the RF channel. It is also necessary to establish when each symbol starts.

Table 7.3 Synchronization requirements per technology

Technology	Frequency accuracy	Time accuracy
GSM	5×10^{-8}	not required*
CDMA	5×10^{-8}	1 μs GPS (10 μs holdover)
UMTS (FDD)	5×10^{-8}	not required
UMTS (TDD)	5×10^{-8}	2.5 μs
WiMAX/LTE (FDD)	5×10^{-6}	not required
WIMAX/LTE (H-FDD)	5×10^{-6}	1 μs GPS (25 μs holdover)
WiMAX/LTE (TDD)	5×10^{-6}	1 μs GPS (25 μs holdover)

Note: * GSM and UMTS network base stations get their synchronization from TDM T1/E1 lines that they use for backhaul. However, for cost reduction purposes, many networks use Ethernet/IP connections for their backhauls, which do not provide the required timing information.

Wireless systems use a cellular structure, in which a central node concentrates the communications of users within a cell. This central node receives different names depending on the technology, it can be called AP (Access Point), BS (Base Station), BTS (Base Transceiver Station) or eNodeB (evolved Node B). The users can be called CL (Client), SB (Subscriber) or U (User).

Each technology uses its own nomenclature, and when a technology is described the nomenclature used in the technology is applied, otherwise the most common nomenclatures are used throughout the book.

7.8.1 Frequency and Time Synchronization

A wireless channel access control can be distributed or centralized:

- In a distributed system, each user monitors the channel and sends information when the channel is free. This may lead to conflicts that are resolved through a contention protocol. Several precautions are taken to minimize possible air conflicts, including a type of structured access using RTS/CTS (Request to Send/Clear to Send) messages, in which case the AP takes a role of access controller. This is the case of Wi-Fi, in which Access Point and Clients use the same procedure to access the channel.
- In a centralized system, an entity, the central node of a cell, has control of the channel and allocates time slots for users to send and receive their information. The initial user access is allowed only at specific time slots in which a contention protocol is used. Centralized systems resort to structured access, defining access time frames and within them access time slots.

In a wireless system, information is sent from different locations, which use independent frequency oscillators. Receivers have to adjust their oscillators to the incoming signal frequency to recover the base band signal. Next they need to identify the start of the symbol stream, so the carried information can be correctly detected.

In a distributed system, information can arrive at any time and the receiver requires a long time to adjust. In this case, preambles, several symbols long, are used to adjust the receiver.

In centralized systems, information is sent at regular intervals and users can continuously tune their oscillators to the central node frequency. The timing is also pre-defined, so only a minor adjustment is required, which is achieved by a single symbol preamble in each frame.

The receiver is always looking for a preamble to find out when a new transmission or frame is starting and it uses that to adjust its oscillator. Preambles also provide information about the remaining content of the signal.

7.8.2 RF Channel Equalization and Reference Signals (Pilot)

Once the Baseband waveform is recovered at the receiver, it has to be adjusted to compensate for the distortion introduced by the RF channel. Information about the RF channel response has to be gathered frequently as it varies with time.

As described in Chapter 5, the RF channel causes the following distortions:

- *Path phase-shift for different frequencies*: Each path has a propagation time and renders different phase shifts for different frequencies.
- *Multi-path phase-shift and amplitude distortion*: The combination of multi-path signals results in phase shift and amplitude distortion. The amplitude distortion is called fading and can be very severe for small durations of time. Diversity techniques are used to get signal information during fade periods.

- *Frequency change due to movement*: Also results in a change of phase.
- *Distortion caused by multi-paths of previous symbols*: This effect is called ISI (inter-symbol interference) and may be limited to a fraction of a symbol or affect several symbols. This extent is defined by the symbol duration and the time difference between the significant paths.
- *Noise addition*: Thermal noise is added to the signal in the receiver. A low noise front-end receiver is used to amplify the signal and minimize the effect of this additive noise in the remaining stages.
- *Waveform distortion*: Non-linearity in the receiver or frequency deviations can further distort the signal.
- *Interference from other sources*: Interference can come from other cells using the same frequencies.

Each one of the above distortions requires special attention from the technology, equipment and network designers. Inter-symbol interference (ISI) can be avoided by the technology design through the use of a cyclic prefix. Phase shift and amplitude distortions can be corrected by the equipment designer using RF channel equalization techniques. Noise addition can be reduced by the use of low noise front-end amplifiers; whereas waveform distortion can be lessened by a careful design. The remaining issues have to be addressed by the network designer; some options to deal with them include the use of advanced antenna systems to counteract fading, proper specification of the duration of the cyclic prefix, and optimization of network resources to minimize interference.

To be corrected, the distortion caused by the RF channel has to be first characterized and its effect on the system must be estimated. This is done by transmitting, together with the data, a matrix of scattered reference signals that carry known information. Reference signals are also called pilots and a training sequence and a preamble can be considered as reference signals also.

The OFDM signal described previously has the orthogonality of the I and Q components assured by using frequencies that are multiples of a base, independently of the phase of each frequency. The mixed I+Q waveform has its orthogonality defined by the use of sine and cosine waveforms, and is sensitive to phase variations of its components. The next sections describe how this property is used in the information extraction.

7.8.3 Information Extraction

We will exemplify here the channel equalization procedure using pilots in an OFDM signal. Pilots carry known information by the receiver and this allows the correction of the channel. Let's assume a four sub-carrier OFDM carrier, in which the first and last sub-carrier will carry pilot information (digit 1) and the two middle sub-carriers will carry the data (1010). In this case the data is modulated using QPSK or a higher modulation scheme, so the I and Q channels are used to carry data information (two bits per sub-carrier in case of QPSK). The pilot uses BPSK, and only the I channel carries its information. This is illustrated in Figure 7.22 to Figure 7.24.

An example of RF channel distortion, considered in this exercise is shown in Figure 7.25.

At the receiver, due to phase distortion, the FFT will detect pilot information on the Q signal also. The receiver will then adjust the phase of the received signal for the first and last sub-carriers until the FFT gives a zero result. At this point the phase shift is compensated. Phase shifts at each sub-carrier should be different, so an interpolation (average, linear or other) is applied to the other sub-carriers. At this moment, the phase shift for each sub-carrier is known and will be applied to all I and Q sub-carriers. The next step is to equalize the amplitudes using the I amplitude received at the pilot carrying sub-carriers.

Pilots carry a Pseudo Random Binary Sequence (PRBS) and these are applied to sub-carriers according to specific patterns, so equalization can be done in frequency and time. The distance between the pilots depends on the flatness of the channel, so the interpolation used is still valid to adjust the intermediate sub-carriers. This is illustrated in Figure 7.26 and Figure 7.27.

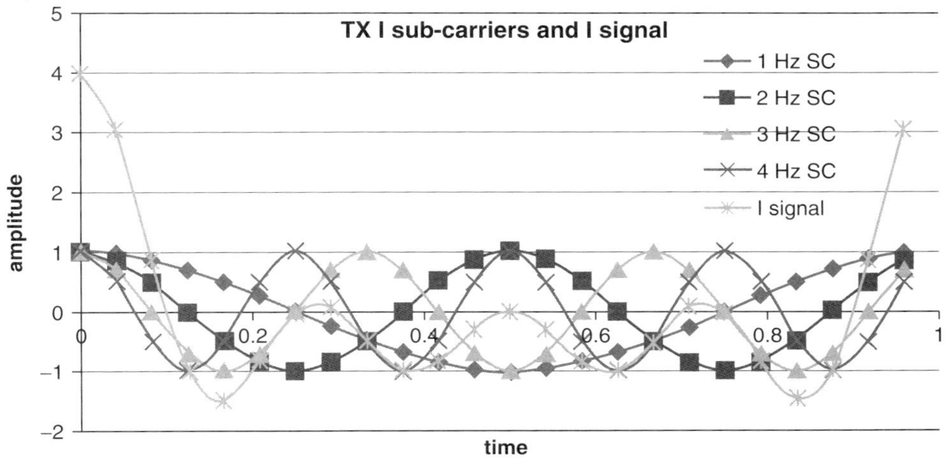

Figure 7.22 Transmit I sub-carriers and composite signal, I signal.

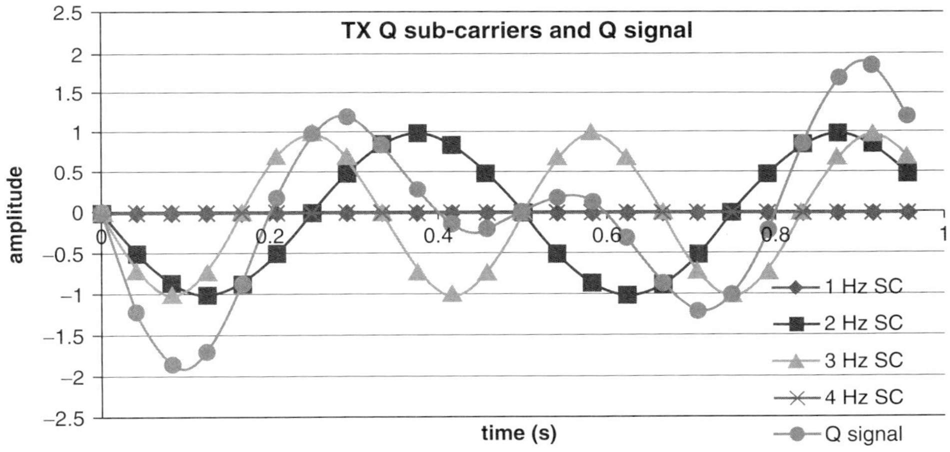

Figure 7.23 Transmit I sub-carriers and composite signal, Q signal.

Once the received waveform has been equalized, the FFT can be calculated and Table 7.4 shows the results for the sub-carriers that carry data. The original data (1010) was properly retrieved, as 0 corresponds to -1 in an NRZ code.

7.9 Error Correction Techniques

In 1927, Harry Nyquist established that the maximum number of independent pulses (f_p) that can be transmitted in a channel with bandwidth (B) was defined by Equation (7.2).

$$f_p = 2B \qquad (7.2) \text{ Nyquist's signaling rate theorem}$$

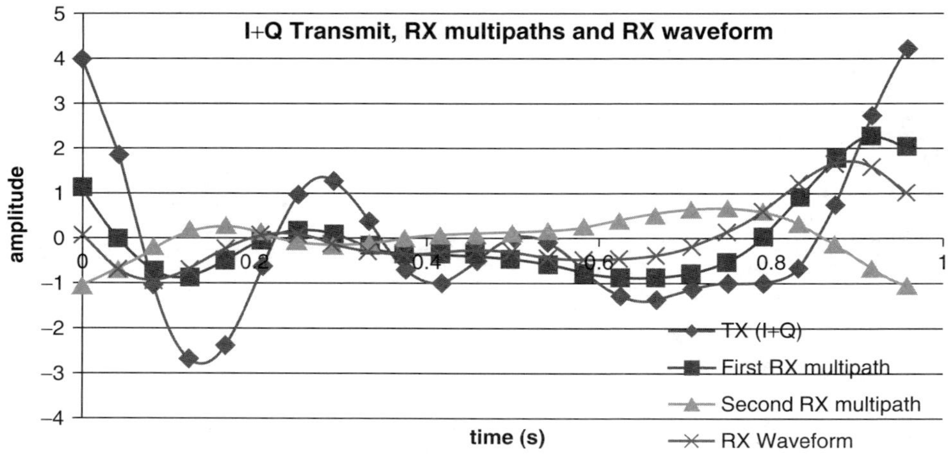

Figure 7.24 I+Q transmit signal, received multipaths and received composed waveform.

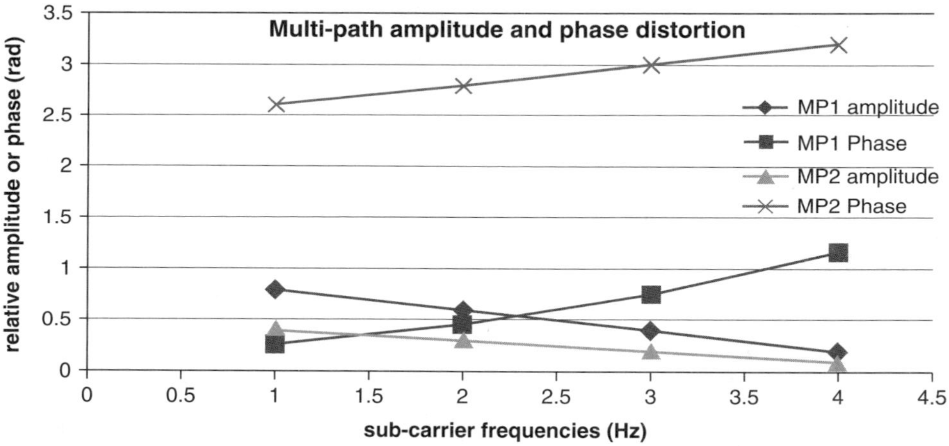

Figure 7.25 Multipath amplitude and phase distortion example.

At the same time, Ralph Hartley defined the maximum signaling data rate that can be sent and received reliably over a communications channel. He defined initially the maximum number (M) of distinct pulses of amplitude (A) that can be detected by a receiver with a discernible threshold (ΔV), and, from it, he calculated the maximum data rate that a channel can pass (R) (Equations (7.3) and (7.4)), but he did not make the connection with noise. The R value corresponds to the channel capacity (C).

$$M + 1\frac{A}{\Delta V} \quad \text{(7.3) Hartley's receiver sensitivity}$$

$$R = f_p \log_2(M) \quad \text{(7.4) Hartley's data signaling rate law}$$

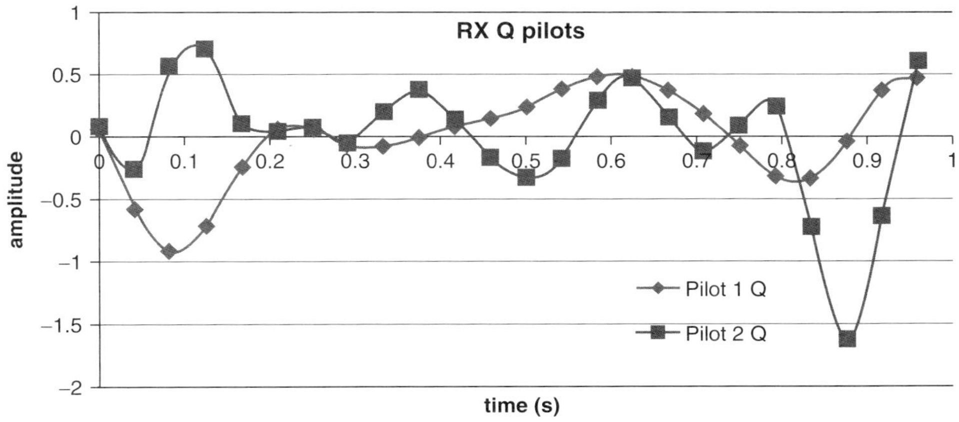

Figure 7.26 Received Q pilots.

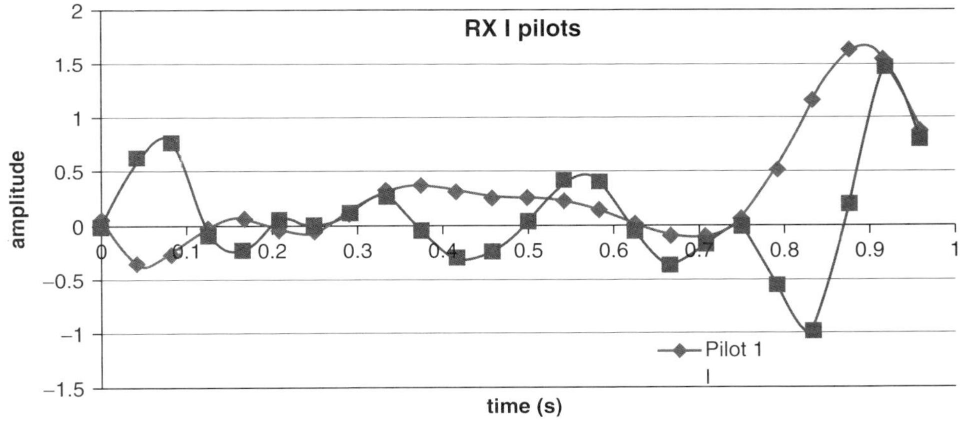

Figure 7.27 Received I pilots.

Table 7.4 DFFT detection values

	DFFT detection	
Sub-carrier	sum	symbol
f2 I	6.544087	1
f2 Q	−4.70484	−1
f3 I	3.095497	1
f3 Q	−4.46556	−1

In 1948, Claude Shannon made the connection between the threshold established by Hartley and noise. He defined noise and signal power at the receiver respectively by Equations (7.5) and (7.6). The V value is the voltage and the (R) value is the characteristic impedance of the circuit.

$$N = \frac{V_N^2}{R} \qquad \text{(7.5) Noise}$$

$$S = \frac{V_s^2}{R} \qquad \text{(7.6) Signal}$$

The total received power was defined by Equation (7.7).

$$V_T^2 = V_S^2 + V_N^2 \qquad \text{(7.7) Signal + noise voltage}$$

Using Hartley's thresholds concept, he established that the maximum number of discernible thresholds (b), could be defined by Equation (7.8).

$$2^b = \frac{V_T}{V_N} = \sqrt{\frac{V_T^2}{V_N^2}} = \sqrt{1 + \frac{S}{N}} \qquad \text{(7.8) Maximum number of discernible thresholds (exponent)}$$

The same equation can be expressed as Equation (7.9).

$$b = \log_2 \left(1 + \frac{S}{N}\right)^{\frac{1}{2}} \qquad \text{(7.9) Maximum number of discernible thresholds}$$

Replacing the Nyquist independent pulse value and Shannon's value for the discernible threshold into Hartley's law, we get the channel capacity C expressed in Equation (7.10).

$$C = B \log_2 \left(1 + \frac{S}{N}\right) \qquad \text{(7.10) Shannon's channel capacity}$$

According to Shannon's equation, the amount of information that can be transmitted through a channel depends on the SNIR. Interference and noise cause errors in the digital signal that limit channel capacity. Assuming that the SNIR cannot be further improved, the elimination of errors is often done through error correcting codes. These codes are added to provide redundancy to the digital signal. The amount of redundancy defines the main characteristics of a code.

The total elimination of errors would imply an extreme application of redundancy, which leads to an extremely large error correction code. Thus, when the number of errors is small, it becomes more efficient to use as a variation of the redundancy method: the Automatic Repeat Request (ARQ). This method repeats transmitted data only when requested by the receiver. The CRC (Cyclic Redundancy Check) added to the MPDU (MAC Protocol Data Unit) is used in the ARQ process to identify errors. Additional error detection codes can also be applied in the receiver itself.

HARQ (Hybrid ARQ) is a variation of ARQ that adds Forward Error Correction (FEC) bits to the method. WiMAX uses two types of HARQ (Hybrid ARQ): Type I, or Chase Combining, and Type II, or Incremental Redundancy.

In Chase Combining, the same data is resent upon request, that is, the information transmitted is a copy of the original data. The receiver uses all previously received versions to improve the chance of decoding. The WiMAX Forum mandates implementation of Chase Combining with use of Convolutional Turbo Codes (CTC) for both upstream and downstream.

In Incremental Redundancy, the code rate and puncturing pattern are changed from one transmission to the next, increasing the chance of a successful decoding, that is, the information transmitted is now different from the original data because of the change in the bits added by the FEC process. The implementation of this type of HARQ is made optional by the WiMAX Forum.

If the maximum specified number of retransmissions is reached, the packet is dropped and a higher level layer has to request its retransmission.

Error correction techniques reduce spectral efficiency by presenting an overhead that diminishes data throughput, hence should be considered in throughput calculations. This overhead varies with the error rate, so an average number should be estimated (10% overhead is a reasonable assumption).

7.10 Resource Allocation and Scheduling

Digital data arrives at the transmitter in packets, which are queued to be transmitted. Packets coming from a user application are defined as a service flow. Resources have to be allocated to queue the packets and send them through the wireless link. Resources are finite and it is necessary to check their availability, before accepting a new session. Resource allocation considers average resource consumption before deciding if enough resources are available and a session can be accepted.

Resource allocation is done according to scheduling algorithms, which consider the requirements of each packet, based on their required QoS (Quality of Service).

Data packets appear in bursts, and although, in average, there are enough resources for all of them, packets may have to be scheduled in time, thus there is certain delay when transmitting them. The content of some packets is not affected by delays, but others, however, are sensitive to it. Scheduling is a complex task and different criteria may be used to implement it. Each vendor develops their own algorithms as technology standards do not cover this aspect.

An efficient scheduler complies with acceptable delays, maximizes throughput and provides an equitable service to all users, offering all of them similar shares of the resources. Some of these objectives, however, may conflict and the way these conflicts are arbitrated is defined by the scheduling algorithm. This chapter next describes the main scheduling algorithms.

7.10.1 FIFO (First In, First Out)

A regular queue uses a FIFO (First In, First Out) scheduler also known as FCFS (First Come, First Served) algorithm. This is the simplest algorithm to implement, but it does not consider important aspects such as:

- packet sensitivity to delays
- packet size
- data tonnage from specific sources
- throughput optimization

A FIFO algorithm is acceptable for Best Effort (BE) flows.

7.10.2 Generalized Processor Sharing (GPS)

A Generalized Processor Sharing (GPS) approach was developed to share the capacity of momentarily congested communication links in an efficient, flexible and fair manner, but it cannot be implemented practically as it considers infinitesimal packet sizes. It is only used as a comparison with other algorithms.

7.10.3 Fair Queuing (FQ)

For other flows than Best Effort, a common algorithm is Fair Queuing (FQ) that balances the throughput of each flow using a bitwise round-robin approach. The packet with the earliest finish time is the one selected for transmission. The calculation of the finish time is computationally intensive and needs to be redone each time a new packet arrives.

7.10.4 Max-Min Fairness (MMF)

A Max-Min Fairness (MMF) algorithm is a simple implementation of FQ and privileges the smallest packets first, by equally dividing resources between all flows. As soon as one flow is satisfied, the resources are again split between the remaining flows. The same applies if a new flow arrives.

These algorithms are fair in terms of balancing throughputs but do not consider the impact of delay on the packet transmission.

7.10.5 Weighted Fair Queuing (WFQ)

Weighted Fair Queuing (WFQ) is a generalization of FQ that assigns weights (priorities) for each flow. Theses weights can be used to implement a Quality of Service (QoS) policy in relation to delays that can provide a guaranteed data rate.

None of the above algorithms considers spectrum efficiency, which is essential in wireless systems, hence the Fairly Shared Spectrum Efficiency (FFSE) algorithm, which can be used as a combined measure of fairness and spectrum efficiency. This is achieved by balancing the wireless throughput while offering a minimum data flow or scheduling priority. User prioritization can be assigned considering the formula in Equation (7.11).

$$P = \frac{T^d}{R^s}$$ (7.11) Scheduling prioritization in FFSE

where:

P = flow prioritization.
T = potential data rate achievable in the time slot.
R = historical average of this flow.
d = scheduler data fairness.
s = scheduler spectrum fairness.

When $d = 1$ and $s = 0$, we have the round-robin approach. When $d = 0$ and $s = 1$, we maximize spectrum usage.

7.11 Establishing Wireless Data Communications

This section describes the general characteristics of establishing a wireless data communication session, as it is different from traditional voice communication sessions (calls) and understanding it is important for a network designer. Detailed implementations are described in the chapters that deal with each specific technology. We will describe here general characteristics of the establishment of a wireless data communication session. Specific implementations will be described in the technology chapters.

Wireless data communications are more complex than voice and follow specific phases:

- Choose carrier and synchronize.
- Associate and Authenticate.

- Wait for event.
- Transmit or receive data.
- Wait for event/Check carriers for possible handoff.

Figure 7.28 represents a wireless connection. It is composed of data equipment and wireless equipment on each side. Depending on the technology, the air connection can be framed (have a defined time frame) or be frameless.

Frameless medium (air) access is done using contention, which happens for each packet. Mechanisms are implemented to minimize contention, through special short messages (RTS Request To Send and CTS Clear To Send).

In framed connections, one of the ends (AP or Base Station) takes the role of controlling access to the air, although the initial subscriber access is still contention based, but restricted to specific frame locations.

7.11.1 Data Transmission

The data equipment (PC, camera, router) sends a TCP/IP message (layers 4 and 3) using an Ethernet connection (layers 2 and 1). The wireless equipment is basically composed of a processor, usually an SBC (Single Board Computer), a radio and an antenna system.

The processor receives the packet through its Ethernet port, reads the Ethernet MAC, analyzes the QoS information, and adds the wireless MAC (layer 2). The message is then placed in one of several buffers according to its QoS. A transmission slot request message is sent to the radio and the message is placed in the air at the specified time and location.

7.11.2 Data Reception

The radio sends all received messages to the processor, where it is analyzed to check the destination. When data is received, the MAC is removed and an Ethernet MAC is added. The packet is placed in a buffer to be transmitted. When the medium is clear, it is sent to the Ethernet port.

7.11.3 Protocol Layers

When data is transferred from one user to another, it passes through several protocol layers. These layers interconnect at SAPs (Service Access Points).

An SDU (Service Data Unit) is a block of data received by a protocol layer. In the transmit direction, the protocol layer adds an overhead to the SDU and creates a PDU (Protocol Data Unit). This PDU

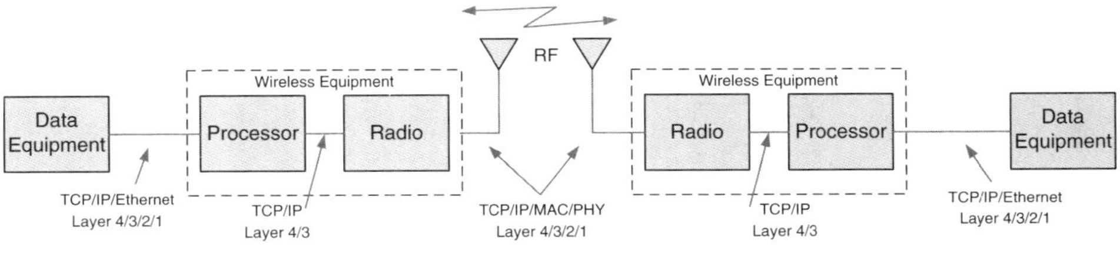

Figure 7.28 Wireless connection block diagram.

then becomes an SDU for the lower layer, as illustrated in Figure 7.29. In the receive direction, the reverse is done, and the protocol overhead is stripped, so the PDU received by a layer is exactly the same that was sent by its equivalent layer at the other end. Each protocol layer ignores the existence of the lower layers.

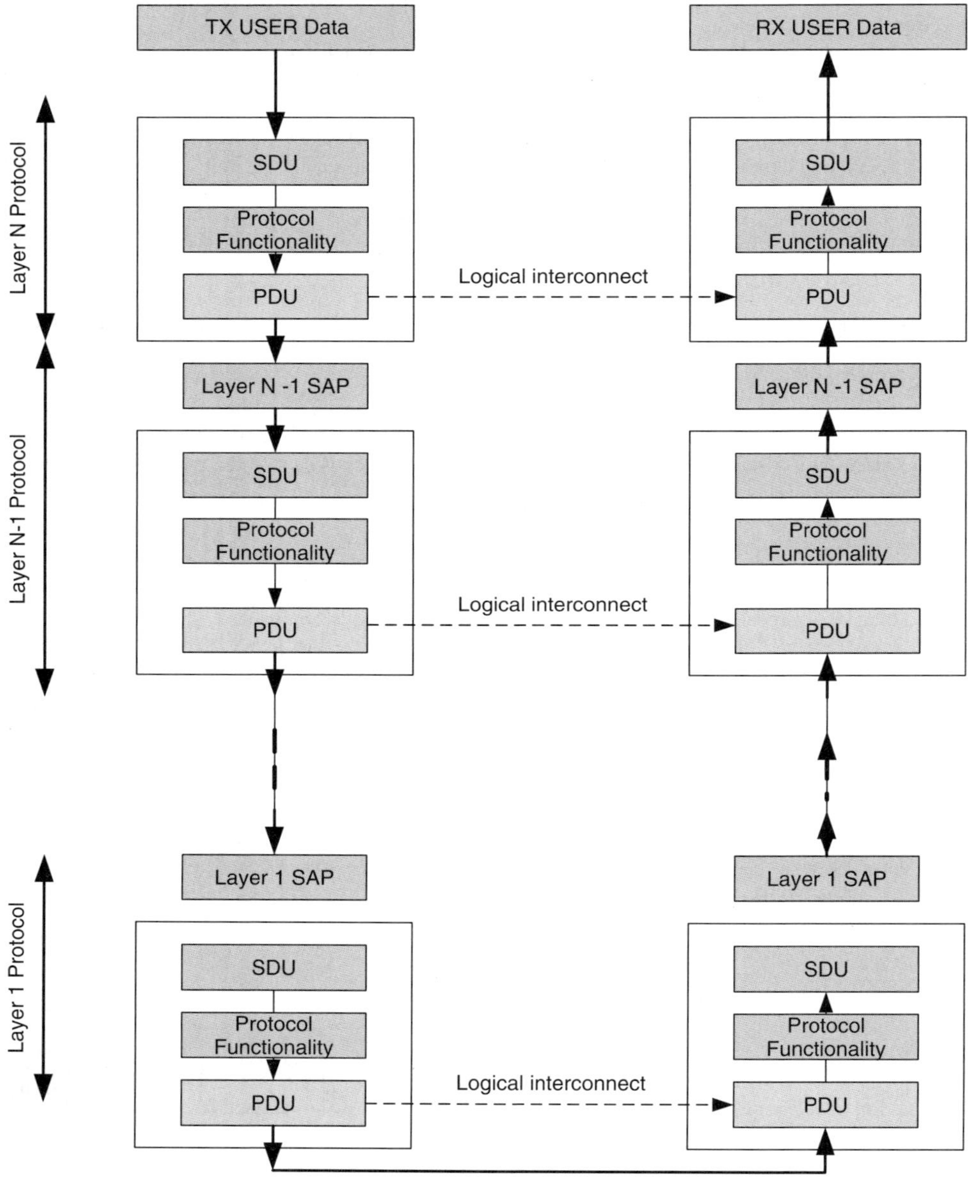

Figure 7.29 Service and protocol data units within different layers.

7.11.4 Wireless Communication Procedure

Figure 7.30 shows the phases of a wireless connection procedure in general terms, and is applicable to all the technologies covered in this book. According to Figure 7.30, in the first phase, the Access Point (AP) or Base Station (BS) broadcasts messages establishing the rules to joining the network.

The user wireless device then searches the bands it is programmed for, looking for compatible broadcasts and learning how to access the network. It chooses the strongest signal and proceeds to registration.

The first access has to be done on a contention basis, regardless of whether it is a framed or frameless technology. This is used to synchronize user transmissions with the framing structure.

Next, users exchange several messages to associate ("I want to join your group of active users"), authenticate ("here are my credentials, please check them"), generate encryption keys ("this is how we will encrypt our data exchange") and exchange capabilities ("this is what I can do and this is what you can do").

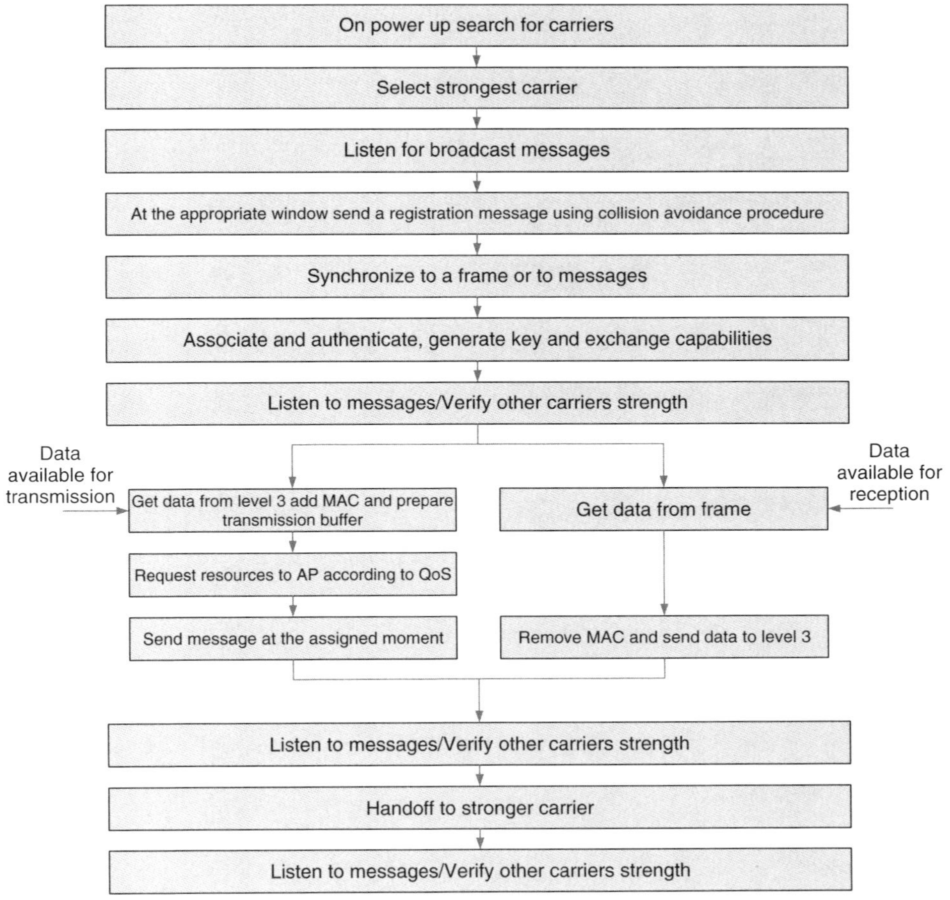

Figure 7.30 Wireless connection procedure.

Once this phase is done, the user listens to messages and periodically adjusts its timing (if required by the technology).

When data is available for transmission, the user gets the data through an Ethernet port, strips the Ethernet MAC and adds the wireless MAC. The data is then stored in a buffer accessible by the radio to wait its turn to be transmitted.

In framed technologies the radio requests the assignment of resources according to the QoS specified for the data. The QoS is generally mapped to the data protocol used by the packet (TCP/IP, ICMP, UDP and others). In a frameless network the random back-off is adjusted according to its priority, so low priority messages wait longer and the data is transmitted according to timer expirations.

Transmitted data is recovered by the receive side radio and delivered to a buffer in the processor. The processor strips the wireless MAC and inserts an Ethernet MAC before sending the data to its destination.

In the idle state the wireless equipment continues to listen to messages in the air and periodically checks the signal strength of its carrier and other available carriers. As network-wise it is more productive to use the best possible carrier as this increases the network throughput capacity, so when a better carrier is detected, a handover is done.

8

OFDM Implementation

The typical implementation of an OFDM equipment used in a WiMAX network is presented in this chapter but the antenna signal processing is not included here, as it is described separately, in Chapter 10. OFDM implementation in an LTE network is similar and the small deviations are described in Chapter 12.

8.1 Transmit Side

Once a wireless data session is established, the timing and the modulation to be used are determined by the scheduling algorithm. The block of bits to be transmitted is then processed. This block usually arrives through an Ethernet connection with a protocol overhead, to which the wireless protocol overhead is added.

The transmit side block diagram is shown in Figure 8.1. It can be divided into four blocks: bit processing, symbol processing, digital IF processing and carrier modulation.

8.1.1 Bit Processing

Bit processing is divided into three stages: randomization, forward error correction coding, and interleaving; each of them is described next.

8.1.1.1 Randomization

The bit sequence is combined with a pseudo-random sequence, to break any long sequence of 1 s or 0 s. A polynomial is used to generate a Pseudo Random Bit Sequence (PRBS), which is mixed with the actual data using an XOR circuit.

8.1.1.2 FEC (Forward Error Correction) Encoding

Redundant data is added to the bit sequence to allow for error correction. The k information bits from the source become an n bit codeword (n,k). The coding ratio R is k/n and represents the fraction of the information contained in each codeword. The addition of FEC increases the bandwidth requirement, in lieu of a higher SNR (signal to noise ratio) requirement.

LTE, WiMAX and WLAN Network Design, Optimization and Performance Analysis, First Edition.
Leonhard Korowajczuk.

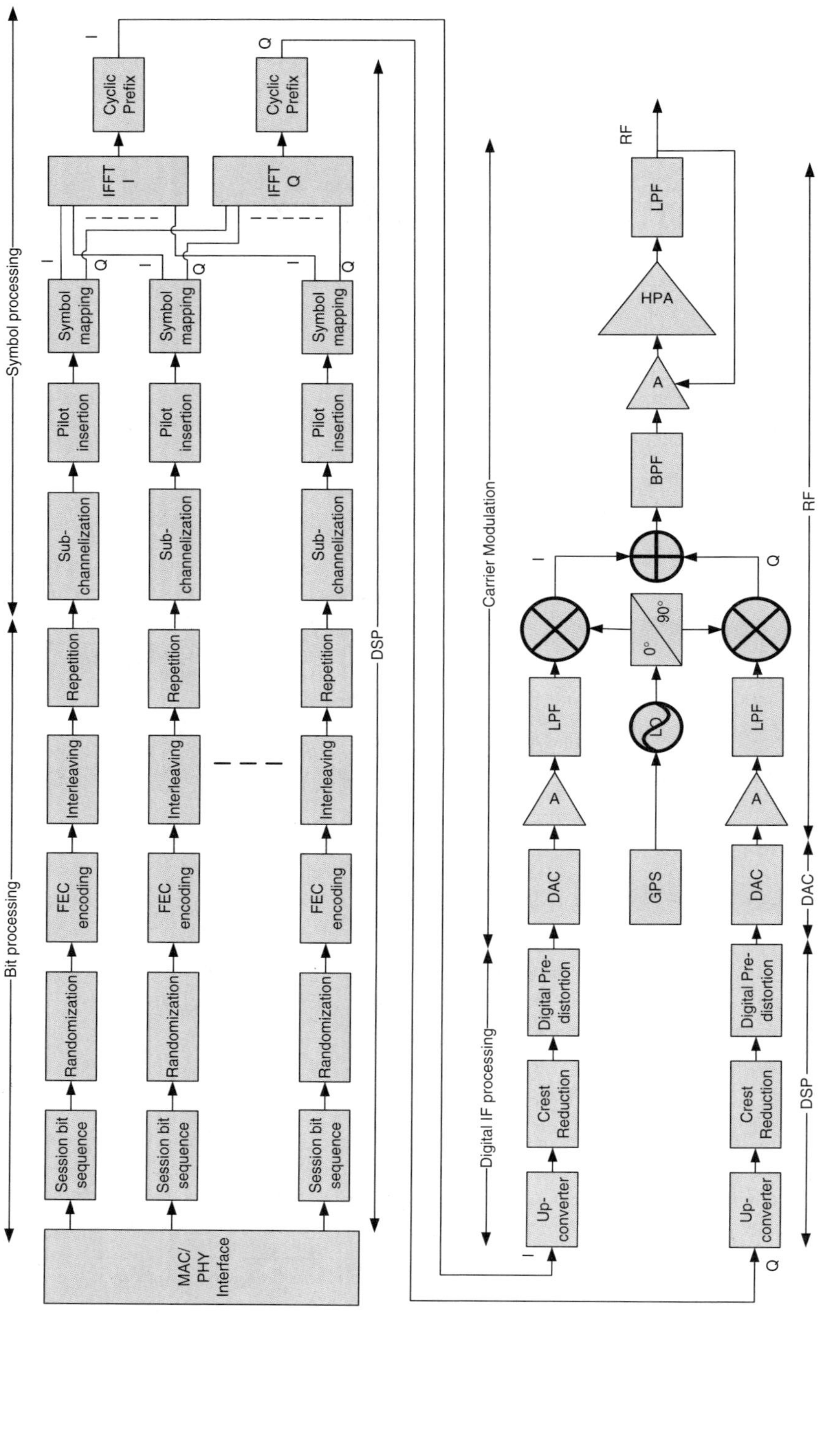

Figure 8.1 OFDM transmit block diagram.

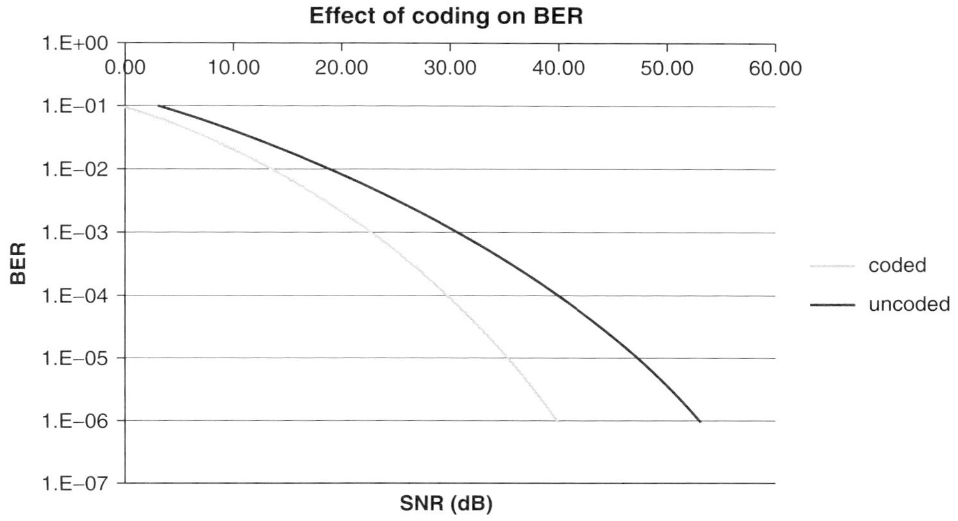

Figure 8.2 Effect of coding on BER.

The performance of FEC codes is statistical and has to be expressed in terms of BERs (Block Error Rates). The error performance of a digital communication system has a waterfall shape, that is, the performance improves with the increase of SNR. As explained above, FEC coding reduces this SNR requirement in exchange for a higher bandwidth. This reduction in SNR is called coding gain; the use of the word "gain", however, is a bit deceptive as the information throughput is in fact reduced for the new SNR required. The coding gain is usually larger for higher BERs, so it may be better to target larger BERs, to maximize information throughput. Today it is common to work with BER of up to 10^{-2}. Figure 8.2 shows the effect of the coding gain on the BER.

The main FEC codes used in broadband systems are:

- Concatenated Reed-Solomon Convolutional Code (RS-CC)
- Block Turbo Coding (BTC)
- Convolutional Turbo Code (CTC)
- Low-Density Parity-Check (LDPC)

The overall strategy of error correction is to eliminate errors in steps, with the goal of optimizing total throughput. Receivers work with relatively high error rates, which are reduced by the FEC codes; residual errors are left to be corrected at higher levels (MAC, Medium Access Level). Anything left after that is treated by retransmitting messages (ARQ, Automatic Repeat Request).

8.1.1.3 Interleaving

When errors affect a sequence of bits, the use of just the few FEC code bits would not be enough to correct these errors. For this reason, bits are grouped in blocks and mixed in a way so that sequential bits fall in different FEC codes. The block is the size of the channel or set of channels allocated to a session.

8.1.2 Symbol Processing

Symbol processing is divided into five stages: sub-channelization, pilot insertion, symbol mapping, iDFFT, and cyclic prefix; each of them described next.

8.1.2.1 Sub-Channelization

OFDM sub-carriers are grouped in sub-channels, which are then assigned to transmit the data. Considering that the modulation to be used in the session was previously defined, the number of subchannels assigned can be determined by calculating the number of symbols required to transmit the source data.

8.1.2.2 Pilot Insertion

Chapter 5 showed that sub-carriers have path losses that vary with frequency and time. Pilots are used to equalize sub-carrier power levels, and are applied to specific sub-carriers. Pilots use pre-defined bit sequences, so they can be easily identified in the receive side and be used to adjust the values of remaining sub-carriers. Wi-Fi uses training sequences instead of pilots.

8.1.2.3 Symbol Mapping

After pilots are mapped onto the bit stream, the complete bit stream is mapped to the remaining sub-channel sub-carriers. These sub-carriers are sine waves that are harmonically related and have to be sampled with different rates, to follow the Shannon sampling theorem.

Each sub-carrier is represented by two sine waves, one designated I (in-phase), and another Q (in-quadrature), as explained in Chapter 4. In this way, the proposed amplitude and phase constellation relationship desired for each symbols is achieved when both sine waves are combined. This combination is done by adding the digital samples of the sine waves. The analog sine waves are never generated, as their sample values are previously stored in the DSP (Digital Signal Processor) memory and just have to be retrieved when combining the signals.

8.1.2.4 iDFFT (Inverse Discrete Fast Fourier Transform)

It is essential for an OFDM transmission that the information be transmitted at a pre-established symbol duration (T), which will be considered to space the sub-carriers ($\Delta f = 1/T$) in frequency, assuring their orthogonality.

The mapping above is used to digitally generate the samples of the combined (sum) signal. Initially the sampling frequency is calculated at twice the number of sub-carriers times the aliasing factor (covered in Chapter 4). The sample times for the symbol duration are calculated and for each sample time the values of each sub-carrier symbol are calculated and accumulated into the sample value. After all the samples have been calculated, the output signal is digitally defined.

Each sub-carrier is composed of an I and a Q component and is defined by a sine or cosine with a phase shift. Chapter 4 showed the composed signal of several sub-carriers forming an OFDM symbol. We can represent this signal by its samples made at the Nyquist–Harley–Shannon rate.

This composed signal is the iDFFT of a series of sub-carrier symbols. We can use a DSP to calculate the value of each sub-carrier symbol at the sampling moment and add all of them. This sum represents the sampling of the combined signal. An N sub-carrier OFDM signal requires 2000 samples and a total of $2N^2$ operations per symbol. However, the majority of the sine/cosine waveforms have several periods within the symbol duration, and have the same values at each period, so the number of operations can be reduced to N*log N.

A 256 sub-carriers' OFDM signal requires 512 samples and, consequently, 130,000 calculations in the duration of 1 symbol. If we assume symbol duration of 100 μs, the DSP must be able to perform 1400 MIPS. This number can be drastically reduced by applying the FFT algorithm to 14 MIPS. The results are two streams of 512 samples each per symbol, which will be up-converted and combined with the OFDM carrier.

8.1.2.5 Cyclic Prefix

The cyclic prefix is added by allowing the iFFT to run for an additional number of samples equivalent in time to the required multipath margin.

8.1.3 Digital IF Processing

The digital IF processing is divided into three stages: up converter, crest factor reduction, digital pre-distortion; each of them described next.

8.1.3.1 Up Converter

I and Q signals can be up-converted to an IF (Intermediate Frequency), which can be handled by a DAC (Digital to Analog Converter). It is possible to combine the I and Q streams at this stage or combine them at the final stage. Typically the IF frequency is around 244 MHz.

At lower carrier frequencies it is possible to use a zero-IF (no IF is used and the base band is modulated directly into the carrier) circuit and the I and Q signals are digitally modulated directly by the carrier frequency. This solution is preferred as its single stage is less susceptible to amplitude and phase distortion than the two-stage IF solution.

In both cases the received waveforms are over-sampled through interpolation to space the aliases and allow more room for the filter to work. As an example a 5 MHz baseband waveform is generated with 11.424 MSPS (Million Samples per Second), the resulting IF waveform is oversampled, through interpolation, with a factor of 8, to 91.392 MSPS.

8.1.3.2 Crest Factor Reduction (CFR)

The Peak to Average Power, described previously, requires large back-offs in the High Power Amplifier (HPA). High power peaks seldom occur and it is more beneficial to limit them to a specified value, then to over-dimension the power amplifier. Allowing the peak to saturate the amplifier increases ACLR (Adjacent Channel Leakage Ratio) and increases EVM (Error Vector Magnitude), clipping it reduces ACLR. This is a not a simple operation, though, as at the signal offered to the HPA is the IQ sum, which was not complete yet, also the samples may not coincide with the peaks, so the sampling interpolation should be done before detecting the peaks. There are many techniques to implement CFR (Crest Factor Reduction) and some are very complex. Clipping peaks reduces ACLR, but the EVM is increased. This procedure is illustrated in Figure 8.3.

8.1.3.3 Digital Pre-Distortion

High Power Amplifiers (HPAs) need high linearity to comply with ACLR and EVM requirements. HPAs are designed to work up to the 1 dB compression point, implying on distortion of high-power

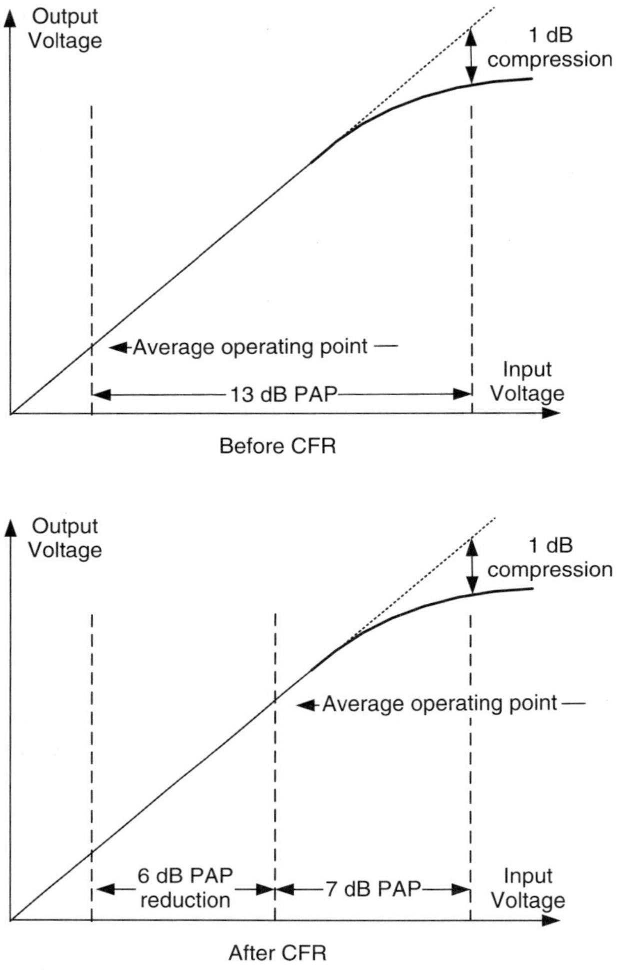

Figure 8.3 Crest reduction.

signals. To compensate this, a pre-distortion is applied to the signal, so that the output continues linearly. A look-up table (LUT) that has the characteristic of the amplifier is used to pre-distort the input signal. More advanced implementation use an adaptive look-up table (ALUT), which is adjusted by comparing the output with the input.

8.1.4 Carrier Modulation

It is divided into 12 stages: digital to analog converter (DAC), pre-amplifier, low-pass filter, GPS receiver, local oscillator, carrier generator, RF modulator, combiner, band-pass filter, amplifier, high power amplifier, once more, low-pass filter; each of them described next.

8.1.4.1 Digital to Analog Converter (DAC)

DACs take the waveform samples and generate an analog signal. This analog signal has the baseband components and its aliases. For this reason, modern DACs create additional samples, by interpolating existing samples (up to eight new samples is usual) and, consequently, separating the aliases from the baseband. A filter can then eliminate those aliases more easily.

8.1.4.2 Pre-Amplifier

DACs generate low power signals, so a pre-amplifier is used to elevate the signal above the noise floor, for subsequent operations.

8.1.4.3 Low-Pass Filter

A low-pass filter reduces the aliases' power, keeping the baseband signal.

8.1.4.4 GPS Receiver

The carrier frequency must be extremely stable, which is achieved through synchronization with a master signal. The only practical solution available today for this synchronization is the use of a Global Navigation Satellite System (GNSS). Table 8.1 shows the status and characteristics of several GNSS systems, some existent today, some to be released in the upcoming years.

Additionally, two regional satellite systems are being planned:

- IRNSS (Indian Region Navigational System): planned 2012, with 3 GEO and 4 MEO satellites.
- QZSS (Quasi Zenith Satellite System): planned for 2013 with 3 satellites.

The standard IEEE 1588 proposes a synchronization system over the Internet, but it has not been adopted yet.

For synchronization, WiMAX specifies the use of GPS, whereas LTE does not recommend any specific synchronization solution.

8.1.4.5 Local Oscillator

The local oscillator has to be extremely stable including in terms of phase jitter. Many issues in the real system can come from the deterioration of the local oscillator.

Table 8.1 Global navigation satellite systems (GNSS)

System	Country	Coding	Orbit (km)	Satellites	Frequency (MHz)	Deployment
GPS	US	CDMA	20,200	24+	1227.6; 1575.42	1993
GLONASS	Russia	FDMA/CDMA	19,100	20	1602	2010
GALILEO	Europe	CDMA	23,222	27+	1176.45; 1278.75	2014
COMPASS	China	CDMA	21,150	30+	1207.14; 1268.52; 1561.098; 1575.42	2011

8.1.4.6 Carrier Generator

Sine and cosine waveforms are generated from the local oscillator. Once again, phase and frequency stability are essential for a good performance.

8.1.4.7 RF Modulator

This stage modulates the baseband signal with the carrier.

8.1.4.8 Combiner

After the baseband signal is modulated, I and Q waveforms are combined in this stage.

8.1.4.9 Band-Pass Filter

Aliases of the baseband created during the DAC are filtered at this stage.

8.1.4.10 Amplifier

A pre-amplifier is then used to bring the signal to the specified level. The output power level is usually adjusted by this amplifier.

8.1.4.11 High Power Amplifier

The HPA is generally a constant gain amplifier. WiMAX HPAs operate in between 20 and 40 dBm. The maximum transmitted power allowed is regulated by local administrations in each country.

8.1.4.12 Low-Pass Filter

Finally, again a low pass filter removes any spurious signals coming from the HPA.

8.2 Receive Side

The received signal is initially demodulated and digitized. All the processing from there onward is digital. The receive side is divided into four blocks: carrier demodulation, digital IF processing, symbol processing and bit processing.

8.2.1 Carrier Demodulation

The eight stages of carrier demodulation (band-pass filter, low noise amplifier, phase locked loop, local oscillator, demodulator, low-pass filter, amplifier, and digital-to-analog converter) are described next.

8.2.1.1 Band-Pass Filter (BF)

The signal received from the antenna is passed though a band-pass filter to remove any out-of-band signal that may saturate the LNA or generate interference.

8.2.1.2 Low Noise Amplifier (LNA)

A low noise amplifier is a special amplifier that has a very low intrinsic noise, usually of the order of few dB. This amplification allows the remaining components to work with a higher level signal, and consequently is less susceptible to the components' intrinsic (thermal) noise.

8.2.1.3 Phase Locked Loop (PLL)

The circuit locks to the symbol transitions and consequently into the transmitter clock.

8.2.1.4 Local Oscillator

The local oscillator generates the carrier frequencies in phase and in quadrature.

8.2.1.5 Demodulator

The demodulator combines the received signal with the carrier to obtain the IF or base band signal, depending on the architecture.

8.2.1.6 Low-Pass Filter

This filter separates the baseband signal.

8.2.1.7 Amplifier

Additional amplification and gain adjustment are done before the signal is sent to the next stage.

8.2.1.8 Analog to Digital Converter (ADC)

The received analog signal is digitized, by sampling and quantizing. Figure 8.4 shows the OFDM sequence.

8.2.2 Digital IF Processing

It is composed of one main stage: digital down-converter.

8.2.2.1 Digital Down-Converter (DDC)

The digitized signal is decimated (re-sampled or interpolated) to obtain the baseband signal. Additional samples may be generated, through interpolation, to separate aliases from the baseband signal and help the filter action. A FIR (Finite Impulse Response) filter is used to separate the baseband.

8.2.3 Symbol Processing

Symbol processing is divided into eight stages: replace cyclic prefix, fast Fourier transform, pilot extraction, OFDMA ranging, channel estimation, equalization, de-sub-channelization, symbol de-mapping; each of the stages is described next.

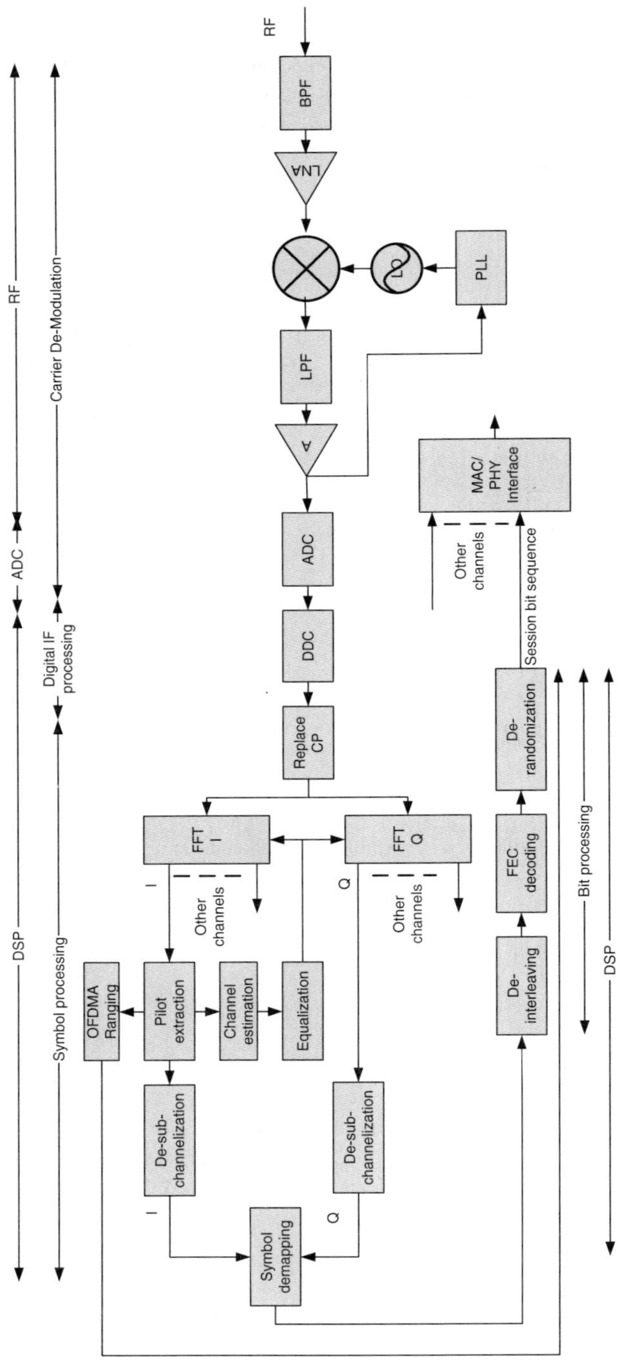

Figure 8.4 OFDM receive block diagram.

8.2.3.1 Replace Cyclic Prefix

The samples from the cyclic prefix replace the initial samples, which may be compromised by multipath.

8.2.3.2 Fast Fourier Transform

The received signal is composed of symbols modulated on orthogonal frequencies. The existence of each frequency can be detected through self-correlation.

The practical implementation of the FFT is illustrated in Figures 8.5, 8.6 and 8.7. Four symbols (10,01,10,11) are modulating subcarriers f1 (1 Hz), f2 (2 Hz), f3 (3 Hz) and f5 (5 Hz). Carrier f4 (4 Hz) is not used.

Figures 8.5 and 8.6 show I and Q waveforms. Figure 8.7 shows the combined I+Q waveform, which is the input to the FFT. The waveforms are shown over two cycles, just to show that the

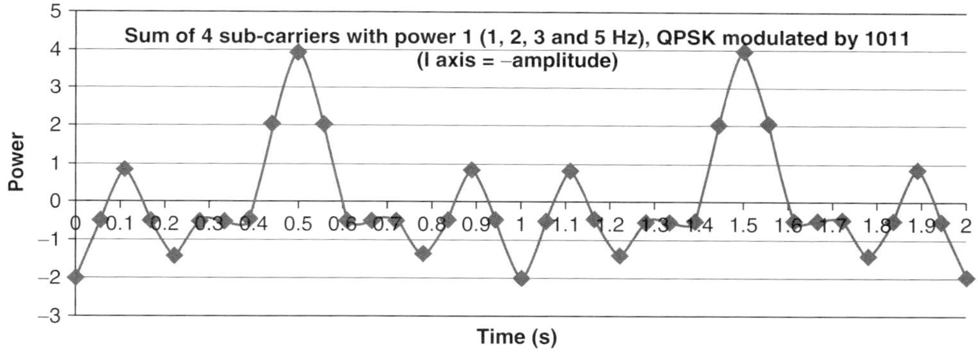

Figure 8.5 Sum of I sub-carriers.

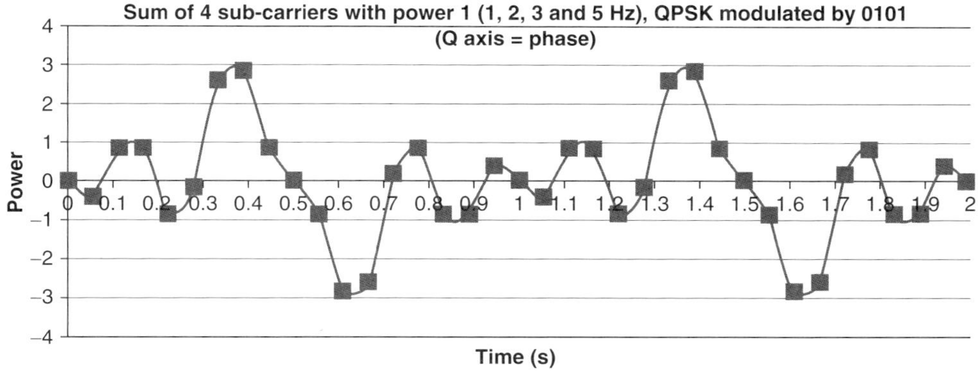

Figure 8.6 Sum of Q sub-carriers.

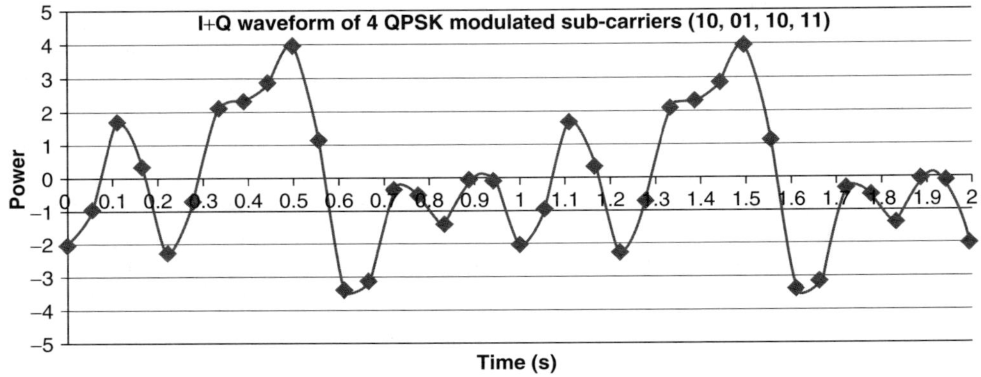

Figure 8.7 Sum of I+Q sub-carriers.

Table 8.2 Sum of I and Q sub-carriers

FFT	f_1	f_2	f_3	f_4	f_5
I = sum((I+Q)*cos(f_n*t))	−9	9	−9	5E-15	−9
Q = sum((I+Q)*sin(f_n*t))	9	−9	9	0.00	−9

combination of the multiple frequencies is cyclical. In practice, the symbol duration is equal to one cycle of the lowest frequency.

The function of the FFT is to find out which frequencies are present in the waveform. This is done by multiplying the samples of the combined waveform (I+Q) with the corresponding sine and cosine values of each frequency at that instant in time, and adding them over one complete cycle. This calculates the cross-correlation between different frequencies and the auto-correlation between the same frequencies.

The sub-carrier frequencies are multiples of the lowest frequency and are orthogonal, so the cross-correlation is zero, the same applies to I and Q signals, which are orthogonal also. The auto-correlation results in a positive or negative sum, depending on the original signal.

Table 8.2 gives the sums obtained for the different sub-carriers in the example above. As can be seen, I gives 1,0,1,1. Let's remember that the sub-carrier four is not used in this sub-channel. Q gives, 0,1,0,1. So, the FFT decoded bit stream is 10011011.

8.2.3.3 Pilot Extraction

The above example illustrated QPSK, but for higher modulations, there is a need to distinguish between four amplitude levels. Besides, the received signal can be contaminated by noise, signals from other sites and distortion. It is necessary to calibrate the thresholds corresponding to the amplitudes. This is achieved through the use of pilots.

Pilots are pre-defined signals that modulate pre-defined sub-carriers. In WiMAX, pilot sub-carriers use the most robust modulation BPSK 1/2 and their power level is 3 dB higher than regular data sub-carriers. This means that information is sent only on the I waveform.

8.2.3.4 OFDMA Ranging

Pilots are used to verify and adjust the timing of received signals. Pilot detection is periodically done within a three symbols' sliding window and information is sent to the mobile to adjust its transmission timing, so reception is maximized for the middle symbol.

8.2.3.5 Channel Estimation

Pilots are strategically distributed in frequency and time, so they can be used to dynamically estimate the channel response. This is used for equalization and also by antenna system algorithms.

8.2.3.6 Equalization

Channel variation can be adjusted in time and frequency using pilot detection information. This significantly improves data symbol detection.

8.2.3.7 De-Sub-Channelization

Once the channel is adjusted, the FFT information becomes more reliable and it is possible to detect the presence and amplitude of the signals transmitted.

8.2.3.8 Symbol De-Mapping

The information extracted by the FFT is then used to un-map the symbols of each sub-channel, obtaining the original bit stream as a result, sometimes with some errors.

8.2.4 Bit Processing Stages

Bit processing is divided into four stages: de-interleaving, FEC encoding, de-randomization, and MAC/PHY interface; each stage is described next.

8.2.4.1 De-Interleaving

This process restores the symbols' order, so they can be sent to error correction.

8.2.4.2 FEC Encoding

The error correction code verifies the most probable symbol that was sent, based on previous and subsequent symbols. In this process puncturing is reversed and the bits that were eliminated are replaced.

8.2.4.3 De-Randomization

The decoded bits are de-randomized using the same sequence used in the randomization process.

8.2.4.4 MAC/PHY Interface

This is the interface that sends bits from the PHY layer to the MAC layer for further processing.

9

Wireless Communications Network (WCN)

9.1 Introduction

A wireless communications network has to integrate with other networks, mainly the Internet and PSTN (Public Switched Telephone Network). This network can be divided into two parts: wireless access network and core network.

Figure 9.1 shows a block diagram with the main components of a WCN. The wireless part comprises what is called the wireless access network, while the remaining part comprises what is called the core network.

9.2 Wireless Access Network

Wireless constitutes only the access to a much larger communication network. Wireless access is performed by fixed, nomadic or mobile subscriber stations (SS) and Radio Base Stations (RBS). These two elements concentrate all the wireless access up to layer 2 and are specified by IEEE for WLAN (802.11), WiMAX (802.16) and by 3GPP for LTE.

Contrary to common belief, the wireless access does not implement traffic control, as information is always sent at the maximum possible throughput for each connection. This way, the limited RF resources are used to its maximum capacity. Traffic control, characterized by limiting users' usage to their subscription plan, is done in the core network.

9.2.1 Subscriber Wireless Stations (SWS)

The nomenclature used to identify Subscriber Wireless Stations (SWSs) varies by application and technology. Some common names are: SS (Subscriber Station), MS (Mobile Station), and CPE (Customer Premise Equipment). Sometimes they are referred to simply as terminals. SWSs can be fixed, nomadic or mobile. They can be installed on a rooftop, desktop, or be portable. They can be used by a single user or multiple users. They can use single or multiple antennas. SWS antennas can be omni or directional.

The SWS functionality is implemented in specialized chipsets, based on DSPs and manufactured by a few specialized companies, in lieu of the traditional semiconductor vendors. This handful of

LTE, WiMAX and WLAN Network Design, Optimization and Performance Analysis, First Edition.
Leonhard Korowajczuk.

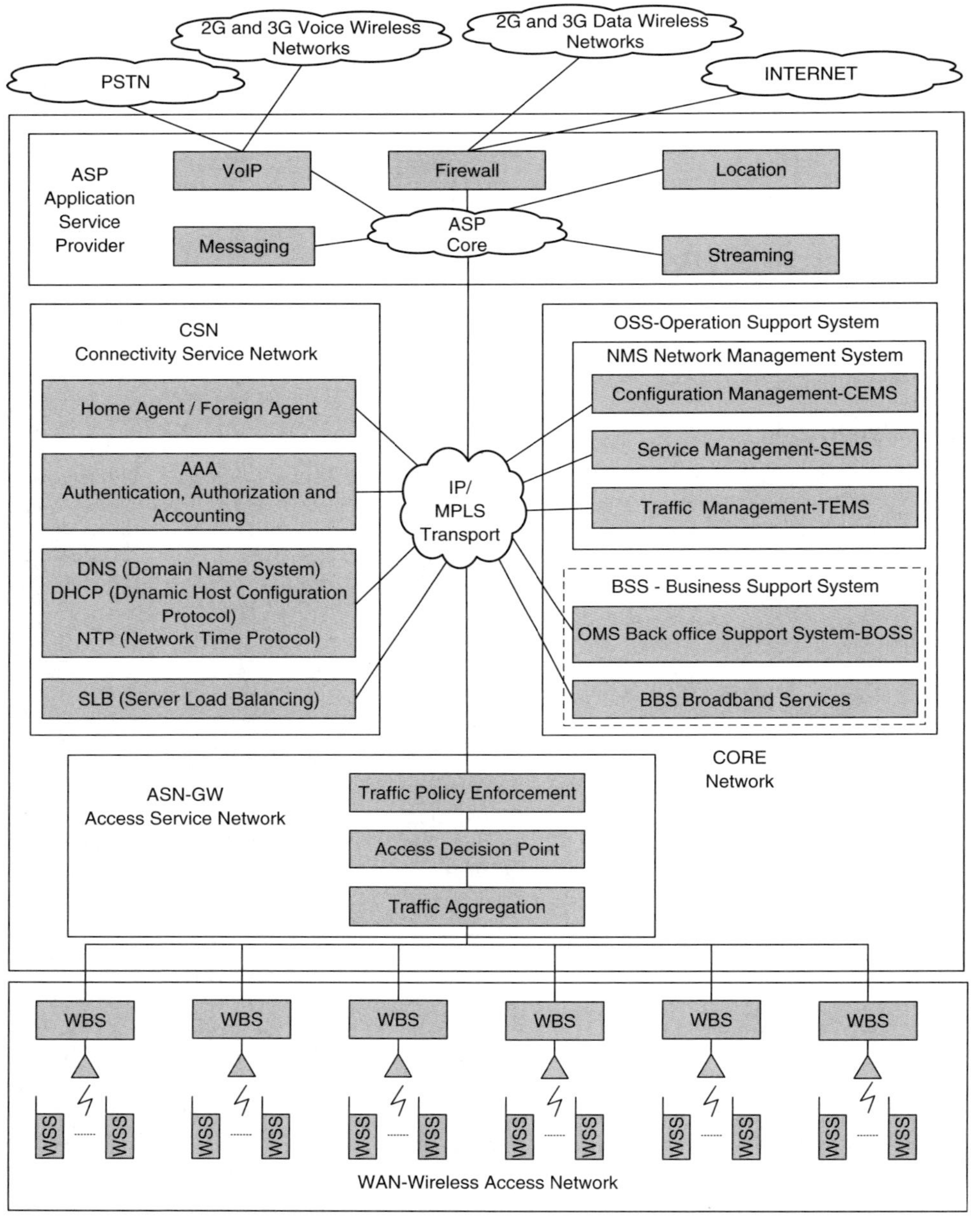

Figure 9.1 Wireless communication network.

companies developed the semiconductors and software required to make them operational. Integrators then use these chipsets and software to create their own products. Some of the chipset vendors are:

- WLAN: Atheros.
- WiMAX: Arraycom, Beceem, Fujitsu, Picochip, Runcom and Sequans.
- LTE: Altair, BitWave, Comsys, Infeon, Qualcomm, Samsung, ST-Ericsson, Wavesat.

Examples of integrators are: Intel, Ericsson, Motorola, Lucent and phone manufacturers.

9.2.2 Wireless Base Stations (WBS)

Wireless base stations provide a wireless service footprint, through a combination of antenna height, antenna pattern, transmitted power and received interference. WBSs are known by many names, which vary by technology. Some of the names are: BTS (Base Terminal Station), BS (Base Station), NB (node B), and eNB (Evolved Node B).

A WBS can have one or more radiation directions, each one forming a sector. Sectors can have one or more radios, each using one RF channel.

A radiation footprint is known as cell, and this applies to the footprint of radios from a sector or to the footprint of the whole WBS. This leads to some confusion, but today the trend is to call the sector footprint a cell.

A WBS footprint can be classified according to its size: macro, micro, pico and femto. These classifications came about with the traffic increase and the need to reduce cell size to cope with it. The WBS of different footprints can be overlaid, so several femto, pico and micro cells can be inside the footprint area of a macro cell. There are no rigid limits to classify cells, but femto cells are generally room wide, pico cells would cover street segments, micro cells cover street intersections, while macro cells cover from one block to very large areas

A WBS can be a standalone or be part of a large wireless array that covers a whole metropolitan area. In the first case it may be more economical to include core functionalities in the WBS, whereas in the second case it is certainly more economical to concentrate core functions in a separate element.

WLAN (802.11) implementations are extremely simple and include little routing functionality. WiMAX (802.16 WiMAX Forum extension) allows for three different integration levels (centralized, distributed and mixed ASN Access Service Network), while LTE strictly defines the WBS (eNB) functionality, without giving much flexibility.

9.3 Core Network

This part of the network manages the wireless communications and interconnects to other networks. It is loosely specified by entities, such as the WiMAX Forum and 3GPP, but a lot of flexibility is left for vendor implementation.

The core network hardware is made of servers and routers that run dedicated and commercial software. Specific functions can be performed by several servers to provide the required processing capacity, implying that processing load control and balancing should be implemented.

Core network tasks can be divided into the following services: Access Service Network (ASN), Connectivity Service, Application Service, and Operational Service.

9.3.1 Access Service Network (ASN)

This service controls the wireless access network and provides the interface between this and the other services. It can be divided into three main parts: traffic aggregation, wireless access decision point, and traffic policy enforcement. The implementation of each part can be split into many different functions and can be packaged into different profiles as described in Section 9.2.1.4.

9.3.1.1 Traffic Aggregation

WBS traffic is routed to an aggregation point, where it is distributed to the core network. Regular switches and routers are used for this task.

9.3.1.2 Wireless Access Decision Point

The decision point implements functions such as: Radio Resource Management (RRM), Scheduling, and Mobility Management (MM), including paging and wireless QoS. The wireless QoS is obtained from determinations done at the Traffic Policy Enforcement level and used to properly perform resource allocation and scheduling.

Mobility functions require interactions with other WBSs, which is done by sending control packets directly to the WBSs (if available) or through the core network.

9.3.1.3 Traffic Policy Enforcement

The traffic policy implements Service Flow Management (SFM), Traffic Shaping (also known as Bandwidth Management), IP Routing and IP QoS.

Internet QoS is specified in 802.1q and 802.1p, but they are usually not implemented by applications, as an approach of over-dimensioning its capacity is assumed in wireline networks. This over-dimensioning cannot be done in wireless and the QoS has to be implemented by the ASN at IP level.

The first Internet QoS implementations used an approach of Integrated Services (IntServ) in which the RSVP (Resource Reservation Protocol) is used to pre-allocate network resources. This approach proved to be inefficient and unreliable, as many reserved connections were poorly used and some were never released.

In the second Internet QoS approach, Differentiated Services (DiffServ), service types get different priorities and are sent to different queues. This queuing mechanism can be used to control user traffic throughput (tonnage) in the short and long term. Two algorithms can be applied.

- *Token bucket*: Virtual tokens are periodically added to each queue, each token corresponding to a certain tonnage. A packet is only sent if there are enough tokens to transmit it. This procedure controls traffic bursts, whereas the periodicity and the amount of token added control the long-term tonnage.
- *Leaky bucket*: A virtual counter is incremented each time a packet is sent and is decremented over time. When the counter exceeds a specified limit, the packet is dropped. The counter increment and the decrement timing can also be used to control long-term tonnage.

Internet applications usually do not identify the IP packet CoS (Class of Service), although the ToS (Type of Service) field exists in the IP header for this finality.

Figure 9.2 shows the IP packet format. The Traffic Policy Layer uses the ToS (Type of Service) field to generate QoS information for the Access Decision Point, which can then appropriately schedule the packet transmission. Packets sent from subscriber stations should come with this information, if not, it will have to be added to be used so the Internet network can route the packets.

The QoS information is mainly derived from the packet protocol, which usually is a good indication of QoS requirements. In Peer to Peer (P2P) applications, this is difficult to do, as the protocol

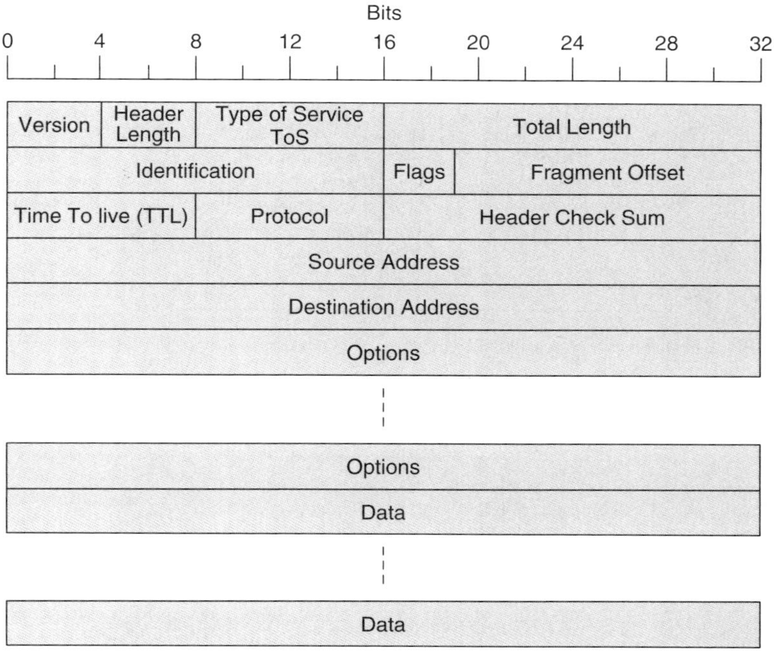

Figure 9.2 IP packet format.

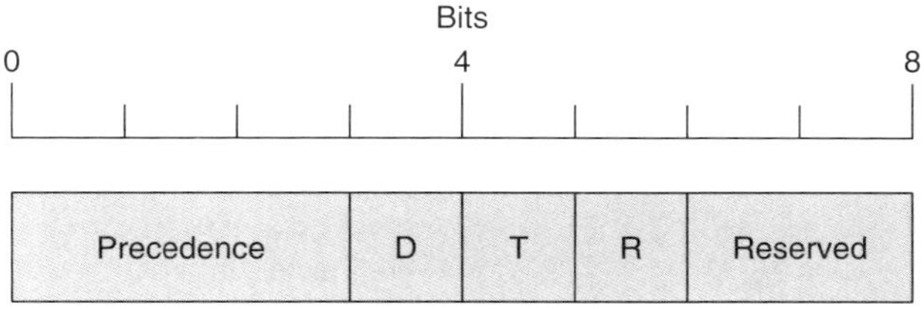

Figure 9.3 IP ToS (Type of Service).

information is buried inside the packet. P2P communications require additional effort to be properly characterized, typical examples are torrents sent between two users.

Figure 9.3 shows the ToS (Type of Service) field in detail. The precedence field defines the priority; D defines delay; T, throughput; and R, reliability. Zero indicates a normal value, whereas "one" indicates an enhanced value, although the specific values themselves are not specified. Table 9.1 defines the Service Priority field values.

The Protocol field is the one used to identify the type of message protocol used. Table 9.2 gives the values used for some of the most common ones.

The 802.1p specification adds a CoS (Class of Service) field in the Ethernet MAC. Although this specification was never published, the three-bit field was included in the 802.1q (VLAN tagging)

Table 9.1 Type of Service priority field

Precedence value	Priority level
000	Routine
001	Priority Level
010	Immediate
011	Flash
100	Flash Override
101	Critical
110	Internetwork Control
111	Network Control

Table 9.2 Protocol types

Protocol	Value
ICMP	1
IGMP	2
TCP	6
UDP	17

Table 9.3 User priority in 802.1q (Ethernet MAC)

User priority	Traffic type
000	Best Effort
001	Background
010	Excellent Effort
011	Critical Applications
100	Video <100 ms latency
101	Voice <10 ms latency
110	Internetwork Control
111	Network Control

specification. This field is called Priority Code Point (PCP) and specifies user packet priority as shown in Table 9.3.

9.3.1.4 ASN Profiles

The functions described above can be packaged in many ways, each technology with its own partitioning. Vendors implement technologies differently, with the implementation even varying from one software release to the next.

To implement its core elements, WLAN uses modules already available for wireline Internet networks.

The WiMAX Forum specifies three ASN implementation profiles:

- Profile A has the ASN functionality is performed by an ASN-GW.
  - Handover control is done in the ASN-GW.
  - Radio Resource Control is done in the ASN-GW, but Radio resource management is done in the WBS.
  - Mobility management is ASN centered.
- Profile B incorporates the ASN-GW inside the WBS.
- Profile C has the ASN functionality mapped between the ASN-GW and the WBS.
  - Handover control is done in the WBS.
  - Radio Resource Control and Radio Resource Management are done in the WBS, but messages between WBSs are sent through the ASN-GW.
  - Mobility management is ASN centered.

WiMAX implementations of the ASN function are described in Chapter 13. The remaining elements are implemented into servers using regular Internet software.

LTE follows the 3GPP guidelines which divide the ASN functionality between:

- PDN (Packet Data Network) which implements the Traffic Policy Enforcement.
- S-GW (Serving Gateway) which implements the Wireless Access Decision Point.
- MME (Mobility Management Entity) which implements the mobility functions of the Wireless Access Decision Point, including HA (Home Agent) and FA (Foreign Agent) functionalities described next in the Connectivity Service.

LTE implements the Access Decision Point (ADP) functionality into the S-GW (Service–Gateway) and the mobility part into the MME (Mobility Management Entity), both described in Chapter 14. Regular software modules used in Internet implementations are also used to implement the remaining parts of the core network.

Vendors are constantly upgrading their products and adding new features. Usually, solutions from different vendors are used within the same network, as each one specializes in parts of the total solution. Due to a large overlap that may exist between solutions, the WAN (Wireless Access Network) vendor is usually the one in charge of integrating all solutions.

9.3.2 Connectivity Service

This service provides Internet connectivity to the different network elements and establishes and tracks subscriber connectivity.

It comprises the following functionalities:

- *Home Agent*: stores information about home subscribers.
- *Foreign Agent*: stores information about roaming subscribers.
- *AAA*: stores information about Authentication, Authorization and Accounting.
- *DNS*: translates the Domain Name System.
- *DHCP*: provides temporary IPs through the Dynamic Host Configuration Protocol.
- *NTP*: provides the network time through the Network Time Protocol.
- *SLB*: assesses Service Load Balancing for WBS and Servers.

9.3.3 Application Service

This service provides application-specific functions, including:

- *VoIP*: translates VoIP packets into circuit-switched signals and vice versa.
- *Streaming*: Buffers streaming information for timely delivery.
- *Messaging*: Stores short messages for delivery and retrieval.
- *Firewall*: provides interconnections to the Internet through a firewall.

9.3.4 Operational Service

This service is the operator's interface to the network, and provides configuration, operation and maintenance interfaces.

The OSS/BSS- Operation Support System/Business Support System performs all activities required to manage the network, from customer maintenance, to network maintenance/performance and billing.

Telecom management elements are defined in ITU M.3000 and are mapped to their main functionalities in Figure 9.4.

The OSS performs the NMS (Network Management System) function, which is divided into Service, Configuration and Traffic Management.

- *Service management*: defines the service characteristics (data rate, QoS ...) to be offered for each customer group. The actual function is implemented in routers and other devices, but the operator interface is done here.

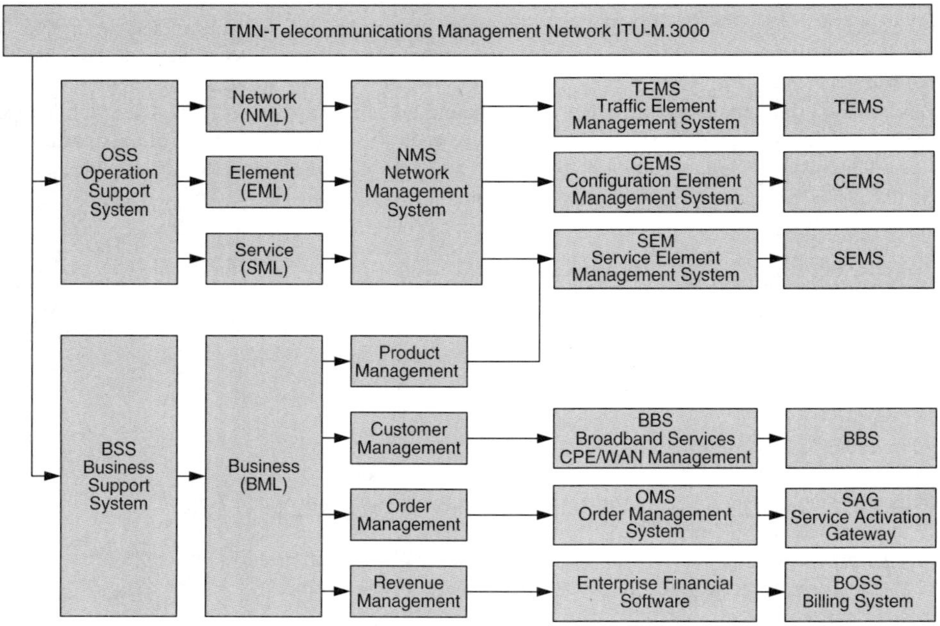

Figure 9.4 Network management components.

- *Configuration management*: defines all network elements and tracks their performance.
- *Traffic management*: gathers network traffic data and prepares statistics.

The BSS (Business Support System) function is divided into Product, Customer, Order and Revenue Management

- *Product management*: product is understood in this context as the service offered to the customers.
- *Customer management*: here all customer data and features are stored, some software packages offer to customers, access to this data.
- *Order management*: includes troubleshoot tickets and other controllable activities. Some vendors include real time network and CPE monitoring.
- *Revenue management*: includes the actual billing system, including issuing bills.

The nomenclature used in the TMN is fluid and varies from vendor to vendor, as products offered may implement only parts of the required functionality. Figure 9.4 represents only one possible naming arrangement.

10

Antenna and Advanced Antenna Systems

10.1 Introduction

Antennas are mechanical elements that transform varying electrical signals into an electro-magnetic field that transfers the signal energy into the space surrounding the antenna, which propagates away from the antenna. Antennas also capture the energy from surrounding electro-magnetic fields and transform it into varying electrical signals. The former perform a signal transmit function and the latter a signal receive function.

The transmitted electro-magnetic field, also known as an RF (radio frequency) wave, propagates in many directions and is reflected, refracted and diffracted by obstacles in its path. This splits the original energy into many components that travel through different paths. A receive antenna receives several of those paths, and its sum creates a unique signal. The received signal is subject to fading as the different wave components either add constructively or cancel each other.

Two receive antennas receive different wave components and the resulting received signals are not exactly the same. This can be explored to enhance the extraction of the transmitted information, by analyzing the signals received from two or more antennas.

The similarity between two received signals is defined by its cross-correlation, which defines how different the signals are from one another. This cross-correlation is similar to the orthogonality test explained in Chapter 4, and can be calculated by a sliding dot product.

A high cross-correlation indicates that the signals are very similar, and were subject to similar fading. A low cross-correlation indicates that the signals are different, and did not suffer similar fading. Receive antennas are said to be correlated if they receive similar signals and uncorrelated if the signals are significantly different.

Antenna correlation can be minimized by spatial or angular separation. In the first case, antennas are spatially separated, with a 10 wavelength separation used as rule of thumb. In the second case, antennas are pointed in slightly different directions, with 30° used as a rule of thumb. There are many factors that can affect correlation in real life, as the existence of line-of-sight and alternative paths, and implementation constraints further reduce the possibility of getting uncorrelated signals from antennas. This difficulty increases even further with the increase in the number of receive antennas. It is hard for a designer to estimate antenna correlation, so field measurements are highly recommended. Many throughput claims of the new technologies are based on unrealistic correlations.

LTE, WiMAX and WLAN Network Design, Optimization and Performance Analysis, First Edition.
Leonhard Korowajczuk.

10.2 Antenna Basics

Antennas play an important role in a wireless connection, and lately this role has been extended by digitally processing received signals from multiple antennas. Besides the traditional use of antennas, new digital techniques allow us to explore the fact that different antennas receive signals from different paths and thus can extract different information from the received signals. If the signals received by the antennas are not correlated, the information from one antenna can complement the other. There are several ways that the information can be extracted and processed, each one better applicable to a specific environment. Broadband technologies are designed to allow the use of these new technologies to increase spectrum efficiency at a relatively modest increase in cost.

Antennas interface the electrical signal with the irradiated signal. An isotropic antenna generates the signal equally in all directions of space so the total available power is equally distributed and each direction receives a small fraction of it. Antennas can be built to concentrate the available power in some directions while sending very little to other directions. An antenna that concentrates the power on a specific plane (generally parallel to ground) is called an omni antenna, whereas if it concentrates the power on a part of this plane, it is called a directional antenna. The direction (angle in relation to the true north) to which the power is concentrated is called the azimuth. There can also be an inclination in relation to the horizontal plane, known as the antenna tilt, which is expressed by the angle in relation to the horizontal plane; if this angle is towards the ground, it is called a downtilt, otherwise it is an uptilt.

Antennas are devices that interface the radio equipment with the RF channel, by transforming electrical energy into RF energy, and vice versa, as illustrated in Figure 10.1.

Isotropic antennas are theoretical antennas that irradiate energy outwards equally in all directions. This energy is spread over the surface of an ever expanding sphere, centered on the antenna that expands at the speed of light. The energy density drops with the increase of the sphere area, which is given by Equation (10.1), where d is the distance from the antenna.

$$A = 4\pi d^2 \qquad \text{(10.1) Area of a sphere}$$

The power density is expressed by Equation (10.2) in W/m^2, where, P_t is the transmitted power.

$$\text{Power} = P_t/(4\pi d^2) \qquad \text{(10.2) Power density of an isotropic antenna}$$

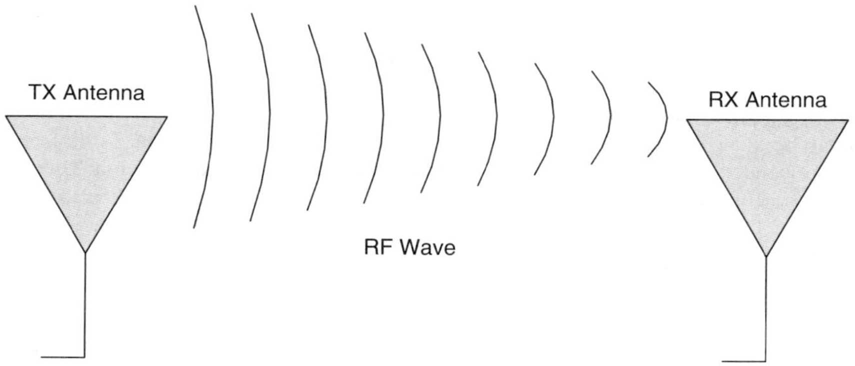

Figure 10.1 RF energy transmission.

An isotropic antenna is defined as having a unitary gain and its effective aperture is calculated for a distance of $\lambda/4\pi$. The isotropic antenna area is then given by Equation (10.3).

$$A_i = \frac{\lambda^2}{4\pi} \qquad \text{(10.3) Area of an isotropic antenna at } \lambda/4\pi$$

Isotropic antennas are theoretical as one cannot build an antenna that radiates equally in all directions. A real antenna concentrates the irradiated power into a smaller region of space, and this area is called the antenna's effective aperture, A_e. The ratio of the effective aperture of an antenna and the area of an isotropic antenna defines the antenna gain, defined in Equation (10.4).

$$G = A_e/A_i = \frac{4\pi A_e}{\lambda^2} \qquad \text{(10.4) Antenna gain}$$

The transmitted power density is then defined by Equation (10.5).

$$\text{Power density} = P_t G_t/(4\pi d^2) \qquad \text{(10.5) Power density of a directional antenna}$$

The receive antenna captures the energy on its effective area A_r, as defined in Equation (10.6).

$$P_r = \frac{P_t G_t}{4\pi d^2} A_r = \frac{P_t G_t}{4\pi d^2} \frac{G_r \lambda^2}{4\pi} \qquad \text{(10.6) Received power by an antenna}$$

This leads us to the Friis Transmission equation (Harald T. Friis, 1893–1976), expressed in Equation (10.7) with values expressed in watt/meter.

$$P_r = P_t G_t G_r (\lambda/4\pi d)^2 \qquad \text{(10.7) Friis Transmission equation}$$

where G_t and G_r are respectively transmit and receive antenna gains above the isotropic antenna gain. The same equation is expressed in dB in Equation (10.8).

$$\begin{aligned} P_r = {} & 10\log(P_t) + 10\log(G_t) + 10\log(G_r) + 20\log(\lambda) \\ & - 21.984 - 20\log(d) \qquad \text{(10.8) Friis Transmission equation (dB)} \end{aligned}$$

Reducing the antenna effective area increases antenna gain, as the power is applied over a smaller area. Antennas have the same gain for transmission and reception.

10.3 Antenna Radiation

Antennas radiate through electric (E-field) and magnetic (H-field) fields. An electric field is defined by the force exerted on a unit charge (1 Coulomb) and is expressed in Newton/Coulomb, which is equivalent to V/m. An electric field points out from a positive charge and into a negative charge, as illustrated in Figure 10.2.

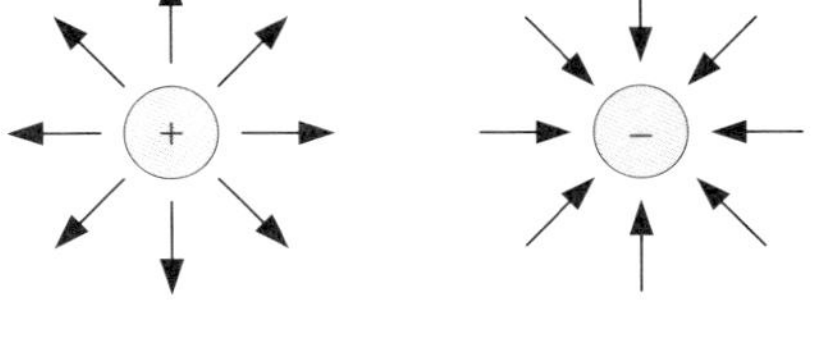

Figure 10.2 Electric field.

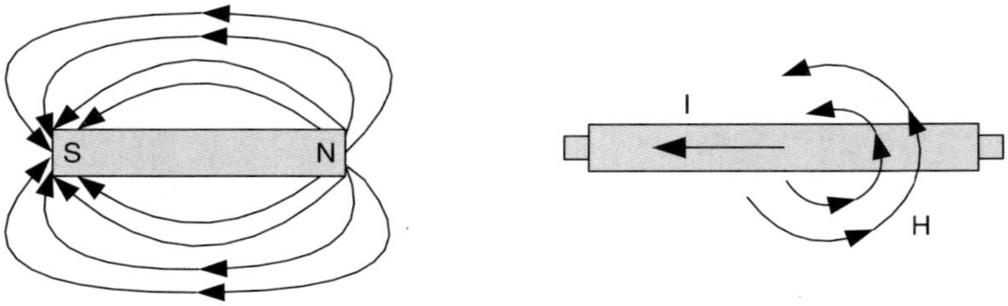

Figure 10.3 Magnetic field.

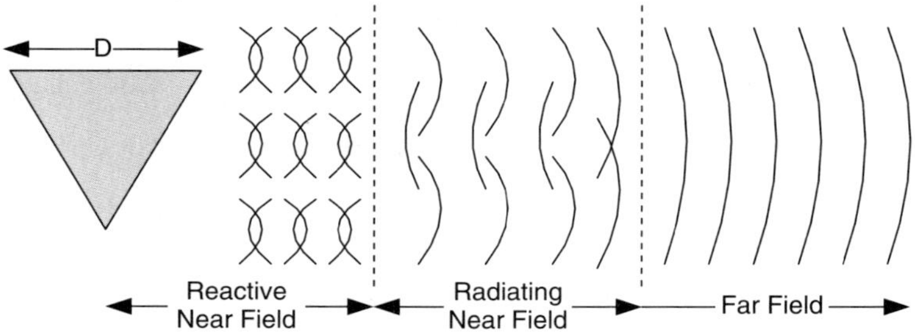

Figure 10.4 Antenna radiation fields.

A magnetic field is a vector that flows between magnetic dipoles and is expressed in Ampere/m. It flows from the positive (North) side to the negative (South) side of a dipole. The magnetic field wraps around a wire in which current is flowing (right-hand thumb rule) and is orthogonal to the electric field, as illustrated in Figure 10.3. It is the interaction of the E-field with the H-field in space that allows for RF wave propagation.

Antenna radiation can be divided into three regions, called fields, as illustrated in Figure 10.4.

10.3.1 Reactive Near Field (Reactive Region)

This is the field in the immediate vicinity of the antenna, where the field is not fully organized yet. It is a reactive field as the E-field and the H-field are 90° out of phase to each other. The boundary of this region is defined by Equation (10.9), where D is the physical antenna size, d is the distance from the antenna and λ is the wavelength.

$$d = 0.62\sqrt{\frac{D^2}{\lambda}} \qquad \text{(10.9) Reactive near field}$$

10.3.2 Radiating Near Field (Fresnel Region)

The reactive field starts to subside and the radiating field starts to emerge in this region. The size of the region is defined by Equation (10.10). Depending on D and λ this region may not exist.

$$0.62\sqrt{\frac{D^3}{\lambda}} < d < \frac{2D^2}{\lambda} \qquad \text{(10.10) Radiating near field}$$

10.3.3 *Far Field (Fraunhofer Region)*

In this region the radiating field is established with the E-field and H-field orthogonal and in-phase with each other. This field starts at a distance defined by Equation (10.11).

$$d > \frac{2D^2}{\lambda} \quad \text{(10.11) Far field}$$

10.4 Antenna Types

The isotropic antenna is a theoretical concept, although there are practical implementations that use six dipoles to approximate the isotropic characteristic. These antennas are used for measurements only. Some practical antennas are described next.

10.4.1 *Dipole (Half Wave Dipole)*

The simplest and most common practical directional antenna configuration is the dipole antenna developed by Heinrich Rudolph Hertz (1857–1894) in 1886. It consists of two center-fed elements, as illustrated in Figure 10.5.

Dipole antenna fields are presented in Figure 10.6. The horizontal lines represent the electrical field, while the vertical circles show the magnetic field.

The dipole gain in relation to the theoretical isotropic antenna is shown in Table 10.1. The most common configuration is the $^1\!/_2$ wavelength dipole.

The dipole input impedance varies with the total length as shown in Figure 10.7. The free space impedance is 376.7Ω.

An isotropic antenna radiates energy equally in all directions, but it is physically impossible to build, as the antenna would need to be a point floating in space radiating energy. As this is impractical, the closest real antenna to that is the dipole antenna.

The increase in power due to the concentration of energy by the antenna into a single beam is specified as a ratio in relation to the power obtained with an isotropic antenna, in which case this ratio is referred to as dBi. This ratio can be also expressed in dBd, that is, the ratio considers a dipole antenna instead of the isotropic one. A gain expressed in dBd is 2.15 lower than when it is expressed in dBi, although both express the same signal level in dBm. Designers should verify the unit required by the planning tool, to express the antenna gain correctly.

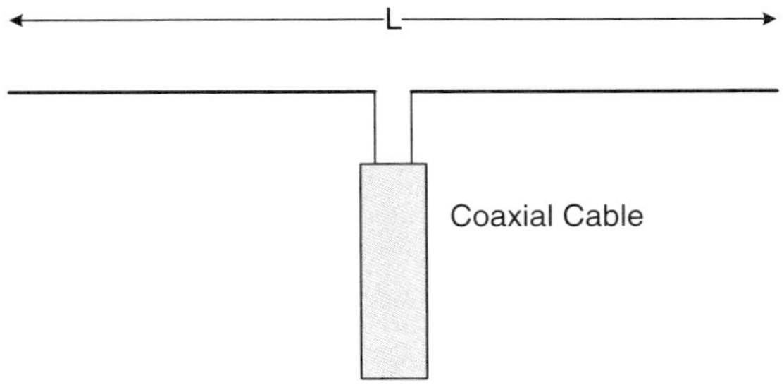

Figure 10.5 Dipole antenna.

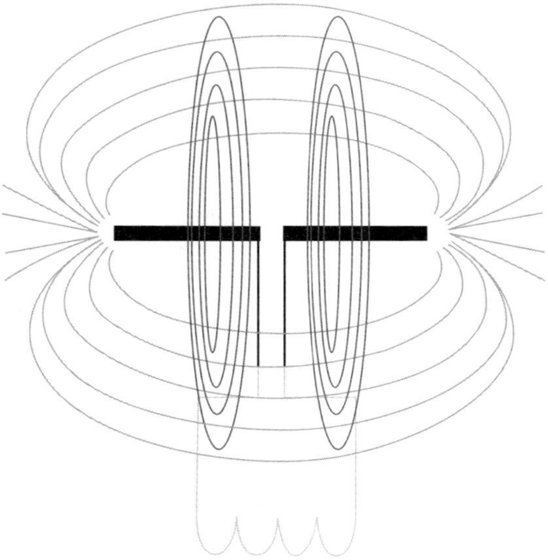

Figure 10.6 Dipole antenna fields.

Table 10.1 Isotropic antenna dipole gain

Dipole length (L) (wavelengths)	Linear gain	Gain (dB)
$\ll 0.5$	1.5	1.76
0.5	1.64	2.15
1	1.8	2.55
1.5	2	3.01
2	2.3	3.62
3	2.8	4.47
4	3.5	5.44
8	7.1	8.51

10.4.2 Quarter Wave Antenna (Whip)

The quarter wave antenna has similar properties to the dipole antenna, but because it only has one of the elements of the dipole, it requires a ground plane to provide a return path for the electric field. The earth can be used as a ground plane for the return of the electric field, but in some cases it has poor conductivity or is distant from the antenna, so an artificial ground plane has to be provided. This ground plane can be any metallic surface (as a vehicle roof), or radial wires placed at the antenna base. The diameter of this ground plane should be at least $\lambda/2$. This antenna is illustrated in Figure 10.8.

10.4.3 Omni Antenna

An omni antenna has a uniform pattern in one of the planes and a directional pattern in the other, plane, as illustrated in Figure 10.9. Antenna patterns are explained in detail in Section 10.5.2.

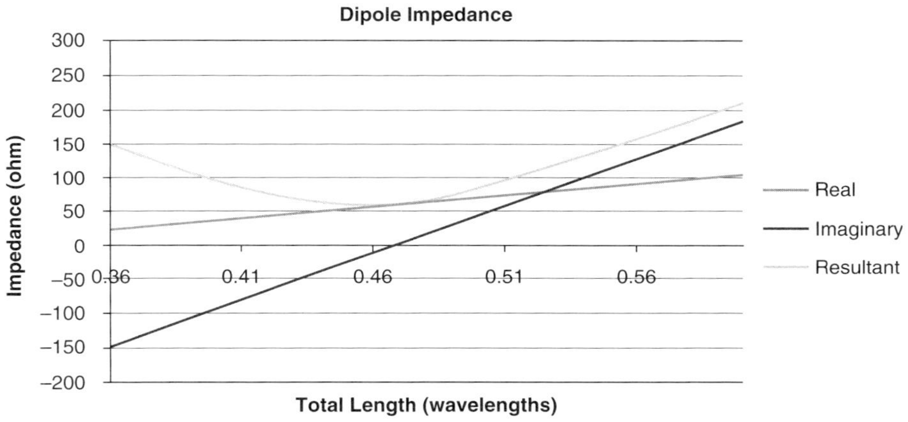

Figure 10.7 Dipole input impedance.

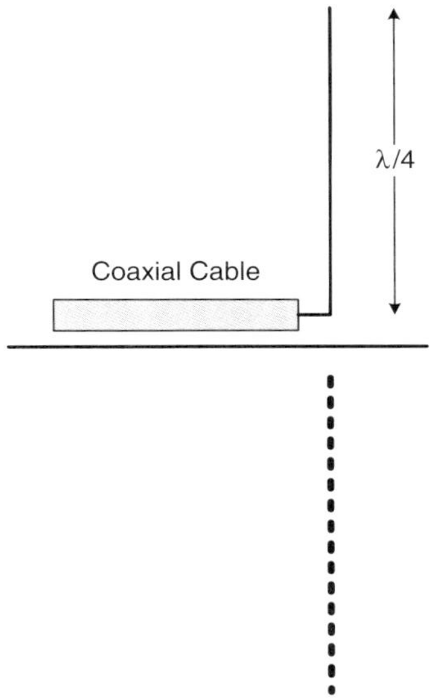

Figure 10.8 Whip antenna.

Antenna patterns present the antenna characteristics only in vertical and horizontal planes, but the antenna pattern is three-dimensional. Specialized algorithms should be used by propagation tools to emulate the antenna pattern in 3D, as illustrated in Figure 10.10.

10.4.4 Parabolic Antenna

The gain of a dipole can be increased by reflecting energy towards it. This is done using a metallic parabolic reflector that concentrates the energy on a dipole located at the focal point of the reflector.

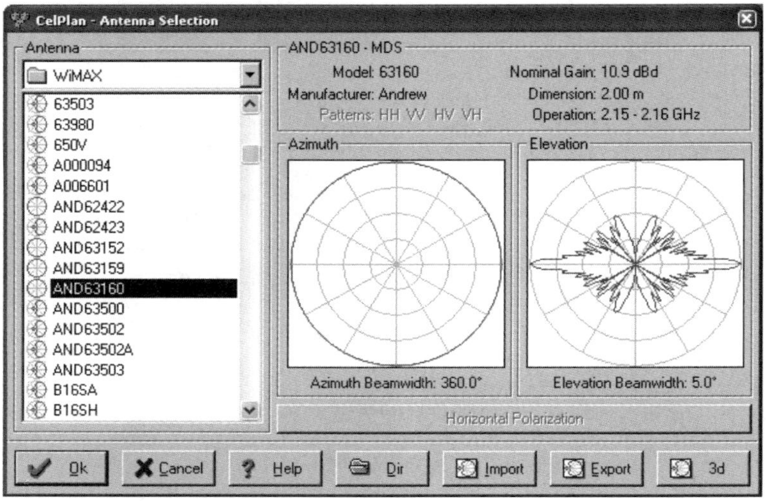

Figure 10.9 Omni antenna sample.

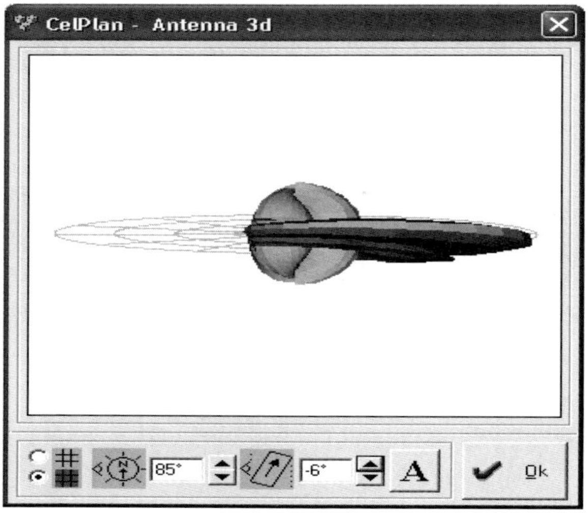

Figure 10.10 3D representation of a directional antenna.

The structure that captures the energy is called the antenna feed. Parabolic reflectors can be cylindrical (having the parabolic form only in one plane) or a dish (which have a paraboloid format). The parabolic antenna with an axial feed is illustrated in Figure 10.11.

A second reflector can be added to the parabolic antenna, which allows the feeder to be placed in the center of the reflector, as illustrated in Figure 10.12. This antenna is known as a Cassegrain antenna.

Parabolic antennas have high gain, are very directional (narrow beam) and have a narrow bandwidth.

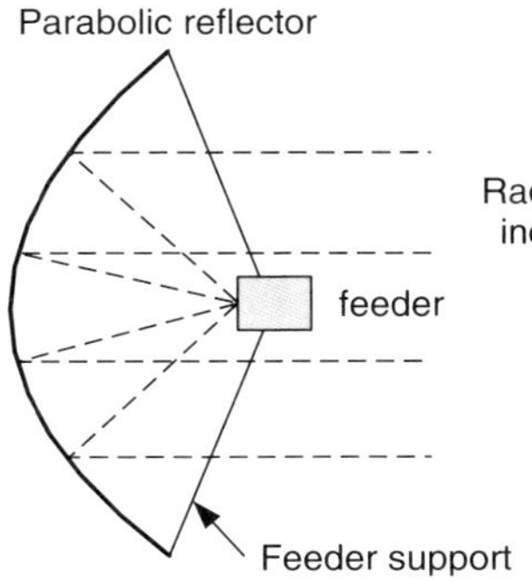

Figure 10.11 Axial parabolic antenna (cylindrical or dish).

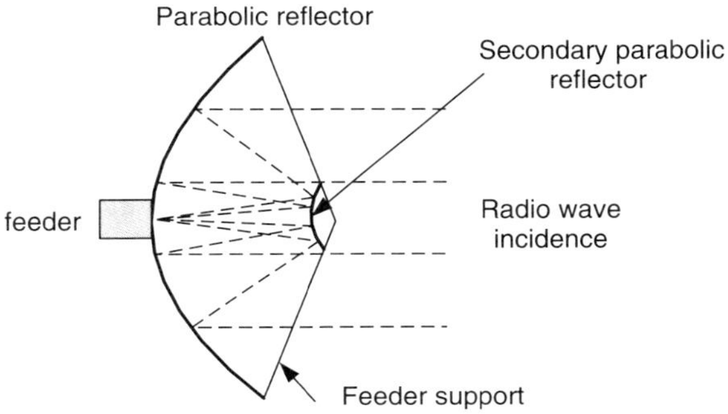

Figure 10.12 Cassegrain parabolic antenna.

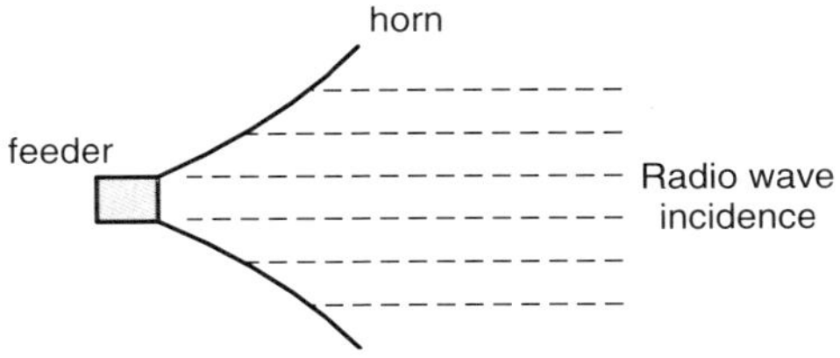

Figure 10.13 Horn antenna.

10.4.5 Horn Antenna

Horn antennas have a metal horn-shaped waveguide in front of the feeder to guide the radio waves energy towards it, illustrated in Figure 10.13. Horn antennas have a high gain and work well over a large bandwidth.

10.4.6 Antenna Type Comparison

Table 10.2 gives the gain and effective area for different antenna types.

Table 10.2 Gain and effective aperture for antennas at different frequencies

Antenna	Power gain	Power gain (dB)	Power gain for A = 1 m² frequency (MHz)				Effective area	Effective area (cm²) for A = 1 m² frequency (MHz)			
			900	1800	2400	3500		900	1800	2400	3500
Wavelength ($\lambda = c/f$) meter			0.33	0.17	0.13	0.09		0.33	0.17	0.13	0.09
Isotropic	1	0.000	1	1	1	1	$\lambda^2/4\pi$	88.4	22.1	12.4	5.8
Small dipole or loop	1.5	1.761	1.5	1.5	1.5	1.5	$1.5\lambda^2/4\pi$	132.6	33.2	18.7	8.8
Half-wave dipole	1.64	2.148	1.64	1.64	1.64	1.64	$1.64\lambda^2/4\pi$	145.0	36.3	20.4	9.6
Horn with mouth area A	$10A/\lambda^2$		90	360	640	1361	0.81 A	0.81	0.81	0.81	0.81
Parabola with face area A	$7A/\lambda^2$		7	7	7	7	0.56 A	0.56	0.56	0.56	0.56

10.5 Antenna Characteristics

Antennas can be defined by their impedance matching, radiation pattern, and polarization. Each of these characteristics is described in the following sections.

10.5.1 Impedance Matching

To maximize the power transfer and avoid reflections, source, line and load impedances should be matched as illustrated in Figure 10.14. For complex impedances (defined by a real and imaginary part) the matching should follow Equation (10.12). The symbol * indicates a complex conjugate pair, in which the imaginary part of one has an opposite sign of the other.

$$Z_{\text{load}} = Z_{\text{source}}^{*} \qquad \text{(10.12) Impedance matching}$$

Impedance mismatching is expressed by the Reflection Coefficient (RC or Γ) and expresses the amplitude of the reflected wave (V_r) relative to the incident wave (V_i). Γ is a complex number that describes the magnitude and phase shift of the reflection, and is defined in Equation (10.13). When

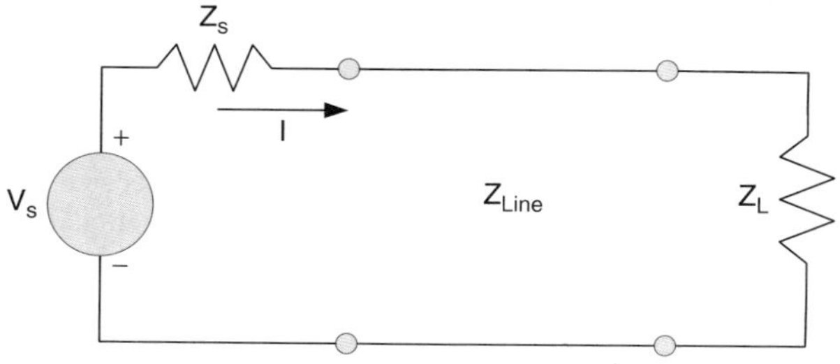

Figure 10.14 Impedance matching.

$\Gamma = -1$ the line is short circuited, $\Gamma = 0$ represents a perfect match, and $\Gamma = 1$ indicates that the line is open.

$$\Gamma = \frac{V_r}{V_i} = \frac{Z_L - Z_S}{Z_L + Z_S} \quad \text{(10.13) Reflection coefficient}$$

The magnitude of Γ is defined by the variable ρ, defined in Equation (10.14).

$$\rho = |\Gamma| \quad \text{(10.14) Reflection coefficient magnitude}$$

The incident and reflected waves interfere with each other causing, as a result, a stationary waveform over the line, which is characterized by nodes (when the minimum magnitude is reached) and antinodes (when the maximum amplitude is reached).

Node and antinode values are calculated by Equations (10.15) and (10.16).

$$V_{max} = V_i + V_r = V_i + \rho V_i = V_i(1 + \rho) \quad \text{(10.15) Standing wave node voltage}$$

$$V_{min} = V_i + V_r = V_i - \rho V_i = V_i(1 - \rho) \quad \text{(10.16) Standing wave antinode voltage}$$

The VSWR (Voltage Standing Wave Ratio) can then be represented by Equation (10.17).

$$VSWR = \frac{V_{max}}{V_{min}} = \frac{1 + \rho}{1 - \rho} \quad \text{(10.17) VSWR}$$

Many problems found in wireless systems are due to defective antennas or transmission lines. The measurement of VSWR is the best way to find out if the antenna is properly matched.

Another way to express matching is through Return Loss or Reflection Loss (RL), which is expressed in dB by Equation (10.18), where P_r is the reflected power and P_i is the incident power.

$$RL(dB) = 10 \log_{10} \frac{P_r}{P_i} \quad \text{(10.18) Return loss}$$

The Return Loss is the negative of the reflection coefficient in dB, as shown in Equation (10.19).

$$RL(dB) = -20 \log_{10} |\Gamma| \quad \text{(10.19) Return loss and reflection coefficient}$$

Acceptable values of the Reflection Coefficient are 0.1 to 0.25. The relationship between RL, Γ, and VSWR is shown in Table 10.3.

10.5.2 Antenna Patterns

The antenna radiation is expressed through its patterns, measured on two orthogonal planes, illustrated in Figure (10.15). The azimuth plane is the plane parallel to ground and the elevation plane is perpendicular to ground.

Figure 10.16 illustrates antenna patterns for a directional antenna. The H-plane (identified by the letter H) is also called the azimuth plane and its pattern is presented in polar form. The main lobe is at zero degrees and points right. The external circle represents the antenna gain and the inner circles are spaced in −10 dB increments.

The E-plane (identified by the letter V) is also called elevation plane and its pattern is presented in polar form. The main lobe is at zero degrees and points right. The external circle represents the antenna gain and the inner circles are spaced in −10 dB increments.

Antennas can be positioned horizontally or vertically and this changes their patterns. The positioning of the antennas defines its polarization. Transmit and receive antennas should use the same polarization to maximize the transfer of energy. The usual polarization is vertical to vertical (VV), in which transmit

Table 10.3 Impedance mismatching coefficients

RL	Γ	VSWR
20	0.10	1.2222
18	0.13	1.288
16	0.16	1.3767
14	0.20	1.4985
12	0.25	1.6709
10	0.32	1.925
8	0.40	2.3229
6	0.50	3.0095
4	0.63	4.4194
2	0.79	8.7242
0	1.00	–

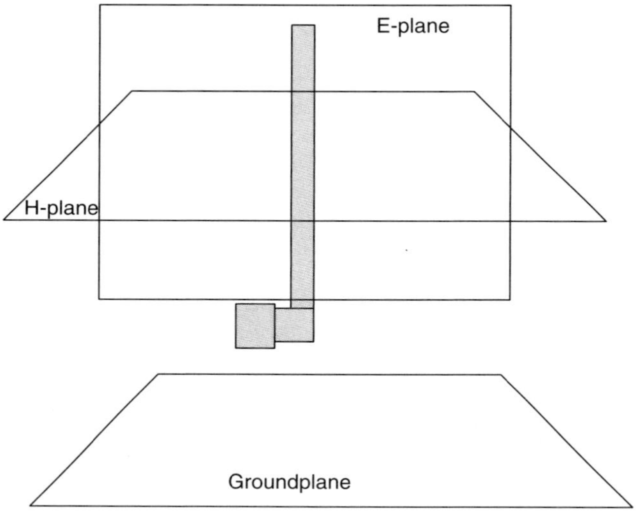

Figure 10.15 Antenna pattern planes.

and receive antennas are vertically polarized. Alternatively, a pair of horizontally (HH) polarized antennas can be used.

Antenna polarization can be used to minimize interference between antenna pairs, and in this case it is important to know the cross-polarization gain of a cross-polarized antenna pair. HV and VH patterns give the pattern when cross-polarization is used. The cross-polarization gain (in reality, a negative gain or loss) is measured in laboratory conditions. In real life, radiowaves have their polarization affected by reflections and the cross-polarization loss is decreased.

In the antenna patterns below, the first capital letter indicates the transmitter polarization, as the receiver is always set for the azimuth position and normalized to a unitary gain when determining antenna patterns. This is illustrated in Figures 10.16, 10.17 and 10.18. Antenna beamwidth is defined by the points were the peak gain is attenuated by 3 dB.

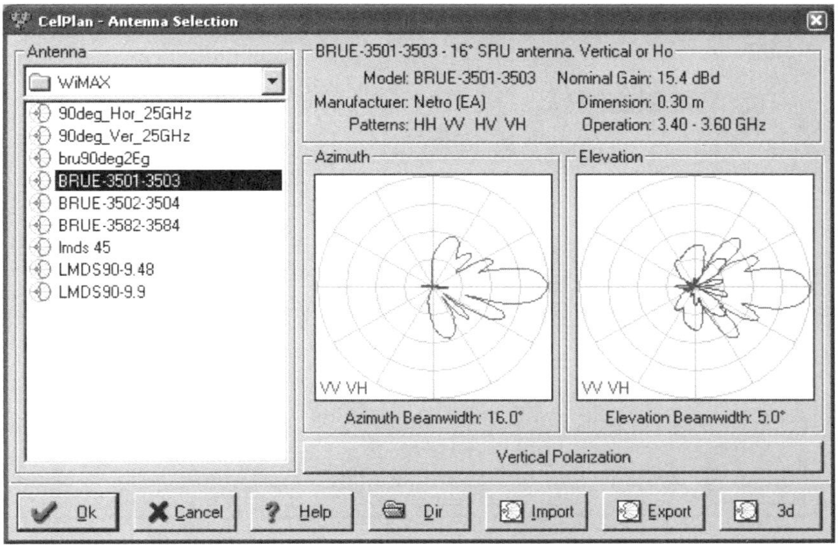

Figure 10.16 Vertical polarization directional antenna pattern sample.

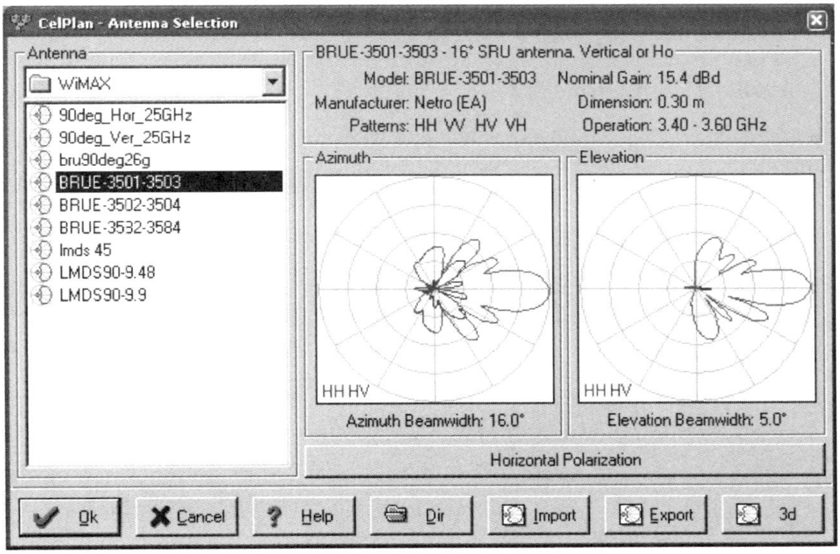

Figure 10.17 Horizontal polarization directional antenna pattern sample.

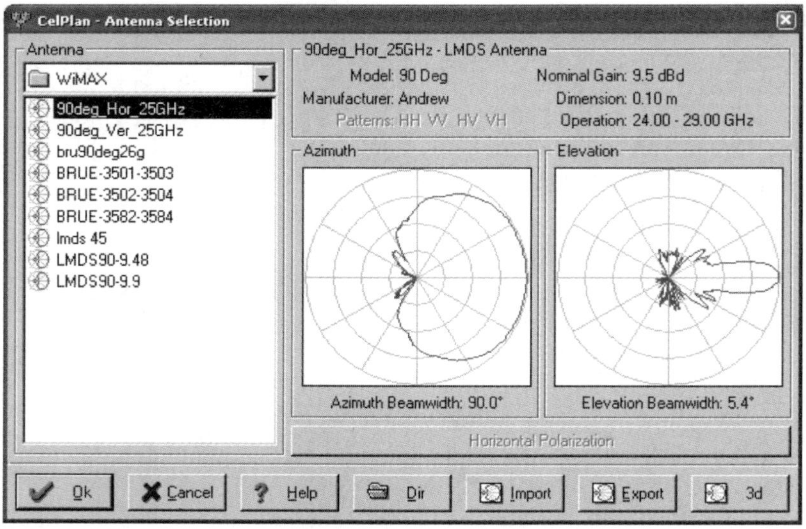

Figure 10.18 Directional antenna pattern sample.

These patterns from the previous figures represent the antenna in two dimensions only, but real antennas are tri-dimensional, thus it is important that a design tool be capable of generating the three-dimensional pattern when doing predictions, as shown in Figure 10.19.

10.5.3 Antenna Polarization

Antenna polarization is defined by the orientation of its E-field (electric field). This polarization can assume three different forms: linear, circular or elliptical.

10.5.3.1 Linear Polarization

In linear polarization the electric field oscillates on the same direction of the propagation and the magnetic field is perpendicular to it. This is illustrated in Figure 10.20 and expressed by Equation (10.20). When the field oscillation is parallel to the ground, the field is described as horizontally polarized, when it is perpendicular to ground, it is described as vertically polarized.

$$\boldsymbol{E} = \cos\left(2\pi f\left(t - \frac{x}{c}\right)\right)\widehat{\boldsymbol{y}} \qquad \text{(10.20) Linearly polarized antenna}$$

10.5.3.2 Circular Polarization

In circular polarization, the E-field has two components 90° out of phase with each other; the resultant field changes direction (rotates) as it propagates and is defined by Equation (10.21).

$$\boldsymbol{E} = \cos\left(2\pi f\left(t - \frac{x}{c}\right)\right)\widehat{\boldsymbol{y}} + \sin\left(2\pi f\left(t - \frac{x}{c}\right)\right)\widehat{\boldsymbol{z}} \qquad \text{(10.21) Circular polarized antenna}$$

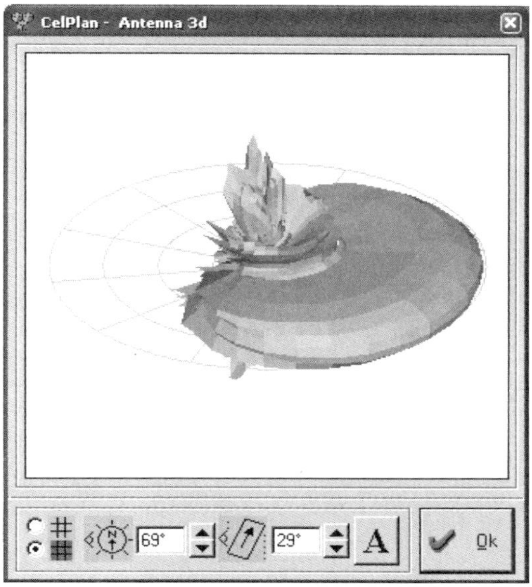

Figure 10.19 3D Representation of directional antenna.

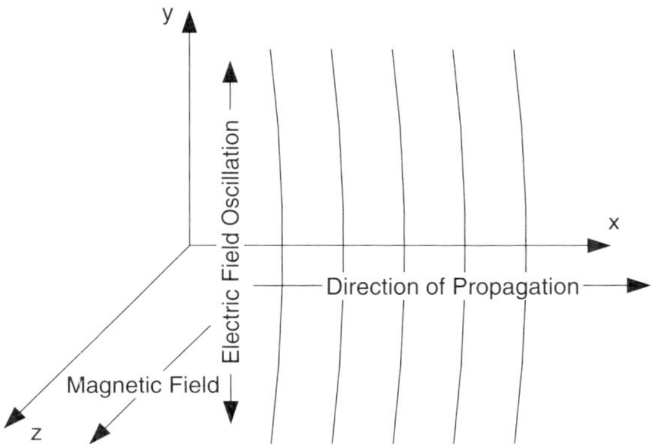

Figure 10.20 Linear polarization.

10.5.3.3 Elliptical Polarization

Elliptical polarization is similar to circular polarization, with the difference that the two electric field components do not have the same amplitude.

10.5.4 Cross-Polarization

Theoretically, a vertically polarized antenna transmits and receives vertically polarized fields only and will not communicate with a horizontally polarized antenna and vice versa. In real life antennas, some of the cross-polarized energy can be detected by the antenna, as explained in the antenna patterns above. Besides, RF waves change polarity when they are reflected or diffracted, so we have to consider that there will be a Polarization Loss Factor (PLF), defined by the rotation angle between the antenna polarities, which is defined in Equation (10.22).

$$PLF = \cos^2 \varphi \qquad (10.22)\ \text{Polarization loss factor}$$

This factor can be applied also when a pair of antennas is not orthogonal to each other. This is the case when transmit antennas are cross-polarized at 45 degrees from vertical, while the receive antenna is vertical. In this case the angle between transmit and receive antennas is 45 degrees. Table 10.4 gives the PLF for different angles.

Advanced antenna systems require several antennas and an economical way is to use different polarizations in the same casing. Cross-polarized antenna pairs (transmit and receive) will receive relatively well uncorrelated signals.

A common configuration is to have two orthogonal antennas polarized 45 degrees from the vertical, as illustrated in Figure 10.21. The receive antennas can be vertically oriented in which case, according to Table 10.4, they will incur a PLF loss of 3 dB, or can be cross-polarized also, receiving well uncorrelated signal each.

Table 10.4 Polarization loss factor

Angle		PLF	
Degrees	Radian	linear	dB
0	0.00	1.00	0
10	0.17	0.97	−0.133
20	0.35	0.88	−0.54
30	0.52	0.75	−1.249
40	0.70	0.59	−2.315
45	0.79	0.50	−3.01
50	0.87	0.41	−3.839
60	1.05	0.25	−6.021
70	1.22	0.12	−9.319
80	1.40	0.03	−15.21
90	1.57	0.00	–

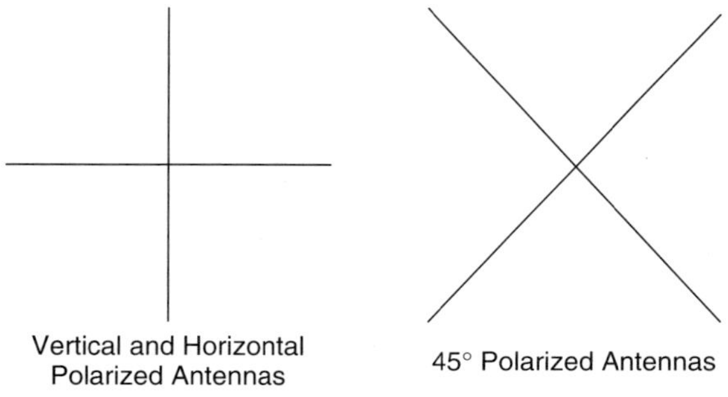

Figure 10.21 Cross-polarized antennas.

10.5.5 Antenna Correlation or Signal Coherence

Multiple antenna systems are said to have correlated antennas if the signals received by the receive antennas are coherent, that is, are similar to each other. When the received signals are differentiated (non-coherent), the antennas are said to be uncorrelated.

Figure 10.22 illustrates the channel (H) between two transmit and receive antennas. This channel is represented by four virtual connections (h_{nn}) for each delayed path. The matrix that defines the virtual connections of each delayed complex channel path is given by Equation (10.23).

$$\boldsymbol{H} = \begin{pmatrix} h_{11} & h_{12} \\ h_{21} & h_{22} \end{pmatrix}$$ (10.23) Matrix H of a complex channel path

The correlation between antennas is a function of the local scattering and is a function of the Angular Spread (AS), Angle of Arrival (AoA) and Direction of Travel (DoT). This correlation is not constant, varying significantly over a geographical area.

The received signal **r** is expressed in Equation (10.24).

$$\boldsymbol{r} = \begin{vmatrix} r_1 \\ r_2 \end{vmatrix} = \begin{pmatrix} h_{11} & h_{12} \\ h_{21} & h_{22} \end{pmatrix} \begin{vmatrix} x_1 \\ x_2 \end{vmatrix} + \begin{vmatrix} n_1 \\ n_2 \end{vmatrix}$$ (10.24) Output complex signal for a single delayed path

The correlation between antennas can be expressed in terms of signal (ρ) observed at each antenna element, as represented in Equation (10.25), where E is the expectancy function.

$$\rho = \frac{E(h_{11}h_{12}^*)}{\sqrt{E(h_{11}h_{11}^*)}\sqrt{E(h_{12}h_{12}^*)}}$$ (10.25) Output complex signal for a single delayed path

This correlation can be normalized assuming the expectation expressed in Equation (10.26).

$$E(h_{11}h_{11}^*) = 1$$ (10.26) Expectancy normalization

Thus the transmit correlations for the antenna pair are given by Equations (10.27) and (10.28).

$$\rho_{tx} = E(h_{11}h_{12}^*)$$ (10.27) Correlation between transmit antennas as measured at antenna 1

$$\rho_{tx} = E(h_{22}h_{21}^*)$$ (10.28) Correlation between transmit antennas as measured at antenna 2

The receive correlations are given by Equations (10.29) and (10.30).

$$\rho_{rx} = E(h_{11}h_{21}^*)$$ (10.29) Correlation between receive antennas as measured at antenna 1

$$\rho_{rx} = E(h_{22}h_{12}^*)$$ (10.30) Correlation between receive antennas as measured at antenna 2

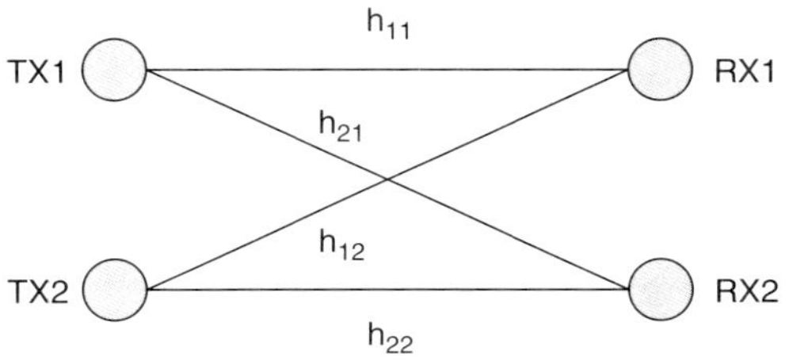

Figure 10.22 2 × 2 Antenna configuration.

Finally the correlation matrix for a 2 × 2 multi antenna channel model can be written as a Kronecker product of two simplified correlation matrixes as indicated in Equation (10.31).

$$R = R_{tx} \otimes R_{rx} \qquad \text{(10.31) Correlation matrix}$$

where R_{tx} and R_{rx} are specified in Equations (10.32) and (10.33).

$$R_{tx} = \begin{bmatrix} 1 & \rho_{tx} \\ \rho_{tx}^* & 1 \end{bmatrix} \qquad \text{(10.32) Transmit correlation matrix}$$

$$R_{rx} = \begin{bmatrix} 1 & \rho_{rx} \\ \rho_{rx}^* & 1 \end{bmatrix} \qquad \text{(10.33) Receive correlation matrix}$$

ITU Advanced Antenna Models replace R_{tx} by R_{eNB}, ρ_{tx} by α and ρ_{rx} by β, as shown in Equation (10.34).

$$R_{eNB} = \begin{pmatrix} 1 & \alpha \\ \alpha^* & 1 \end{pmatrix} R_{UE} = \begin{pmatrix} 1 & \beta \\ \beta^* & 1 \end{pmatrix} \qquad \text{(10.34) eNB and UE antenna correlation}$$

The 2 × 2 correlation matrix is then defined by Equation (10.35).

$$R_{spatial} = R_{eNB} \otimes R_{UE} = \begin{bmatrix} 1 & \alpha \\ \alpha^* & 1 \end{bmatrix} \otimes \begin{bmatrix} 1 & \beta \\ \beta^* & 1 \end{bmatrix}$$

$$= \begin{bmatrix} 1 & \beta & \alpha & \alpha\beta \\ \beta^* & 1 & \alpha\beta^* & \alpha \\ \alpha^* & \alpha^*\beta & 1 & \beta \\ \alpha^*\beta^* & \alpha^* & \beta^* & 1 \end{bmatrix} \qquad \text{(10.35) Spatial antenna correlation}$$

The α and β values represent the different channel types and range from 0 to 1. Example values for low, medium and high correlation are shown in Table 10.5.

Practical implementations of high and medium correlation are illustrated in Figure 10.23. A low correlation is not illustrated as it is hard to achieve. Some authors claim that separations above 10 λ in NLOS situations could be considered as low correlation.

10.6 Multiple Antennas Arrangements

A wireless system provides communications between Base Stations (BS) and Subscriber Stations (SS) or Mobile Stations (MS). Base stations allow the installation of multiple antennas with large separations between them. Subscriber stations are more restricted in this sense and mobile stations even more so.

Multiple antenna systems are classified according to the number of antennas at the transmitter (RF channel inputs) and receiver (RF channel outputs). The use of multiple antennas implies in addition

Table 10.5 ITU correlation factors for different antenna configurations

	Low correlation	Medium correlation	High correlation
α	0	0.3	0.9
β	0	0.9	0.9

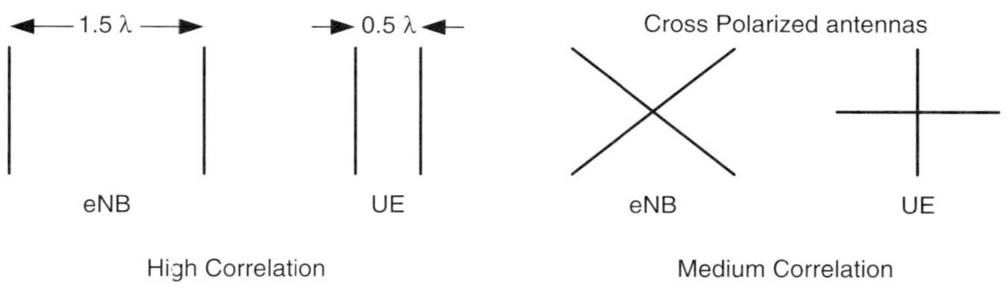

Figure 10.23 ITU antenna configurations for different correlations.

extra processing on one or both sides of the RF channel. It must be also noted that each direction (downlink and uplink) may be configured with different solutions.

The nomenclature used to classify the channels refers to the channel (air) between transmit and receive antennas. The input (In) defines how many signals are sent through the air, while the output (Out) defines how many signals are received from the air.

10.6.1 SISO (Single In to Single Out)

Traditionally, a wireless link has one transmit and one receive antenna, which can be described as a SISO (Single In signal and Single Out signal) configuration in relation to the wireless channel. In SISO, multipath signals are received by the antenna with the combined signal being subject to fading, which should be compensated using the techniques already described in previous chapters. A SISO link is illustrated in Figure 10.24.

Multiple antenna techniques rely on the existence of different paths between antennas to eliminate fading. Signals with similar fading characteristics are said to be coherent, whereas signals with different fading are not coherent (or diverse).

Co-located antennas have to be optimized to provide the desired signal diversity (no coherence), which can be done by adjusting the position or azimuth angle of the antennas. Spacing of at least $\lambda/2$ (half a wavelength) and an angular shift of at least 1/8 of the antenna beamwidth may optimize signal diversity.

Optimizing the antenna's position is not sufficient to assure non-coherent signals, as the number of LOS components in relation to indirect components has a large influence on signal coherence. LOS paths tend to be coherent, whereas non-LOS paths tend to be non-coherent. The amount of LOS

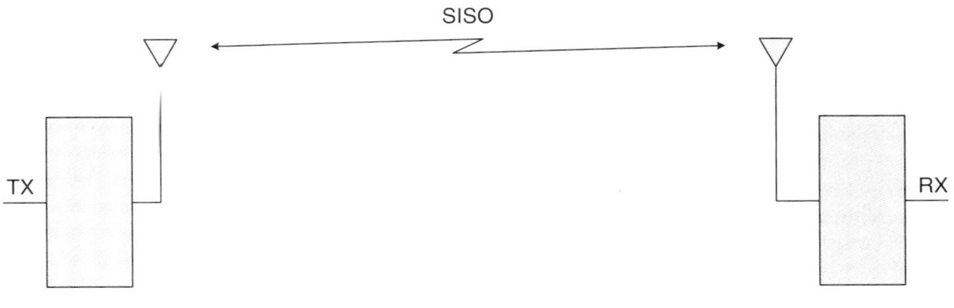

Figure 10.24 SISO configuration – one transmit and one receive antenna.

present in a multipath signal is defined in a Ricean distribution by the k factor, which is defined as the ratio of the dominant component's signal power over the (local-mean) scattered components power.

- Signals have high coherence. High k factor: >10; multipath signal follows a Gaussian distribution.
- Signals have medium coherence. Medium k factor: $10 < k > 2$; multipath signal follows a Ricean distribution.
- Signals have low coherence. Low k factor: $k < 2$; multipath signal follows a Rayleigh distribution.

The k factor can be estimated on a pixel basis from the geographical data (topography and morphology) by RF prediction tools.

The techniques described next apply to both DL and UL directions, although it is difficult to use multiple antennas in mobile phones, and, if implemented, the antennas are not fully uncorrelated.

10.6.2 SIMO (Single In to Multiple Out)

Adding additional receive antennas creates alternative paths that will receive different multipath components and consequently be subject to different fading instances. The difference in the resulting signals is defined by the correlation factor.

This set-up is called a SIMO (Single In signal and Multiple Out signals) configuration, and is also known as Receive Diversity. It is illustrated in Figure 10.25. The multiple output signals from the RF channel have to be combined, so a single signal is sent to the receiver. In Receive Diversity the receiver learns information about the channel by analyzing known transmissions, such as preamble and pilots. This is the method usually used in 2 G cellular networks.

There are three basic techniques with which the signals can be combined before the receiver:

- Selection Combining (SC): the strongest signal is always selected.
- Equal Gain Combining (EGC): the signal are simply added together (phase correction may be applied).
- Maximal Ratio Combining (MRC): the signals are added together weighted by their SNR.

These receive diversity methods are described in more detail in Section 10.7. Receive diversity improves the overall SNR with the number of antennas, but for this to happen, the paths should be non-coherent and the combining device optimal. During the network design, the path correlation should be considered as an efficiency factor to be applied to the combining device gain.

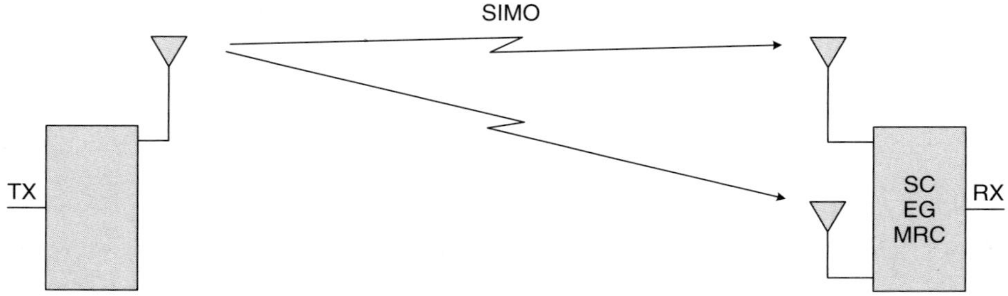

Figure 10.25 SIMO configuration – receive diversity.

10.6.3 MISO (Multiple In to Single Out)

Adding additional transmit antennas creates alternative paths that will have different multipath components and, consequently, be subject to different fading instances. The amount of difference in the resulting signals is defined by the coherence factor.

Multiple transmit antenna configurations are called MISO (Multiple-In signal and Single-Out signal), and are also known as Transmit Diversity, illustrated in Figure 10.26. The receive antenna receives signals coming from two or more transmit antennas, so, in principle, it receives n times the power, n being the number of transmit antennas. This gain in power is known as the array gain and can be positive or negative (representing a loss) depending on the signal coherence. The signal coherence is a direct function of the antenna correlation.

There are two modes of transmitting schemes for MISO: open loop, and closed loop. In open loop, the transmitter does not have any information about the channel. In closed loop, the receiver sends Channel State Information (CSI) to the transmitter on a regular basis. The transmitter uses this information to adjust its transmissions using one of the methods below:

- *Transmit Channel Diversity*: The transmitter evaluates periodically the antenna that gives the best results and transmits on it. Only one antenna transmits at a time.
- *Linear Diversity Coding*: A linear pre-coding is applied at the transmitter and a post-coder is used at the receiver, both using the CSI information. Both antennas transmit simultaneously.

Additional transmit diversity can only be obtained by replacing sets of multiple symbols by orthogonal signals and transmitting each orthogonal symbol on a different antenna. Alamouti proposed an orthogonal code for two symbol blocks in a method called Space Time Block Coding (STBC), which is specified for use in WiMAX, and described in Section 10.8.3.1. Transmit diversity does not improve overall SNR, but reduces it by averaging the fading over two symbols.

The most common types of transmit diversity methods are described in Section 10.8.

10.6.4 MISO-SIMO

The transmit and receive diversity methods described above can be combined, as illustrated in Figure 10.27.

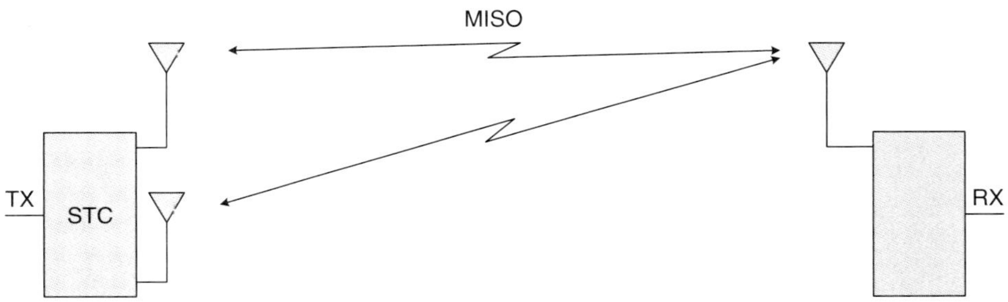

Figure 10.26 MISO configuration – transmit diversity.

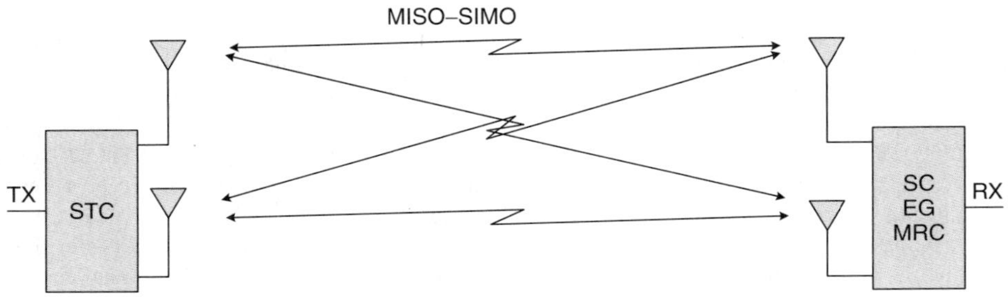

Figure 10.27 MISO-SIMO – receive and transmit diversities combined.

10.6.5 MIMO (Multiple In to Multiple Out)

Spatial multiplexing (SM) or MIMO (Multiple In signal and Multiple Out signal) are improvements over the previous solution that uses simultaneous transmit and receive diversity. The use of transmit and receive diversity increases the robustness of the channel, the use of spatial multiplexing trades this robustness for capacity as explained next.

In spatial multiplexing, each antenna transmits different data, so each receiver receives copies of different streams of data. A matrix is then assembled to relate each transmit signal to each receive antenna; as long as this matrix has a number of unique values equal or larger than the number of transmitted streams, it is possible to mathematically decode the data.

In principle, the throughput can be multiplied by the number of transmit antennas, but this can lead to confusion when trying to understand the gain given by MIMO techniques. This multiplication of throughput does not actually happen in real life because the channels are not completely orthogonal and interfere with each other.

The SNIR requirement at each receive antenna is higher than if only one transmission was being performed, which implies choosing a modulation scheme with a lower throughput. So although the nominal throughput of the link is multiplied by the number of antennas, it is reduced due to the higher SNIR requirement (which forces use of a lower modulation scheme). The final result is between both values and depends largely of the antenna correlation. Spatial Multiplexing (SM) is illustrated in Figure 10.28.

In open loop, the transmitter is unaware of the channel, but the receiver can recover some channel information using one of the methods below. Spatial Multiplexing is described in more detail in Section 10.10.

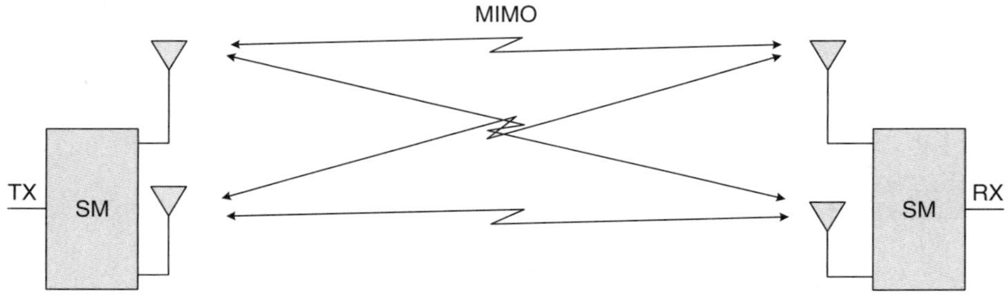

Figure 10.28 MIMO – spatial multiplexing.

- *Maximum Likelihood Detection (MLD)*: the decoder looks for the maximum likelihood vector over several symbols. This implies examining multiple possibilities that increase exponentially with the number of modulation levels.
- *Linear Detectors (LD)*: the decoder applies the inverse of the channel to amplify it, trying to remove the channel distortions.
- *Interference Cancellation (BLAST Bell Labs Layered Space Time)*: this technique adds another level of randomness by circulating the data through the different antennas, providing space diversity additionally to time diversity.
- *SVD (Single Value Decomposition) pre-coding and post-coding*: the diagonalization done in BLAST can be done by applying the channel knowledge.
- *Linear Pre-Coding and Post-Coding (LPCP)*: channel knowledge decomposes the channel model in a set of parallel channels and more power is directed to the channels with more gain. The reverse operation is done at the receiver.

10.6.6 Adaptive MIMO Switching (AMS)

Transmit diversity presents a higher throughput than spatial multiplexing at low SNIR levels, whereas spatial multiplexing results in a higher throughput than transmit diversity for high SNIR levels. This technique automatically chooses the solution that gives the best throughput at each location.

10.6.7 Uplink MIMO (UL-MIMO)

This is also a Spatial Multiplexing (SM) technique but used in the upstream only, as illustrated in Figure 10.29. It is also known as Uplink Collaborative MIMO. Because transmissions arrive from different locations they are non-coherent between themselves, thus providing a better performance than the one obtained in the downlink spatial multiplexing, although the self-interference issues continue to be present.

10.7 Receive Diversity

A radio link is subject to quality variations that degrade its throughput capacity. Adding an additional receiver provides a diverse path that may complement the information received by the first receiver and, consequently, help in recovering throughput. Receive diversity techniques are of the SIMO type.

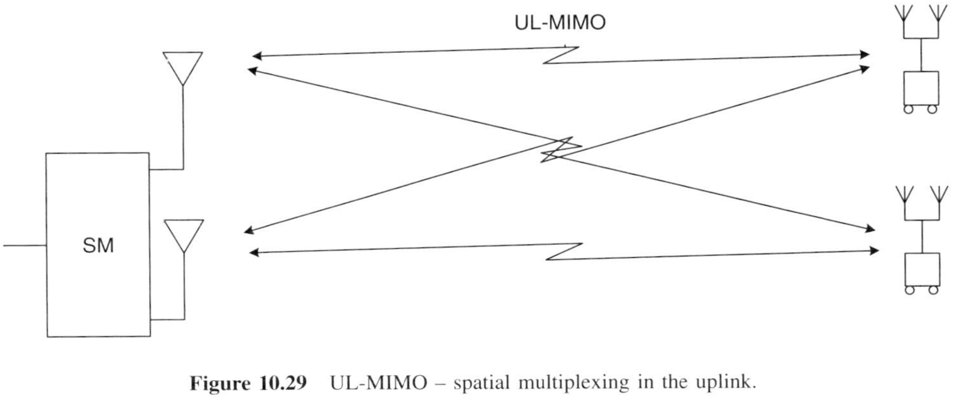

Figure 10.29 UL-MIMO – spatial multiplexing in the uplink.

To better understand receive diversity, assume that a transmitter sends a signal **s**, which travels through two different RF channels specified by Equations (10.36) and (10.37).

$$h_0 = \alpha_0 e^{j\theta_0} \quad \text{(10.36) RF channel 1}$$

$$h_1 = \alpha_1 e^{j\theta_1} \quad \text{(10.37) RF channel 2}$$

The received signals have noise added to them in Equations (10.38) and (10.39).

$$r_0 = sh_0 + n_0 \quad \text{(10.38) RF channel 1 with noise}$$

$$r_1 = sh_1 + n_1 \quad \text{(10.39) RF channel 2 with noise}$$

This combination of signal and noise will then be received at the other end of the communication channel. There are three main methods commonly used to benefit from this received diversity, described next.

10.7.1 Equal Gain Combining (EGC)

Equal Gain Combining is illustrated in Figure 10.30. The signals received from both branches are combined, as expressed in Equation (10.40). Each branch should have its own LNA, to avoid combiner loss.

$$r = s(h_0 + h_1) + n_0 + n_1 \quad \text{(10.40) Equal Gain Combining}$$

For coherent channels (identical or nearly identical), there is no real gain as the signal and the noise rise together by 3 dB, that is, even though there was an increase in the signal (added twice), there was exactly the same increase in noise, thus it results in the same as receiving in only one antenna, that is, no diversity.

For non-coherent channels there is a significant gain when fading occurs in one channel and not in the other. This solution is the easiest type of receive diversity to implement, but the benefit only happens when the probability of fading overlap between the channels is small. The use of different polarizations for the Rx antennas helps to maximize this benefit.

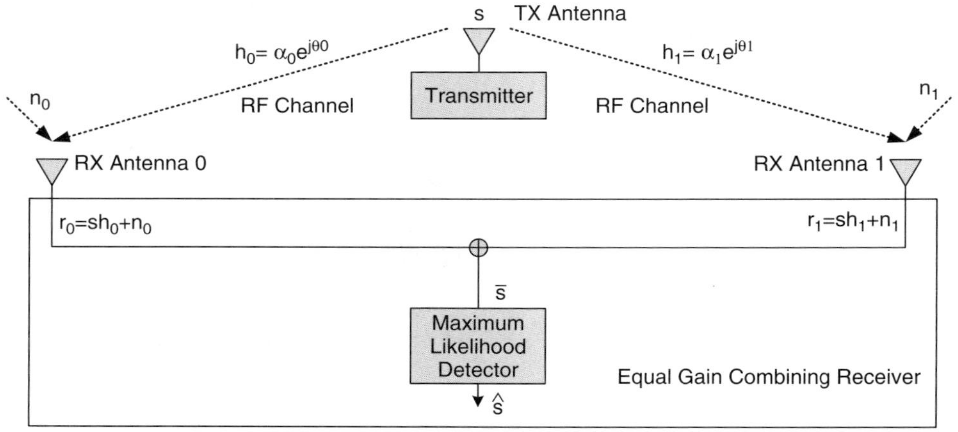

Figure 10.30 Equal gain combining receiver.

10.7.2 Diversity Selection Combining (DSC)

Diversity Selection Combining is another variation of SIMO and is illustrated in Figure 10.31 and expressed by Equation (10.41). In this technique, each branch analyzes known transmitted information (such as pilots) and informs its Signal to Noise Ratio (SNR) to the diversity switch, which chooses the branch with the best ratio. This is a technique commonly adopted by WLAN systems.

$$r = sh_0 + n_0 \quad \text{or} \quad r = sh_1 + n_1 \qquad \text{(10.41) Diversity Selection Combining}$$

This method should only be used when the channel coherence time is much longer than the symbol duration, so the best branch chosen from pilot analysis will hold for several symbols (at least until the next channel assessment); otherwise, by the time the switch is made from one antenna to the other, the channel might have changed already, thus affecting the SNR at each antenna. There is no gain if the channels are coherent and for non-coherent channels the gain can be significant if the fading does not coincide in both channels.

10.7.3 Maximal Ratio Combining (MRC)

Another SIMO technique is Maximal Ratio Combining, illustrated in Figure 10.32 and expressed by Equations (10.42) to (10.44). In this method, the phase and gain of each branch is optimally adjusted prior to combining the signals. This requires a good knowledge of each channel, which is derived from pilot analysis. This method estimates each channel independently and then multiplies the signal by the convoluted channel. This allows the recovery of the best possible signal, although the noise can be a major impairment, as it is also amplified with the faded signal. The method performs well when the signals are quite above noise level. It is a more complex and expensive method to implement,

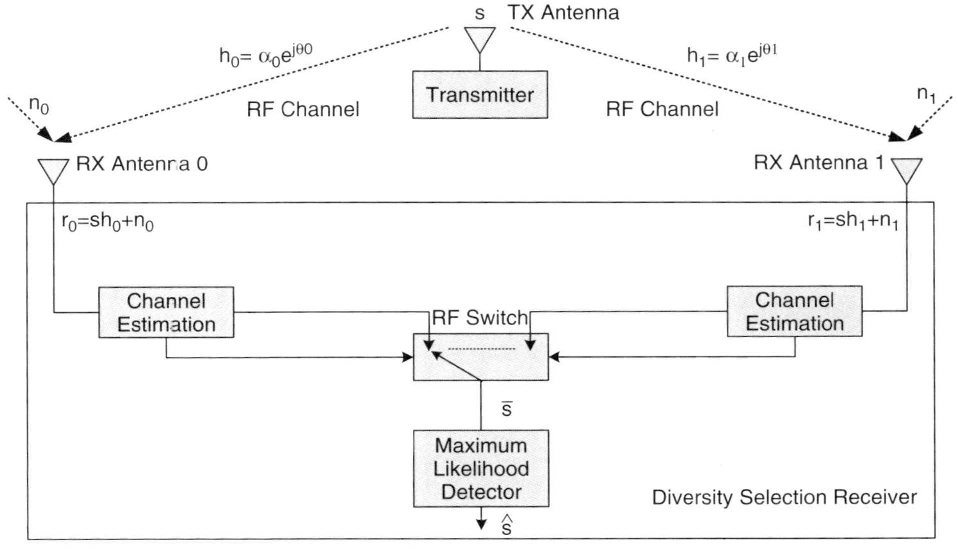

Figure 10.31 Diversity selection receiver.

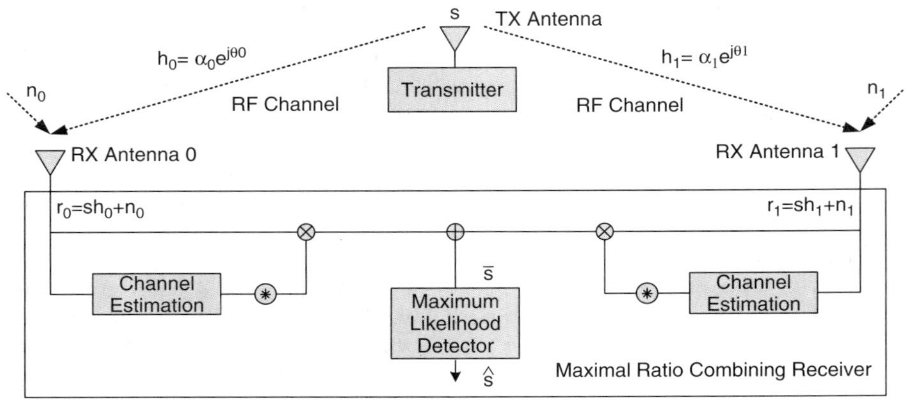

Figure 10.32 Maximal ratio combining receiver.

than the two previous receive diversity options. It is usually used in the BS but not in MSs.

$$\tilde{s} = h_0^* r_0 + h_1^* r_1 \quad \text{(10.42) Received signal}$$

$$\tilde{s} = h_0^*(sh_0 + n_0) + h_1^*(sh_1 + n_1) \quad \text{(10.43) Received signal with noise}$$

$$\tilde{s} = (\alpha_0^2 + \alpha_1^2)s + h_0^* n_0 + h_1^* n_1 \quad \text{(10.44) Maximal Ratio Combining}$$

In this technique, in case of fading in one of the branches, a maximum gain is applied to it, which also increases the noise level. As an example, assume that the signals are being received at −60 dBm and that the noise floor is at −90 dBm. A fading of 30 dB implies a 30 dB gain for this branch and, consequently, the noise reaches −60 dBm, drastically reducing the SNR. A possible solution in this case would be to drop the signal that requires a higher gain, that is, a mix of MRC and DSC.

10.7.4 Maximal Likelihood Detector (MLD)

Also a type of SIMO, this detector verifies the distance between the received signal and the possible constellation values, by calculating the Euclidian distance between the received signal and the possible constellation states. The Euclidean distance is the "ordinary" distance between constellation points of a modulation and the received value. Equation (10.45) gives the distance between two vectors (Euclidean distance).

$$d^2(x, y) = (x - y)(x^* - y^*) \quad \text{(10.45) Euclidean distance}$$

The reconstituted signal $\tilde{s}$ has to be compared to the different constellation values, using Equation (10.46).

$$d^2(s_0, s_i) \leq d^2(s_0, s_k), \qquad \forall i >< k \quad \text{(10.46) Constellation distance}$$

The number of possible states can be very high, when high modulations are used, and the detection is done over multiple combinations of s symbols. Several sub-optimal methods were developed, including zero forcing, minimum mean square error, decision feedback and sphere detectors.

- *Zero forcing (ZF) detectors* invert the channel matrix and have small complexity but perform badly at low SNR.

- *Minimum Mean Square Error (MSSE) detectors* reduce the combined effect of interference between the channels and noise, but require knowledge of the SNR, which can only be roughly estimated at this stage.
- *Decision Feedback (DF) receivers* make the decision on one symbol and subtract its effect to decide on the other symbol. This leads to error propagation.
- *Sphere detectors (SD)* reduce the number of symbols to be analyzed by the ML detector, by performing the analysis in stages. It may preserve the optimality while reducing complexity.

10.7.5 Performance Comparison for Receive Diversity Techniques

In Table 10.6, we have a set of sub-carriers (rows) and a sequence of symbols in time (columns). Pilots are represented by P and data by D. Fade can be detected when pilots are measured, between two pilot symbols, fade can be estimated.

The DSC method has to wait for a pilot analysis to be done, so it can choose the best signal. EGC, on the other hand, always adds the two signals, but the faded signal does not contribute much. MRC adjusts the signal levels and phases before adding them, but also amplifies the received noise. The bottom three rows of Table 10.6 compare the decisions taken by the different algorithms.

For indoor applications, where the signal is well above the noise level, MRC is the best solution. For outdoor applications, DSC provides the best results, as long as the channel coherence time is large when compared to the symbol time and EGC is the best compromise when the coherence time is on the order of the symbol duration.

Figure 10.33 shows typical gains provided by each method for different channel correlations (negligible, low, medium and high correlation). Actual values may vary according to the implementation and the environment. A network designer should evaluate the receiver algorithm and the kind of the environment to derive his own values.

10.8 Transmit Diversity

The addition of more transmitters creates new multipaths and may increase the detection options in a method known as transmit diversity, which is of the MISO type. Transmit diversity can be represented by a matrix that relates data sent in the antennas to a certain sequence of symbols. In

Table 10.6 Receive detector performance comparison

	Symbol n	Symbol n+1	Symbol n+2	Symbol n+3	Symbol n+4	Symbol n+5	Symbol n+6	Symbol n+7	Symbol n+8	Symbol n+9	Symbol n+10
sub-carrier n	P	D	P	D	P	D	P	D	P	D	P
sub-carrier n+1	D	D	D	D	D	D	D	D	D	D	D
sub-carrier n+2	D	D	D	D	D	D	D	D	D	D	D
sub-carrier n+3	P	D	P	D	P	D	P	D	P	D	P
Pilot Ch1	fade		ok		ok		fade		ok		fade
Pilot Ch2	ok		ok		fade		fade		ok		ok
DSC		2	2	2	2	1	1	1	1	1	2
EGC	2	1+2	1+2	1+2	1	1+2	1+2	1+2	1+2	1+2	2
MRC	1+2	1+2	1+2	1+2	1+2	1+2	1+2	1+2	1+2	1+2	1+2

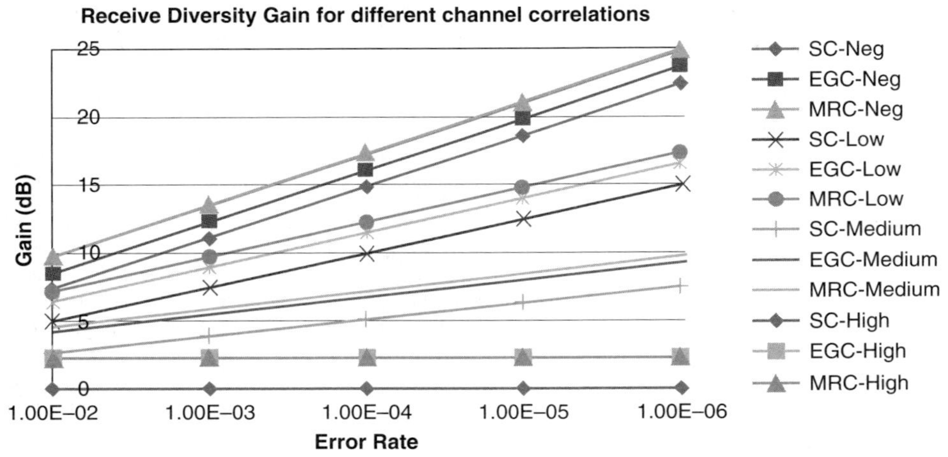

Figure 10.33 Maximal ratio combining receiver.

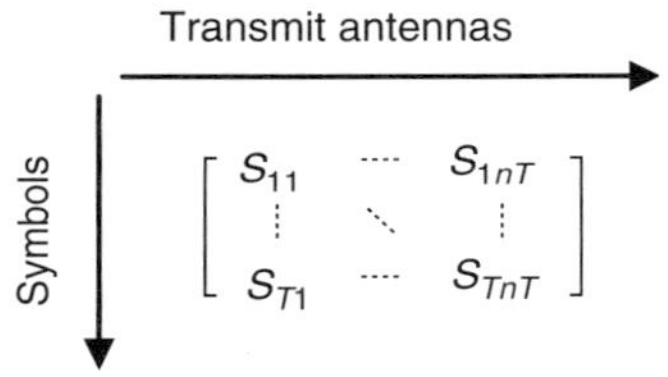

Figure 10.34 Transmit diversity matrix.

the usual representation, there are T time slots and n_T transmit antennas, with s_{ij} representing a modulated symbol. The T length represents the transmit diversity block size, which is illustrated in Figure 10.34.

This matrix is defined by a code rate that expresses the number of symbols that can be transmitted on the course of one block. A block that encodes k symbols has its code rate defined by Equation (10.47).

$$r = \frac{k}{T} \qquad \text{(10.47) Matrix code rate}$$

The three most common types of transmit diversity techniques are described in the following sections.

10.8.1 Receiver-Based Transmit Selection

Receiver-Based Transmit Selection, a type of MISO, is illustrated in Figure 10.35 and expressed in Equation (10.48). In TDD systems, transmit and receive directions use the same channel. When channels vary very slowly, it is reasonable to assume that the best channel used for the receive side would be the best one to transmit as well. However, this only holds true for channels with a coherence time larger than the frame time.

$$r = s_0 h_0 + n \quad \text{or} \quad r = s_1 h_1 + n \qquad \text{(10.48) Received based transmit selection}$$

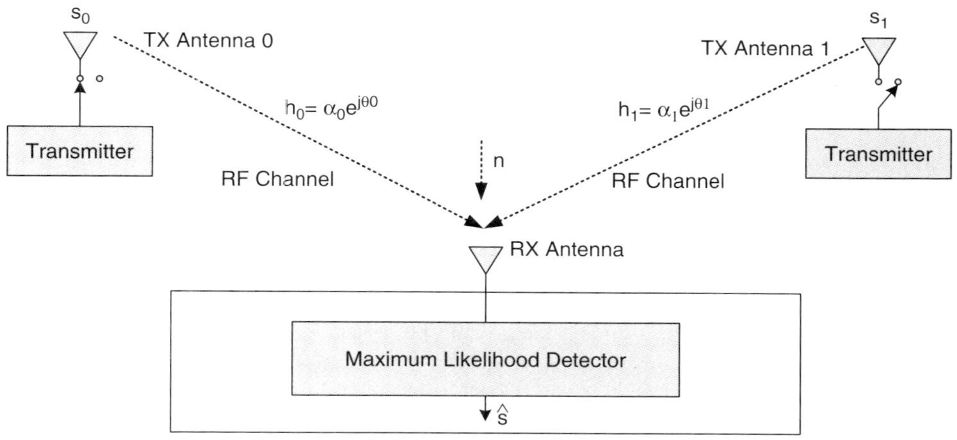

Figure 10.35 Receive-based transmit selection.

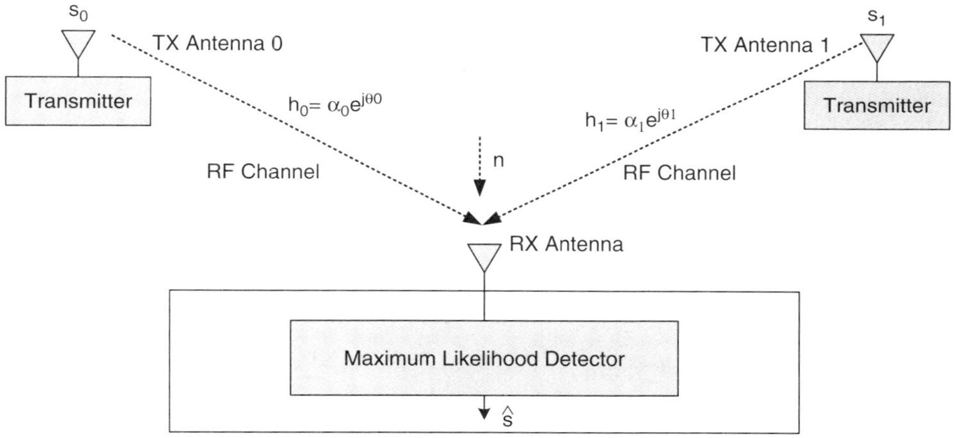

Figure 10.36 Transmit redundancy.

10.8.2 Transmit Redundancy

Another type of MISO, Transmit Redundancy is illustrated in Figure 10.36 and expressed in Equation (10.49). In this method, transmission is made on both channels all the time, which requires both signals to arrive at the antennas at the same time, that is, the circuits should be designed to avoid different delays in the path to the antennas and cable lengths should be exactly the same.

In coherent channels (similar channels), the received signal will increases by 3 dB, whereas in non-coherent channels, the extra path reduces multipath fading but also becomes a source of interference. Designers should configure antennas to obtain uncorrelated channels but with a low dispersion between them. An example would be to use directional antennas with azimuth angle diversity (pointing antennas to slightly different angles) or use different antenna polarities. It is important for designers to analyze the sources of multipath before deciding on the best deployment strategy.

$$r = s_0 h_0 + s_1 h_1 + n \quad \text{(10.49) Received signal transmit redundancy}$$

10.8.3 Space Time Transmit Diversity

Additional schemes of SIMO that add a time component to the space diversity provided by multiple antennas have also been proposed. Some examples are Delay Diversity and Space Time Trellis. Both methods rely on creating additional multipath and are complex to implement.

A simpler method was proposed by Siavash Alamouti. In his proposition, the multipath is delayed by a full symbol, and then a conjugate value is sent to cancel the reactive part of the signal. This technique is easy to implement, but requires the channel to remain stable over a period of two symbols. This means that the coherence time should be larger than two symbols. This method is called Space-Time Block Coding (STBC or STC), also known as Matrix A and is described next.

10.8.3.1 Space Time Block Code: Alamouti's Code (Matrix A)

In this technique, each transmission block is made of two symbols in time. Each antenna sends the information as depicted in Table 10.7. The operations applied over the information were carefully chosen to cancel the unwanted information at each antenna. Thus, even though different information is sent by each antenna on one symbol, the same information is repeated over the next symbol, therefore, this is still considered a diversity scheme.

This is the only type of code that can reach a coding rate of 1. In the WiMAX standard, this matrix is referred to as Matrix A. Equation (10.50) shows how the matrix is built. BS support of this method is mandatory in the WiMAX and LTE standards.

$$X = \begin{bmatrix} s_0 & s_1 \\ -s_1^* & s_0^* \end{bmatrix} \quad \text{(10.50) Matrix A}$$

The received signal for the first symbol (0) and the second symbol (1) are shown in Equations (10.51) and (10.52).

$$r_0 = h_0 s_0 + h_1 s_1 + n_0 \quad \text{(10.51) First symbol received signal}$$

$$r_1 = -h_0 s_1^* + h_1 s_0^* + n_1 \quad \text{(10.52) Second symbol received signal}$$

We have now two RF channels present and to be able to detect them, alternate pilots should be sent by each antenna, so the receiver can estimate the channels independently. This scheme only works if the channels are approximately constant over the period of two symbols (coherence time should be larger than two symbols). Once the channels are estimated, the original signals can be obtained by a simple combination of the received signals and the estimated channel responses. The output signals

Table 10.7 Alamouti's Matrix A

Alamouti	Antenna 0	Antenna 1
Symbol 0	S_0	S_1
Symbol 1	$-S_1{}^*$	$S_0{}^*$

are presented in Equations (10.53) and (10.54).

$$\breve{s}_0 = h_0^* r_0 + h_1 r_1^* = (\alpha_0^2 + \alpha_1^2) s_0 + h_0^* n_0 + h_1 n_1^* \qquad (10.53)\text{ Space Time Block code}-s_0$$

$$\breve{s}_1 = h_1^* r_0 + h_0 r_1^* = (\alpha_0^2 + \alpha_1^2) s_1 - h_0 n_1^* + h_1^* n_0 \qquad (10.54)\text{ Space Time Block code}-s_1$$

This scheme has the same drawback as the MRC receiver, as the noise is amplified when one of the signals fades. The same solution suggested for MRC can be applied here. The Space Time Block Code is illustrated in Figure 10.37.

10.9 Transmit and Receive Diversity (TRD)

Transmit diversity can be mixed with MRC to provide a fourth order diversity MIMO scheme (2 × 2) as can be seen in Figure 10.38. The procedure is defined by Equations (10.55) to (10.60).

$$r_0 = h_0 s_0 + h_1 s_1 + n_0 \qquad (10.55)\text{ TRD received signal 0}$$

$$r_1 = -h_0 s_1^* - h_1 s_0^* + n_1 \qquad (10.56)\text{ TRD received signal 1}$$

$$r_2 = h_2 s_0 + h_3 s_1 + n_2 \qquad (10.57)\text{ TRD received signal 2}$$

$$r_3 = -h_2 s_1^* - h_3 s_0^* + n_3 \qquad (10.58)\text{ TRD received signal 3}$$

$$\breve{s}_0 = h_0^* r_0 + h_1 r_1^* + h_2^* r_2 + h_3 r_3^*$$
$$= (\alpha_0^2 + \alpha_1^2 + \alpha_2^2 + \alpha_3^2) s_0 + h_0^* n_0 + h_1 n_1^* + h_2^* n_2 + h_3 n_3^* \qquad (10.59)\text{ TRD output signal 0}$$

$$\breve{s}_1 = h_1^* r_0 - h_0 r_1^* + h_3^* r_2 - h_2 r_3^*$$
$$= (\alpha_0^2 + \alpha_1^2 + \alpha_2^2 + \alpha_3^2) s_1 - h_0 n_1^* + h_1^* n_0 - h_2 n_3^* + h_3^* n_2 \qquad (10.60)\text{ TRD output signal 1}$$

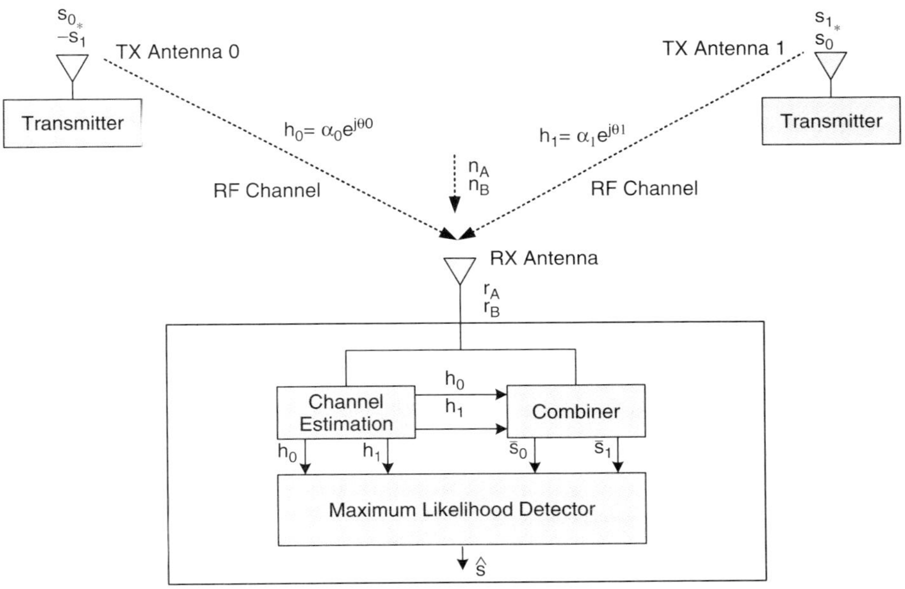

Figure 10.37 Matrix A MIMO.

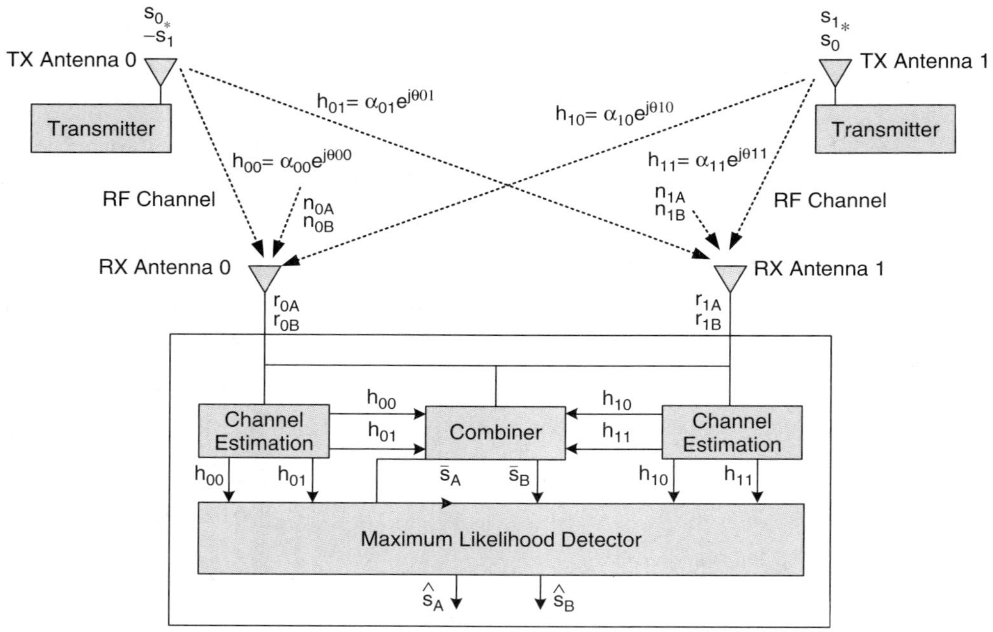

Figure 10.38 Transmit and receive diversity.

10.10 Spatial Multiplexing (Matrix B)

It is also possible to increase network capacity by sending different information from each transmit antenna. This is the case of Spatial Multiplexing (also known as Matrix B, or commonly referred to, albeit incorrectly, as MIMO B). In this technique, there is no diversity as the information transmitted by each antenna is different.

Each channel response is estimated using alternate pilots for each transmitter. It is defined in the WiMAX standard as Matrix B and is illustrated in Figure 10.39. Matrix B symbol allocation is shown in Equation (10.61).

It is possible to increase network capacity by sending different information from each transmit antenna. There is no diversity in this scheme. Each channel response is estimated using alternate pilots for each transmitter. It is defined in the WiMAX standard as Matrix B and is illustrated in Figure 10.39. Matrix B symbol allocation is shown in Equation (10.61).

$$X = [s_1 \quad s_2] \tag{10.61) Matrix B}$$

The received signals are expressed in Equations (10.62) to (10.64).

$$r_0 = h_{00}s_0 + h_{10}s_1 + n_0 \tag{10.62) Matrix B receive 0}$$

$$r_1 = h_{01}s_0 + h_{11}s_1 + n_1 \tag{10.63) Matrix B receive 1}$$

$$\begin{pmatrix} r_0 \\ r_1 \end{pmatrix} = \begin{pmatrix} h_{00} & h_{10} \\ h_{01} & h_{11} \end{pmatrix} \begin{pmatrix} s_0 \\ s_1 \end{pmatrix} + \begin{pmatrix} n_0 \\ n_1 \end{pmatrix} \tag{10.64) Matrix B receive signal}$$

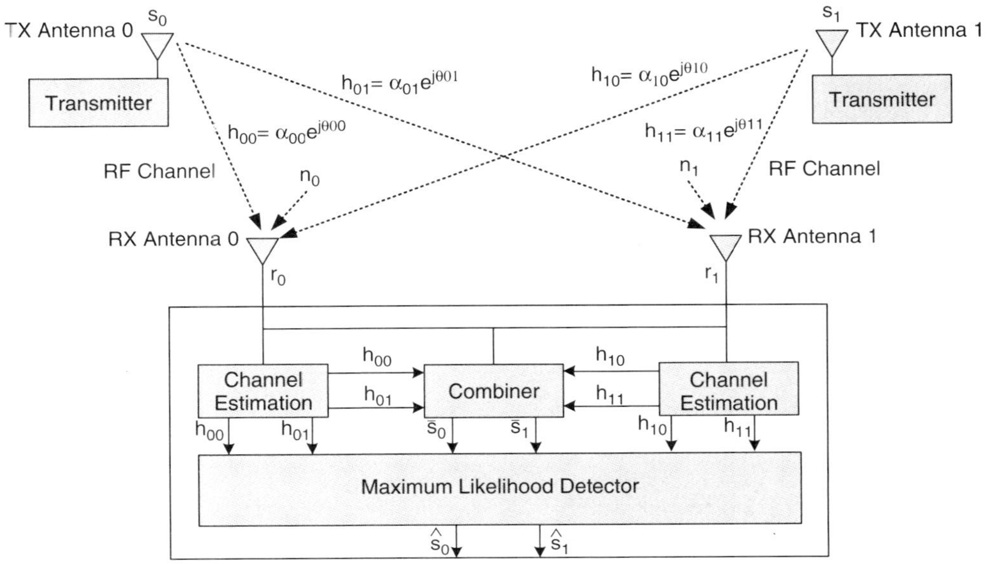

Figure 10.39 Matrix B MIMO.

The Maximum Likelihood Detector (MLD) has to consider possible combinations of s_0 and s_1, which could be a large number. The total number of combinations for 64QAM, 16QAM and 4 QAM is 14,512. This number can be reduced to 1,152 combinations if a quadrant approach is used. In this approach, instead of checking all possible combinations, quadrants are tested first, eliminating the rejected quadrant combinations. The MLD algorithm is shown in Equation (10.65).

$$D(s_0, s_1) = \{|r_0 - h_{00}s_0 - h_{21}s_2|^2 + |r_1 - h_{12}s_0 - h_{11}s_1|^2\} \quad (10.65) \text{ Maximum likelihood detector}$$

Table 10.8 shows the best combinations for the various numbers of antennas.

Table 10.8 MIMO type depending on number of antennas

	Number of RX antennas			
Number of TX antennas	1	2	3	4
1	Baseline	Downlink: MRC Uplink: Collaborative MIMO	MRC	MRC
2	STC (Matrix A)	2xSMX (Matrix B) STC+ 2xMRC (Matrix A)	2xSMX (Matrix B) STC+3xMRC (Matrix A)	STC+ 4xMRC (Matrix A)
4	STC (Matrix A)	2xSMX (Matrix B) STC+ 2xMRC (Matrix A)	2xSMX (Matrix B) STC+3xMRC (Matrix A)	4xSMX (Matrix C)

10.11 Diversity Performance

The performance of the different diversity methods is derived from the literature and we display average values here. Figures 10.40 to Figure 10.44 show the SNR required for different levels of Bit Error Rate depending on the number of antennas and modulation scheme.

Figures 10.45 to Figure 10.48 show gain provided by different MIMO techniques depending on the desired Bit Error Rate (BER).

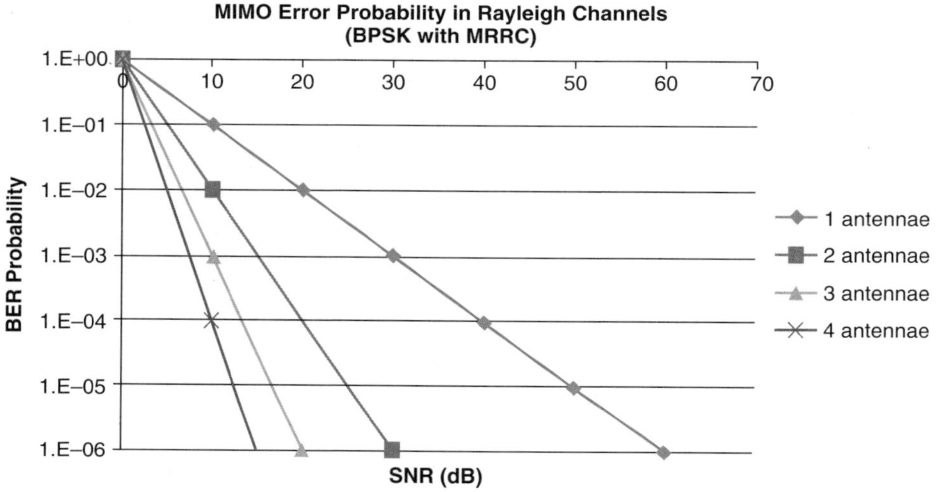

Figure 10.40 MIMO error probability in a Rayleigh channel.

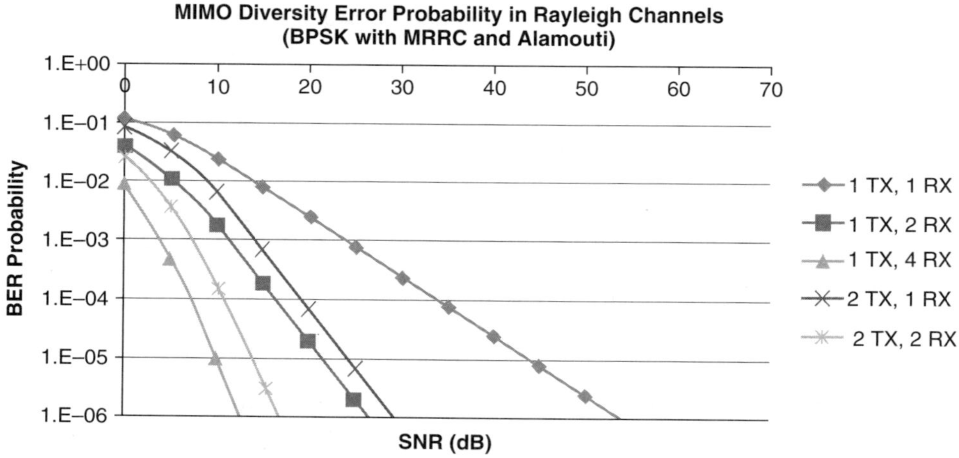

Figure 10.41 MIMO Diversity error probability in a Rayleigh channel.

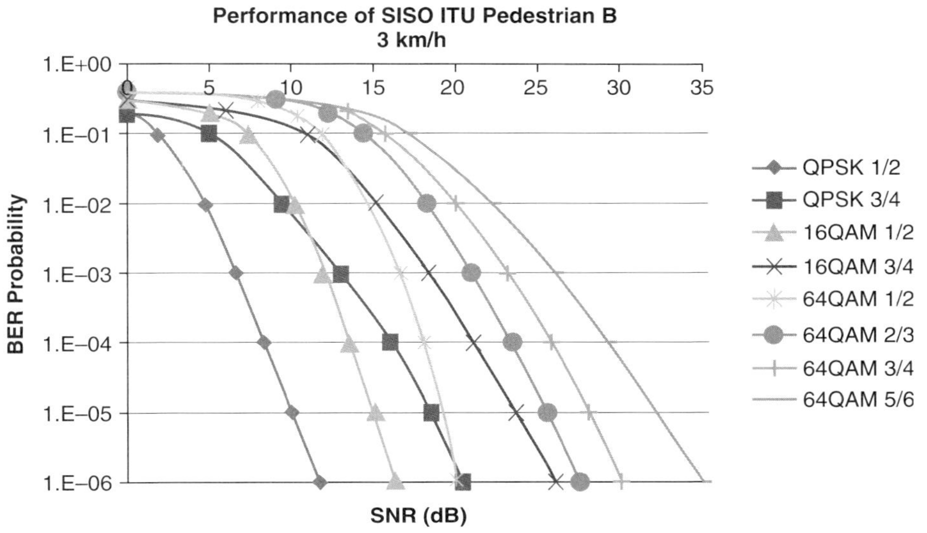

Figure 10.42 Performance of SISO ITU for Pedestrian B.

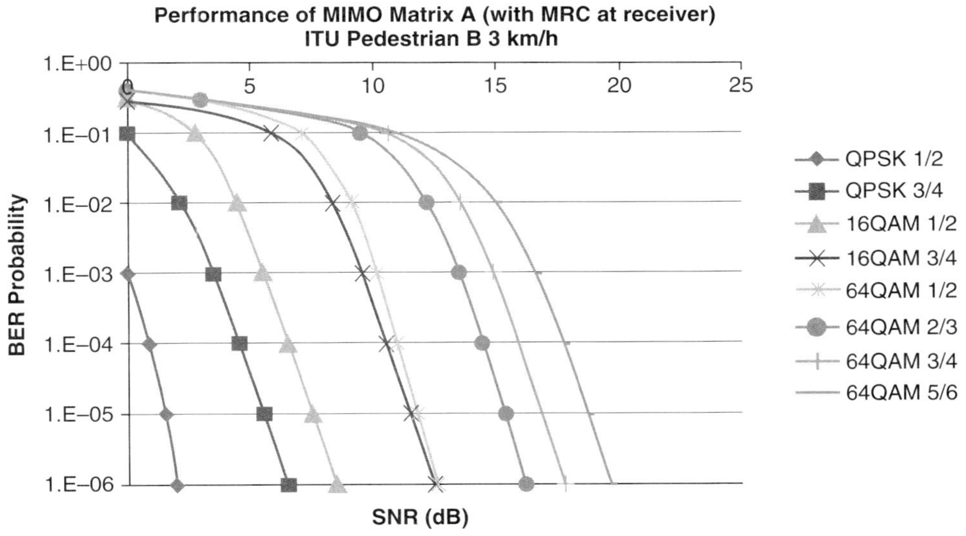

Figure 10.43 Performance of MIMO Matrix A.

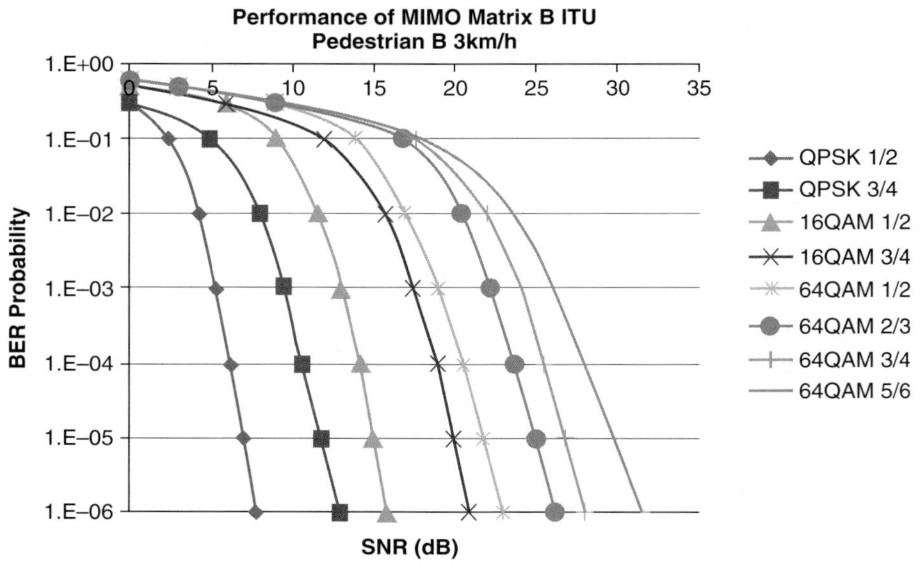

Figure 10.44 Performance of MIMO Matrix B.

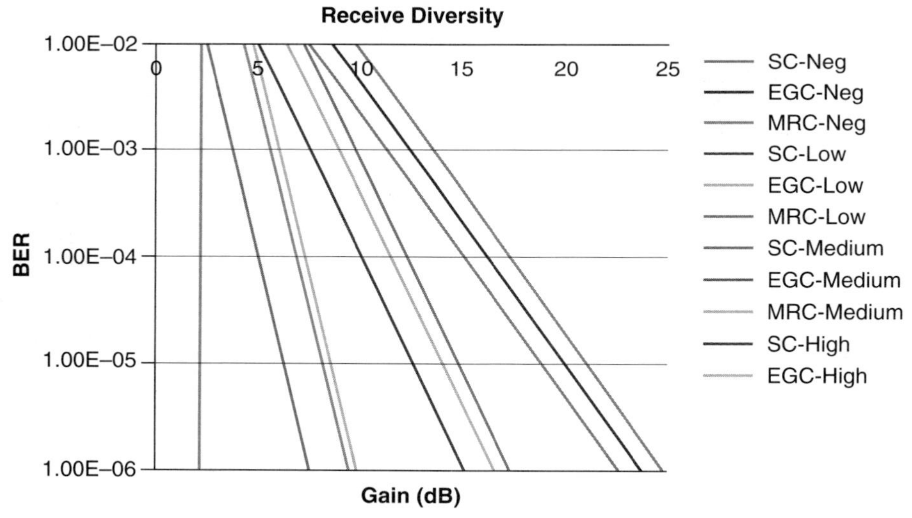

Figure 10.45 Performance of receive diversity technique.

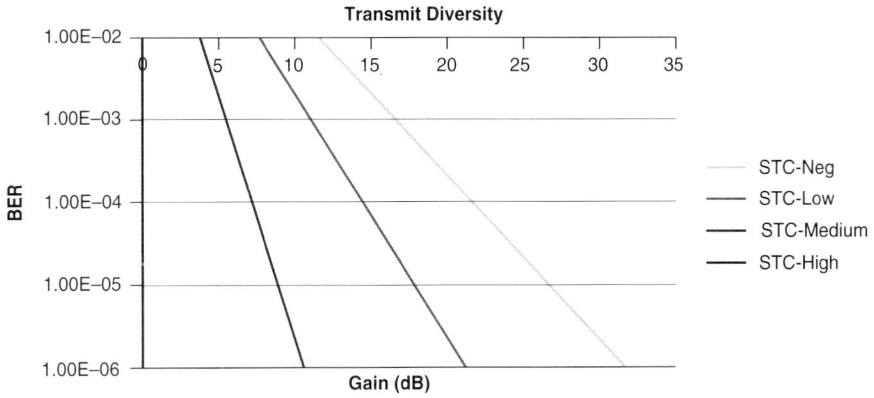

Figure 10.46 Performance of transmit diversity technique.

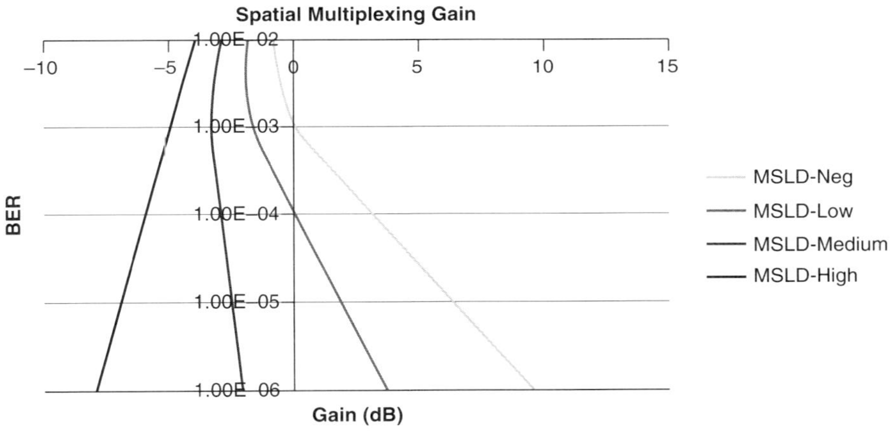

Figure 10.47 Performance of Spatial Multiplexing Gain.

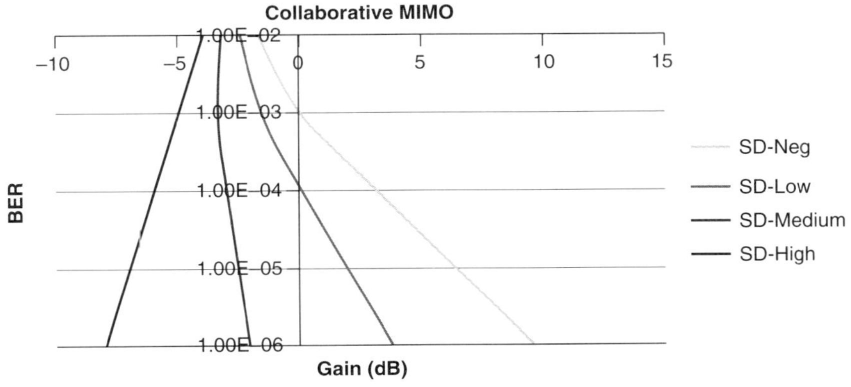

Figure 10.48 Performance of collaborative MIMO.

10.12 Antenna Array System (AAS), Advanced Antenna System (AAS) or Adaptive Antenna Steering (AAS) or Beamforming

Advanced antenna systems can be built by multiple elements which are fed with different signal phases and can generate nulls and poles at certain directions. This feature is used to reinforce signals and cancel interference. Two of the main methods, direction of arrival and antenna steering, are described next.

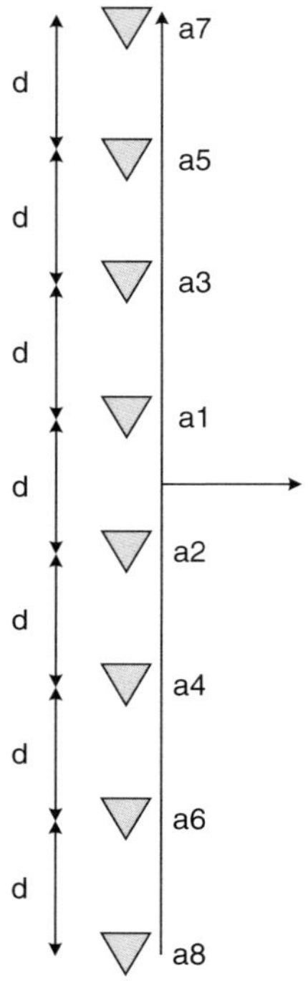

Figure 10.49 Array (linear) of antennas.

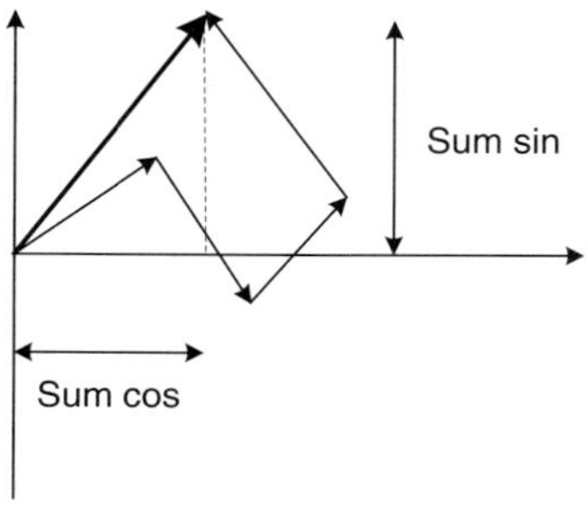

Figure 10.50 Pattern calculation for array of antennas.

- *Direction of Arrival (DoA) Beamforming* is done by detecting the direction with which the signal and interferers arrive, reinforcing the first one and canceling the others. The maximum number of cancelled signals is equal to the number of antenna elements minus one. The reinforcement is usually of the order of few dB and the canceling is not complete either. As different implementations have large variations, these parameters have to be specified at the design time based on the equipment used.
- *Antenna Steering or Beamforming* is used to direct the signal transmission or reception towards the desired signal or away from interferers. The concept is based on the combination of signals from an array of antennas. These arrays are also known as smart antennas and Figure 10.49 illustrates an array (linear) of eight antennas, spaced by a distance d.

Although the array is made up of omni antennas, it has a directional pattern that can be calculated by adding the signal received from each antenna at a certain distance, as illustrated in Figure 10.50. This combination results in the pattern of Figure 10.51 for eight antennas separated by $\lambda/2$.

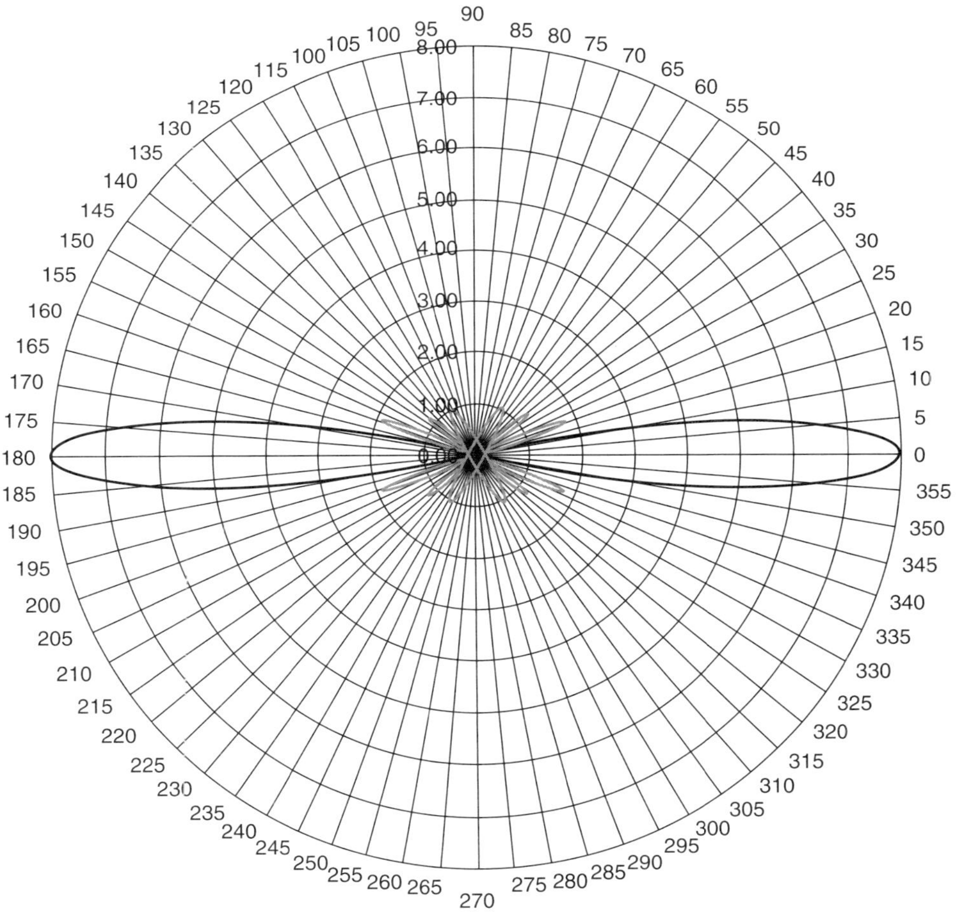

Figure 10.51 Antenna pattern for 8 antennas separated by $\lambda/2$.

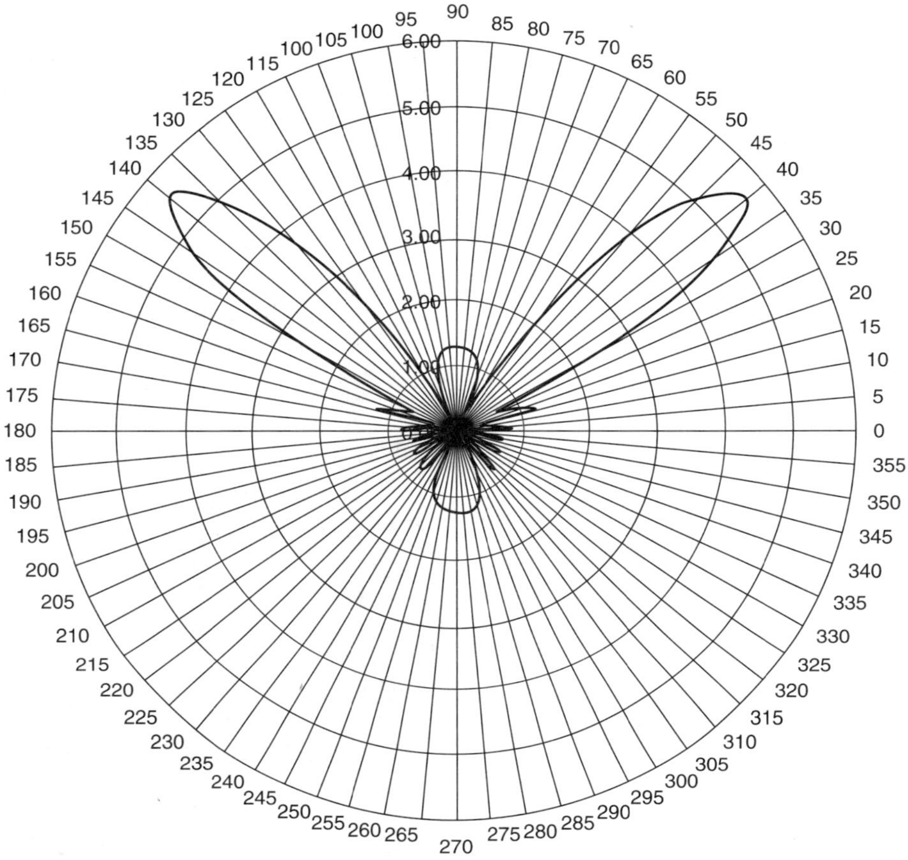

Figure 10.52 Modified antenna pattern.

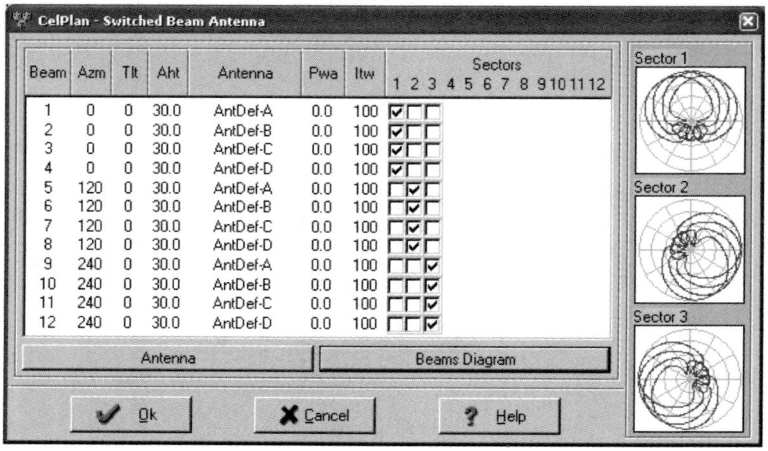

Figure 10.53 Static beamforming (switched beam antenna).

This pattern can be modified by changing the phases of the signals to each antenna, as shown in Figure 10.52.

This pattern forming can be static or adaptive. The static beamforming is known as switched beam and provides beams in pre-defined directions that can be turned on or off. Figure 10.53 shows an example of a dialog box for a configuration of this type of system in a planning tool.

The adaptive beamforming uses the information of symbols received by the array, to define the desired pattern. The signal phases to different antennas are adjusted dynamically. The array antennas can be distributed in a line, forming linear arrays, or on a plane, forming planar arrays.

The number of elements in the array defines how many directions can be chosen simultaneously. This number is equal to N-1 directions, where N is the number of elements in the array. The phases can be adjusted to enhance the reception for the direction or cancel the signal from it.

11

Radio Performance

11.1 Introduction

Network performance is ultimately defined by the radio capability to recover the original information. A radio is made of RF hardware and a signal processing hardware and software. The RF hardware has a defined SNR (Signal to Noise Ratio) to be able to extract the information from the received signal. Typical examples are a SINAD (Signal to Interference Noise And Distortion) of 12 dB for FM signals.

The signal processing hardware and software pre-process the information before transmitting it to improve the chances of recovery, by using error correction codes, interleaving and scrambling. All this results in different SNR requirements for different environments and different error rates, for each possible throughput. Due to this, radio performance has to be estimated for all possible operating conditions. In this chapter we cover how this estimation can be done. This methodology was derived from the CelPlanner software developed by CelPlan Technologies.

Basic radio performance can be defined by its Receive Sensitivity, which is the minimum input signal that results in an output with a desired signal to noise ratio and is defined by Equation (11.1).

$$S_i = k(T_t + T_{RX})B\frac{S_0}{N_0} \qquad \text{(11.1) Receiver sensitivity}$$

where:

S_i = Sensitivity in Watt.
k = Boltzmann's constant (1.38×10^{-23} J/K).
T_t = Source thermal noise at input (290°K).
T_{RX} = Equivalent noise temperature increment of the receiver (typically 4°K for the BS and 12°K for the mobile).
B = Bandwidth (Hz).
S_0/N_0 = Minimum required SNR at output.

This equation can be divided into three parts:

1. the input RF noise;
2. the receiver circuit thermal noise;
3. the required Signal to Noise Ratio.

LTE, WiMAX and WLAN Network Design, Optimization and Performance Analysis, First Edition.
Leonhard Korowajczuk.

Table 11.1 RF noise for different bandwidths

Bandwidth (Hz)	RF Noise (dBm)
1	−174
10	−164
100	−154
1 K	−144
10 K	−134
30 K	−129
100 K	−124
200 K	−121
1 M	−114
1.5 M	−112
5 M	−107
10 M	−104
20 M	−101
40 M	−98

11.2 Input RF Noise

The input RF signal noise that is dependent of the antenna's environment temperature, is generally assumed as 290°K (17°C, 88°F), and defined by kT_tB. The input RF noise expressed in dBm is given by Equation (11.2). Table 11.1 gives thermal noise values for different bandwidths.

$$N_{\text{RFdbm}} = -174 + 10\log B \qquad \text{(11.2) Input RF signal noise}$$

11.3 Receive Circuit Noise

The receive circuit noise is of thermal origin and defined by $kT_{RX}B$, and it expresses the increase in noise caused by circuit components. Typical receive circuit noise values are 4°K for the BS and 12°K for mobile radios. This noise increase can be expressed in dB and then is known as Noise Figure (NF). Noise figure can be measured by the ratio of the input noise to the circuit noise where the SNR (Signal to Noise Ratio) is being measured, as shown in Equation (11.3).

$$NF = SNR_{\text{in}}/SNR_{\text{out}} \qquad \text{(11.3) Receive circuit noise}$$

Noise figures are published by manufacturers for active components, and for passive stages the noise figure is equal to its attenuation. When many stages are cascaded, the total noise figure can be calculated from noise figures and gains of each stage by the use of the Friis formula (Harold T. Friis, 1883–1976) shown in Equation (11.4).

$$NF = NF_1 + \frac{NF_2 - 1}{G_1} + \frac{NF_3 - 1}{G_1G_2} + \frac{NF_4 - 1}{G_1G_{2G_s}} + \cdots + \frac{NF_n - 1}{G_1G_2\cdot\cdots G_n - 1} \qquad \text{(11.4) Noise figure}$$

11.4 Signal to Noise Ratio

A received signal does not have a constant power value and can only be defined by a statistical distribution. When line of sight (LOS) is present, the distribution tends to be Gaussian, and when no line of sight is available, it tends to follow a Rayleigh distribution. Noise typically has a Gaussian distribution.

SNR distribution, if it is to be properly expressed, should have an average value and a statistical distribution associated with it. It is difficult to determine the resultant type of distribution, but in practical terms we will have to add to the noise many interfering signals and based on the Central Limit Theorem we can assume that the distribution is Gaussian, due to the large number of independent contributors to the noise.

Conventionally, SNR values are expressed by their average value assuming a Gaussian distribution, although the standard deviation of the distribution is rarely mentioned. In our designs we have measured standard deviation values between 6 and 10 dB.

11.4.1 Modulation Constellation SNR

The SNR requirement is directly connected to the modulation constellation used and the amplitude distance between the symbols. A certain average SNR value will cause an amount of wrong detections and it is possible to map SNR average values to Bit Error rates (BER).

For BPSK, a signal with a Gaussian distribution has the BER given by Equation (11.5).

$$P_b = \frac{1}{2} erfc \left(\sqrt{\frac{E_b}{N_0}} \right) \qquad \text{(11.5) BER probability for BPSK}$$

where:

P_b = Probability of receiving one bit in error.
E_b = Bit energy.
N_0 = Noise energy.

This equation can also be expressed in terms of SNR as expressed in Equation (11.6).

$$\text{SNR} = \frac{E_B R}{N_0 B} \qquad \text{(11.6) SNR}$$

where:

B = Bandwidth.
R = Data Rate.

B/R represents the spectral efficiency and for a value of 1 we have the BER probability expressed by Equation (11.7).

$$P_b = \frac{1}{2} erfc \left(\sqrt{\frac{S}{N}} \right) \qquad \text{(11.7) BER probability in terms of SNR}$$

Figure 11.1 shows the BER curve for BPSK modulation in an AWGN channel and in a Rayleigh channel. Both curves have similar SNR requirement for high BER values (e.g. 10^{-1}), but this requirement increases significantly for low BER values (e.g. 10^{-5}).

11.4.2 Error Correction Codes

Error correction codes can be applied to reduce BER by adding redundant information to the data. Suppose that a network can accept an error rate of 10^{-4}, which will be corrected by repetition at

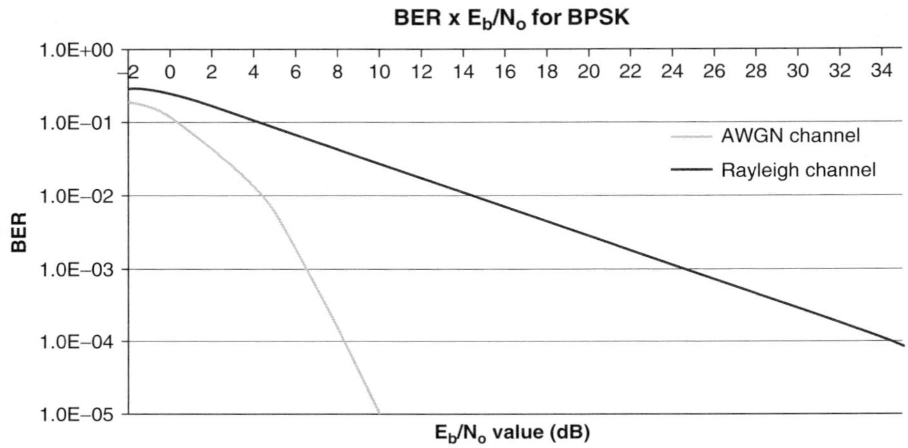

Figure 11.1 E_b/N_0 requirement for different BER for BPSK modulation.

higher levels, like TCP/IP. For a Rayleigh channel it will require, per Figure 11.1, an SNR of 34 dB. An error correction could allow the radio to work at a BER of 10^{-2}, by correcting the number of errors and reducing the BER to 10^{-4}, requiring an SNR of 14 dB. The lower SNR requirement would allow a higher modulation scheme to be used to compensate for the throughput loss by the error correction code overhead. Changing the modulation scheme from BPSK to 64QAM increases the throughput 6 times, while an error correction code of ½ (data to data plus error correction code ratio), reduces it by 2 times. From the graph it can be seen that error correction codes are more advantageous in Rayleigh channels than in Gaussian channels.

The advantage of error correction codes is that they reduce the number of retransmissions and fix errors instantly on decoding. They also allow the use of higher modulations with acceptable BER. The disadvantage of error correcting codes is that they represent an overhead and thus reduce data throughput. The final throughput obtained is what determines if their use is advantageous.

This improvement cannot be easily established as it varies with the type of channel and the SNR value. A practical rule of thumb is that they are more efficient at high SNR levels and for non-line of sight situations.

The maximum system throughput capacity was calculated by Shannon in 1948, based on ideas proposed by Nyquist and Hartley, and is shown in Equation (11.8).

$$C = B\log_2\left(1 + \frac{S}{N}\right) \quad \text{(11.8) Shannon capacity}$$

where;

C = Channel capacity in bits/s.
B = bandwidth in Hz.
N = total noise in W or V2.
S/N = Signal to Noise Ratio (SNR) or Carrier to Noise Ratio (CNR) expressed in linear units.

Table 11.2 gives the channel capacity for different bandwidths and SNR values. It can be approximated by Equation (11.9).

$$C = 0.32 * \text{SNR}(dB) * B \quad \text{(11.9) Approximate channel capacity value}$$

Table 11.2 Shannon's channel capacity

	Shannon's capacity (bit)							
	SNR (dB)							
Bandwidth in Hz	5	10	15	20	25	30	35	40
1	1.6	3.3	4.9	6.6	8.3	9.9	11.6	13.2
10	17	33	50	66	83	100	116	133
100	166	332	498	664	830	997	1,163	1,329
1,000	1,661	3,322	4,983	6,644	8,305	9,966	11,627	13,288
10,000	16,610	33,219	49,829	66,439	83,048	99,658	116,267	132,877
100,000	166,096	332,193	498,289	664,386	830,482	996,578	1,162,675	1,328,771
1,000,000	1,660,964	3,321,928	4,982,892	6,643,856	8,304,820	9,965,784	11,626,748	13,287,712
10,000,000	16,609,640	33,219,281	49,828,921	66,438,562	83,048,202	99,657,843	116,267,483	132,877,124

The most common error correction codes used today are Forward Error Correction codes (FEC). They have this name because they add redundant data in advance of the error occurrence. They can be classified into two types:

- *Convolutional codes:* process the information on a bit-by-bit basis and are most suitable to be implemented in hardware. A Viterbi decoder (1967) is usually used to implement them and is an optimum decoder.
- *Block codes:* process information on a block-by-block basis. Turbo codes and Low Density Parity-check codes (LDPC) are two options of block codes and can provide nearly optimal efficiency.

Shannon also calculated the maximum capacity that can be achieved by an error correction code for different BERs. This value can be calculated by the formula in Equation (11.10) and is presented in Table 11.3 for a channel with 10 MHz bandwidth:

$$R(p_b) = \frac{C}{1 - p_b \log p_b} \qquad (11.10)\ \text{Maximum channel data rate}$$

where:

$R(p_b)$ = Maximum data rate that can be achieved in the channel.
C = Channel capacity for the desired bandwidth and available SNR.
p_b = Bit error rate probability (ratio of bits with error to total number of bits).

Table 11.3 Shannon's capacity for different received BER

	Maximum capacity for a 10 MHz bandwidth for different BER							
	Bandwidth in Hz							
BER	5	10	15	20	25	30	35	40
0.1	14,554,772	29,109,545	43,664,317	58,219,089	72,773,862	87,328,634	101,883,407	116,438,179
0.01	16,215,266	32,430,533	48,645,799	64,861,066	81,076,332	97,291,599	113,506,865	129,722,131
0.001	16,552,797	33,105,594	49,658,391	66,211,188	82,763,984	99,316,781	115,869,578	132,422,375
0.0001	16,602,279	33,204,557	49,806,836	66,409,114	83,011,393	99,613,671	116,215,950	132,818,229
0.00001	16,608,738	33,217,476	49,826,214	66,434,952	83,043,690	99,652,427	116,261,165	132,869,903
0.000001	16,609,534	33,219,067	49,828,601	66,438,134	83,047,668	99,657,202	116,266,735	132,876,269

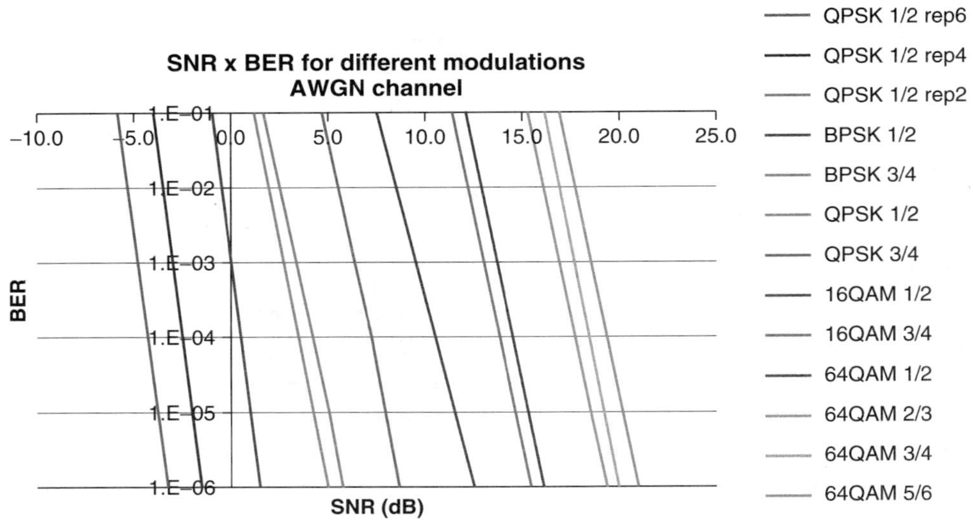

Figure 11.2 SNR requirement for different BER for various modulations in an AWGN channel.

Figure 11.2 shows typical SNR required for each BER at different modulations for an AWGN (Additive White Gaussian Noise) channel, using a Convolutional Turbo Code. Those are theoretical values and practical values can vary with the actual implementation.

The SNR requirement in a Rayleigh channel for different BER and different modulations is shown in Figure 11.3. The increase in SNR requirement from a BER of 10^{-1} to 10^{-6} is about 30 dB.

There are many tradeoffs that can be applied at this point, such as increasing the operational BER and correcting errors using an error correction code. The addition of the error correction code decreases the throughput, but this can be compensated for by the use of a higher modulation due to a smaller SNR required.

The combination of modulation and coding rate is called a modulation scheme. The ratio used to express the coding rate represents the number of data bits over the total number of bits, that is, it indicates the impact of the error correction codes in the throughput.

Table 11.4 gives some examples of possible combinations of modulation and coding rates for the same SNR. Figures 11.1–11.3 are used to get the required SNR and corresponding error rates. Table 11.4 is divided into four sets for comparison purposes.

- In the first set, the channel is Rayleigh and we assume a SNR of 14 dB. This SNR results in an error rate of 10^{-2} for BPSK modulation (from Figure 11.1). Using a coding rate of $1/2$ the BER can be reduced to 10^{-3} (from Figure 11.3), but the throughput drops by 50%.Using QPSK with a coding rate of $1/2$, will keep the error rate at 10^{-2} and provide the same output as without any coding. So, there is no advantage or disadvantage in using error correction.
- In the second set, a SNR = 34 dB was considered and using the same reasoning as above, an advantage is obtained in using coding, as the throughput can be increased more than threefold.
- In the third set, the channel is AWGN and coding reduces throughput.
- The fourth set uses also AWGN and provides a throughput gain around 1.5 times.

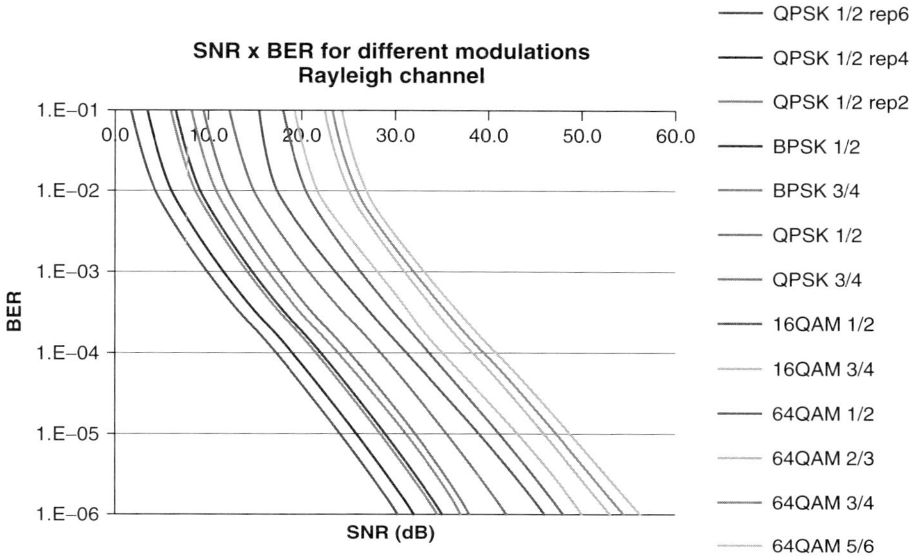

Figure 11.3 SNR requirement for different BER for various modulations in a Rayleigh channel.

Table 11.4 Comparison of modulation schemes

Channel	SNR (dB)	Modulation	Coding Rate	BER	Normalized throughput
Rayleigh	14	BPSK		10^{-2}	1
Rayleigh	14	BPSK	1/2	10^{-3}	0.5
Rayleigh	14	QPSK	1/2	10^{-2}	1
Rayleigh	34	BPSK		10^{-4}	1
Rayleigh	34	BPSK	1/2	10^{-6}	0.5
Rayleigh	34	QPSK	1/2	5×10^{-6}	1
Rayleigh	34	16QAM	1/2	5×10^{-5}	2
Rayleigh	34	64QAM	1/2	10^{-4}	3
AWGN	5	BPSK		10^{-2}	1
AWGN	5	QPSK	1/2	10^{-6}	0.5
AWGN	5	QPSK	3/4	10^{-1}	0.75
AWGN	10	BPSK		10^{-5}	1
AWGN	10	BPSK	1/2	10^{-7}	0.5
AWGN	10	16QAM	1/2	10^{-3}	2

The modulation scheme has to be chosen at the transmitter, so the channel performance has to be evaluated before that decision. New wireless broadband technologies use many modulation schemes, and the choice is automatically done on a package-by-package basis, based on a SNR threshold table, configured by the user or embedded in the radio software code. This procedure is called AMC (Adaptive Modulation and Coding).

11.4.3 SNR and Throughput

Figure 11.4 shows the items that influence Throughput (left side) and SNR (right side). The first item is the modulation scheme (modulation and error correction coding), which is determined by the SNR and required BER and defines the maximum throughput. Speed and permutations change fading characteristics, thus affecting the SNR requirements.

Automatic Repeat reQuest (ARQ) or Hybrid ARQ (HARQ) is a layer 2 procedure that allows error correction by resending messages that did not receive an acknowledgement within a certain time frame. It may be advantageous to specify each modulation scheme at higher BER levels and correct the errors using ARQ.

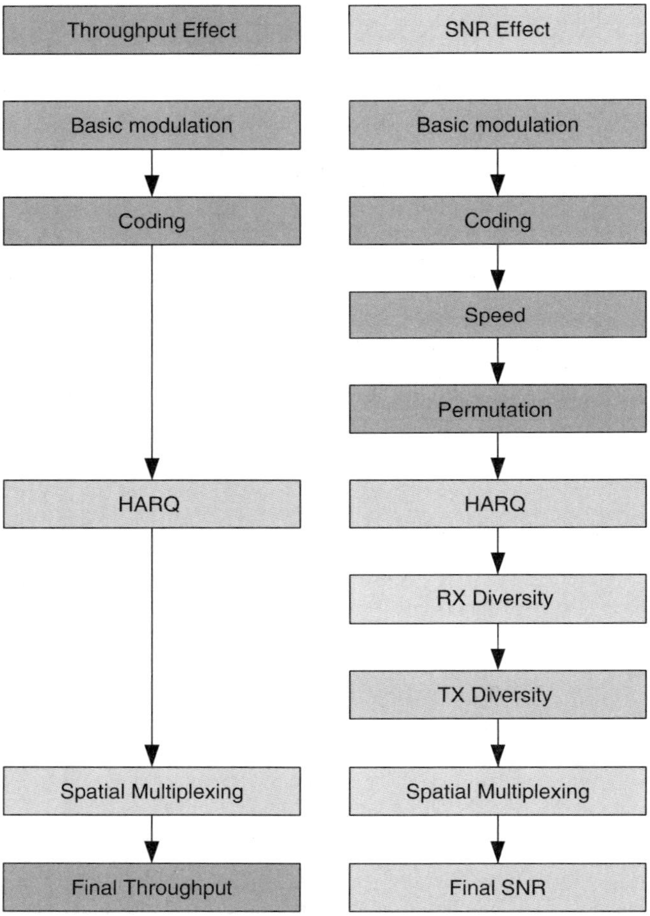

Figure 11.4 Throughput calculation in WiMAX systems.

The ideal BER threshold depends on the average message length. For example, a 10^{-4} BER for average message lengths of 60 Bytes means that 1 in 20 messages will be received in error. The retransmission of this message increases the traffic by slightly more than 5%.

Finally, the antenna algorithms (diversity or spatial multiplexing) applied to the transmission and reception, affect the SNR and, consequently, the throughput.

11.5 Radio Sensitivity Calculations

During the design we must be able to calculate the network's throughput. This requires estimating channel characteristics and the effects of each of the techniques available for recovering the transmitted signal.

Next, a step by step approach to calculate the throughput of a wireless connection over different channel types is presented. The calculation should be done for all available modulation schemes, different BERs (10^{-1} to 10^{-6}) and different channel fading environments as defined by ranges of the k parameter in the Ricean distribution, listed below.

- Rayleigh: $k<=1$
- Ricean: $1 < k<=2$
- Ricean: $2 < k<=10$
- Gaussian: $k > 10$

Impairments or gains due to the techniques used in WiMAX and LTE are considered in terms of SNR and throughput. The final result is obtained from the sum of gains and losses for each technique applied.

All of this is summarized in a set of tables presented as Tables 11.5 to 11.14. Values vary with implementations from vendor to vendor, so actual values should be obtained from vendors or from measurements.

We next illustrate the tables used by CelPlanner as implemented by CelPlan Technologies. These tables are essential to calculate properly the SNR and from it the radio sensitivity. The tables required to get the desired value are:

- Modulation Scheme SNR
- FEC Algorithms Gains
- Mobility Effect
- Permutation Effect
- HARQ Effect
- Improvement Reduction Factor for Antenna Systems
- Receive Diversity
- Transmit Diversity
- Spatial Multiplexing

This results in thousands of possible combinations and the design tool has to calculate the actual value on a prediction pixel-by-pixel basis. A table summarizing the values should be available for the designer, so it can evaluate the sensitivity numbers used by the design tool. An example of such a table is shown in Figure 11.8 on p. 308.

11.5.1 Modulation Scheme SNR

Table 11.5 Static fading, CTC, no permutation, required SNR in dB

Fading Type	Rayleigh k<=1						Ricean 1<k<=2						Ricean 2<k<=10						AWGN k>10					
BER Scheme	10^{-1}	10^{-2}	10^{-3}	10^{-4}	10^{-5}	10^{-6}	10^{-1}	10^{-2}	10^{-3}	10^{-4}	10^{-5}	10^{-6}	10^{-1}	10^{-2}	10^{-3}	10^{-4}	10^{-5}	10^{-6}	10^{-1} 1.E-01	10^{-2} 1.E-02	10^{-3} 1.E-03	10^{-4} 1.E-04	10^{-5} 1.E-05	10^{-6} 1.E-06
QPSK 1/2 rep 6	1.72	4.28	9.76	17.22	24.32	30.22	−0.78	1.09	4.91	10.05	14.95	19.05	−3.28	−2.09	0.07	2.89	5.59	7.89	−5.78	−5.28	−4.78	−4.28	−3.78	−3.28
QPSK 1/2 rep 4	3.50	6.06	11.54	19.00	26.10	32.00	1.00	2.87	6.69	11.83	16.73	20.83	−1.50	−0.31	1.85	4.67	7.37	9.67	−4.00	−3.50	−3.00	−2.50	−2.00	−1.50
QPSK 1/2 rep 2	6.50	9.06	14.54	22.00	29.10	35.00	4.00	5.87	9.69	14.83	19.73	23.83	1.50	2.69	4.85	7.67	10.37	12.67	−1.00	−0.50	0.00	0.50	1.00	1.50
BPSK 1/2	6.50	9.06	14.54	22.00	29.10	35.00	4.00	5.87	9.69	14.83	19.73	23.83	1.50	2.69	4.85	7.67	10.37	12.67	−1.00	−0.50	0.00	0.50	1.00	1.50
BPSK 3/4	9.20	12.66	19.16	27.92	36.30	43.40	6.70	9.27	13.87	20.01	25.87	30.83	4.20	5.89	8.59	12.11	15.43	18.27	1.70	2.50	3.30	4.20	5.00	5.70
QPSK 1/2	9.50	12.06	17.54	25.00	32.10	38.00	7.00	8.87	12.69	17.83	22.73	26.83	4.50	5.69	7.85	10.67	13.37	15.67	2.00	2.50	3.00	3.50	4.00	4.50
QPSK 3/4	12.20	15.66	22.16	30.92	39.30	46.40	9.70	12.27	16.87	23.01	28.87	33.83	7.20	8.89	11.59	15.11	18.43	21.27	4.70	5.50	6.30	7.20	8.00	8.70
16QAM 1/2	15.30	17.32	23.60	30.98	39.00	46.00	12.70	14.38	18.90	24.15	29.83	34.83	10.10	11.44	14.20	17.33	20.67	23.67	7.50	8.50	9.50	10.50	11.50	12.50
16QAM 3/4	19.20	21.76	28.50	36.26	44.58	51.80	16.60	18.57	23.33	28.77	34.59	39.67	14.00	15.39	18.17	21.29	24.59	27.53	11.40	12.20	13.00	13.80	14.60	15.40
64QAM 1/2	19.70	21.24	26.36	33.12	39.82	46.60	17.17	18.46	22.14	26.91	31.65	36.43	14.63	15.68	17.92	20.71	23.47	26.27	12.10	12.90	13.70	14.50	15.30	16.10
64QAM 2/3	22.40	25.08	30.84	38.24	45.58	53.00	20.03	22.09	26.19	31.39	36.55	41.77	17.67	19.09	21.55	24.55	27.53	30.53	15.30	16.10	16.90	17.70	18.50	19.30
64QAM 3/4	22.70	26.04	31.96	39.52	47.02	54.60	20.50	22.99	27.21	32.51	37.78	43.10	18.30	19.95	22.45	25.51	28.54	31.60	16.10	16.90	17.70	18.50	19.30	20.10
64QAM 5/6	23.00	27.00	33.08	40.80	48.46	56.20	20.97	23.90	28.22	33.63	39.01	44.43	18.93	20.80	23.36	26.47	29.55	32.67	16.90	17.70	18.50	19.30	20.10	20.90

11.5.2 FEC Algorithm Gains

Table 11.6 Coding factor in relation to CTC (dB)

Fading type	Rayleigh k<=1						Ricean 1<k<=2						Ricean 2<k<=10						AWGN k>10					
BER	10^{-1}	10^{-2}	10^{-3}	10^{-4}	10^{-5}	10^{-6}	10^{-1}	10^{-2}	10^{-3}	10^{-4}	10^{-5}	10^{-6}	10^{-1}	10^{-2}	10^{-3}	10^{-4}	10^{-5}	10^{-6}	10^{-1}	10^{-2}	10^{-3}	10^{-4}	10^{-5}	10^{-6}
CTC	0	0	0	0	0	0	0	0	0	0	0	0	0	0	0	0	0	0	0	0	0	0	0	0
CC	−2.5	−2.5	−2.5	−2.5	−2.5	−2.5	−2.5	−2.5	−2.5	−2.5	−2.5	−2.5	−2.5	−2.5	−2.5	−2.5	−2.5	−2.5	−2.5	−2.5	−2.5	−2.5	−2.5	−2.5
LDPC	−0.5	−0.5	−0.5	−0.5	−0.5	−0.5	−0.5	−0.5	−0.5	−0.5	−0.5	−0.5	−0.5	−0.5	−0.5	−0.5	−0.5	−0.5	−0.5	−0.5	−0.5	−0.5	−0.5	−0.5
BTC	0.5	0.5	0.5	0.5	0.5	0.5	0.5	0.5	0.5	0.5	0.5	0.5	0.5	0.5	0.5	0.5	0.5	0.5	0.5	0.5	0.5	0.5	0.5	0.5
ZCC	−0.5	−0.5	−0.5	−0.5	−0.5	−0.5	−0.5	−0.5	−0.5	−0.5	−0.5	−0.5	−0.5	−0.5	−0.5	−0.5	−0.5	−0.5	−0.5	−0.5	−0.5	−0.5	−0.5	−0.5

11.5.3 Mobility Effect

Table 11.7 AMC symbol, no permutation for different channels, SNR improvement in dB

Fading type	Rayleigh k<=1						Ricean 1<k<=2						Ricean 2<k<=10						AWGN k>10						
BER Schemes	10^{-1}	10^{-2}	10^{-3}	10^{-4}	10^{-5}	10^{-6}	10^{-1}	10^{-2}	10^{-3}	10^{-4}	10^{-5}	10^{-6}	10^{-1}	10^{-2}	10^{-3}	10^{-4}	10^{-5}	10^{-6}	10^{-1}	10^{-2}	10^{-3}	10^{-4}	10^{-5}	10^{-6}	Speed (km/h)
Speed=0 km/h																									
QPSK 1/2 rep 6	0	0	0	0	0	0	0	0	0	0	0	0	0	0	0	0	0	0	0	0	0	0	0	0	0
QPSK 1/2 rep 4	0	0	0	0	0	0	0	0	0	0	0	0	0	0	0	0	0	0	0	0	0	0	0	0	0
QPSK 1/2 rep 2	0	0	0	0	0	0	0	0	0	0	0	0	0	0	0	0	0	0	0	0	0	0	0	0	0
BPSK 1/2	0	0	0	0	0	0	0	0	0	0	0	0	0	0	0	0	0	0	0	0	0	0	0	0	0
BPSK 3/4	0	0	0	0	0	0	0	0	0	0	0	0	0	0	0	0	0	0	0	0	0	0	0	0	0
QPSK 1/2	0	0	0	0	0	0	0	0	0	0	0	0	0	0	0	0	0	0	0	0	0	0	0	0	0
QPSK 3/4	0	0	0	0	0	0	0	0	0	0	0	0	0	0	0	0	0	0	0	0	0	0	0	0	0
16QAM 1/2	0	0	0	0	0	0	0	0	0	0	0	0	0	0	0	0	0	0	0	0	0	0	0	0	0
16QAM 3/4	0	0	0	0	0	0	0	0	0	0	0	0	0	0	0	0	0	0	0	0	0	0	0	0	0
64QAM 1/2	0	0	0	0	0	0	0	0	0	0	0	0	0	0	0	0	0	0	0	0	0	0	0	0	0
64QAM 2/3	0	0	0	0	0	0	0	0	0	0	0	0	0	0	0	0	0	0	0	0	0	0	0	0	0
64QAM 3/4	0	0	0	0	0	0	0	0	0	0	0	0	0	0	0	0	0	0	0	0	0	0	0	0	0
64QAM 5/6	0	0	0	0	0	0	0	0	0	0	0	0	0	0	0	0	0	0	0	0	0	0	0	0	0
Speed=3 km/h																									
QPSK 1/2 rep 6	−1	1	0	1	2	3	−1	1	0	1	2	3	−1	1	0	1	2	3	−1	1	0	1	2	0	3
QPSK 1/2 rep 4	−1	1	0	1	2	3	−1	1	0	1	2	3	−1	1	0	1	2	3	−1	1	0	1	2	0	3
QPSK 1/2 rep 2	−1	1	0	1	2	3	−1	1	0	1	2	3	−1	1	0	1	2	3	−1	1	0	1	2	0	3
BPSK 1/2	−1	1	0	1	2	3	−1	1	0	1	2	3	−1	1	0	1	2	3	−1	1	0	1	2	0	3
BPSK 3/4	−1	1	0	1	2	3	−1	1	0	1	2	3	−1	1	0	1	2	3	−1	1	0	1	2	0	3
QPSK 1/2	−1	1	0	1	2	3	−1	1	0	1	2	3	−1	1	0	1	2	3	−1	1	0	1	2	3	3
QPSK 3/4	−1	1	0	1	2	3	−1	1	0	1	2	3	−1	1	0	1	2	3	−1	1	0	1	2	3	3
16QAM 1/2	−1	1	0	1	2	3	−1	1	0	1	2	3	−1	1	0	1	2	3	−1	1	0	1	2	3	3
16QAM 3/4	−1	1	0	1	2	3	−1	1	0	1	2	3	−1	1	0	1	2	3	−1	1	0	1	2	3	3
64QAM 1/2	−1	1	0	1	2	3	−1	1	0	1	2	3	−1	1	0	1	2	3	−1	1	0	1	2	3	3
64QAM 2/3	−1	1	0	1	2	3	−1	1	0	1	2	3	−1	1	0	1	2	3	−1	1	0	1	2	3	3
64QAM 3/4	−1	1	0	1	2	3	−1	1	0	1	2	3	−1	1	0	1	2	3	−1	1	0	1	2	3	3
64QAM 5/6	−1	1	0	1	2	3	−1	1	0	1	2	3	−1	1	0	1	2	3	−1	1	0	1	2	3	3

Speed = 30 km/h																									
QPSK 1/2 rep 6	−2	−1	0	1	2	4	−2	−1	0	1	2	4	−2	−1	0	1	2	4	−2	−1	0	1	2	0	30
QPSK 1/2 rep 4	−2	−1	0	1	2	4	−2	−1	0	1	2	4	−2	−1	0	1	2	4	−2	−1	0	1	2	0	30
QPSK 1/2 rep 2	−2	−1	0	1	2	4	−2	−1	0	1	2	4	−2	−1	0	1	2	4	−2	−1	0	1	2	0	30
BPSK 1/2	−2	−1	0	1	2	4	−2	−1	0	1	2	4	−2	−1	0	1	2	4	−2	−1	0	1	2	0	30
BPSK 3/4	−2	−1	0	1	2	4	−2	−1	0	1	2	4	−2	−1	0	1	2	4	−2	−1	0	1	2	0	30
QPSK 1/2	−2	−1	0	1	2	4	−2	−1	0	1	2	4	−2	−1	0	1	2	4	−2	−1	0	1	2	4	30
QPSK 3/4	−2	−1	0	1	2	4	−2	−1	0	1	2	4	−2	−1	0	1	2	4	−2	−1	0	1	2	4	30
16QAM 1/2	−2	−1	0	1	2	4	−2	−1	0	1	2	4	−2	−1	0	1	2	4	−2	−1	0	1	2	4	30
16QAM 3/4	−2	−1	0	1	2	4	−2	−1	0	1	2	4	−2	−1	0	1	2	4	−2	−1	0	1	2	4	30
64QAM 1/2	−2	−1	0	1	2	4	−2	−1	0	1	2	4	−2	−1	0	1	2	4	−2	−1	0	1	2	4	30
64QAM 2/3	−2	−1	0	1	2	4	−2	−1	0	1	2	4	−2	−1	0	1	2	4	−2	−1	0	1	2	4	30
64QAM 3/4	−2	−1	0	1	2	4	−2	−1	0	1	2	4	−2	−1	0	1	2	4	−2	−1	0	1	2	4	30
64QAM 5/6	−2	−1	0	1	2	4	−2	−1	0	1	2	4	−2	−1	0	1	2	4	−2	−1	0	1	2	4	30
Speed = 120 km/h																									
QPSK 1/2 rep 6	−4	−2	−1	0	1	2	−4	−2	−1	0	1	2	−4	−2	−1	0	1	2	−4	−2	−1	0	1	0	120
QPSK 1/2 rep 4	−4	−2	−1	0	1	2	−4	−2	−1	0	1	2	−4	−2	−1	0	1	2	−4	−2	−1	0	1	0	120
QPSK 1/2 rep 2	−4	−2	−1	0	1	2	−4	−2	−1	0	1	2	−4	−2	−1	0	1	2	−4	−2	−1	0	1	0	120
BPSK 1/2	−4	−2	−1	0	1	2	−4	−2	−1	0	1	2	−4	−2	−1	0	1	2	−4	−2	−1	0	1	0	120
BPSK 3/4	−4	−2	−1	0	1	2	−4	−2	−1	0	1	2	−4	−2	−1	0	1	2	−4	−2	−1	0	1	0	120
QPSK 1/2	−4	−2	−1	0	1	2	−4	−2	−1	0	1	2	−4	−2	−1	0	1	2	−4	−2	−1	0	1	2	120
QPSK 3/4	−4	−2	−1	0	1	2	−4	−2	−1	0	1	2	−4	−2	−1	0	1	2	−4	−2	−1	0	1	2	120
16QAM 1/2	−4	−2	−1	0	1	2	−4	−2	−1	0	1	2	−4	−2	−1	0	1	2	−4	−2	−1	0	1	2	120
16QAM 3/4	−4	−2	−1	0	1	2	−4	−2	−1	0	1	2	−4	−2	−1	0	1	2	−4	−2	−1	0	1	2	120
64QAM 1/2	−4	−2	−1	0	1	2	−4	−2	−1	0	1	2	−4	−2	−1	0	1	2	−4	−2	−1	0	1	2	120
64QAM 2/3	−4	−2	−1	0	1	2	−4	−2	−1	0	1	2	−4	−2	−1	0	1	2	−4	−2	−1	0	1	2	120
64QAM 3/4	−4	−2	−1	0	1	2	−4	−2	−1	0	1	2	−4	−2	−1	0	1	2	−4	−2	−1	0	1	2	120
64QAM 5/6	−4	−2	−1	0	1	2	−4	−2	−1	0	1	2	−4	−2	−1	0	1	2	−4	−2	−1	0	1	2	120

11.5.4 Permutation Effect

Table 11.8 Symbol permutation factor (dB)

Fading type	Rayleigh k<=1						Ricean 1<k<=2						Ricean 2<k<=10						AWGN k>10					
BER	10^{-1}	10^{-2}	10^{-3}	10^{-4}	10^{-5}	10^{-6}	10^{-1}	10^{-2}	10^{-3}	10^{-4}	10^{-5}	10^{-6}	10^{-1}	10^{-2}	10^{-3}	10^{-4}	10^{-5}	10^{-6}	10^{-1}	10^{-2}	10^{-3}	10^{-4}	10^{-5}	10^{-6}
Schemes																								
Downlink																								
DL-PUSC	1	1	1	1	1	1	1	1	1	1	1	1	1	1	1	1	1	1	1	1	1	1	1	1
DL-FUSC	−0.5	−0.5	−0.5	−0.5	−0.5	−0.5	−0.5	−0.5	−0.5	−0.5	−0.5	−0.5	−0.5	−0.5	−0.5	−0.5	−0.5	−0.5	−0.5	−0.5	−0.5	−0.5	−0.5	−0.5
DL-OFUSC	0	0	0	0	0	0	0	0	0	0	0	0	0	0	0	0	0	0	0	0	0	0	0	0
DL-TUSC	−0.5	−0.5	−0.5	−0.5	−0.5	−0.5	−0.5	−0.5	−0.5	−0.5	−0.5	−0.5	−0.5	−0.5	−0.5	−0.5	−0.5	−0.5	−0.5	−0.5	−0.5	−0.5	−0.5	−0.5
DL-TUSC-2	0.5	0.5	0.5	0.5	0.5	0.5	0.5	0.5	0.5	0.5	0.5	0.5	0.5	0.5	0.5	0.5	0.5	0.5	0.5	0.5	0.5	0.5	0.5	0.5
DL-AMC-1-6	0	0	0	0	0	0	0	0	0	0	0	0	0	0	0	0	0	0	0	0	0	0	0	0
DL-AMC-2-3	0	0	0	0	0	0	0	0	0	0	0	0	0	0	0	0	0	0	0	0	0	0	0	0
DL-AMC-3-2	0	0	0	0	0	0	0	0	0	0	0	0	0	0	0	0	0	0	0	0	0	0	0	0
Uplink																								
UL-PUSC	−0.5	−0.5	−0.5	−0.5	−0.5	−0.5	−0.5	−0.5	−0.5	−0.5	−0.5	−0.5	−0.5	−0.5	−0.5	−0.5	−0.5	−0.5	−0.5	−0.5	−0.5	−0.5	−0.5	−0.5
UL-OPUSC	0.5	0.5	0.5	0.5	0.5	0.5	0.5	0.5	0.5	0.5	0.5	0.5	0.5	0.5	0.5	0.5	0.5	0.5	0.5	0.5	0.5	0.5	0.5	0.5
UL-AMC-1-6	0	0	0	0	0	0	0	0	0	0	0	0	0	0	0	0	0	0	0	0	0	0	0	0
UL-AMC-2-3	0	0	0	0	0	0	0	0	0	0	0	0	0	0	0	0	0	0	0	0	0	0	0	0
UL-AMC-3-2	0	0	0	0	0	0	0	0	0	0	0	0	0	0	0	0	0	0	0	0	0	0	0	0

11.5.5 HARQ Effect

Table 11.9 HARQ SNR improvement in dB

Fading type	Rayleigh k<=1						Ricean 1<k<=2						Ricean 2<k<=10						AWGN k>10					
BER	10^{-1}	10^{-2}	10^{-3}	10^{-4}	10^{-5}	10^{-6}	10^{-1}	10^{-2}	10^{-3}	10^{-4}	10^{-5}	10^{-6}	10^{-1}	10^{-2}	10^{-3}	10^{-4}	10^{-5}	10^{-6}	10^{-1}	10^{-2}	10^{-3}	10^{-4}	10^{-5}	10^{-6}
All Schemes	0	0	0	0	0	0	0	0	0	0	0	0	0	0	0	0	0	0	0	0	0	0	0	0
Type 1																								
QPSK 1/2 rep 6	0.13	0.08	0.04	0	0	0	0.13	0.08	0.04	0	0	0	0.13	0.08	0.04	0	0	0	0.13	0.08	0.04	0	0	0
QPSK 1/2 rep 4	0.19	0.13	0.06	0	0	0	0.19	0.13	0.06	0	0	0	0.19	0.13	0.06	0	0	0	0.19	0.13	0.06	0	0	0
QPSK 1/2 rep 2	0.38	0.25	0.13	0	0	0	0.38	0.25	0.13	0	0	0	0.38	0.25	0.13	0	0	0	0.38	0.25	0.13	0	0	0
BPSK 1/2	0.38	0.25	0.13	0	0	0	0.38	0.25	0.13	0	0	0	0.38	0.25	0.13	0	0	0	0.38	0.25	0.13	0	0	0
BPSK 3/4	0.56	0.38	0.19	0	0	0	0.56	0.38	0.19	0	0	0	0.56	0.38	0.19	0	0	0	0.56	0.38	0.19	0	0	0
QPSK 1/2	0.75	0.5	0.25	0	0	0	0.75	0.5	0.25	0	0	0	0.75	0.5	0.25	0	0	0	0.75	0.5	0.25	0	0	0
QPSK 3/4	1.125	0.75	0.38	0	0	0	1.13	0.75	0.38	0	0	0	1.13	0.75	0.38	0	0	0	1.125	0.75	0.375	0	0	0
16QAM 1/2	1.125	0.75	0.38	0	0	0	1.13	0.75	0.38	0	0	0	1.13	0.75	0.38	0	0	0	1.125	0.75	0.375	0	0	0
16QAM 3/4	1.5	1	0.5	0	0	0	1.5	1	0.5	0	0	0	1.5	1	0.5	0	0	0	1.5	1	0.5	0	0	0
64QAM 1/2	1.625	1.25	0.88	0.5	0.13	0	1.63	1.25	0.88	0.5	0.13	0	1.63	1.25	0.88	0.5	0.13	0	1.625	1.25	0.875	0.5	0.125	0
64QAM 2/3	1.875	1.5	1.13	0.75	0.38	0	1.88	1.5	1.13	0.75	0.38	0	1.88	1.5	1.13	0.75	0.38	0	1.875	1.5	1.125	0.75	0.375	0
64QAM 3/4	2.5	2	1.5	1	0.5	0	2.5	2	1.5	1	0.5	0	2.5	2	1.5	1	0.5	0	2.5	2	1.5	1	0.5	0
64QAM 5/6	3.13	2.50	1.88	1.25	0.63	0.00	3.13	2.50	1.88	1.25	0.63	0.00	3.13	2.50	1.88	1.25	0.63	0.00	3.13	2.50	1.88	1.25	0.63	0.00
Type 2																								
QPSK 1/2 rep 6	0.50	0.33	0.17	0	0	0	0.50	0.33	0.17	0	0	0	0.50	0.33	0.17	0	0	0	0.50	0.33	0.17	0	0	0
QPSK 1/2 rep 4	0.56	0.38	0.19	0	0	0	0.56	0.38	0.19	0	0	0	0.56	0.38	0.19	0	0	0	0.56	0.38	0.19	0	0	0
QPSK 1/2 rep 2	0.75	0.50	0.25	0	0	0	0.75	0.50	0.25	0	0	0	0.75	0.50	0.25	0	0	0	0.75	0.50	0.25	0	0	0
BPSK 1/2	0.75	0.50	0.25	0	0	0	0.75	0.50	0.25	0	0	0	0.75	0.50	0.25	0	0	0	0.75	0.50	0.25	0	0	0
BPSK 3/4	0.94	0.63	0.31	0	0	0	0.94	0.63	0.31	0	0	0	0.94	0.63	0.31	0	0	0	0.94	0.63	0.31	0	0	0
QPSK 1/2	1.125	0.75	0.38	0	0	0	1.13	0.75	0.38	0	0	0	1.13	0.75	0.38	0	0	0	1.125	0.75	0.375	0	0	0
QPSK 3/4	1.5	1	0.5	0	0	0	1.5	1	0.5	0	0	0	1.5	1	0.5	0	0	0	1.5	1	0.5	0	0	0
16QAM 1/2	2.25	1.5	0.75	0	0	0	2.25	1.5	0.75	0	0	0	2.25	1.5	0.75	0	0	0	2.25	1.5	0.75	0	0	0
16QAM 3/4	2.25	1.5	0.75	0	0	0	2.25	1.5	0.75	0	0	0	2.25	1.5	0.75	0	0	0	2.25	1.5	0.75	0	0	0
64QAM 1/2	2	1.5	1	0.5	0	0	2	1.5	1	0.5	0	0	2	1.5	1	0.5	0	0	2	1.5	1	0.5	0	0
64QAM 2/3	2.625	2	1.38	0.75	0.13	0	2.63	2	1.38	0.75	0.13	0	2.63	2	1.38	0.75	0.13	0	2.625	2	1.375	0.75	0.125	0
64QAM 3/4	3.25	2.5	1.75	1	0.25	0	3.25	2.5	1.75	1	0.25	0	3.25	2.5	1.75	1	0.25	0	3.25	2.5	1.75	1	0.25	0
64QAM 5/6	3.88	3.00	2.13	1.25	0.38	0.00	3.88	3.00	2.13	1.25	0.38	0.00	3.88	3.00	2.13	1.25	0.38	0.00	3.88	3.00	2.13	1.25	0.38	0.00

11.5.6 *Improvement Reduction Factor for Antenna Systems*

Table 11.10 Improvement reduction factor

Fading type		Factor
Rayleigh	k<=1	1
Ricean	1<k<=2	0.6
Ricean	2<k<=10	0.35
AWGN	k>10	0.2

11.5.7 *Receive Diversity*

Table 11.11 RX Diversity, Rayleigh, improvement in dB

Antenna correlation	Small						Low						Medium						High					
BER	10^{-1}	10^{-2}	10^{-3}	10^{-4}	10^{-5}	10^{-6}	10^{-1}	10^{-2}	10^{-3}	10^{-4}	10^{-5}	10^{-6}	10^{-1}	10^{-2}	10^{-3}	10^{-4}	10^{-5}	10^{-6}	10^{-1}	10^{-2}	10^{-3}	10^{-4}	10^{-5}	10^{-6}
$N = 2$																								
SC	3.75	7.50	11.25	15.00	18.75	22.50	2.50	5.00	7.50	10.00	12.50	15.00	1.25	2.50	3.75	5.00	6.25	7.50	0.00	0.00	0.00	0.00	0.00	0.00
EGC	4.88	8.63	12.38	16.13	19.88	23.63	4.01	6.51	9.01	11.51	14.01	16.51	3.13	4.38	5.63	6.88	8.13	9.38	2.26	2.26	2.26	2.26	2.26	2.26
MRC	6.01	9.76	13.51	17.26	21.01	24.76	4.76	7.26	9.76	12.26	14.76	17.26	3.51	4.76	6.01	7.26	8.51	9.76	2.26	2.26	2.26	2.26	2.26	2.26
$N = 3$																								
SC	5.00	10.00	15.00	20.00	25.00	30.00	3.33	6.67	10.00	13.33	16.67	20.00	1.67	3.33	5.00	6.67	8.33	10.00	0.00	0.00	0.00	0.00	0.00	0.00
EGC	6.79	11.79	16.79	21.79	26.79	31.79	5.72	9.05	12.39	15.72	19.05	22.39	4.65	6.32	7.98	9.65	11.32	12.98	3.58	3.58	3.58	3.58	3.58	3.58
MRC	8.58	13.58	18.58	23.58	28.58	33.58	6.91	10.25	13.58	16.91	20.25	23.58	5.25	6.91	8.58	10.25	11.91	13.58	3.58	3.58	3.58	3.58	3.58	3.58
$N = 4$																								
SC	5.63	11.25	16.88	22.50	28.13	33.75	3.75	7.50	11.25	15.00	18.75	22.50	1.88	3.75	5.63	7.50	9.38	11.25	0.00	0.00	0.00	0.00	0.00	0.00
EGC	7.88	13.51	19.13	24.76	30.38	36.01	6.76	10.51	14.26	18.01	21.76	25.51	5.64	7.51	9.39	11.26	13.14	15.01	4.52	4.52	4.52	4.52	4.52	4.52
MRC	10.14	15.77	21.39	27.02	32.64	38.27	8.27	12.02	15.77	19.52	23.27	27.02	6.39	8.27	10.14	12.02	13.89	15.77	4.52	4.52	4.52	4.52	4.52	4.52

11.5.8 Transmit Diversity

Table 11.12 TX Diversity, Rayleigh, improvement in dB

Antenna correlation	Small						Low						Medium						High					
BER	10^{-1}	10^{-2}	10^{-3}	10^{-4}	10^{-5}	10^{-6}	10^{-1}	10^{-2}	10^{-3}	10^{-4}	10^{-5}	10^{-6}	10^{-1}	10^{-2}	10^{-3}	10^{-4}	10^{-5}	10^{-6}	10^{-1}	10^{-2}	10^{-3}	10^{-4}	10^{-5}	10^{-6}
$N = 2$																								
STC/STBC	6.55	11.55	16.55	21.55	26.55	31.55	4.37	7.70	11.03	14.37	17.70	21.03	2.18	3.85	5.52	7.18	8.85	10.52	0.00	0.00	0.00	0.00	0.00	0.00
TSD	5.97	10.97	15.97	20.97	25.97	30.97	3.98	7.31	10.65	13.98	17.31	20.65	1.99	3.66	5.32	6.99	8.66	10.32	0.00	0.00	0.00	0.00	0.00	0.00
LDP	5.46	10.46	15.46	20.46	25.46	30.46	3.64	6.97	10.31	13.64	16.97	20.31	1.82	3.49	5.15	6.82	8.49	10.15	0.00	0.00	0.00	0.00	0.00	0.00
$N = 3$																								
STC/STBC	8.22	14.88	21.55	28.22	34.88	41.55	5.48	9.92	14.37	18.81	23.25	27.70	2.74	4.96	7.18	9.41	11.63	13.85	0.00	0.00	0.00	0.00	0.00	0.00
TSD	7.64	14.30	20.97	27.64	34.30	40.97	5.09	9.53	13.98	18.42	22.87	27.31	2.55	4.77	6.99	9.21	11.43	13.66	0.00	0.00	0.00	0.00	0.00	0.00
LDP	7.12	13.79	20.46	27.12	33.79	40.46	4.75	9.19	13.64	18.08	22.53	26.97	2.37	4.60	6.82	9.04	11.26	13.49	0.00	0.00	0.00	0.00	0.00	0.00
$N = 4$																								
STC/STBC	9.05	16.55	24.05	31.55	39.05	46.55	6.03	11.03	16.03	21.03	26.03	31.03	3.02	5.52	8.02	10.52	13.02	15.52	0.00	0.00	0.00	0.00	0.00	0.00
TSD	8.47	15.97	23.47	30.97	38.47	45.97	5.65	10.65	15.65	20.65	25.65	30.65	2.82	5.32	7.82	10.32	12.82	15.32	0.00	0.00	0.00	0.00	0.00	0.00
LDP	7.96	15.46	22.96	30.46	37.96	45.46	5.31	10.31	15.31	20.31	25.31	30.31	2.65	5.15	7.65	10.15	12.65	15.15	0.00	0.00	0.00	0.00	0.00	0.00

11.5.9 *Spatial Multiplexing*

Table 11.13 Spatial Multiplexing DL, Rayleigh, improvement in dB

Antenna correlation	Small						Low						Medium						High					
BER	10^{-1}	10^{-2}	10^{-3}	10^{-4}	10^{-5}	10^{-6}	10^{-1}	10^{-2}	10^{-3}	10^{-4}	10^{-5}	10^{-6}	10^{-1}	10^{-2}	10^{-3}	10^{-4}	10^{-5}	10^{-6}	10^{-1}	10^{-2}	10^{-3}	10^{-4}	10^{-5}	10^{-6}
$N = 2$																								
MLD-SD	−1.6	−0.8	0	3.2	6.4	9.6	−2.07	−1.87	−1.67	0.13	1.93	3.73	−2.53	−2.93	−3.33	−2.93	−2.53	−2.13	−3	−4	−5	−6	−7	−8
BLAST	−1.4	−0.7	−0.7	0.35	1.05	2.1	−1.93	−1.80	−2.13	−1.77	−1.63	−1.27	−2.47	−2.90	−3.57	−3.88	−4.32	−4.63	−3	−4	−5	−6	−7	−8
SIC-MUD	−1.2	−0.6	−0.6	0.3	0.9	1.8	−1.80	−1.73	−2.07	−1.80	−1.73	−1.47	−2.40	−2.87	−3.53	−3.90	−4.37	−4.73	−3	−4	−5	−6	−7	−8
SVD	−1.5	−0.75	0	3	6	9	−2.00	−1.83	−1.67	0.00	1.67	3.33	−2.50	−2.92	−3.33	−3.00	−2.67	−2.33	−3	−4	−5	−6	−7	−8
LPD	−1.6	−0.8	0	3.2	6.4	9.6	−2.07	−1.87	−1.67	0.13	1.93	3.73	−2.53	−2.93	−3.33	−2.93	−2.53	−2.13	−3	−4	−5	−6	−7	−8
$N = 3$																								
MLD-SD	−2.4	−1.2	0	4.8	9.6	14.4	−2.60	−2.13	−1.67	1.20	4.07	6.93	−2.80	−3.07	−3.33	−2.40	−1.47	−0.53	−3	−4	−5	−6	−7	−8
BLAST	−2.1	−1.05	−1.05	0.53	1.58	3.15	−2.40	−2.03	−2.37	−1.65	−1.28	−0.57	−2.70	−3.02	−3.68	−3.83	−4.14	−4.28	−3	−4	−5	−6	−7	−8
SIC-MUD	−1.8	−0.9	−0.9	0.45	1.35	2.7	−2.20	−1.93	−2.27	−1.70	−1.43	−0.87	−2.60	−2.97	−3.63	−3.85	−4.22	−4.43	−3	−4	−5	−6	−7	−8
SVD	−2.25	−1.125	0	4.5	9	13.5	−2.50	−2.08	−1.67	1.00	3.67	6.33	−2.75	−3.04	−3.33	−2.50	−1.67	−0.83	−3	−4	−5	−6	−7	−8
LPD	−2.4	−1.2	0	4.8	9.6	14.4	−2.60	−2.13	−1.67	1.20	4.07	6.93	−2.80	−3.07	−3.33	−2.40	−1.47	−0.53	−3	−4	−5	−6	−7	−8
$N = 4$																								
MLD-SD	−3.2	−1.6	0	6.4	12.8	19.2	−3.13	−2.40	−1.67	2.27	6.20	10.13	−3.07	−3.20	−3.33	−1.87	−0.40	1.07	−3	−4	−5	−6	−7	−8
BLAST	−2.8	−1.4	−1.4	0.7	2.1	4.2	−2.87	−2.27	−2.60	−1.53	−0.93	0.13	−2.93	−3.13	−3.80	−3.77	−3.97	−3.93	−3	−4	−5	−6	−7	−8
SIC-MUD	−2.4	−1.2	−1.2	0.6	1.8	3.6	−2.60	−2.13	−2.47	−1.60	−1.13	−0.27	−2.80	−3.07	−3.73	−3.80	−4.07	−4.13	−3	−4	−5	−6	−7	−8
SVD	−3	−1.5	0	6	12	18	−3.00	−2.33	−1.67	2.00	5.67	9.33	−3.00	−3.17	−3.33	−2.00	−0.67	0.67	−3	−4	−5	−6	−7	−8
LPD	−3.2	−1.6	0	6.4	12.8	19.2	−3.13	−2.40	−1.67	2.27	6.20	10.13	−3.07	−3.20	−3.33	−1.87	−0.40	1.07	−3	−4	−5	−6	−7	−8

11.5.10 Spatial Multiplexing

Table 11.14 Spatial Multiplexing DL, Rayleigh, improvement in dB

Antenna correlation	Small						Low						Medium						High					
BER	10^{-1}	10^{-2}	10^{-3}	10^{-4}	10^{-5}	10^{-6}	10^{-1}	10^{-2}	10^{-3}	10^{-4}	10^{-5}	10^{-6}	10^{-1}	10^{-2}	10^{-3}	10^{-4}	10^{-5}	10^{-6}	10^{-1}	10^{-2}	10^{-3}	10^{-4}	10^{-5}	10^{-6}
$N = 2$																								
MLD-SD	−1.6	−0.8	0	3.2	6.4	9.6	−2.07	−1.87	−1.67	0.13	1.93	3.73	−2.53	−2.93	−3.33	−2.93	−2.53	−2.13	−3	−4	−5	−6	−7	−8
BLAST	−1.4	−0.7	−0.7	0.35	1.05	2.1	−1.93	−1.80	−2.13	−1.77	−1.63	−1.27	−2.47	−2.90	−3.57	−3.88	−4.32	−4.63	−3	−4	−5	−6	−7	−8
SIC-MUD	−1.2	−0.6	−0.6	0.3	0.9	1.8	−1.80	−1.73	−2.07	−1.80	−1.73	−1.47	−2.40	−2.87	−3.53	−3.90	−4.37	−4.73	−3	−4	−5	−6	−7	−8
SVD	−1.5	−0.75	0	3	6	9	−2.00	−1.83	−1.67	0.00	1.67	3.33	−2.50	−2.92	−3.33	−3.00	−2.67	−2.33	−3	−4	−5	−6	−7	−8
LPP	−1.6	−0.8	0	3.2	6.4	9.6	−2.07	−1.87	−1.67	0.13	1.93	3.73	−2.53	−2.93	−3.33	−2.93	−2.53	−2.13	−3	−4	−5	−6	−7	−8
$N = 3$																								
MLD-SD	−2.4	−1.2	0	4.8	9.6	14.4	−2.60	−2.13	−1.67	1.20	4.07	6.93	−2.80	−3.07	−3.33	−2.40	−1.47	−0.53	−3	−4	−5	−6	−7	−8
BLAST	−2.1	−1.05	−1.05	0.53	1.58	3.15	−2.40	−2.03	−2.37	−1.65	−1.28	−0.57	−2.70	−3.02	−3.68	−3.83	−4.14	−4.28	−3	−4	−5	−6	−7	−8
SIC-MUD	−1.8	−0.9	−0.9	0.45	1.35	2.7	−2.20	−1.93	−2.27	−1.70	−1.43	−0.87	−2.60	−2.97	−3.63	−3.85	−4.22	−4.43	−3	−4	−5	−6	−7	−8
SVD	−2.25	−1.125	0	4.5	9	13.5	−2.50	−2.08	−1.67	1.00	3.67	6.33	−2.75	−3.04	−3.33	−2.50	−1.67	−0.83	−3	−4	−5	−6	−7	−8
LPP	−2.4	−1.2	0	4.8	9.6	14.4	−2.60	−2.13	−1.67	1.20	4.07	6.93	−2.80	−3.07	−3.33	−2.40	−1.47	−0.53	−3	−4	−5	−6	−7	−8
$N = 4$																								
MLD-SD	−3.2	−1.6	0	6.4	12.8	19.2	−3.13	−2.40	−1.67	2.27	6.20	10.13	−3.07	−3.20	−3.33	−1.87	−0.40	1.07	−3	−4	−5	−6	−7	−8
BLAST	−2.8	−1.4	−1.4	0.7	2.1	4.2	−2.87	−2.27	−2.60	−1.53	−0.93	0.13	−2.93	−3.13	−3.80	−3.77	−3.97	−3.93	−3	−4	−5	−6	−7	−8
SIC-MUD	−2.4	−1.2	−1.2	0.6	1.8	3.6	−2.60	−2.13	−2.47	−1.60	−1.13	−0.27	−2.80	−3.07	−3.73	−3.80	−4.07	−4.13	−3	−4	−5	−6	−7	−8
SVD	−3	−1.5	0	6	12	18	−3.00	−2.33	−1.67	2.00	5.67	9.33	−3.00	−3.17	−3.33	−2.00	−0.67	0.67	−3	−4	−5	−6	−7	−8
LPP	−3.2	−1.6	0	6.4	12.8	19.2	−3.13	−2.40	−1.67	2.27	6.20	10.13	−3.07	−3.20	−3.33	−1.87	−0.40	1.07	−3	−4	−5	−6	−7	−8

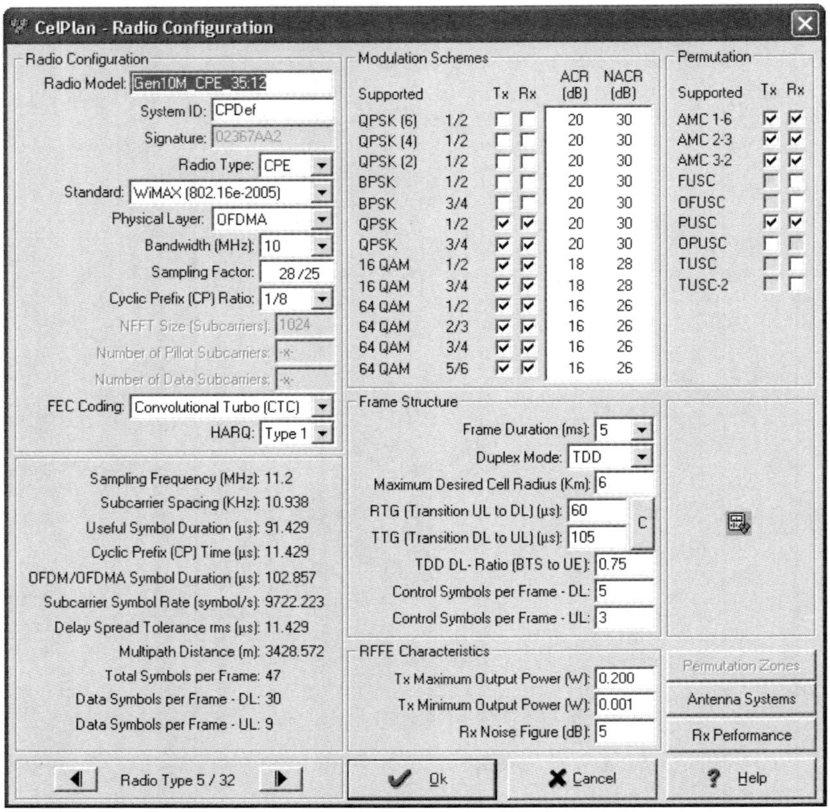

Figure 11.5 General radio parameters configuration dialogue.

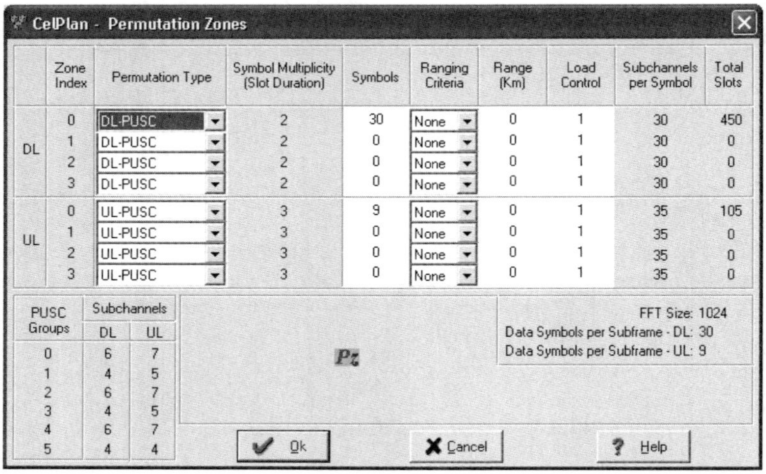

Figure 11.6 Radio zones configuration dialogue.

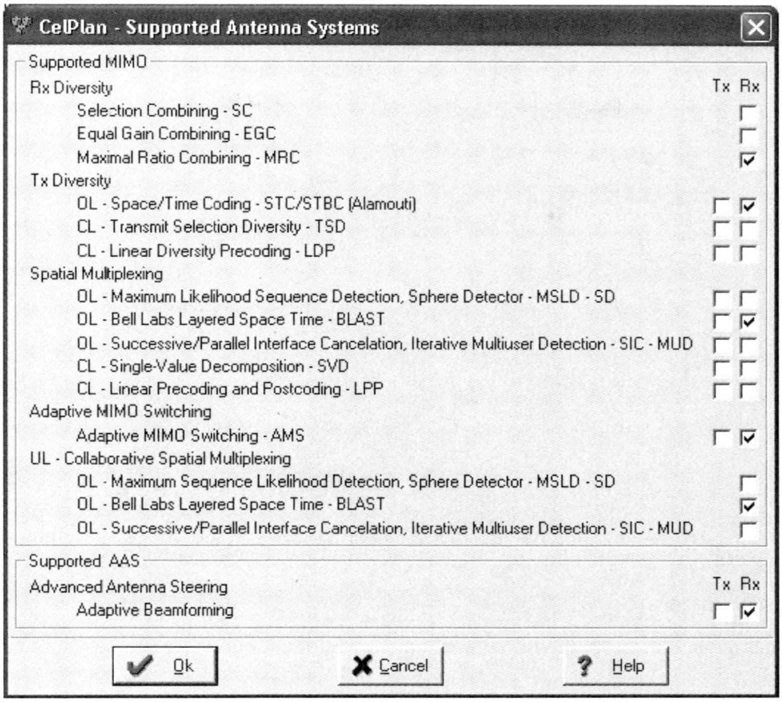

Figure 11.7 Radio antenna systems configuration dialogue.

11.6 Radio Configuration

Besides defining radio sensitivity, designers must also configure other radio characteristics such as technology, and modulation and permutation schemes supported. This radio set-up should be done for each type of Base Station and CPE radio used in the network. An example of configuration is shown in Figure 11.5 to Figure 11.7.

Due to the huge number of combinations, more than one outcome is possible and the one that provides the highest throughput should be selected.

To calculate the throughput of a connection between a Base Station Radio and a User Radio the following steps should be followed:

- Check the radios' compatibility, as more than one radio may be offering service at the pixel. One way of performing quickly this check in the design tool is to create a signature parameter that reflects the features supported by the radio. Radios with the same signature are considered compatible.
- Calculate propagation channel loss to find the average received power and draw a value from the statistical distribution for the pixel.
- Apply the same criteria above for other environmental losses.
- Sum both values to obtain the instantaneous path loss for each transmit radio.
- Calculate throughput for each compatible modulation scheme, considering mobility, permutation, HARQ type, service BER target and the different possibilities of Antenna Systems.
- The combination that provides the highest throughput should be selected.

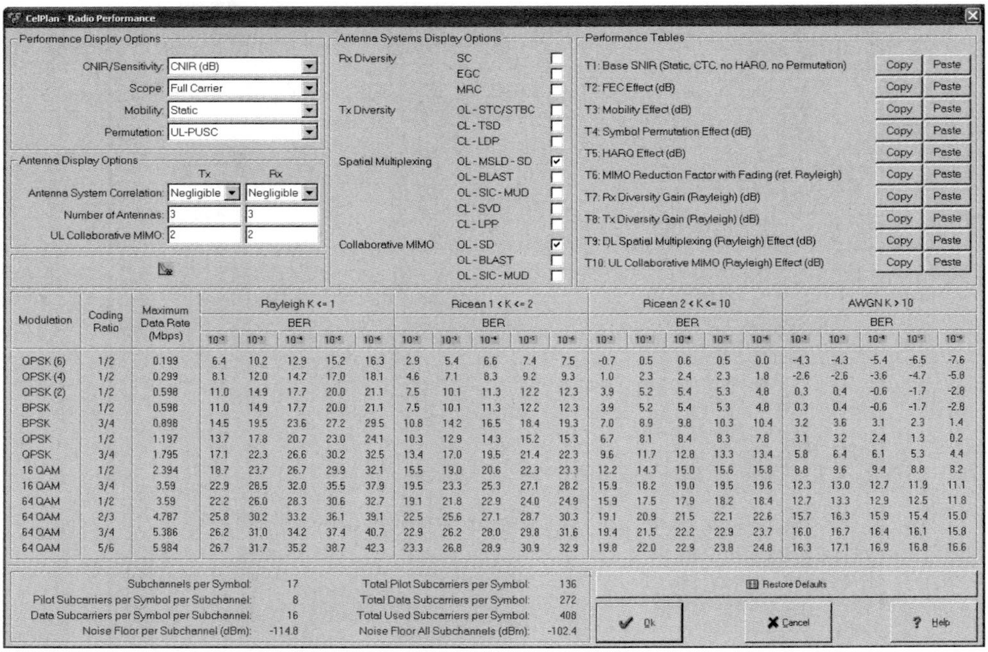

Modulation	Coding Ratio	Maximum Data Rate (Mbps)	Rayleigh K <= 1 BER					Ricean 1 < K <= 2 BER					Ricean 2 < K <= 10 BER					AWGN K > 10 BER				
			10^{-2}	10^{-3}	10^{-4}	10^{-5}	10^{-6}	10^{-2}	10^{-3}	10^{-4}	10^{-5}	10^{-6}	10^{-2}	10^{-3}	10^{-4}	10^{-5}	10^{-6}	10^{-2}	10^{-3}	10^{-4}	10^{-5}	10^{-6}
QPSK (6)	1/2	0.199	6.4	10.2	12.9	15.2	16.3	2.9	5.4	6.6	7.4	7.5	-0.7	0.5	0.6	0.5	0.0	-4.3	-4.3	-5.4	-6.5	-7.6
QPSK (4)	1/2	0.299	8.1	12.0	14.7	17.0	18.1	4.6	7.1	8.3	9.2	9.3	1.0	2.3	2.4	2.3	1.8	-2.6	-2.6	-3.6	-4.7	-5.8
QPSK (2)	1/2	0.598	11.0	14.9	17.7	20.0	21.1	7.5	10.1	11.3	12.2	12.3	3.9	5.2	5.4	5.3	4.8	0.3	0.4	-0.6	-1.7	-2.8
BPSK	1/2	0.598	11.0	14.9	17.7	20.0	21.1	7.5	10.1	11.3	12.2	12.3	3.9	5.2	5.4	5.3	4.8	0.3	0.4	-0.6	-1.7	-2.8
BPSK	3/4	0.898	14.5	19.5	23.6	27.2	29.5	10.8	14.2	16.5	18.4	19.3	7.0	8.9	9.8	10.3	10.4	3.2	3.6	3.1	2.3	1.4
QPSK	1/2	1.197	13.7	17.8	20.7	23.0	24.1	10.3	12.9	14.3	15.2	15.3	6.7	8.1	8.4	8.3	7.8	3.1	3.2	2.4	1.3	0.2
QPSK	3/4	1.795	17.1	22.3	26.6	30.2	32.5	13.4	17.0	19.5	21.4	22.3	9.6	11.7	12.8	13.3	13.4	5.8	6.4	6.1	5.3	4.4
16 QAM	1/2	2.394	18.7	23.7	26.7	29.9	32.1	15.5	19.0	20.6	22.3	23.3	12.2	14.3	15.0	15.6	15.8	8.8	9.6	9.4	8.8	8.2
16 QAM	3/4	3.59	22.9	28.5	32.0	35.5	37.9	19.5	23.3	25.3	27.1	28.2	15.9	18.2	19.0	19.5	19.6	12.3	13.0	12.7	11.9	11.1
64 QAM	1/2	3.59	22.2	26.0	28.3	30.6	32.7	19.1	21.8	22.9	24.0	24.9	15.9	17.5	17.9	18.2	18.4	12.7	13.3	12.9	12.5	11.8
64 QAM	2/3	4.787	25.8	30.2	33.2	36.1	39.1	22.5	25.6	27.1	28.7	30.3	19.1	20.9	21.5	22.1	22.6	15.7	16.3	15.9	15.4	15.0
64 QAM	3/4	5.386	26.2	31.0	34.2	37.4	40.7	22.9	26.2	28.0	29.8	31.6	19.4	21.5	22.2	22.9	23.7	16.0	16.7	16.4	16.1	15.8
64 QAM	5/6	5.984	26.7	31.7	35.2	38.7	42.3	23.3	26.8	28.9	30.9	32.9	19.8	22.0	22.9	23.8	24.8	16.3	17.1	16.9	16.8	16.6

Figure 11.8 Receiver performance table.

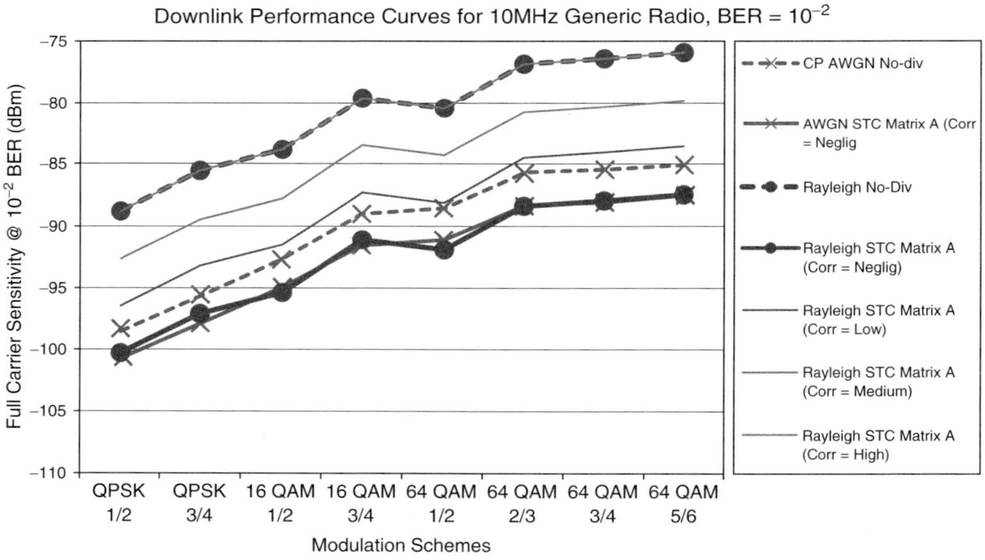

Figure 11.9 Downlink performance for a generic radio with 10 MHz bandwidth.

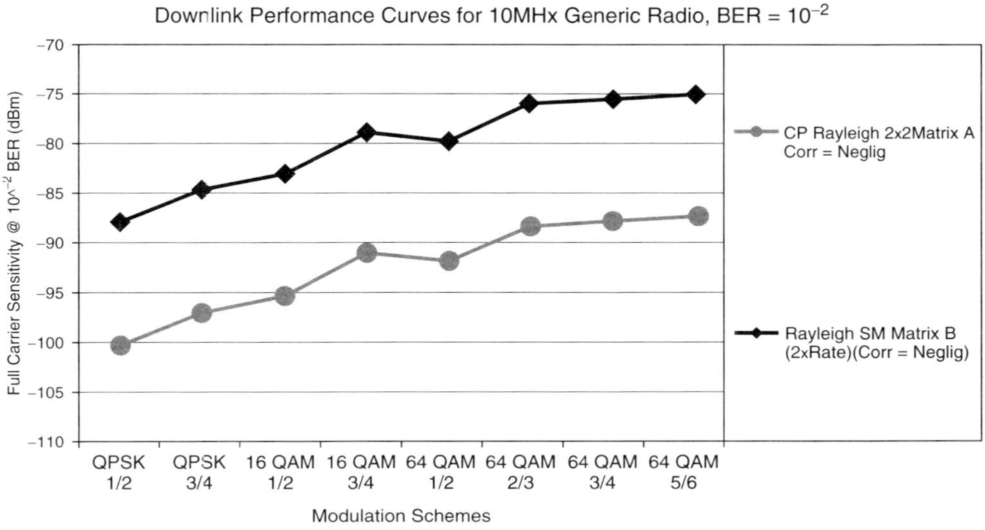

Figure 11.10 Downlink performance for a generic radio with 10 MHz bandwidth (detail).

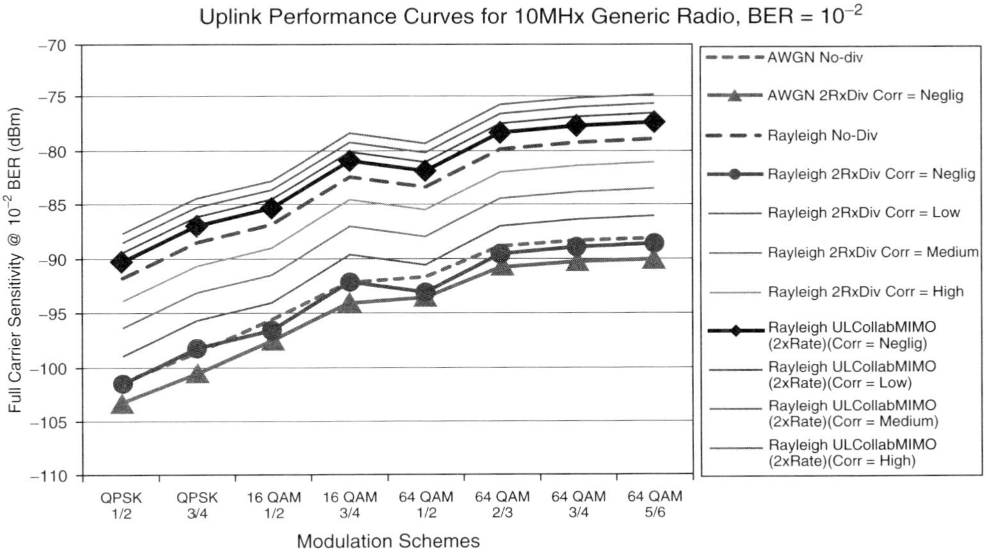

Figure 11.11 Uplink performance for a generic radio with 10 MHz bandwidth.

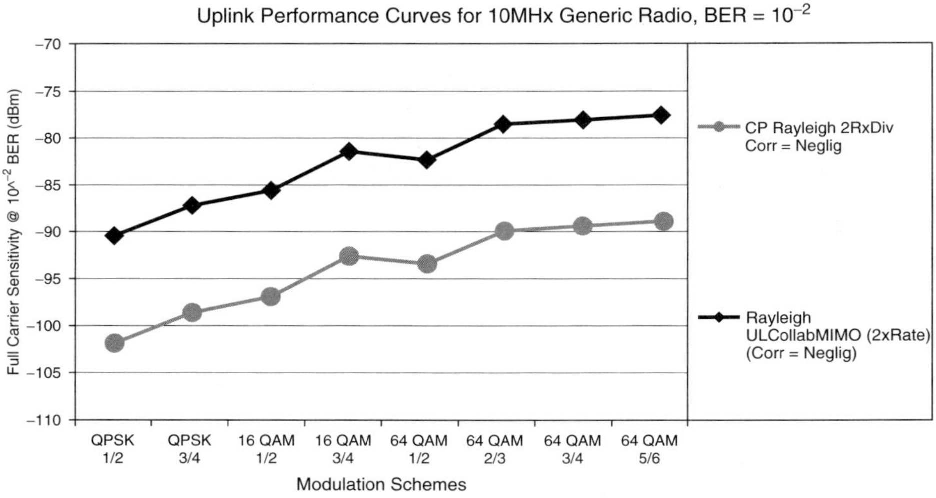

Figure 11.12 Uplink performance for a generic radio with 10 MHz bandwidth (detail).

Figure 11.8 illustrates a dialogue that provides the throughput and required SINR (Signal to Interference and Noise Ratio) for the modulation schemes used in WiMAX in different fading environments for different Error Rates. This table is illustrative and reflects the content of Tables 11.5 to 11.14 configured above. The actual values should be calculated in real time during each prediction, considering statistical variations of all parameters involved.

All these performance variations should be considered by the Traffic Simulation process. Figure 11.9 to Figure 11.12 show performance curves for different scenarios.

12

Wireless LAN

12.1 Standardization

The popularization of computers at home created the need to interconnect several devices, such as desktops, switches, laptops, printers, cable modems, and so on. WLAN (Wireless Local Area Network) became attractive as it did not require wires and gave mobility to laptops, but to be implemented it required the RF spectrum.

The RF spectrum is regulated by government agencies, requiring that users obtain licenses to use it. Industries, universities and radio amateurs were assigned unlicensed bands for use in laboratories and short range communications, as is the case of the ISM (Industrial, Scientific and Medical Equipment) bands and the U-NII (Unlicensed National Information Infrastructure) band. These bands were used in microwave ovens, car alarms, wireless phones and other similar devices. Over time, regulators conceded that these bands could also be used for short range broadband communications. ISM bands are listed in Table 12.1 and UNII bands in Table 12.2. UNII bands are subject to DFS (Dynamic Frequency Selection) because they are shared with radar operations and have to switch frequencies if a radar signal is detected.

Communication protocols available in the early 1980s were oriented towards voice and not appropriate for wireless broadband data LANs. Efforts to specify a new protocol started in the late 1980s and were carried out by the IEEE (Institute of Electrical and Electronics Engineers) in the USA and ETSI (European Telecommunications Standards Institute) in Europe. These specifications focused on indoor data transmission and were oriented towards data packets, using IP (Internet Protocol). The idea was to wirelessly emulate the successful Ethernet links.

The first specification done for a wireless LAN was released by ETSI in 1996 (HiperLAN/1), and used a single FSK/GMSK (Frequency Shift Keying with Gaussian Minimum-Shift Keying) carrier with a special layer for Channel Access and Control (CAC). It was capable of reaching an air data rate of 10 Mbps, for a 50 m range.

The first IEEE specification was released in 1997 and specified the use of CCK (Complementary Code Keying) with DSSS (Direct Sequence Spread Spectrum) or Frequency Hopping, and CSMA/CA (Carrier Sense Multiple Access/Collision Avoidance) as the medium access method. The air data rate was 1 or 2 Mbps and it was specified for the 2.4 GHz band. The air data rate includes the Error Correction Code, but does not include reference signals, control signals and channel access overhead, which may reduce the effective data throughput to 20% of the initial air data rate.

Within the IEEE, several study groups were created to examine specific issues related to the WLAN implementation, each group identified by 802.11 followed by a letter (e.g. a, b, d, e ...). Some groups

LTE, WiMAX and WLAN Network Design, Optimization and Performance Analysis, First Edition.
Leonhard Korowajczuk.

Table 12.1 ISM band

ISM band as per ITU-R		
Frequency (MHz)		
From	To	Bandwidth
6.765	6.795	0.030
13.553	13.560	0.007
26.957	27.283	0.326
40.66	40.70	0.04
433.05	433.92	0.87
902	928	26
2400	2500	100
5725	5875	150
24,000	24,250	250
61,000	61,500	500
122,000	123,000	1000
244,000	246,000	2000

Table 12.2 U-NII band

Frequency (MHz)				
From	To	Bandwidth	Power (W)	DFS
5150	5250	100	0.05	no
5250	5350	100	0.25	yes
5470	5725	255	0.25	yes
5725	5825	100	1.00	no

released their specifications ahead of others, so the release order does not match the letter sequence. This standard evolved and in 1999 a new specification (802.11b) was released, improving on the original one and reaching an air data rate of 11 Mbps.

It was clear though that further throughput increase would not be possible using single carrier solutions, due to the symbol duration becoming shorter than the multipath spread. This same amendment included then a new technology (802.11a), which used OFDM (Orthogonal Frequency Division Multiplexing), to increase the air data rate to 54 Mbps. This technology was specified for use in the 5 GHz band.

In 2000, ETSI issued a new specification (HiperLAN/2) very similar to 802.11a. In June 2003, the OFDM technology specified by 802.11a for the 5.4 GHz was extended to the 2.4 GHz band and became known as 802.11g. In 2007, IEEE released a comprehensive specification named IEEE Std 802.11-2007, which included the amendments of groups a, b, d, e, g, h, i and j.

The latest IEEE-issued specification is amendment 802.11n, entitled "Enhancement for higher throughput". It adds several features that may increase throughput, although claims of air speeds of 600 Mbps are greatly exaggerated.

Even the most detailed specifications are not sufficient to warranty inter-operability between different manufacturers. A global association WEA (Wi-Fi Alliance) was formed to promote WLAN's growth and assure inter-operability through a certification program. This trade group established the

Table 12.3 802.11 releases

Release	Protocol	Band	Modulation	Bandwidth	MAC
Jun 97	legacy	2.4	DSSS	20	CSMA/CA
Sep 99	b	2.4	DSSS	20	CSMA/CA
Sep 99	a	5	OFDM	20	CSMA/CA
Jun 03	g	5	OFDM	20	CSMA/CA
Mar 07	a,b,d,e,g,h,i,j	2.4/4	DSSS/OFDM	5,10,20	CSMA/CA
Oct 09	n	2.4/5	OFDM	40,20,10,5	CSMA/CA

Table 12.4 HiperLAN releases

Release	Protocol	Band	Modulation	Bandwidth	MAC
Jun 96	1	2.4	FSK, GMSK	20	CAC
Feb 00	2	2.4	OFDM	20	Dynamic TDMA

Table 12.5 IEEE WLAN 2.4 GHz unlicensed channels

Channel	Center Frequency (MHz)	USA	World
1	2412	yes	yes
2	2417	yes	yes
3	2422	yes	yes
4	2427	yes	yes
5	2432	yes	yes
6	2437	yes	yes
7	2442	yes	yes
8	2447	yes	yes
9	2452	yes	yes
10	2457	yes	yes
11	2462	yes	yes
12	2467	no	yes
13	2472	no	yes

trademark Wi-Fi for 802.11 certified products. A summary of 802.11 and HiperLAN releases is given respectively in Table 12.3 and Table 12.4.

WLAN 802.11 became very widespread and, today, all laptops are 802.11b/g enabled and operate in the 2.4 GHz band. The use of the 5 GHz band is still incipient and is mostly used more for professional applications. Some vendors altered the standard specifications to extend the operation of 802.11g/a for outdoor applications, such as, Citywide MAN (Metropolitan Area Networks), Point to Point backhaul and Public Safety Video Surveillance.

WLAN channels specified in the IEEE Standard are listed in Table 12.5 for the 2.4 GHz unlicensed band, in Table 12.6 for the 3.6 GHz licensed band, in Table 12.7 for the 4.9 GHz Public Safety licensed band and in Table 12.8 for the 5 GHz unlicensed bands.

Table 12.6 IEEE WLAN 3.6 GHz unlicensed channels

USA				
		Bandwidth (MHz)		
Channel	Center Frequency (MHz)	5	10	20
131	3657.5	x		
132	3660.0		x	
132	3662.5	x		
133	3665.0			x
133	3667.5	x		
134	3670.0		x	
135	3672.5	x		
135	3675.0			
136	3677.5	x		
136	3680.0		x	
137	3682.5	x		
137	3685.0			x
138	3687.5	x		
138	3690.0		x	

Table 12.7 IEEE WLAN 4.9 GHz licensed channels

Public Safety USA				
	Bandwidth (MHz)			
Center Frequency (MHz)	Comments	5	10	20
4942.5	25/500 mW	1		
4945.0	50/1000 mW		11	
4947.5	Public Safety 25/500 mW	2		
4950.0	Public Safety 50/1000 mW			
4952.5	Public Safety 25/500 mW	3		
4955.0	Public Safety 50/1000 mW		13	
4955.0	100/2000 mW			21
4957.5	Public Safety 25/500 mW	4		
4960.0	Public Safety 50/1000 mW			
4962.5	Public Safety 25/500 mW	5		
4965.0	Public Safety 50/1000 mW		15	
4967.5	Public Safety 25/500 mW	6		
4970.0	Public Safety 50/1000 mW			
4972.5	Public Safety 25/500 mW	7		
4975.0	Public Safety 50/1000 mW		17	
4975.0	Public Safety 100/2000 mW			25
4977.5	Public Safety 25/500 mW	8		
4980.0	Public Safety 50/1000 mW			
4982.5	Public Safety 25/500 mW	9		
4985.0	Public Safety 50/1000 mW		19	
4987.5	Public Safety 25/500 mW	10		

Table 12.8 IEEE WLAN 5 GHz unlicensed channels

USA					Europe		
			Bandwidth (MHz)			Bandwidth (MHz)	
Channel	Center frequency (MHz)	Comments	20	40	Comments	20	40
36	5180	40 mW	x	x	200 mW	x	x
40	5200	40 mW	x	*	200 mW	x	*
44	5220	40 mW	x	x	200 mW	x	x
48	5240	40 mW	x	*	200 mW	x	*
52	5260	200 mW	x	x	200 mW	x	x
56	5280	200 mW	x	*	200 mW	x	*
60	5300	200 mW	x	x	200 mW	x	x
64	5320	200 mW	x	*	200 mW	x	*
100	5500	indoor DFS 200 mW	x	x	1000 mW	x	x
104	5520	indoor DFS 200 mW	x	*	1000 mW	x	*
108	5540	indoor DFS 200 mW	x	x	1000 mW	x	x
112	5560	indoor DFS 200 mW	x	*	1000 mW	x	*
116	5580	indoor DFS 200 mW	x	x	1000 mW	x	x
120	5600	disabled	x	*	1000 mW	x	*
124	5620	disabled	x	x	1000 mW	x	x
128	5640	disabled	x	*	1000 mW	x	*
132	5660	disabled	x	x	1000 mW	x	x
136	5680	indoor DFS 200 mW	x	*	1000 mW	x	*
140	5700	200 mW	x		1000 mW	x	
149	5745	800 mW	x	x			
153	5765	800 mW	x	*			
157	5785	800 mW	x	x			
161	5805	800 mW	x	*			
165	5825	1000 MW	x				
x	in use						
*	alternative use						

The remaining sections of this chapter refer to the 802.11 standard in its OFDM version (802.11a/g/n), as it is the protocol that provides the highest throughput and should replace all other 802.11 implementations over time.

12.2 Architecture

There are two possible implementations of a WLAN (Wireless Local Area Network) foreseen in the standard:

- *IBSS (Independent Basic Service Set)*: This is an ad-hoc implementation, in which all elements have the same hierarchy and functionality: an element is called an STA (Station). STAs can establish communication directly between themselves, and also perform DS (Distribution Service), by relaying information to other STAs. An IBSS is illustrated in Figure 12.1.
- *InfraBSS (Infrastructure Basic Service Set)*: This implementation requires an infrastructure element that concentrates all communications and performs the DS to STAs and portals. This infrastructure element is called AP (Access Point) and accumulates the function of a STA and a DS. The DS

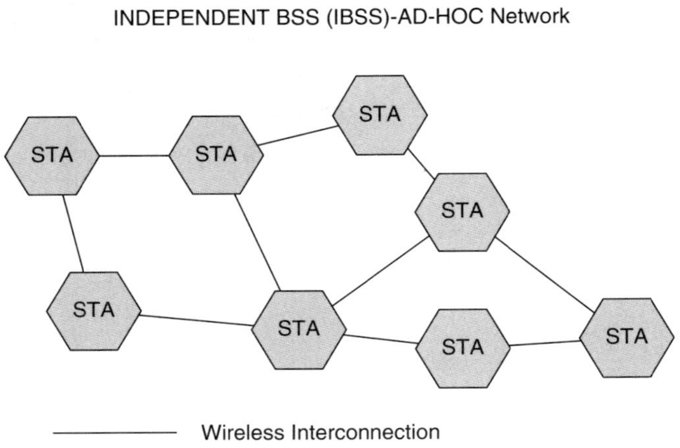

Figure 12.1 Independent BSS (IBSS), ad-hoc network.

function can also be used to interconnect different InfraBSSs forming an ESS (Extended Service Set). This set-up allows the WLAN to grow without restrictions. An InfraBSS is illustrated in Figure 12.2.

The wireless interconnections are done using RF channels and protocols (standards) define how this is done. This definition covers basically layers 1 and 2 of the OSI stack.

The Physical layer (PHY) establishes how the RF channel should be modulated so the signal can be recovered to provide the desired data throughput.

The Medium Access Layer (MAC) establishes how the data is packaged and the channel access procedures. It also defines link management and control.

Summarizing:

- An STA (Station) is any device that has 802.11 compliant PHY (Physical Layer) and MAC (Medium Access Control) interface to the WM (Wireless Medium).
- An AP (Access Point) is an entity that has station functionality and provides access to the distribution services, via the wireless medium.
- The BSS (Basic Service Set) is a set of member stations that may communicate between themselves.

The complete set of architectural services provided by 802.11 is:

- Authentication/De-authentication
- Association/Disassociation/Re-association
- Distribution/Integration
- Privacy
- MSDU (MAC Service Data Unit) delivery

12.3 The IEEE Std 802.11-2007

All the amendments to the 802.11-1999 Standard were incorporated in the 802.11-2007 Standard. These amendments are: a, b, d, e, g, h, i and j.

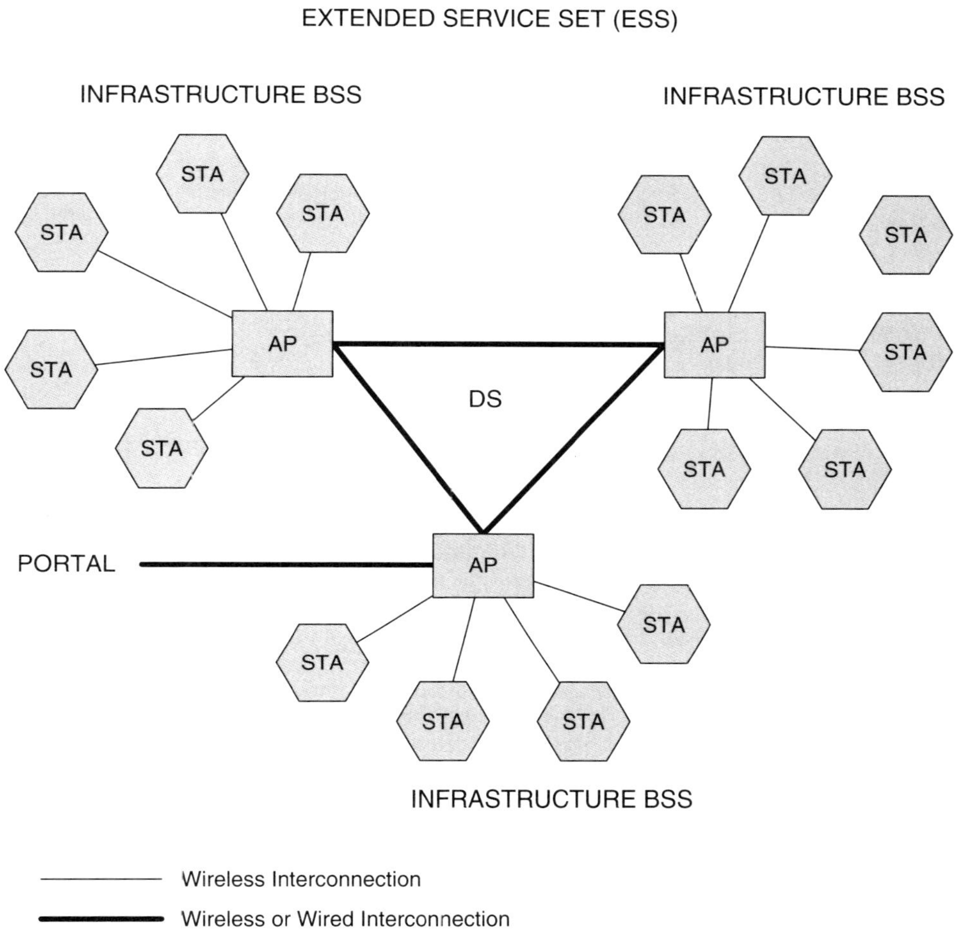

Figure 12.2 Infrastructure BSS (InfraBSS).

- 802.11a: OFDM operation in the 5 GHz band.
- 802.11b: Incorporates Spread Spectrum 1 and 2 Mbit/s modes and adds 5.5 and 11 Mbit/s modes in 2.4 GHz band.
- 802.11d: Adds radio transmitter parameters to AP transmissions so an STA can comply within the regulatory domain (country specific).
- 802.11e: Enhances MAC to support LAN applications with Quality of Service (voice, audio, video ...).
  - Introduces Hybrid Coordination Function (HCF), which specifies two QoS mechanisms.
  - Enhanced Distributed Channel Access (EDCA).
  - HCF Controlled Channel Access (HCCA).
  - Eight User Priorities (UP) are supported.
- 802.11g: OFDM operation in the 2.4 GHz band.

- 802.11h: Enables regulatory acceptance of 5 GHz in Europe.
  - Improvements in Transmit Power Control (TPC).
  - Dynamic Channel Selection (DCS) in presence of radar signals.
  - Channel energy measurements and reporting.
- 802.11i: 802.11-1999 Std specified WEP (Wired Equivalent Privacy) and SKA (Shared Key Authentication).
  - AES-CCMP (Advanced Encryption Standard CTR (Counter Mode) with CBC-MAC (Cipher Block Chaining- Message Authentication Code) Protocol) based on AES-128 in CCM (CTR with CBC-MAC) mode.
  - TKIP(Temporal Key Integrity Protocol): used to upgrade WEP deployments.
  - PMK: Pair-wise Master Key.
  - WPA (Wi-Fi Protected Access) encryption.
- 802.11j: Enable regulatory acceptance of 802.11 in Japan.
  - Addition of 10 MHz and 5 MHz channels.
  - Inclusion of 4.9 GHz for Public Safety.

12.3.1 Physical (PH) Layer

The OFDM implementation considers an FFT of 64 bits, regardless of the channel bandwidth. The OFDM carrier has 52 non-null sub-carriers and one central carrier. No transmission is done at the central carrier. The 52 sub-carriers are composed of 48 data carriers and 4 pilot carriers. The durations described next apply to a 20 MHz bandwidth, while Table 12.16 on p. 334 gives the durations for 10 and 5 MHz channels.

In 802.11, sub-carriers are always spaced by 312.5 kHz, which implies a symbol duration of 3.2 μs. The Cyclic Prefix (or guard interval) has always 0.8 μs (25% of symbol duration).

Packets are mapped to sub-carriers and sent as soon as access to the RF channel is allowed and synchronization information is provided by training sequences, sent in front of each data packet. This procedure is called PLCP (Physical Layer Convergence Procedure) and is illustrated in Figure 12.3. The basic PHY transmission unit is a PSDU (PLCP Service Data Unit), which is composed of:

- PLCP Preamble: contains two sets of training sequences used for synchronization (SYNC).
  - The first set called SYNC (Synchronization) or short training sequence has 10 symbols (alternating 1+j and −1−j between the used sub-carriers) of 0.8 μs duration each, over 12 uniformly spaced sub-carriers. It provides AGC (Automatic Gain Control) convergence, diversity selection, timing acquisition and coarse frequency acquisition.
  - The second set called SFD (Start Frame Delimiter) or long training sequence has 2 symbols of 3.2 μs duration, over the 52 non-null sub-carriers (transmitting a +1, +1,−1,−1 pattern on them), This long sequence is preceded by a guard interval (cyclic prefix) of 1.6 μs. This set is used for channel estimation and fine frequency acquisition.

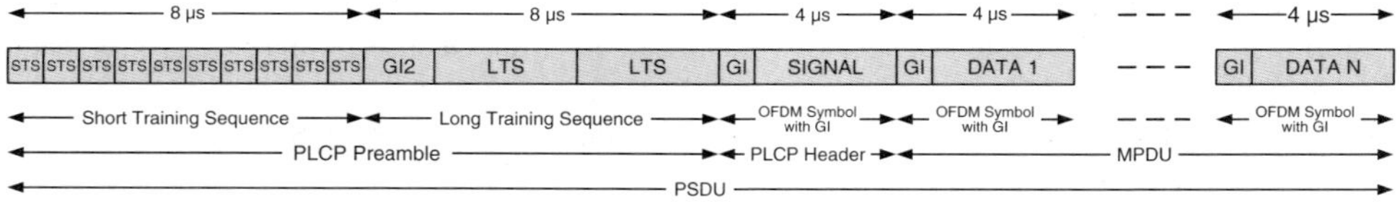

Figure 12.3 Physical Layer Convergence Procedure (PLCP).

- PLCP Header: has duration of 1 OFDM symbol and carries information about the modulation rate and the PSDU length in octets. A parity bit is also included. This header is sent using BPSK ½ modulation.
- PSDU (PLCP Service Data Unit): composed of a 16 bit Service field (used for scrambler initialization with some bits reserved for future use) and scrambled data. The data bits are combined with a 127 pseudo-random sequence to produce the scrambled data sequence. The sequence is then encoded by a Convolutional encoder according to the rate established for the connection (1/2, 2/3 or 3/4) and a two-stage block interleaver that maps adjacent bits to non-adjacent sub-carriers. Sub-carriers −21, −7, +7 and +21 are used by pilots which are BPSK modulated by a 127 pseudo-random sequence.
- Tail and padding bits: The 6 tail bits are all zero, so the Convolutional encoder is returned to a zero state. The padding bits are used to complete the number of data bits to a multiple of 48.

A PSDU (PLCP Service Data Unit) represents the complete information to be transmitted at a time in the wireless media. It carries the PLCP Preamble, PLCP Header and MPDU (MAC Protocol Data Unit). The MPDU carries the IP data packet, which has a 20 bytes header and a data part that ranges from 0 to 65,535 bytes.

The IP header has the following fields:

- Version (4 bits)
- Header length (4 bits)
- Differentiated Services (8 bits)
- Total Length (16 bits)
- Identification (16 bits)
- Flags (3 bits)
- Fragment offset (13 bits)
- Time to live (8 bits)
- Protocol (8 bits)
- Header Checksum (16 bits)
- Source Address (32 bits)
- Destination Address (32 bits)

The complete OFDM carrier is represented in Figure 12.4. The sub-carrier at the center of the OFDM carrier frequency is numbered as zero and is left unused, due to possible leakages of the RF carrier at this frequency. There are 26 active sub-carriers on each side of it, numbered from −26 to +26. Twelve subcarriers are used for the Short Training Sequence (STS) and carry the 10 SYNC symbols, with total duration of 8 μs. The Long Training Sequence (LTS) is transmitted in all active sub-carriers and carries the two SFD (Start Frame Delimiter) symbols with total duration of 8 μs, including a guard band of 0.8 μs. This Guard Band (GB) is also known as Guard Interval (GI) and represents the OFDM cyclic prefix. All other symbols use the 52 subcarriers and have a guard band of 0.8 μs and data duration of 3.2 μs. The first symbol carries the PLCP Header, also known as SIGNAL. The remaining symbols carry the MAC PDU.

12.3.2 Medium Access Control (MAC) Layer

The medium here is the RF channel and messages should be transmitted through it from STAs to STAs or Portals, using the DS in InfraBSS. The access to the RF channel is done using Coordination Functions (CF).

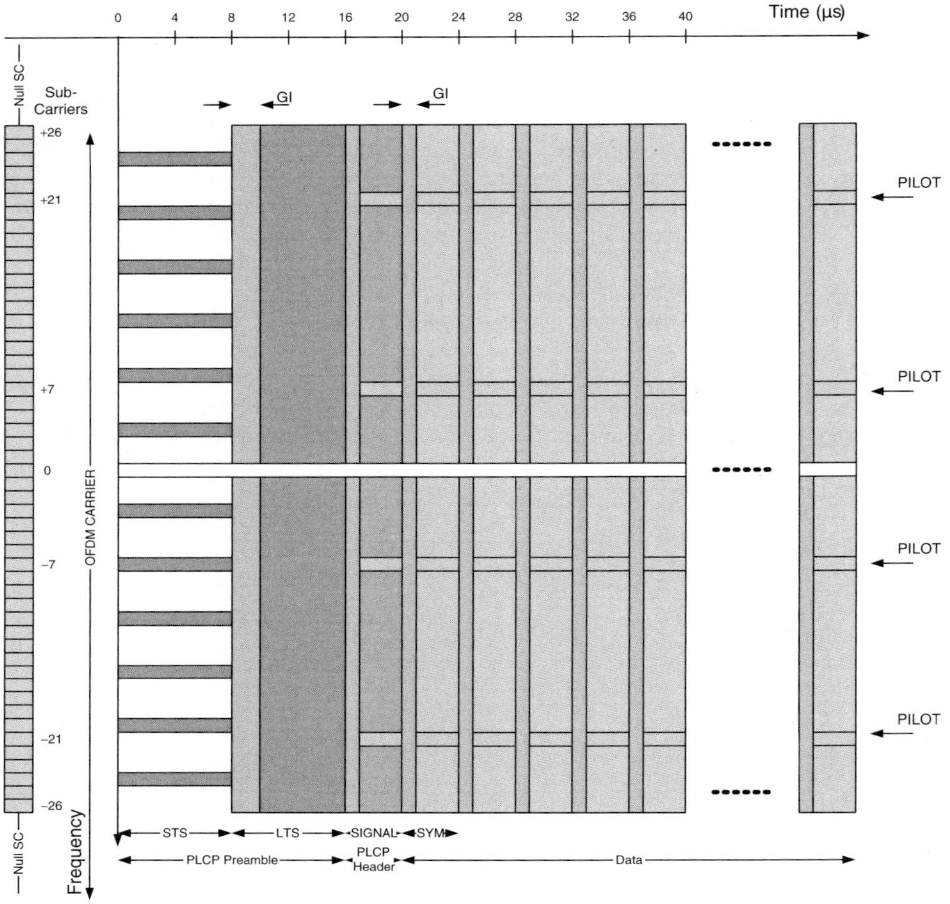

Figure 12.4 Physical Layer (PHY).

The MAC layer defines the messages that can be interchanged in the wireless media to provide the desired functionalities. These messages are called frames and are divided in three groups:

- *Management Frames*: used to manage STAs and overall timing.
- *Data Frames*: used to send the actual data.
- *Control Frames*: used to control the medium access and frames inter-exchange.

The general MAC frame format is represented in Figure 12.5.

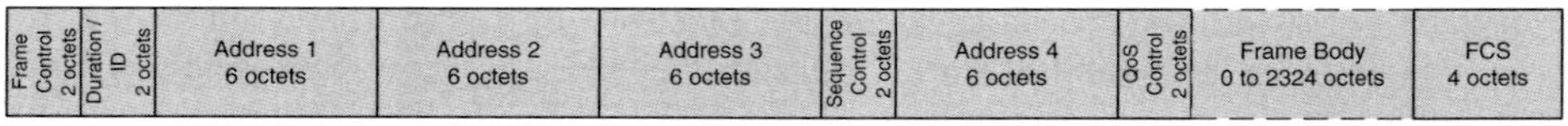

Figure 12.5 Medium Access Control (MAC) frame format.

Table 12.9 MAC address configuration

To	From	Address 1 (RA)	Address 2 (TA)	Address 3	Address 4
STA	STA	RA = DA	TA = SA	BSSID	N/A
STA	AP	RA = DA	TA = BSSID or AP STA	SA	N/A
AP	STA	RA = BSSID or AP STA	TA = SA	DA	N/A
AP	AP	RA	TA	DA	SA

Key:
STA Station
AP Access Point
SA Source Address
DA Destination Address
TA Transmitting STA Address
RA Receiving STA Address
BSSID Basic Service Set ID
N/A Not Available (field is omitted)

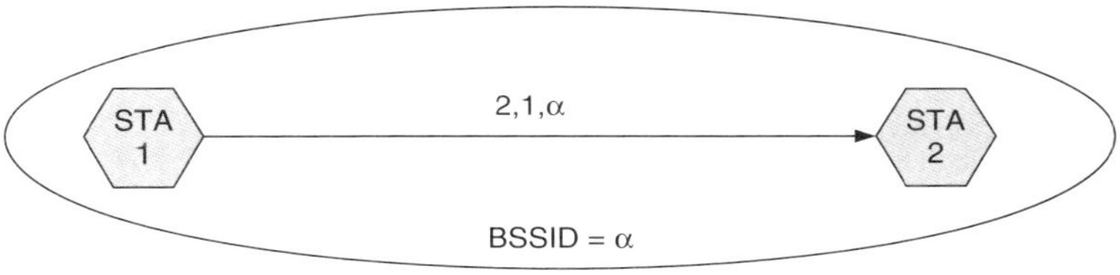

Figure 12.6 STA to STA addressing (IBSS).

12.3.2.1 Distribution System

The MAC address supports different configurations of the wireless networks as presented in Table 12.9. The address fields carry the MAC ID of each element. We should remember that each AP is also an STA, being the AP responsible for the routing function.

Address 1 has the receiving STA address, which is also the destination address. When the receiving side is an AP in the DS function the BSSID field receives the address of the STA contained in the AP. Address 2 has the transmitting STA address, which is also the source address. When the transmitting side is an AP in the DS function the BSSID field receives the address of the STA contained in the AP. Address 3 is the original source or the final destination address. When no AP is involved, the BSSID is sent. When the communication is between APs, the final destination address is sent. Address 4 is used only in WDS (Wireless Distribution System) mode and it carries the initial source address.

Figure 12.6 shows the address fields for an IBSS network. It applies also for management messages between an AP and its STAs as shown in Figure 12.7. Figure 12.8 shows the address fields for the messages sent to transfer data between two APs, using WDS.

12.3.2.2 Management Frames

The following management frames are available:

- *ACK*: This message returns an acknowledgement to a received message and has 14 octets. When an ACK is not received within a specified time, the message is resent, up to a specified maximum number of repetitions.

- *RTS/CTS*: An STA cannot listen to the air messages when it is transmitting, so if a collision occurs, it is only discovered by the lack of an ACK message. This is very upsetting for long messages. To avoid this issue, the long message can be preceded by a short one, called RTS (Request to Send), and complemented by another message called CTS (Clear to Send). This reduces the repetition time in case of a collision. Collisions are common, however, when there are hidden nodes in the network, which is when some STAs in a BSS cannot hear the transmission of all other STAs, allowing one STA to start transmitting when another one is already transmitting. The RTS frame has 20 octets, whereas the CTS frame has 14 octets. RTS and CTS should always be used if there is the possibility of hidden nodes occurring in the system.
- *PS-Pol*: This message is sent by an STA in power save (PS) mode to request data from an AP. This message has 20 octets.
- *CF-End*: This message is transmitted to announce the end of the Contention Free (CF) period. This message has 20 octets.
- *CF-End+CF-ACK*: This message is transmitted to announce the end of the Contention Free (CF) period, but stating that there are still pending ACKs. This message has 24 octets.
- *Block ACK Req*: This message requests an ACK for a block of messages.
- *Block ACK*: This message includes ACKs for a block of messages. This message has 152 octets.

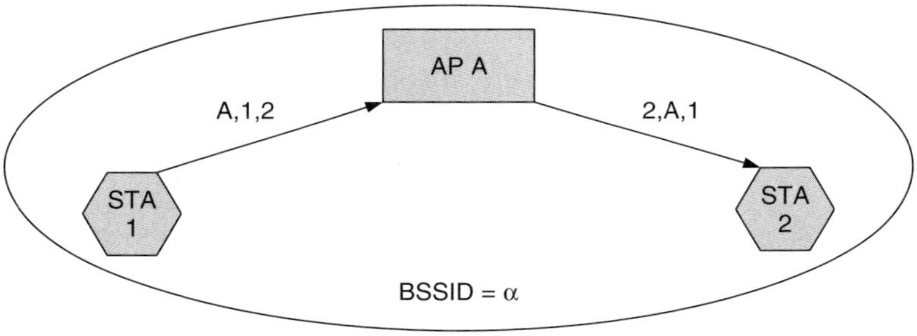

Figure 12.7 STA to STA addressing (InfraBSS).

Figure 12.8 STA to STA addressing through WDS.

Table 12.10 Maximum MPDU duration for best channel conditions

Maximum MPDU duration for best conditions (64 QAM 3/4)	
Minimum MPDU data octets	2324
Data to (FEC + Data) rate	0.75
(Header + tail) MAC octets	36
Total Octets(Data + FEC + MAC)	3135
64QAM modulation sub-carriers	522.4
Required OFDM symbols	10.9
PHY overhead duration (μs)	16.0
OFDM symbol duration (μs)	4
Total MPDU duration (μs)	60

Table 12.11 Minimum MSDU duration for best channel conditions

Minimum MPDU duration for best conditions (64 QAM 3/4)	
Maximum MPDU data octets	30
Data to (FEC + Data) rate	0.75
(Header + tail) MAC octets	36
Total Octets(Data + FEC + MAC)	76
64QAM modulation sub-carriers	12.7
Required OFDM symbols	0.3
PHY overhead duration (μs)	16.0
OFDM symbol duration (μs)	4
Total MPDU duration (μs)	20

12.3.2.3 Data Frames

The data frame follows the general frame format, so it has an overhead of 36 octets and may contain up to 2324 octets of data. The longest data message corresponds to 11 OFDM carriers (44 μs) when using 64QAM 2/3 or 49 sub carriers (490 μs) when using QPSK ½. Tables 12.10 and 12.11 give MPDU duration when transferring the maximum and minimum number of octets, using 64 QAM ¾.

Tables 12.12 and 12.13 give the MPDU duration when transferring the maximum and minimum number of octets, using QPSK 1/2.

IP data packets of up to 10 bytes have the same duration of 20 μs, regardless of the modulation used. The largest packets have duration of 60 μs for 64QAM and 212 μs for QPSK.

Sending large data frames is an efficient procedure as it minimizes overhead, but at the same the FER (Frame Error Rate) increases. There is an optimum MPDU size that maximizes throughput for each environment. It is possible to specify the maximum MPDU size to be transmitted, so the original MPDU is fragmented into smaller frames. Fragmented frames from the same MPDU are transmitted in sequence and are timed not to be interrupted by other transmissions.

When a frame does not get an ACK within a certain time window, the MPDU is retransmitted up to a specified number of times. During this time window, duplicates may be received, in this case, the MAC procedure drops the repeated frames.

Table 12.12 Maximum MPDU duration for worst channel conditions

Maximum MPDU duration for worst conditions (QPSK 1/2)	
Minimum MPDU data octets	2324
Data to (FEC + Data) rate	0.5
(Header + tail) MAC octets	36
Total Octets(Data + FEC + MAC)	4684
QPSK modulation sub-carriers	2342
Required OFDM symbols	48.8
PHY overhead duration (μs)	16
OFDM symbol duration (μs)	4
Total MPDU duration (μs)	212

Table 12.13 Minimum MPDU duration for worst channel conditions

Minimum MPDU duration for worst conditions (QPSK 1/2)	
Maximum MPDU data octets	30
Data to (FEC + Data) rate	0.5
(Header + tail) MAC octets	36
Total Octets(Data + FEC + MAC)	96
QPSK modulation sub-carriers	48
Required OFDM symbols	1
PHY overhead duration (μs)	16
OFDM symbol duration (μs)	4
Total MPDU duration (μs)	20

12.3.2.4 Control Frames

Control Frames have variable lengths, depending on the information to be carried. The average length of each message is 30 octets. The following are the most common messages sent in the control frame:

- *Beacon*: This is a message that defines an access coordination period. It provides basic information for STAs. The Beacon frame contains the TSF (Timing Synchronization Function) timer, beacon interval, capability information, SSID (Service Set Identifier), PHY parameter set, IBSS parameter set and TIM (Traffic Indication Map).
- *Disassociation*: This message informs that the STA is disassociated.
- *Association Request*: This message asks for an association.
- *Association Response*: This message informs if an association request was successful or not.
- *Probe Request*: In this message, information about the WLAN is requested by the STA.
- *Probe Response*: The AP response to a probe request.
- *Authentication*: This message provides data for authentication.
- *De-authentication*: This message informs that the STA was de-authenticated.
- *Action*: This message informs different actions.

12.3.3 RF Channel Access

The RF channel access function is called Coordination function and it can be distributed or have a coordinator (point).

12.3.3.1 Distributed Coordination Function (DCF)

In DCF, the coordination is distributed between all stations. This is the primary RF channel access method and is done using CSMA/CA (Carrier Sense Multiple Access with Collision Avoidance). In this method a station wishing to transmit has to first listen to the channel for a predetermined amount of time so as to check for any activity on the channel. If the channel is sensed "idle," then the station is permitted to transmit. If the channel is sensed as "busy," the station has to defer its transmission. DCF periods are illustrated in Figure 12.9, where the Beacon message defines the start of each period.

Because 802.11 is a TDD (Time Division Duplex) system, only one transmission can happen at a time, and both the signal travel time and Transmit (Tx) to Receive (Rx) switching should be considered in the system timing. Several timers called IFS (Inter-Frame Space) are specified with this purpose (Figure 12.10):

- *SIFS (Short Inter Frame Space)*: This is the nominal time that the MAC and PHY require to receive the last symbol of a frame at the air interface, process the frame, and respond on the air interface, with the first symbol of the earliest possible response frame. This time is used between subsequent transmissions related interactions of a given STA, meaning that it includes all messages from RTS until the ACK. The SIFS calculation is calculated by the formula below and its typical value is 10 μs.

$$\text{SIFS} = \text{RX processing delay}(3\,\mu\text{s}) + \text{RxTxturnaround time}(5\,\mu\text{s})$$

- *ST (Slot Time)*: The timing of different Stations can be out of phase by the maximum link propagation time (e.g. NAV timing). The Slot Time is used to compensate for this difference, when there are interactions between different STAs. The ST is used to define the PIFS, DIFS and CBP times. The typical ST time is 20 μs, but it does vary with the maximum link distance. The formula below is used to calculate ST.

$$\text{ST} = \text{CCA(Clear Channel Assessment) time } (10\,\mu\text{s}) + \text{RxTx turnaround time } (5\,\mu\text{s})$$
$$+ \text{ air propagation time } (1\,\mu\text{s for every } 300\,\text{m}) + \text{RX processing delay } (3\,\mu\text{s})$$

- *PIFS (Point Inter Frame Space)*: This timing is used by STAs working under PCF (Point Coordination Function), described in Section 12.3.3.2. PIFS typical value is 30 μs and is calculated by the following formula:

$$\text{PIFS} = \text{SIFS} + \text{ST}$$

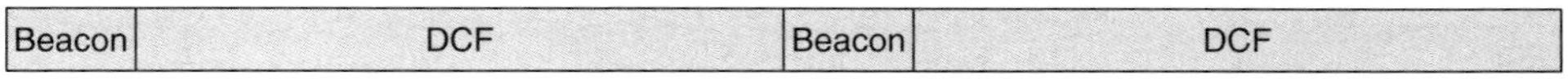

Figure 12.9 Distributed Coordination Function (DCF).

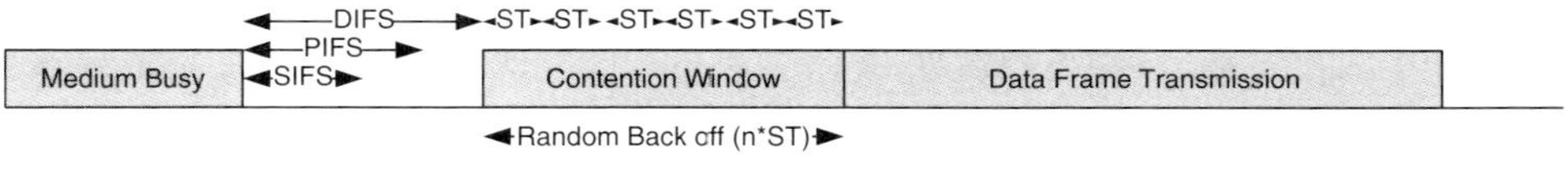

Figure 12.10 Inter-frame spacing.

- *DIFS (Distributed Inter Frame Space)*: This timing is used by STAs working under DCF (Distributed Coordination Function), described in Section 12.3.3.1. Its typical value is 40 μs.

$$\text{DIFS} = \text{SIFS} + 2\text{ST}$$

- *AIFS (Arbitration Inter Frame Space)*: This is used by QoS STAs to transmit all data frames.

$$\text{AIFS} = \text{SIFS} + \text{DIFS}$$

- *EIFS (Extended Inter Frame Space)*: This timing should be used to access the wireless medium when the previous frame did not receive an ACK.

$$\text{EIFS} = \text{SIFS} + \text{DIFS} + \text{ACK}$$

- *RBT (Random Back-off Time)*: An STA desiring to initiate a data transfer invokes the Carrier Sense (CS) mechanism to determine the busy/idle state of the medium. If the medium is busy, the STA shall defer until the medium is continuously idle for a DIFS period. The STA generates then a random delay (back-off), which is a multiple of ST and if the medium is still idle it initiates transmission.

The random back-off depends on the number of active STAs expected to be active at the same time and is defined by a Contention Window (CW). The CW is specified by a minimum (CW_{min}) and maximum value (CW_{max}), which can be set to the values, 7, 15, 31, 63, 127 and 255.

A random number between 0 and CW is chosen by each STA. The duration of the random back-off time is given by this random number multiplied by Slot Time. The Slot Time is used to assure that if one STA starts transmitting in one of the slots there will be enough time for the signal to reach all STAs. If a conflict occurs (two STA transmitted and no ACK was received), the CW_{min} value is increased to the next value.

The CW starts at CW_{min} and goes to the next value when an unsuccessful transmission happens. It returns to CW_{min} after a successful transmission. CW_{min} and CW_{max} should be chosen according to the conflict probability acceptable between STAs.

The CW counter is decremented at each ST that the medium is free. When someone else takes the medium, the counting stops and is reinitiated for the next CW opportunity.

12.3.3.2 Collision Avoidance Procedure

The procedure for collision avoidance is described next and is illustrated in Figure 12.11.The CCA (Clear Channel Assessment) functionality plays a key role in this process, but there is no reliable way

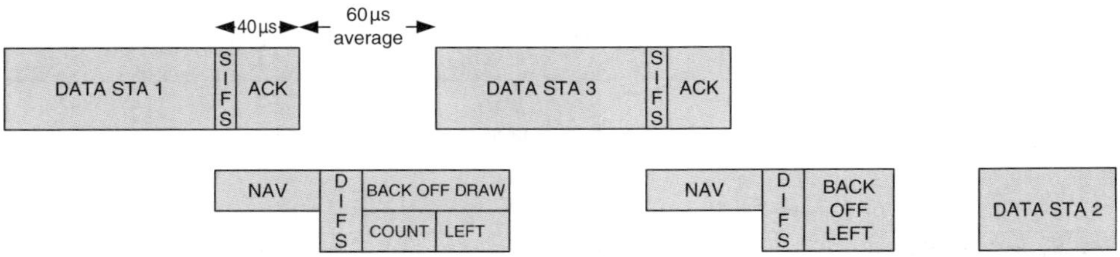

Figure 12.11 Collision avoidance procedure.

of detecting activity on the wireless medium just by listening to it with the Carrier Sense (CS), due to the possible existence of hidden nodes. This is circumvented by providing the duration of each transmission in the MAC message. This duration is indicated using the NAV (Network Allocation Vector) field in the MAC message.

The STA sends this indication when sending a message, so other STAs can time the end of the transmission (including an allowance for the ACK message). If hidden nodes exist, the RTS/CTS mechanism should be used and a NAV is sent in each message. This circumvents the hidden node issue as all STAs must be able to listen to AP messages before transmitting. This procedure is illustrated in Figure 12.12. The usage of NAV does not synchronize perfectly the STAs and the difference can be as long as the RF propagation time from the AP to the edge of the cell (1 μs every 300 meters).

12.3.3.3 Point Coordination Function (PCF)

In PCF, a Point Coordinator (PC) is established at the BSS to coordinate data transmissions by STAs that have requested it. In this case, the period between Beacon messages alternates between PCF and DCF, as illustrated in Figure 12.13:

- STAs indicate to a Point Coordinator (PC) if they are capable of responding to a poll and whether they want to be polled.
- A PC retains control of the medium in the CFP (Contention Free Period) by gaining access to the channel prior to STAs in DCF mode.
- The PC issues polls to STAs on the polling list.
- The PC ends PCF when timer expires or all transmissions are done.

12.3.4 Power Management

Stations (STAs) can be in active or Power Save (PS) mode. The AP holds any frames to STAs in PS mode until it is time to send a DTIM (Delivery Traffic Indication Message) in a Beacon frame. The STA is informed that there is a frame for it in the DTIM message and remains in active mode; otherwise it goes into power save mode until the next DTIM.

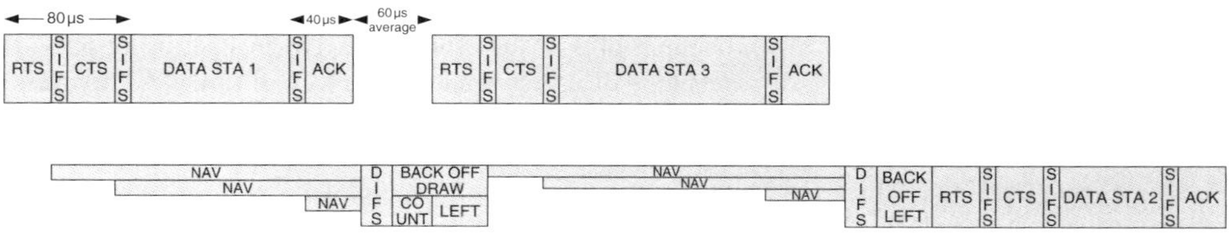

Figure 12.12 Collision avoidance procedure with RTS and CTS.

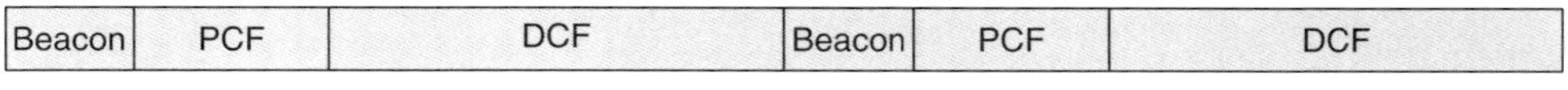

Figure 12.13 Point coordination function.

12.4 Enhancements for Higher Throughputs, Amendment 5: 802.11n-2009

The medium access procedure of 802.11 is very inefficient and extensive work was done to increase the protocol efficiency, while keeping compatibility with existing systems. The main focus of these efforts is the n amendment, but amendments k, y and w are also included in this specification:

- *802.11k*: Interfaces for higher layers for radio network measurements.
- *802.11n*: Support HT (High Throughput) rate at 20 MHz.
- *Spectral efficiency* of 3 bit/sec/Hz.
- *Alamouti coding* (Space Time Block Coding).
- *Turbo coding*.
- *802.11y*: Enables high powered units to operate in the 3650 to 3700 MHz band.
- *802.11w*: Increases security of management frames.

The compatibility with existing STAs reduced the number of options available to increase throughput. After many proposals were considered, the following new features were chosen:

- Use of 40 MHz bandwidth.
- Use of MIMO (Multiple Input Multiple Output), including:
  - Receiver Diversity using Selection Combining.
  - Receiver Diversity using Maximum Ratio Combining (MRC).
  - Channel Aware Transmitter Diversity using Channel State Information (CSI).
  - Channel Unaware Transmitter Diversity using Space Time Block Code (STBC), Alamouti algorithm.
- Spatial Multiplexing with up to four antennas and other antenna algorithms.
- Additional FEC options (Binary Convolutional Code with 5/6 ratio and LDPC (Low-Density Parity Check).
- Changes in the Cyclic Prefix (12.5% and 50% option).
- Reduced Inter-Frame Space.
- Increased data length per frame, using MAC aggregation and block acknowledgement.

Claims of 600 Mbps are made in the standard. This extraordinary (but elusive) throughput is calculated based on the following steps:

1. The calculation starts with the air throughput of 54 Mbps for a 20 MHz channel using 64QAM ¾. This rate applies only to a low percentage of users which are very close to the AP (typically 5% of AP total area coverage).
2. The throughput is increased to 58.5 Mbps by adding four data sub-carriers to the original 48 data sub-carriers. The extra sub-carriers increase the interference between adjacent channels and may even reduce the overall throughput, by requiring a lower throughput modulation scheme.
3. The use of a FEC of 5/6 instead of ¾ increases the throughput to 65 Mbps. This lower FEC ratio requires an even better SNR and will reduce further the area in which this signal can be used.
4. A reduction of the Guard interval to 12.5% of the symbol results in symbol duration of 3.5 μs, thus increasing the throughput to 72.2 Mbps. This reduces the maximum multipath spread to 100 m.
5. Next, the 20 MHz channel is replaced by a 40 MHz channel, taking the throughput to 144.4 MHz. This is a valid claim, but double the spectrum usage.
6. Use of spatial multiplexing with four antennas, increases the throughput to 577.8 Mbps. For this to happen, the antennae signals received in the four antennas should be totally uncorrelated, which

does not happen in real life deployments. In some cases the use of multiple antennas can even reduce the throughput.

7. Finally the use of shorter IFS (Inter-Frame Space), frame aggregation and block ACK takes the throughput to 600 Mbps, as the claimed improvement for these additional changes is 4%.

Some of these claims apply only to very short distance indoor applications, and other can only provide improvements outdoor, so it does not make sense to consider both effects together.

Real life indoor applications (a TV set connected to a PC) with a 40 MHz channel in the 5 GHz band and a single user get a maximum air throughput of 150 Mbit/s.

In real life outdoor applications, where many users are distributed over the whole area, a 40 MHz channel is not recommended (due to excessive spectrum use) and many of the new features would have to be disabled to provide the required communication distance, in which case hardly any throughput gain will be noticed (see throughput calculations below).

Additionally there is a protocol overhead required to provide compatibility between the 2003 (non-HT) configurations and the new High Throughput (HT) configurations. This further reduces the throughput for 802.11n. The HT configuration requires additional overhead with the increase of the number of antennas, undermining the gain by deploying multiple antennas.

The 802.11n-2009 took many years in preparation, so vendors issued much pre-n equipment which, of course, lacks compatibility between them. The final standard assumed three different PHY (Physical Layer) configurations and five different PLCP (Physical Layer Convergence Procedure) configurations.

12.4.1 Physical Layer

The PHY configurations are:

- *Non HT (NHT)*: This configuration is equivalent to the 802.11-2003 specifications.
- *HT Mixed (HTM)*: This configuration supports legacy and new HT configurations.
- *HT Green Field (HTGF)*: This configuration does not provide compatibility with the legacy equipment, and should better benefit from the new features.

Each of these configurations can be implemented in a 20 MHz and a 40 MHz channel. The 40 MHz channel can be implemented as a single channel or duplicate (bonding) of two 20 MHz channels. In this latter case, data is split between two 20 MHz channels and then reassembled.

Previously implemented 10 MHz and 5 MHz channels are not contemplated in 802.11n.

The PHY also supports from 1 to 4 spatial antenna streams. This implementation is shown in the block diagrams in Figures 12.14 and 12.15. The mandatory implementation in the standard is the support for two streams in the AP and one stream in the STA.

The CSD (Cyclic Shift Delay) is inserted to assure a minimum shift between the spatial streams. The first stream is used as reference; the second stream has a delay of −400 ns, the third a delay of −200 ns and the fourth a delay of −600 ns.

The HT physical layer supports several modulation schemes that consider the bandwidth and the number of spatial streams. They are divided into two blocks:

- *EQM (Equal Modulation Schemes)*: in this block all spatial streams use the same modulation scheme. These modulation schemes require that the adopted MS (Modulation Scheme) must be the lowest of all streams. Numbered from 0 to 32 and shown below for 20 and 40 MHz.
- *UEQM (UnEqual Modulation Schemes)*: in this block each spatial scheme uses a different modulations scheme. These modulation schemes allow that spatial streams with better SNR get better modulation schemes than the ones with a worst SNR. Numbered from 33 to 76.

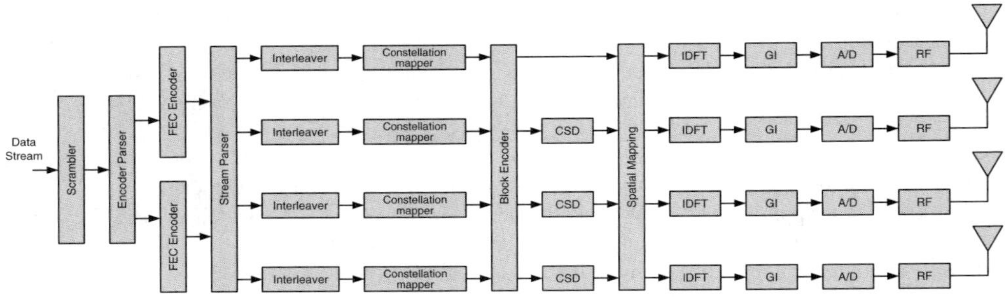

Figure 12.14 TX MIMO block diagram.

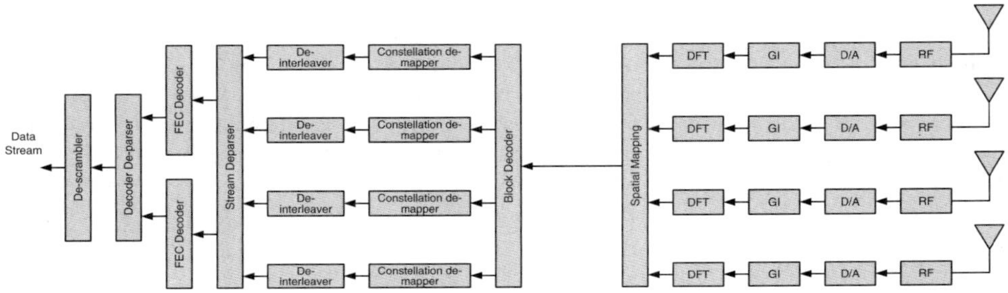

Figure 12.15 RX MIMO block diagram.

Modulation Coding Scheme (NCS) indexes for 20 MHz and 40 MHz are presented in Tables 12.14 and 12.15.

12.4.2 MAC Layer

12.4.2.1 PLCP (Physical Layer Convergence Procedure)

Possible PLCP configurations are:

- *Non-HT (NHT)*: compatible with the legacy equipment.
- *Non-HT duplicate (NHTD)*: data is sent to two adjacent 20 MHz legacy channels and re-assembled again.
- *HT Mixed (HTM)*: legacy and HT products are supported. The PSDU has legacy and HT training and signal sequences.
- *HT Greenfield (HTG)*: only HT products are supported.
- *MCS-32*: devised for noisy environments, supports only one stream and uses as modulation scheme only BPSK 1/2. It is specified for 40 MHz, using duplicated 20 MHz channels.

The legacy, HT Mixed and HT Greenfield PSDUs are shown in Figures 12.16, 12.17 and 12.18.

Table 12.14 Modulation indexes for 20 MHz

MCS Index	Modulation	FEC Rate	CBW (MHz)	Spatial Streams	Bits per single sub-carrier	Coded sub-carriers	Pilots	Number of coded bits per OFDM Symbol	Number of data bits per OFDM Symbol	Data Rate for GI = 0.8 (Mbps)	Data Rate for GI = 0.4 (Mbps)
0	BPSK	1/2	20	1	1	56	4	52	26	6.5	7.2
1	QPSK	1/2	20	1	2	56	4	104	52	13.0	14.4
2	QPSK	3/4	20	1	2	56	4	104	78	19.5	21.7
3	16QAM	1/2	20	1	4	56	4	208	104	26.0	28.9
4	16QAM	3/4	20	1	4	56	4	208	156	39.0	43.3
5	64QAM	1/2	20	1	6	56	4	312	156	39.0	43.3
6	64QAM	3/4	20	1	6	56	4	312	234	58.5	65.0
7	64QAM	5/6	20	1	6	56	4	312	260	65.0	72.2
8	BPSK	1/2	20	2	1	56	4	104	52	13.0	14.4
9	QPSK	1/2	20	2	2	56	4	208	104	26.0	28.9
10	QPSK	3/4	20	2	2	56	4	208	156	39.0	43.3
11	16QAM	1/2	20	2	4	56	4	416	208	52.0	57.8
12	16QAM	3/4	20	2	4	56	4	416	312	78.0	86.7
13	64QAM	1/2	20	2	6	56	4	624	312	78.0	86.7
14	64QAM	3/4	20	2	6	56	4	624	468	117.0	130.0
15	64QAM	5/6	20	2	6	56	4	624	520	130.0	144.4
16	BPSK	1/2	20	3	1	56	4	156	78	19.5	21.7
17	QPSK	1/2	20	3	2	56	4	312	156	39.0	43.3
18	QPSK	3/4	20	3	2	56	4	312	234	58.5	65.0
19	16QAM	1/2	20	3	4	56	4	624	312	78.0	86.7
20	16QAM	3/4	20	3	4	56	4	624	468	117.0	130.0
21	64QAM	1/2	20	3	6	56	4	936	468	117.0	130.0
22	64QAM	3/4	20	3	6	56	4	936	702	175.5	195.0
23	64QAM	5/6	20	3	6	56	4	936	780	195.0	216.7
24	BPSK	1/2	20	4	1	56	4	208	104	26.0	28.9
25	QPSK	1/2	20	4	2	56	4	416	208	52.0	57.8
26	QPSK	3/4	20	4	2	56	4	416	312	78.0	86.7
27	16QAM	1/2	20	4	4	56	4	832	416	104.0	115.6
28	16QAM	3/4	20	4	4	56	4	832	624	156.0	173.3
29	64QAM	1/2	20	4	6	56	4	1248	624	156.0	173.3
30	64QAM	3/4	20	4	6	56	4	1248	936	234.0	260.0
31	64QAM	5/6	20	4	6	56	4	1248	1040	260.0	288.9

Table 12.15 Modulation indexes for 40 MHz

MCS Index	Modulation	FEC Rate	CBW (MHz)	Spatial Streams	Bits per single sub-carrier	Coded sub-carriers	Pilots	Number of coded bits per OFDM Symbol	Number of data bits per OFDM Symbol	Data Rate for GI = 0.8 (Mbps)	Data Rate for GI = 0.4 (Mbps)
0	BPSK	1/2	40	1	1	114	6	108	54	13.5	15.0
1	QPSK	1/2	40	1	2	114	6	216	108	27.0	30.0
2	QPSK	3/4	40	1	2	114	6	216	162	40.5	45.0
3	16QAM	1/2	40	1	4	114	6	432	216	54.0	60.0
4	16QAM	3/4	40	1	4	114	6	432	324	81.0	90.0
5	64QAM	1/2	40	1	6	114	6	648	324	81.0	90.0
6	64QAM	3/4	40	1	6	114	6	648	486	121.5	135.0
7	64QAM	5/6	40	1	6	114	6	648	540	135.0	150.0
8	BPSK	1/2	40	2	1	114	6	216	108	27.0	30.0
9	QPSK	1/2	40	2	2	114	6	432	216	54.0	60.0
10	QPSK	3/4	40	2	2	114	6	432	324	81.0	90.0
11	16QAM	1/2	40	2	4	114	6	864	432	108.0	120.0
12	16QAM	3/4	40	2	4	114	6	864	648	162.0	180.0
13	64QAM	1/2	40	2	6	114	6	1296	648	162.0	180.0
14	64QAM	3/4	40	2	6	114	6	1296	972	243.0	270.0
15	64QAM	5/6	40	2	6	114	6	1296	1080	270.0	300.0
16	BPSK	1/2	40	3	1	114	6	324	162	40.5	45.0
17	QPSK	1/2	40	3	2	114	6	648	324	81.0	90.0
18	QPSK	3/4	40	3	2	114	6	648	486	121.5	135.0
19	16QAM	1/2	40	3	4	114	6	1296	648	162.0	180.0
20	16QAM	3/4	40	3	4	114	6	1296	972	243.0	270.0
21	64QAM	1/2	40	3	6	114	6	1944	972	243.0	270.0
22	64QAM	3/4	40	3	6	114	6	1944	1458	364.5	405.0
23	64QAM	5/6	40	3	6	114	6	1944	1620	405.0	450.0
24	BPSK	1/2	40	4	1	114	6	432	216	54.0	60.0
25	QPSK	1/2	40	4	2	114	6	864	432	108.0	120.0
26	QPSK	3/4	40	4	2	114	6	864	648	162.0	180.0
27	16QAM	1/2	40	4	4	114	6	1728	864	216.0	240.0
28	16QAM	3/4	40	4	4	114	6	1728	1296	324.0	360.0
29	64QAM	1/2	40	4	6	114	6	2592	1296	324.0	360.0
30	64QAM	3/4	40	4	6	114	6	2592	1944	486.0	540.0
31	64QAM	5/6	40	4	6	114	6	2592	2160	540.0	600.0
32	BPSK	1/2	20 × 2	1	1	56 × 2	4 × 2	52 × 2	26 × 2	6.0 × 2	6.7 × 2

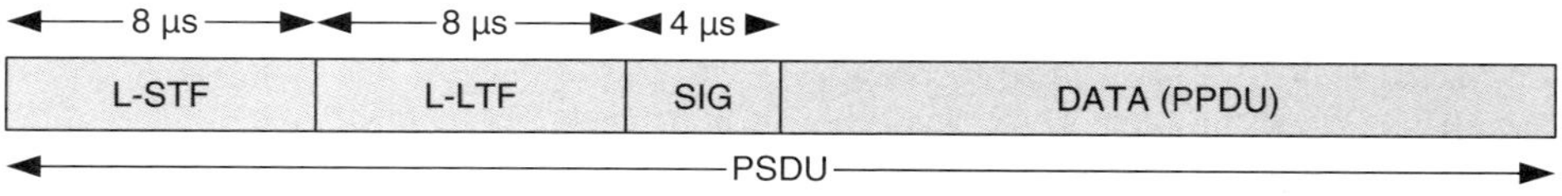

Figure 12.16 Legacy PSDU.

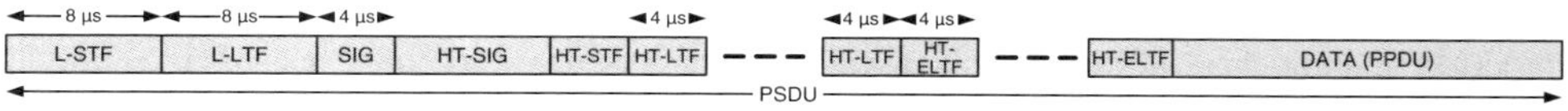

Figure 12.17 HT Mixed PSDU.

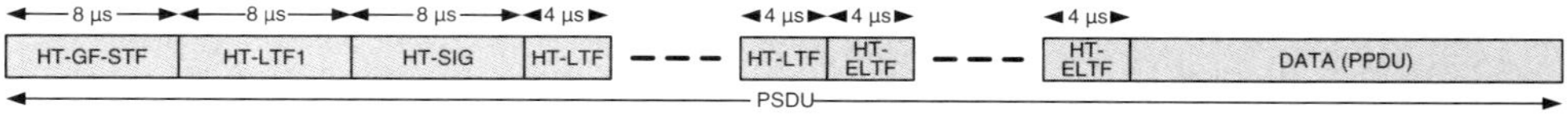

Figure 12.18 HT greenfield PSDU.

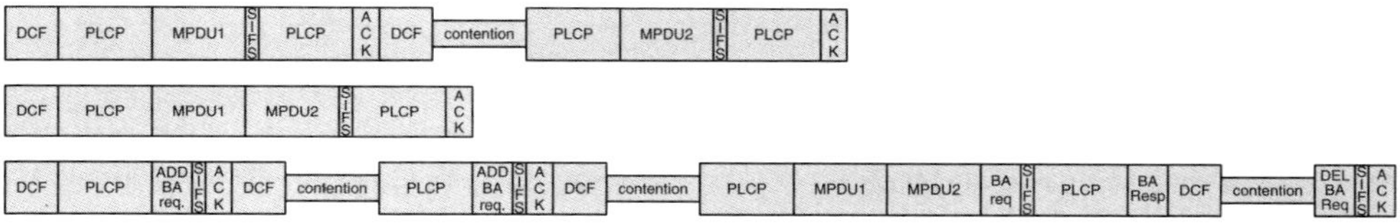

Figure 12.19 Frame aggregation.

12.4.2.2 Frame Aggregation

Aggregating several frames to the same receiver minimizes the number of ACKs sent and in principle reduces the overall transmission time as shown in Figure 12.19, but the protocol requires the establishment of a Block ACK (BA) session followed by a deletion, which may increase the overall time.

12.5 Work in Progress

There are still several groups working on future enhancements as described below:

- *802.11f*: Specify the information to be exchanged by AP to support Distribution Service (DS) function.
  - Enable implementation of Distribution Systems containing APs of different vendors.
- *802.11p*: WAVE (Wireless Access for Vehicular Environment).
  - Oriented towards ITS (Intelligent Transportation Systems).
  - Communication between Road Side Units and vehicles and directly between vehicles.
- *802.11r*: Reduce connectivity time when transitioning between BTSs.
  - Enable QoS when roaming.

- *802.11s*: Implement ESS (Extended Service Set) using WDS (Wireless Distribution System).
  - Standardize 802.11 four address format.
- *802.11t*: Enable testing, comparison and deployment planning of 802.11 devices.
- *802.11u*: Enable working with other networks.
- *802.11v*: Enable centralized management of attached stations.

12.6 Throughput

The effective throughput of an 802.11 network depends on the type of implementation, bandwidth, average packet size and number of users. Throughput can be specified for different conditions and generally for marketing reasons an unrealistic throughput is announced.

We can characterize the throughput in the air and for different scenarios of average packet size and number of users and this is shown in Table 12.16. The spectral efficiency is shown in Table 12.17.

Table 12.16 WLAN general parameters

		Wi-Fi 802.11-2007			802.11n-2009	
		OFDM			HT MF	HT GF
Maximum Throughput (Mbit/s)	# users	11a/g			11n	
Bandwidth (MHz)		5.0	10.0	20.0	20.0	40.0
Maximum rate considering access - Mbit/s	air	13.50	27.00	54.00	60.00	135.00
(1packet = 32 B, ST = 10 μs, DIFS = 30 μs, SIFS = 10 μs)	1	0.85	1.03	1.16	1.26	1.36
TDD ratio = 0.5 PER = 1% Control Frame overhead = 20%	5	0.51	0.69	0.76	0.82	0.89
Link length = 1000 m Uplink: RTS/CTS	10	0.45	0.51	0.54	0.59	0.63
Maximum rate considering access - Mbit/s	air	13.50	27.00	54.00	60.00	135.00
(1packet = 64 B, ST = 10 μs, DIFS = 30 μs, SIFS = 10 μs)	1	1.58	1.97	2.26	2.45	2.65
TDD ratio = 0.5 PER = 1% Control Frame overhead = 20%	5	0.97	1.33	1.49	1.61	1.75
Link length = 1000 m Uplink: RTS/CTS	10	0.85	0.99	1.07	1.16	1.26
Maximum rate considering access - Mbit/s	air	13.50	27.0	54.0	60.0	135.0
(1packet = 128 B, ST = 10 μs, DIFS = 30 μs, SIFS = 10 μs)	1	2.77	3.63	4.29	4.65	5.03
TDD ratio = 0.5 PER = 1% Control Frame overhead = 20%	5	1.77	2.13	2.86	3.10	3.36
Link length = 1000 m Uplink: RTS/CTS	10	1.27	1.87	2.08	2.25	2.44
Maximum rate considering access - Mbit/s	air	13.50	27.0	54.0	60.0	135.0
(1packet = 512 B, ST = 10 μs, DIFS = 30 μs, SIFS = 10 μs)	1	6.34	9.75	13.33	14.44	15.64
TDD ratio = 0.5 PER = 1% Control Frame overhead = 20%	5	3.91	6.44	8.04	8.71	9.44
Link length = 1000 m Uplink: RTS/CTS	10	3.65	4.73	5.55	6.01	6.51
Maximum rate considering access - Mbit/s	air	13.50	27.00	54.00	60.00	135.00
(1packet = 1024 B, ST = 10 μs, DIFS = 30 μs, SIFS = 10 μs)	1	8.08	13.56	20.52	22.23	24.08
TDD ratio = 0.5 PER = 1% Control Frame overhead = 20%	5	5.70	8.17	13.37	14.48	15.69
Link length = 1000 m Uplink: RTS/CTS	10	4.40	7.61	9.72	10.53	11.41
Maximum rate considering access - Mbit/s	air	13.50	27.0	54.0	60.0	135.0
(1packet = 2048 B, ST = 10 μs, DIFS = 30 μs, SIFS = 10 μs)	1	9.37	16.86	28.09	30.43	32.97
TDD ratio = 0.5 PER = 1% Control Frame overhead = 20%	5	6.42	11.76	16.72	18.11	19.62
Link length = 1000 m Uplink: RTS/CTS	10	6.18	9.01	15.57	16.87	18.27

Table 12.17 Spectrum efficiency

		Wi-Fi 802.11-2007			802.11n-2009	
		OFDM			HT MF	HT GF
Spectral efficiency (bits/Hz)	# users	11a/g			11n	
Bandwidth (MHz)		5.0	10.0	20.0	20.0	40.0
(1packet = 128 B, ST = 10 µs, DIFS = 30 µs, SIFS = 10 µs)	1	0.6	0.4	0.2	0.2	0.1
TDD ratio = 0.5 PER = 1% Control Frame overhead = 20%	5	0.4	0.2	0.1	0.2	0.1
Link length = 1000 m Uplink: RTS/CTS	10	0.3	0.2	0.1	0.1	0.1
(1packet = 2048 B, ST = 10 µs, DIFS = 30 µs, SIFS = 10 µs)	1	1.9	1.7	1.4	1.5	0.8242
TDD ratio = 0.5 PER = 1% Control Frame overhead = 20%	5	1.3	1.2	0.8	0.9	0.5
Link length = 1000 m Uplink: RTS/CTS	10	1.2	0.9	0.8	0.8	0.5

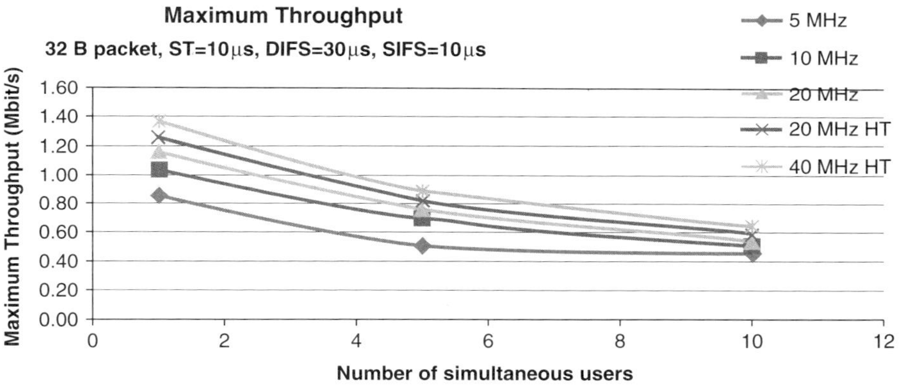

Figure 12.20 Maximum throughput for 32-byte data packet.

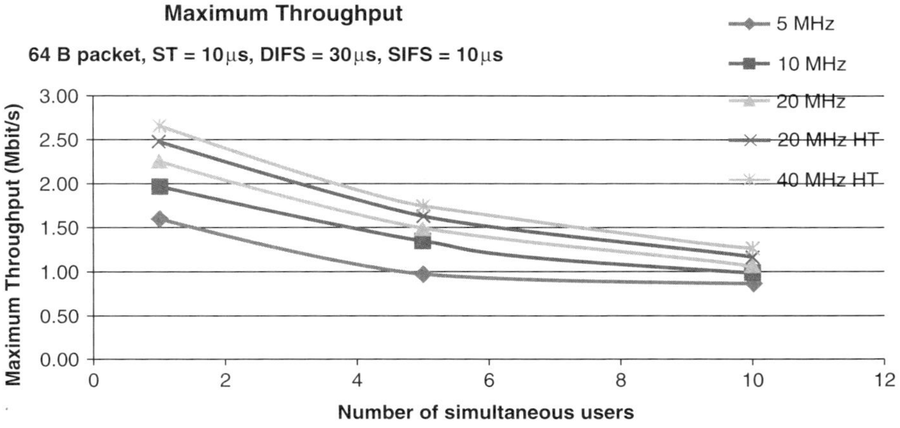

Figure 12.21 Maximum throughput for 64-byte data packet.

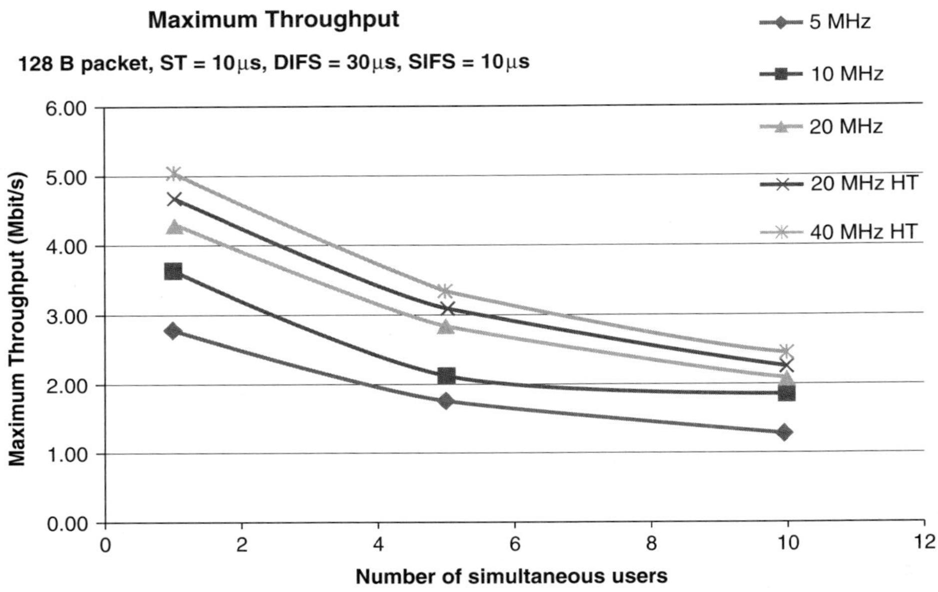

Figure 12.22 Maximum throughput for 128-byte data packet.

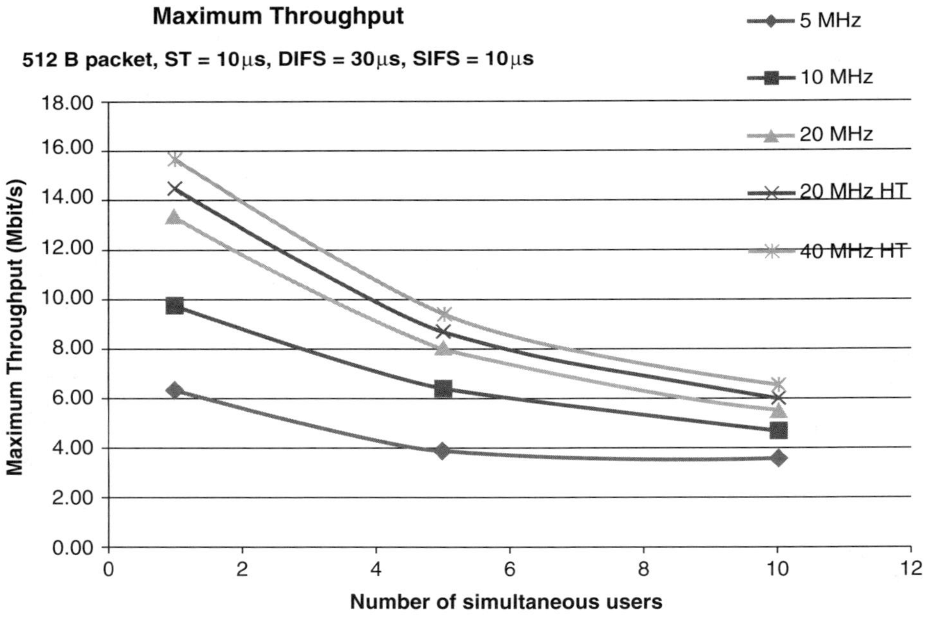

Figure 12.23 Maximum throughput for 512-byte data packet.

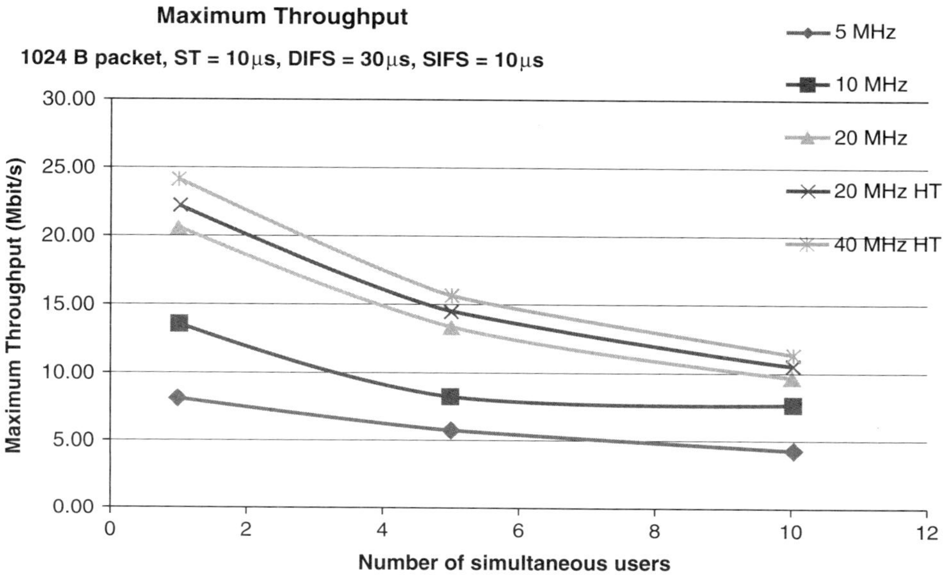

Figure 12.24 Maximum throughput for 1024-byte data packet.

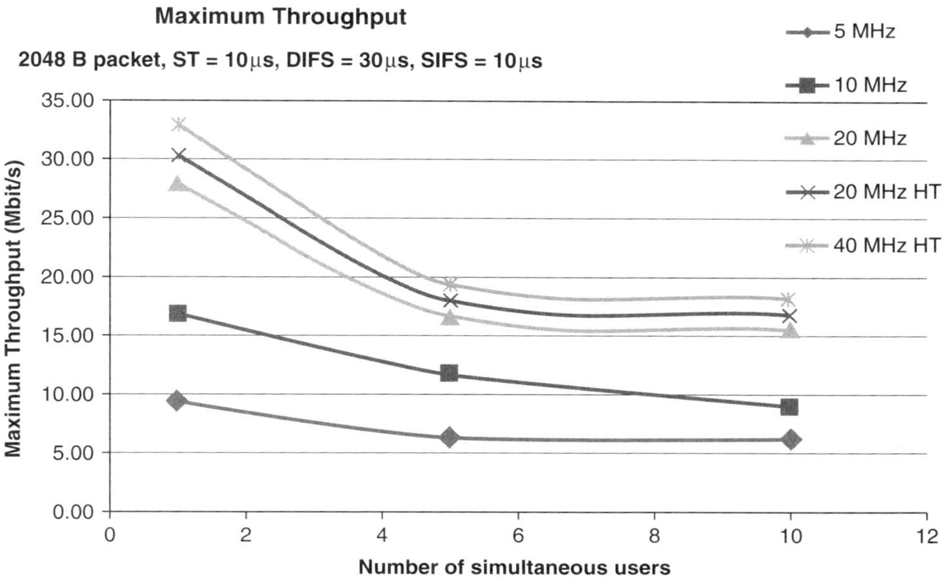

Figure 12.25 Maximum throughput for 2048-byte data packet.

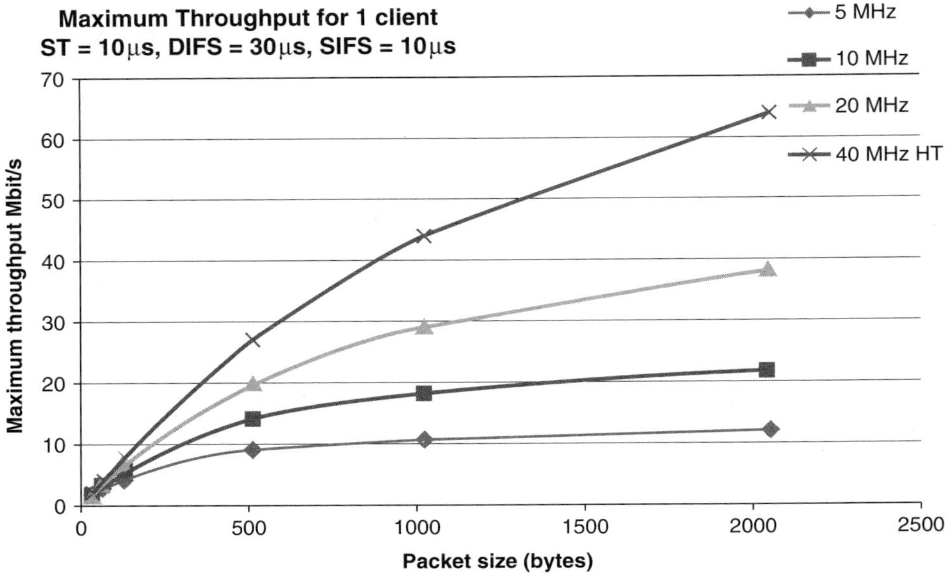

Figure 12.26 Maximum throughput for 1 client.

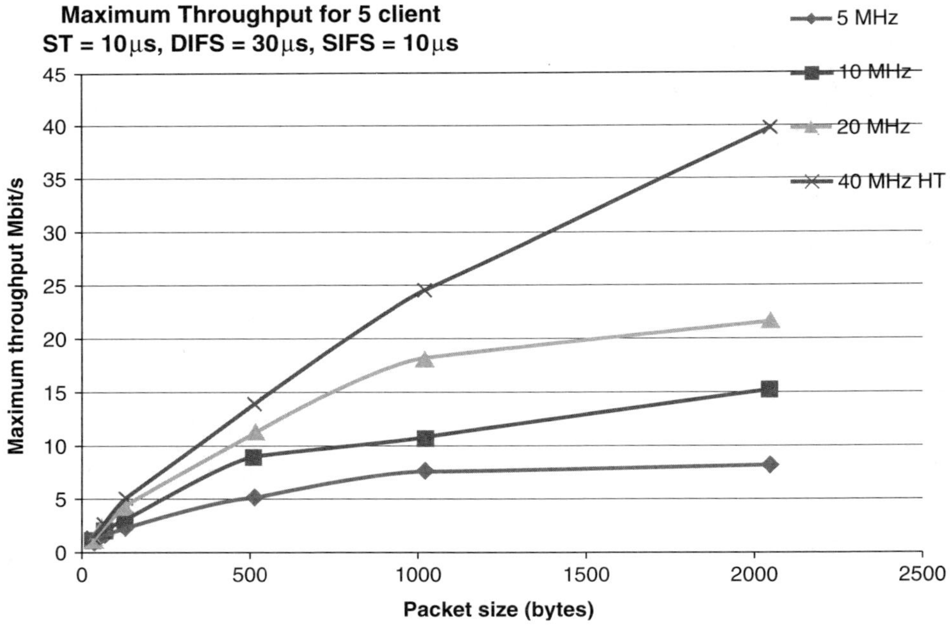

Figure 12.27 Maximum throughput for 5 clients.

The spectral efficiency does not go above 2.4 bit/Hz and is achieved only for a 5 MHz channel. The goal of 802.11n was to achieve 3 bit/Hz, but this is only achieved on paper. In real life, the best it can achieve is 1.6 bit/Hz with 1 user only. Throughput increase due to Space Time Coding (STC) was not accounted for as it varies significantly, so an additional 20% efficiency could be considered in the best case.

Figures 12.20 to 12.25 show the throughput for packets of 64 Bytes to 1024 Bytes of data, for the different bandwidths and varying number of users.

Figures 12.26 and 12.27 show the throughput for 1 and 5 clients.

13

WiMAX

The design of a WiMAX (Worldwide Interoperability for Microwave Access) wireless network requires vast knowledge of the technology, a detailed market and network representation, and powerful tools to design the network and provide optimized solutions. WiMAX is the first solution conceived to support IP data efficiently and be capable of providing wireless high speed data to wide areas, improving spectrum efficiency over previous technologies. Certain aspects of the WiMAX technology (e.g. security functions, roaming, IP network architecture) that do not affect network design are not the topic of this book and, hence, are only briefly discussed in this material.

13.1 Standardization

13.1.1 The WiMAX Standards

The need for a wireless solution that could address higher data rates became clear with the deployment of IEEE 802.11a/g networks (OFDM-based) that provided 54 Mbps IP-based communication mainly in indoor environments. This technology, although very successful, presents limitations in terms of multipath performance and is very inefficient when multiple users are present, due to its conflict-based access mechanism. WiMAX was developed with this experience in mind and also to accountfor issues observed with 3G and 3.5G technologies.

The WiMAX standard was developed by the IEEE in several phases. It is based on the work done by IEEE-802 LAN/MAN Standards Committee (LMSC), which was created in February 1980 to define standards for Local and Metropolitan Area Networks (LANs and MANs).

More than twenty Working Groups (WGs) and Technical Advisory Groups (TAGs) were created within this committee to standardize different aspects of LANs and MANs. Some of the original groups are inactive and some are in hibernation but most of the groups are still active, adding to and modifying standards. This material focuses on the information in the documents of Working Group 802.16. Detailed information on this and other groups can be found at www.ieee802.org.

The original goal of the Broadband Wireless Access (BWA) Working Group, 802.16, was to specify a point-to-point broadband technology to be used above 11 GHz. Today, the IEEE states: "This standard specifies the air interface, including the medium access control layer (MAC) and physical layer (PHY), of combined fixed and mobile point-to-multipoint broadband wireless access (BWA) systems providing multiple services" (IEEE, 2008). The use of the IEEE standard is voluntary. The idea behind it is to enable worldwide deployment of the technology by providing guidelines for interoperability.

LTE, WiMAX and WLAN Network Design, Optimization and Performance Analysis, First Edition.
Leonhard Korowajczuk.

Work by the 802.16 group proceeds today on different projects. When a project is approved, it is assigned a suffix letter, for example, 802.16d. A new project is usually developed by a Sub-Group (SG) or Task Group. Certain projects are published as amendments to the standard while others involve a full revision of the whole standard.

Table 13.1 briefly describes existing projects within the 802.16 group. Terminated projects indicate projects that were completed and became an active standard.

Even though the idea behind the standard is to allow interoperability among operators, because the IEEE standards aim to cover all possible applications of a given technology, the 802.16 standard supports such a multitude of configuration options and features that it could lead to non-compatible networks if vendors choose different implementation options.

13.1.2 The WiMAX Forum

This possibility of non-compatible equipment being created led to the creation of the WiMAX Forum, a non-profit organization, created to disseminate and further standardize the WiMAX technology.

The IEEE standards define only the physical layer (PHY) and the medium access layer (MAC), but this is not enough to assure network inter-operability. The WiMAX Forum took the responsibility for creating guidelines for an end-to-end WiMAX network architecture, covering roaming and integration with other networks (WLAN and 3G). Those guidelines are based on IETF RFCs, IEEE standards, DOCSIS (Data Over Cable Service Interface Specification) security protocols and the 3GPP IMS (3rd Generation Partnership Project IP Multimedia Subsystem) fixed and mobile convergence protocol.

The Forum took on also the task of specifying sets of minimum features (profiles) that should be supported by all vendors, making the WiMAX equipment interoperable. These profiles define mandatory and optional parameters for implementation in the base stations and CPEs. A certification program was established to certify equipment that fulfills the requirements of each profile.

Certification test profiles were issued in two groups: wave 1 (mainly fixed stations features) and wave 2 (mobile station features). Both wave1 and wave2 certified products are already available.

The WiMAX Forum organizes events called Plugfests, in which vendors can interconnect their equipment to perform interoperability tests. Additional information about these events and the WiMAX Forum can be obtained at www.wimaxforum.org.

13.1.3 WiMAX Advantages

The major improvements of WiMAX over traditional 3G technologies are the following (each of these points is discussed further):

- fully packet switched (IP based)
- Orthogonal Frequency Division Multiplexing (OFDM)
- Orthogonal Frequency Division Multiple Access (OFDMA)
- Time Division Duplexing (TDD)
- multi-level adaptive modulation (up to 64QAM)
- stronger error correction techniques
- designed for advanced antenna systems

WiMAX supports shared rates up to 70 Mbps, ranges up to 10 km and leverages the IP protocol used today by wired networks, thus being able to directly interconnect to them.

Table 13.1 WiMAX standards

Project	Description
P802.16m	Currently in pre-draft stage. It addresses the "Advanced Air Interface" and is an amendment to projects 802.12-2004 and 802.16e-2005. This project should result in the creation of WiMAX Release 2 (WiMAX 2.0)
Draft P802.16h	Draft amendment under development, addressing "Improved Coexistence Mechanisms for License-Exempt Operation"
Draft P802.16i	Draft amendment under development, addressing "Management Plane Procedures and Services"
Draft P802.16j	Draft amendment under development, addressing "Multihop Relay Specification"
Draft P802.16Rev2	Draft under development for the IEEE Standard for Local and Metropolitan Area Networks – Part 16: Air Interface for Broadband Wireless Access Systems (this draft consolidates Standards 802.16e-2005, 802.16-2004 – including Cor1-2005, 802.16f-2005, and 802.16g-2007). When this revision is complete it will supersede Std 802.16-2004 and all related amendments and corrigenda.
Std 802.16-2004	Active standard. Revision of IEEE Std 802.16 (including 802.16-2001, 802.16c-2002, and 802.16a-2003). This standard was developed by the Task Group d under the name P802.16-REVd. The IEEE Std 802.16-2004 revision added the 2–11 GHz range and supported a fixed FFT of 256 for all bandwidths and an FFT of 2048 for the 20 MHz bandwidth, but it only targeted fixed subscribers. The shortcomings of this standard were corrected in IEEE Std. 802.16-2004/Cor1 (2005).
Std 802.16g-2007	Active amendment to 802.16, it addresses "Management Plane Procedures and Services"
Std 802.16f-2005	Active amendment to 802.16 developed by the Network Management Task Group, it addresses the "Management Information Base"
Std 802.16e-2005	Active amendment to 802.16 developed by Task Group e, it addresses "Physical and Medium Access Control Layers for Combined Fixed and Mobile Operation in Licensed Bands" (including mobility and fast handover.) This group was working in parallel to IEEE Std. 802.16-2004 since 2002, aiming to develop a mobile version of the standard. This became IEEE Std 802.16e/Cor1 amendment and corrigendum that supports fixed and mobile communications in bands below 6 GHz.
Std 802.16-2004/Cor1-2005	Active corrigendum to 802.16-2004, published along with 802.16e-2005, developed by the Maintenance Task Group
Std 802.16.2-2004	Active revision of 802.16.2-2001 developed by Task Group 2, it addresses the "Coexistence of Fixed Broadband Wireless Access Systems"
Std 802.16/Conformances	Conformance01-2003: developed by Task Group C; IEEE Standard for Conformance to IEEE 802.16 – Part 1: Protocol Implementation Conformance Statements for 10–66 GHz WirelessMAN-SC™ Air Interface; Conformance02-2003: developed by Task Group C; IEEE Standard for Conformance to IEEE 802.16 – Part 2: Test Suite Structure and Test Purposes (TSS&TP) for 10–66 GHz WirelessMAN-SC™; Conformance03-2004: developed by Task Group C; IEEE Standard for Conformance to IEEE 802.16 – Part 3: Radio Conformance Tests (RCT) for 10–66 GHz WirelessMAN-SC Air Interface 10–66 GHz WirelessMAN-SC™ Air Interface;

(*continued overleaf*)

Table 13.1 *(continued)*

Project	Description
	Conformance04-2006: developed by Conformance Task Group; IEEE Standard for Conformance to IEEE 802.16 – Part 4: Protocol Implementation Conformance Statement (PICS) Pro-forma for Frequencies below 11 GHz
Std 802.16k-2007	Active amendment of 802.1D (previously amended by 802.17a), developed by the Network Management Task Group, it addresses "Media Access Control (MAC) Bridges – Bridging of 802.16"
Draft P802.16d	Terminated project. This project concluded in 2004 with the release of 802.16-2004.
Draft P802.16-2004/Cor2	Terminated project.
Std 802.16-2001	Superseded by 802.16-2004
Std 802.16a-2003	Superseded by 802.16-2004
Std 802.16c-2002	Superseded by 802.16-2004
Std 802.16.2-2001	Superseded by 802.16-2004

13.1.4 WiMAX Claims

Marketing claims for all technologies are usually exaggerated, mainly for new standard releases; specially as marketing competition increases among LTE, WiMAX, and WLAN. Even though the claims are based on real features of the technologies, they refer to very specific situations for each parameter, are not valid simultaneously, and most likely will not be achieved in real situations, but it is important to register them.

For the 802.16 e version claims are made of 144 Mbit/s in the downstream and 35 Mbit/s in the upstream, with a spectral efficiency of 3.7 bit/s/Hz. For 802.16 m claims reach 1 Gbit/s for fixed users.

13.2 Network Architecture

WCS architecture is covered in Chapter 9. The WiMAX architecture is illustrated in Figure 13.1. It can be divided into ASN (Access Service Network), CSN (Connection Service Network), ASP (Application Service Provider) and OSS (Operation Support System).

Interfaces between the architectural blocks are defined in Figure 13.2.

- R1: is the air interface between the SS and the BTS and is described in detail in this chapter.
- R2: is a logical interface between SS and CSN (a direct physical connection between both does not exist). It is associated with Authentication, Service Authorization, IP host configuration management and mobility management.
- R3: is the interface between ASN and CSN. It supports AAA, policy management and mobility management.
- R4: consists of protocols between ASNs to coordinate mobility management.
- R5: consists of protocols for communications between CSNs.
- R6: consists of protocols for communications between BTS and ASN.
- R8: consists of protocols for communications between BTSs.

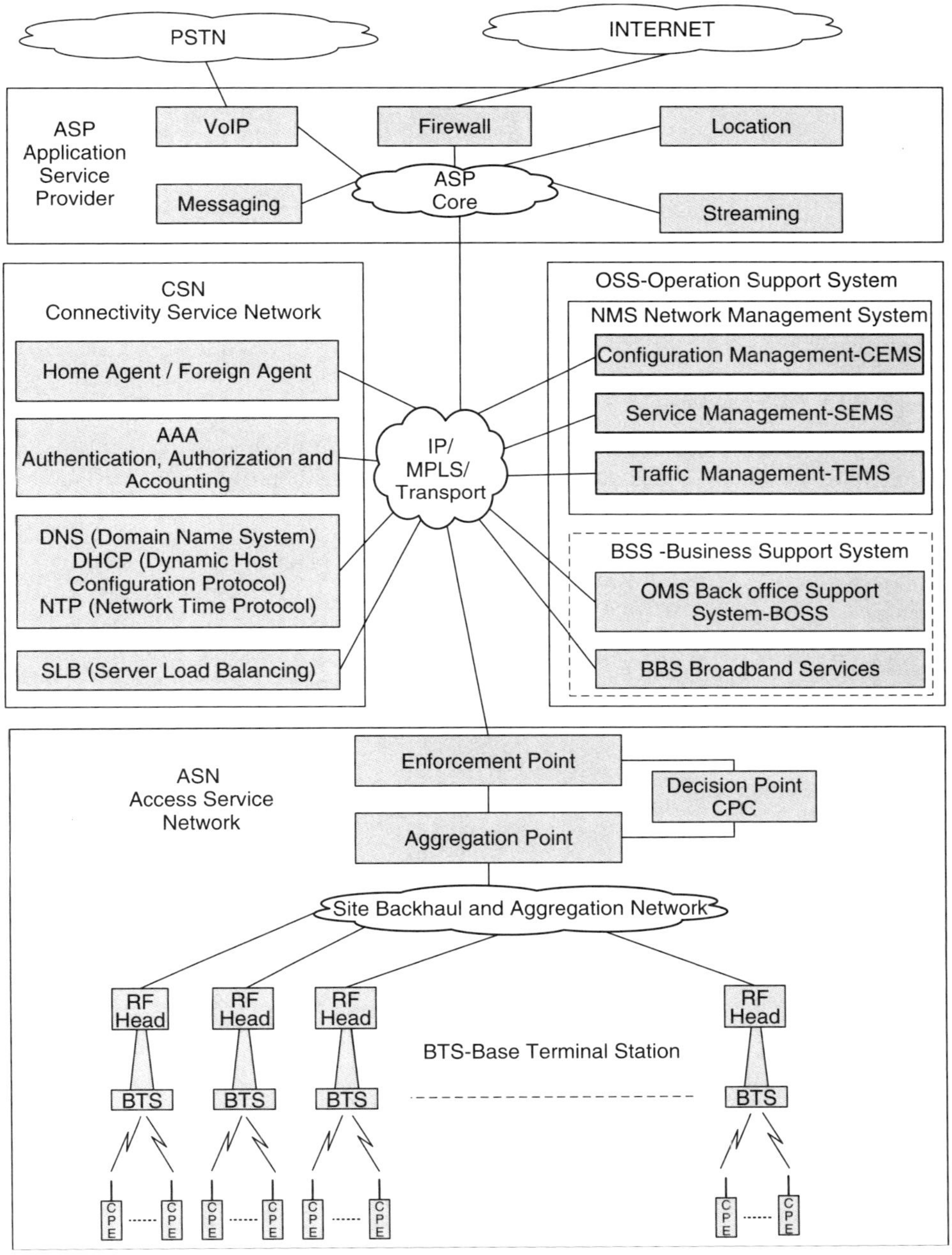

Figure 13.1 WiMAX network architecture.

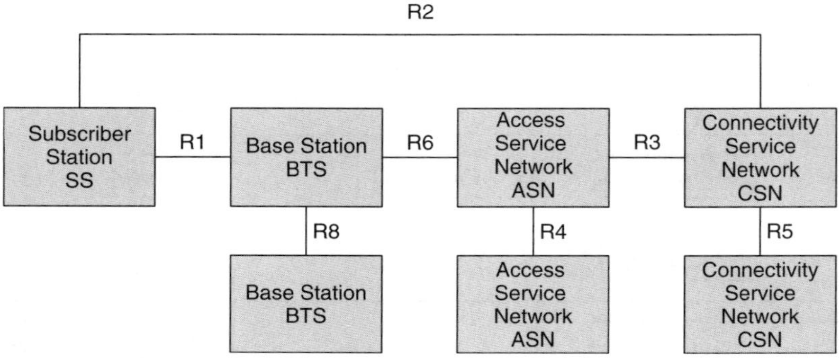

Figure 13.2 WiMAX interfaces.

13.2.1 ASN (Access Service Network)

The ASN provides subscriber access to the network and includes CPEs, Access Points (AP) or Base Stations (BTS), and the ASN gateway (GW). The ASN-GW aggregates BS traffic and implements functions that are common to all base stations. It also interfaces with the other elements of the network. Its main functions here are:

- wireless access
- aggregation point
- enforcement point
- decision point

The main ASN components are described next.

13.2.1.1 Base Station (BS)

The main Base Station characteristics are:

- indoor or outdoor
- number of sectors: single or multiple
- one or more Ethernet ports
- integrated or separate RF GPS
- tower-top or ground-based
- integrated or separate RF Head and antenna
- MIMO support
- beamforming support

13.2.1.2 RF Head (RFH)

The RF head can be separated from the Base Station, and mounted close to the antenna to lower losses. Its features are:

- It is typically outdoor.
- It supports two or more antennas.
- The output power at each antenna is typically between 2 W and 5 W.

13.2.2 CPE

The main characteristics of a CPE are:

- fixed or portable
- stand-alone or embedded
- indoor or outdoor
- data only, or data and VoIP support (1 or more channels)

13.2.3 ASN-GW (Access Service Network Gateway)

The ASN-GW aligns and synchronizes over-the-air and IP subscriber management, QoS policy enforcement, and mobility as a single centralized entity. ASN-GW hardware is made up of computers, routers and switches.

The ASN-GW main functions are the following:

- First hop IP router for the set of subscribers under its domain.
- Acts as a foreign agent for mobile nodes.
- Operates as the EP (Enforcement Point) edge router.
- Is the connectivity point to the AAA to authenticate the subscriber and retrieve attributes describing the subscriber's authorized set of capabilities.
- Supports idle mode.
- Supports paging.

The following capabilities are of particular importance in the ASN-GW:

- MPLS (Multi Protocol Label Switching):
  - It is a protocol-agnostic data-carrying mechanism, in which data packets are assigned labels. Packet-forwarding decisions are made based on label contents, without examining the packet itself. This creates end-to-end circuits across any type of transport medium, using any protocol.
- Traffic Engineering:
  - RSVP (Resource Reservation Protocol RFC 3209), LDP (Label Distribution Protocol RFC 3036, 3478), Layer 2 VPN (Virtual Private Network) Transport Independent, multicast and others.
- Quality of Service:
  - Packet classification (RFC 2474, 2475, 2597, 2598).
  - IntServ (Integrated Services) or DiffServ (Differentiated Services):
    - ACL (Access Control List) and TCL (Transit Control List), ingress policing.
    - BGP (Border Gateway Protocol) attribute-based QoS; class-based ingress policing and egress shaping.
    - priority queuing and EDRR (Enhanced Deficit Round-Robin).
    - RED (Random Early Detection) and WRED (Weighted RED).
    - MPLS (Multi Protocol Label Switching LSP (Label Switched Path RFC 3270).
    - ATM (Asynchronous Transfer Mode) queuing per user.
- Layer 2 Tunnel Protocol:
  - LNS (Layer 2 Tunnel Protocol Network Server) L.
  - LAC (Layer 2 Tunnel Protocol Access Concentrator).
- Routing Protocols:
  - BGP-4 (Border Gateway Protocol RFC 1771).

  - IS-IS (Intermediate System to Intermediate System RFC 1195 and ISO/IEC10589).
  - OSPF (Open Shortest Path First RFC 2328), RIP (Routing Information Protocol RFC 2453), VRRP (Virtual Router Redundancy Protocol RFC 2338).
- Multicast Protocols:
  - PIM-SM (Protocol Independent Multicast Sparse Mode RFC 2362 + IETF Draft), PIM-DM (Protocol Independent Multicast- Dense Mode IEFT Draft), IGMP (Internet Group Management Protocol RFC 3376), SSM (Source Specific Multicast RFC 3569), MBGP (Multi Protocol BGP RFC 2858), MSDP(Multicast Source Delivery Protocol RFC 3618).

13.2.3.1 Decision Point

The decision point is responsible for the following functions:

- Local Key Distribution function
- Mobility (paging) controller
- QoS Policy Decision Point
- Subscriber Access Control

13.2.3.2 Router/FA (Foreign Agent) or Enforcement Point

This is responsible for the functions below:

- IP traffic routing
- subscriber authentication
- policy management
- QoS capacities at subscriber level
- Layer 2 tunnels terminations
- support for Multiple Foreign Agents in the same platform
- Multi-bind Interface support
- IP in IP and GRE (Generic Routing Encapsulation) tunnels
- support for Overlapping IP addresses
- timestamp-based replay protection
- MAC-based forwarding
- Registration Revocation support
- Reverse Tunnel Support
- mobile IP statistics/counter support

13.2.3.3 Switch or Aggregation Point

This is responsible for the Switch layer2.

13.2.4 CSN (Connectivity Service Network)

The CSN performs centralized functions such as address translation, local and foreign subscribers' tracking, authentication, and call records' storage. These functions are all software controlled and are computer-intensive. Some functionality is spread to other computers in the ASN or OSS/BSS, to use existing processing capability.

This system is responsible for the connectivity to the subscriber and its main functions are described below.

13.2.4.1 Home Agent (HA)/Foreign Agent (FA)

Both an HA and FA are required for mobility:

- A home agent stores information about mobile nodes whose permanent home address is in the home agent's network.
- A foreign agent stores information about mobile nodes visiting its network. Foreign agents also advertise care-of addresses, which are used by Mobile IP.

13.2.4.2 AAA (Authentication, Authorization, and Accounting)

The following functions are provided by AAA:

- session control per subscriber;
- customer self-service;
- service administration;
- RADIUS (Remote Authentication Dial In User Service) accounting interface for unified billing and accounting;
- enhanced security services such as parental control, and time of day content filtering;
- personal firewall and intrusion detection services;
- location lock, restricting subscriber location;
- enhanced QoS control based on dynamic subscriber profile and service requirements to assist with optimal delivery of on-demand services such as video-on-demand;
- management of resale (wholesale and retail) of network resources through third parties.

The EAP (Extensible Authentication Protocol) method used for authentication is EAP-TTLS (EAP-Tunneled Transport Layer Security) where a certificate is stored on the Server and a CHAP (Challenge Handshake Authentication Protocol) password authentication is used. CHAP ensures that the password is not sent between ISP and CPE in the clear.

RADIUS (Remote Authentication Dial In User Server) client should be implemented for eventual dial-in customers.

One or more computer platforms may be used and in this case. Server Load Balancing (SLB) should be used to divide the traffic between the platforms.

13.2.4.3 DNS (Domain Name Server)/DHCP (Dynamic Host Control Protocol)/ NTP (Network Time Protocol) Server

The DNS translates domain names into numerical identifiers to address networking equipment. The DHCP provides IP addresses of computers in the network and their configuration. Both functions can be provided by or more servers and in this case SLB should be used.

13.2.5 OSS/BSS (Operation Support System/Business Support System)

The OSS/BSS performs all activities required to manage the network, from customer maintenance to network maintenance/performance and billing.

13.2.5.1 NMS (Network Management System)

An NMS should support the following functions:

- configuration management
- customer management
- performance management
- fault management
- security management
- policy management

Policy management is seldom offered by a single product and product functionalities overlap.

13.2.5.2 Element Management System (CEMS)

A CEMS should support the following functions:

- configuration management
- fault management
- performance management

13.2.5.3 Service Element Management System (SEMS)

A SEMS should support the following functions:

- define services
- define and view subscriber accounts
- define Access Service offerings
- define Lawful Intercept Service offerings
- Video Service offerings
- Web-based login
- Wi-Fi and WiMAX access
- dynamic service selection
- scheduled time of day-based access
- duration-based access based on total time online
- volume-based access with download limits
- invalid PPP login redirect
- define Bandwidth Service offerings
- define IP redirect Service offerings
- dynamic service selection
- captive portal/redirect services
- tiered bandwidth
- bandwidth on demand
- Video-on-Demand
- URL filtering
- dynamic traffic prioritization

Some of these services also create new billing models, helping to diversify the subscriber base and increase service penetration. For example, operators can charge lower prices for services that are only offered at certain times of the day or only allow a specific amount of content to be downloaded.

Key benefits and characteristics include:

- *Value added services*: delivers a range of new services to increase average revenue per subscriber.
- *Customer self-service*: reduces operating costs by permitting subscribers to self-select the services they desire through a Web portal.
- *Cooperative proxy support*: coordinates with existing RADIUS, Lightweight Directory Access Protocol (LDAP) and Extensible Authentication Protocol (EAP).
- *Extensible Authentication Protocol (EAP)*: serves and leverages existing RADIUS infrastructure to provide an integrated mechanism for Authentication, Authorization and Accounting (AAA), and service delivery.
- *Open OSS integration*: extends a standardized SOAP (Simple Object Access Protocol)/XML (eXtensibled Markup Language) gateway to accept service activation and change requests from Web portals and back-office OSSs.
- *Single subscriber sign-on*: provides subscribers with a single identity across client (PPP Point-to-Point Protocol) or clientless (DHCP) access methods including support for multiple end devices and multiple network transports.
- *Flexible Service Provisioning*: utilizes an open service definition model permitting the creation of new, value-added services, using third party application service providers and application vendors.
- *Cooperative proxy support*: coordinates with existing RADIUS, Lightweight Directory Access Protocol (LDAP).
- *Authentication Protocol (EAP)*: serves and leverages existing RADIUS infrastructure to provide an integrated mechanism for Authentication, Authorization, and Accounting (AAA) and service delivery.
- *Open OSS integration*: extends a standardized SOAP/XML gateway to accept service activation and change requests from Web portals and back office OSSs.
- *Field-proven scalability*: delivers a redundant architecture featuring clustering and load balancing that can grow to support millions of subscribers.

13.2.5.4 Traffic Element Management System (TEMS)

A TEMS should support the following functions:

- Administration Manager
- Topology Manager
- MPLS Manager
- Device Configuration Manager
- Access Control List Manager
- Rate Limiting Manager
- MAC Filter Manager
- Event Manager
- Configuration Manager
- Service Director
- Report Manager
- Change Manager
- RF Monitoring Manager

13.2.5.5 BBS (Business Broadband Services) Automated Service Initiation

Device management provides the following services:

- individual or *en masse* updates to device software and configurations;
- configuration and management of advanced services such as VPNs, gaming, security, VoIP, and IPTV;
- real-time CPE monitoring and troubleshooting;
- customer self-service provisioning and troubleshooting;
- automated business logic, including provisioning, monitoring, and analysis;
- multi-vendor CPE support, allowing management of third-party devices;
- powerful device management for simplified, reliable management of all configurable devices at the network edge;
- sophisticated policy management for consistent policy implementation across different CPEs;
- field-proven scalable and secure architecture, scaling from a few to millions of devices using highly secure administration.

13.2.5.6 SAG (Service Activation Gateway)

This is a software product that provides a single point interface into the network for provisioning and activating user (subscriber) services. It interfaces software from different vendors. Its functions are:

- provisioning AAA
- Order Management System (OMS)
- CPE provisioning
- AAA provisioning
- VoIP provisioning
- Voice Mail Server provisioning
- DHCP provisioning

13.2.5.7 BOSS (Back Office Support System)

This package overlaps with the previous offerings and it can be deployed in parallel or as a replacement. It has the following features:

- Customer care, fulfillment, and billing with service creation and Management
  - Customer care
  - billing
  - reporting
  - mediation
  - prepaid rating and charging
  - customer portal
  - prepaid cards
  - order management
  - administration and security
  - MIS (Management Information System) reporting
  - delivery of bills to postpaid subscribers
  - workflow definition
  - Executive Views

- Interconnect billing
  - billing
  - reporting
  - mediation
- Trouble Ticket Manager
  - customer trouble tickets
  - network trouble tickets
  - tools, searches and alerts
  - TTM (Trouble Ticket management) configuration
- Fraud Manager
- Network Asset Management

13.2.6 ASP (Application Service Provider)

Provides support to applications such as VoIP, messaging, streaming, location, and firewall. These features are generally not offered by equipment vendors; they are, instead, implemented by service providers.

13.2.6.1 VoIP (Voice over IP) Application and gateway to PSTN (Public Service Telephone Network)

This provides connectivity with the telephone network, including VoIP translation.

13.2.6.2 Messaging Application

This provides storage and forwarding of messages.

13.2.6.3 Firewall and Internet Access

This is provided by specialized routers.

13.2.6.4 Location Application

This is required in some countries, for emergency calls (e.g. 911 in the USA).

13.2.6.5 Streaming Application

This provides streaming support for distribution and broadcasting.

13.3 Physical Layer (PHY)

It is easier to understand OFDM by first considering the Frequency Division Multiplex (FDM) part of it. FDM is a transmission technique in which numerous signals (channels) are transmitted simultaneously on a single communication line over multiple frequencies. Each signal is assigned a different frequency range (subcarrier) within the main channel; a spacing (guard band) is placed between these subcarriers to avoid signal overlap. This guard band is usually larger than the information channel.

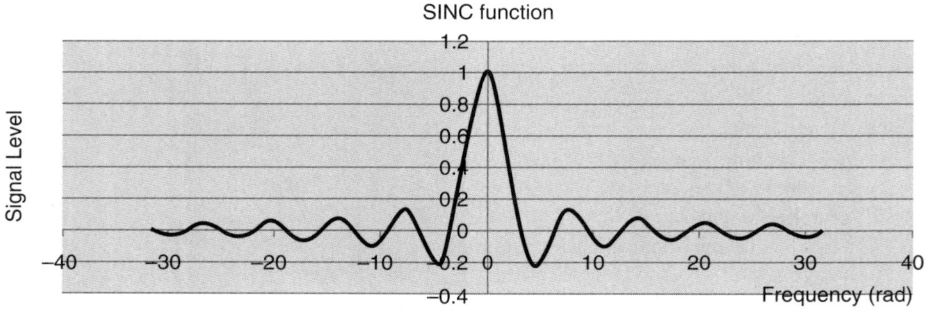

Figure 13.3 Spectrum of a frequency modulated by digital signal.

A single frequency modulated by a digital signal results in a spectral distribution defined by the SINC function is explained in Chapter 4 and shown in Section 4.2. The nulls spacing of the SINC signal is directly proportional to the symbol rate of the digital signal. The total signal occupies a large bandwidth due to the existence of successive peaks and zeros, causing interference to adjacent channels. As only the first peak is enough to convey the information, FDM technology filters the frequencies above and below it, while OFDM brings channels close together so their peaks coincide with the nulls of the other channels, minimizing interference.

Equations (13.1) and (13.2) define the OFDM carrier separation and data rate as a function of the symbol duration.

$$T_s = \frac{1}{\Delta f} \qquad \text{(13.1) OFDM carrier separation (nulls spacing)}$$

where:

T_s = Symbol duration.
Δf = Frequency step between sub-carriers.

$$\text{Data Rate} = \frac{1}{T_s} \qquad \text{(13.2) OFDM data rate}$$

The SINC function is defined by Equation (13.3):

$$\text{sinc}(\pi f) = \frac{\sin \pi f}{\pi f} \qquad \text{(13.3) SINC function}$$

Interference exists because the center frequency of one channel is affected by the side bands of the adjacent channels. Non-orthogonal technologies try to mitigate this problem by increasing the distance between channel frequencies and by using filters to better isolate adjacent channels.

Orthogonal technologies solve this interference issue by bringing frequency channels closer together. This apparently contradictory approach exploits the fact that the modulation of a single frequency by a digital signal results in a very specific spectrum that has peaks and zeros with defined frequency spacing. Placing channel peaks at the zeros of adjacent channels avoids the interference between frequencies, thus providing increased capacity and higher spectrum efficiency. For this concept to work, the distance between the orthogonal frequencies should be an integer multiple of the base

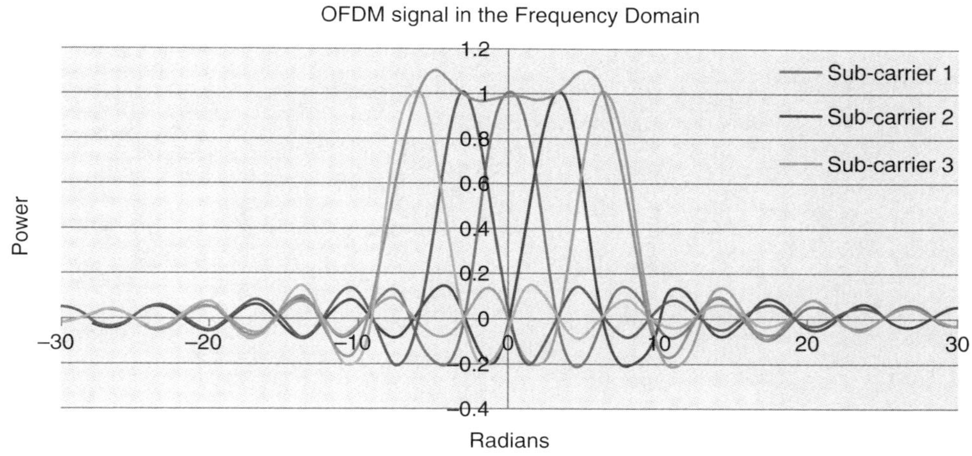

Figure 13.4 OFDM signal with five sub-carriers shown in the frequency domain.

frequency. For example, for a base frequency of 10 kHz, orthogonal frequencies occur at 20 kHz, 30 kHz, 40 kHz, etc.

OFDM is basically a Frequency Division Multiplex (FDM) technology that applies this orthogonality concept to define the distance between adjacent carriers. An OFDM carrier uses subcarrier frequencies that are integer multiples of the spacing between zero level occurrences, which are inversely proportional to the transmitted symbol duration.

A high data rate signal requires a large bandwidth for transmission and consequently the symbol duration becomes very short. This makes it susceptible to severe multipath (ISI Inter Symbol Interference), in which each symbol interferes with many subsequent symbols.

The OFDM concept can be further extended by splitting the high data rate signal into several lower rate streams, each modulating an OFDM sub-carrier. Figure 13.4 shows an OFDM carrier with five frequencies (sub-carriers) where the interference level at each subcarrier peak (sampling moment) is zero. The line represents the composite spectrum.

Figure 13.5 shows the same five subcarriers and their composite signal (line) depicted in time domain. The use of subcarriers allows OFDM to break a wideband channel into many narrowband channels without losing spectral efficiency. Each subcarrier can be modulated by a lower data rate, increasing symbol time and diminishing signal bandwidth. This makes the detection of each subcarrier much easier because the system operates in a region where the RF channel can be considered flat. The fact that there is no interference between adjacent subcarriers allows OFDM to eliminate guard bands between them.

Traditional 2G and 3G technologies use a single fixed channel bandwidth whereas WiMAX allows the use of multiple channel bandwidths, varying from 1.25 MHz to 20 MHz, hence permitting network configuration to be adjusted according to the services provided and spectrum availability.

An OFDM carrier is defined in three domains: frequency, time, and power, as illustrated in Figure 13.6. In Figure 13.6, null subcarriers work as lower and upper guard bands and do not transmit data. The DC subcarrier is the subcarrier whose frequency corresponds to the RF center frequency of the base station (carrier) and also does not transmit any data, due to possible carrier signal leakage at this frequency.

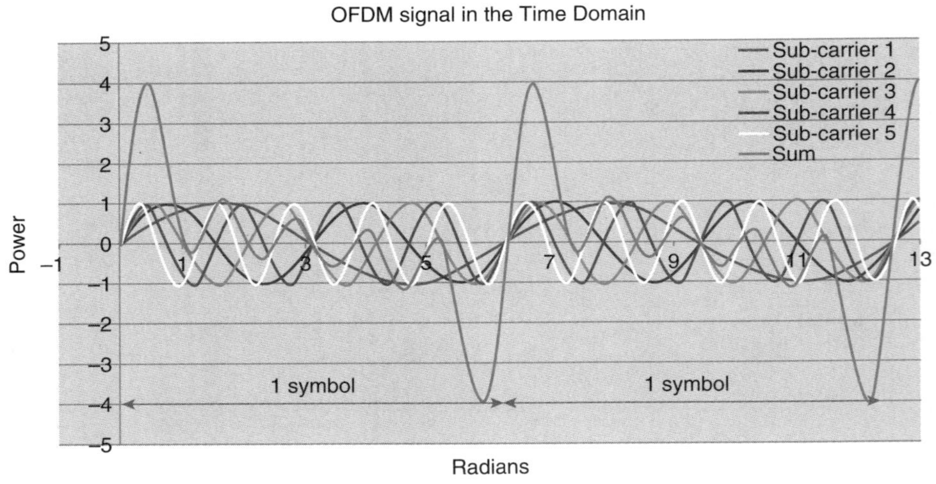

Figure 13.5 OFDM signal with five sub-carriers shown in the time domain.

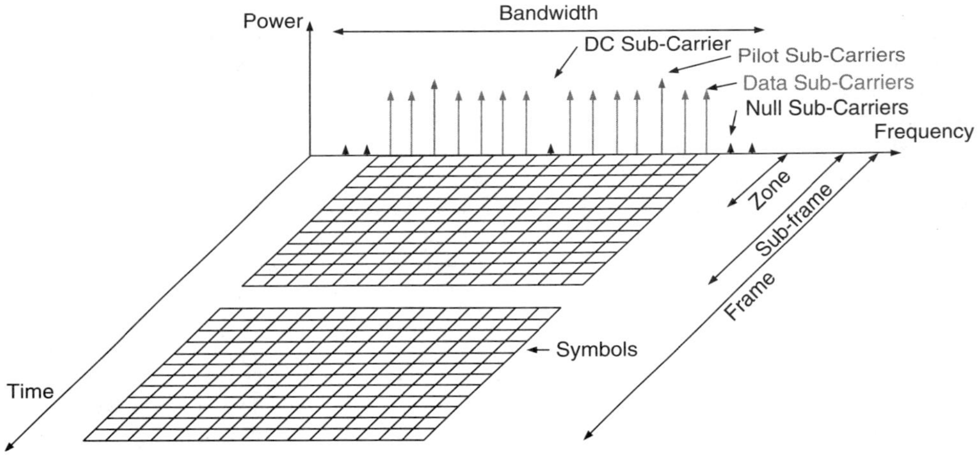

Figure 13.6 OFDM carrier represented in frequency, time, and power domains.

13.3.1 OFDM Carrier in Frequency Domain

An OFDM carrier is composed of several subcarriers, spaced at regular intervals. Each subcarrier carries a fraction of the data, thus their individual data rate is low but the total throughput is high. The higher the number of subcarriers, the better the tolerance to multipath spread, but the larger the processing power required.

Then, assuming a bandwidth of 10 MHz and a multipath spread of 10 μs, what would be the best partition of the bandwidth in subcarriers? For now, let's ignore guard bands. A channel with

Table 13.2 Calculation of number of subcarriers

Alternative	1	2	3	4	5
Subcarrier bandwidth (MHz)	10	1	0.1	0.01	0.001
Symbol Duration (μs)	0.1	1	10	100	1000
Multipath Spread (μs)	10	10	10	10	10
Multipath Affected Symbols	100	10	1	0.1	0.01
Throughput per subcarrier (Msps)	10	1	0.1	0.01	0.001
Number of subcarriers in 10 MHz	1	10	100	1000	10000
Total Throughput (Msps)	10	10	10	10	10

a bandwidth of 10 MHz can accommodate 10 Msps (Millions of symbols per second). Table 13.2 analyzes five different alternatives for determining the number of subcarriers.

In the first alternative, a single subcarrier is used and the multipath overlap covers 100 symbols, that is, signals from the previous 100 symbols are all mixed up and the recovery of the desired signal becomes impossible. This also makes alternative 2 unfeasible, as the signals of 10 symbols still interfere on the current symbol. This leaves alternatives 3, 4, and 5 with, respectively, 100, 1000, and 10000 subcarriers with overlaps of 1 symbol or less.

Alternative 3 still affects the second symbol (one symbol overlap), which is not as bad as the prior two alternatives, but still makes recovery of the original signal difficult. In alternative 4, however, the interference happens on the first 10% (0.1 multipath overlap) of the symbol, which can be completely eliminated with the use of a technique called the Cyclic Prefix, described in Section 13.3.2.1.

Considering that alternative 4 can solve the overlap problem, there is no need for option 5, which would further increase symbol duration and narrow subcarrier bandwidth. Thus, the WiMAX standard selected alternative number 4.

Also, frequency inaccuracies limit the carrier separation to around 10 kHz, because a 1 GHz signal with 10^{-6} precision, results in a 1 kHz deviation, which is $^{1}/_{10}$ of the separation.

Besides, an analog signal has to be sampled at twice the bandwidth, and per Nyquist–Shannon sampling theorem, the required sampling results in a 40 Msps for a 20 MHz bandwidth, which is the maximum number of samples that can be processed with today's digital signal processors. This means that solutions with a higher bandwidth and a higher number of subcarriers would have performance issues.

13.3.1.1 Subcarriers

Figure 13.7 shows an OFDM carrier, composed of several subcarriers. Not all subcarriers can be used for data transmission, because, even though OFDM technology allows optimal filtering of adjacent frequencies within the carrier, channels adjacent to the OFDM carrier itself still have to be protected. About 18% of the total subcarriers are left as guard band (approximately half at each side of the OFDM carrier); this is still much better than traditional technologies that may use anywhere from 20–66% of the spectrum for this purpose.

The central subcarrier also is not used for data transmission. It is called the DC subcarrier and is left unmodulated to avoid interference from carrier leakage.

Both guard band and DC carriers are considered null carriers, that is, there is no transmitted energy on them, allowing the signal from the data carriers to fade naturally, thus avoiding interference in adjacent channels.

Although each subcarrier operates over a flat channel, the same cannot be said about the composite spectrum of all sub-carriers, because such a wide flat channel would not be possible, hence pilot

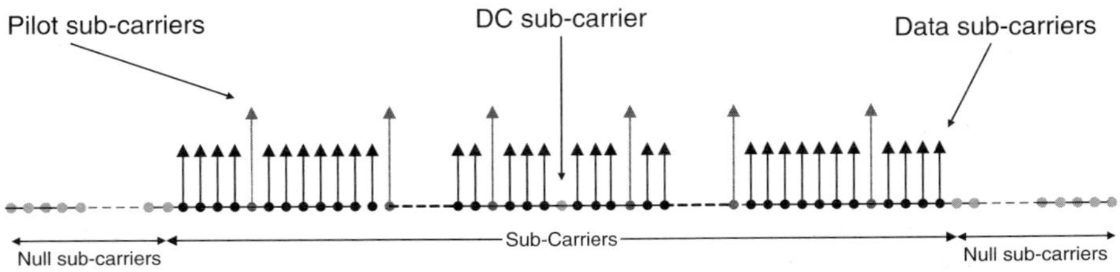

Figure 13.7 OFDM carrier and sub-carriers.

subcarriers are used to help equalize the gains across the band and in time. These pilot subcarriers are distributed throughout the bandwidth and carry known patterns of data, allowing for an amplitude equalization of the band, so all subcarriers are received with the same level.

Pilot subcarrier symbols carry known information that can be used to equalize a channel. The receiver has to experiment with different equalizations until the pilot data is properly detected. This equalization setting can then be used to extract the unknown data information sent on adjacent subcarriers. The equalization process in WiMAX is simpler than in other technologies because the multipath distortion is eliminated by the use of Cyclic Prefixes. The task left is the equalization of the signal levels of the different subcarriers.

Three main types of equalizers can be used for pilot detection: linear equalizer, non-linear equalizer, and Maximum Likelihood Sequence Detection (MLSD). The standard does not select a specific methodology and the decision of which method to use is left to the vendor's discretion.

In linear equalizers, the signal goes through a filter that simulates the inverse of the channel. It is simple to implement but not very efficient and can even increase the noise floor. Non-linear equalizers estimate previously decoded symbols' interference and subtracts it from the signal. It is subject to error propagation.

MLSD examines all possible combinations for a sequence of symbols and chooses the most likely one. Its complexity increases with the constellation size and time delay.

The use of pilots represents an overhead to the system, but is required to improve reception. Around 11% of the subcarriers are pilots in the downlink and 33% in the uplink. The best trade-off between number of pilots and data subcarriers varies with the channel environment, thus different data allocation schemes were envisaged to deal with different channel types.

Once null subcarriers, DC subcarrier, and pilot subcarriers are discounted, only about 60% of the total number of subcarriers is left for actual data transmission.

13.3.1.2 Subchannelization

Data and pilot subcarriers can be grouped into subchannels, which represent the smallest unit for data allocation. Subchannels are spread over the entire spectrum and their use is controlled by the subchannel index. In WiMAX, a subchannel has always 48 symbols of data, plus pilots.

The concept of subchannelization is used in the uplink of 802.16–2004 and both in the uplink and downlink of 802.16e. On the downlink, base stations allocate subchannels to subscribers based on their data requirements and channel conditions, as this allows them to transmit with lower modulation schemes to subscribers with poor channel quality, and higher modulation schemes to subscribers with good SNR.

On the uplink, subscribers can only use subchannelization if the base station acknowledges that it is capable of decoding subchannels.

13.3.2 OFDM Carrier in Time Domain

13.3.2.1 Symbol and Cyclic Prefix

The smallest significant allocation unit in the time domain is the symbol; its duration is related to the subcarrier spacing and defines the data rate.

The number of subcarriers and their spacing varies with the channel bandwidth; the definition of these parameters is a trade-off between multipath protection and design cost and complexity, that is, a higher number of subcarriers avoids inter-symbol interference (ISI) caused by delay spread, but it increases the system's cost and complexity due to the need for higher processing power.

Two different approaches can be taken to tackle this issue:

- As the bandwidth varies, keep the total number of subcarriers constant but vary the symbol duration.
- As the bandwidth varies, keep the symbol duration constant but vary the total number of subcarriers.

In both cases, the bandwidth corresponds to the number of subcarriers multiplied by the inverse of the symbol duration.

WiMAX OFDM uses the first approach, with symbol duration varying from 8 to 128 μs; WiMAX OFDMA, or scalable OFDMA (SOFDMA), uses the second approach, with a fixed symbol duration of 102.86 μs, which corresponds to a subcarrier spacing of 10.94 kHz.

A significant problem in digital systems is the possibility of Inter-Symbol Interference (ISI), caused by the time spread between the arrival of the first and last received multipath signals (delay spread). Part of the received symbol waveform at the receiver is corrupted by multipath spread of the previous symbol. There are two ways of dealing with ISI: increase the time between two consecutive symbols or use equalization (calculating the impact of ISI in each symbol).

The delay spread varies depending on the frequency being used as well as on the terrain and relative speed of the transmitter and receiver. Increased symbol duration improves the delay spread immunity with the use of a Cyclic Prefix (CP) to eliminate ISI. CP is the repetition of the last part of the received waveform (data), which is added to the beginning or end of the data payload. This causes the symbol duration to be extended. As long as the CP duration corresponds to the maximum expected delay spread, the ISI issue is eliminated.

This technique only works for OFDM systems because of their long symbol durations. Other technologies cannot take advantage of this concept due to their short symbol duration when compared with their system delay spread (e.g. 3.69 μs symbol duration for GSM systems versus 18 μs of delay spread).

The OFDM signal is produced by a block called the IFFT (Inverse Fast Fourier Transform). A Forward FFT multiplies a signal numerous times by complex exponentials over the range of frequencies. It then sums each product and plots the results in what is called the spectrum, which is the signal representation in the frequency domain. The IFFT/FFT converts this spectrum back to time domain signal by again multiplying it by a succession of sinusoids.

The IFFT is used in OFDM systems because it can generate the time-domain signal at once, instead of converting one carrier at a time and then adding them all up. The results of applying IFFT and FFT are the same and, if you apply both blocks, the output is identical to the input. Thus, if IFFT is applied at one end to generate the OFDM signal, the reverse, FFT, has to be applied to the other end to recover the original input.

The symbol extension for the cyclic prefix is done by extending the IFFT duration at the transmitter. Its cyclic waveform is illustrated in Figure 13.8. This extension is possible because the FFT and the transmission of the symbols in the wireless medium are two independent processes, thus the duration of the actual transmitted symbol can be longer than the original symbol duration presented to the FFT.

At the receiver, the last part of the received waveform replaces the initial part perfectly (Figure 13.8), the symbol is trimmed to its original duration, and the FFT can be performed with the right symbol

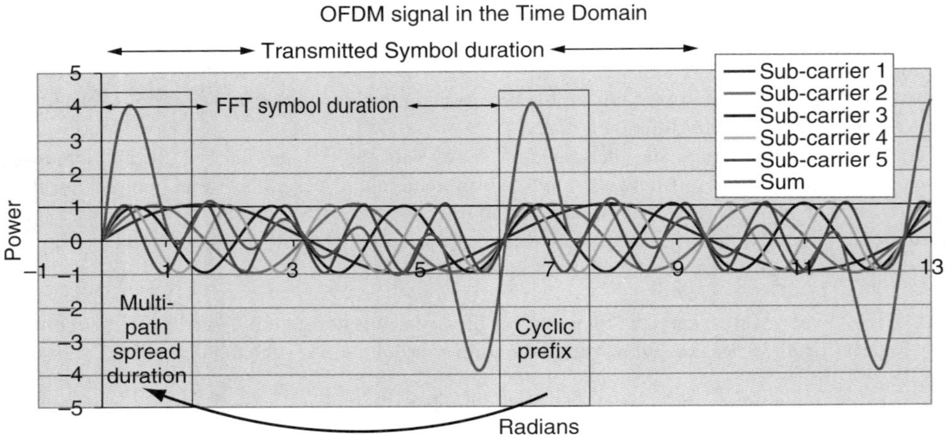

Figure 13.8 Cyclic waveform of IFFT.

Table 13.3 Maximum multipath spread distance for OFDMA symbol fractions

Symbol fraction	Multipath spread distance (m)
1/4	6857.14
1/8	3428.57
1/16	1714.29
1/32	857.14

duration. In fact, no replacement is necessary as the FFT function is cyclical and it is sufficient that a whole period is present. In summary, the symbol duration is only increased between the transmitter and the receiver.

This symbol extension is specified in the standard as $^1/_4$, $^1/_8$, $^1/_{16}$ or $^1/_{32}$ of the symbol time to provide flexibility for different environments. The WiMAX Forum narrows this down, specifying a mandatory implementation of the $^1/_8$ fraction, making optional the implementation of other fractions. The maximum multipath spread for each Cyclic Prefix duration is shown in Table 13.3.The drawback of the cyclic prefix addition is that it introduces overhead into the system, representing a loss in time efficiency of about 12.5% for the $^1/_8$ fraction.

13.3.2.2 Constant OFDM

IEEE Std. 802.16–2004, WiMAX OFDM, specifies a constant number of 256 subcarriers for all bandwidths and its OFDM parameters are presented in Table 13.4. Observe that the sampling bandwidth increases in relation to the nominal bandwidth to improve the implementation of the anti-aliasing filter, as sampling done for the nominal bandwidth rate would require a square (very sharp) filter.

Table 13.4 OFDM parameters of IEEE Std. 802.16-2004, WiMAX OFDM

IEEE Std. 802.16-2004	OFDM parameters				
Bandwidth (MHz)	1.75	3.5	7	14	28
Sampling factor	8/7	8/7	8/7	8/7	8/7
Sampling factor	1.14	1.14	1.14	1.14	1.14
Sampling bandwidth (MHz)	2	4	8	16	32
Sampling period (ns)	500	250	125	63	31
FFT size	256	256	256	256	256
Subcarrier frequency spacing (kHz)	7.813	15.63	31.25	62.5	125
Useful symbol time (μs)	128	64	32	16	8
Guard ratio	1/8	1/8	1/8	1/8	1/8
Cycle prefix duration (μs)	16	8	4	2	1
OFDM symbol time (μs)	144	72	36	18	9

Table 13.5 OFDM parameters of IEEE Std. 802.16e, WiMAX OFDMA

IEEE Std. 802.16e	OFDM parameters			
Bandwidth (MHz)	1.25	5	10	20
Sampling factor	28/25	28/25	28/25	28/25
Sampling factor	1.12	1.12	1.12	1.12
Sampling Bandwidth MHz)	1.4	5.6	11.2	22.4
Sampling period (ns)	714	179	89	45
FFT size	128	512	1024	2048
Subcarrier frequency spacing (kHz)	10.9375	10.9375	10.9375	10.9375
Useful symbol time (μs)	91.43	91.43	91.43	91.43
Guard ratio	1/8	1/8	1/8	1/8
Cycle prefix duration (μs)	11.43	11.43	11.43	11.43
OFDM symbol time (μs)	102.86	102.86	102.86	102.86

13.3.2.3 Scalable OFDM

As explained previously, in IEEE Std. 802.16e, WiMAX OFDMA, the spacing between carriers is proportional to the data rate. The first specification of the technology used a fixed number of carriers regardless of the bandwidth, which resulted in different performances for different spectrum bandwidths. To solve this issue, IEEE Std. 802.16e specified a scalable number of subcarriers that varies with the bandwidth, keeping the spacing between subcarriers constant, as shown in Table 13.5. Because of the scalability in the number of subcarriers, this technology is known as Scalable OFDMA (SOFDMA).

13.3.2.4 Duplexing

OFDM was originally intended as a single signal transmission, but some kind of multiple access technique had to be combined with it to allow for multiple users. In any communication network, two directions of transmission (duplexing) must be accommodated: from the BS to the SSs or MSs, defined as downlink, and from SSs or MSs to the BS, defined as uplink. To deal with this, the 802.16

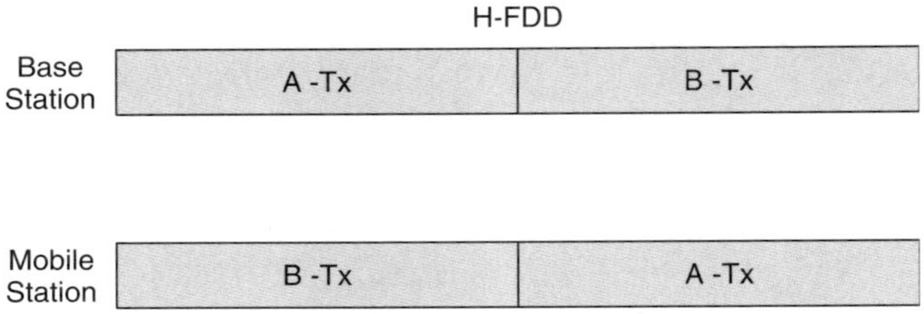

Figure 13.9 H-FDD time allocation of a frequency channel.

standard includes two duplexing techniques: Frequency Division Duplexing (FDD) and Time Division Duplexing (TDD).

In a full duplex system both directions (transmit and receive) are available to the user simultaneously, which can be accomplished by either FDD or TDD. In a half duplex system, users can only use one direction at a time, that is, they either transmit or receive information; one example of such a system is H-FDD (Half Duplex FDD).

In FDD mode, both BS and SS transmit at the same time in different frequencies, one for the downlink and one for the uplink. To allow isolation of these frequencies in the receiver, FDD requires a significant separation between both frequencies (45 MHz minimum separation is typical).

In H-FDD, the BS and SS also use different frequencies, but they do not transmit at the same time. This avoids the frequency separation issue but uses frequencies only 50% of the time, allocating the remaining 50% to another cell. This time allocation is illustrated in Figure 13.9 for two cells A and B.

Both FDD and its H-FDD variation use a fixed duration frame for downlink and uplink transmission. This works well for symmetrical services that require the same data rate for both directions, however, it is not suitable for applications such as the Internet, in which MSs and SSs offer an asymmetrical demand to the network, requiring different throughput for the uplink and downlink. Even though asymmetrical frequency bands could be considered in FDD, this not easily done, as it would require resources to be idle in the direction with less traffic. For such applications, the TDD mode is more efficient because it offers adaptive distribution of frame duration for uplink and downlink transmission. One advantage of FDD, however, is the fact that users do not need to wait for their assigned subframe, thus reducing system latency.

In TDD, separate transmission times are allocated for downlink and uplink transparently to users. This is illustrated in Figure 13.10. The allocation cycle is defined by a frame period, which is divided into a downlink subframe and an uplink subframe. The length of these subframes, in turn, is a multiple of the symbol duration.

The two subframes can have different durations to accommodate asymmetrical traffic in downlink (DL) and uplink (UL). The ratio between the uplink subframe duration and the downlink subframe duration is the TDD ratio, which must be unique for the whole network, that is, all BSs use the same ratio; this uniqueness avoids interference between downlink and uplink signals. A typical ratio is 60%, that is, 50% more downlink than uplink, but this value should be adjusted by network designers based on the desired services configuration.

The split in DL and UL adds some inefficiency to TDD transmission as the TDD ratio represents an average value of network throughput demand; this may cause the system to be under-utilized at moments in which the demand differs from this average. This loss in efficiency is estimated to be between 5% and 10%.

Another advantage of TDD systems is that both uplink and downlink directions share the same RF conditions of the channel (unlike FDD systems, because of the use of distinct frequencies for

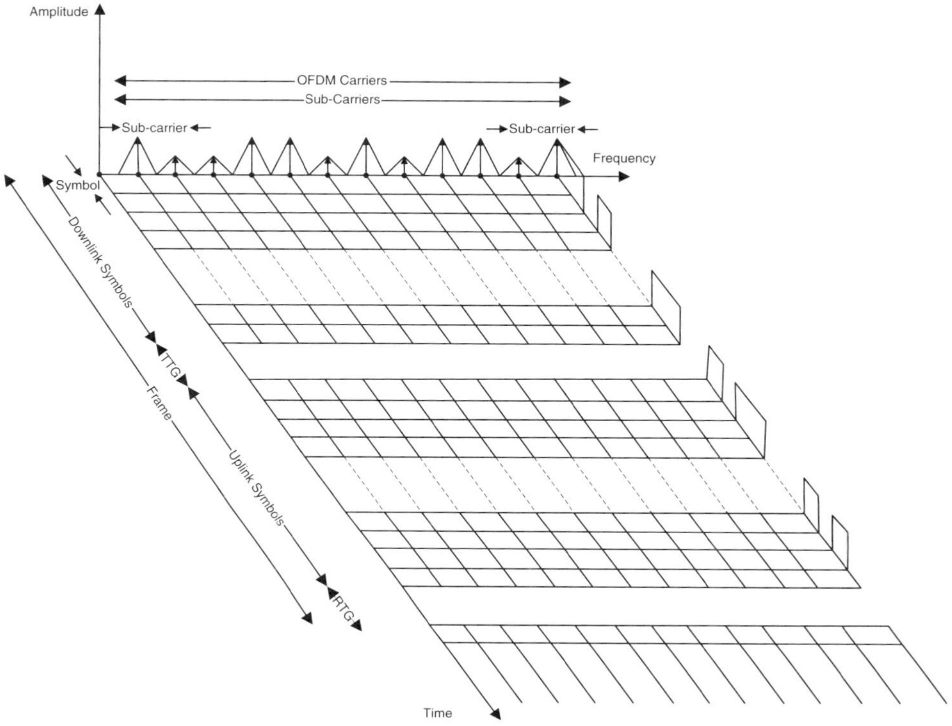

Figure 13.10 TDD transmission in OFDM.

each direction), which allows measurement data collected in the uplink to be used for RF tuning of the downlink.

13.3.2.5 Frame and Synchronization

As previously explained, a WiMAX TDD network requires frame level synchronization. Frame timing is controlled by the APs (Access Points) also called Base Stations (BS). Downlink frames are synchronized at the base station by GPS, time and synchronization signals are extracted at the receiver. Figure 13.11 shows subframes of various base stations to illustrate the need for synchronization.

The received downlink frame timing is used to synchronize the uplink frame transmission at the SS as adjustment information is sent by the BS to the SS to adjust its timing. There are two synchronization functions: timing synchronization and frequency synchronization.

- Timing synchronization is less critical than in other technologies because the symbol time is larger and the equalization is easier (the multipath spread is a fraction of the symbol time).
- Frequency synchronization, however, is more critical than in other technologies because the subcarriers are very close to each other.

On the SSs side, the frame synchronization requires SSs to have knowledge of the time it takes for a signal to travel from the BS to their location; thus indirectly obtaining the distance between

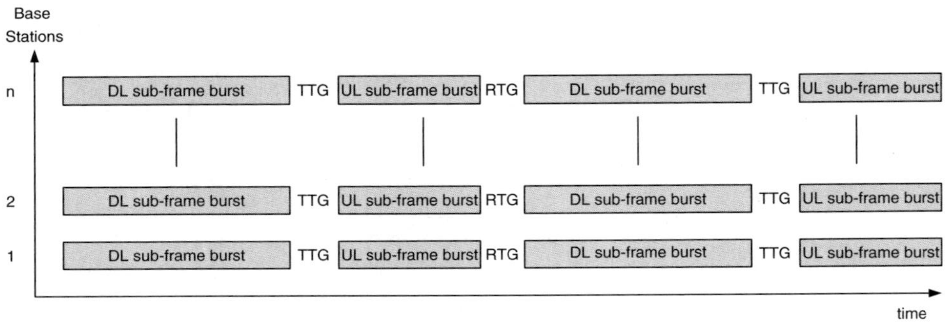

Figure 13.11 DL and UL subframes of multiple base stations.

themselves and the BS. This is achieved through a process called ranging. Ranging is performed over one to two consecutive symbols by the BS. In this process, the BS evaluates by how much the SS timing is skewed. This information is then sent to the SS in a message.

There are four ranging procedures in WiMAX:

- The Initial Ranging is requested by the SS when accessing a new cell (2 symbols duration).
- The Periodic Ranging is requested periodically to maintain synchronization (1 symbol duration every 250 ms to 5 s).
- The Bandwidth Request is performed when the SS needs to request UL resources (1 symbol duration).
- The Handover, which is requested during the handover procedure to the target cell (2 symbols duration).

Ranging opportunities are presented periodically and do not appear in all frames.

As mentioned before, WiMAX divides time into continuous frames and although it allows several duplexing modes, it focused initially on TDD (Time Division Duplexing), which requires frame synchronization (TDD ratio) at the network level.

In WiMAX, base stations control the time access by defining the beginning of downlink subframes. Only a small portion of the uplink subframe is left for autonomous access. Base stations are synchronized between themselves using GPS or a similar system. Uplink frame access is synchronized using the ranging technique.

The WiMAX frame is the heart of the technology, as it optimizes the access to the airways in an unprecedented way. IP data comes in chunks of data (packets) that vary in size from few bytes to some MB, with an average value of a few KB. The traditional conflict-based protocol is very inefficient as proved in IEEE Std. 802.11 deployments, in which frames are not used, and instead, all network elements fight for the RF channel. The WiMAX approach is more efficient, as frames indicate when one is to use the RF channel, thus significantly reducing the overhead for conflict and packet synchronization.

This is only possible because of global frame synchronization throughout the network, allowing all network elements to know the moment data is available or should be sent. Only minor adjustments are required at the start of each frame. BSs have a guaranteed time to send data and SS/MSs are granted their time by the BSs. SS/MSs adjust their timing periodically (through ranging), so the data they send arrives at the right moment at the BS. Any minor misalignments are absorbed by the Cyclic Prefix.

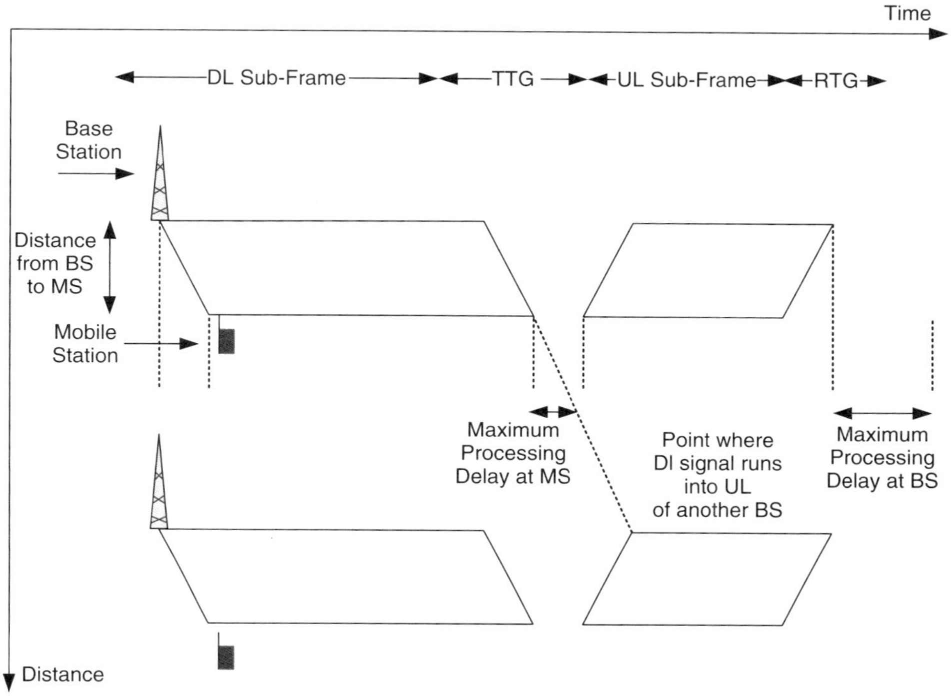

Figure 13.12 Transmission of DL and UL subframes in TDD mode.

Several frame sizes are specified for the WiMAX technology: 2.0, 2.5, 4.0, 5.0, 8.0, 10.0, 12.5 and 20.0 ms. The WiMAX Forum mandates support for 5 ms frames, all others are optional.

In TDD mode, the radio switches between transmit and receive according to the moment in time. A CPE initially receives the DL subframe and then switches to transmit mode at the appropriate moment in the UL subframe.

BSs can only specify the start of the UL subframe after all SSs/MSs receive the DL subframe, process the data, switch to transmit mode, and the transmitted data is received at the BS. Figure 13.12 shows the transmission process. A time gap happens between the DL subframe and the UL subframe due to the propagation and processing delays involved in the transmission process. This gap is called the Transmit Transition Gap (TTG) and it should be greater than twice the maximum network propagation delay between any BS and the most distant SS/MS plus their processing time. A similar time gap, called the Receive Transition Gap (RTG), exists between the UL subframe and the DL subframe and it should be greater than the BS processing time, as illustrated in Figure 13.12.

For example, a given 5 ms frame can accommodate 47 symbols (102.86 μs * 47 = 4834.42 μs). Assuming a processing time of 40 μs at the end of each subframe (DL and UL), 85.58 μs are left as the roundtrip propagation delay, or 42.79 μs each way. Considering that the speed of light can be approximated to 3.00×10^8 m/s, this delay indicates that the SS can be as far as 12.8 km away from the BS. The TTG in the example of Figure 13.12 is 125.58 μs long (85.58 μs propagation delay plus 40 μs processing delay at the SS), which gives a distance of 37.6 km for the downlink of a BS not to interfere with the UL frame of another BS.

The TTG and RTG add gaps in the communication process, using additional resources of the system, but they are essential as a mechanism to avoid interference. In the example of Figure 13.12, these gaps use about 3.3% of the total resources.

13.3.2.6 Zones

It may be necessary to differentiate between different SSs/MSs communicating with a BS in terms of resource allocation (for interference averaging, for example) and this is achieved by further dividing subframes into zones.

Zones are configured according to the application being used and the number of frequencies available in the system, which define the reuse criteria. The ranging procedure, explained previously, allows SSs and the BS to determine how far apart they are and this can be used with the concept of zones to limit certain applications to specific areas of the cell.

Zones have time and geographical area connotations. Resources can be assigned during specific frame zones to CPEs close to the BS and during other frame zones to CPEs at the edge of the cell. Zones can be used to determine sub-channelization schemes, use of segmentation, and antenna system applications.

13.3.3 OFDM Carrier in the Power Domain

13.3.3.1 Modulation

To be transmitted, the source information has to be modulated onto the OFDM carrier. WiMAX is capable of using different modulations in different subcarriers.

In the modulation process, sets of bits are combined into symbols and assigned to carrier states (phase x energy) forming a constellation. Constellations can be represented in polar form, showing the phase and magnitude in the same diagram, as illustrated in Figure 13.13. The distance between constellation points represents how much noise the modulation can accommodate.

Figure 13.14 shows QPSK, 16 QAM, and 64QAM modulations superimposed. The levels of each modulation are such that the average power of each one is the same. Additional details about modulation schemes are given in Chapter 4.

WiMAX supports the following modulations: BPSK (not used for data), QPSK, 16-QAM, and 64-QAM. These modulations are combined with coding algorithms, code rates, and repetitions to form what are known as modulation schemes.

In terms of coding algorithms, WiMAX supports both convolutional and convolutional turbo. WiMAX radios can use 25 different code rates, the main ones being $^1/_3$, $^1/_2$, $^2/_3$, $^3/_4$, $^5/_6$ and repetition rates of x2, x4, and x6. However, not all combinations of these parameters are allowed. The standard specifies 27 possible combinations, some of which are optional and not implemented by all vendors.

The pilot subcarriers are modulated by pre-defined sequences with boosted BPSK, which has a higher power level than other modulations, whereas data subcarriers are modulated by the highest modulation scheme possible in each connection, through what is known as Adaptive Modulation and Coding (AMC).

A BPSK symbol can carry 1 bit of data, whereas a QPSK symbol can carry 2 bits, a 16-QAM symbol 4 bits, and a 64-QAM symbol, 6 bits. Forward Error Correction Codes add an additional overhead of between 17% (for $^5/_6$ rate coding) and 67% (for $^1/_3$ rate coding).

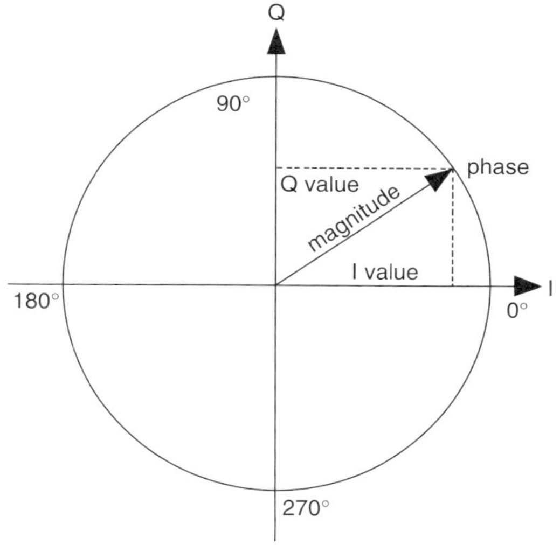

Figure 13.13 Polar and rectangular constellation diagram.

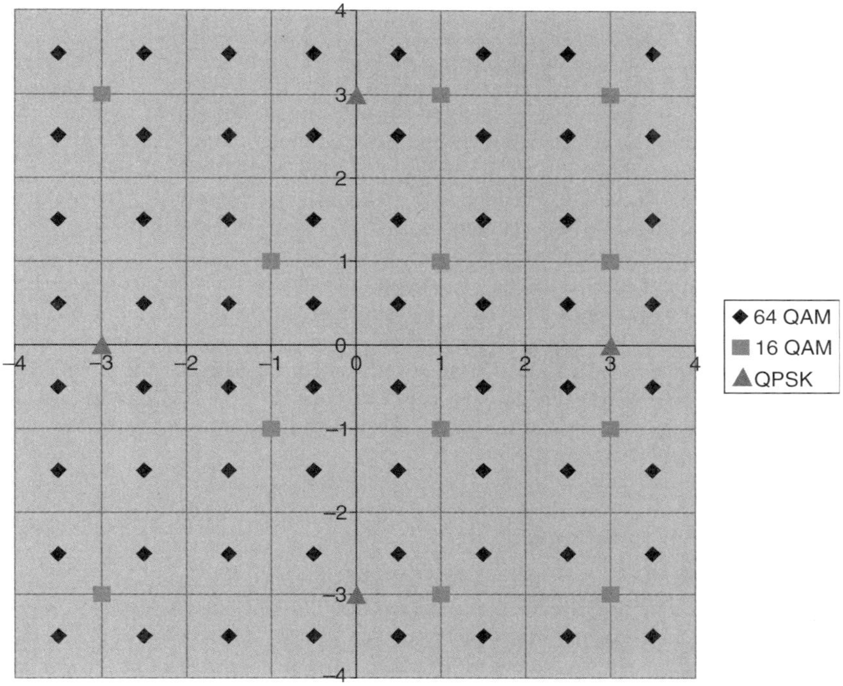

Figure 13.14 Representation of QPSK, 16-QAM, and 64-QAM modulations.

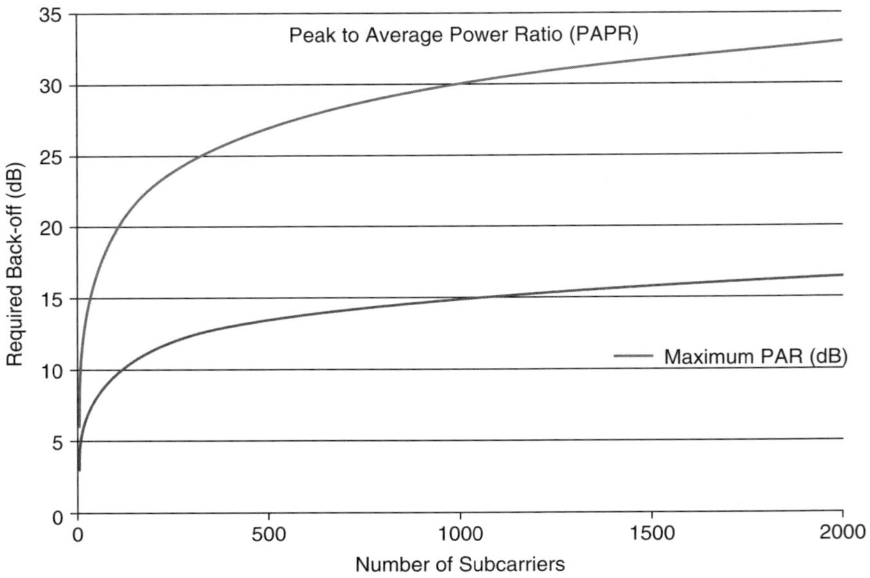

Figure 13.15 Peak to Average Power Ratio (PAPR) in WiMAX.

13.3.3.2 OFDM Peak to Average Power Ratio (PAPR)

The OFDM signal is the sum of several subcarriers that are modulated by different modulation schemes. The sum of these levels has a distribution centered on an average level with a diminishing probability of reaching very high levels. Theoretically, for 1000 OFDM subcarriers, the peak signal can be close to 30 dB above average, although with an infinitesimal probability. The use of a pseudo-random code to scramble the data helps to randomize it and reduce the probability that all subcarriers have the same symbol at the same time.

Nevertheless when a high level is reached, it saturates the amplifier and results in a distortion. This implies that a large back-off has to be used in the amplifier, increasing its size and cost. Besides, the amplifier gain decreases for high levels even when clipping the signal. The amplifier response is improved through a digital distortion of the input signal that compensates for the compression caused by it (Digital Pre-Distortion). The PAPR issue could be considered as impairment during network design and a performance loss added to reflect this. Figure 13.15 shows the required back-off in the power amplifier.

13.3.3.3 Transmit Power and Power Control

Power level is a variable that impacts the reach of a WiMAX network. The limits for transmit power are regulated by local or national agencies, such as the FCC in the United States. It is important that the BS power be adapted to the maximum power of the SS and MS devices to provide a balanced link. The transmit power of a BS radio has to be typically 10 dB higher than the SS/MS power.

The pilot and the preamble power are BPSK modulated and boosted 2.5 dB above the average value specified for other modulations. SSs/MSs transmit power is classified by the WiMAX Forum into four distinct power classes: 20 dBm, 23 dBm, 27 dBm and 30 dBm.

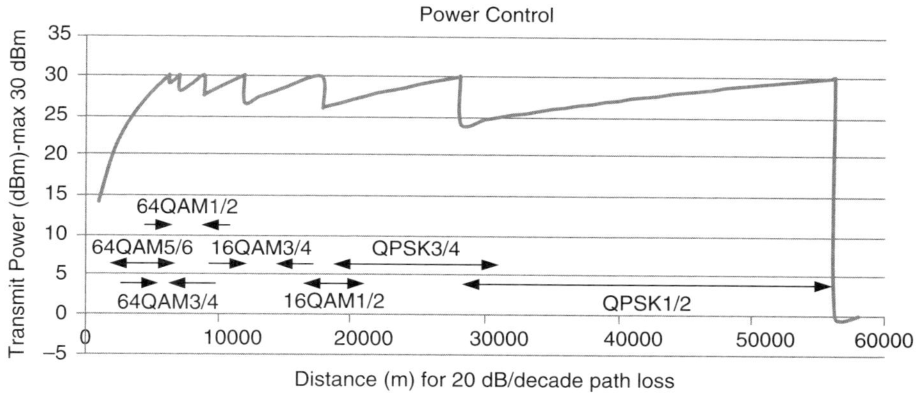

Figure 13.16 Variation of transmitted power with distance for a 20 dB/decade path loss.

Power control can be used in both the downstream and upstream with a dynamic range of up to 50 dB, although this only applies when the highest modulation rate is reached. Figure 13.16 shows the variation of transmitted power with distance, for a 20 dB/decade path loss. Distances displayed in the graph change significantly with the path–loss slope. It is important to note that lowest modulation rates cover most of the distance and consequently the cell area.

13.3.3.4 Power Operation Modes

SS/MS devices can operate in three different modes:

- *Normal mode*: the device is constantly monitoring downlink messages and updating its ranging values.
- *Sleep mode*: the device goes into sleep mode and periodically wakes and updates its ranging values.
- *Idle mode*: the device unregisters and goes into sleep mode. It must be paged for incoming data and re-register again for outgoing data. This mode is useful on situations where the device is moving (e.g. highways) to avoid excessive registrations.

13.4 Multiple Access OFDMA

A 10 MHz WiMAX frame, if totally filled, can carry close to 20,000 data symbols every 5 ms or 3.12 Ms/s. The throughput ranges between 3.1 Mbps to 15.6 Mbps. This is a very large amount of data that few users would require.

To maximize carrier usage, it is necessary to split the resources dynamically between several users. This technique is called Multiple Access and exists in many technologies. In WiMAX, the Orthogonal Frequency Division Multiplex (OFDM) becomes Orthogonal Frequency Division Multiple Access (OFDMA).

The first WiMAX specification 802.16d did not provide Multiple Access, so the whole carrier was allocated at least for the duration of a frame to a single user. This made sense as the first deployments were point-to-point oriented, but it became very wasteful for point-to-multipoint networks.

For point-to-multipoint networks, specification 802.16e was developed using OFDMA. It divided the carrier into groups of sub-carriers in the frequency domain and groups of symbols in the time domain, forming slots. Those slots could be defined on a frame basis and were attributed to users on a frame or multi frame basis.

13.5 WiMAX Network Layers

The main aspects of the WiMAX technology were discussed in the previous sections. This section analyzes how all of it comes together.

13.5.1 The PHY Layer

The Physical Layer (PHY) is responsible for the physical connection of the two network elements communicating with each other (i.e. MS/SS and BS). The PHY is responsible for all physical characteristics of the transmission (e.g. bit processing and modulation). In WiMAX, the PHY can be OFDM or OFDMA.

The PHY block diagram presented in Figure 13.17 shows the operations performed in the downlink and uplink. It is divided into the following functional blocks:

- Bit processing
- Symbol Processing
- Digital IF processing
- Analog/Digital Conversion
- Carrier modulation

The basic signal transformations are the following:

13.5.1.1 Transmit Direction

- Data is divided into blocks (multiples of 48 bits later mapped to sub-channels).
- Data is combined with a pseudo-random sequence to avoid long sequences of identical values (minimize PAR issues).
- Data is coded using a Forward Error Correction (FEC) code.
- Data rate is adjusted through repeating or puncturing.
- Data is interleaved (data segments are inter-exchanged in position) to improve error correction by separating blocks of data subject to prolonged fading.
- Data is mapped to subcarrier symbols according to the selected permutation scheme and pilots are inserted.
- Signal is converted from the frequency domain to time domain through an IFFT (Inverse Fourier Transform) operation.
- The Cyclic Prefix is added, extending the symbol duration.
- Signal is up-converted.
- The crest factor is reduced and the signal is pre-distorted to compensate for power amplifier response.
- Signal is converted from digital to analog.
- Signal modulates transmit carrier.

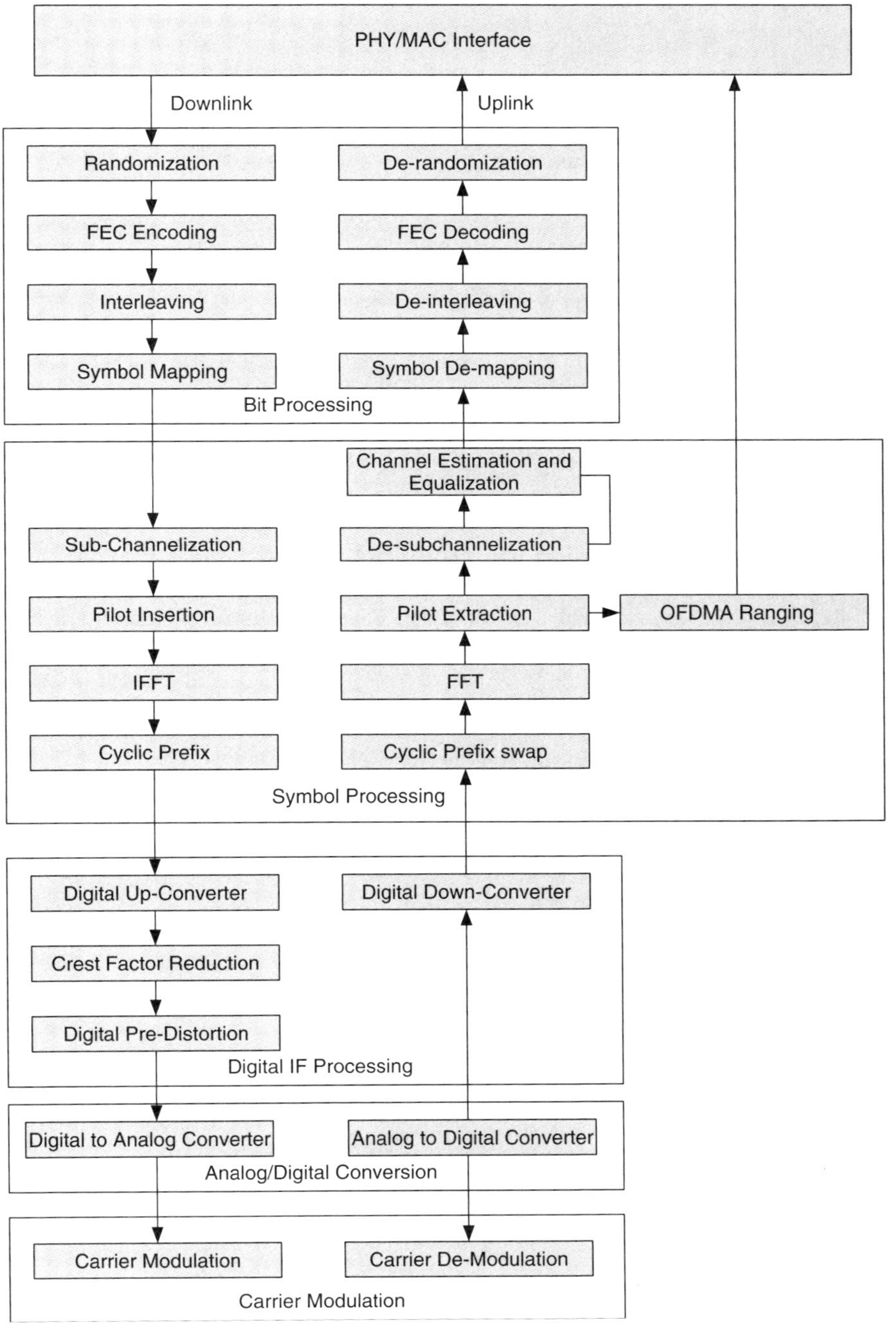

Figure 13.17 PHY block diagram.

13.5.1.2 Receive Direction

- Signal is demodulated.
- Signal is sampled and converted from analog to digital.
- Signal is down-converted.
- Signal tail replaces the Cyclic Prefix.
- Signal is converted from time domain to frequency domain using an FFT (Fast Fourier transform) operation.
- Pilot is extracted and ranging operation is performed.
- Sub-channels are extracted.
- Channel estimation is made from pilots, and signal is equalized.
- Data is extracted and symbols are de-mapped.
- De-interleaving is performed.
- FEC decoding is performed.
- De-randomization is performed.
- Data blocks are recovered from sub-channels.

13.5.2 The MAC (Data) Layer

The Open Systems Interconnection Reference Model (OSI) describes communications networks as a group of seven layers: Application Layer, Presentation Layer, Session Layer, Transport Layer, Network Layer, Data Link Layer, and Physical Layer. Each of the layers represents protocols that perform similar functions, receiving services and information from the layer below it and providing services and information to the layer above it.

From the seven layers present in the OSI model, only two are described in the 802.16 standard: the Data Link Layer and the Physical Layer (PHY). The Data Link Layer is subdivided into the Logical Link Control (LLC) and the Media Access Control (MAC) Layers; the first, LLC, often applying the IEEE 802.2 standard whereas the latter, MAC, is described in detail in the 802.16 standard.

The MAC layer itself is further subdivided into three sub-layers: the Convergence Sub-layer (CS), the Common Part Sub-layer (CPS), and the Security Sub-layer (SS). Figure 13.18 illustrates these layers.

The WiMAX Convergence Sub-layer interfaces with upper layers through the following user data protocols: IP (Internet Protocol) v4 and v6 and Ethernet. IPv4 and IPv6 implementation is mandatory

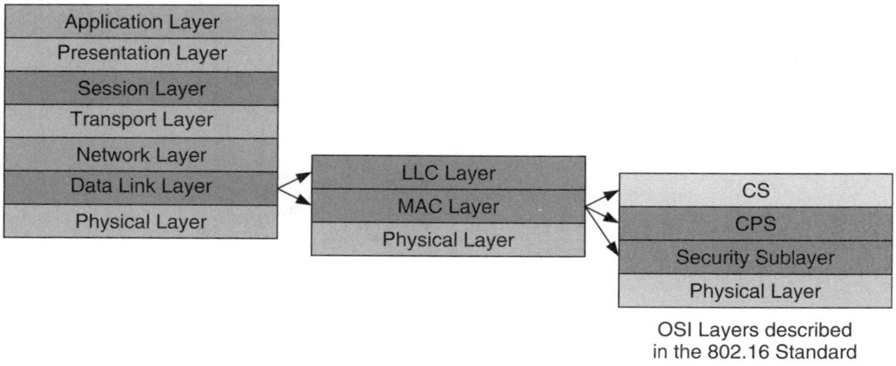

Figure 13.18 OSI layers, and the layers and sub-layers included in the 802.16 standard.

per the WiMAX Forum Mobile System Profile, whereas Ethernet support is optional. This sub-layer is also responsible for payload header suppression and packet classification.

The Common Part Sub-layer (CPS) is responsible for network entry and initialization, connection management, ARQ/HARQ use, and mobility-related tasks. Each of these features is described in more detail in this section.

One of the main functions of the Security Sub-layer (SS) is to take care of data encryption.

13.5.2.1 Data Unit

When data is transferred from one user to another, it passes through several protocol layers. These layers interconnect at SAPs (Service Access Points).

An SDU (Service Data Unit) is a block of data received by a protocol layer. In the transmit direction, the protocol layer adds an overhead to the SDU and creates a PDU (Protocol Data Unit). This PDU then becomes an SDU for the lower layer, as illustrated in Figure 13.19. In the receive direction, the reverse is done and the protocol overhead is stripped, so the PDU received by a layer is exactly the same that was sent by its equivalent layer at the other end. It is as if each protocol layer ignored the existence of the lower layers.

13.5.2.2 MAC-PDU

The MAC Protocol Data Unit (MAC-PDU) is composed of a fixed length header, a variable length payload, and an optional CRC (Cyclic Redundancy Code). The MAC-PDU can have a maximum length of 2048 bytes. A generic wireless MAC-PDU is represented in Figure 13.20. The top part of Figure 13.20 shows the header in detail and indicates the length (in bits) of each component. The bottom part shows the complete PDU and the length (in bytes) of each component.

In Figure 13.20, HT indicates the Header Type, which is used to indicate whether a payload follows (type 0) or not (type 1). The payload is not present in signaling MAC-PDUs (e.g. a bandwidth request). For 802.16e systems, the HT is always set to 0 in the downlink, because the payload is always present.

EC (Encryption Control) indicates if the content is encrypted (1) or not (0); the header itself is never encrypted. When this field is set to 1, the Encryption Key Sequence (EKS) becomes relevant and provides the index of the Traffic Encryption Key (TEK) and Initialization Vector (IV) used for encryption of the payload. If EC is set to 0 (payload not encrypted), the two EKS bits are reserved for other purposes.

The 6 bits reserved for Type describe any sub-headers and special payload types contained in the payload (e.g. fragmented PDUs).

ESF stands for Extended Sub-header Field and indicates whether sub-headers are present after the generic header (1). If they are absent, this value is set to 0. The CI (CRC Indicator) follows the same principle: 1 if CRC is present, 0 if it is absent.

One reserved bit (RSV) is also present in the header. Reserved bits are usually set to 0 for transmission and ignored in the reception. These bits are usually saved to support new functions.

The length field (LEN) indicates, in bytes, the total length of the MPDU, including the header, sub-header, payload, and CRC.

CID stands for Connection Identifier. In traditional wireless technologies, users are assigned channels that are kept allocated during the span of a connection. In WiMAX, users are instead assigned a set of CIDs, which is a 16-bit long parameter that uniquely identifies the connection between the BS and the SS and is linked to the messages sent. CIDs eliminate the need for specialized channels and allow free resources to be allocated dynamically. CIDs can be used for both downlink and uplink or only in one direction. CIDs are classified into different categories, such as basic connection, primary management, user data transport, and broadcast and are kept during the whole session.

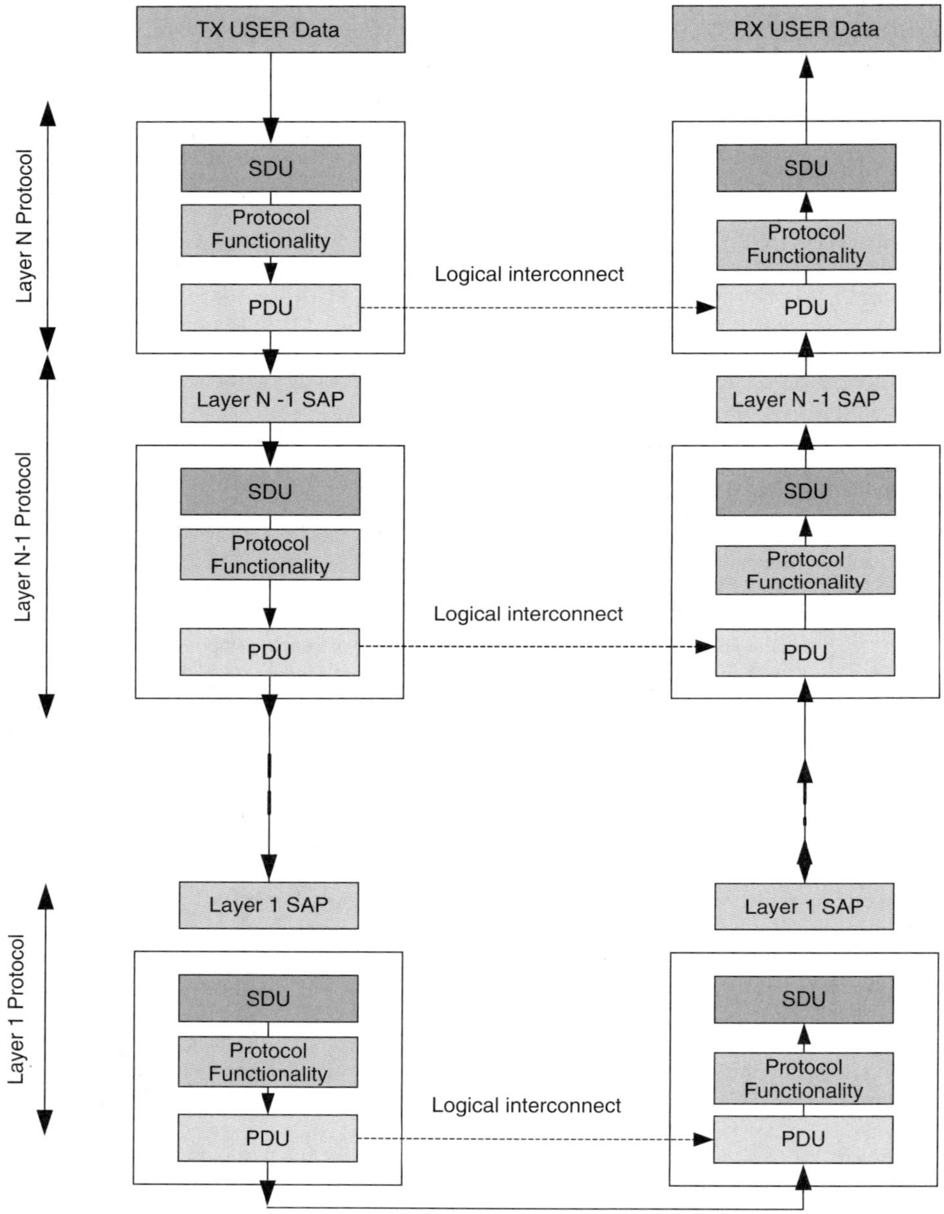

Figure 13.19 Service and protocol data units within different layers.

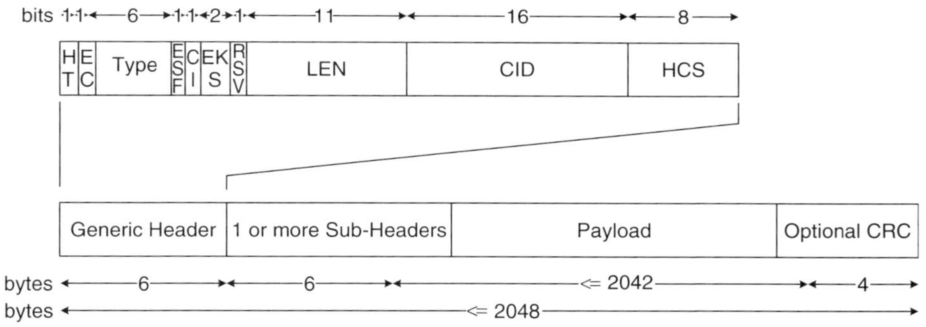

Figure 13.20 Generic wireless MAC-PDU.

Basic Connection CIDs are mandatory and used for MAC control messages, for example, Downlink Burst Profile Change Request and Response messages (DBPC-REQ/RSP).

Primary Management CIDs are also mandatory and used to send MAC control messages that are not sent through basic connection CIDs, such as dynamic service flow messages (e.g. Dynamic Service Flow Addition Request DSA-REQ).

When CIDs are related to user data transport connections, they identify the service flow to which the PDU belongs. A service flow is a flow of packets that matches a given Quality of Service (QoS) and exists for both downlink and uplink. The flows may exist even when not actually carrying traffic and are identified by a 32-bit long parameter called the Service Flow Identifier (SFID). An independent CID is needed for each active service flow (i.e. different QoS requirements).

The existence of SFIDs allows the dynamic management of service flows with messages that can add, change, or delete existing flows in the system without affecting all network users.

Broadcast CIDs carry broadcast data messages such as channel descriptors (UCD/DCD) and mapping (DL-MAP and UL-MAP).

The 8 bits of Header Check Sum (HCS) are mandatory and used to detect transmission errors in the header of the MPDU.

The generic header of an MPDU has a fixed length, the payload, however, has a variable length because it may contain a single SDU, a fraction of an SDU or multiple SDUs. In the second case, a fragmentation header (FH) is used to describe the SDU. On the other hand, when multiple SDUs are present, one or more packing sub-headers (PS) are inserted after the generic header, or immediately before each SDU. The indication of presence and location of these sub-headers is given by the Type in the generic header.

The variable payload size, the fact that one or more sub-headers may be present after the generic header, and the optional use of CRC (Cyclic Redundancy Check) make for a variable-length MPDU, which provides more efficient data transmission, for example, multiple MPDUs can be transmitted in a single burst to save PHY overhead.

Signaling messages do not use payloads, and do not require any sub-headers, thus they have a slightly different format. Figure 13.21 shows, as an example, the MAC-PDU used in a bandwidth

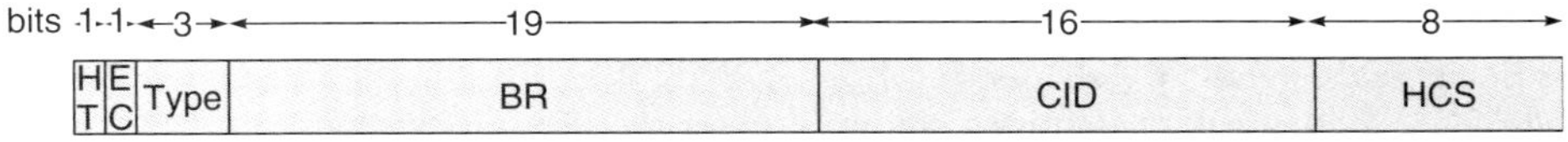

Figure 13.21 Bandwidth request MAC-PDU.

request. This type of message is sent only after the SS already has a transport connection set-up. The BR field indicates the bandwidth request message, that is, the amount of uplink resources (given in bytes) being requested by the SS. The Type field determines if the requested bandwidth is incremental (000) or aggregate (001), in which case the message should be combined with previous requests. The CID in this message specifies the connection for which the request is being made.

The MAC-PDU is an overhead to the transmitted user data and has to be considered in network dimensioning because it diminishes data throughput. This overhead varies with the message size, thus an average number should be estimated based on the size of the packets being transmitted. Additionally, protocol control messages also add to the overhead and must be considered in throughput calculations. For a mix of services, a 5% overhead can be considered an average.

13.5.3 Error Correction

According to Shannon's equation, the amount of information that can be transmitted through a channel depends on the SNIR. Interference and noise cause errors in the digital signal limiting channel capacity. Assuming that the SNIR cannot be improved further, the elimination of errors is often done through error correcting codes. The codes are added to provide redundancy to the digital signal. The amount of redundancy defines one of the characteristics of the code.

The total elimination of errors would imply continuously applied redundancy, which may lead to an extremely large error correction code. Thus, when the number of errors is small, it becomes more efficient to use a variation of this redundancy method: the Automatic Repeat Request (ARQ). This method repeats transmitted data only when requested by the receiver. The CRC added to the MPDU is a form of ARQ. Additional error detection codes can also be applied in the receiver itself.

HARQ (Hybrid ARQ) is a variation of ARQ that adds Forward Error Correction (FEC) bits to the method. WiMAX uses two types of HARQ (Hybrid ARQ): Type I, or Chase Combining, and Type II, or Incremental Redundancy.

In Chase Combining, the same data is resent upon request, that is, the information transmitted is a copy of the original data. The receiver uses all previously received versions to improve the chance of decoding. The WiMAX Forum mandates implementation of Chase Combining with use of Convolutional Turbo Codes (CTC) for both upstream and downstream.

In Incremental Redundancy (IR), the code rate and puncturing pattern are changed from one transmission to the next, increasing the chance of a successful decoding, that is, the information transmitted is now different from the original data because of the change in the bits added by the FEC process. The implementation of this type of HARQ is made optional by the WiMAX Forum.

If the maximum specified number of retransmissions is reached, the packet is dropped and a higher level layer has to request its retransmission.

Error correction techniques reduce spectral efficiency by presenting an overhead that diminishes data throughput, hence should be considered in throughput calculations. This overhead varies with the error rate, so an average number should be estimated (10% overhead is a reasonable assumption).

13.5.4 Frame Description

As described previously, in WiMAX, frames indicate when one is to use the RF channel. Frames are divided into two subframes: the downlink subframe (Figure 13.24, on p. 380) and the uplink subframe (Figure 13.25, on p. 381). For transmission, subchannels are grouped, on a subframe basis, into data units in the size required by each user.

The downlink subframe starts with a preamble spread over the first symbol of the OFDM carrier. The preamble is a pre-defined bit pattern (PN code) broadcast to the system that allows determination of FFT size, frame synchronization, and cell and segment identification.

The preamble is followed by a detailed map of the contents of both the DL and UL subframes, which allows direct access to data in the subframes, eliminating the need of preambles for each data packet.

The preamble and mapping information extend for about 3 OFDMA symbols in the DL and approximately 2 OFDMA symbols in the UL, which represents an additional 10% use of the available resources.

13.5.4.1 Downlink Subframe

As explained previously, a time gap, known as Transmit Transition Gap (TTG), happens between the DL subframe and the UL subframe due to the propagation and processing delays involved in the transmission process. A similar time gap exists between the UL subframe and the next DL subframe and is called the Receive Transition Gap (RTG)

A downlink subframe always starts RTG μs after the last UL subframe. The start of the new subframe is further confirmed by the preamble symbol, which is a PN code spread over the first symbol subcarriers. The preamble carries information of the IDCell and segment number of the sector.

The IDCell is an integer that identifies the cell during messaging so one user cell does not get data destined for another. There are 32 IDCell codes (ranging from 0 to 31) that can be attributed to three-sectored cells (each sector defined by a segment). Omni cells have 18 IDCell codes. In practice, however, the codes can be applied to any configuration.

The standard defines 114 preamble PN codes for each FFT size: thus, considering the five possible FFT sizes (128, 512, 1024, and 2048 for 802.16e and 256 and 2048 for 802.16d), there are 570 unique codes in total.

Preamble PN codes haves indexes from 0 to 113. Considering that there are 32 possible IDCell values (from 0 to 31) and 3 segments (0, 1, 2), only 96 different combinations are possible (codes 0 to 95); the remaining combinations (codes 96 to 113) are used for omni cells, which have 18 possible IDCell values.

After the preamble starts the first zone is obligatorily a PUSC zone (explained in Section 13.7.3.3), so it can support segmentation. The first information to be mapped in that zone is the FCH (Frame Control Header) that uses two symbols and defines the location and size of the DL MAP and the UL MAP.

The FCH contains the DL Frame Prefix with the subchannel bitmap and details about the DL MAP message (length, indication of type of coding and repetition coding). Thus, the FCH is essential for the SSs to decode the DL MAP. The subchannel map indicates which subchannels are used by the segment. Transmission of the FCH is done in QPSK 1/2 with four repetitions using CTC. An exception is made for FFTs size 128, in this case, the FCH uses only 1 symbol and is not repeated four times. Figure 13.22 illustrates the structure of the FCH for FFTs size 128, and Figure 13.23 illustrates the structure for other FFT sizes.

Several data slots using the same channel Modulation and Coding Scheme (MCS) are called a burst. DL and UL MAPS map, respectively, all downlink and uplink data bursts. When a base station assigns a burst to an SS, it also informs the MCS used for the burst through a number called the Interval Usage Code (IUC).

In the downlink, this code is the DIUC (Downlink IUC): in the uplink, it is the UIUC (Uplink IUC). Certain values of these codes are defined in the standard (e.g. UIUC 0 indicates Fast-Feedback channel), others are defined by the operator (e.g. DIUC codes 0–12 and UIUC codes 1–10).

The DL MAP has variable length and describes the structure of the rest of the frame. It is the DL MAP that defines who uses each part of the frame and its associated DIUC (Downlink Interval Usage

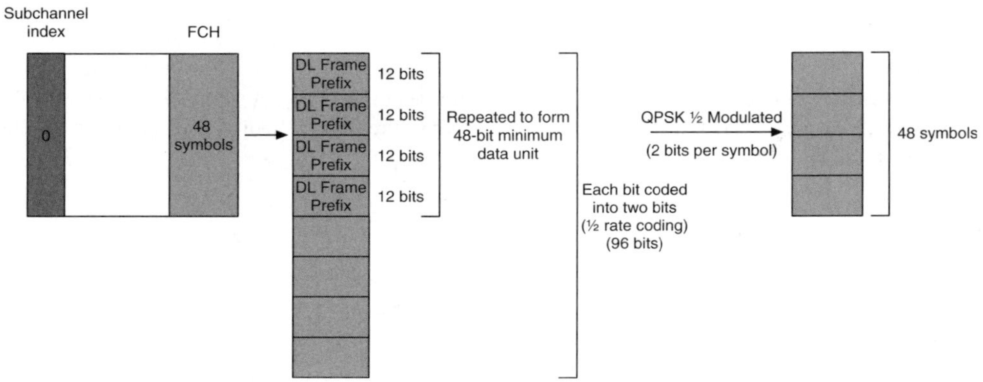

Figure 13.22 FCH description for FFT size 128.

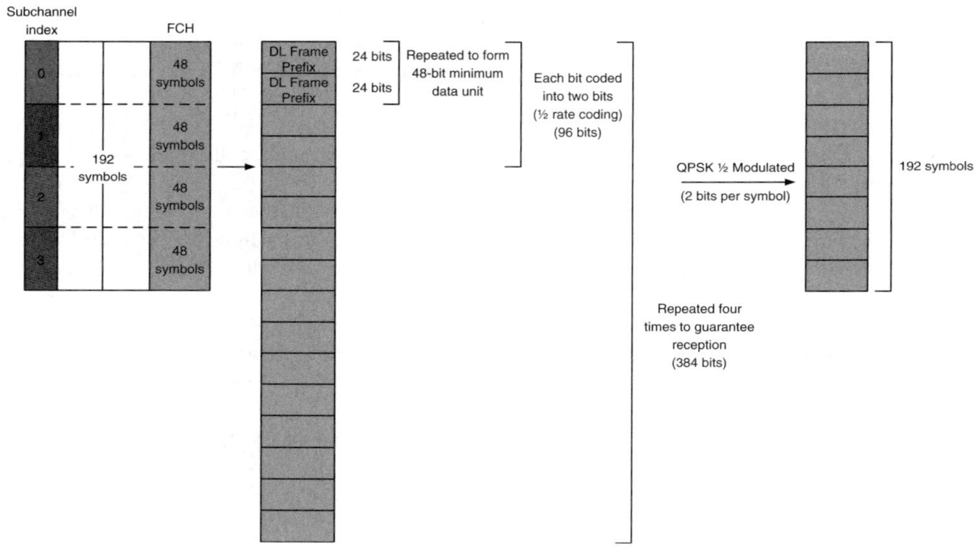

Figure 13.23 FCH description for other FFT sizes.

Code). The DL MAP message also contains the base station ID, the length of the DL subframe, and the starting time of each DL data burst.

As with the FCH, the DL MAP is encoded using $1/2$ rate convolutional coding and transmitted using QPSK modulation. The code rate is specified in the FCH.

Similarly to the DL MAP, the UL MAP also has variable length; however, whatever the length is, it is defined in the DL MAP along with its modulation and coding scheme (through the DIUC). The downlink mapping information always pertains to the current frame, in which the DL MAP message is located.

The UL MAP contains the length of the uplink subframe, and the Allocation Start Time of each UL data burst (with its associated UIUC). The Allocation Start Time is referenced from the beginning of the downlink frame and may be of such length that the UL MAP may be referencing a future frame and not the current one.

After the UL MAP, the downlink frame may contain Channel Descriptor (CD) messages. Not every downlink frame includes these but the interval between two consecutive CD messages should not be longer than 10 s.

The length and MCS of the Downlink Channel Descriptor (DCD) are defined in the DL MAP. Some of the information contained in the DCD is the downlink center frequency, total frame length, TTG and RTG, hand-off parameters, DIUC values and their associated operator-defined MCSs.

The DL MAP also defines the length and MCS of the Uplink Channel Descriptor (UCD). Some of the information contained in the UCD is the uplink center frequency, ranging parameters, UL_PermBase (Uplink Permutation Base), and UIUC values and their associated operator-defined MCSs.

DL bursts are always rectangular sections of the frame, as illustrated in Figure 13.24, with the width of the rectangle being the number of subchannels and the length being the number of allocated slots. For this reason, a padding area is added between the CDs and data burst blocks. Within each burst, data is allocated downward from left to right as shown at the bottom of Figure 13.24.

13.5.4.2 Uplink Subframe

The uplink subframe mapping is done in the downlink subframe, thus all SS/MSs know when to send data. The uplink subframe starts TTG μs after the downlink subframe. If there is absolutely no uplink traffic in the system (e.g. broadcast only), there are no UL subframes.

The first zone of the uplink subframe also uses uplink PUSC permutation (described in Section 13.7.3.4), with IDCell as its permutation base. In this zone, a three-symbols wide area is periodically allocated for ranging operations. Part of this region is used for initial and handover ranging, and part for bandwidth request and periodic ranging.

Other regions that may be present in the UL subframe are used for HARQ and ACK (Acknowledge) messages and for the CQICH (Channel Quality Indicator Channel), which allows SSs to feedback information on the state of the channel to the BS.

This division of the uplink frame into periodic regions optimizes the sending of these small but frequent messages. The location of the regions may change depending on each system implementation.

In the UL subframe, data bursts do not have to maintain a rectangular shape as in the DL, instead data units (slots) are allocated horizontally from left to right, then downward as illustrated in Figure 13.25. Each data burst goes all the way to the zone boundary before being extended to the next symbol. Within the slots, data is allocated downward, and then left to right (indicated at the bottom of Figure 13.25). Additional zones are transmitted when the current zone has being completely used.

13.5.5 Resource Management

Resource management is the function in the BS that allocates resources to the transmission of data. During this process, each active user is assigned a Connection ID (CID) to identify its sessions. Each CID is 16 bit long.

The SDUs (Service Data Units) are classified into Service Flows, which establish the following QoS parameters: packet error rate, latency, and jitter throughput. The packet error rate is provided by the allocated modulation schemes. Latency and jitter throughput are provided by the scheduling mechanisms.

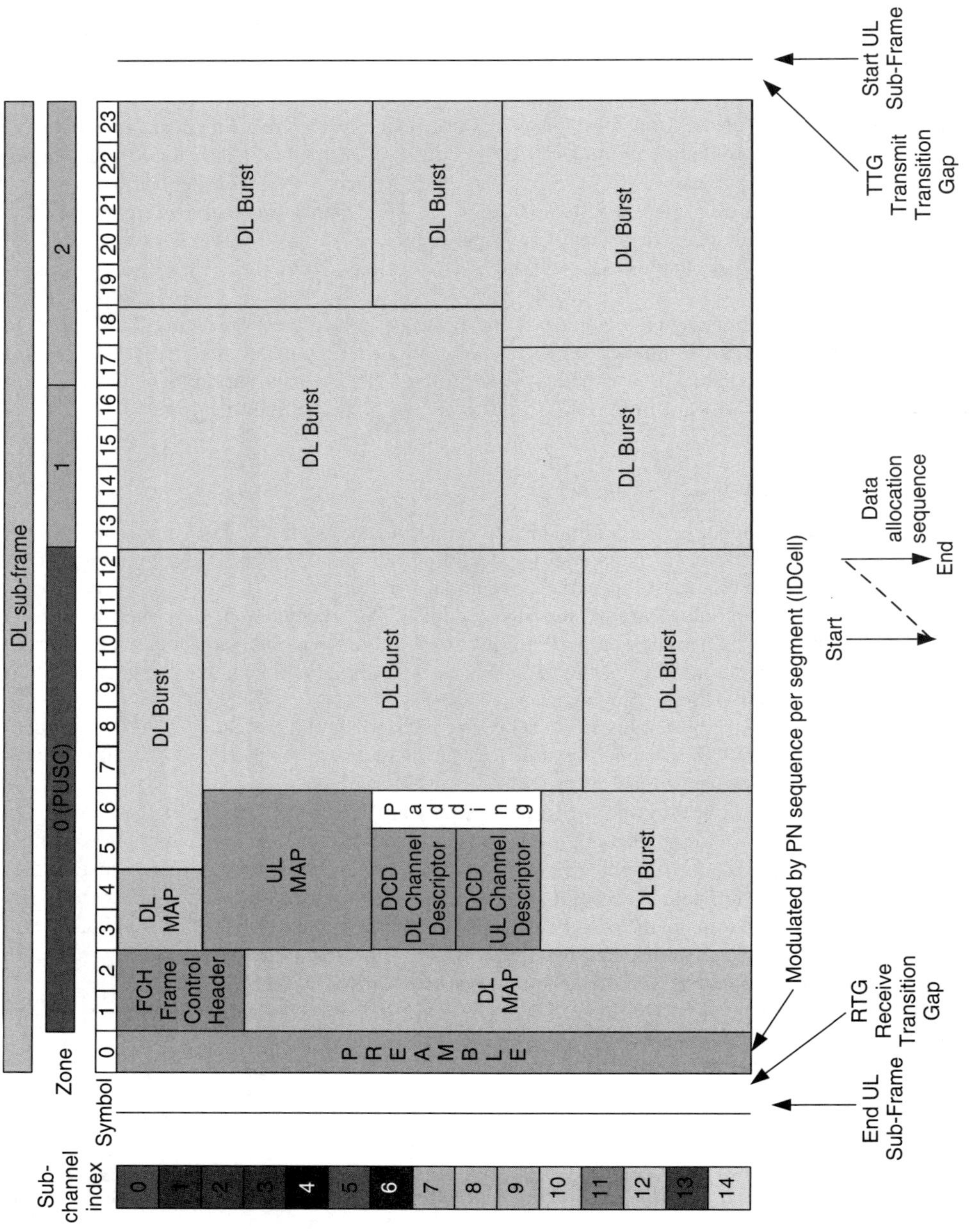

Figure 13.24 Downlink subframe.

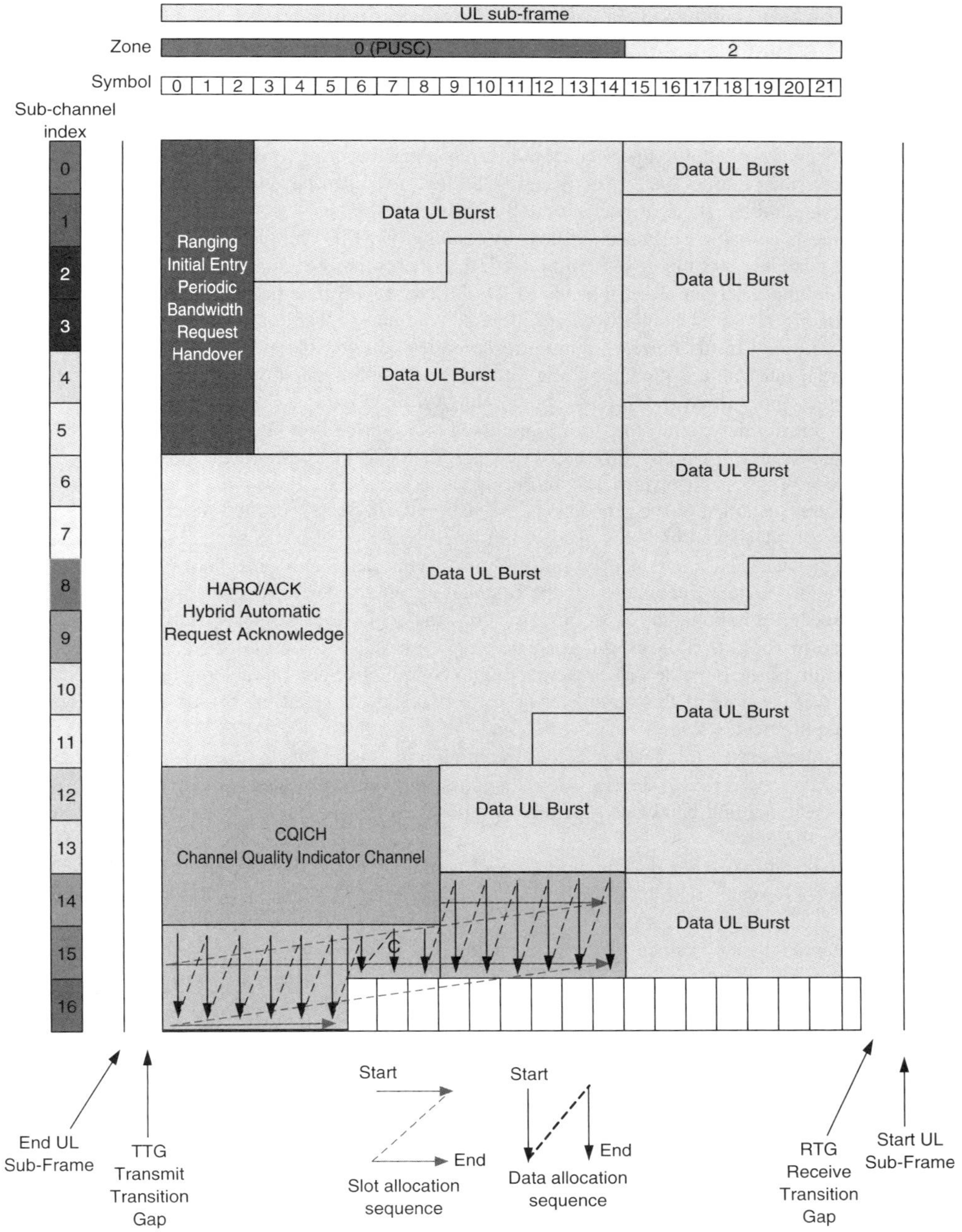

Figure 13.25 Uplink subframe.

13.5.5.1 Service Flows and QOS Categories

Service flows may be provisioned but not necessarily active. Such flows are called provisioned service flows and are assigned an SFID by the network. They do not, however, have a CID.

The Quality of Service (QoS) is a parameter that appears within the encoding of every service flow, which can be of one of the following three types: provisioned, admitted, or active. A provisioned service flow may be directly activated or go through a two-step process in which it is first admitted and later activated. An admitted service flow has resources assigned to its QoS parameter set but it is in a transient state, not completely activated. Admitted flows can be activated and later deactivated returning to the admitted stage or back to the provisioned stage.

An SS requests activation of a service flow by sending its SFID and QoS parameters to the BS. If resources are available and the flow is authorized, the BS responds by mapping the flow to a CID. If the BS is activating a service flow, it sends SFID, CID, and QoS parameters to the SS.

The QoS parameter set is a subset of the following parameters: traffic priority, maximum sustained traffic rate, maximum traffic burst, minimum reserved traffic rate, tolerated jitter, maximum latency, unsolicited grant interval, unsolicited polling interval, and vendor-specific QoS parameters. Not all of these parameters apply to all traffic classes in a WiMAX network.

The traffic priority determines the priority assigned to a service flow. It is only used when all other QoS parameters of two (or more) different flows are identical; in that case, the flow with the higher priority is given preference (lower delay, buffering rights).

The maximum sustained traffic rate defines the peak rate (in bits/s) for the service flow; the parameter considers the SDUs and includes overhead inflicted by MAC. The BS and SS control the service flow to, and on average, maintain it within this limit. This parameter only lists a threshold for the service flow and does not guarantee that the rate can be obtained. The decision of what to do with flows that overstep this boundary (e.g. delay or drop the service) is vendor-specific. This parameter can be omitted or set to 0, indicating that no maximum rate has been established.

The maximum traffic burst describes the maximum continuous burst of data that the network should support for a service flow. If this parameter is set to 0, it indicates that the service flow has no burst reservation requirements.

The minimum reserved traffic rate specifies the minimum rate (in bits/s) reserved for a service flow. If the minimum reserved rate of the BS is less than the bandwidth requested for the connection, the BS can allocate the remaining bandwidth for other purposes. If this parameter is omitted, no bandwidth is reserved for the flow.

The tolerated jitter specifies the maximum delay variation for the connection. It represents a commitment only for services type UGS or ertPS; otherwise it should be set to '000000', indicating no commitment.

The maximum latency defines the maximum interval between the entry of a packet at the CS and the transmission of its corresponding SDU. This parameter represents a service commitment and should be guaranteed by the BS/SS. Omission of the maximum latency, or a value of 0, indicates no commitment. BSs/SSs do not have to guarantee the maximum latency for flows that exceed their minimum reserved rate.

The unsolicited grant and polling intervals define, respectively, the time between successive data grant and polling opportunities for a service flow. The grant interval applies to flows type UGS and ertPS, whereas the polling interval applies to rtPS flows.

The scheduler classifies data according to the following five QoS categories, which, in turn, define how PHY resources are allocated:

- *UGS (Unsolicited Grant of Service)*: in this category, the scheduler periodically allocates a fixed amount of data in the downlink and uplink. This category represents real time applications with strong jitter constraints. For services like this, once the connection is started, a connection ID for a specific rate is defined and kept throughout the call. A typical example of this type of service is voice without silence suppression and T1/E1 transport.

- *rtPS (real time Polling Service)*: in this category, the scheduler periodically verifies the amount of data to be allocated in the downlink and, by unicast messaging, the amount of data to be allocated in the uplink. As in UGS, this category also represents real time applications but with moderate jitter requirements; it provides for variable rate resources up to the guaranteed rate. A typical example is Video on Demand (VoD).
- *ertPS (enhanced real time Polling Service)*: in this category, the scheduler periodically allocates a fixed amount of data in the downlink and uplink, but messages can be sent during this allocation to adjust the amount of data to be allocated next. This category builds on top of rtPS and UGS by representing real time applications with strong jitter constraints but providing for variable rate service. It typically applies to voice with silence suppression.
- *nrtPS (non real time Polling Service)*: in this category, the scheduler verifies, with low periodicity, the amount of data to be allocated in the downlink and, by less frequent unicast messaging, the amount of data to be allocated in the uplink, but SSs can request more frequent allocations by contention-based polling. This category represents delay-tolerant data streams comprising variable-sized data packets for which a minimum data rate is required. A typical example is web browsing for gold consumers (guaranteed minimum throughput).
- *BE (Best Effort)*: in this category, the scheduler allocates data only when no other allocation is required, the amount of data being limited to resource availability. SSs can only request allocation using contention-based polling. Only a maximum data rate is defined for this type of services, no delay or jitter constraints are considered. Data streams are handled on a space-available basis. A typical example is web browsing for regular consumers (no guaranteed minimum throughput).

Resource management variations occur when there is congestion, as different courses of action can be taken. Service degradation or service denial can be applied when required and vary with the complexity of the scheduler.

At each frame, the different queues are analyzed and assigned to the downstream and upstream subframes in data bursts. This assignment is displayed in the DL-MAP and UL-MAP.

Scheduling algorithms are not specified in the WiMAX standard and its implementation is left to the vendor's discretion. WiMAX schedulers have a much more complex job than the those users of previous technologies, thus simpler implementations are expected at the beginning but the performance of schedulers should evolve as field experience is gathered. Section 7.10 describes the most common scheduling algorithms, which are used in existing WiMAX equipment.

13.5.5.2 Downlink Resource Allocation

In the downlink, the SSs listen to all DL MAP messages to determine which allocations relate to them. Within the mapping message, an SS looks for its CIDs to determine whether to process a given data burst or not.

If a data burst is to be processed, the SS uses the DIUC provided in the DL MAP to check, in the DCD, which modulation and coding scheme to use. To allocate the actual data burst, the BS provides offset values for the OFDMA symbol and subchannel, which indicate the beginning of the burst. Because downlink bursts are always rectangular in shape, the end of the burst is determined by providing its dimensions, that is, number of OFDMA symbols and subchannels. Figure 13.26 illustrates this concept.

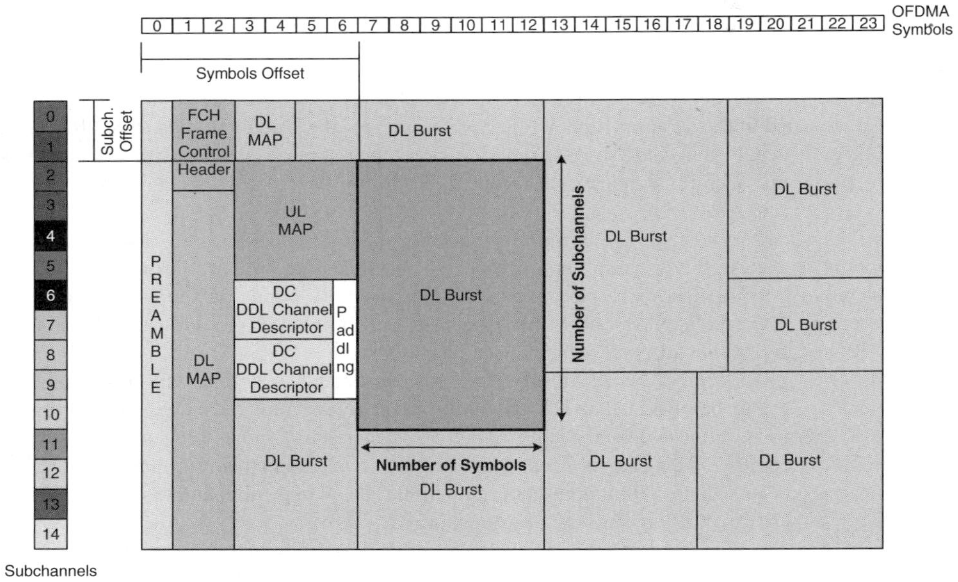

Figure 13.26 Downlink data burst allocation.

13.5.5.3 Uplink Resource Allocation

Because data bursts in the uplink do not follow a specific pattern, for uplink resource allocation, SSs have to read the Allocation Start Time to find the beginning of allocations and then continuously read resource allocations. Then they have to continuously read resource allocations because UL data bursts are continuously allocated one after the other. The UL MAP message provides the duration of each burst in number of slots. Figure 13.27 illustrates the concept.

13.5.5.4 Zone Configuration

A zone can be configured by defining distance ranges from the BS, so SSs outside a certain range can be excluded of a zone for load control, allowing interference reduction by averaging. Zones can also define use of different segmentation and permutation schemes (explained in Section 13.7.3).

Figure 13.28 shows a sample planning tool dialog with several parameters that can be configured for each zone. Note that based on the number of symbols defined per zone, the total number of slots available can be automatically calculated by the tool because the slot duration and number of subchannels per symbol of each zone are defined by the selected permutation scheme.

13.6 WiMAX Operation Phases

BSs are constantly broadcasting network messages and SSs are listening to their messages. A BS can address a specific SS by broadcasting a message to it (paging or data burst). The frame preamble is sent even if there is no traffic to be sent in the frame.

The SS has then to do a ranging operation to adjust its timing and power before accessing the network. Registered SSs are only allowed to receive messages and send replies back in time intervals

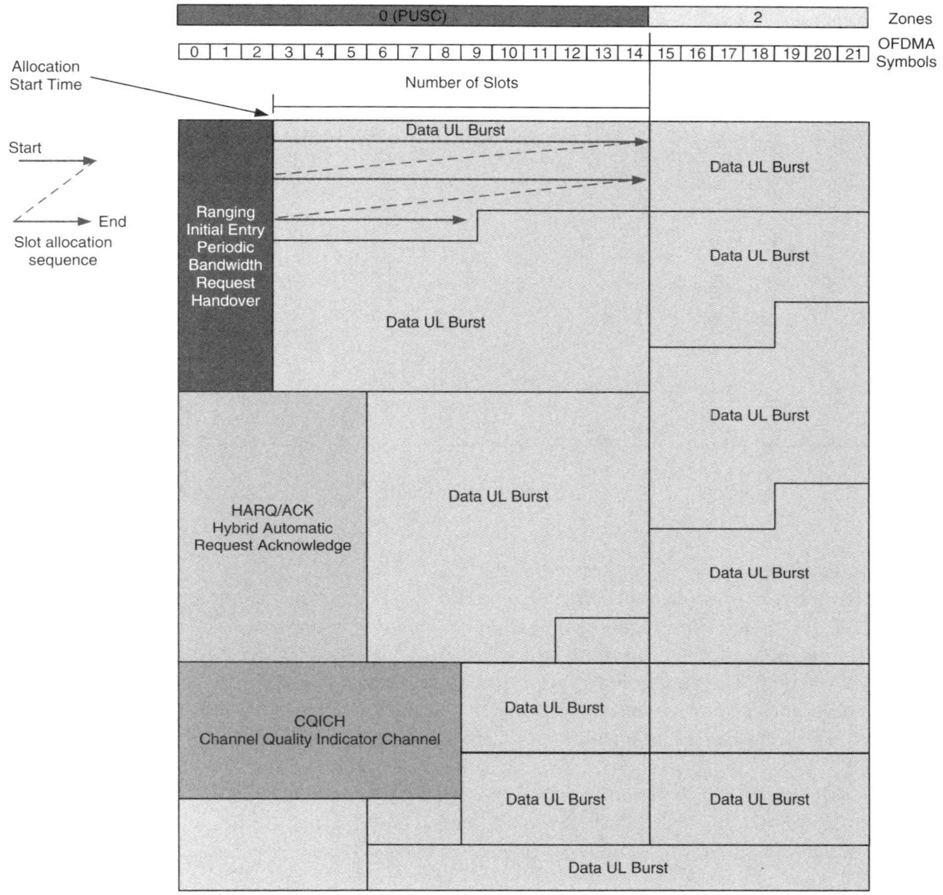

Figure 13.27 Uplink data burst allocation.

defined by the BS. When an SS wants to initiate a data exchange, it has to send its intent through a BW request ranging operation during a specific time interval. The message sent is a PN code with 144 bits that is formed by a BS-specific ID and the SS ID. Multiple SSs can access the same ranging opportunity and be distinguished by their unique PN codes.

Several connections can be simultaneously established between a BS and an SS and they are identified by a Connection Identifier (CID).

The following is a list of network operation phases:

- *Broadcast Message*: the BS periodically broadcasts the following messages: DL Channel Descriptor (DCD), UL Channel Descriptor (UCD), DL Medium Access Protocol (DL-MAP), UL Medium Access Protocol (UL-MAP).
- *Initial Ranging*: the SS sends a ranging code request at a specific interval. The BS acknowledges with timing and power adjustment. Ranging is repeated periodically according to UL-MAP scheduling.
- *Basic Capability Exchange*: the SS sends its basic capabilities and BS confirms its receipt.

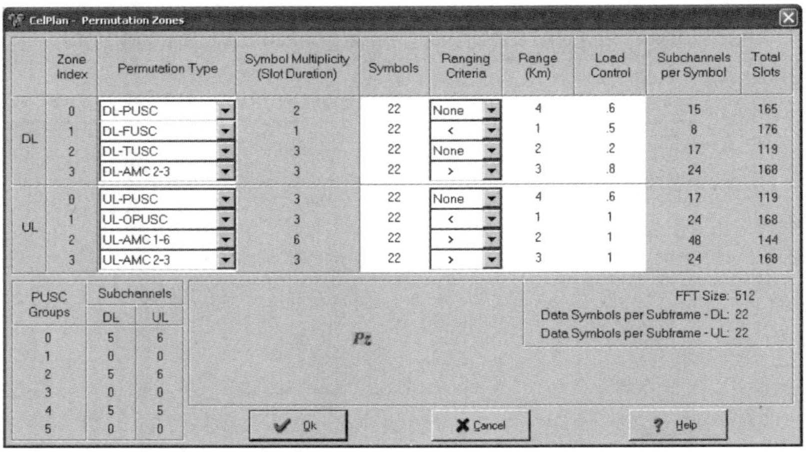

Figure 13.28 Configuration of zones within DL and UL subframes.

- *Registration message*: the SS registers with the BS.
- *Service Flow set-up for user data*: After registration the BS informs SS of service flow availability to the AAA (Authentication, Authorization and Accounting) server.
- *Authentication and key generation*: the SS receives pre-provisioned service flows to connect directly to the AAA server. The encryption key generation procedure follows.
- *Transfer data in the DL direction*: the BS informs in the DL map that a data burst is being sent to a specific SS in the frame, its location and CID identifier. The SS reads the DL map and fetches the data.
- *Transfer data in the UL direction*: the SS makes a BW request through a ranging access at the specified frame interval. The BS sends the UL map specifying the location in the frame where the UL data burst should be transmitted and later fetches this UL data burst.
- *Paging*: is done to locate an SS in idle mode that has already registered in the paging zone. Paging can be only done at specific intervals.
- *Handover*: the BS allocates scan intervals for the SS. During these intervals the SS does a cell scanning operation to locate handover candidates.

13.7 WiMAX Interference Reduction Techniques

The use of large bandwidth increases spectral efficiency in terms of data throughput. The previous statement is correct for WiMAX systems only when considering an isolated cell; in a network with multiple cells, however, the large bandwidth of a WIMAX carrier reduces the total number of carriers available (due to limited spectrum) and interference becomes a major issue, affecting data throughput. The following paragraphs describe the two main techniques for reducing this interference in WiMAX networks.

13.7.1 Interference Avoidance and Segmentation

In the interference avoidance technique, resources are split between users, so they do not interfere with each other. This can be achieved by segmenting a carrier, a concept known as segmentation.

WiMAX standard IEEE Std. 802.16e identifies up to three segments for each carrier. This leads to a practical increase in resources, because the carrier is now multiplied by three; however, it also implies a threefold reduction of throughput capacity (not considering the reduction by interference). Segmentation also improves carrier adjacency interference. To aid receivers to tune to a specific subset of subcarriers, additional pilots are added within each segment further reducing throughput.

Interference avoidance should be planned by using optimization tools considering the segments as a resource multiplier and accounting for carrier adjacency interference reduction.

13.7.2 Interference Averaging and Permutation Schemes

Interference is not distributed evenly along the network or along the subcarriers. When a fixed allocation of subcarriers is used, some connections suffer severe interference while others suffer very little. This is even more noticeable at light traffic loads.

In interference averaging, there is a pseudo-random use of resources, so the interferers and the interfered rotate and the overall interference is averaged between all users. With less interference, error correction codes (e.g. FEC) are more effective and more capable of restoring system capacity.

Interference averaging should be used with care and be assessed by strong design tools that can verify network performance beforehand, because a network can go from an acceptable to disastrous performance just by a slight increase in interference.

Interference averaging can only be properly evaluated through dynamic traffic simulation, when the varying network interference is taken into consideration.

The pseudo-random distribution of subcarriers, or permutation schemes, can be achieved in many ways and, to give users more flexibility, several subcarrier permutation schemes were included in the standards and are known as Pilot and Data Allocation Schemes (PDAS).

A permutation scheme defines the set of pilots and data subcarriers that carry data and how they are mapped into sub-channels. The pilot content is known and is used to estimate the RF channel; this knowledge is then used to extract the data.

Permutation schemes vary with pilot to data subcarriers ratio and the method of pseudorandom distribution of subcarriers. Table 13.6 lists the different permutation schemes and their corresponding pilot to data ratio. Each of these schemes is described in Section 13.7.3.

Permutation schemes are used to introduce frequency diversity in the WiMAX network by allocating subcarriers to sub-channels. Two allocation techniques are defined in the standard: Distributed Subcarrier Allocation (DSCA) and Adjacent Subcarrier Allocation (ASCA).

In DSCA, the subcarriers of a sub-channel are randomly spread throughout the channel, hence maximizing the frequency diversity effect. This approach reduces the chance of using the same subcarrier

Table 13.6 Pilot to data ratio of different permutation schemes

Subcarrier ratio	Pilot/Data
FUSC	0.108
OFUSC	0.125
PUSC DL	0.167
PUSC UL	0.500
OPUSC UL	0.125
AMC	0.125

in neighbor sectors; however, the fact that the subcarriers are distributed in the whole bandwidth makes channel estimation harder to achieve. This is the choice technique for situations where the RF channel characteristics change quickly, such as in a mobile environment.

In ASCA, on the other hand, the subcarriers are adjacent to each other within a sub-channel. This technique looks for the best subcarriers available (higher SNIR) targeting increased performance. The use of adjacent subcarriers makes channel estimation easier and they are a perfect fit for Adaptive Antenna Systems (AAS); it does, however, require fairly stable environment conditions, hence it has mostly been used for stationary or low-speed applications (e.g. nomadic).

The following section describes each of the permutation schemes defined in the standard.

13.7.3 Permutation Schemes

As described in the previous section, Pilot and Data Allocation Schemes make use of permutation to reduce network interference. These schemes are combined with different sub-channelization and segmentation techniques to address different situations. All of these combinations, however, provide sub-channels with 48 subcarrier data symbols; the variations are the depth in symbols, the permutation type and the number of pilots. Often, when the term permutation scheme is used, it implies this combination of pilot allocation, data permutation type, sub-channelization and segmentation.

The following is a brief explanation of the permutation schemes included in the standard:

- *FUSC (Full Usage of Sub-channels)*: formulated for the downlink for full allocation of sub-channels, this scheme uses DSCA but does not address the much smaller requirements of the uplink.
- *OFUSC (Optional Full Usage of Sub-channels)*: variation of FUSC, in which pilot subcarriers are evenly spaced by eight data subcarriers.
- *PUSC-DL (Partial Usage of Sub-channels)*: formulated for partial usage of sub-channels, allowing adjacent sectors to use different subcarriers and, consequently, avoiding interference. This scheme uses DSCA and addresses the uplink lower throughput requirement. Because this segmentation of sub-channels is not obligatory in this scheme, PUSC can fully replace FUSC. This scheme, however, requires a large number of pilots for channel equalization leading to a reduction in throughput.
- *PUSC-UL*: developed to support the small data units expected in the uplink direction. This scheme uses DSCA.
- *OPUSC (Optional Partial Usage of Sub-channels)*: alternative to PUSC-UL, also uses DSCA but requires fewer pilots and, consequently, increases throughput, though at a performance cost.
- *TUSC (Tiled Usage of Sub-channels)*: also uses DSCA and was developed to pair with PUSC-UL in the downlink, because FUSC and PUSC-DL do not pair with PUSC-UL.
- *OTUSC (Optional Tiled Usage of Sub-channels)*: also uses DSCA and is the downlink scheme created to pair with uplink's OPUSC.
- *AMC (Adjacent Mapping of Carriers)*: scheme for both the downlink and uplink using ASCA; this scheme is used to provide a contiguous set of subcarriers, which is required by AAS systems.

The mobile WiMAX standard defines the following schemes as mandatory:

- PUSC, FUSC, and AMC for the downlink.
- PUSC and AMC for the uplink.

The general sequence presented in Table 13.7 is followed to perform sub-channelization for each sub-channelization scheme.

Table 13.7 Sub-channelization sequence

Permutation scheme	Sequence
FUSC	pilot allocation subcarrier permutation subcarrier mapping to sub-channel
PUSC-DL	grouping of subcarriers in clusters of 14 subcarriers cluster permutation cluster mapping to segments and pilot allocation data subcarrier permutation to sub-channels within a segment
PUSC-UL	subcarrier permutation grouping of subcarriers in tiles of 9 subcarriers and pilot allocation tile mapping to slots mapping of data subcarriers to sub-channels
AMC	grouping of subcarriers in bins of 9 subcarriers and pilot allocation mapping bins to sub-channels

13.7.3.1 FUSC

FUSC (Full Usage of Sub-channels) permutation is optional and can only be used in the downlink after the first zone. In this permutation, the first step is to logically remove guard and DC subcarriers (null subcarriers). FUSC uses two types of pilots: constant pilots that keep a fixed position along all zones and variable pilots, which change their position from one symbol to the next. The latter moves 6 subcarriers positions from one symbol to the next. After the null subcarriers are removed, a set of constant pilots is mapped. Next, variable pilots per zone are mapped according to the number of symbols in the zone.

Subcarriers are then renumbered (permutated) using an algorithm based on the PermBase variable assigned to the cell. The PermBase is an integer (ranging from 0 to 31) in the permutation formula that guarantees that two sectors using the same frequency band, but with different PermBase values, will be using a different set of subcarriers.

Finally, the remaining subcarriers are mapped into sub-channels. Each sub-channel is assigned 48 subcarriers. The duration of a sub-channel is one symbol.

Figure 13.29 illustrates the above procedure for a 5 MHz channel with 512 subcarriers. Subcarriers (frequency axis) are illustrated in the vertical and symbols (time axis) on the horizontal.

The left part of Figure 13.29 indicates the logical transformations that subcarriers undergo to form the sub-channels. The objective of these transformations is to provide carrier permutation to average interference. The right part of Figure 13.29 shows the sub-frame and data allocation procedure, which consists of the allocation of a data burst to a data unit composed of several sub-channels. The concept of preamble and the data allocation sequence, presented in Figure 13.29, is described in Section 13.5.4.

Table 13.8 summarizes the main FUSC permutation characteristics for different bandwidths.

13.7.3.2 OFUSC

OFUSC is a variation of FUSC in which some null subcarriers are traded for pilot subcarriers to improve equalization at a loss in adjacent channel interference. In OFUSC, there are no variable pilots, instead, pilot subcarriers are evenly spaced by eight data subcarriers. Table 13.9 summarizes the main OFUSC permutation characteristics for different bandwidths.

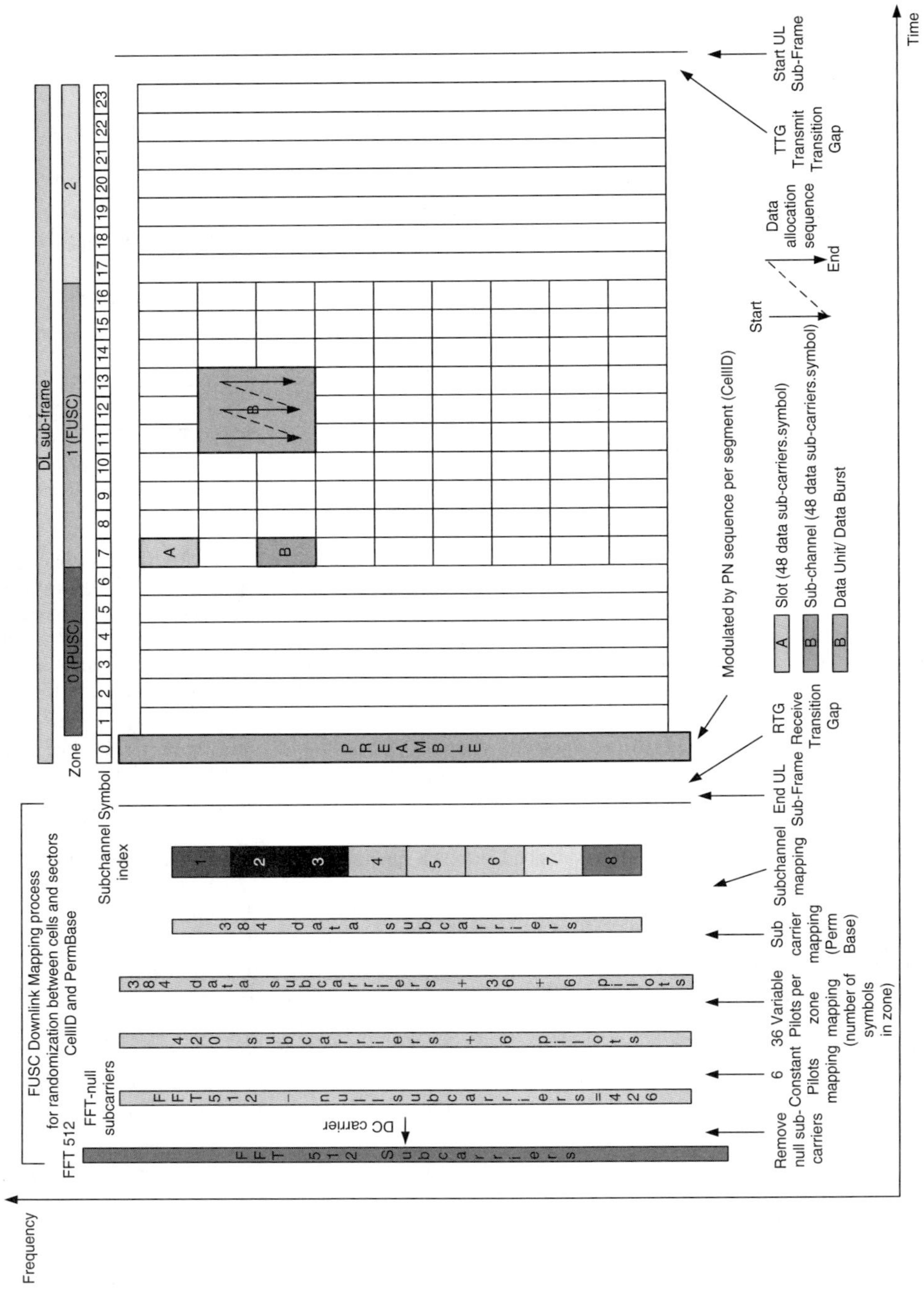

Figure 13.29 Description of FUSC permutation scheme for a 5 MHz carrier.

Table 13.8 Main characteristics of FUSC permutation

	DL-FUSC			
Bandwidth (MHz)	1.25	5	10	20
Total subcarriers	128	512	1024	2048
Lower frequency guard subcarriers (left)	11	43	87	173
Higher frequency guard subcarriers (right)	10	42	86	172
Total guard subcarriers	21	85	173	345
DC subcarriers	1	1	1	1
Number of used subcarriers	106	426	850	1702
Number of variable pilots of set 0	3	18	35	71
Number of variable pilots of set 1	3	18	35	71
Number of constant pilots of set 0	2	3	6	12
Number of constant pilots of set 1	2	3	6	12
Total number of pilot subcarriers	10	42	82	166
Number of data subcarriers	96	384	768	1536
Number of symbols per sub-channel	1	1	1	1
Number of data subcarriers symbols per sub-channel	48	48	48	48
Number of sub-channels	2	8	16	32
Max DL-FUSC symbols per frame	2688	10,752	21,504	43,008
Max DL-FUSC Meg symbols per second	0.54	2.15	4.30	8.60
Subcarrier permutation	PermBase	PermBase	PermBase	PermBase

Table 13.9 Main characteristics of OFUSC permutation

	DL-OFUSC			
Bandwidth (MHz)	1.25	5	10	20
Total subcarriers	128	512	1024	2048
Lower frequency guard subcarriers (left)	10	40	80	160
Higher frequency guard subcarriers (right)	9	39	79	159
Total guard subcarriers	19	79	159	319
DC subcarriers	1	1	1	1
Number of used subcarriers	109	433	865	1729
Total number of pilot subcarriers	12	48	96	192
Number of data subcarriers	96	384	768	1536
Number of symbols per sub-channel	1	1	1	1
Number of data subcarriers symbols per sub-channel	48	48	48	48
Number of sub-channels	2	8	16	32
Max DL-OFUSC symbols per frame	2688	10,752	21,504	49,152
Max DL-OFUSC Mega symbols per second	0.54	2.15	4.30	9.83
Subcarrier permutation	PermBase	PermBase	PermBase	PermBase

13.7.3.3 PUSC-DL

PUSC is always used in the first zone in the downlink direction. This is the zone where the frame mapping information resides. Section 13.5.4 describes frame mapping in detail. This permutation scheme can be used also in the next zones, if they exist.

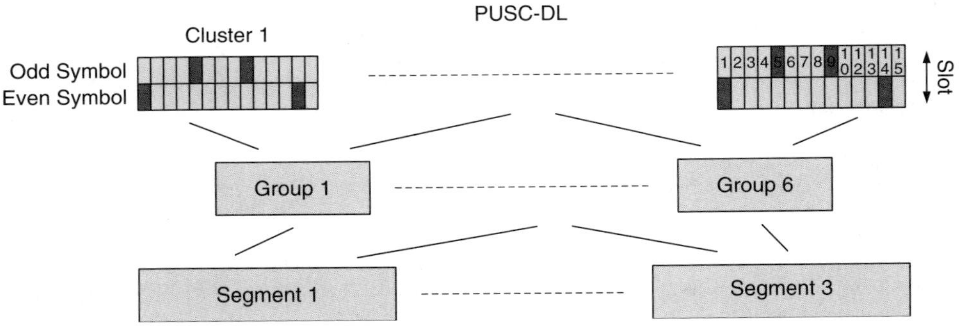

Figure 13.30 Pilot allocation in PUSC-DL.

In PUSC-DL, the first step is to logically remove null subcarriers (guard and DC). The remaining subcarriers are grouped into clusters with 14 sequential subcarriers each. In every cluster, two subcarriers are designated as pilots (odd symbols use the 5th and 9th sub carriers, while even symbols use the 1st and 13th, as illustrated in Figure 13.30).

Clusters are re-numbered (permutated) according to a pre-defined sequence. A fixed mapping is used on the first zone and a PermBase mapping in the next zones.

A pair of clusters over two symbols forms a slot. Slots are allocated to six groups according to pre-defined numbers, so groups may have different sizes. Data subcarriers are remapped within each group following an IDCell-based mapping for the first zone or a PermBase mapping for the next zones.

The IDCell is an integer that identifies the cell during messaging so one cell does not get data destined for another. There are 32 IDCell codes (ranging from 0 to 31) that can be attributed to sectored cells. Omni cells have 18 IDCell codes.

The use of IDCell as permutation criteria in the first zone is required because the receiver does not know which PermBase is being used until it reads the mapping information of the frame. Because the use of IDCell may result in conflicts between same cell sectors, the mapping information is allocated per segment, that is, it applies an interference avoidance technique to solve the issue (see Section 13.7).

After remapping of the subcarriers within the groups, each sub-channel is then assigned two slots from a group. Considering that each cluster within the slots has 2 pilots, and that a slot has two clusters, that gives a total of 48 data symbols per sub-channel (2 slots), as illustrated in Figure 13.31.

Groups are then assigned to 1 to 6 segments. During the design, these segments are assigned to cells or sectors. Having different group sizes gives the flexibility to provide higher capacity to some sectors in relation to others. Assigning different segments to neighbor sectors eliminates interference between them, but reduces sector throughput.

Figure 13.31 illustrates a 5 MHz PUSC-DL carrier with an FFT of 512. Subcarriers (frequency axis) are illustrated in the vertical and symbols (time axis) on the horizontal. The main characteristics of the permutation are presented in Table 13.10. The left part of Figure 13.31 indicates the logical transformations that the subcarriers undergo to form sub-channels. The objective of these transformations is to provide carrier permutation to average interference. The right part of Figure 13.31 indicates the sub-frame and the data allocation procedure. The data allocation procedure allocates a data burst to a data unit, which, in its turn, is composed of several sub-channels.

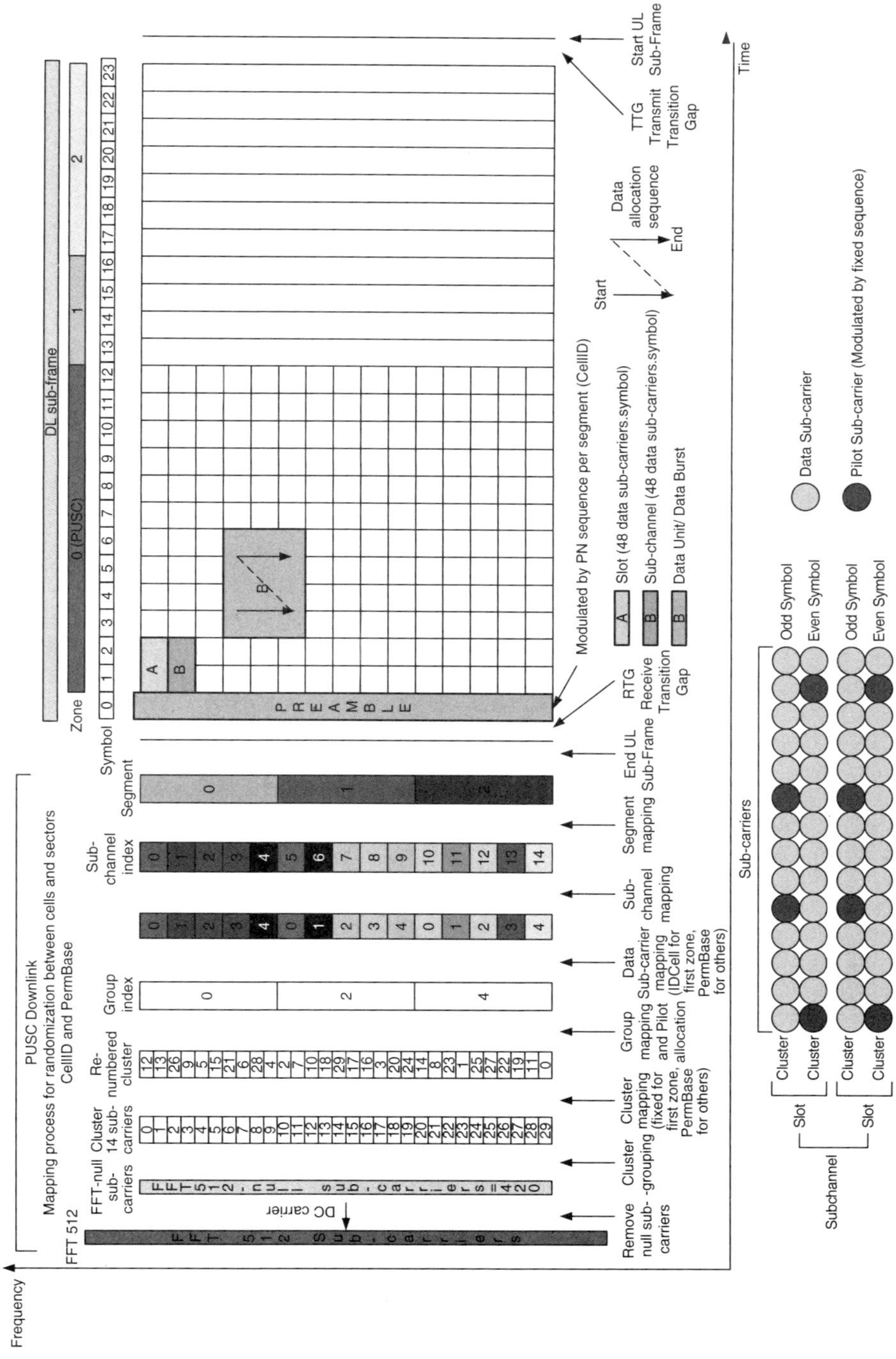

Figure 13.31 Description of PUSC-DL permutation scheme for a 5 MHz carrier.

Table 13.10 Main characteristics of PUSC-DL permutation

	DL-PUSC			
Bandwidth (MHz)	1.25	5	10	20
Total subcarriers	128	512	1024	2048
Lower frequency guard subcarriers (left)	22	46	92	184
Higher frequency guard subcarriers (right)	21	45	91	183
Total guard subcarriers	43	91	183	367
DC subcarriers	1	1	1	1
Number of used subcarriers	84	420	840	1680
Number of subcarriers per cluster per symbol	14	14	14	14
Number of pilots per cluster per symbol	2	2	2	2
Number of data subcarriers per cluster per symbol	12	12	12	12
Number of clusters per symbol	6	30	60	120
Number of symbols per sub-channel	2	2	2	2
Number of data subcarriers symbols per sub-channel	48	48	48	48
Number of clusters per sub channel	4	4	4	4
Number of sub-channels	3	15	30	60
Max DL-PUSC symbols per frame	2016	10,080	20,160	40,320
Max DL-PUSC Mega symbols per second	0.40	2.02	4.03	8.06
PUSC utilization factor	1/3	1/3	1/3	1/3
Max DL-PUSC Mega symbols per second	0.13	0.67	1.34	2.69
Subcarrier permutation (IDCell in the first zone and Perm Base in others)	IDCell PermBase	IDCell PermBase	IDCell PermBase	IDCell PermBase

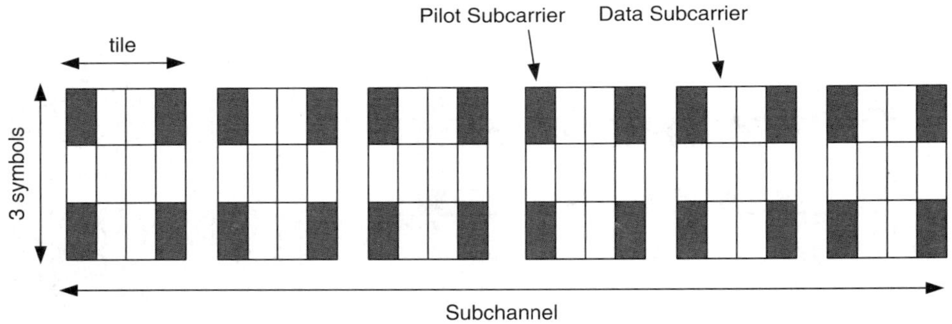

Figure 13.32 Pilot allocation in PUSC-UL.

13.7.3.4 PUSC-UL

In PUSC-UL, the first step is to remove logically null subcarriers; the remaining subcarriers are then mapped into tiles using PermBase-based permutation. Every 4 contiguous data subcarriers form a tile over 3 symbols. Pilots are allocated in each tile as shown in Figure 13.32. In each tile, one out of every three subcarriers is a pilot.

Tiles are mapped into slots, with 6 tiles per slot. The slots define the sub-channel index. The tile mapping to sub-channels is done using PermBase. Sub-channels are also formed by 6 tiles chosen according to a pre-defined permutation. Each tile has 4 pilots, so each sub-channel has 24 pilots and 48 data subcarriers. The high number of pilots provides a strong basis for data recovery.

Figure 13.33 illustrates a 5 MHz PUSC-UL carrier with an FFT of 512. Subcarriers (frequency axis) are illustrated in the vertical and symbols (time axis) on the horizontal. The main characteristics of the permutation are presented in Table 13.11.

The left part of Figure 13.33 indicates the logical transformations that the subcarriers undergo to form sub-channels. The objective of these transformations is to provide carrier permutation to average interference. The right part of Figure 13.33 indicates the sub-frame and the data allocation procedure. The data allocation procedure allocates a data burst to a data unit, which, in its turn, is composed of several sub-channels.

13.7.3.5 OPUSC-UL

Very similar in concept to PUSC-UL, this scheme presents a few variations. Only three subcarriers are grouped together to form a tile. The scheme is, however, able to provide a larger throughput due to the use of fewer pilots. This reduction in the number of pilots makes channel detection more difficult, hence this scheme is appropriate only for "well-behaved" channels. Table 13.12 summarizes the main characteristics of the permutation and Figure 13.34 illustrates the pilot allocation method.

13.7.3.6 TUSC

This scheme is used in the downstream to work with PUSC-UL. The scheme has the same structure as the upstream PUSC, thus the pilot information received can be used to beamform the return signal.

13.7.3.7 OTUSC

OTUSC is a scheme similar to TUSC, but pairing with OPUSC-UL.

13.7.3.8 AMC (Adjacent Mapping of Sub-Carriers)

In AMC, the first step is to remove logically null (guard) sub-carriers. Bins are then formed by creating groups of 9 contiguous subcarriers, the middle one being the pilot. Six bins are then joined to form a slot. This can be done in four different ways: 1 bin over 6 symbols, 2 bins over 3 symbols, 3 bins over 2 symbols or 6 bins over 1 symbol. Each of these ways is commonly represented as NxM, where N is the number of bins and M is the number of symbols, for example, AMC 2×3 stands for AMC with 2 bins over 3 symbols. The bins are mapped to sub-channels using the IDCell as the permutation base. Regardless of the NxM combination used, the total number of data subcarriers per sub-channel is always 48. This is illustrated in Figure 13.35.

AMC can be used both in the downstream and upstream. Its use of contiguous subcarriers makes this the only permutation scheme available for use with Advanced Antenna Systems (AAS). Only the downstream is illustrated here, as the upstream is similar.

Figure 13.36 illustrates the mapping procedures and pilot distribution for a 5 MHz carrier with an FFT of 512 using AMC 2×3. Subcarriers (frequency axis) are illustrated in the vertical and symbols (time axis) on the horizontal.

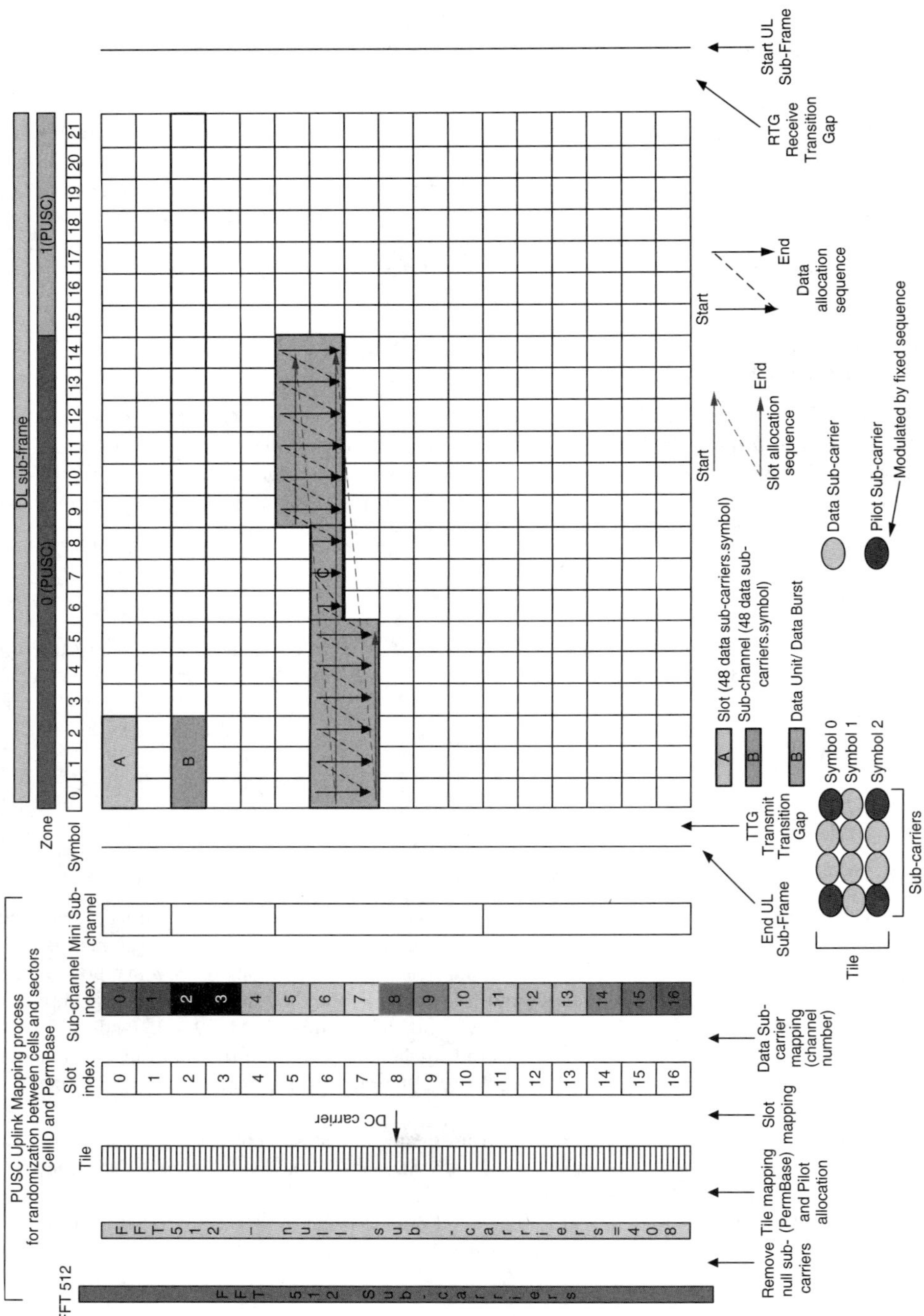

Figure 13.33 Description of PUSC-UL permutation scheme for a 5 MHz carrier.

Table 13.11 Main characteristics of PUSC-UL permutation

	PUSC-UL			
Bandwidth (MHz)	1.25	5	10	20
Total subcarriers	128	512	1024	2048
Lower frequency guard subcarriers (left)	16	52	92	184
Higher frequency guard subcarriers (right)	15	51	91	183
Total guard subcarriers	31	103	183	367
DC subcarriers	1	1	1	1
Number of used subcarriers	96	408	840	1680
Number of symbols per sub-channel	3	3	3	3
Number of subcarriers per tile	4	4	4	4
Number of pilots per tile	4	4	4	4
Number of data subcarriers per tile	8	8	8	8
Number of tiles	24	102	210	420
Number of data subcarriers symbols per sub-channel	48	48	48	48
Number of tiles per sub channel	6	6	6	6
Number of sub-channels	4	17	35	70
Max UL-PUSC symbols per frame	1024	4352	8960	17,920
Max UL-PUSC Mega symbols per second	0.20	0.87	1.79	3.58
PUSC utilization factor	1/3	1/3	1/3	1/3
Max UL-PUSC Mega symbols per second	0.07	0.29	0.60	1.19
Subcarrier permutation	yes	yes	yes	yes

Table 13.12 Main characteristics of OPUSC-UL permutation

	OPUSC-UL			
Bandwidth (MHz)	1.25	5	10	20
Total subcarriers	128	512	1024	2048
Lower frequency guard subcarriers (left)	10	40	80	160
Higher frequency guard subcarriers (right)	9	39	79	159
Total guard subcarriers	19	79	159	319
DC subcarriers	1	1	1	1
Number of used subcarriers	108	432	864	1728
Number of symbols per sub-channel	3	3	3	3
Number of subcarriers per tile	3	3	3	3
Number of pilots per tile	1	1	1	1
Number of data subcarriers per tile	8	8	8	8
Number of tiles	36	144	288	576
Number of data subcarriers symbols per sub-channel	48	48	48	48
Number of tiles per sub channel	6	6	6	6
Number of sub-channels	6	24	48	96
Max UL-OPUSC data symbols per frame	1536	6144	12,288	24,576
Max UL-OPUSC data Mega symbols per second	0.31	1.23	2.46	4.92
PUSC utilization factor	1/3	1/3	1/3	1/3
Max UL-OPUSC data Mega symbols per second	0.10	0.41	0.82	1.64
Subcarrier permutation	yes	yes	yes	yes

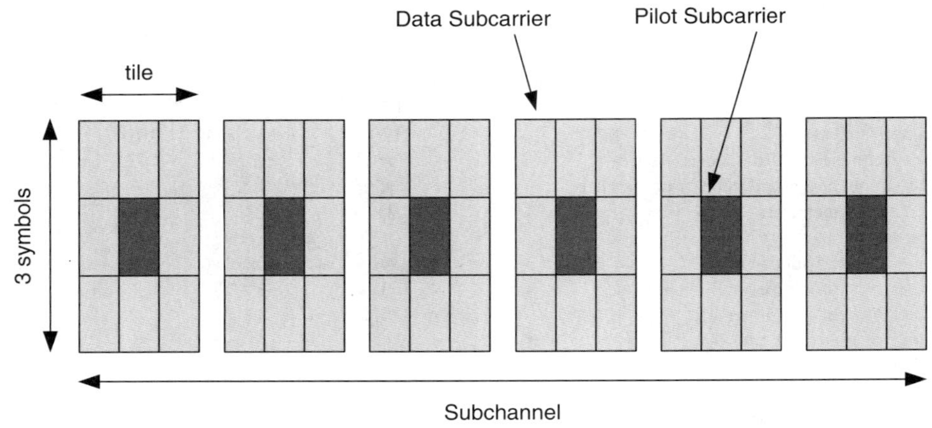

Figure 13.34 Pilot allocation in OPUSC-UL.

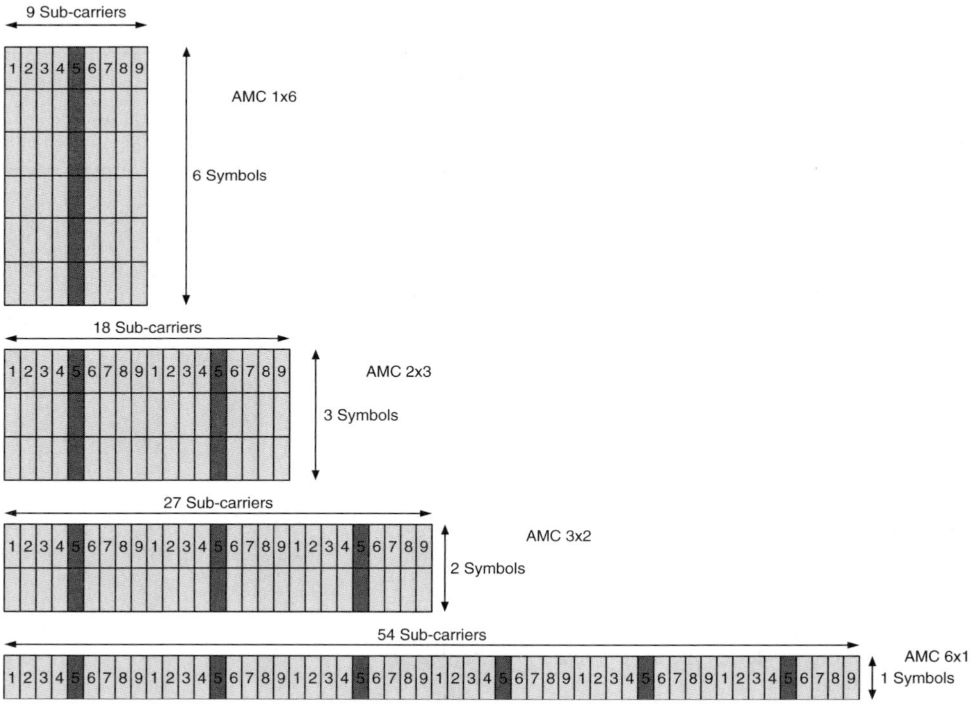

Figure 13.35 Pilot allocation in AMC permutation.

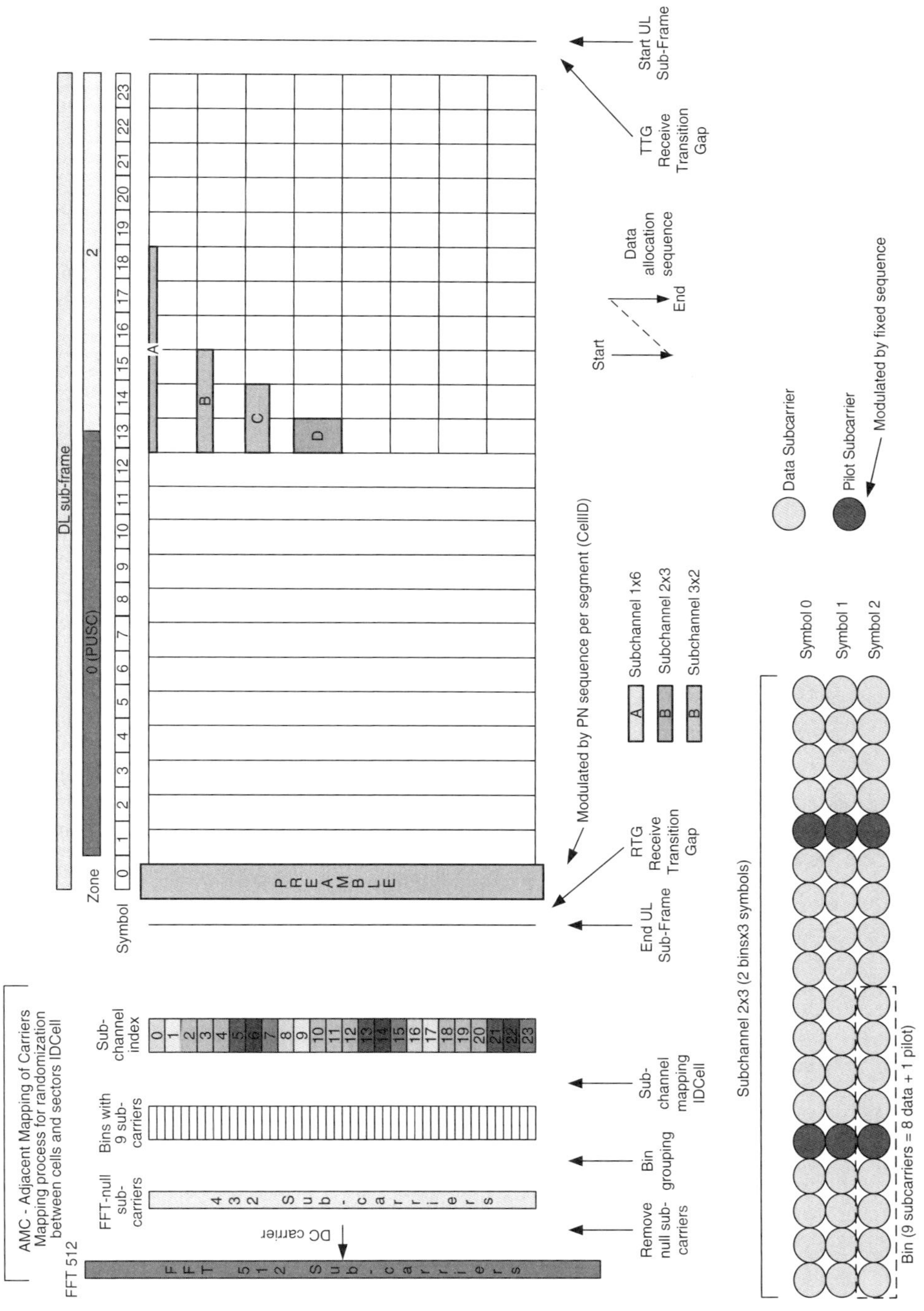

Figure 13.36 Description of AMC 2 × 3 permutation scheme for a 5 MHz carrier.

Table 13.13 Main characteristics of AMC permutation

	DL/UL AMC			
Bandwidth (MHz)	1.25	5	10	20
Lower frequency guard subcarriers (left)	10	40	80	160
Higher frequency guard subcarriers (right)	9	39	79	159
Total guard subcarriers	19	79	159	319
DC subcarriers	1	1	1	1
Number of used subcarriers	108	432	864	1728
Number of data subcarriers per bin	8	8	8	8
Number of pilots per bin	1	1	1	1
Number of bins per slot	6	6	6	6
Number of symbols per slot	1	1	1	1
Number of symbols per sub-channel	1	1	1	1
Number of data subcarriers symbols per sub-channel	48	48	48	48
Number of sub-channels	2	8	16	32
Max AMC DL data symbols per frame	2688	10,752	21,504	43,008
Max AMC DL Mega data symbols per second	0.54	2.15	4.30	8.60
Max AMC UL data symbols per frame	1536	6144	12,288	24,576
Max AMC UL data Mega symbols per second	0.31	1.23	2.46	4.92
Subcarrier permutation	no	no	no	no

The left part of Figure 13.36 shows the logical transformations that the subcarriers undergo to form sub-channels, whereas the right part shows the sub-frame and data allocation procedure. Table 13.13 summarizes the main characteristics of the AMC 2×3 permutations.

13.7.4 Permutation Summary

Tables 13.14 and 13.15 show the data rates obtained with each permutation/sub-channelization scheme. A higher rate does not necessarily represent the best solution, because it may be achieved due to a trade-off with other features that may imply requirement of a higher SNIR. The example used to create these tables considers a frame duration of 5 ms and $^1/_8$ cyclic prefix, with 28 symbols for the downlink and 16 symbols for the uplink (TDD ratio of 0.636).

Table 13.14 Data rate (symbols/frame) for different permutation schemes

	Summary symbol/frame			
Bandwidth (MHz)	1.25	5	10	20
DL-FUSC	2688	10,752	21,504	43,008
DL-OFUSC	2688	10,752	21,504	49,152
DL-PUSC	2016	10,080	20,160	40,320
DL- AMC	2688	10,752	21,504	43,008
UL-PUSC	1024	4352	8960	17,920
UL-OPUSC	1536	6144	12,288	24,576
UL-AMC	1536	6144	12,288	24,576

Table 13.15 Data rate (msymbols/second) for different permutation schemes

	Summary msymbols/second			
Bandwidth (MHz)	1.25	5	10	20
DL-FUSC	0.54	2.15	4.30	8.60
DL-OFUSC	0.54	2.15	4.30	9.83
DL-PUSC	0.40	2.02	4.03	8.06
DL-AMC	0.54	2.15	4.30	8.60
UL-PUSC	0.20	0.87	1.79	3.58
UL-OPUSC	0.31	1.23	2.46	4.92
UL-AMC	0.31	1.23	2.46	4.92

13.8 WiMAX Resource Planning

WiMAX resource planning has some peculiarities that differentiate it from other technologies.

13.8.1 WiMAX Frequency Planning

Several strategies can be used when performing a WiMAX frequency plan:

- Frequency channels can be reserved for point to point connections.
- Frequency channels can be reserved for the cell core coverage.
- Frequency channels can be partially used (segmented).
- Frequency channels can be allocated by zones.
- Frequency channels can be partially loaded when using interference averaging.

A combination of all these concepts can be applied together to a system, as illustrated in Figure 13.37. In this example, three carriers are originally available. Carrier number three is not shown in Figure 13.37 as it is reserved for point-to-point rooftop connections. Carrier two is segmented (a, b, c) and used for coverage close to the cell centers (zoning), whereas carrier one is used in the outskirts of each cell, also using segmentation to avoid interference between adjacent cells.

13.8.1.1 Frequency Reuse and Segmentation Scheme

For the frequency reuse and segmentation schemes (FRSS) discussed next, the following convention has been adopted: FRSS = (Nc, Ns, Nf, Ns), where:

- Nc is the number of cell sites per cluster, that is, number of BTSs needed to consume all available spectrum (frequency channels).
- Ns is the number of sectors per BTS.
- Nf is the number of frequencies available.
- Ns is the number of segments used for each frequency.

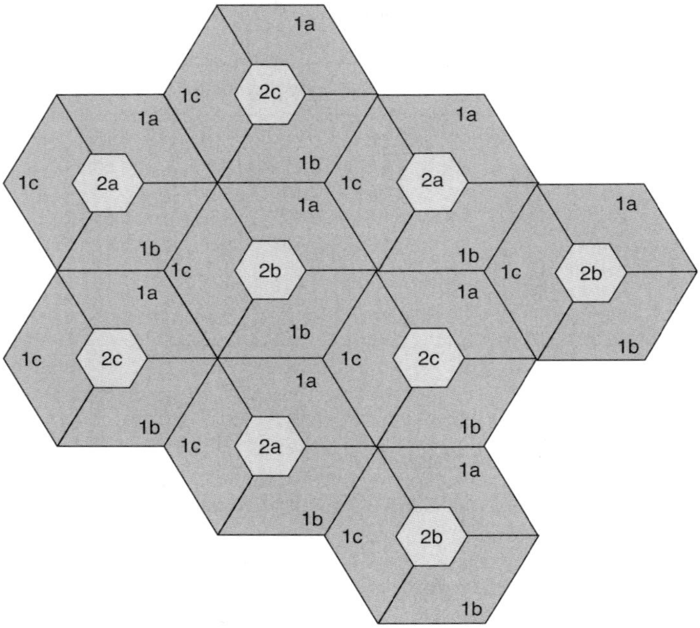

Figure 13.37 Multi-layer frequency plan with segmentation and zoning.

13.8.1.2 Scenario A: One Carrier Available

In this scenario, only one WiMAX frequency is available for the design. The following options could be considered for this implementation.

Option 1: Reuse (1, 3, 1, 1) with Interference Averaging

Considering the FRSS concept described previously (1, 3, 1, 1) indicates that each cluster is formed by only 1 cell, with three sectors. Only 1 frequency channel is available and segmentation is not used (1 segment). This configuration is illustrated in Figure 13.38.

Only under very light load circumstances it is possible for this option to provide service. Such a set-up can only be achieved by the sacrifice of user throughputs, by averaging interference proportional to the load fraction and through the use of different PermBase values for each cell. A resource management control mechanism could even be in place to guarantee that the load would NOT exceed a certain fraction, therefore assuring that this trade-off (throughput for performance) is respected. In such a system, however, under high load circumstances, very low SNIR values would make the system operation prohibitive.

Option 2: Reuse (1, 3, 1, 3) with Orthogonal Segmentation (same PermBase)

This scenario makes use of the segmentation technique, where a segment with one-third of the sub-channels is assigned to each sector. It can be implemented only through the use of PUSC permutation, using the resource of sub-channel groups to define each segment.

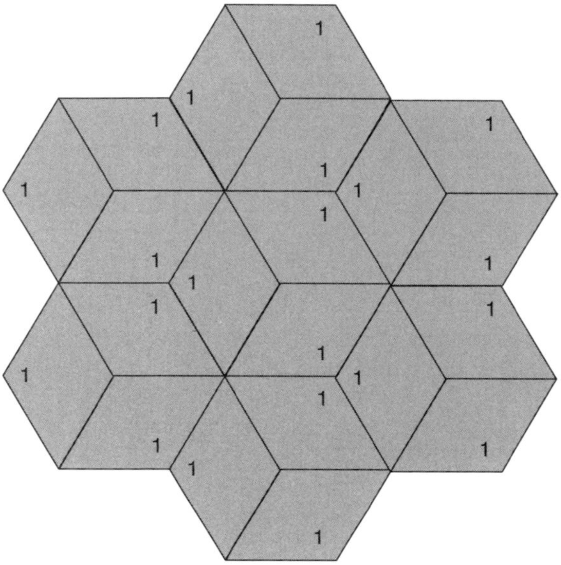

Figure 13.38 Reuse (1, 3, 1, 1).

The use of same PermBase values for all sectors (so segments are orthogonal) allows some sources of interference (e.g. adjacent sectors) to be reduced to zero, while other sources will have full conflict (e.g. all sectors C in all cells will be totally in synch with the same permutation sequences). This configuration is illustrated in Figure 13.39.

Option 3: Reuse (1, 3, 1, 3) with Non-Orthogonal Segmentation (Different Permbase Per Sector)

This scenario corresponds to the same segmentation technique described in option 2 (and same illustration, Figure 13.39), where a segment with $^1/_3$ of the sub-channels is assigned to each sector. However, in this scenario different PermBase values are assigned to each sector. This scheme can also be implemented only through the use of PUSC permutation, with the resource of sub-channel groups to define each segment.

By using different PermBase values, the interference from all sectors is averaged to $^1/_3$ (because the probability of conflict is reduced to $^1/_3$ of what it would be with synchronized sequences). This $^1/_3$ averaging, however, occurs even for the adjacent sectors, which, in some circumstances, may become very restrictive, because the worst geographically located interference sources are not avoided.

Under full load, one can demonstrate that the use of same PermBase values for all sectors (option 2) in a system is a better choice than that of different PermBase values per sector (option 3). However, for lightly or medium loaded circumstances, the optimum solution could be different. Only network simulations that rely on traffic distribution and load per sector allow such comparison and optimization.

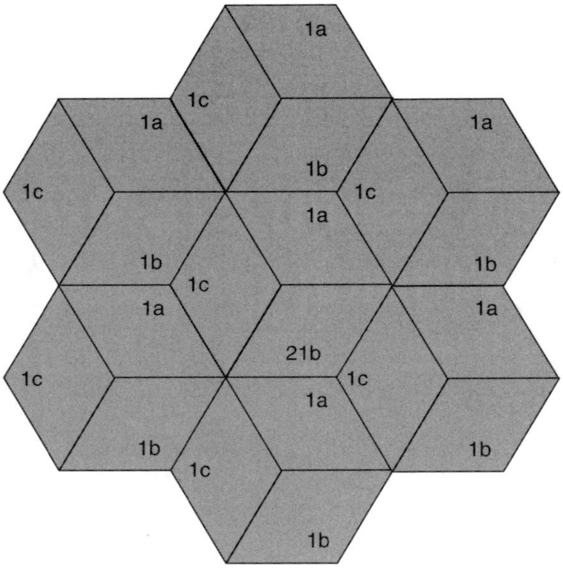

Figure 13.39 Reuse (1, 3 ,1, 3).

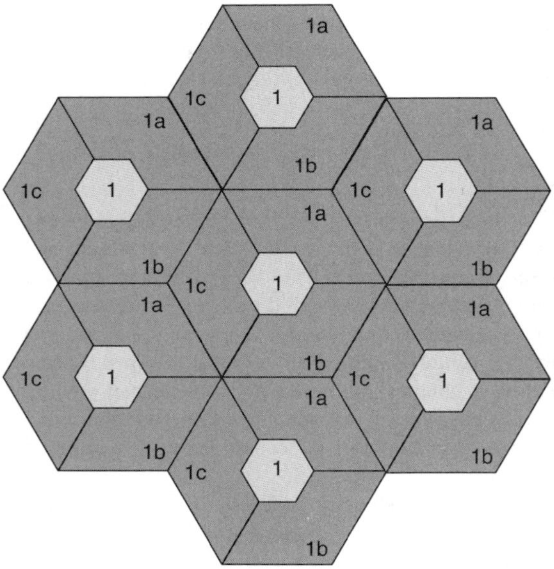

Figure 13.40 Fractional Frequency Reuse (FFR).

Option 4: Fractional Frequency Reuse

This option corresponds to a combination of scenarios (1, 3, 1, 1) and (1, 3, 1, 3), described previously; that is, subscribers in a region closer to the center of cell use FRSS = (1, 3, 1, 1), whereas closer to the edge of the cell, segmentation follows FRSS = (1, 3, 1, 3).

The zoning structure would probably be implemented with a PUSC or FUSC zone for the inner area at different PermBase values, and a segmented PUSC zone for the outer areas, with same PermBase value for all sectors. This configuration is illustrated in Figure 13.40.

13.8.1.3 Scenario B: Three Carriers Available

In this scenario, three WiMAX frequency channels are available for planning. The following options could be considered for this implementation.

Option 1: Reuse (1,3,3,1) with Interference Averaging

This option corresponds to the assignment of one full frequency channels to each sector, therefore using up the full spectrum within each three-sectored BTS. Each sector is configured with a different PermBase value.

This scenario is expected to give equivalent SNIR performance to option 3 discussed in Scenario A, with FRSS = (1, 3, 1, 3), however, it provides three times more throughput, as now each resource (frequency channel) corresponds to a full 10 MHz channel. This configuration is illustrated in Figure 13.41.

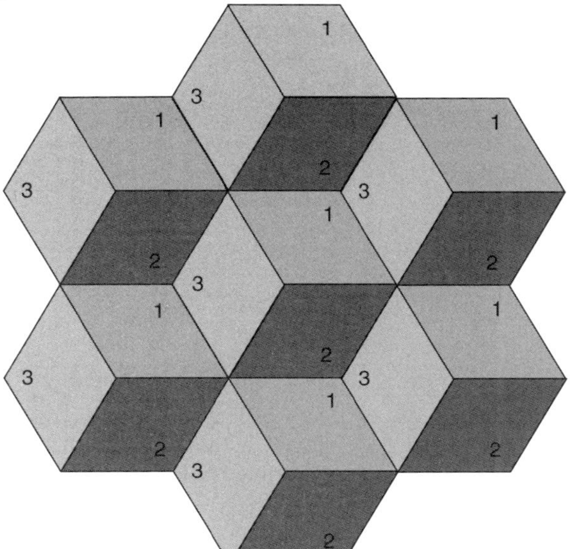

Figure 13.41 Reuse (1, 3 ,3, 1).

Option 2: Reuse (3, 3, 3, 3) with Orthogonal Segmentation (Same PermBase)

This option corresponds to assigning 1/3 of the full carrier (one segment) to each sector, therefore reusing the full spectrum in three three-sectored BTS clusters. All sectors are configured with the same PermBase value.

This scenario provides much higher SNIR performance than option 1 of Scenario B, but at the cost of reducing the channel bandwidth to one third. However, the better operating SNIR condition allows for the use of higher modulation schemes during link adaptation, which might end up compensating for, or even exceeding, performance in terms of final throughput. It is only possible to fully compare these two options by simulating the system at different load fractions. This configuration is illustrated in Figure 13.42.

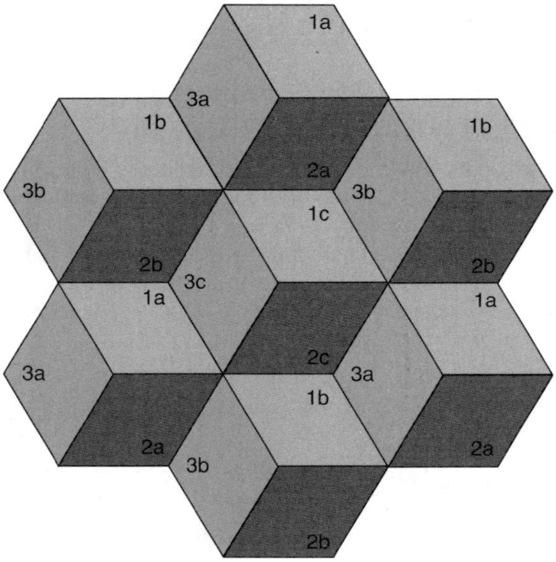

Figure 13.42 Reuse (3, 3, 3, 3).

13.8.2 WiMAX Code Planning (Cell Identification)

As in other technologies, WiMAX cells have to be identified during messaging so one cell does not get data destined for another. This is done through a variable called IDCell. There are 32 IDCell codes (numbered 0 to 31) that can be used with sectored cells and 18 codes for omni cells. IDCell is explained in detail in Section 13.5.4.1.

IDCell planning should distribute the IDCell index so same index cells have the minimum possible interaction (signal overlap).

13.8.2.1 Permutation Base Planning

Additional permutation, independent of the IDCell, may be required for other zones; this is achieved by using DL_PermBase and UL_PermBase variables.

DL_PermBase is a parameter used in the generation of the permutation sequence for both PUSC and FUSC schemes. It is an integer value that may vary from 0 to 31 for sectored cells and 0 to 17 for omni cells.

UL_PermBase is a parameter used in the generation of the permutation sequence for PUSC scheme in the uplink. It is an integer value that may vary from 0 to 69.

The DL_PermBase variable is equal to the IDCell in the first downlink zone.

13.8.2.2 When to Use Fixed or Variable DL_PermBase

To use interference avoidance (orthogonalization) among segments (i.e. different groups of sub-channels from the same carrier are assigned to different sectors), the same PermBase should be used in all sectors. Example of application. Consider a given WiMAX 10 MHz channel, using PUSC permutation, for which we assign sub-channel groups 0 and 1 (with 6 and 4 sub-channels, respectively) to all sectors A, groups 2 and 3 to all sectors B, and groups 4 and 5 to sectors C. This type of arrangement is sometimes referred to as segmentation and provides for an actual reuse of 1x3, where the unit being reused corresponds to a segment of $^1/_3$ of the full channel capacity. The only way to guarantee that these groups of sub-channels will not have coincident subcarriers at the same time is to assign the same PermBase to all of them.

However, when the idea is to average interference by generating random conflict probability, different PermBase parameters should be used for each sector. Example of application. Consider a given WiMAX 10 MHz channel, using PUSC permutation, where we assign all sub-channel groups from 0 to 5 to all sectors. This type of arrangement corresponds to a reuse of 1x1 (sometimes considered feasible in WiMAX systems). The only way for such systems to possibly work once one reaches cell edges would be in circumstances where the system traffic load is low (i.e. only a fraction of the sub-channels is active at a time), because this allows interference to be averaged down proportionally to this load fraction. The technique used to diminish the probability of conflict among different sectors is to assign different pseudo-random permutation sequences to each sector, that is configure different PermBase values.

13.8.3 Tips for PermBase Resource Planning

On the first PUSC zone of the downlink the DL_PermBase is not configurable, as it is mandatory by the standard for this variable to have the same value as the parameter IDCell (also used in the Preamble for cell identification).

On the remaining zones, the DL_PermBase is configurable by network management and may be set to a common value for all sectors in the system, or may vary per sector, depending on the desired application.

In the permutation formulas, the variable PermBase may generate the same permutation sequence. For instance, in a 1024 FFT system using FUSC (where 16 sub-channels are available per symbol), both PermBase = 0 and PermBase = 16 lead to the same sequence.

Therefore the plan should consider the minimum value between the number of sub-channels and the PermBase. The number of sub-channels for each permutation scheme is presented in Table 13.16.

13.8.4 Spectrum Efficiency

This chapter analyzed, for each domain (frequency, time and power), how the carrier resources were used and what percentage of the resources was allocated for data and to assure data integrity (overhead). We divided it into structural and coding overhead, shown in Table 13.17 and Table 13.18.

Table 13.16 Number of sub-channels per permutation scheme

Permutation name	128	512	1024	2048
DL-FUSC	2	8	16	32
DL-OFUSC	2	8	16	32
DL-PUSC	3	15	30	60
DL-TUSC	4	17	35	70
DL-TUSC2	6	24	48	96
DL-AMC-1-6	12	48	96	192
DL-AMC-2-3	6	24	48	96
DL-AMC-3-2	4	16	32	64
UL-PUSC	4	17	35	70
UL-OPUSC	6	24	48	96
UL-AMC-1-6	12	48	96	192
UL-AMC-2-3	6	24	48	96
UL-AMC-3-2	4	16	32	64

Table 13.17 Structural overheads

Carrier overhead	Percentage
Guard Bands	18
Pilot DL and UL	23
Cyclic Prefix	13
TDD partition	5
TDD gap	3
OFDMA preamble and mapping	10
Total structural overhead	72
Available for data	28

Table 13.18 Coding overheads

Data overhead	Minimum (%)	Maximum (%)
Coding	17	50
MAC overhead	3	5
HARQ	10	15
Total	30	70
Available for data	20	8

There is another overhead directly related to the data retrieval, called the coding overhead. Table 13.18 lists this overhead for the minimum and maximum coding overhead. As we can see, only 8–20% of the carrier is available for the actual data, which is more than many other technologies can claim. There is still plenty of room for new ideas to optimize spectrum usage, so we can rest assured that there will be new technology generations in the future.

14

Universal Mobile Telecommunication System – Long Term Evolution (UMTS-LTE)

14.1 Introduction

In Europe, the main standardization bodies involved in wireless specifications are the ITU (International Telecommunications Union), the ETSI (European Telecommunications Standardization Institute), and the 3GPP/3GPP2 (3rd Generation Partnership Project). The IMT2000 (International Mobile Telecommunications for beyond year 2000) is a set of standards created for mobile communications to meet ITU specifications.

The 3GPP is the GSM (Global System for Mobile communications) evolution standardization group within IMT2000. The 3GPP is a partnership of several partner members that are standardization organisms in different countries: ARIB (Association of Radio Industries and Businesses) from Japan, ATIS (Alliance for Telecommunications Industry Solutions) from North America, CCSA (China Communications Standards Association), ETSI, TTA (Telecommunications Technology Association) from Korea and TTC (Telecommunications Technology Committee) from Japan.

3GPP2 is the standardization group for cdma2000 (code division multiple access 2000) within ITU's IMT2000 project and is a partnership of ARIB/TTC (Japan), CCSA (China), TIA (Telecommunications Industry Association) from North America, and TTA (Korea).

The first generation of wireless technologies was analog, being the main technologies TACS (Total Access Communication System) in Europe and AMPS (Advanced Mobile Phone Service) in the USA. Data transmission, at the time relied on the use of modems.

The second wireless generation was divided between two technologies: CDMA (Code Division Multiple Access) in the USA and GSM (Global System for Mobile communications) in Europe. CDMA uses spread spectrum technology and code division multiplexing. It was derived from military technology used in World War II. It was initially optimized for voice and, even though data transmission was possible, that was only at low speed.

LTE, WiMAX and WLAN Network Design, Optimization and Performance Analysis, First Edition.
Leonhard Korowajczuk.

GSM is a circuit switched TDM (Time Division Multiplex) designed for voice, and it had its phase 1 specifications ready in 1990 and was commercially launched in 1992. GSM phase 2 specifications were ready in 1995. Data transmission started with SMS (Short Message Service) using HSCSD (High-Speed Circuit Switched Data) over GSM voice circuits. The first standard to consider packet data was GPRS (General Packet Radio Service), improved later to EDGE (Enhanced Data rates for GSM Evolution), but both still relied on wireless channels designed for voice. The GSM (formerly Groupe Spécial Mobile) technology and its evolution were initially specified by CEPT (Conférence des administrations Europeans des Posts et Télécommunications), then by ETSI (European Telecommunications Standard Institute) and today the specifications are the responsibility of the 3GPP.

The first data transmissions were done over dial-up lines using HSCSD (High-Speed Circuit Switched Data), which claimed speeds of 57.6 kbps, even though 9.6 kbps was a more realistic number.

The first packet data solution was GPRS, deployed in 2001, claiming 57 to 114 kbps, with 14.4–28.8 kbps more realistic numbers. It was followed by EDGE in 2002 which claimed 384 kbps, being a 56.6 kbps a more realistic number.

It became clear by then that a new standard was required to provide channels appropriate for packet data. The third wireless generation tried to address this data demand.

Meanwhile, CDMA systems evolved to IS-2000 1 × RTT (Information System-2000 Single Carrier Radio Transmission Technology) and then to a data only solution called EVDO (EVolution Data Optimized).

The GSM branch decided to change to cdma (code division multiple access) technology for its third generation, using as radio technology WCDMA (Wideband Code Division Multiple Access) and the solution was adopted under the umbrella of UMTS (Universal Mobile Telecommunication System).

UMTS was initially defined for voice, but had the potential to add channels optimized for packet data. Data specific implementations were done with the release of HSDPA (High Speed Downlink Packet Access) for downlink data in Release 5, and HSUPA (High Speed Uplink Packet Data) for uplink data in Release 6. Both directions were consolidated in HSPA (High Speed Packet Data) and HSPA+ (HSPA evolution introduced in Release 7). These technologies claim speeds from 384 kbps to 2 Mbps, with more realistic numbers in the range of 56.6 kbps to 256 kbps.

HSPA (High Speed Packet Access) technology, although theoretically able to deliver high data rates, is limited in the distance in which data can be delivered, as symbol duration decreases with the data rate increase and becomes much smaller than the multipath distance, making data retrieval difficult. Another deficiency of this technology is that it is based on the original voice architecture instead of being an all IP technology.

The multipath issue was solved by the OFDM (Orthogonal Frequency Division Multiplex) technology. This was confirmed by the success achieved with WLAN (Wireless Local Area Network 802.11) and the subsequent introduction of WiMAX (Worldwide Interoperability for Microwave Access 802.16) technology. Both technologies were standardized by the IEEE (Institute of Electrical and Electronics Engineers) from North America.

The 3GPP concluded then that it would be left behind and started work on its fourth generation standard also using OFDM, their first technology was designed primarily for data, voice being carried as packet data. When this work started, WLAN had been in existence for nearly a decade and WiMAX was already being deployed.

The new technology was nicknamed LTE (Long Term Evolution) and, to differentiate it from WiMAX, a set of requirements was established to target some of the weak points of the other OFDM technology. The idea for LTE was also to keep some compatibility with legacy infrastructure.

The main objectives specified for the LTE technology were:

- Co-existence with legacy standards while evolving towards an all-IP network.
- Spectral bandwidth from 1.4 to 20 MHz.

- Spectral efficiency three to four times better than HSPA release 6 in the downlink (4 bit/s/Hz) and 2 to 3 times better than HSPA release 6 in the uplink (bit/s/Hz). Table 14.1 lists the goals for different scenarios of LTE and LTE-A (LTE-Advanced).
- Latency smaller than 5 ms for small packets.
- High mobility (up to 120 km/h).
- Cell range up to 5, 30 and 100 km.
- Downlink based on OFDMA (Orthogonal frequency Division Multiple Access).
- Uplink based on SC-OFDMA (Single-Carrier OFDMA).
- FDD (Frequency Division Duplex) and TDD (Time Division Duplex).
- Bandwidth scalability and multi-band operation.
- A new Evolved Packet Core (EPC).
- Simplified architecture and mobility management.
- Interworking with existing UTRAN (UMTS Terrestrial Radio Access Network) systems and other non 3GPP systems.

These objectives have been highly promoted by marketers in a competition with other technologies, such as WiMAX. Soon the claims were raised and Table 14.2 shows some of the more recent claims.

The announced claims can be achieved only in very special conditions and have little relation to the actual throughputs expected in a real network. It is left to the network designer to bring reality to the table and tell operators and investors that the actual throughput will be closer to one-tenth of the advertised value.

Many other claims, such as Reuse of 1 and SON (Self-Organizing Networks) are also being overstated and their limitations not explained. Other claims, such as cell radius, have been sacrificed for a larger throughput.

This kind of marketing has caused a lot of harm to the industry and it is time for the community to establish some guidelines for these excessive claims. Meanwhile, it is the network designer's task

Table 14.1 LTE spectral efficiency objectives

ITU spectral efficiency objectives (bit/s/Hz)				
	Downlink		Uplink	
ITU Scenario	LTE	LTE- A	LTE	LTE- A
Indoor Hotspot	3.0	5.0	2.3	4.0
Urban Micro	2.6	3.1	1.8	2.2
Urban Macro	2.2	2.8	1.4	1.8
Rural Macro	1.1	2.6	0.7	2.0

Table 14.2 LTE marketing claims

	Channel bandwidth (MHz)	Downlink throughput (Mbit/s)	Uplink throughput (Mbit/s)	Symbol duration (μs)	Maximum multipath delay spread (m)
HSDPA	5	14.4	0.384	0.26	781
HSUPA	5	14.4	5.75	0.26	781
HSPA+	5	26	11.5	0.26	781
LTE	20	300	75	66.67	5000

to explain to customers the technology limitations and many times these explanations are received with disbelief, as the customer tends to believe in the claims presented by vendors and the press. Sometimes the only way to make customers understand real life performance of the technology is to educate them, explaining the theory behind the different implementations of the technology. This was one of the motivations behind this book.

We illustrate the issue with an example: cars can do 40 km per liter of petrol and can do 300 km/hour, but not both at the same time. We can even say that a car can make 100 km per liter, without stating that the measurement was made downhill. The same applies to some of the latest wireless technology claims.

14.2 Standardization

3GPP standards are in a state of constant evolution and this is reflected in the specifications, which are classified by topic into series (numbered from 01 to 99). The specifications follow a general format 3GPP TS SS.NNN vR.XX.YY, where TS stands for Temporary Specification, SS defines the specification series, NNN the number of the specification inside the series, v stands for version, R for release, XX indicates a specification change and YY an editorial change.

The 3GPP aggregates documents versions into Releases (Rel) to consolidate evolution stages and provide developers with a consistent and stable platform for development. Releases have stages as defined below:

- Stage 1: The desired service is described from the user's point of view.
- Stage 2: A logical analysis is done, breaking the problem in functional elements and information flows.
- Stage 3: Documentation of the functionality and protocols.

Table 14.3 gives the 3GPP 3G Releases. 3GPP did not upgrade LTE to a 4G technology status, although, in the literature, we find references to LTE as 3.5G and 4G. LTE Advanced was recently accepted by 3GPP as a 4G technology. In this book, this preference is to consider LTE in general as a 4G technology, due to the significant differences from previous technologies. Actually the real differences between LTE and LTE Advanced are minimal.

The above standards do not refer to LTE, as this name is not recognized by 3GPP, although used by the industry. The term used is UMTS E-UTRA (Universal Mobile Telecommunications System Evolved Universal Terrestrial Radio Access).

Table 14.3 3GPP 3G standards evolution

3 GPP standard	Release date	Commercial introduction	Main technology introduced
Rel-99	Mar 00	2003	UMTS
Rel-4	Mar 01	none	
Rel-5	Oct 02	2006	HSDPA, IMS
Rel-6	Dec 04	2007	HSUPA
Rel-7	Dec 06	2009	HSPA+
Rel-8	Dec 08	2011 (forecast)	LTE, EPC, IMS+
Rel-9	Dec 09	2012 (forecast)	LTE TDD, MBMS
Rel-10	Mar 11	2013 (forecast)	LTE-A, SON
Rel-11	Dec 12	2015 (forecast)	TBD

This book is updated to Release 8/9 and describes the features planned for Releases 10 and 11. These releases, however, are still changing and new versions are being issued.

A summary of the specifications and features introduced in each release is given next.

14.2.1 Release 8 (December 2008)

- LTE Radio Access Modes
- Transmission Bandwidth
- Peak data rates

14.2.2 Release 9 (December 2009)

- SON (Self-Organizing Networks)
- MBMS (Multimedia Broadcast/Multicast Service)
- Multi RAT (Radio Access Technology) Internetworking and System Reselection
- Inter RAT Handover

14.2.3 Release 10 (March 2011)

Release 10 is known as LTE Advanced (LTE-A), it should be released in March 2011 and includes:

- Backward compatibility with LTE.
- Bandwidth extension (up to five 20 MHz carriers).
- Spectrum aggregation (aggregation of non-contiguous channels, even from different bands).
- MIMO enhancements.
  - Uplink spatial multiplexing (up to 4 layers).
  - Downlink spatial multiplexing (up to 8 layers).
  - Enhanced downlink multi-user MIMO.
  - Combined non-code-book-based beam-forming and spatial multiplexing.
- Relay techniques (coverage extension and data rate, multi-frequency repeater-like function).
  - Creates new cell and self-backhauls to donor cell.
- Coordinated multipoint reception and transmission.
  - Scheduling coordination.
  - Joint transmission and reception.
  - Reduces inter-cell interference and improves signal strength.

14.2.4 Release 11 (December 2012)

- Advanced IP Interconnection of Services
- System improvements to Machine Type Communications
- Network-provided location information for IMS

14.2.5 LTE 3GPP Standards

3GPP standards can be found at: http://www.3gpp.org/ftp/Specs/html-info/36-series.htm. Table 14.4 list the standards issued to this date. Each standard has several versions and all of them are available at the above site.

Table 14.4 3GPP LTE (EPS) Standards

TS 36.101	E-UTRA: User Equipment (UE) radio transmission and reception
TS 36.104	E-UTRA: Base Station (BS) radio transmission and reception
TS 36.106	E-UTRA: FDD repeater radio transmission and reception
TS 36.113	E-UTRA: Base Station (BS) and repeater Electro Magnetic Compatibility (EMC)
TS 36.124	E-UTRA: Electromagnetic compatibility (EMC) requirements for mobile terminals and ancillary equipment
TS 36.133	E-UTRA: Requirements for support of radio resource management
TS 36.141	E-UTRA: Base Station (BS) conformance testing
TS 36.143	E-UTRA: FDD repeater conformance testing
TS 36.171	E-UTRA: Requirements for Support of Assisted Global Navigation Satellite System (A-GNSS)
TS 36.201	E-UTRA: LTE physical layer; General description
TS 36.211	E-UTRA: Physical channels and modulation
TS 36.212	E-UTRA: Multiplexing and channel coding
TS 36.213	E-UTRA: Physical layer procedures
TS 36.214	E-UTRA: Physical layer; Measurements
TS 36.300	E-UTRA and E-UTRAN: Overall description; Stage 2
TS 36.302	E-UTRA: Services provided by the physical layer
TS 36.304	E-UTRA: User Equipment (UE) procedures in idle mode
TS 36.305	E-UTRA: Stage 2 functional specification of User Equipment (UE) positioning in E-UTRAN
TS 36.306	E-UTRA: User Equipment (UE) radio access capabilities
TS 36.307	E-UTRA: Requirements on User Equipments (UEs) supporting a release-independent frequency band
TS 36.314	E-UTRA: Layer 2 - Measurements
TS 36.321	E-UTRA: Medium Access Control (MAC) protocol specification
TS 36.322	E-UTRA: Radio Link Control (RLC) protocol specification
TS 36.323	E-UTRA: Packet Data Convergence Protocol (PDCP) specification
TS 36.331	E-UTRA: Radio Resource Control (RRC); Protocol specification
TS 36.355	E-UTRA: LTE Positioning Protocol (LPP)
TS 36.401	E-UTRAN: Architecture description
TS 36.410	E-UTRAN: S1 layer 1 general aspects and principles
TS 36.411	E-UTRAN: S1 layer 1
TS 36.412	E-UTRAN: S1 Signalling transport
TS 36.413	E-UTRAN: S1 Application Protocol (S1AP)
TS 36.414	E-UTRAN: S1 data transport
TS 36.420	E-UTRAN: X2 general aspects and principles
TS 36.421	E-UTRAN: X2 layer 1
TS 36.422	E-UTRAN: X2 Signalling transport
TS 36.423	E-UTRAN: X2 Application Protocol (X2AP)
TS 36.424	E-UTRAN: X2 data transport
TS 36.440	E-UTRAN: General aspects and principles for interfaces supporting Multimedia Broadcast Multicast Service (MBMS) within E-UTRAN
TS 36.441	E-UTRAN: Layer 1 for interfaces supporting Multimedia Broadcast Multicast Service (MBMS) within E-UTRAN
TS 36.442	E-UTRAN: Signalling Transport for interfaces supporting Multimedia Broadcast Multicast Service (MBMS) within E-UTRAN
TS 36.443	E-UTRAN: M2 Application Protocol (M2AP)
TS 36.444	E-UTRAN: M3 Application Protocol (M3AP)
TS 36.445	E-UTRAN: M1 data transport
TS 36.455	E-UTRA: LTE Positioning Protocol A (LPP a)

Table 14.4 (*continued*)

TS 36.508	E-UTRA and EPC: Common test environments for User Equipment (UE) conformance testing
TS 36.509	E-UTRA and EPC: Special conformance testing functions for User Equipment (UE)
TS 36.521-1	E-UTRA: User Equipment (UE) conformance specification; Radio transmission and reception; Part 1: Conformance testing
TS 36.521-2	E-UTRA: User Equipment (UE) conformance specification; Radio transmission and reception; Part 2: Implementation Conformance Statement (ICS)
TS 36.521-3	E-UTRA: User Equipment (UE) conformance specification; Radio transmission and reception; Part 3: Radio Resource Management (RRM) conformance testing
TS 36.523-1	E-UTRA and EPC: User Equipment (UE) conformance specification; Part 1: Protocol conformance specification
TS 36.523-2	E-UTRA and EPC: User Equipment (UE) conformance specification; Part 2: Implementation Conformance Statement (ICS) proforma specification
TS 36.523-3	E-UTRA and EPC: User Equipment (UE) conformance specification; Part 3: Test suites
TR 36.800	UTRA and E-UTRA: Extended UMTS/LTE 800 Work Item Technical Report
TR 36.805	E-UTRA: Study on minimization of drive-tests in next generation networks
TR 36.806	E-UTRA: Relay architectures for E-UTRA (LTE-Advanced)
TR 36.807	E-UTRA: User Equipment (UE) radio transmission and reception
TR 36.808	E-UTRA: Carrier Aggregation Base Station (BS) radio transmission and reception
TR 36.810	UTRA and E-UTRA: UMTS/LTE in 800 MHz for Europe
TR 36.814	E-UTRA: Further advancements for E-UTRA physical layer aspects
TR 36.815	Further Advancements for E-UTRA; LTE-Advanced feasibility studies in RAN WG4
TR 36.821	Extended UMTS/LTE 1500 work item technical report
TR 36.902	Evolved Universal Terrestrial Radio Access Network (E-UTRAN); Self-configuring and self-optimizing network (SON) use cases and solutions
TR 36.903	E-UTRA: Derivation of test tolerances for multi-cell Radio Resource Management (RRM) conformance tests
TR 36.912	Feasibility study for Further Advancements for E-UTRA (LTE-Advanced)
TR 36.913	Requirements for further advancements for Evolved Universal Terrestrial Radio Access (E-UTRA) (LTE-Advanced)
TR 36.921	E-UTRA: FDD Home eNode B (HeNB) Radio Frequency (RF) requirements analysis
TR 36.922	E-UTRA: TDD Home eNode B (HeNB) Radio Frequency (RF) requirements analysis
TR 36.931	RF requirements for LTE Pico NodeB
TR 36.938	Evolved Universal Terrestrial Radio Access Network (E-UTRAN); Improved network controlled mobility between E-UTRAN and 3GPP2/mobile WiMAX radio technologies
TR 36.942	E-UTRA: Radio Frequency (RF) system scenarios
TR 36.956	E-UTRA: Repeater planning guidelines and system analysis

14.3 Frequency Bands

3GPP standardized frequency bands in which the technology could be deployed worldwide. Those bands were divided into FDD and TDD. It is questionable, however, whether a single phone would be able to operate in all these bands and in both FDD and TDD modes. Besides, in some countries, more flexibility is given to operators, even allowing them to operate TDD in FDD-assigned bands.

The 3GPP also standardized the bandwidths that can be used in each band. This standardization is shown in Table 14.5.

The 3GPP specified a universal channel numbering defined as ARFCN (Absolute Radio Frequency Channel Number). In this system, channels are numbered in 100 KHz steps within a band, with ranges

Table 14.5 LTE FDD and TDD bands

FDD band	Downlink (MHz)		Channel range	Uplink (MHz)		Channel range	Duplex	Bandwidth (MHz)						Usage areas
	From	To		From	To			1.4	3	5	10	15	20	
1	2110	2170	0-599	1920	1980	18000-18599	FDD			Y	Y	Y	Y	UMTS
2	1930	1990	600-1199	1850	1910	18600-19199	FDD	Y	Y	Y	Y	Y	Y	Americas
3	1805	1880	1200-1949	1710	1785	19200-19949	FDD	Y	Y	Y	Y	Y	Y	GSM1800
4	2110	2155	1950-2399	1710	1755	19950-20399	FDD	Y	Y	Y	Y	Y	Y	Americas
5	869	894	2400-2649	824	849	20400-20649	FDD	Y	Y	Y	Y			Americas
6	875	885	2650-2749	830	840	20650-20749	FDD			Y	Y			Japan and Australia
7	2620	2690	2750-3449	2500	2570	20750-21449	FDD			Y	Y			UMTS
8	925	960	3450-3799	880	915	21450-21799	FDD	Y	Y	Y	Y			GSM900
9	1845	1880	3800-4249	1750	1785	21800-22149	FDD			Y	Y	Y	Y	Japan and Australia
10	2120	2170	4250-4749	1710	1770	22150-22749	FDD			Y	Y	Y	Y	Americas
11	1476	1501	4750-4949	1428	1453	22750-22949	FDD			Y	Y			Japan and Australia
12	728	746	5000-5179	698	716	2300-23179	FDD	Y	Y	Y	Y			USA
13	746	756	5180-5279	777	787	23180-23279	FDD			Y	Y			USA
14	758	768	5280-5379	788	798	23280-23379	FDD			Y	Y			USA
17	734	746	5730-5849	704	716	23730-23849	FDD			Y	Y			USA
33	1900	1920	36000-36199	1900	1920	36000-36199	TDD	Y	Y	Y	Y	Y	Y	UMTS
34	2010	2025	36200-36349	2010	2025	36200-36349	TDD	Y	Y	Y	Y	Y		UMTS
35	1850	1910	36350-36949	1850	1910	36350-36949	TDD	Y	Y	Y	Y	Y	Y	Americas
36	1930	1990	36950-37549	1930	1990	36950-37549	TDD	Y	Y	Y	Y	Y	Y	Americas
37	1910	1930	37550-37749	1910	1930	37550-37749	TDD	Y	Y	Y	Y	Y	Y	Americas
38	2570	2620	37750-38249	2570	2620	37750-38249	TDD	Y	Y	Y	Y	Y	Y	UMTS
39	1880	1920	38250-38649	1880	1920	38250-38649	TDD	Y	Y	Y	Y	Y	Y	China
40	2300	2400	38650-39649	2300	2400	38650-39649	TDD	Y	Y	Y	Y	Y	Y	China

assigned for each band in the downlink and uplink. The channel name follows the central carrier of the band; channels that follow outside of the band due to the selected bandwidth should not be used.

This numbering scheme does not have a simple relation to the frequency used and does not provide a numbering for additional bands or extensions. The 100 kHz step is also restrictive as a Resource Block has 180 kHz, while the possible step would be 200 kHz.

14.4 Architecture

With the increase in data traffic it became clear that the existing network architecture was not optimized for packet switching and did not provide adequate response times.

In contrast, the standard IP-based architecture adopted by WiMAX had the required performance and was much easier to implement. The 3GPP decided, then, to stay midway, simplifying the existing architecture, but keeping some of its structure. The next section describes the existing GERAN (GSM/Edge Radio Access Network)/UTRAN architecture and the new EPS (Evolved Packet System) architecture.

14.4.1 GSM and UMTS Architectures

The 3GPP architecture for the GSM/UMTS network (Release 99 to Release 7) is shown in Figure 14.1. The simplified diagram of Figure 14.4 represents, at the top, the circuit switched equipment used in a GSM network, GERAN and CS CORE, and at the bottom, represents the UMTS packet switched equipment used in UMTS, UTRAN and PS CORE. The shared CORE is used by both technologies. The interfaces between the equipment are identified in each connection.

The meaning of each acronym is described next.

AuC	Authentication Center
BSC	Base Station Controller
BTS	Base Terminal Station
CN	Core Network
CORE	Core Network
CS	Circuit Switched
EIR	Equipment Identify Register
GERAN	GSM Edge Radio Access Network
GGSN	Gateway GPRS Support Node
GMSC	Gateway MSC
GSM	Global System for Mobile Communications
HLR	Home Location Register
MSC	Mobile Switching Center
NB	Node B
PS	Packet Switching
RNC	Radio Network Controller
SGSN	Serving GPRS Support Node
UMTS	Universal Mobile Telecommunication System
UTRAN	Universal Terrestrial Radio Access Network

On the GSM side of the network, the BTSs are very simple; the BSC carries the entire burden. The fact that the BSC has to be "consulted" all the time affects the speed of the system; this approach is acceptable for voice but is not good for data transmissions.

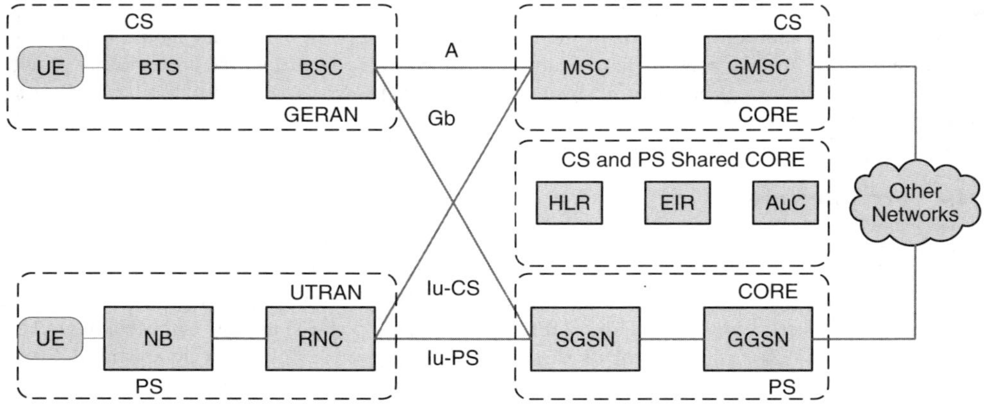

Figure 14.1 Simplified 3GPP GSM and UMTS network architecture.

In UMTS, the UTRAN has the Node B as its Base Station and the RNC as its main controller. The node B was made as simple as possible to lower equipment costs and the intelligence was concentrated in the RNC, including all mobility-related functions. This means that all messages have to reach the RNC, be processed, and return to the Node B, increasing latency. It was soon found that this was a limiting factor in the support of real time data, such as VoIP (Voice over IP) and video. Thus, it became clear that this architecture was adequate for circuit switching, but did not perform well with packet switching and was one of the limitations of the HSDPA technology.

WiMAX adopted the IP architecture drastically improving the system performance and reducing cost. The 3GPP decided then to change its core architecture, formulating a new architecture described in Section 14.4.2.

14.4.2 EPS Architecture

The new system architecture was proposed to cope with an all IP network, called Evolved Packet System (EPS). This system is divided into core network and access network.

The core network is called System Architecture Evolution (SAE) and is based on the Evolved Packet Core (EPC). The access network is based on a Long Term Evolution (LTE) and relies on the Evolved Universal Mobile Telecommunication System Terrestrial Radio Access Network (E-UTRAN).

The terms SAE and LTE are being replaced, respectively by EPC and E-UTRAN. LTE is being used in the marketing world to designate the EPS, but the term was dropped from the 3GPP technical specifications. A simple sketch of this architecture is illustrated in Figure 14.2. The Evolved Packet System (EPS) architecture is detailed in Figure 14.3.

The RNC (Radio Network Controller) was eliminated and its functionality was split between the eNB (Enhanced Node B) and a Mobility Management Entity (MME). The eNB took over all the functions that do not require information from other cells, whereas the MME took over the functions

Figure 14.2 EPS architecture elements.

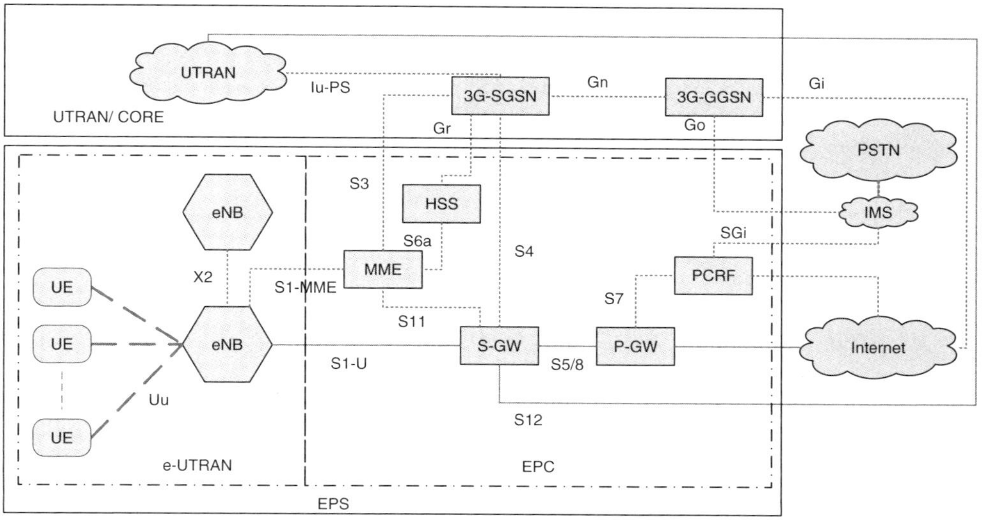

Figure 14.3 EPS (LTE) detailed architecture.

that required information from more than one cell. The interfaces between equipments are identified in each connection.

As the transmission is of data only, there is no need for soft handover, thus simplifying the architecture.

The list of acronyms used in Figure 14.3 is presented next:

AS	Access Stratum
eNB.	evolved Node B
EPC	Evolved Packet Core
EPS	Evolved Packet System
E-UTRAN	Evolved-UTRAN
GGSN	GPRS Gateway Support Node
HSS	Home Subscriber Server
IMS	IP Multimedia Subsystem
LTE	Long Term Evolution
MME	Mobility Management Entity
NAS	Non Access Stratum
PCRF	Policy Control and Charging Rules Function
PDN	Packet Data Network
P-GW	PDN Gateway
PSTN	Public Switched Telecommunications Network
SGSN	Serving GPRS Support Node
S-GW	Serving Gateway
UE	User Equipment
UTRAN	Universal Terrestrial Radio Access Network

The next sections describe each of the architecture elements.

14.4.3 eNodeB (eNB)

This is the equivalent of a Base Station and performs the E-UTRAN functions. The protocols between the UE and the eNB are called Access Stratum (AS) protocols. The eNBs are interconnected to their neighbors, so they can exchange information directly with each other. Its main functionalities are:

- Radio resource management functions such as radio bearer control, radio admission control, scheduling, dynamic allocation of resources and radio mobility control.
- Radio data encryption.
- Header compression to reduce IP overhead for small packets (e.g. voice).
- Connectivity with EPC elements.

14.4.4 Mobility Management Entity (MME)

The MME processes the signaling between the UE and the core network. The protocols used between UE and the CN are called Non-Access Stratum (NAS). The main functionalities supported by the MME are:

- Bearer Management: establishment, maintenance and release of bearers.
- Connection Management: establishment and security of connections between the UE and the network.

14.4.5 Serving Gateway (S-GW)

The S-GW serves as the local mobility anchor for data bearers when the UE moves from one eNB to another, when it is in idle state or during inter-working with other 3 GPP networks. It temporarily stores downlink data while the UE is paged. It also performs administrative tasks such as data tonnage management.

14.4.6 Packet Data Network Gateway (PDN-GW or P-GW)

This gateway is responsible for the UE IP allocation and QoS (Quality of Service) enforcement. It separates the downlink IP packets into different bearers according to Traffic Flow Templates (TFT). It also serves as mobility anchor when inter-working with other non 3 GPP networks (e.g. cdma2000 and WiMAX).

14.4.7 Policy Control and Charging Rules Function (PCRF)

This function is responsible for the QoS policy and instructs the P-GW which class identifiers to use according to each UE subscription profile.

14.4.8 Home Subscriber Server (HSS)

The HSS holds the databases of subscribers' public and private identities, security variables, and location information. It contains the HLR (Home Location Register), EIR (Equipment Identity Register), and AuC (Authentication Center).

14.4.9 IP Multimedia Sub-System (IMS)

The IMS is responsible for the interconnection of IP packets with other networks, such as GPRS, DSL or WiMAX). It interfaces with other servers using CAMEL (Customized Applications for Mobile network Enhanced Logic), OSA (Open Service Access), and SIP (Session Initiation Protocol).

14.4.10 Voice over LTE via Generic Access (VoLGA)

The VoLGA Forum was created by several vendors to provide operators with ways to deliver mobile voice and messaging services over LTE networks, based on the existing 3GPP Generic Access Network (GAN).

VoLGA aim is to make GSM and UMTS circuit switched (CS) voice service accessible to packet switched (PS) E-UTRAN and UTRAN users. This is done by the addition of a VANC (VoLGA Access Network Controller) to the E-UTRAN architecture, which uses the 3GPP Generic Access Network (GAN) for generic IP access from CS services.

14.4.11 Architecture Interfaces

The architecture interfaces are described next:

- Access Stratum Interfaces: The term "Stratum" is used by the 3GPP and stands for layer. The access stratum refers to the layer of interfaces that connects the eNB.
- Uu: is the radio interface between UE and eNB.
- S1-U: connects the UE to the EPC (S-GW) in the user plane. It uses the IP protocols (GTP (GPRS Tunneling Protocol)/UDP (User Datagram Protocol)/IP (Internet Protocol)), instead of the SS7 (Signaling System 7) protocol used in GSM and UMTS.
- X2: interconnects eNBs (pair-wise) directly and uses the same protocols as the S1 interface. The following functionalities are implemented over this interface:
  - initiation
  - mobility
  - load balancing
  - interference management
  - UE historical information
  - Non Access Stratum Interfaces: Theses interfaces carry the messages between the core elements.
- S1-MME: connects the UE to the EPC (S-GW) in the user plane. It uses the IP protocol (SCTP (Stream Control Transmission Protocol)/IP (Internet Protocol)) instead of the SS7 protocol used in GSM and UMTS. This interface supports SON (Self-Organizing Network) functions. The following functionalities are implemented over this interface:
  - UE initiation
  - context management
  - bearer management
  - paging
  - mobility
  - load management
- S11: interconnects MME and S-GW.

- S6a: interconnects MME and HSS.
- S12: interconnects S-GW with legacy UTRAN for Inter-RAT (Radio Access Technology) handover.
- S5/8: interconnects S-GW to P-GW and uses GTP (GPRS Tunneling Protocol).
- S7: interconnects P-GW with PCRF.
- S3: interconnects MME with 3G-SGSN for Inter-RAT (Radio Access Technology) handover.
- Gr: interconnects HSS to 3G-SGSN for Inter-RAT (Radio Access Technology) handover.
- Go: interconnects IMS to 3G-GGSN.
- SGi: interconnects PCRF to IMS.

Figure 14.4 shows how the different EPS blocks are interconnected. The core functionality uses regular processors (dedicated or commercial PCs) interconnected by routers. The architecture is such that more than one processor can be used for a function if this is required for capacity or reliability.

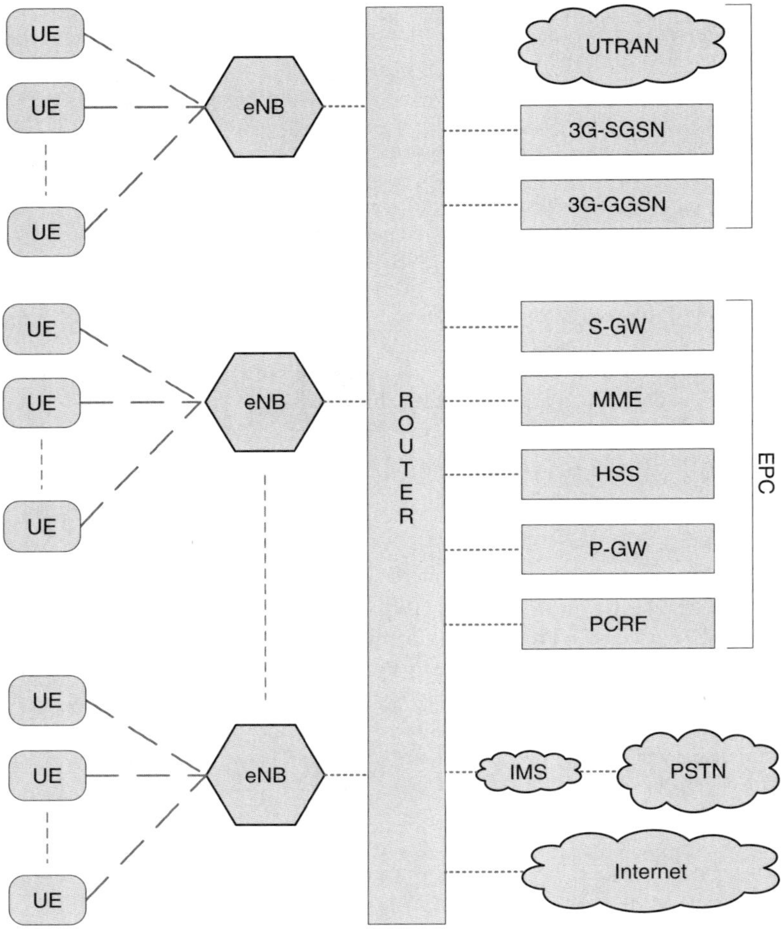

Figure 14.4 LTE architecture.

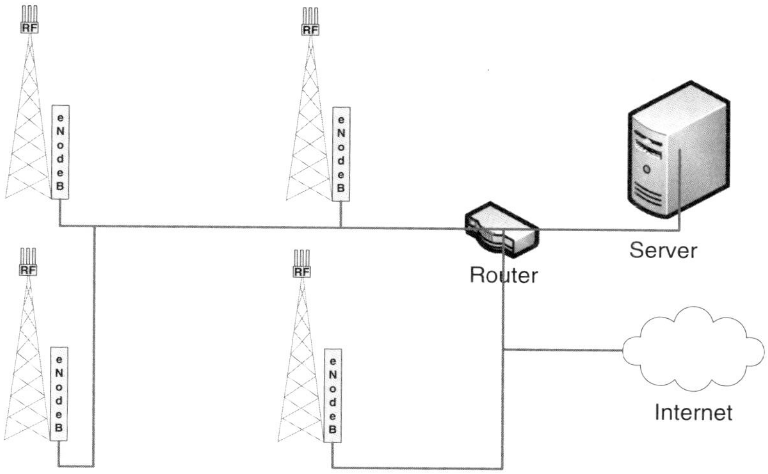

Figure 14.5 LTE components' interconnection.

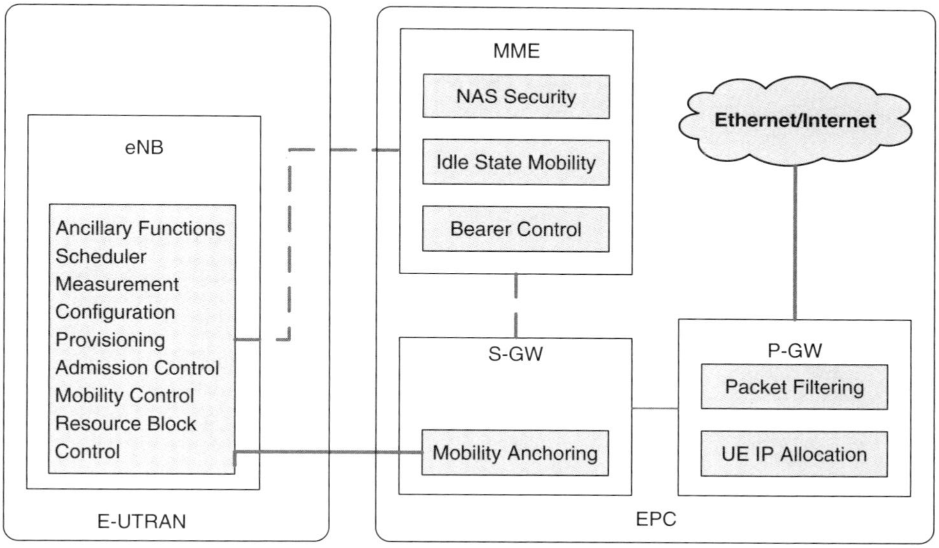

Figure 14.6 LTE functionality distribution.

This more flexible architecture allows the use of commercial components and lowers the overall cost of the product when compared to UMTS architecture. It also allows hardware to be updated more easily with the evolution of computers.

The EPC functionality resides on servers, and one or more servers can be used to accommodate processing requirements. The actual interconnection is extremely simple as shown in Figure 14.5.

Figure 14.6 shows the LTE functionality distribution between the main elements. UE connects to eNB and then to the MME processor that present itself first (if there is more than one). MME provides the UE with an IP and encryption parameters. In this architecture, the eNB assumes the majority of the functions of the RNC in UMTS architecture. The MME is responsible for the bearer control and security. The S-GW performs mobility anchoring during handover and the P-GW provides IP routing and implements packet filtering (including throughput) according to the user's profile.

14.5 Wireless Message Flow and Protocol Stack

Information in this new architecture is IP-based and consequently structured in packets. These packets have variable sizes and different birth rates. The different elements of the architecture have to establish connection between them and also with the external world, assuring at the same time the integrity of the information. IP messages are manipulated by a protocol stack with several levels, each with its own functionality. Figure 14.7 illustrates the layers of the wireless protocol stack between the UE and eNB.

14.5.1 Messages

Messages are IP information received from the UEs, from the external world or generated internally to control the inter-working of the architecture. Messages can be classified as:

- Broadcast Messages or Dedicated Messages
- Control or Traffic Messages
- Downlink, Uplink or bi-directional Messages

The main message types are:

- System Information
- Paging
- LTE PHY Frame Content
- Random Access
- User Message structure
- User application information
- HARQ
- Measurements

For the messages to be sent through the different elements message bearers, message channels and auxiliary signals have to be established. The message flow is illustrated in Figure 14.8. The codes used in Figure 14.8 are:

ARQ	Automatic Repeat reQuest
BCCH	Broadband Control Channel
BCH	Broadcast Channel
CCCH	Common Control Channel
DCCH	Dedicated Control Channel
DL	Downlink
DTCH	Dedicated Traffic Channel
GTP	GPRS Tunneling Protocol
HARQ	Hybrid Automatic Repeat reQuest
IP	Internet Protocol
MAC	Medium Access Control
MCCH	Multicast Control Channel
MCH	Multicast Channel

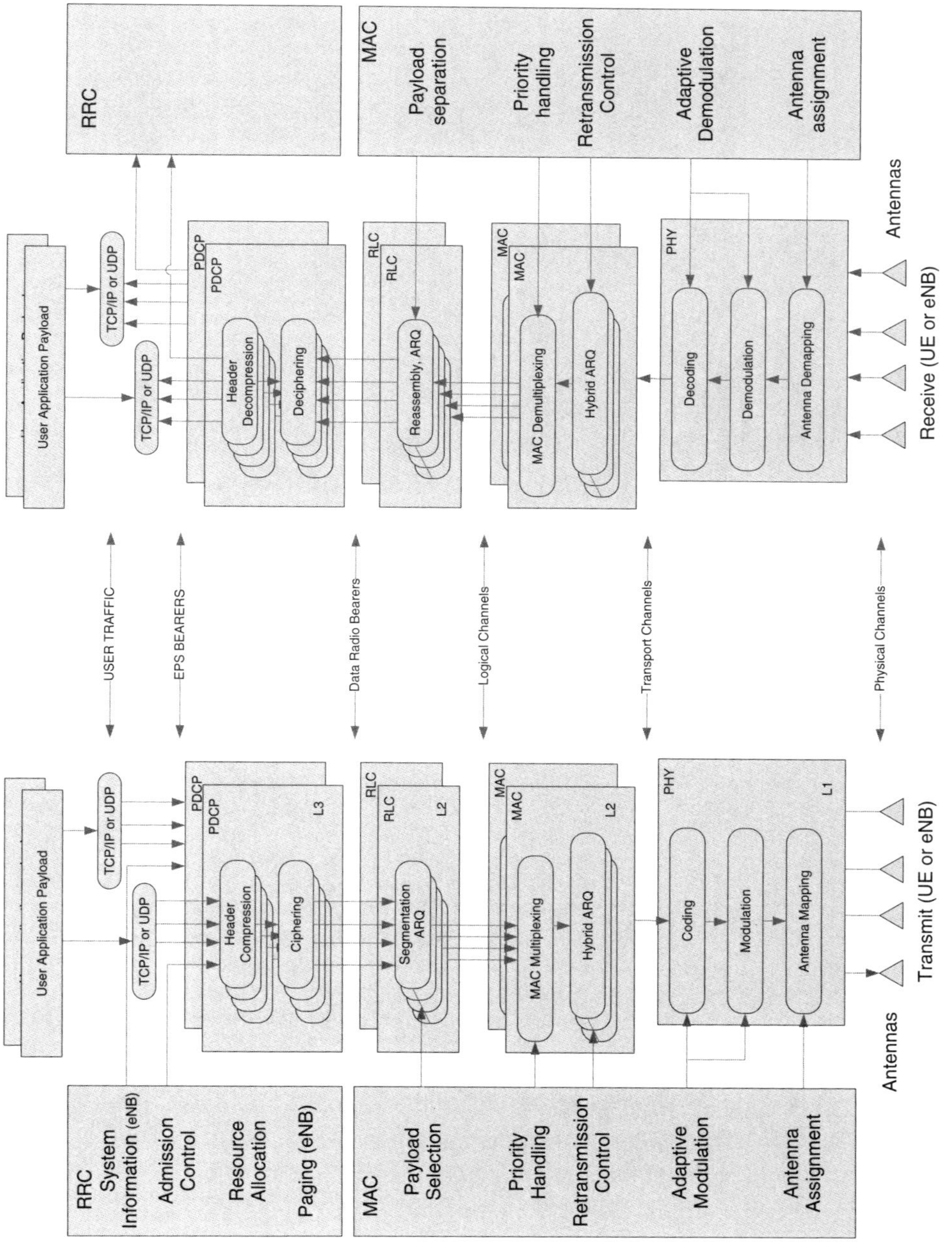

Figure 14.7 LTE message flow and protocol stack.

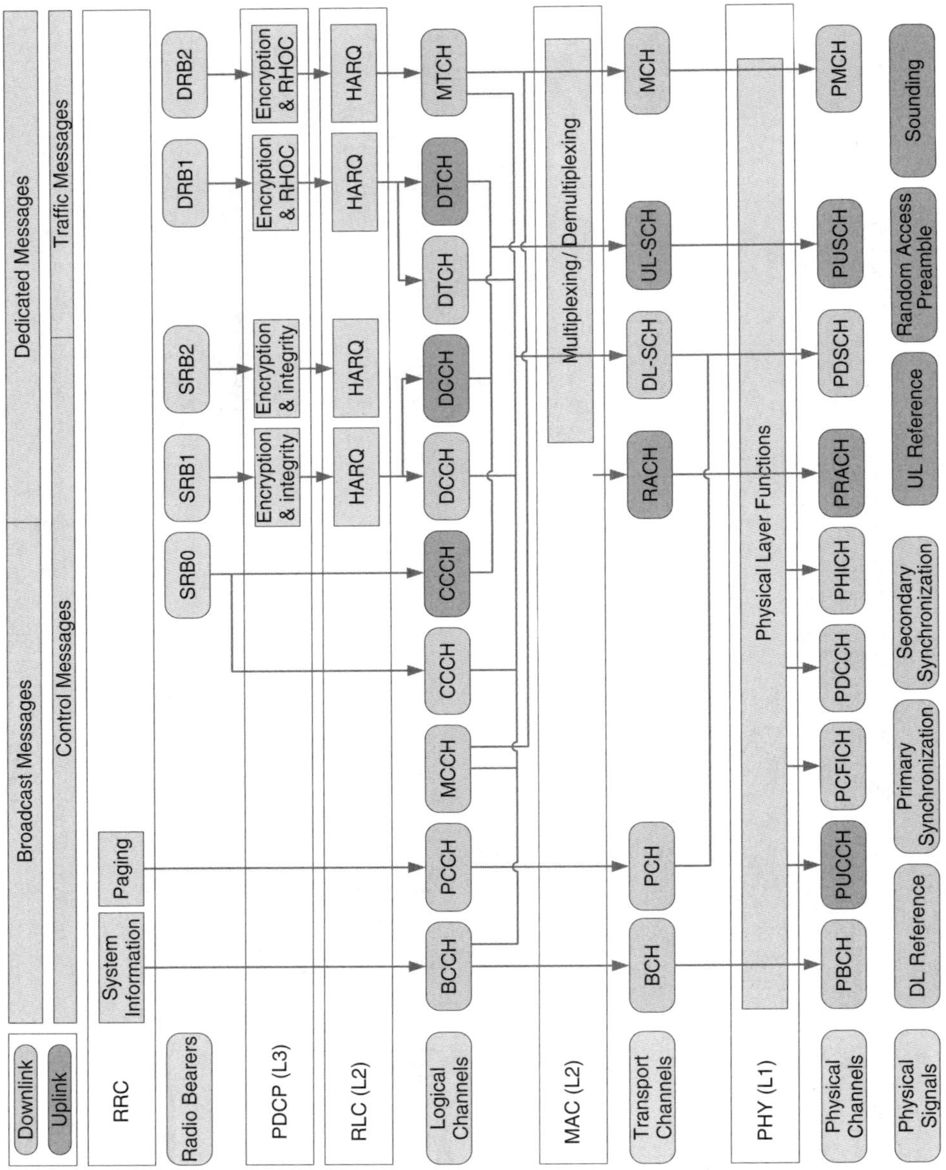

Figure 14.8 RRC states.

MTCH	Multicast Traffic Channel
PBCH	Physical Broadcast Channel
PCCH	Paging Control Channel
PCFICH	Physical Control Format Indicator Channel
PDCCH	Physical Downlink Control Channel
PDCP	Packet Data Convergence Protocol
PDSCH	Physical Downlink Shared Channel
PHICH	Physical Hybrid ARQ Indicator Channel
PHY	Physical Layer
PMCH	Physical Multicast Channel
PRACH	Physical Random Access Channel
PUCCH	Physical Uplink Control Channel
PUSCH	Physical Uplink Shared Channel
RACH	Random Access Channel
RLC	Radio Link Control
RRC	Radio Resource Control
SCH	Synchronization Channel
SCTP	Stream Control Transmission Protocol
SRB	Signaling Radio Bearer
UDP	User Datagram Protocol
UL	Uplink

14.5.2 Protocol Layers

The Protocol Stack is a set of software modules responsible for specific tasks in the connection establishment and message handling process.

User data cannot be sent directly through the wireless system. Several control messages have to be created to allow the UE to properly connect to the wireless network. Due to the variability of the wireless media, the integrity of the data has to be verified. Data correction by higher layers would be too slow to be acceptable, so the wireless layers have to provide their own data correction and integrity tests. This is done through a protocol stack in which each layer is responsible for specific functions.

Data that is passed from one protocol layer to the next is called SDU (Service Data Unit) and once it is encapsulated by that protocol layer, it is called PDU (Protocol Data Unit).

14.5.2.1 RRM Radio Resource Management

The RRM is responsible for the following functions:

- Radio bearer control: establishment, release and configuration of bearer radio resources.
- Radio admission control: administration of bearer allocation.
- Resource allocation: scheduling and allocation of resource for wireless communication.
- Load balancing: performance of intra-eNB and inter-eNB traffic re-distribution.
- Intra and Inter-cell interference coordination: management of resource blocks to avoid conflict between neighbor cells using same or neighbor frequencies.
- Mobility control: management of mobility during idle and connected modes.
- Inter-RAT (Radio Access Technology) mobility management: management of resources during inter-RAT handover.

14.5.2.2 RRC Radio Resource Control

The RRC is responsible for the System Information (MIB and SIB), creation of SRB1 and DRB for each UE, transmit and receive of NAS messages, initial security activation, paging support, and measurement support.

The PDCP (Packet Data Convergence Protocol) layer carries out the header compression and decompression of the IP protocol. This is important to reduce overhead, mainly for short messages, such as the ones used in VoIP. A VoIP packet has between 50 and 200 Bytes, whereas the IP header has 20 Bytes, representing a significant overhead and throughput reduction.

Compressing the header increases efficiency in general, and is essential for applications that use small packets. The header compression used by PDCP is based on the RoHC (Robust Header Compression) standard RFC3095 or RFC 2507 or RFC1144, developed in 2001 by the IETF (Internet Engineering Task Force).

The RoHC is based on the fact that there is a significant redundancy in the header fields of consecutive packets of the same stream. After the first header is sent, the remaining fields seldom change, a fact that is exploited by the protocol. The protocol uses the packet Sequence Number (SN) to establish variations and only sends changed values from one IP packet to the next. IP fields have different compression mechanisms specified to optimize performance and increase robustness. Compression effectiveness can reach 90%, reducing the overhead to around 2 Bytes per packet.

The UE RRC has the following states:

- RRC Idle: The characteristics of this state are:
  - UE is located in the same Tracking Area (TA). TA is one of several areas that the network is divided for paging purposes.
  - Cell re-selection (done when so instructed by the actual cell).
- Paging without feedback.
- S1 connection is not established.
- UE measures but does not report: RSRP (Reference Signal Received Power), RSCP (Received Signal Code Power) and RXLEV (Receive Level).
- DCCH and DTCH are not established.
- RRC Connected: the characteristics of this state are listed below and illustrated in Figure 14.9:
  - S1 connection is established to MME and X2 to neighbor eNB.
  - UE performs initial and periodic Random Access.
  - UE performs neighbor cell measurements and reports to eNB.
  - Data may be exchanged.

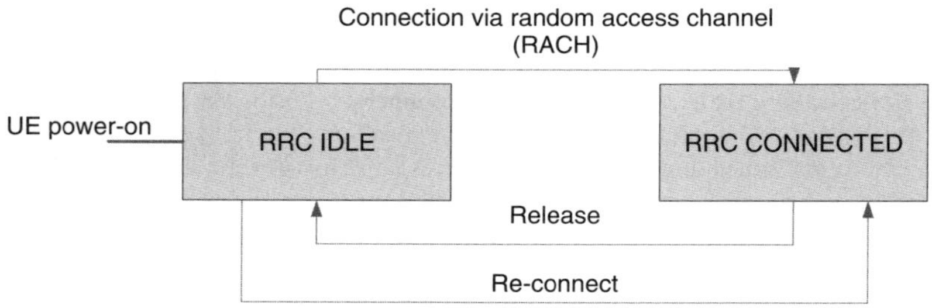

Figure 14.9 LTE message flow.

14.5.2.3 RLC Radio Link Protocol

This layer adjusts the SDU (Service Data Unit) to the size required by the next layer. This is done by segmentation or concatenation of the SDU data. The RLC implements ARQ (Automatic Repeat Request) for messages that are sent in an acknowledgement mode. RLC uses SDU numbering to guarantee in-sequence delivery to higher levels. The information is gathered into Logical Channels, which are then sent to the MAC layer.

14.5.2.4 MAC Medium Access Protocol

This layer multiplexes data and control bearers and adds a CRC (Cyclic Redundancy Code) to assure data integrity. The amount of added redundancy targets a packet rate of approximately 10%, and a HARQ procedure is used to correct the packets received in error, thus maximizing the throughput. The information is gathered into Transport Channels, which are then sent to the PHY layer.

14.5.2.5 PHY Physical Layer

This layer is responsible for mapping the data onto the OFDM frame, performing FEC (Forward Error Correction), and AMC (Adaptive Modulation and Coding). The data is mapped onto Resource Blocks (12 sub-carrier × 1 symbol).

The PHY is also responsible for the multi-antenna processing, such as layer mapping and pre-coding, transmit diversity (Space Frequency Block Coding), closed or open loop MIMO spatial multiplexing and closed loop beam forming.

The relationship between the downlink channels is shown in Figure 14.10 and the relationship between the uplink channels is shown in Figure 14.11

14.5.3 Message Bearers

LTE is an All IP (AIP) technology, and this means that all user data is IP-based. When a user is connected to an external PDN (Packet Data Network), a connection is established through the LTE network and it is always on, until the user disconnects from the PDN. The bearer is an information transmission path of defined capacity, delay, bit error rate, etc. The bearer definition is done at the network level. It represents a logical association between a UE and a PDN (Public Data Network).

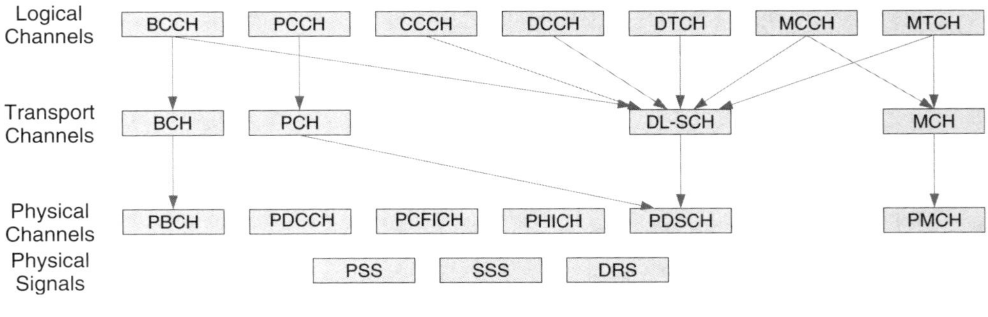

Figure 14.10 Downlink channel relationship.

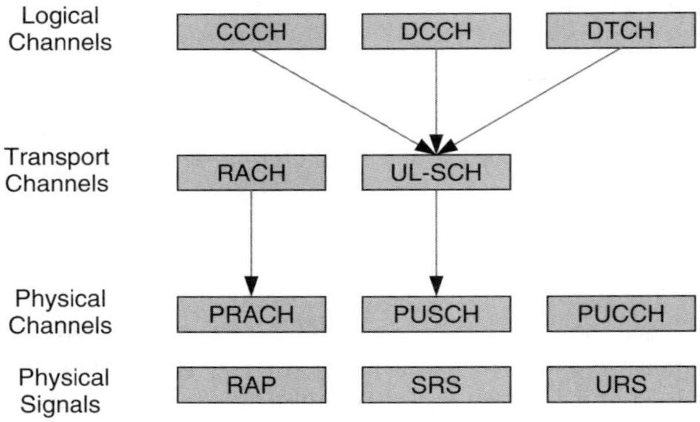

Figure 14.11 Uplink channel relationship.

The bearer is a carrier of information with a given QoS (Quality of Service) associated with it, which defines quality characteristics of the service. An analogy can be made with a mail carrier (FEDEX, UPS, which are the bearers) and the different shipment options (overnight priority, overnight, next day, second day, ground) represent QoS. Temporary addresses are associated to each bearer to indicate the origin and destination of the data.

14.5.3.1 Quality of Service (QoS)

The possible QoS associated to a bearer are specified by nine QCI (QoS Class Identifier), as shown in Table 14.6.

14.5.3.2 Bearers

To exchange information between a UE and a PDN specific bearers have to be established. A UE has a default bearer established when it joins a network, and additional dedicated bearers as need arises, up to a total of 3 control information bearers and 11 data bearers. Each of the bearers is transmitted

Table 14.6 QCI categories

QCI	Resource type	Priority	Packet delay budget (ms)	Packet error loss rate	Example of service
1	GBR	2	100	10^{-2}	Conversational voice
2	GBR	4	150	10^{-3}	Conversational video (live streaming)
3	GBR	5	300	10^{-6}	Non conversational video (buffered streaming)
4	GBR	3	50	10^{-3}	Real time gaming
5	Non-GBR	1	100	10^{-6}	IMS signaling
6	Non-GBR	7	100	10^{-3}	Voice, video (live streaming)
7	Non-GBR	6	300	10^{-6}	Video (buffered streaming)
8	Non-GBR	8	300	10^{-6}	TCP based (www, e-mail, FTP, P2P, file sharing)
9	Non-GBR	9	300	10^{-6}	

GBR- Guaranteed Bit Rate

over a different message channel. Section 14.5.4 describes these channels of the downlink and uplink in more detail.

The control information bearers are:

- SRB0 Signaling Radio Bearer 0 RRC messages over CCCH. It is used for RRC connection request, Reject, Reestablishment, Reestablishment request and Reestablishment reject.
- SRB1 Signaling Radio Bearer 1 NAS (Non-Access Stratum) messages over the DCCH, uses acknowledged mode. It is used for UE capability inquiry or information, RRC connection set-up complete, reconfiguration, security mode command, measurement report, and mobility commands.
- SRB2 Signaling Radio Bearer 2 high priority RRC messages over DCCH, uses acknowledged mode. It is used for DL and UL information transfer.
- Data bearers identified by DRB (Data Radio Bearer), numbered from 1 to 11. Control and data packets are sent to different bearers according to their QoS requirements. User packets are analyzed according to their origin and destination addresses and their port number to define their QoS requirement according to a TFT (Traffic Flow Template) associated with a bearer.

14.5.4 Message Channels

Channels can be divided into: logical channels, transport channels, and physical channels.

14.5.4.1 Logical Channels

Logical channels carry specific information between processes. They are represented by four-letter acronyms, in which the first two letters represent the type of information being carried and last two are always CH (meaning Channel). Logical channels carry the messages between the RLC and MAC layers. The logical channels are:

- Downlink Control Channels:
  - BCCH Broadcast Control Channel: broadcasts system information periodically, respecting DRX (Discontinuous Reception), which preserves UE battery life. It contains the MIB (Master Information Block), which has a defined structure to allow any UE to decode the data. Additional system information is carried by the CCCH.
  - PCCH Paging Control Channel: notifies the UE of an incoming call, and it also considers DRX. It informs when there is updated system information in the BCCH.
  - MCCH Multicast Control Channel: provides control information related to the reception of the MBMS (Multimedia Broadcast/Multicast Service).
  - CCCH Common Control Channel: provides control information before a link is established between the UE and eNB. It is used until a DCCH is established.
  - DCCH Dedicated Control Channel: this channel is used after an RRC connection is established and the access (contention) has been successfully resolved. It is always associated with a DTCH.
- Uplink Control Channels:
  - CCCH Common Control Channel: provides control information before a link is established between the UE and eNB. It is used until a DCCH is established.
  - DCCH Dedicated Control Channel: this channel is used after an RRC connection is established and the access (contention) has been successfully resolved. It is always associated with a DTCH.
- Downlink Traffic Channels:
  - DTCH Dedicated Traffic Channel: it is used to transmit user data. UM-RLC (Unacknowledged Mode Radio Link Control) is used with time-critical services like VoIP and AM-RLC (Acknowledged Mode Radio Link Controller) for data that requires low BER (10^{-6} or lower).

- MTCH Multicast Traffic Channel: it is used for point-to-multi-point data transmission.
- Uplink Traffic Channels:
  - DTCH Dedicated Traffic Channel: it is used to transmit user data. UM-RLC (Unacknowledged Mode Radio Link Control) is used with time-critical services, such as VoIP and AM-RLC (Acknowledged Mode Radio Link Controller) for data that requires low BER (10^{-6} or lower).

Transport channels add the MAC frame to the logical messages, allowing for message integrity analysis. They use three- or four-letter acronyms, in which the two last letters always CH mean Channel. The first and second letters (if existent) define the type of information being carried.

Transport channels carry the messages between the MAC and PHY layers. The transport channels are:

- Downlink:
  - BCH Broadcast Channel: it is used transport the BCCH channel.
  - PCH Paging Channel: it is used to transport the PCCH channel.
  - MCH Multicast Channel: it is used to transport the MCCH. It supports SFN (Single Frequency Network) by combining signals that arrive from multiple cells transmissions.
  - DL-SCH Downlink Shared Channel: Transports all the remaining channels.
- Uplink:
  - RACH Random Access Channel: Transmits information about the Random Access Channel and how the UE can connect to it to obtain timing advance information.
  - UL-SCH Uplink Shared Channel: it is used to transport uplink control and traffic channels.

14.5.4.2 Physical Channels

Physical channels carry the transport channel messages between wireless or wired hardware. They are identified by four- to six-letter acronyms in which the two last letters are always CH (meaning Channel). The first two to four letters define the type of information being carried. These channels are defined by their signal format, frequency and so on. The physical channels are:

- Downlink:
  - PBCH Physical Broadcast Channel: transmits the broadcast channel BCH. It is sent over 72 central sub-carriers on symbols 0 to 3 of slot 1, using QPSK modulation.
  - PCFICH Physical Control Format indicator Channel: transmits the size of the PDCCH in OFDM symbols.
  - PDCCH Physical Downlink Control Channel: specifies the transport channel format, allocates resources and HARQ information.
  - PHICH Physical HARQ acknowledgement Indicator Channel: acknowledges UL transmissions.
  - PDSCH or DL-SCH Physical Downlink Shared Channel: carries data to the UE (payload).
  - PMCH Physical Multicast Shared Channel: carries the downlink payload.
- Uplink:
  - PRACH Physical Random Access Channel: contention-based random access channel, used to adjust the UE timing. It occupies 72 sub-carriers (5 RB) by 2 slots deep. Once the random access is successful, the message is transmitted in the UL-SCH.
  - PUCCH Physical Uplink Control Channel:-acknowledges DL transmissions, sends CQI (Channel Quality Indicator) data, MIMO feedback and makes requests for UL transmissions.
  - PUSCH or UL-SCH Physical Uplink Shared Channel: carries the UE data (payload).

14.5.5 Physical Signals

- Downlink:
  - PSS Primary Synchronization Signal: is used for synchronization and defines 3 groups of cell IDs.
  - SSS Secondary Synchronization Signal: is used for synchronization and defines 168 cells for each of the groups defined by PSS.
  - DRS Downlink Reference Signal: provides a reference signal (pilot) for the demodulation of downlink data.
- Uplink:
  - URS Uplink Reference Signal: provides a reference signal (pilot) for the demodulation of uplink data.
  - RAP Random Access Preamble: is part of the PRACH message and uses a Zadoff-Chu sequence.
  - SRS Sounding Reference Signal: is sent every second slot in the second and one before last RBs of the bandwidth and also as the last symbol in a DL-SCH transmission. It is used to access the UE channel quality and timing.

14.6 Wireline Message Flow and Protocol Stacks

Each layer in the message protocol from one device connects to the same layer in the next device, so the lower and upper layers became transparent. This is illustrated in Figures 14.12 and 14.13. Figure 14.12 illustrates the message exchange at the E-UTRAN level, whereas Figure 14.13 illustrates the control plane message exchange.

The wireline connections use regular IP protocols or the legacy GPRS protocol:

- L1: Ethernet PHY.
- L2: Ethernet MAC.
- UDP/IP: User Datagram Protocol/Internet Protocol (RFC 768).
- IP: Internet Protocol (RFC 791).
- GTP-U: GTP (GPRS- General Packet Radio Service Tunneling Protocol) User Plane (3GTS 29.060).
- SCTP: Stream Control Transmission Protocol (RFC 2960).
- S1-AP: S1 Interface Application Part Protocol (3GTS 36.413).

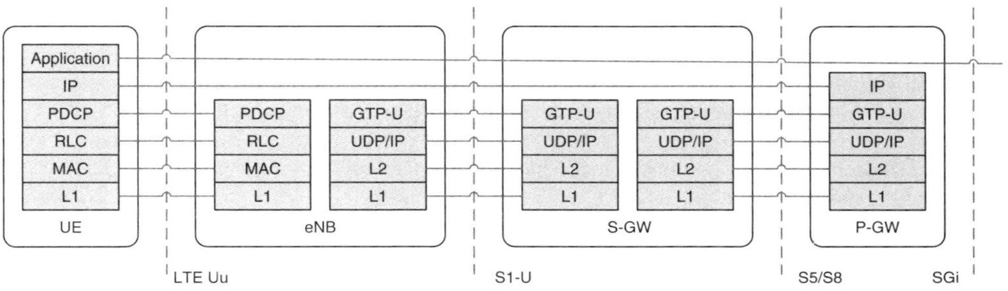

Figure 14.12 E-UTRAN message exchange.

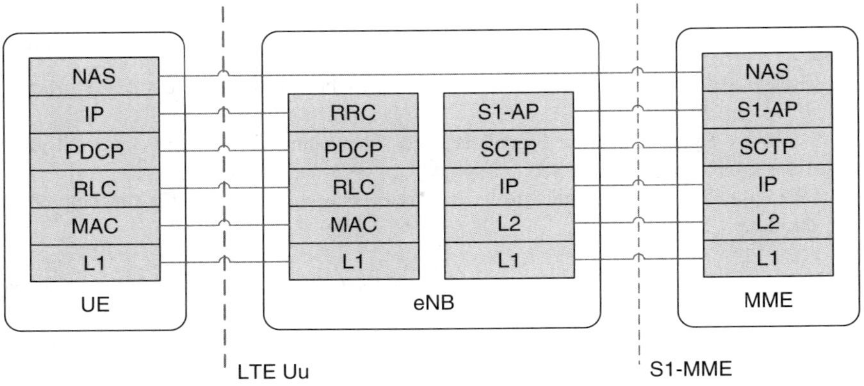

Figure 14.13 Control plane message exchange.

14.7 Identifiers

Identifiers are labels used to identify a network or a network element. The following are the main identifiers used by LTE.

- Network level:
  - PLMN (Public Land Mobile Network) ID: ID assigned to the PLMN.
  - Bearer ID: ID assigned by the MME to a bearer.
  - MMEI (Mobility Management Entity Identifier): identifies the entity that manages mobility.
  - Cell ID (eNB): is assigned to the eNB sector and identifies it uniquely in the system. The CellID should be planned in such a way that co-channels sectors with the same ID do not overlap. There are 168 groups of 3 CellIDs (considering three sector sites).
  - TAI (Tracking Area Identity): identifies the tracking area and is composed of a TAC (Tracking Area Code), a MNC (Mobile Network Code) and an MCC (Mobile Country Code).
- UE level:
  - IMSI (International Mobile Subscriber Identity): permanent number assigned by the service provider to identify the subscriber. The IMSI is stored in the USIM (Universal Subscriber Identity Module), which is contained in the UICC (Universal Integrated Circuit Card). The UICC is the smart card used in GSM, UMTS and now LTE phones. It is also stored in the network's HSS (Home Subscriber Server).
  - IMEI (International Mobile Equipment Identity): permanently identifies the wireless phone and is assigned by the phone manufacturer. It is used in black lists to ban stolen phones. It is stored in the HSS.
  - GUTI (Global Unique Terminal Identifier): dynamic identity assigned by the MME (Mobility Management Entity) as long as the UE is registered with the network EPC. It is stored in the UE and MME.
  - IP address (Internet Protocol): dynamic identity assigned by the PGW (Packet Data Network Gateway) while the UE is registered with the EPC. It is stored in the UE and PGW.
  - C-RNTI Cell Radio Network Temporary Identity: UE temporary identity assigned by eNB while the UE is connected to the eNB. It is stored in the eNB and UE.

14.8 HARQ Procedure

To understand HARQ functioning we need to understand the Forward Error Correction (FEC) method used in LTE. This method is based on turbo coding.

14.8.1 Turbo Code

Turbo coding was created in 1993 by Berrou, Glavieux and Thitimajshima and achieves a nearly Shannon throughput. The code is based on combining two regular convolutional codes of a transport block, one applied to the regular block (systematic bits) and another to an interleaved version of the same block, as illustrated in Figure 14.14. Thus some bits that were weakly decoded in the original block may have a better decoding in the interleaved block and this allows the code to select the strongest decoding for each bit on the receive side.

The turbo code circuit receives a payload of k input bits, creates two sets of m parity bits, generating an output of $k + 2m = n$ bits. The code rate is then defined by the ratio k/n.

LTE defines 188 block sizes varying from 40 to 6144 bits. The generation of parity bits uses a convolutional code with a ratio of 1/2 (1 payload and 1 redundancy), resulting in a final ratio of 1/3 (1 payload and 2 redundancy). This output code is called the mother code. All mother codes generated in LTE have a coding ratio of 1/3. The coding scheme required to transmit the data does require coding rates that varies from 1/3 to 4/5. The transport block adds a 24 bit (3 bytes) CRC to each block and this allows to verify if a block was received in error or not.

Lower code rates can be obtained using a technique called puncturing. Lower redundancy rates are obtained simply by eliminating bits from the code. Table 14.7 shows the rates obtained when puncturing is done over a set of 8 bits. Zeros in the puncturing vector represent redundancy bits that are suppressed, and ones represent actual redundancy bits. Table 14.7 illustrates puncturing for 4 bits of data and 8 bits of redundancy. A 0 corresponds to a punctured bit (discarded) and a 1 corresponds to a bit that is kept.

Data blocks received in error will be resent through the ARQ process in the RLC layer, but the ARQ presents a significant overhead and delay, so its use should be minimized.

Reducing the number of ARQs requires working with a low BLER, but this is in conflict with the desire of increasing the modulation types. It was found that it would be advantageous to work

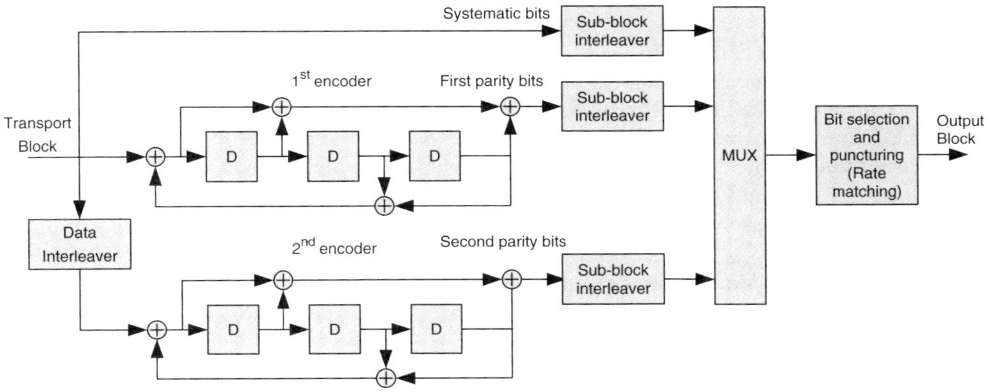

Figure 14.14 Turbo code encoder.

Table 14.7 Turbo code rate and typical puncturing table

Code rate		Puncturing pattern							
1	1.00	0	0	0	0	0	0	0	0
4/5	0.80	0	0	0	0	0	0	1	0
2/3	0.67	0	0	0	0	0	0	1	1
4/7	0.57	0	0	0	0	0	1	1	1
1/2	0.50	1	0	0	0	0	1	1	1
4/9	0.44	1	0	1	0	0	1	1	1
2/5	0.40	1	0	1	0	1	1	1	1
4/11	0.36	1	1	1	0	1	1	1	1
1/3	0.33	1	1	1	1	1	1	1	1

with higher BLER levels to increase the overall throughput, as long as the error correction procedures represented a lower overhead. For this reason, it was decided to implement a more optimized ARQ procedure at transport block level, which would solve the bulk of the block errors.

14.8.2 Incremental Redundancy

LTE makes use of the Incremental Redundancy (IR) methodology to transmit user data. In this method data is transmitted with the lowest rate that produces BLER (Block Error Rate) of about 10%. Blocks received in error are retransmitted using a technique called HARQ (Hybrid ARQ).

There are three types of HARQ:

1. *Type 1 Packet Combining or Chase Combining (CH)*: in this case, the energy of the retransmitted packet is added to the original packet, improving the chances of detection. This method uses the same redundancy for good and back packets.
2. *Type 2 Full Incremental Redundancy (FIR)*: this method gradually decreases the coding rate, by sending additional bits of the punctured code, not sent previously. These bits are then combined with the previous one to form a more powerful error correction code. This method may not use any redundancy for good packets and only adds redundancy incrementally to correct bad received packets.
3. *Type 3 Partial Incremental Redundancy (PIR)*: this method resends the packet with increased redundancy, so the new retransmission has a better probability of being decoded by itself or it can be chase combined with the previously received packets to increase the energy of the bits. This method allows lower redundancy for good packets and adds redundancy to retransmit bad packets. Incremental Redundancy relies on Rate Compatible Punctured Turbo Codes (RCPT) or Rate Compatible Punctured Convolutional Codes (RCPC) in which puncturing is done by removing bits from a mother code, so all punctured versions have common bits.

Figures 14.15 to 14.16 show expected BLER to SNR curves for different Modulations and Coding Schemes (MCS) and HARQ methods. It can be seen that HARQ types 2 and 3 require less SNR than type 1.

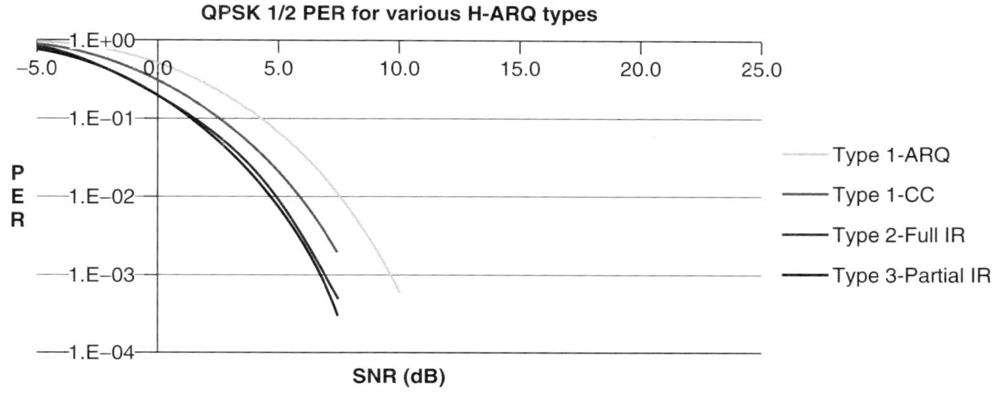

Figure 14.15 PER × SNR × H-ARQ (QPSK ½).

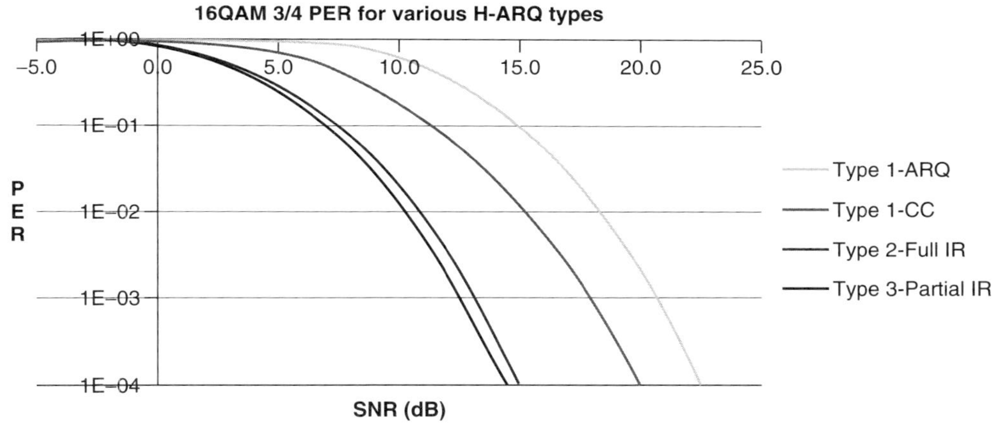

Figure 14.16 PER × SNR × HARQ (16QAM ¾).

Figures 14.18 to 14.20 show SNR to throughput relations for different MCS and HARQ methods. It can be seen that HARQ methods 2 and 3 give better results than method 1.

The use of HARQ is advantageous in terms of throughput, but it adds delays to the information delivery. The maximum number of retransmissions is configured by the RRC and varies between 1 and 28, with 5 being a typical number. To minimize delays, new data can be transmitted during the HARQ process and it is up to the RLC layer to put the received blocks in sequence. It is even possible to have up to 8 blocks in HARQ at the same time.

LTE works using an N-Stop and Wait protocol, that is, after transmitting a data block, an a ACK or NACK is expected, meanwhile all the context is stored in memory. Other N (up to 8) processes can run in parallel, to expedite the delivery of data.

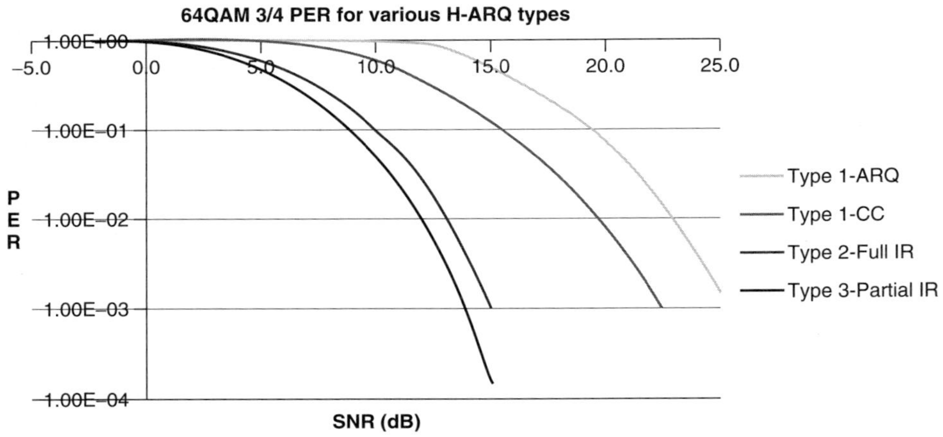

Figure 14.17 PER × SNR × HARQ (64QAM ¾).

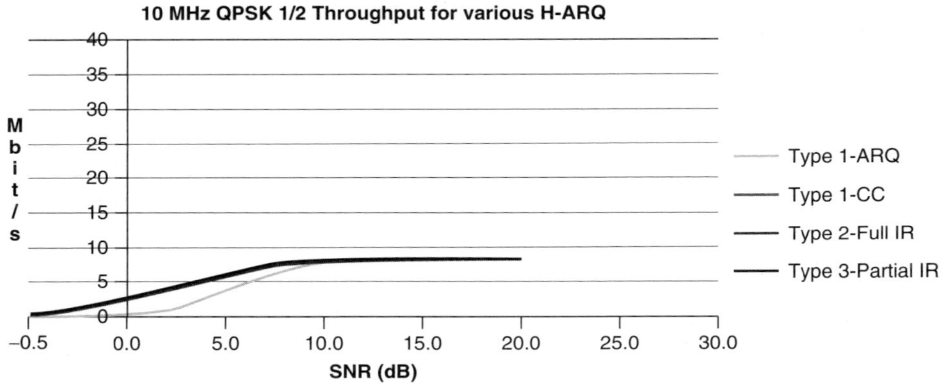

Figure 14.18 Throughput × SNR × HRQ (QPSK ½).

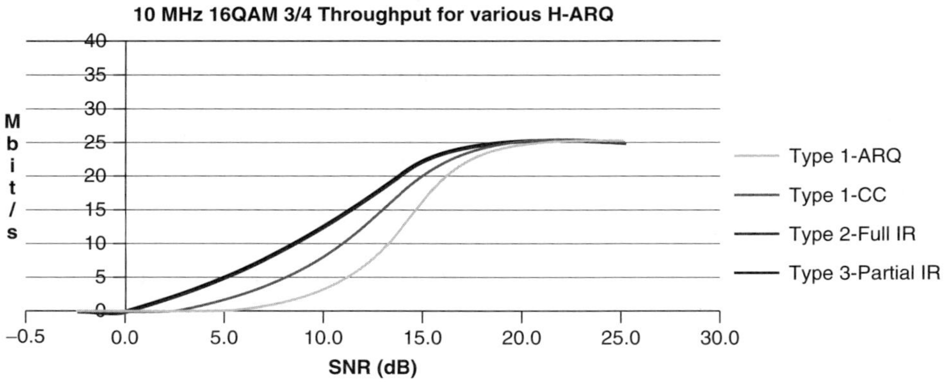

Figure 14.19 Throughput × SNR × HRQ (16QAM ¾).

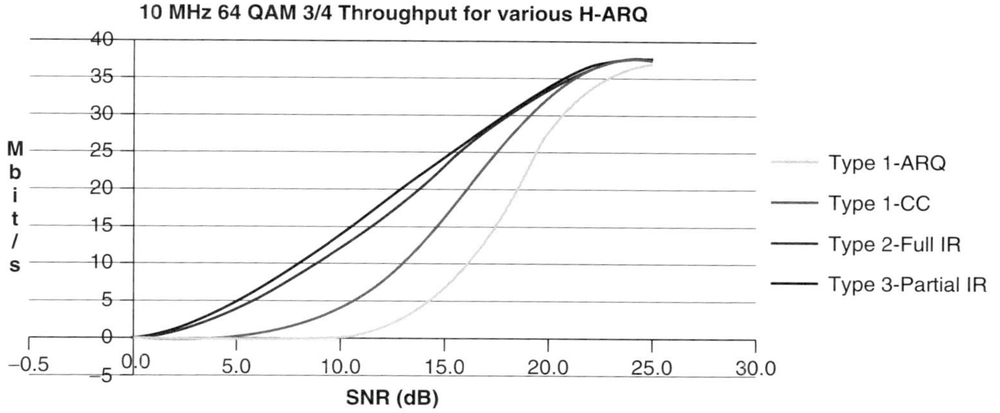

Figure 14.20 Throughput × SNR × HRQ (64QAM ¾).

14.9 Scrambling Sequences

Wireless signals mix in the air, so there is a need for code sequences that can be uniquely identified even when mixed with other sequences. This means that the codes should have a low correlation. When this is applied to shifted versions of the same code, it is called low auto correlation. Ideally, these codes should have constant amplitude. Codes that satisfy this property are called CAZAC (Constant Amplitude Zero Auto Correlation) codes. One set of sequences that has CAZAC properties are the Zadoff-Chu sequences.

A Zadoff–Chu (ZC) sequence is a complex-valued mathematical sequence which, when applied to radio signals, gives rise to an electromagnetic signal of constant amplitude. A Zadoff–Chu sequence that has not been shifted is known as a "root sequence". Cyclic shifted versions of the sequence do not cross-correlate with each other when the signal is recovered at the receiver.

The sequence then exhibits the useful property in which cyclic-shifted versions of it remain orthogonal to one another, provided that each cyclic shift, when viewed within the time domain of the signal, is greater than the combined propagation delay and multi-path delay-spread of that signal between transmitter and receiver.The sequences themselves are not orthogonal to each other.

In the frequency domain, this ZC sequence can be expressed by Equation (14.1):

$$X_{ZC}^{u}(k) = e^{-j\pi u \frac{k(k+1)}{M_{ZC}}} \qquad \text{(14.1) Zadoff-Chu sequence}$$

where:

$MZC =$ sequence length (preferably a prime number).

$U =$ sequence index, represented by integers prime to MZC.

14.10 Physical Layer (PHY)

LTE PHY uses a framed OFDM (Orthogonal Division Multiplex) structure and the same frame arrangement is used for FDD and TDD, downlink and uplink as illustrated in Figure 14.21.

The downlink access is done using OFDMA (Orthogonal Division Multiple Access) while the uplink uses DFT-S-OFDMA (Discrete Fourier Transform-Shared-Orthogonal Division Multiple

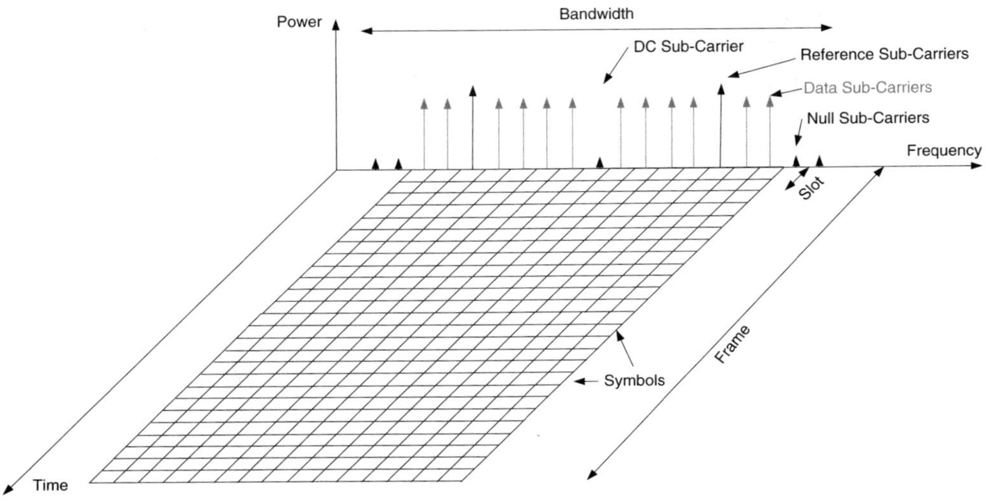

Figure 14.21 OFDMA composition.

Access) also known as SC-OFDMA (Single Carrier-OFDMA). A frame is 10 ms long and the sub-carrier spacing is 15 kHz.

14.10.1 PHY Downlink

Figure 14.22 shows a block diagram for a downlink (eNB to UE) connection. The configuration shown is for full 2 × 2 MIMO. In the downlink, regular OFDMA is used. On the transmit side two Transport Blocks (TBs) are prepared for transmission, assuming spatial multiplexing. A CRC is added to the TB to allow for integrity checking. The TB is segmented or appended into Coding Blocks (CB) and another CRC is added for HARQ purposes. A turbo encoder or a convolutional encoder is used to encode the data at $^{1}/_{3}$ coding rate. The output coding rate is matched to the desired coding rate by puncturing. The data is then mapped to the modulation constellation, according to the estimated path requirement for a 10% BER (Block Error Rate). A pre-coder is applied according to the antenna configuration chosen for the path. The signal is then mapped directly onto the sub-carriers in one or more Resource Blocks (RBs). Each Resource Block is a set of 12 sub-carriers by one symbol. The Reference Signal (RS), known also as pilot is inserted in specific sub-carriers. An IFFT (Inverse Fourier Transform) is then applied to the sub-carrier in the transmission bandwidth. Finally, a CP (Cyclic Prefix) is inserted according to the sector configuration (normal or extended CP). The signal is then sent to each of the antenna's RF equipment.

On the receive side, the signal is received by two RF equipments and the CP is removed. An FFT (Fast Fourier Transform) is applied to the whole bandwidth and the sub-carriers are extracted. The sub-carrier content is then de-mapped from the constellation states to a sequence of bits. The MIMO receiver equalizes and processes these bits, sending them to the two receive branches. The RBs are de-mapped and a Likelihood Receiver Generator (LRG) sends the most likely streams to the turbo decoder for error correction. The CRC is then checked and, if it fails, an HARQ request is sent. The information is stored until received correctly or it will be discarded, up to 8 HARQ tries are done. The data is finally re-assembled or disassembled and each TBs CRC is checked. A failed CRC triggers an ARQ request.

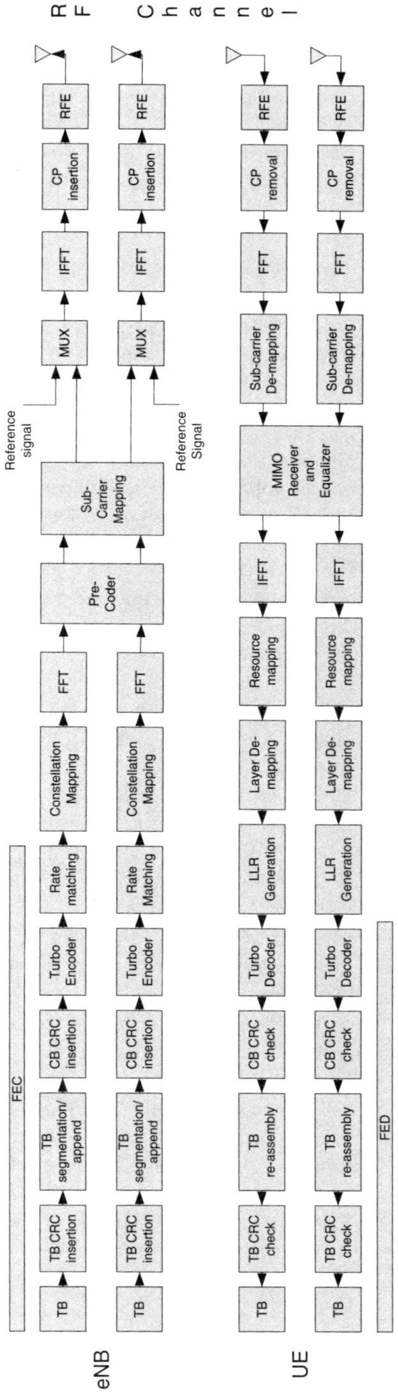

Figure 14.22 Downlink PHY block diagram for 2 × 2 MIMO.

This procedure assures that the number of errors passed to the higher layers is small, although the RF part works at a 10% BER. The advantage of this procedure is a higher throughput and the drawback is processing time jitter. Applications that require a low delay have to target lower BERs.

14.10.1.1 Peak to Average Power Ratio (PAPR)

The OFDM signal is made by the sum of many sub-carriers and the resulting signal can present high peaks. The ratio between the average and the peak power ratio is called the PAPR (Peak to Average Power Ratio).

If all 1000 sub-carriers were identical, the resulting signal would peak at 30 dB above the power level of a single sub-carrier. The probability of this happening is infinitesimal, but peaks of 10 to 15 dB have a larger probability of occurring. It is shown in the literature that a 7 dB PAPR clipping almost does not affect the BER performance, whereas a 3 dB PAPR clipping affects it by 2 to 3 dB.

14.10.2 PHY Uplink

Figure 14.23 shows a block diagram for an uplink (UE to eNB) connection. The configuration shown is for full 2×2 MIMO. In the uplink, DFT-S-OFDMA is used. On the transmit side, two Transport Blocks (TBs) are prepared for transmission, assuming spatial multiplexing. A CRC is added to the TB to allow for integrity checking. The TB is segmented or appended into Coding Blocks (CB) and another CRC is added for HARQ purposes. A turbo encoder or a convolutional encoder is used to encode the data at $1/3$ coding rate. The output coding rate is matched to the desired coding rate by puncturing. The data is then mapped to the modulation constellation, according to the estimated path requirement for a 10% BER (Block Error Rate).

The signal is then sequenced in time and passed through an FFT (Fast Fourier Transform) that decomposes it into frequencies that coincide with the sub-carriers. A pre-coder is applied according to the antenna configuration chosen for the path. The signal is then mapped directly to the sub-carriers in one or more Resource Blocks (RBs). Each Resource Block has a set of 12 sub-carriers by one symbol. The Reference Signal (RS), also known as a pilot, is inserted into specific sub-carriers. An IFFT (Inverse Fourier Transform) is then applied to the sub-carrier in the transmission bandwidth. Finally a CP (Cyclic Prefix) is inserted according to the sector configuration (normal or extended CP). The signal is then sent to each of the antenna's RF equipment.

On the receive side, the signal is received by two RF equipment and the CP is removed. An FFT (Fast Fourier Transform) is applied to the whole bandwidth and the sub-carriers are extracted. The sub-carrier content is then de-mapped from the constellation states onto a sequence of bits. The MIMO receiver equalizes and processes these bits, sending them to the two receive branches. An IFFT is applied to restore the original signal, which has the modulated signals aligned in time. The RBs are de-mapped and a Likelihood Receiver Generator (LRG) sends the most likely streams to the turbo decoder for error correction. The CRC is then checked and, if it fails, an HARQ request is sent. The information is stored until received correctly or it is discarded, and up to 8 HARQ tries are done. The data is finally re-assembled or disassembled and each TBs CRC is checked. A failed CRC triggers an ARQ request.

The decision to use the DFT-S-OFDMA was done to reduce the PAPR, but in practice the reduction is of about 2.5 dB for QPSK and about 0.5 dB for 64QAM. On average we can say that a reduction of 1.5 db can be expected. This gain is lost by a similar loss in frequency diversity.

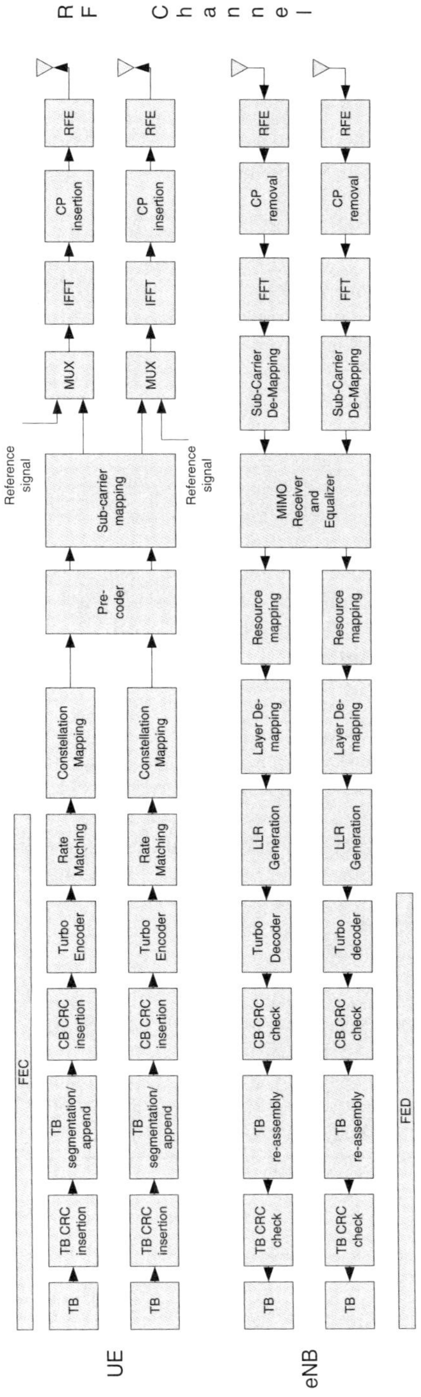

Figure 14.23 Uplink PHY block diagram for 2 × 2 MIMO.

14.11 PHY Structure

LTE uses a framed structure defined by the eNB, for the downlink and uplink. The same basic structure is valid for both directions but the channel allocation is different for each. The uplink frame timing is defined at the eNB, so the UE has to adjust its timing for its signal to arrive at the right moment at the eNB. The sub-carrier spacing is 15 kHz with six standardized bandwidths: 1.4, 3, 5, 10, 15, and 20 MHz

As mentioned in Section 14.10, in the frequency domain, the Physical Layer uses:

- OFDMA (Orthogonal Frequency Division Multiple Access) in the downlink.
- DFT-S-OFDMA (Discrete Fourier Transform- Spread- OFDMA) in the uplink. The spread is done over multiples of 12 sub-carriers and is erroneously called SC-OFDMA (Single Carrier-OFDMA). In reality, the modulation is still applied to each sub-carrier, but because each set of sub-carriers represent a single waveform, it was termed a single carrier.

Null carriers are used to reduce neighbor channel interference, resulting in 90% usage efficiency with the exception of the first bandwidth, as shown in Table 14.8. The sub-carriers are grouped in groups of 12 consecutive sub-carriers.

In the time domain, the 3GPP choose a base time unit of Ts = 1/(15,000*2048) = 32.55 ns to define the different PHY durations.

LTE defines a TTI (Transmission Time Interval) as a frame of 10 ms (307,200 Ts) that represents the transmission periodicity. This frame is further divided into 10 sub-frames, each sub-frame with two slots of 0.5 ms (15,360 Ts). The frame layout is shown in Figure 14.24.

The 15 KHz sub-carrier spacing implies in symbol duration of:

$$\text{Tsymbol} = 1/15{,}000 = 66.666 \text{ us}$$

Consequently, a symbol has a duration of 2048 Ts.

Two Cyclic Prefixes (CPs) are possible, one for distances up to 1.406 km (4.687 µs, $1/16^{\text{th}}$ of symbol duration or 144 Ts) and another for distances up to 5 km (16.666 µs, $1/4^{\text{th}}$ of symbol duration or 512 Ts).

The shorter CP allows 7 symbols per slot, but when using this shorter version, the first CP has to be extended to 5.2 us (160 Ts) to adjust for the frame size. The longer CP allows for 6 symbols per

Table 14.8 Channel bandwidth

Channel bandwidth (MHz)	1.4	3	5	10	15	20
Transmission bandwidth (MHz)	1.08	2.7	4.5	9	13.5	18
Bandwidth efficiency (%)	77.1	90.0	90.0	90.0	90.0	90.0
FFT size	128	256	512	1024	1536	2048
Number of used sub-carriers	72	180	300	600	900	1200
Number of sub-carrier groups	6	15	25	50	75	100

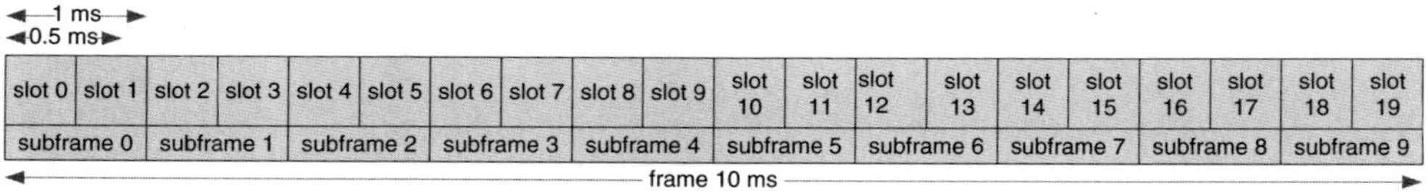

Figure 14.24 FDD frame in the time domain.

slot. The slot structure with the short CP is shown in Figure 14.25 and the structure with the long CP is shown in Figure 14.26.

In principle, both CPs can be used in a network, but they should be uniform per sector (cell) basis. Different CPs could be used for the downlink and uplink. The CP duration has to accommodate the multipath spread and timing deviations present mainly in the uplink.

A Resource Element (RE) is the smallest unit to which data can be assigned and is defined as one sub-carrier wide with one symbol duration. Depending on the modulation, one to six bits can be assigned to a resource element.

The smallest unit that can be scheduled for transmission is a Resource Block (RB), which is 12 sub-carriers wide (180 kHz) and has one slot duration (6 or 7 symbols, depending of the cyclic prefix used). Demodulation Reference Signals use RB resources in the downlink, reducing the number of REs available for data. Based on these parameters, Table 14.9 shows the information capacity for different bandwidths.

A Resource Block with the normal CP has 84 Resource Elements, as shown in Figure 14.27 and a Resource Block with the extended CP has 72 Resource Elements, as shown in Figure 14.28.

The location of the Physical Channels in the frame is defined in the next sections.

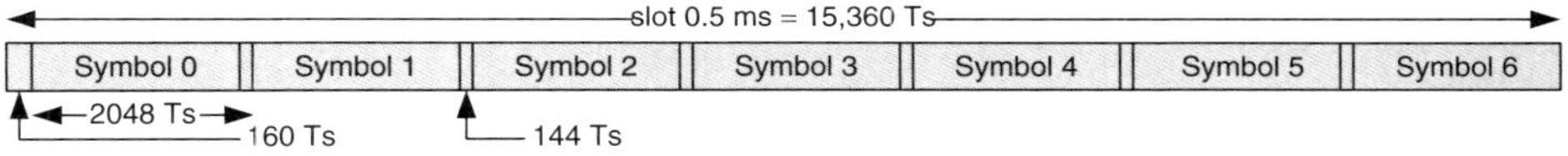

Figure 14.25 Slot structure with short CP.

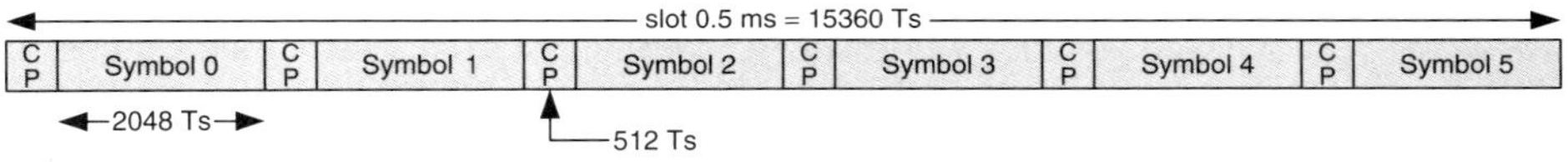

Figure 14.26 Slot structure with long CP.

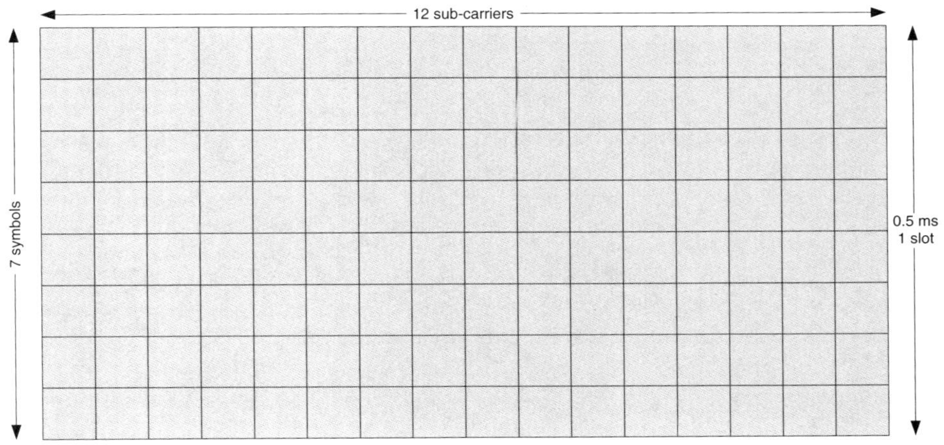

Figure 14.27 Resource block with short CP.

Table 14.9 RF channel bandwidth and information capacity

Channel bandwidth (MHz)	1.4	3	5	10	15	20
Transmission bandwidth (MHz)	1.08	2.7	4.5	9	13.5	18
Number of used sub-carriers	72	180	300	600	900	1200
Number of sub-carrier groups	6	15	25	50	75	100
			Normal CP			
Number of Resource Blocks per frame	120	300	500	1000	1500	2000
Number of Resource Elements per frame	10,080	25,200	42,000	84,000	126,000	168,000
Number of Resource Elements per second	1,008,000	2,520,000	4,200,000	8,400,000	12,600,000	16,800,000
Minimum Throughput with no overhead (Mbit/s)	2.02	5.04	8.40	16.80	25.20	33.60
Maximum Throughput with no overhead (Mbit/s)	6.05	15.12	25.20	50.40	75.60	100.80
			Extended CP			
Number of Resource Blocks per frame	120	300	500	1000	1500	2000
Number of Resource Elements per frame	8640	21,600	36,000	72,000	108,000	144,000
Number of Resource Elements per second	864,000	2,160,000	3,600,000	7,200,000	10,800,000	14,400,000
Minimum Throughput with no overhead (Mbit/s)	1.73	4.32	7.20	14.40	21.60	28.80
Maximum Throughput with no overhead (Mbit/s)	5.18	12.96	21.60	43.20	64.80	86.40

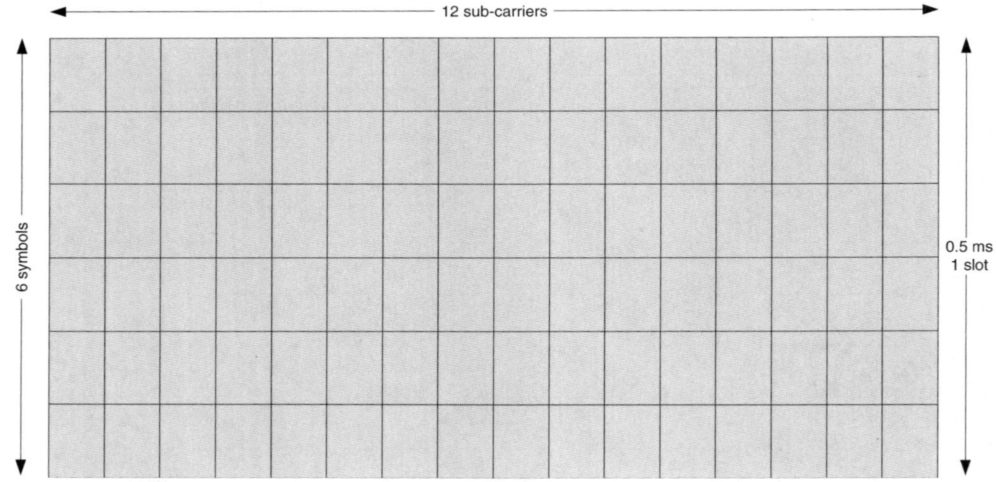

Figure 14.28 Resource block with long CP.

14.11.1 Downlink Physical Channels

The downlink frame is presented in Figure 14.37 on p. 456.

14.11.1.1 PSS (Primary Synchronization Signal)

The PSS has 72 central sub-carriers reserved at symbol 6 (or 5 if extended CP is being used) of slots 0 and 10 of a frame. This signal, however, uses only 62 sub-carriers, so it can be detected by a 64 bit FFT. It uses one of three ZC (Zadoff-Chu) sequences of length 63, with roots M = 25, 29, and 34. This allows the receiver to synchronize to the frame and detect the frame identity (0, 1, or 2, respectively to the roots). Each ZC sequence is trimmed to 62 bits before use.

14.11.1.2 SSS (Secondary Synchronization Signal)

The SSS has 72 central sub-carriers reserved at symbol 5 (or 4 if extended CP is used) of slot 0 and 10 of a frame. As the PSS, the SSS only uses 62 sub-carriers, so it can be detected by a 64 bit FFT.

The SSS sequence is made by interleaving, in the frequency domain, two BPSK modulate sequences of length 31, called SSCX and SSCY. These sequences are cyclic shifts of a single M-sequence of length 31, and are derived from the physical layer cell identity group. The Physical layer Cell Identity (PCI) can then be derived by combining the PSS and SSS information. There are a total of 3 (0 to 1) × 168 (0 to 167) = 504cell (sector) identities.

An m-sequence is a pseudo-random binary sequence which can be created by a shift register of length n, resulting in a sequence length of 2^n-1. The cyclic shifts (m_0 and m_1) are defined for the 168 cell identities in the specifications.

In SSS each of the two m-sequences are further scrambled by binary scrambling codes of length 3. These codes are also based on the cell ID and are different for each message, as well as for the first and the second occurrences of SSS in a frame.

14.11.1.3 RS (Reference Signal)

Reference Signals are required to estimate the DL channel for coherent detection. Four RSs are sent per Resource Block (RB) for antenna port 0 and 1 and two RSs for ports 2 and 3. The positions reserved in an RB for each antenna are shown in Figure 14.29.

An antenna cannot transmit when a symbol is being used for the RS transmission by another antenna, to allow for a clean channel estimate. A 4 port solution is only considered for slow varying channels, so antennas 2 and 3 do not need as many RS. The channel response is estimated based on the interpolation or average of a set of neighbor RSs.

The RS is a complex signal made by the product of the PSS and SSS signal. Consequently, there are 504 PCIs (Physical-layer Cell Identities) in LTE. Those identities should be planned by the network designer.

Antenna port 4 is used for MBSFN and port 5 for Dedicated Reference Signals (DRS) for beam-forming.

14.11.1.4 PBCH (Physical Broadcast Channel)

The PBCH carries the MIB (Master Information Block) and has a payload of only 24 bits. The MIB informs DL bandwidth, the frame number and the PHICH configuration. Additionally, DL antenna configuration, 40 ms timing and the TX diversity type used for PBCH are also informed.

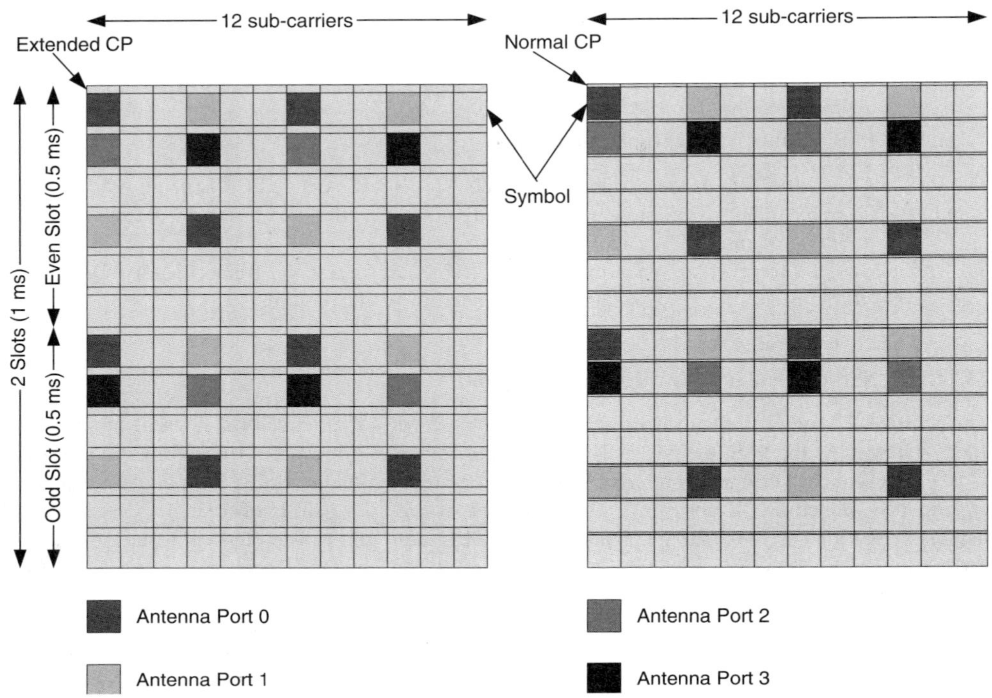

Figure 14.29 Antenna port reference signal allocation.

This channel information is essential for decoding the remaining of the frame, so it is transmitted in a highly reliable form. A 16 bit CRC is added to the payload and a $^1/_3$ convolutional code is then applied, resulting in a total of 120 bits. The modulation used is QPSK, resulting in 60 symbols.

The channel has assigned the first four symbols of slot 1 on 72 sub-carriers centered on the carrier frequency (4 sets of 6 RBs). This implies that the information is always available at the same location regardless of the bandwidth. The channel information fits into one set of RBs, so it has a repetition of 4 times per frame. The PBCH TTI (Transmission Time Interval) is 40 ms, so the channel is repeated again in the three following frames, to a total of 16 repetitions (12 dB gain).

14.11.1.5 PCFICH (Physical Control Format Indicator Channel)

Control channels are allocated to the first 1, 2, or 3 symbols of each sub-frame, across all sub-carriers, as shown in Figure 14.30. The number of symbols is defined by the CFI (Control Format Indicator) which informs the number of PDCCH OFDM symbols per sub-frame and is sent in this channel. Each of the three CFI values has a 32 bit codeword associated to it, resulting in 16 symbols with QPSK modulation. These symbols are spread in REGs (Resource Group Elements) of 4 symbols and their exact location depends on the cell PCI.

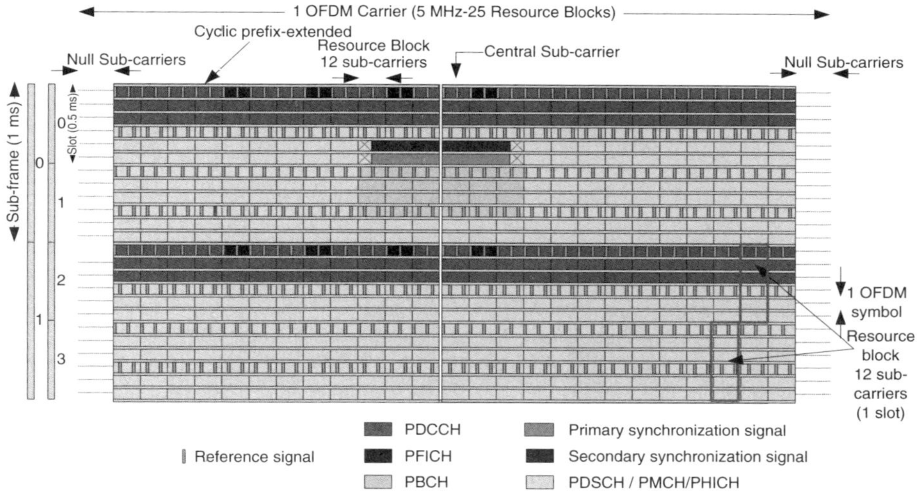

Figure 14.30 PFICH and PDCCH PHY location.

14.11.1.6 PDCCH (Physical Dedicated Control Channel)

The PDCCH carries the Downlink Control Information (DCI) message. This message carries the PDSCH assignment, PUSCH RB grants, modulation and coding scheme, HARQ type, PUSCH power control and CQI request. It also maps the PUSCH in its sub-frame (1 ms).

The DCI message varies in size depending on the number of RBS to be assigned and type of information carried, varying from 26 to 176 bits. This information has 16 bit CRC attached and a $^1/_3$ convolutional code applied to it.

A DCI is mapped to a CCE (Control Channel Element) which comprises four Resource Element Groups (REGs), each containing 8 bits. REGs are QPSK modulated.

14.11.1.7 PHICH (Physical Hybrid ARQ Indicator Channel)

This channel provides the HARQ response to the uplink messages. An ACK (Acknowledgement) is signaled as 1 and a NACK (Not Acknowledged) as 0. The channel information is carried by REGs, supporting from 4 to 8 simultaneous replies. This is achieved by combining the BPSK modulated channel information with an orthogonal Walsh code and with a cell-specific pseudo-random scrambling sequence. To assure a reliable reception the code is repeated 3 times.

14.11.1.8 PDSCH (Physical Downlink Shared Channel)

The information to be sent in this channel follows Figure 14.31. The transport block has a 24 bit CRC (Cyclic Redundancy Code) attached to it. This transport block is then segmented into one or more code blocks, with a minimum size of 40 bits and maximum size of 6144 bits. Another 24 bit CRC is attached to each code block.

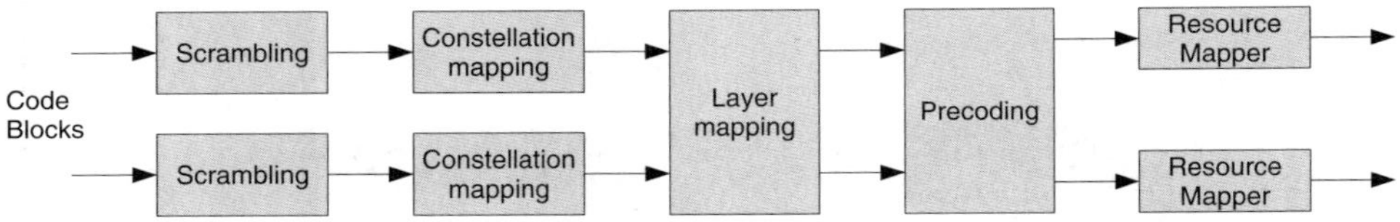

Figure 14.31 PDSCH encoding.

Each code block undergoes turbo coding with a code rate of 1/3. This rate may be above the estimated required rate for the transmission, in this case, the output rate is adjusted by puncturing the code block. Finally, the segmented code blocks are concatenated and sent for transmission.

One or two processed transport blocks can be sent at the same time depending on the antenna algorithm used for transmission. The information to be sent is scrambled with a cell-specific pseudo-random sequence. This improves the reception as information from other cells is descrambled incorrectly and, therefore, appears as uncorrelated noise. The scrambled information is then modulated in QPSK, 16QAM or 64QAM, depending on channel conditions.

The complex symbols are then mapped to 1, 2, 3, or 4 layers, depending on the number of antennas and algorithm used. Precoding is then applied according to the antenna algorithm used: Single Port Transmission, Transmit Diversity, or Spatial Multiplexing, the latter one with CDD (Cyclic Delay Diversity) or not.

CDD applies different delays to the antenna transmission to improve fading response. The precoding alternatives are shown in Figure 14.32. After that, the information is sent to the antennas. The PDSCH is mapped to all remaining empty spots in the PHY.

HARQ retransmissions are sent in this channel after the original transmission if a NACK is received. Up to 8 HARQ processes can be tracked in parallel. Each retransmission should occur 8 ms after the previous one. The retransmissions may be smaller than the original transmission, by sending only additional error correction bits.

14.11.2 Uplink Physical Channels

The uplink frame is shown in Figure 14.38 on p. 457. Note that it represents the uplink frame as received by the eNB. This means that the UE have to adjust their timing prior to transmitting, so the

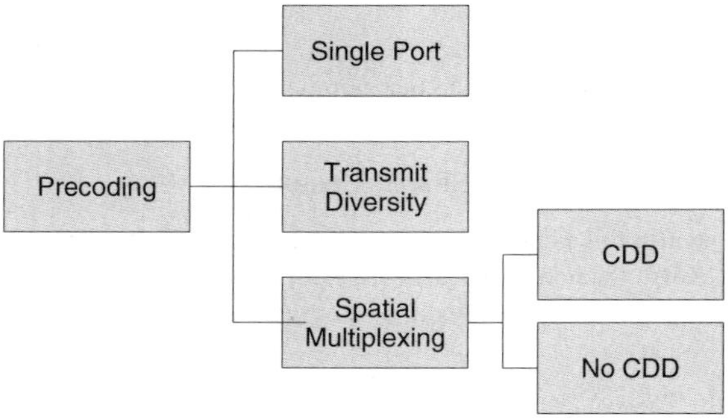

Figure 14.32 Antenna precoding types.

information from different UEs arrives consistently. It is a sound practice to leave symbols in between resource blocks to accommodate any possible variations in timing.

14.11.2.1 DRS (Demodulation Reference Signal)

This signal is used by eNB to estimate the channel and perform coherent demodulation of the PUSCH and PUCCH. The DRS uses one symbol per slot of each Resource Block that transmits data. It can be configured as a Long Block (LB) with one symbol duration or as a Short Block (SB) divided in two half symbols, as illustrated respectively, in Figures 14.33 and 14.34.

The LB DRS has duration of 1 symbol (66.66 μs) and is spread over the channel TBs sub-carriers. The SB DRS has approximately half the duration (25.88 μs for the normal CP and 25 μs for the extended CP), implying that each symbol uses the spacing of two sub-carriers. The short block use does not provide improvement at low and medium speeds, but helps in channel equalization for high speeds (above 120 km/h).

A cell should assign a specific reference signal to its UEs which would individualize their transmission from transmissions from other cells UEs. Uplink DRS are allocated to each cell to be used by the UEs. In principle, each UE uses different sub-carrier within a cell, so there is no conflict between them, with the exception when Uplink Collaborative MIMO (UCM) is used and two UEs transmit using the same RBs. In this case, more than one DRS should be available per cell and they should be orthogonal, so both transmissions can be properly identified.

The number of codes that can be generated according to the number of RBs is given in Table 14.10. The standard requires a minimum of 30 codes to be assigned to different cells, plus additional orthogonal codes to be used within a cell for UCM. A possible choice for these codes was the Zadoff-Chu (ZC) sequences, but not enough sequences can be generated for 1 and 2 RBs. The solution was to do create regular QPSK sequences (non-Zadoff-Chu) for 1 and 2 RBs.

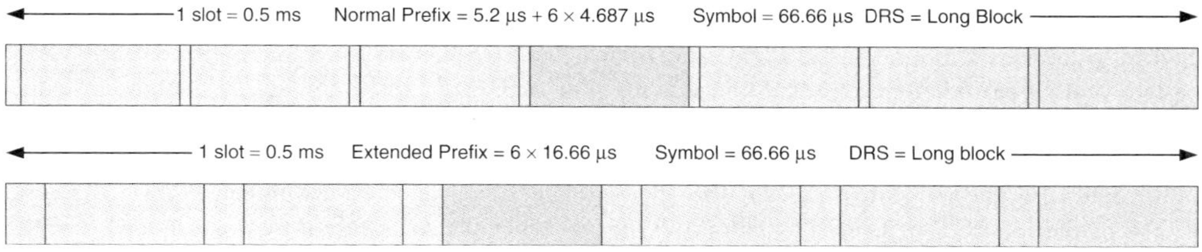

Figure 14.33 Demodulation reference signal location with long block configuration (upstream).

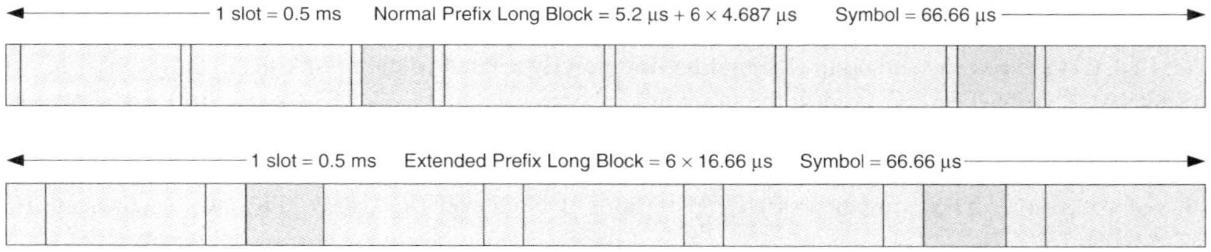

Figure 14.34 Demodulation reference signal location with short block configuration (upstream).

Table 14.10 Number of codes per resource blocks

Resource blocks	Possible ZC codes	QPSK sequences	Allocated sequence groups
1	6	30	30 single
2	12	30	30 single
3	30		30
4	44		30
5	60		30

The Zadoff-Chu sequences form 30 sequence groups, each allocated to a different cell. A group is characterized by its base sequence, but can also have other sequences with equally spaced time shifts, resulting in a set of orthogonal sequences. To keep orthogonality, the shift should be larger than the multipath delay spread. Assuming a delay spread of 1 CP, we get 12 (66.66/5.55) sequences for normal CP and 4 (66.66/16.66) sequences for extended CP. This uses cyclic shifts to get orthogonal sequences for UCM use, but only if using more than 3TBs.

Sequence planning has to be done on a cell (sector) basis, with different sequences assigned to each cell, so possible interference between similar sequences is minimized.

Optionally, sequence hopping can be used. The standard defines 17 hopping configurations, which hop the 30 sequences over the 20 slots of a frame. In this case the 17 hopping patterns should be assigned to the cells in an optimized way. In case of hopping, the shift offset is fixed for all slots.

The DRS overhead is about 14% for the normal CP and 16.6% for the extended CP.

14.11.2.2 SRS (Sounding Reference Signal)

The SRS is used to assess channel quality as an input to sub-carrier scheduling of TBs. It is also used to estimate initial MCS and power control. This signal uses one symbol for the whole UE bandwidth. There are 16 possible configurations of SRS in a frame, from 0 to 10 SRS. The SRS transmission is always in the last symbol of an assigned sub-frame. The SRS can be programmed on a case-by-case basis or with a periodicity between 2 and 320 ms. The SRS size can be configured between 4 and 96 RBs, but it is limited by the PUCCH slots on both sides of the bandwidth and by the UE power. The SRS signal is sent as an SC-OFDMA transmission using QPSK.

Several UEs may need to send an SRS in the same frame and to accommodate this, LTE uses IFDMA (Interleaved Frequency Division Multiple Access). The same reference signals used for DRS are used for SRS, but they are sent every second sub-carrier, so it is possible to interleave sequences from two UEs. The SRS represents an overhead of 7% as the symbol at the end of the sub-frame can seldom be used.

14.11.2.3 PUCCH (Physical Uplink Control Channel)

The PUCCH is used to send control signaling not directly related to the uplink data being sent, such as ACK/NACK to packets received in the downlink, channel quality CQI, MIMO feedback for downlink transmissions and Scheduling Requests (SR).

A PUCCH region is made by an RB on one side of the band followed by another RB on the other side of the frame. The number of PUCCH regions depends on the bandwidth and varies from 1(2 RBs) to 16 (32 RBs).

Several UEs can multiplex their data on a single PUCCH region. This is done by using CDM (Code Division Multiplex). A Zadoff-Chu sequence is assigned to each cell and is modulated with the

UE QAM data. Each UE is assigned a different cyclic shift, so several UEs can transmit in the same PUCCH region. The number of shifts depends on the channel variability, up to 12 shifts are possible for well-behaved channels. Each orthogonal channel can support up to 22 bits.

There are 7 formats specified for PUCCH and they carry HARQ ACK/NACK and CQI information in different combinations.

14.11.2.4 PUSCH (Physical Uplink Shared Channel)

This channel information is processed similarly to the PDSCH.

HARQ retransmissions are sent in this channel after the original transmission if a NACK is received. Up to 8 HARQ processes can be tracked in parallel. Each retransmission should occur 8 ms after the previous one. The retransmissions may be smaller than the original transmission if adaptive HARQ is used.

When control data is scheduled to be transmitted in a sub-frame to which a PUSCH was assigned, it may be transmitted in the PUSCH instead of waiting for the PUCCH. In this case, signaling data is assigned to different resource elements and the multiplexing is done prior to the FFT processing in the FFT-S-OFDMA process.

14.11.2.5 PRACH (Physical Random Access Channel)

The PRACH allows UEs to initially access the system and adjust their timing, so their signal can arrive at the eNB synchronized with the eNB frame timing. The signal transmitted by the UE arrives at the eNB after a trip delay. Table 14.11 gives trip delays for several distances.

The UE does not know how far it is from the eNB, so it sends initially its PRACH transmission as it were co-located with the eNB. This means that the UE transmission arrives a round trip delay later. The PRACH PHY is illustrated in Figure 14.35.

Several UEs may transmit simultaneously aiming to keep a low probability of conflict. A fixed bandwidth of 6 RBs was chosen for the PRACH, as a compromise between overhead and probability of conflict. Four formats were defined with 1, 2, and 3 sub-frame durations.

Format 0 is defined for one sub-frame duration and corresponds to 1 ms. This time has to accommodate the PRACH symbol, its CP (Cyclic Prefix) and the round trip delay. This frame duration should accommodate the PRACH symbol, its CP (1/12th of the symbol), and the GT (Guard Time). It was dimensioned for a cell radius of 7.5 km (100 µs round trip time). Considering that the UE requires about 30 µs to process the signal, the resulting PRACH symbol duration was left with 800 µs. The other formats use 2 ms and 3 ms durations and can support cell radii of up to 50 km.

Table 14.11 Round trip delay for different distances

Distance (km)	Trip time (µs)	Round trip time (µs)
1	3.3	6.7
2	6.7	13.3
5	16.7	33.3
10	33.3	66.7
15	50.0	100.0
20	66.7	133.3
25	83.3	166.7
50	166.7	333.3
100	333.3	666.7

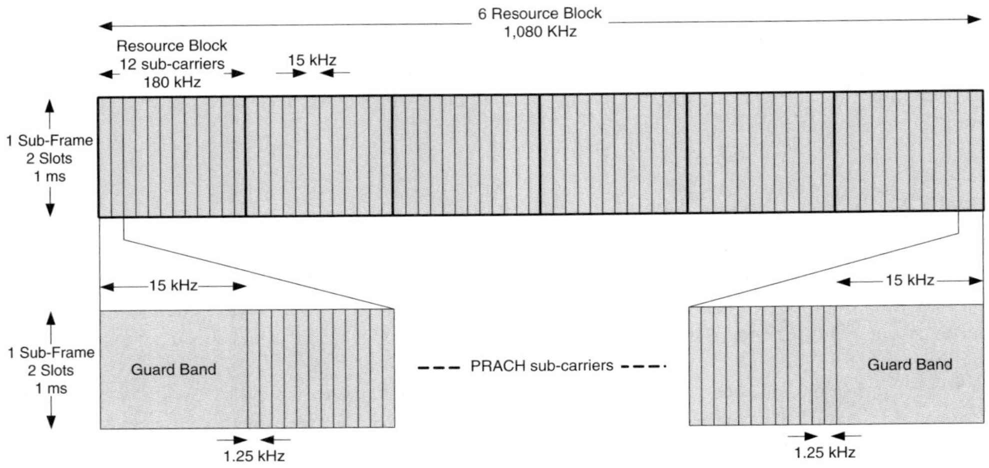

Figure 14.35 PRACH PHY.

The 800 μs PRACH forms only one symbol, so the carrier spacing of this symbol should be 1.25 kHz. This means that the sub-carrier spacing for the PRACH sub-carriers is 1.25 kHz, resulting in 864 sub-carriers in the reserved 1080 kHz bandwidth of 6 RBs. A total of 30 kHz is left as guard band (half on each side), so 839 subcarriers are left for transmitting the signal.

To avoid conflicts between UEs that arrive simultaneously, a large pool of orthogonal codes must be available. This orthogonality is provided by a set of Zadoff-Chu sequences, which are not orthogonal to each other but can be made orthogonal by cyclic shifts. Zadoff-Chu sequences have to have a prime number of elements; LTE chose 839 as the number of sub-carriers to satisfy this requirement.

LTE designers considered that 64 sequences (signatures) should suffice. The sequence transmitted by the UE is called the preamble. The number of orthogonal cyclic shifts that can be applied on one sequence depends on the distance between UE and eNB and varies from 1 shift for 100 km to 64 shifts for 0.7 km. Additional sequences should be used to provide the 64 sequences, although those will not be orthogonal to each other. In practice, this limits the cell size to 6 km.

14.11.3 Downlink PHY Assignments

Resource blocks are assigned symmetrically around the central sub-carrier. The central sub-carrier is not used as it is subject to leakage of the OFDM carrier. PHY assignments are illustrated in Figure 14.36.

The central carrier is not used, as leakage is expected from the modulator circuit and this would affect the quality of the transmitted signal. The remaining sub-carriers are evenly spaced around the sub-carrier, so for bandwidths with even number of sub-carrier groups (1.4, 10 and 20 MHz), the central carrier falls in between the groups, while for bandwidths with an odd number of sub-carriers (3, 5 and 15 MHz), it will fall in the middle of a group.

The physical signals and channels assignment for downlink transmission are described next.

The Reference Signals (RS), are the equivalent of pilots, and are assigned to sub-carrier 0 and 6 in symbols 0 and 3 of each Resource Block. When the short cyclic prefix is used, they are assigned to symbols 0 and 4.

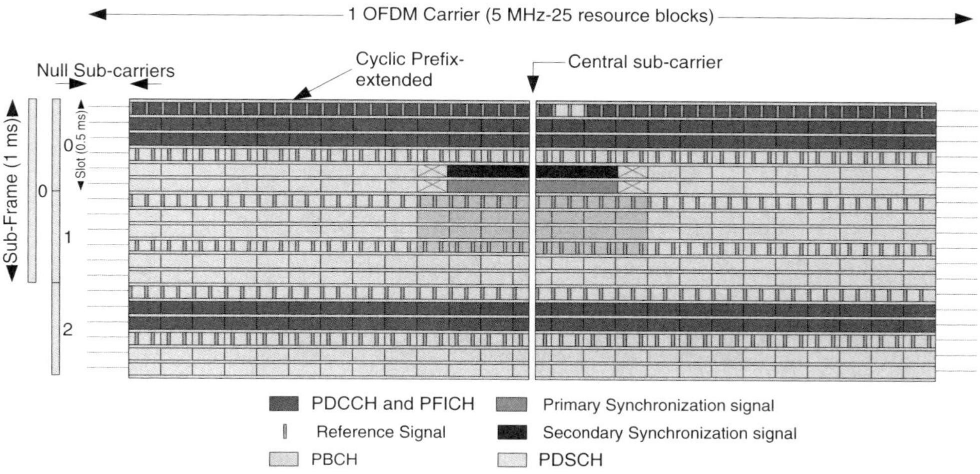

Figure 14.36 Central sub-carrier allocation to RS, PSS, SSS, PDCCH, PFICH, PBSCH and PDSCH.

The Primary Synchronization Signal (PSS) is assigned to the middle 62 sub-carriers (31 on each side of the central carrier) in the last symbol of slot 0. This is done, so a 64 FFT can be used to detect the signal. This signal uses a Zadoff-Chu sequence, which has the same power level but a variable phase. It defines three different cell IDs.

The Secondary Synchronization Signal (SSS) is assigned to the middle 62 sub-carriers (31on each side of the central carrier) in the one before last symbol of the first slot 0. This is done so a 64 FFT can be used to detect the signal. This signal gives one of 168 possible cell identity group numbers.

The PSS and SSS are reassigned on slot 10.

The PBCH is assigned to the middle 84 sub carriers (42 on each side of the central carrier) in the first 4 symbols of the second slot (slot 1, symbols 0, 1, 2 and 3). This channel uses only QPSK modulation.

The PFICH carries the number of OFDM symbols used to carry the PDCCH information in a sub-frame. It is allocated to the first OFDM symbol of every sub-frame and its assignment to sub-carriers follows the Cell ID.

The PDCCH is the physical channel that carries channel allocation and control information. It is allocated to symbols 1 and 2 of the first slot of every sub-frame. The exact number of symbols is specified by the PFICH channel.

The remaining resource elements in each RB can then be allocated to the remaining downlink channels: PDSCH, PMCH and PHICH.

The overall configuration of the PHY frame is shown in Figure 14.37.

14.11.4 Uplink PHY Assignments

As explained in Section 14.10, LTE uses DFT-S-OFDMA in the uplink. The uplink PHY is illustrated in Figure 14.38. In the uplink the central carrier is placed in the middle of the UE transmitting bandwidth, which is always a multiple of 180 kHz. This implies that there will be some leakage in band and may require special filtering of the carrier frequency.

The physical signals and channels assignment for uplink transmission are shown next.

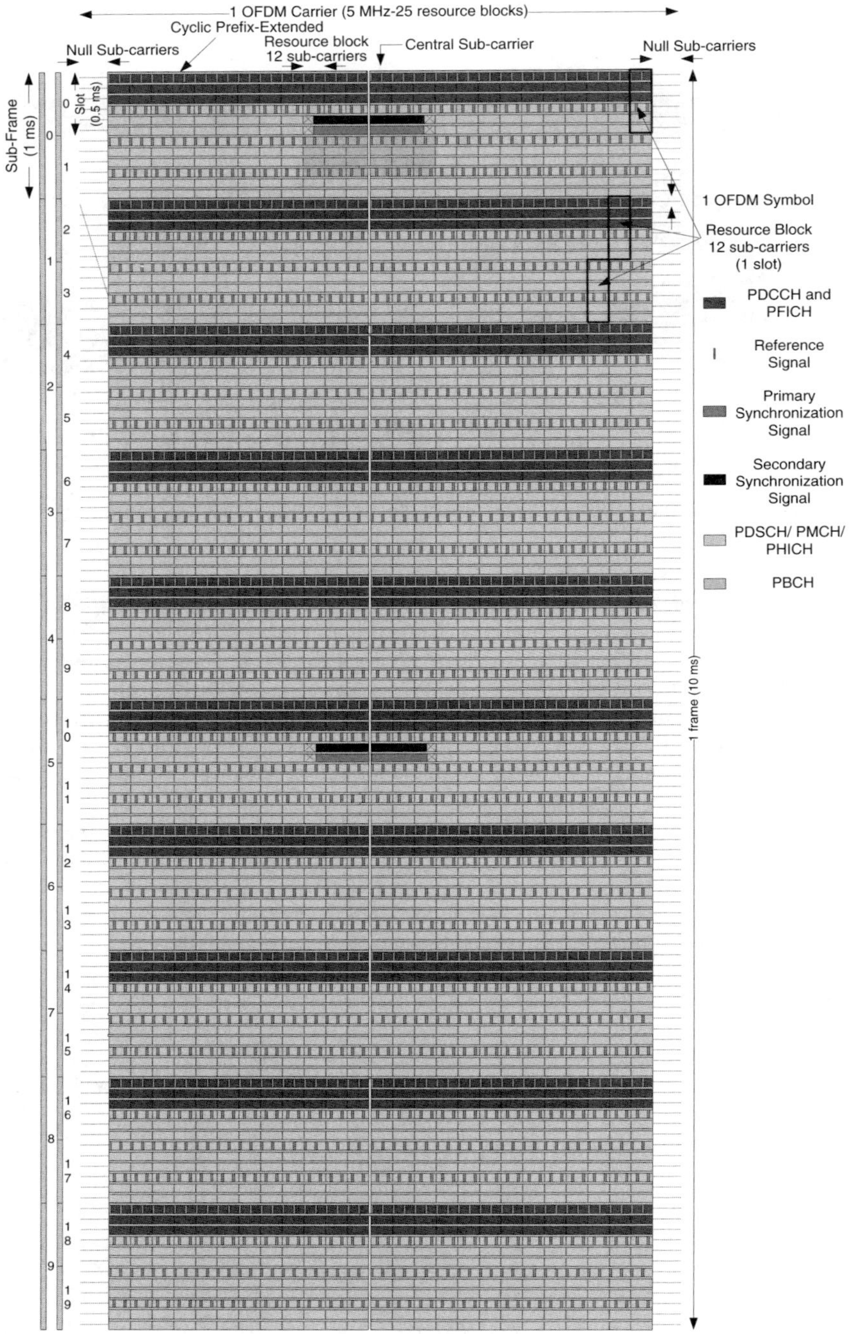

Figure 14.37 LTE PHY frame.

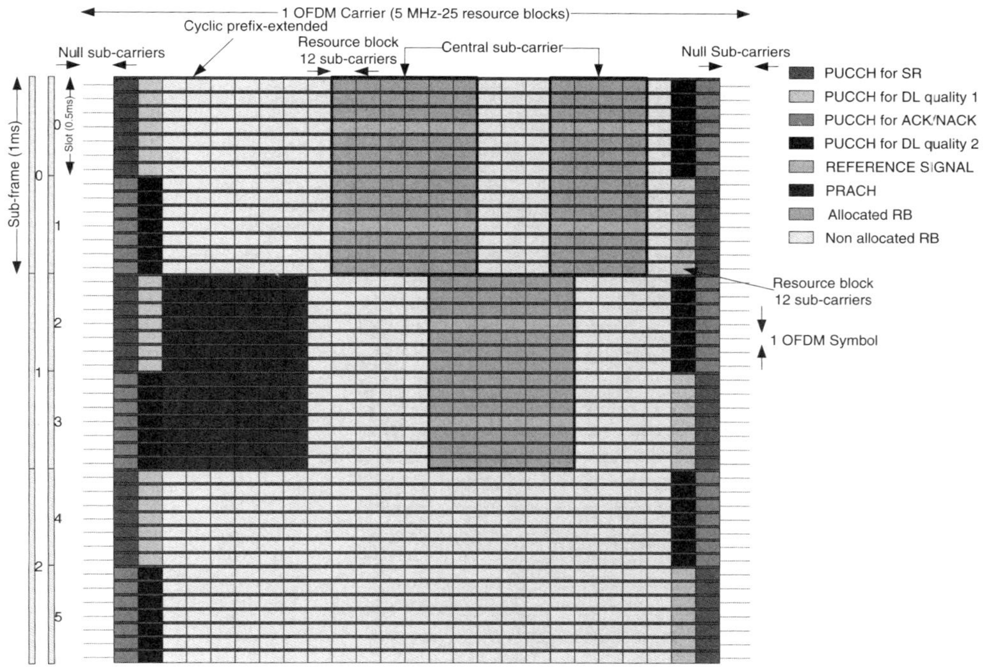

Figure 14.38 Uplink PHY detail.

The demodulation Reference Signal is assigned to the 4th symbol of each slot, when a normal CP is used and in the 3rd symbol when extended CP is used, including in the control channels area. A complete uplink PHY frame is illustrated in Figure 14.39.

14.12 PHY TDD

The TDD LTE frame structure has the same structure as the FDD one, but sub-frame 1 and sub-frame 6 (optionally) are used for changing the TDD direction, as illustrated in Figure 14.40.

There are 7 uplink–downlink configurations, as shown in Table 14.12. Several configurations of the TDD switching frame are defined, as seen in Tables 14.13 and 14.14.

14.13 Multimedia Broadcast/Multicast Service (MBMS)

This is an IP-based broadcast service (downlink only), like mobile TV. The system can support up to sixteen 300 kHz mobile TV channels in a single 5 MHz carrier bandwidth. The transmission is done only on the downlink direction (no uplink) and it can be an isolated system or part of a regular LTE transmission in which a part of the band is used for MBMS.

MBMS will compete with DVB-T (Digital Video Broadcasting – Terrestrial), DVB-H (Digital Video Broadcasting – Handheld) and DMB (Digital Mobile Broadcast), but as it can be deployed in coexistence with regular LTE services, it benefits from the LTE infrastructure and does not require

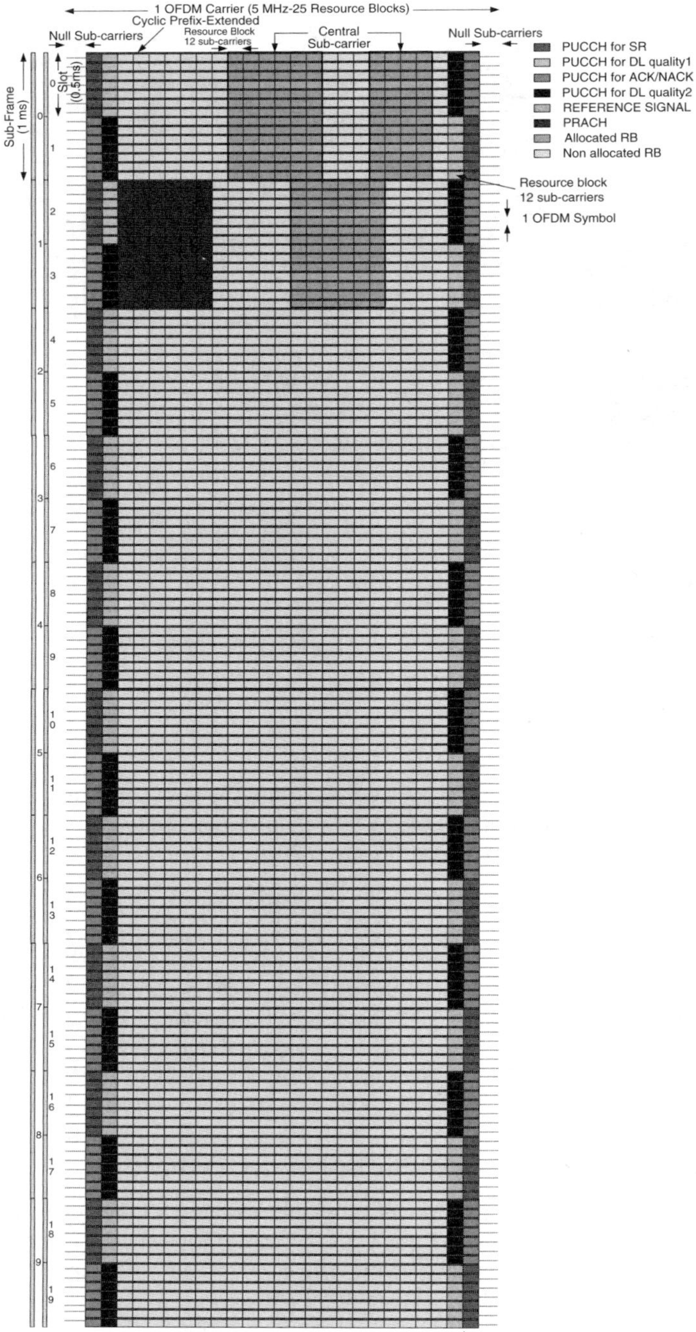

Figure 14.39 Uplink PHY frame.

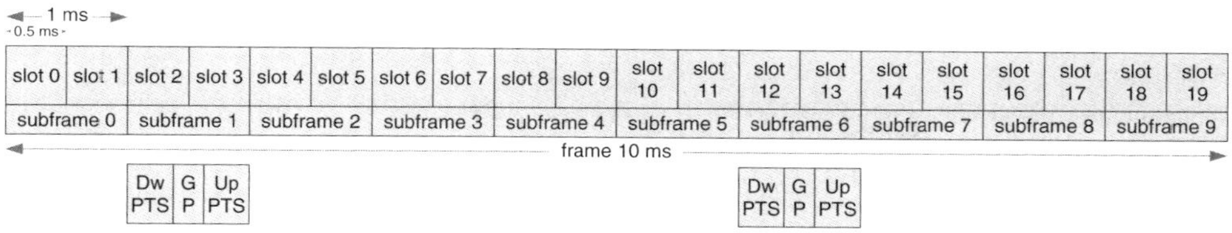

Figure 14.40 TDD frame.

Table 14.12 TDD switching configurations

Up/Down configuration	TDD periodicity (ms)	0	1	2	3	4	5	6	7	8	9	D/U
0	5	D	S	U	U	U	D	S	U	U	U	0.3
1	5	D	S	U	U	D	D	S	U	U	D	1.0
2	5	D	S	U	D	D	D	S	U	D	D	3.0
3	10	D	S	U	U	U	D	D	D	D	D	2.0
4	10	D	S	U	U	D	D	D	D	D	D	3.5
5	10	D	S	U	D	D	D	D	D	D	D	8.0
6	5	D	S	U	U	U	D	S	U	U	D	0.6

Key
D = downlink frame
U = uplink frame
S = TDD switching frame

Table 14.13 TDD switching configurations (normal cyclic prefix)

TDD configuration	DwPTS	GP	UpPTS	DwPTS	GP	UpPTS	DwPTS	GP	UpPTS
		T_s			ms			km	
0	6592	21936	2192	0.21	0.71	0.07	64	214	21
1	19760	8768	2192	0.64	0.29	0.07	193	86	21
2	21952	6576	2192	0.71	0.21	0.07	214	64	21
3	24144	4384	2192	0.79	0.14	0.07	236	43	21
4	26336	2192	2192	0.86	0.07	0.07	257	21	21
5	6592	19744	4384	0.21	0.64	0.14	64	193	43
6	19760	6576	4384	0.64	0.21	0.14	193	64	43
7	21952	4384	4384	0.71	0.14	0.14	214	43	43
8	24144	2192	4384	0.79	0.07	0.14	236	21	43

additional spectrum. Content applications follow a long tail curve and MBMS can be seen as complementary to the technologies listed above, which will satisfy the demand for high applications, while MBMS will better serve the many low usage applications. This is illustrated in Figure 14.41.

MBMS can be implemented as a single or multi-cell transmission, in the latter case, cell contents should be synchronized so the UE can soft-combine the energy of multiple transmitters. This concept is also known as Single Frequency Network (SFN).

Table 14.14 TDD switching configurations (extended cyclic prefix)

TDD Configuration	DwPTS	GP	UpPTS	DwPTS	GP	UpPTS	DwPTS	GP	UpPTS
		T_s			ms			km	
0	7680	20480	2560	0.25	0.67	0.08	75	200	25
1	20480	7680	2560	0.67	0.25	0.08	200	75	25
2	23040	5120	2560	0.75	0.17	0.08	225	50	25
3	25600	2560	2560	0.83	0.08	0.08	250	25	25
4	7680	17920	5120	0.25	0.58	0.17	75	175	50
5	20480	5120	5120	0.67	0.17	0.17	200	50	50
6	23040	2560	5120	0.75	0.08	0.17	225	25	50

Key
DwPTS = Downlink Pilot Time Slot
GP = Guard Period
UpPTS = Uplink Pilot Time Slot

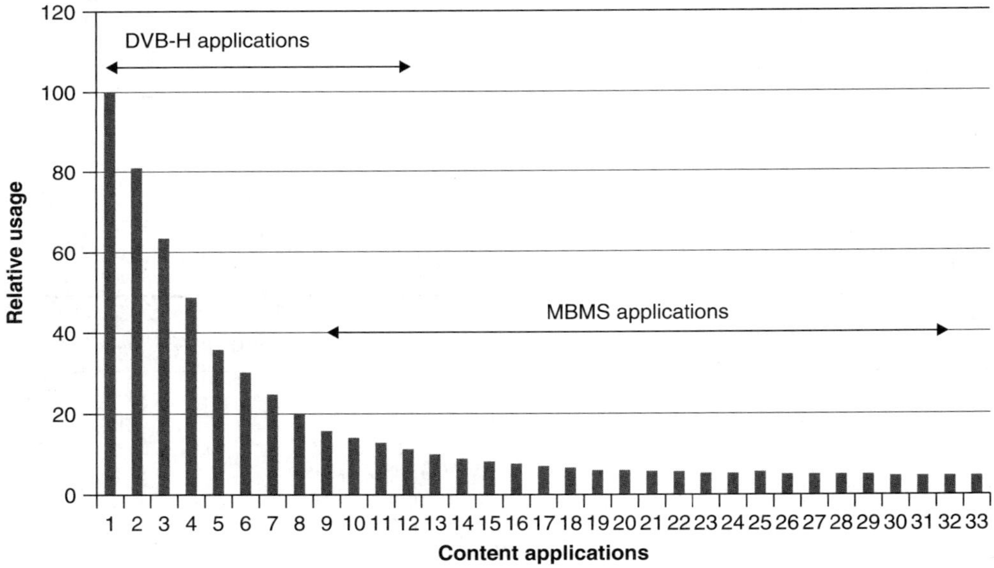

Figure 14.41 Application areas for DVB-H and LTE MBMS.

MBMS cells are foreseen to provide service to large areas, so they should cope with additional multipath. These large cells are subject to long multipath and require an extended CP. To keep the symbol to CP ratio in 1/4, the symbol duration has to be extended, which is achieved by doubling the number of sub-carriers, reducing the sub-carrier spacing to 7.5 kHz and, consequently, obtaining a symbol duration of 133.3 μs. The cyclic prefix then becomes 33.3 μs. This prolonged cyclic prefix is required so the network can be operated as a Single Frequency Network (SFN) and can accept signals from many sites.

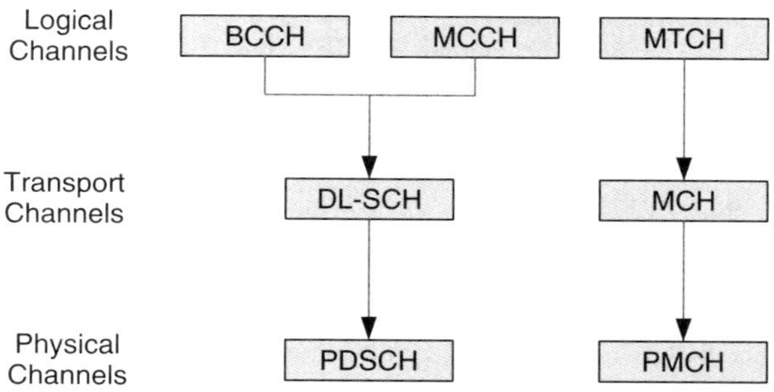

Figure 14.42 MBMS channels.

MBMS uses RS (Reference Signals) more closely spaced than other LTE transmissions. The RS is sent every fourth 7.5 KHz sub-carrier or every other 15 kHz sub-carrier. This improves channel estimates due to the long delay spread possible in this kind of transmission.

In a Multimedia Broadcast Single Frequency Network (MBMS), information is transmitted over the air by multiple time-synchronized cells. Each UE receives the transmission from these cells with different delays, which should all fit inside the CP, including their multipath signals. This improves SNR, although transmissions that fall outside the CP window become interference.

There is no uplink in MBMS, so there is no PDCCH (which would carry the uplink control information). MBMS can send dedicated data to specific users using paging as a notification mechanism.

The method of achieving time synchronization is not defined in the LTE standards, but GPS is the obvious choice. Further synchronization is required at the TB allocation level.

The MBMS frame structure is the same as the one used for regular LTE, but sub-frames 0, 4, and 5 are reserved for unicast transmissions.

MBMS uses the following channels, illustrated in Figure 14.42:

- BCCH (Broadcast Control Channel): logical channel that provides basic network identification and information about the MCCH location on DL-SCH.
- MCCH (Multicast Control Channel): logical channel that carries control information about the downlink channel.
- DL-SCH (Downlink Shared Channel): transport channels that carries BCCH and MCCH.
- PDSCH (Physical Downlink Shared Channel): physical channel that carries the DL-SCH information.
- MTCH (Multicast Transport Channel): logical channels that carries the information to be broadcasted.
- MCH (Multicast Channel): transport channel that carries the MTCH.
- PMCH (Physical Multicast Channel): physical channel that carries the MCH information to be broadcast.

14.14 Call Placement Scenario

This section describes a call placement by a UE to explain how it learns about the frame structure and its data and how it receives and sends information.

LTE uses a framed structure to send information through the wireless channel. The frame structure is partially known by the UE as it is standardized, but its contents have to be defined by the eNB.

The UE has to check on the different frequencies it is programmed to operate for the presence of this frame; for this it looks for two synchronization signals (PSS and SSS). These signals are pseudo-random orthogonal sequences, and are always placed in the same position around the carrier regardless of the bandwidth. Once detected, the UE has the frame synchronization and eNB ID. At this moment the UE is in the RRC_IDLE state.

To decode the remaining of the frame data, the UE has to look for Reference Signals RS (pilots), whose content is known and can be used to establish the channel frequency response for each sub-carrier. This allows the UE to equalize the phase and amplitude of the sub-carriers over frequency and time.

Once a frame is detected and the sub-carriers equalized, the UE has to get information about its content. This information is periodically broadcast by the eNB through the PBCH, which has a known location. This channel carries the Master Information Block (MIB), which carries the System Frame Number (SFN), the downlink bandwidth, the number of antenna ports, and the configuration of the PHICH.

Next, the UE looks for the PCFICH, which is transmitted in a pre-defined location. This channel informs the size of the PDCCH.

The PDCCH informs the mapping of the PCH, DL-SCH in the downlink frame, and the scheduling of the UL-SCH in the upcoming uplink frames, as well as HARQ information. From the downlink mapping, the UE learns the location of the PRACH.

The next step for the UE is to attach to the eNB, so it has to adjust its timing to the eNB timing, which depends on the distance between both. The UE sends a message in the PRACH and waits for the timing advance information coming from the eNB. The PRACH area has to use a contention mechanism as it is not possible to forecast who will be accessing the area. If there is no reply from the eNB, the PRACH message is resent.

Once the timing advance is received, the UE can properly communicate with the eNB, but it still needs to periodically adjust its timing advance and any spontaneous UE access still has to go through the PRACH procedure.

The UE requests the MME to the eNB to be attached to the eNB in question. The MME establishes the required bearers, communication tunnels, and encryption. The UE is then put in the Tracking Area (TA) to which the eNB belongs.

The UE can negotiate a DRX state with the eNB and it is told the periodicity with which it has to wake up and listen for messages.

The UE periodically checks the paging channel to see if it is being paged. When the UE is paged, or has data to be sent, it has to attach to the eNB. It also collects measurements on its frequency and other frequencies to identify neighbors and potential handover candidates.

When the UE has data to be sent, it connects to the eNB by sending a message in the PRACH channel and changing its state to RRC_CONNECT. Next it requests, through the PRACH, a slot to send its data by specifying the instance, the amount of data per frame and the number of frames required to send it. The UE also sends measurement data that it collects for handover decisions.

The RRM in the eNB analyzes the request and the measurement data, allocates RBs (Resource Blocks) and specifies the coding scheme to be used for this specific connection.

It then listens to the PDCCH to learn the scheduled frame and RBs for its transmission. Before sending the data, the UE sends control messages in the PUCCH, using one of the orthogonal codes available.

The data is sent to the PDCP layer where it is assembled into one or more TBs (Transport Blocks) according to the antenna algorithm selected by the RRM. Each TB is scrambled, interleaved, and has a CRC added. This CRC is used for ARQ purposes. The RRM selects the antenna scheme and coding scheme that provides the maximum throughput with a BLER of 10%.

The TB is then sent to the RLC layer which adds the Turbo code with 1/3 rate and then punctures it for the eNB specified rate. It then segments or appends TBs to comply with the maximum MAC size and adds another CRC, which is used for HARQ purposes.

The CBs are sent to the MAC layer which adds the wireless protocol and inserts another CRC, which is used to verify MAC data integrity.

Finally, the PHY layer maps the MAC packets onto the TB sub-carriers specified by the eNB. As the uplink uses DFT-S-OFDMA, it has to map the data in time, perform an FFT and then map the frequency data to the TB sub-carriers. This is done separately for the I and Q components of the modulation. The sub-carriers are then divided, up-converted, modulate the carrier, and are sent to the antennas defined by the antenna algorithm selected.

On the receive side, the received signal is separated into the I and Q components and demodulated. The reverse process is performed and, at each step, the CRC is checked. A packet with a wrong MAC CRC is discarded and resent. A packet with a wrong CB CRC goes through a HARQ process and a packet with a wrong TB CRC undergoes an ARQ process.

Once the data exchange is in process, the antenna scheme and coding scheme are periodically adjusted by the eNB by its own volition or by request from the UE.

Once all the data is sent, the UE sends a message through the PRACH to release the resources, but it continues in the RRC_CONNECTED state until an inactivity timer expires.

During the RRC_CONNECT state, the UE periodically measures the signal from the eNB it is connected and from other eNBs from which it can decode a signal. This information is reported to the eNB and passed to the RRM layer.

An eNB data transmission undergoes a very similar process, but it does a paging first to alert the UE of the imminent transmission of data, assigns RBs and informs the assignment through the PDCCH. The data is then sent on the sub-carriers and symbols of the specified RB. In the case of the eNB, a straight OFDM modulation is done. The rest of the procedure is the same as for the uplink.

14.15 PHY Characteristics and Performance

14.15.1 Transmitter

Some transmitter settings are specified in the LTE standards, but, in general, additional regional restrictions apply.

14.15.1.1 Power Settings

The BS transmit power depends on the regulations of each country for the band used. Typical values are between 32 and 64 W (45 to 48 dBm).

The LTE standard specifies only class 3 UE power as 23 dBm ±2 dB (200 mW). The minimum transmit power is specified as −40 dBm and the off power (transmitter switched off) as −50 dBm.

14.15.1.2 Out of Band (OoB) Emissions

Out of band emissions are defined for several classes and Figures 14.43 and 14.44 show the most common ones for the BS and UE.

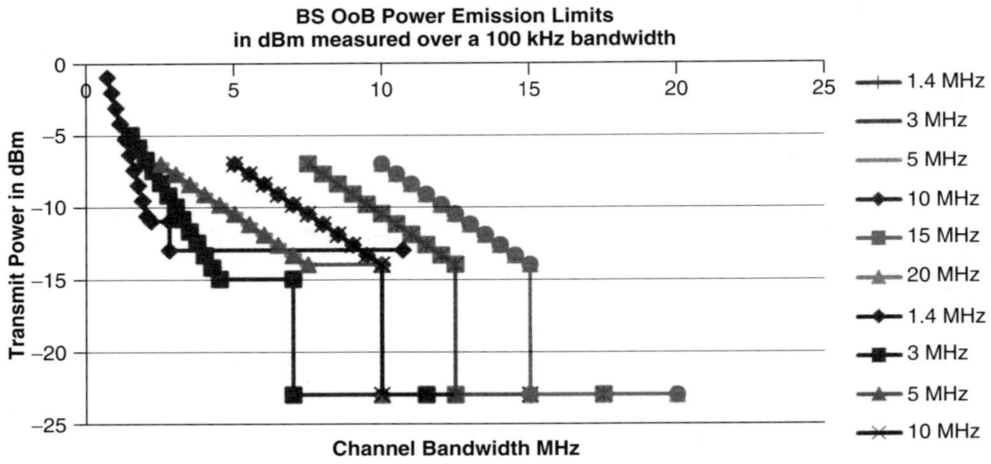

Figure 14.43 BS Out of band emissions.

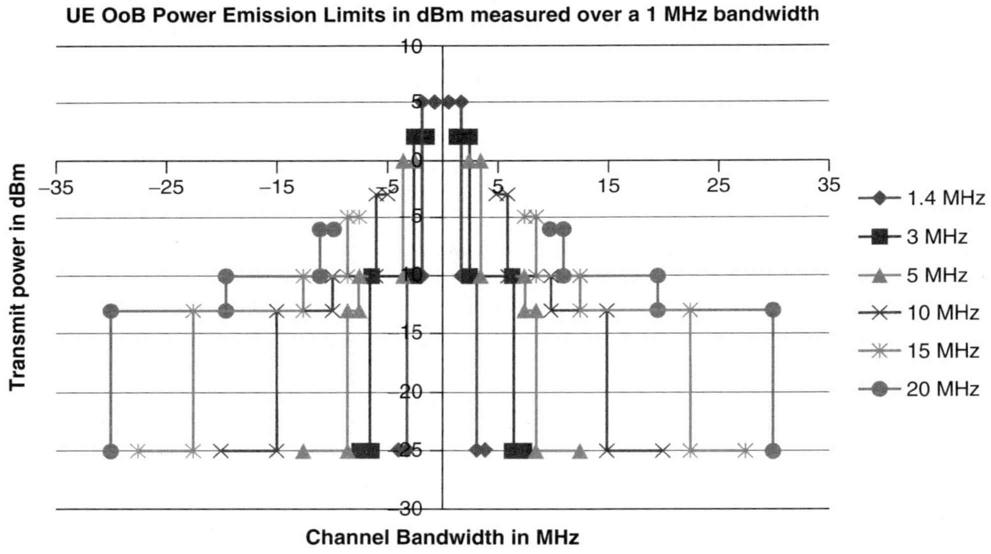

Figure 14.44 UE Out of band emissions.

14.15.1.3 Other

The intermodulation is specified as −40 dBc (dB in relation to the carrier level). The ACLR (Adjacent Channel Leakage Ratio) is specified as the ratio of the mean power on the assigned channel to the mean power of the adjacent channel, and should be higher than 45 dB.

The EVM (Error Vector Measurement) specified for the BS and UE is shown in Table 14.15.

Table 14.15 EVM values for different modulations

Value	(%)
QPSK	17.50
16 QAM	12.50
64 QAM	8

14.15.2 Receiver

The BS receiver usually has a better noise figure than the UE receiver.

14.15.2.1 Sensitivity

The expected receiver sensitivity is shown in Figure 14.45, and considers a 9 dB Noise Figure (NF) for the UE. The receiver sensitivity for the BS should be 5 dB better. There is degradation for FDD depending on the band separation as shown in Table 14.16. LTE standard figures are more lenient by 0 to 4 dB. A designer should obtain actual equipment vendor figures for the final network design.

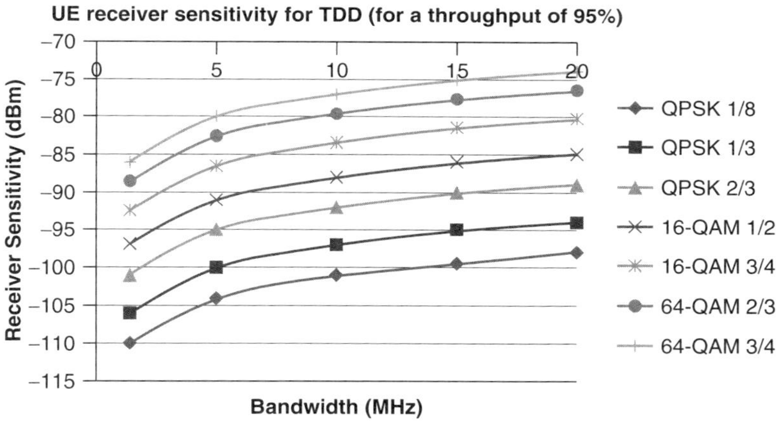

Figure 14.45 UE Receiver sensitivity for TDD.

Table 14.16 Receiver sensitivity decrease

TX to RX separation	Receiver sensitivity decrease
<50 MHz	4 dB
<100 MHz	3 dB
<150 MHz	2 dB
<200 MHz	1 dB

14.15.3 Power Saving

LTE supports power savings at the UE, through DTX and RTX. There are many circuits that consume significant amount of power during idle UE periods and disconnecting them significantly prolongs battery power duration.

The RRC sets a cycle during which UE scheduling and paging information is transmitted, outside this period, the UE can disconnect nearly all of its circuits. When the UE awakes from an RTX cycle, it should listen to the PDCCH, to check for downlink data.

LTE specifies a long RTX for periods in which the UE is idle and a short RTX for periods when periodic data is transmitted, such as VoIP usage.

DTX implies interrupting the transmission of RS by the UE. In principle, the transmitter should only be connected when there is something to transmit from the UE (control or data). The BS interprets the lack of response for a HARQ as a NACK, so the UE does not need to switch on its transmitter to send a NACK.

14.16 Multiple Antennas in LTE

Initial wireless deployments (about 1983) used omni antennas with receive diversity, achieved by separating the antennas by at least 10λ. Around 1990, sector antennas were introduced, with beam widths of 60°, 90°, and 105° and the use of mechanical tilt. This was followed by antennas for 6 sector sites, with 33° and 45° beam widths.

At around 1998, dual polarization antennas were introduced and electrical tilt. Around 2000, electronics were introduced inside antennas with motorized electrical phase shifters. And about 2003, multi band antennas were introduced, allowing the same enclosures to be used to accommodate several antennas. Since then electronics has found its way inside antennas.

The challenge of using multiple antennas is due to the space and cabling required by them. Tower real state is expensive and UEs do not have space to accommodate them. Multiplexing signal on cables and co-packing antennas in the same enclosure became a goal. Modern wireless antennas use multi-column planar arrays, as they optimize tower space.

MIMO technology requires that signals received by antenna pairs be uncorrelated. This requires antennas to be spatially separated, which can be achieved by physical or polarization separation.

- *Physical separation*: There is no guarantee that physical separation of antennas will assure uncorrelated reception, but the probability increases with distance. The industry has followed a rule of thumb of at least 10 λ separation.
- *Polarization separation*: RF signal are orthogonal with each other when they have a 90° cross-polarization. This means that an antenna with one polarization should not pick up signals from an antenna with 90° cross-polarization. In practice, the isolation for a point-to-point link with pencil beam antennas is about 25 dB. In mobile connections the signals change polarization due to reflections and diffractions, so practical cross-polarization values drop between 5 and 10 dB. Polarization depends on the antenna orientation and this is impossible to guarantee for a portable phone which can be used in any position. Using a 45° slant (inclination) gives a 3 dB loss in relation to a vertical or horizontal position. The usage of two cross-polarized antennas with +45° and −45° slants optimizes the reception of portable phones and uncorrelation. Cross-polarization cannot be used in portable phones as cross-polarized antennas are directional.

Further uncorrelation of antenna signals can be obtained by mixing space separation and cross-polarization techniques.

14.16.1 Antenna Configurations

Base Station antennas use several elements to provide high gains without restringing beam width, as there is space to accommodate them. Gains of 12 to 18 dBi are common for three sectored antennas.

UEs have little space for antennas, so reduced size dipoles have to be used, which results in gains between −3 and 0 dBi.

Table 14.17 gives the sizes required for different frequencies and antenna configurations.

LTE standardized the following antenna configurations: 1 × 1, 2 × 2, 4 × 2 and 4 × 4 (this representation means "Number of TX antennas × Number of RX antennas").

Additionally, beamforming technology was also considered. Beamforming requires closely spaced antennas (0.5 λ).

Typical diversity antenna configurations are shown in Figure 14.46. The first configuration has 2 antennas diversity with spatial separation. The second configuration has 2 antennas diversity with cross-polarization. The third configuration supports four diversity antennas using cross-polarization and spatial separation.

Beamforming configurations use a Butler matrix to provide different shifts to the beam. Typical configurations are shown in Figure 14.47.

Multi antenna technology is still in its infancy and it is even hard to separate claims from reality. Many of today's claims are related to peak throughputs that can only be reached with very few users and the results with large number of users can be very disappointing.

The best results can be obtained by conservative configurations such as 2 × 2, as higher configurations can be very disappointing when compared to claims. The technology should evolve further and further improvements will be achieved, but far from the marketing claims made today. The designer should always verify if tests done in the lab will hold with the increase in the number of UEs.

14.16.2 LTE Antenna Algorithms

The LTE protocols were implemented to support existing and envisaged multiple antenna techniques. Figure 14.48 shows some configurations foreseen for LTE.

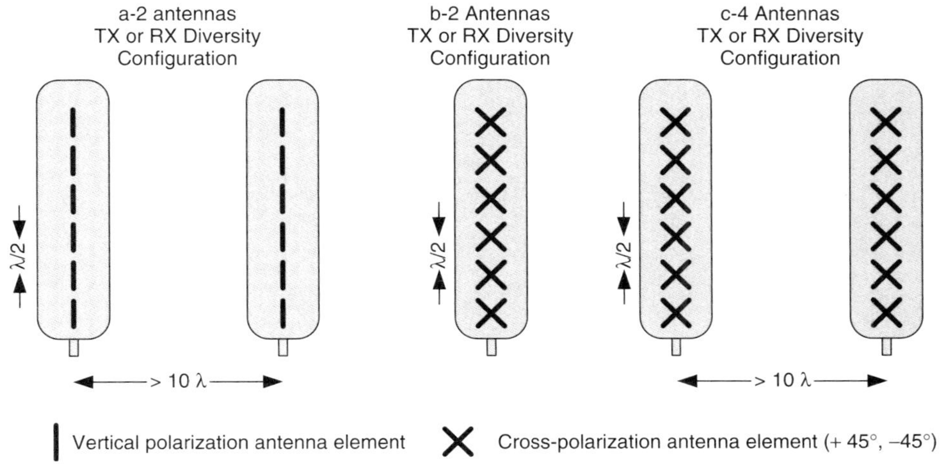

Figure 14.46 Antenna configurations.

Table 14.17 Antenna clusters dimensions

			Base station antennas (≈15 dBi)		Base station antennas (≈8 dBi)		Base station or fixed UE antennas (≈3 dBi)		mobile UE antennas (≈0 dBi)		mobile UE antennas (≈0 dBi)	
Frequency (MHz)	Wavelength (m)	1/2 Wavelength (m)	Dimensioning for 10 cross pol elements and 4 antennas without enclosure		Dimensioning for 5 cross pol elements and 2 antennas without enclosure		Dimensioning for 2 cross pol elements and 2 antennas without enclosure		Dimensioning for 1 cross pol elements and 2 antennas without enclosure		Dimensioning for 1 cross pol elements and 1 antennas without enclosure	
			Height (m)	Width (m)	Height (m)	Width (m)	Height (m)	Width (m)	Height (m)	Width (m)	Height (m)	Width (m)
850	0.35	0.18	1.85	0.71	0.97	0.26	0.44	0.26	0.26	0.26	0.18	0.18
1900	0.16	0.08	0.83	0.32	0.43	0.12	0.20	0.12	0.12	0.12	0.08	0.08
2500	0.12	0.06	0.63	0.24	0.33	0.09	0.15	0.09	0.09	0.09	0.06	0.06
3500	0.09	0.04	0.45	0.17	0.24	0.06	0.11	0.06	0.06	0.06	0.04	0.04
5000	0.06	0.03	0.32	0.12	0.17	0.05	0.08	0.05	0.05	0.05	0.03	0.03

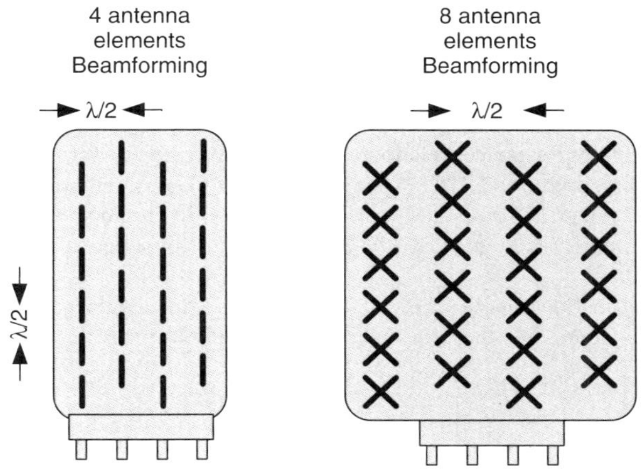

Figure 14.47 Beamforming antenna configuration with 4 and 8 antennas.

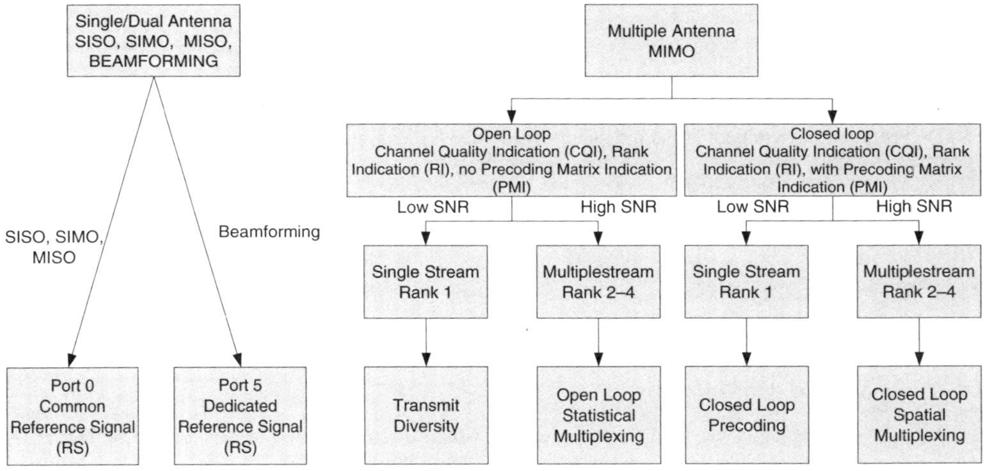

Figure 14.48 Antenna algorithm configurations foreseen for LTE.

Single and dual antenna configurations provide support for SISO, SIMO and MISO configurations. Channel information is obtained from common Reference Signals. Beamforming antennas require dedicated Reference Signals.

MIMO antenna configurations use RS information, Channel Quality Indicators (CQI) and Rank Indication (RI). Open loop MIMO does not get a precise estimation of the RF channel, due to fast channel variations. Closed loop MIMO estimates the channel, thus allowing the construction of a PMI (Precoding Matrix Indication).

14.16.3 Transmit Diversity

Transmit Diversity (TD) is used to reduce the effect of interference and fading due to multipath. The same information is sent in two or more antennas, thus improving the chances of the receiver correctly detecting the signal. Each transmission interferes with the other, so special algorithms are used over a pair of symbols to allow for the cancelation of mutual interference.

LTE uses a similar algorithm to STBC (Space Time Block Codes), but applied in frequency instead of time, called SFBC (Spatial Frequency Block Codes). In STBC a block is formed by a pair of two sequential symbols in time, while in SFBC the block is formed by a pair of two sub-carriers adjacent in frequency.

This is represented in Equation (14.2), where $y^0(1)$ represents symbols transmitted from antenna 0, sub-carrier 1 and $\times$ represents the data to be modulated in each sub-carrier.

$$\begin{bmatrix} y^0(1) & y^0(2) \\ y^1(1) & y^1(2) \end{bmatrix} = \begin{bmatrix} x_1 & x_2 \\ -x_2^* & x_1^* \end{bmatrix} \qquad \text{(14.2) STBC matrix}$$

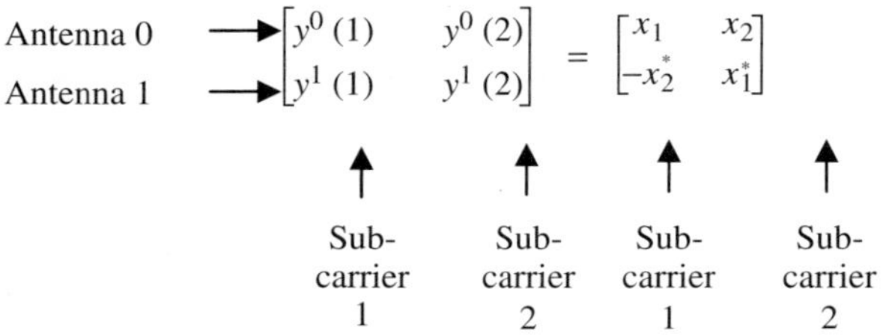

Alamouti codes exist only for 2×2 antenna configurations, so for 4 antennas transmit diversity FSTD (Frequency Switched Transmit Diversity) is used, as shown in Equation (14.3).

$$\begin{bmatrix} y^0(1) & y^0(2)y^0(3) & y^0(4) \\ y^1(1) & y^1(2)y^{10}(3) & y^1(4) \\ y^2(1) & y^2(2)y^2(3) & y^2(4) \\ y^{30}(1) & y^3(2)y^3(3) & y^3(4) \end{bmatrix} = \begin{bmatrix} x_1 & x_2 & 0 & 0 \\ 0 & 0 & x_3 & x_4 \\ -x_2^* & -x_1^* & 0 & 0 \\ 0 & 0 & -x_4^* & -x_3^* \end{bmatrix} \qquad \text{(14.3) 4 antennas transmit diversity}$$

14.16.4 Spatial Multiplexing

In Spatial Multiplexing (SM), N different streams of data are transmitted at the same time and at the same frequency on N transmit antennas and received on N or less receive antennas. The different transmissions interfere with each other, but this can be partially compensated if there is knowledge about the channel between each pair (TX and RX) of antennas.

In the downlink, channel information is obtained from the Reference Signal (RS), as RSs transmission alternates between antennas. In the uplink, this is achieved by the CQI (Channel Quality Indicator) information sent in the CCCH.

A coded transport block from the MAC forms a codeword, which is precoded into a spatial layer, and then assigned to an antenna. The precoding uses a pre-defined codebook, and the best code is selected, based on the available channel knowledge if any, otherwise a unitary mapping is done. The codebook codes try to invert the channel response, to facilitate detection. The final detection is done using a Maximum Likelihood Detector (MLD) which has to test at least 1152 combinations.

LTE defines transmission ranks, which define the number of spatial streams transmitted. Different data (sequential) from the same codeword can be transmitted in different antennas or two different codewords can be transmitted at the same time. Rank 1 defines that only 1 codeword is to be used and the other 3 ranks define which codeword is assigned to which spatial layer.

In the case of open loop spatial multiplexing, Cyclic Delay Diversity (CCD) can be used. In this case, the same set of data is transmitted in different antennas with different delays to force a multipath.

14.16.5 Beamforming

Beamforming Networks (BFNs) do not provide diversity or spatial multiplexing, but they reduce multipath and, potentially, interference.

Phased arrays are used to create narrow beam widths by sending phase-shifted versions of the same signal to several antennas, resulting in a combined narrow beam transmission. The phase shifts can be adjusted to steer the beam. The higher the number of paths, the higher is the directivity. Phased arrays variable phase shifts are digitally generated and this implies that they have to be generated by the radio equipment.

Fixed adaptive phase arrays can be generated at the antenna itself by using multiple beam matrix feeds. These elements use sets of fixed phase couplers and 90° hybrids (directional couplers). Two implementations are possible: Parallel feed Butler matrix and series feed Blass matrix.

14.16.5.1 Parallel Feed Butler Matrix

This is a bidirectional NxN matrix that creates N lobes. An 8×8 matrix generates eight $10°$ lobes between $+60°$ and $-60°$. The signal received from the UE will be strongest at a specific output, determining the direction of arrival. Any transmission for this UE is sent to the same port. This is shown in Figure 14.49. This circuit is easy to implement and has a small loss, but changes its response with frequency.

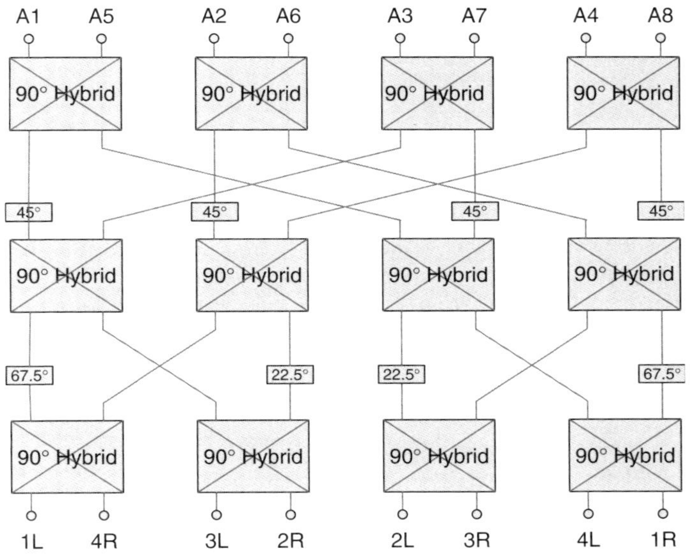

Figure 14.49 Butler matrix circuit.

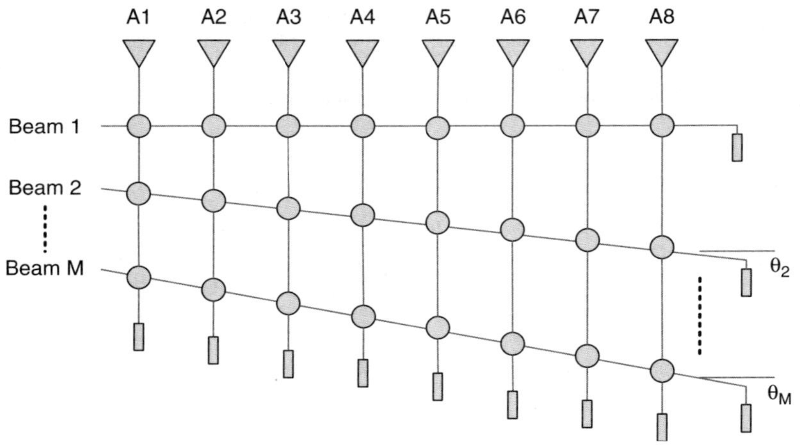

Figure 14.50 Blass matrix circuit.

14.16.5.2 Series Feed Blass Matrix

This is a bidirectional NxM matrix that creates N lobes. An 8×8 matrix generates eight 10° lobes between $+60^\circ$ and -60°. The signal received from the UE will be strongest at a specific output, so any transmission for this UE is sent to the same port. This is shown in Figure 14.50. The circles are directional couplers, while the rectangles are terminations. This circuit requires more components and has more loss, but provides the same response for a broad range of frequencies.

The actual advantage provided by a beamforming circuit depends on each situation. As it diminishes multipath and provides some discrimination against interference, the improvement in SNR can be of several dB. It is up to the designer to evaluate the seriousness of the impairments and establish an SNR improvement due to beamforming, unless the planning tool already considers multiple beams in the analysis.

14.17 Resource Planning in LTE

The number of broadband frequencies (OFDM carriers or RF channels) available is usually small and, in some cases a single frequency channel is available.

LTE networks have to work with a small reuse factor in the network, including reuse of 1, so all frequencies are used in all cells. Low reuses can only be achieved if interference is mitigated through load and scheduling control. LTE specifications do not establish how this can be achieved and leave the decision up to the equipment vendors.

Frequency assignment can be classified into full reuse, hard reuse, fractional reuse, or soft reuse.

14.17.1 Full Reuse

In full reuse, all frequencies can be assigned in every cell, independently of the UE location. In this case cells can only support small loads, and even so, only if mechanisms to avoid or distribute interference are used. LTE requires eNBs to coordinate RBs allocations among themselves. This is done through the X2 interface and the process is called ICIC (Inter-cell Interference Coordination), but once again, the technique used for this implementation is left up to vendors. This means that for quite a while it will not be possible to combine different vendors in the same network, unless planning is done with a design tool and only vendors that fully support the design tool implementation be selected.

This concept allows for a maximum load of around 20% of the cell capacity. This mechanism has to be developed by the vendors and most probably will not work between different vendors. Even so, the control channels and RSs may interfere, but as they have more robust modulation schemes, the areas of interference are reduced. This type of reuse should only be considered if there are no more frequencies available and with the knowledge that service will be affected in some areas.

14.17.2 Hard Reuse

Hard reuse is the traditional planning and assignment of frequencies to cells. A minimum number of 7 frequencies is required if the full carrier is to be used in every cell. It is possible to have a reuse of 3 frequencies, if load and assignment control is used for the user information channels. With a reuse of 3, cells can work at 50% to 70% of their capacity. Even with a reuse of 7, the maximum capacity should be avoided as at least 5% load should be left for handover traffic.

14.17.3 Fractional Reuse

Fractional reuse consists of allocating two groups of frequencies per cell, one group closer to the center of the cell and another on the periphery. The allocation is done according to the UE distance to the eNB, based on the delay adjusted by the ranging procedure.

With this technique, a reuse of 1 becomes a reuse of 2 and a reuse of 3 becomes a reuse of 6. Some load control may be required depending on the design. A cell could be loaded to between 60–80% of its full capacity. This technique is also used in WiMAX networks and is explained in more detail in Section 13.8.

14.17.4 Soft Reuse

LTE does not offer the interference mitigation mechanism used in WiMAX with the permutation schemes (PUSC, FUSC). Instead, it relies on vendors to use the eNB direct connectivity through the X2 interface to perform interference mitigation. This can be done by using fractional reuse as explained in Section 14.17.3, but with variable criteria, such as distance or RS power. The RS power criteria is problematic as indoor cells located near may seem to be located far away, whereas distant cells with direct line of sight may seem closer than they in fact are.

14.18 Self-Organizing Network (SON)

The deployment and operation of broadband networks like LTE require significant investment and expenditures, mainly if the deployment is done without proper planning. The 3GPP specified, in Releases 8 and 9, some procedures to facilitate the deployment of the new networks, making them more plug-and-play like. These features are referred to as SON. Many of the features are inspired in the plug-and-play concept used in personal computers

The main SON goals are:

- *Self-configuration*: plug-and-play detection of the eNB by the gateways and vice versa. Once the eNB is connected to the network it should obtain an IP address from the DHCP/DNS server and automatically find the S-GW and MME to which it should connect. Then it should establish secure tunnels for O&M, S1 and X2 interfaces. After connecting to its dedicated management entity, it downloads the last software version and launches it.

- *Self-configuration of RF transport*: the downloaded software version comes with default parameters, which should be automatically adjusted by the eNB. The list of these parameters is not defined, but the bulk of them come from a prediction tool to which the eNB should be connected.
- *Physical Cell ID (PCI) determination*: the eNB listens to the air and detects all transmissions, which are then sent to the planning tool. The reply will be the eNB PCI.
- *Automatic Neighbor Relations (ANR)*: each eNB starts its operation with the Neighbor Relations provided by the Planning Tool (PT). Then it periodically sends to the planning tool neighbor relations gathered by the UEs. This allows the PT to optimize the ANR and feed it back to the eNBs.
- *Tracking Area Planning*: a wireless market is portioned into non-overlapping Tracking Areas (TA), identified by a TAI (Tracking Area Identifier). These areas are used to facilitate the paging procedure and UEs are required to inform when the best serving eNB has a different TAI of the previous one. The MME requests that paging messages be sent to the UE's registered TA and, if no reply is received, the whole network is paged. The main network-related reason why a UE might not respond to the paging is because the RACH channel was overloaded, meaning that the TA is too large. The goal of this feature is to automatically adjust the TA area. This is much easier done by a planning tool, but feedback from the eNBs can be used to optimize the TAs.
- *Network operation*:
  - *Auto inventory*: each eNB should report its hardware and software components to a centralized database.
  - *Automated software upgrade*: periodically the eNB should check or be notified of software upgrades.
  - *Self-testing and self-healing*: the eNB should do periodical self-diagnostics and in case of failure notify adjoining eNBs, to cover the gap.
- *Network optimization*:
  - *Load balancing*: the eNBs should exchange load and scheduling information, to avoid conflicts. This does not exclude the need of proper frequency planning, as this simple coordination between eNBs, might lead to large capacity reduction to avoid interference.
  - *Handover parameters optimization*: UE's reported measurements could be used by the planning tool to adjust the handover parameters. However, the overhead caused by this kind of reporting can be significant for the benefit it would give, if any.
  - *RF Plan optimization*: the optimization here does not apply to frequencies but to resources, such as TBs. It is more of a co-ordination function that is essential in LTE when low reuse is used.
  - *Interference control*: interference can be avoided by resource (TBs) use co-ordination. The bulk of the planning can be done by using a planning tool, in which resources are divided into groups and each eNB has a group of resources allocated. Fine adjustments can be done live between eNBs, by borrowing resources in overload situations. The resource planning should consider the number of frequencies and resource (TB) groups to form a pool of nearly orthogonal resources that can be distributed according to the traffic requirements of each eNB sector.
  - *Energy savings*: energy savings measures are encouraged but the implementation is left up to vendors.

Although SON has very noble purposes, many of the claims are exaggerated and real benefits are much smaller than promised. One of the short-comings is the lack of an entity that centralizes network data. It is expected that vendors will implement this entity in the OSS and integrate it with a planning tool.

A planning tool will always be required, for the initial network planning and to maintain the network. It can benefit though from real life data collected by eNBs and UEs to update its database.

14.19 RAT (Radio Access Technology) Internetworking

LTE architecture and protocols allow for interconnection of dissimilar networks, including networks based on packet or circuit switching. RAT internetworking foresees network re-selection and handover between e-UTRAN to/from GERAN, UTRAN and even cdma2000, WLAN and WiMAX. To perform this a set of procedures is specified in the standard. Figure 14.51 illustrates handover and re-selection between different RAT networks. Table 14.18 presents the parameters analyzed for the different RAT to decide when to hand over or perform a re-selection.

14.20 LTE Radio Propagation Channel Considerations

To properly model the wireless part of LTE, an extremely complex RF analysis is required, when compared to the 3G technologies. The RF channel has to be modeled for narrow and wide bandwidth and multipath has to be modeled by its components as well as its arrival direction.

The ITU and 3GPP modeling is conceived to test and compare technological solutions and their implementations. This type of modeling does not apply to real networks that consist of thousands of different channel models combined and defined on a per pixel basis. This section covers models conceived for testing and comparison of algorithms and their implementations.

The modeling has to address propagation path loss, shadow (large-scale or slow) fading and multi-path (small-scale or fast) fading. Uniform criteria are established so different solutions can be compared on the same basis. These models can be used as reference for establishing network-wide models, but cannot replace them.

14.20.1 SISO Channel Models

SISO and SIMO models do not require modeling of multipath direction, only of its components.

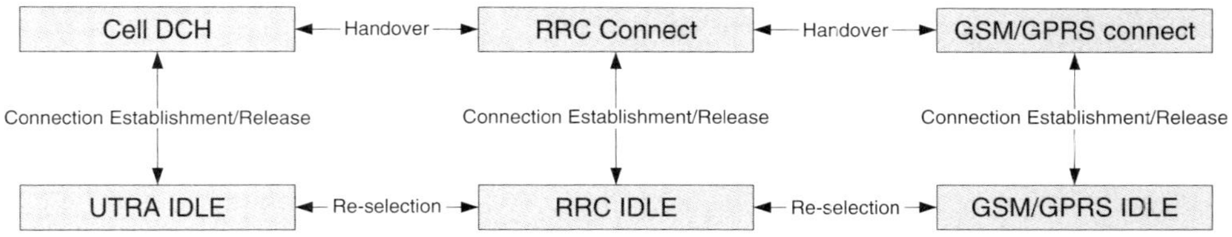

Figure 14.51 Inter-RAT networking.

Table 14.18 Cell search parameters per RAT

Cell search		
LTE	UMTS	GSM
PSS	P-SCH	RRSI
SSS	S-SCH	BSCI ID
PBCH	Primary scrambling code ID	FCCH detection
SFN	System Frame Number	SCH detection

14.20.1.1 ITU Channel Models

The classic ITU multipath model used for UMTS models 3 scenarios and is shown in Table 14.19. It was developed for bandwidths of up to 5 MHz and low relative speeds.

14.20.1.2 4G Extended ITU Channel Models

The classic ITU model was conceived for channels of up to 5 MHz of bandwidth, but LTE channels can have up to 20 MHz, which required an extension of the model, as the number of components increases with the bandwidth.

The models proposed for LTE are shown in Table 14.20. Additionally to multipath considerations were given for speed simulation through the Doppler Effect and this is also shown.

14.20.2 MIMO Channel Models

The previous channel models addressed multipath components and their relative power, but did not address the incoming direction of multipath required for proper MIMO evaluation. Additionally, it was soon found that the multipath pattern was much more complex than what was suggested by previous models.

Propagation between two locations can be divided into several main paths, as illustrated in Figure 14.53 on p. 478, and each of these paths can be further divided in a cluster of multipath.

MIMO theory development was done considering completely independent propagation channels, but correlation between paths changes the picture significantly. Many MIMO tests have been done using independent ITU channel modeling and the result obtained could not be repeated in real life environments. Channel models were then required to simulate, at least partially, a real life environment.

14.20.2.1 ITU Models with Spatial Correlation

The ITU models were then further extended for use with theoretical MIMO studies. Three antenna correlation scenarios are specified by Equation (14.4).

$$R_{eNB} = \begin{pmatrix} 1 & \alpha \\ \alpha^* & 1 \end{pmatrix} R_{UE} = \begin{pmatrix} 1 & \beta \\ \beta^* & 1 \end{pmatrix} \quad (14.4)\ \text{Antenna correlation}$$

Table 14.19 3G ITU channel models

ITU models	ITU pedestrian A		ITU vehicular A		ITU pedestrian B	
Multipath	Relative delay (ns)	Relative power (dB)	Relative delay (ns)	Relative power (dB)	Relative delay (ns)	Relative power (dB)
1	0	0	0	0	0	0
2	110	−9.7	310	−1	200	−0.9
3	190	−19.2	710	−9	800	−4.9
4	410	−22.8	1090	−10	1200	−8
5			1730	−15	2300	−7.8
6			2510	−20	3700	−23.9

Table 14.20 4G Extended ITU channel models

Extended ITU models	Extended Pedestrian A (EPA)		Extended Vehicular A (EVA)		Extended Typical Urban (ETU)	
Multipath	Relative delay (ns)	Relative power (dB)	Relative delay (ns)	Relative power (dB)	Relative delay (ns)	Relative power (dB)
1	0	0	0	0	0	−1
2	30	−1	30	−1.5	50	−1
3	70	−2	150	−1.4	120	−1
4	80	−3	310	−3.6	200	0
5	110	−8	710	−0.6	230	0
6	190	−17.2	1090	−9.1	500	0
7	410	−20.8	1730	−7	1600	−3
8			2510	−12	2300	−5
9				−16.9	5000	−7
Doppler Shift	Shift (Hz)		Shift (Hz)		Shift (Hz)	
1 (2 km/h)	5		5		5	
2 (30 km/h)	70		70		70	
3 (120 km/h)	300		300		300	
4 (350 km/h)	1000		1000		1000	

Table 14.21 ITU correlation factors for different antenna configurations

	Low correlation	Medium correlation	High correlation
α	0	0.3	0.9
β	0	0.9	0.9

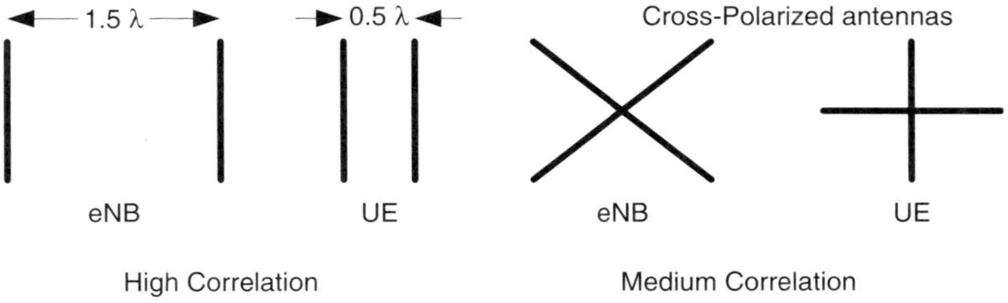

Figure 14.52 ITU antenna configurations for different correlations.

Table 14.21 gives correlation factors for configurations shown in Figure 14.52. No configuration is given for low correlation, as it is seldom achieved in real life systems.

14.20.2.2 SCM (Spatial Channel Model)

The need for spatial consideration became evident when MIMO applications were considered, and the 3GPP specified the SCM (Spatial Channel Model). In this model, six sets of multipath clusters are

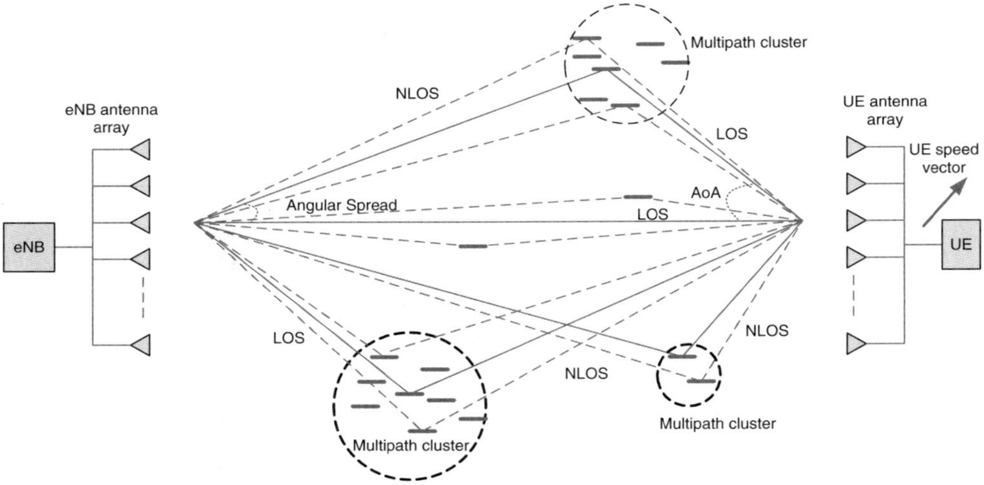

Figure 14.53 Spatial channel model.

Table 14.22 Spatial channel model (SCM)

	Suburban	Urban macro cell	Urban micro cell
Paths	6	6	6
Multipath per path	20	20	20
Total angular spread at eNB	5°	12°	19°
Angular spread per path at eNb	2°	2°	2°
Total angle of arrival at UE	68°	68°	68°
Angle of arrival per path at UE	35°	35°	35°
RMS delay spread (μs)	0.17	0.65	0.25
Shadow fading std (dB)	8	8	4 (LOS); 10 (NLOS)
Path loss model	$31.5 + 35\log_{10}(d)$	$34.5 + 35\log_{10}(d)$	$30.183 + 26\log_{10}(d)$ for LOS
Path loss model			$34.53+38\log_{10}(d)$ for NLOS

considered each with 20 multipaths, as shown in Figure 14.53 The spatial representation can be used bi-directionally, from the eNb to UE and vice versa. Several spatial parameters were then defined for three scenarios as shown in Table 14.22. This was still a very simplistic model to represent real life complexity, but it was quite a step forward in relation to previous models.

14.20.2.3 Winner Channel Model

A consortium of vendors was formed in 2002 as an External Evaluation Group (EEG) to ITU and was called Wireless world INitiative New Radio (WINNER- www.ist-winner.org). This consortium made several contributions to the ITU and 3GPP, one of them was an extension to the SCM, known as WINNER I model or SCME (Spatial Channel Model Extended).

The intention of this model was to extend the channel bandwidth from 5 MHz to 100 MHz, which was achieved by creating midpaths for each path of the SCM. Three midpaths were added, Suburban

Table 14.23 Phase 2 WINNER channel model scenarios

A1	Indoor office
A2	Indoor to outdoor
B1	Urban micro cell
B2	Bad urban microcell
B3	Indoor hotspot
B4	Outdoor to indoor
B5	Stationary feeder
C1	Suburban macro cell
C2	Urban macro cell
C3	Bad urban macro cell
C4	Urban macro outdoor to indoor
D1	Rural macro cell
D2	Moving networks

and Urban Macro scenarios and 4 midpaths for the Urban Micro scenario. Midpath delays were made compatible with SCM path delays, resulting in just 18 or 24 midpaths per path. This brings the total number of multipaths arriving at the receiver to up to 480.

This model did not take into account the actual frequency used and the UE speed, and there was not an agreement on the scenarios, so many alternative scenarios were proposed.

A phase 2, more complete, WINNER model was proposed. This model is not related to any geographical location, so it has to model many different scenarios to provide samples of real life occurrences. Table 14.23 lists the scenarios considered in this model.

Instead of using fixed assignments for each scenario, the model uses randomly generated parameters. The model generation steps are listed below:

1. Step 1 Set up the environment, network layout and antenna array parameters.
2. Step 2 Assign propagation conditions (LOS, NLOS, or OLOS) according to probability.
3. Step 3 Calculate the path loss using the formula in Equation (14.5).

$$PL = A \log_{10} d + B + C \log_{10}\left(\frac{f_c}{5}\right) + X \qquad (14.5)\ \text{Path loss for Winner model}$$

where:
$PL =$ Path loss in dB.
$A =$ Path loss exponent.
$d =$ Distance in meters.
$B =$ Intercept parameter.
$C =$ Path loss frequency dependence.
$f_c =$ Carrier frequency.
$X =$ Environment specific term.

4. Step 4 Generate the large-scale parameters: delay spread, angular spread, Ricean k-factor and shadow fading.
5. Step 5 Randomly generate the delays.
6. Step 6 Generate cluster powers.
7. Step 7 Generate azimuth arrival angles and azimuth departure angles. Optionally, generate elevation angles.

8. Step 8 Randomly couple departure angles with arrival angles within clusters.
9. Step 9 Generate cross-polarization power ratios.
10. Step 10 Draw random initial phases for VV, VH, HV and HH polarizations.
11. Step 11 Generate channel coefficients for each channel. CDL (Clustered Delay Lines) are specified for the scenarios listed above.
12. Step 12 Apply path loss and shadowing for channel coefficients.

14.20.2.4 LTE Evaluation Model

Channel models are also used for evaluating eNB and UE performance, for which simpler models can be used. Fixed TDL (Tapped Delay Line) of SCME were proposed for link and system level simulations. These models are listed in Table 14.24.

Figure 14.54 and Figure 14.55 provide 3GPP evaluation configuration, respectively, for the eNB and UE.

14.20.2.5 Practical Prediction Models

The WINNER phase 2 model shows the difficulty in modeling multipath and other propagation impairments, mainly because it considers one specific scenario at a time. In real life each prediction pixel is subject to a different scenario. The outcome of all multipath is a fading pattern, which can be represented by a Ricean distribution.

This concept is applied in the CelPlan model described in Section 10.6.1, in which a Ricean k factor is calculated on a pixel basis, considering factors such as distance, antenna aperture, LOS, and environment.

Table 14.24 LTE performance evaluation models

Channel model	Scenarios	eNB antennas	UE antennas
SCM-A	Suburban macro	3 sector; 0.5 λ spacing	Handset in talk position
SCM-B	Urban macro (low spread)	6 sector; 0.5 λ spacing	Handset in data position
SCM-C	Urban macro (high spread)	3 sector; 4 λ spacing	Laptop
SCM-D	Urban micro	6 sector; 4 λ spacing	Laptop

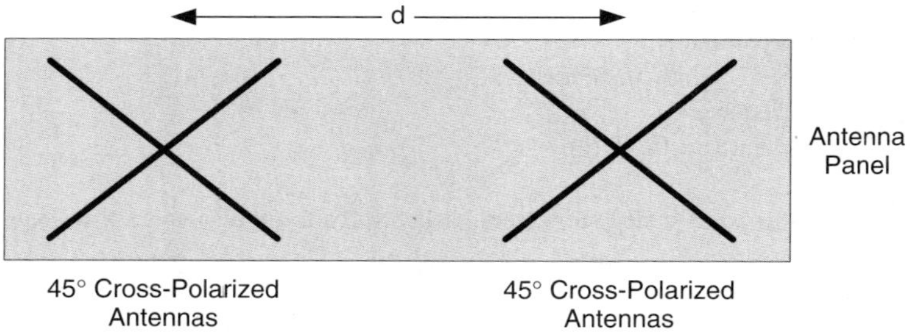

Figure 14.54 eNB antenna model for evaluation purposes.

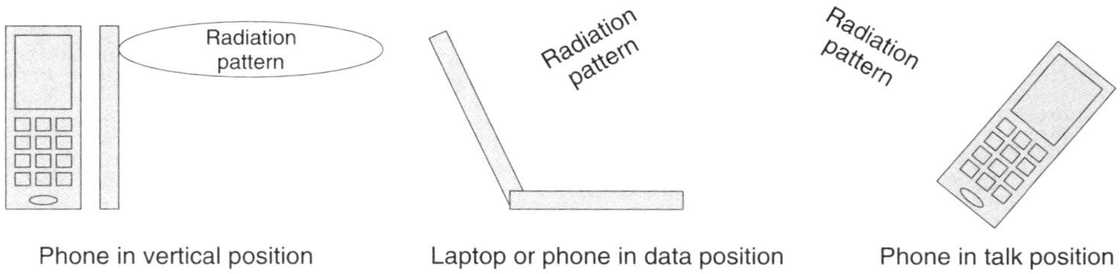

Figure 14.55 UE antenna positioning for evaluation purposes.

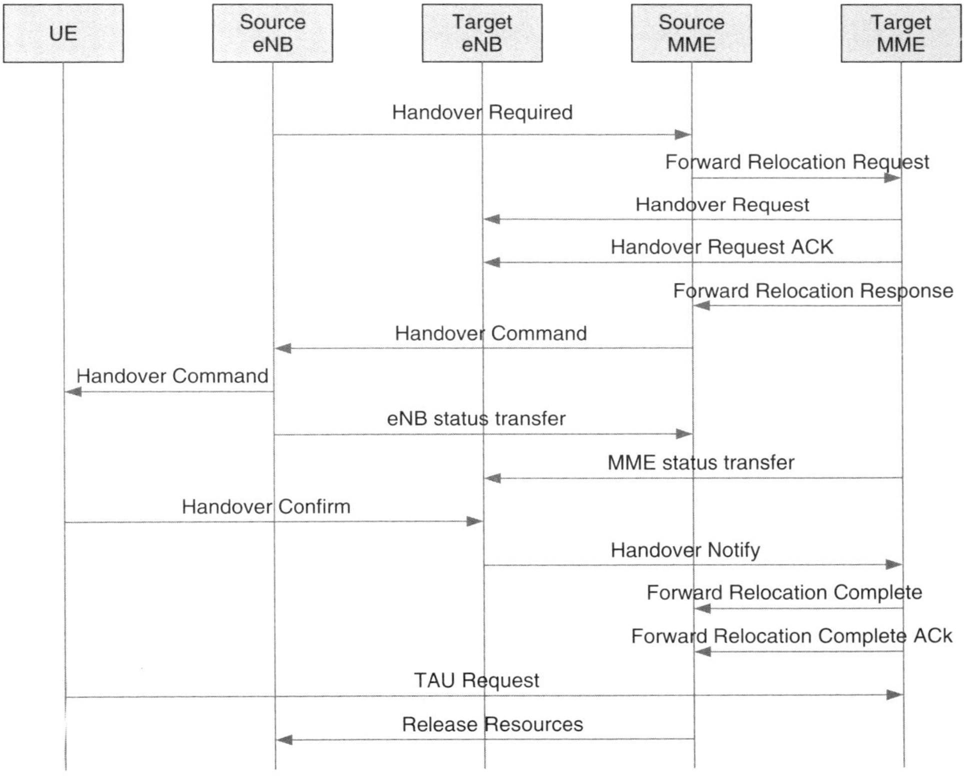

Figure 14.56 Handover messages using S1 interface.

14.21 Handover Procedures in LTE

LTE only uses hard handover, which occurs in between transmissions, but there may be data stored at the eNB to be transmitted. This data has to be sent to the new eNB and this can be done through the X2 (direct) or S1 (via Gateway) interfaces. The X2 protocol is much simpler than the S1 protocol and they are illustrated, respectively, in Figure 14.56 and Figure 14.57.

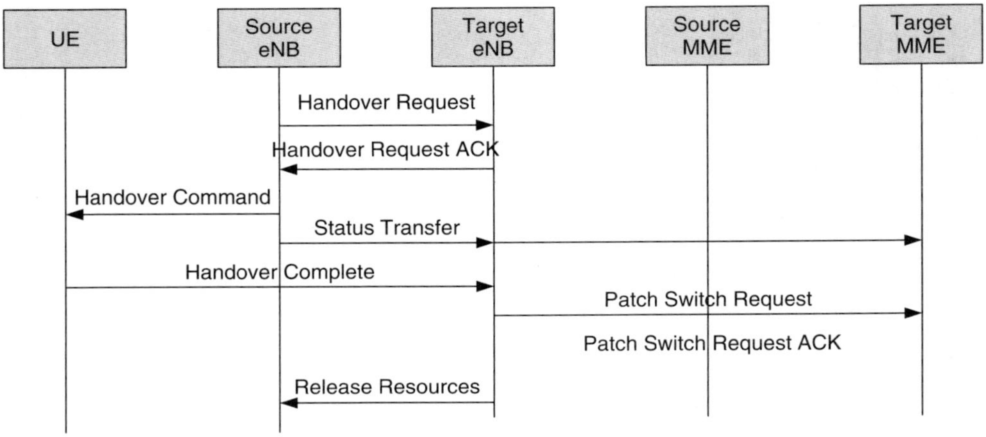

Figure 14.57 Handover messages using X2 interface.

14.22 Measurements

The UE and eNB are required to do measurements at the physical layer and report it to the RRM layer, for handover and resource management purposes.

14.22.1 UE Measurements

The main measurements performed by the UE are listed next:

- Reference Signal Receive Power (RSRP): this measurement is done during the symbols that carry the Reference Signal and are not very precise, with a tolerance of ±6 dB.
- Receive Signal Strength Indicator (RSSI): entire received power in the band.
- Reference Signal Receive Quality (RSRQ): defined by the ratio of the RSRP and carrier RSSI. The accuracy of this measurement is ±4 dB.

Additional measurements are required for inter-RAT handover and cell selection:

- UTRA RSSI: UMTS entire received power.
- UTRA Receive Signal Code Power (RSCP): is done for UMTS signals and corresponds to the power of the cdma signal in the Common Pilot Indicator Channel (CPICH).
- UTRA FDD CPICH E_c/N_0: is the ratio of the CPICH power to the total power density of the channel.
- GSM carrier RSSI: total power of the GSM channel.
- cdma2000 1XRTT Pilot Strength: this pilot is typically 7 dB below the total channel power.
- cdma2000 1xEV-DO Pilot Strength: this pilot is defined in the time domain.

14.22.2 eNB Measurements

- Downlink Reference Signal (RS) TX Power: this power is reported for the antenna connector.
- Received Interference Power: informs the power measured at a non-scheduled RB (Resource Block). The accuracy should be ±4 dB.
- Thermal Noise Power: informs the power measured over the whole bandwidth when there are no signals being transmitted.

14.23 LTE Practical System Capacity

14.23.1 Downlink Capacity

There is a significant difference between the peak throughput that can be achieved by a technology and the real throughput of a network cell. A network designer has to be realistic and consider what this real throughput is. Many network developers have difficulty understanding why the promised peak rates cannot be achieved, mainly if previous budgetary estimates used them.

The calculation of the real throughput starts by calculating the capacity based on the frame structure used for LTE. Table 14.25 calculates the throughput based on sub-carrier allocations, framing and cyclic prefix. The maximum throughput is based on 64QAM, whereas the minimum throughput is based on QPSK.

Next, the control channel overhead is considered in Table 14.26. If MIMO is used, there is twice more RS for two antennas, which further reduces capacity by 2.5%.

In Table 14.27, scheduling and interference inefficiencies are considered. Parameters in the table are estimated for 3 RF channels and soft frequency reuse.

MIMO gain increases in areas where higher modulations can be used; spatial multiplexing may offer increased throughput, depending on the antenna correlations. Assuming cross-polarized antennas, it may be possible to achieve medium correlation, which should result in a throughput gain of about 25%, as shown in Table 14.28.

These tables allow the conclusion that, in real life, the actual LTE downlink throughput per cell (sector) for a 20 MHz bandwidth will be between 4 Mbit/s and 28 Mbit/s. In many cases, the majority of the cell area is served by QPSK and, consequently, the cell average throughput is around 10 Mbit/s. This throughput has then to be divided by the simultaneous users in the cell.

14.23.2 Uplink Capacity

The frame capacity for the uplink is the same as for the downlink. Table 14.29 considers the overhead introduced by the control channel. In this table, only a single RS was considered, corresponding to a single antenna configuration.

Inefficiencies are considered in Table 14.30 and are higher than the ones for the downlink, as the allocation of resources requires communication between UE and eNB.

Finally, MIMO gain is applied in Table 14.31.

These tables conclude that, in real life, the actual LTE uplink throughput per cell (sector) for a 20 MHz bandwidth is between 3 Mbit/s and 26 Mbit/s. In many cases, the majority of the cell area is served by QPSK, consequently giving a cell average throughput of around 8 Mbit/s. This throughput has then to be divided by the simultaneous users in the cell.

Table 14.25 LTE framed throughput per cell

	Normal CP						Extended CP					
Channel bandwidth (MHz)	1.4	3	5	10	15	20	1.4	3	5	10	15	20
Transmission bandwidth (MHz)	1.08	2.7	4.5	9	13.5	18	1.08	2.7	4.5	9	13.5	18
Bandwidth efficiency (%)	77	90	90	90	90	90	77	90	90	90	90	90
FFT size	128	256	512	1024	1536	2048	128	256	512	1024	1536	2048
Number of used sub-carriers	72	180	300	600	900	1200	72	180	300	600	900	1200
Number of sub-carrier groups	6	15	25	50	75	100	6	15	25	50	75	100
Number of Resource Blocks/frame	120	300	500	1000	1500	2000	120	300	500	1000	1500	2000
Number of Resource Elements/rame (thousand)	10	25	42	84	126	168	9	22	36	72	108	144
Number of Resource Elements second (million)	1.01	2.52	4.20	8.40	12.60	16.80	0.86	2.16	3.60	7.20	10.80	14.40
Minimum throughput with no overhead and QPSK (Mbit/s)	2.02	5.0	8.4	16.8	25.2	33.6	1.73	4.3	7.2	14.4	21.6	28.8
Maximum throughput with no overhead and 64QAM (Mbit/s)	6.05	15.1	25.2	50.4	75.6	100.8	5.18	13.0	21.6	43.2	64.8	86.4

Table 14.26 LTE downlink throughput per cell considering overhead

	Normal CP						Extended CP					
Channel bandwidth (MHz)	1.4	3	5	10	15	20	1.4	3	5	10	15	20
Number of sub-carrier groups	6	15	25	50	75	100	6	15	25	50	75	100
Total Resource Elements/frame (thousand)	10	25	42	84	126	168	9	22	36	72	108	144
Reference Signals RE/frame (thousand)	0.24	0.60	1.0	2.0	3.0	4.0	0.24	0.60	1.0	2.0	3.0	4.0
PSS RE/frame (thousand)	0.50	1.26	2.1	4.2	6.3	8.4	0.43	1.08	1.80	3.60	5.40	7.20
SSS RE/frame (thousand)	0.50	1.26	2.1	4.2	6.3	8.4	0.43	1.08	1.80	3.60	5.40	7.20
PBCH RE/frame (thousand)	0.48	1.20	2.0	4.0	6.0	8.0	0.41	1.02	1.70	3.40	5.10	6.80
PDCCH RE/frame (thousand)	2.28	5.70	9.5	19.0	28.5	38.0	1.9	4.8	8.0	16.0	24.0	32.0
PDSCH RE/frame (thousand)	6.07	15.2	25.3	50.6	75.9	101.2	5.2	13.0	21.7	43.4	65.1	86.8
Channel coding overhead (turbo code at $^1/_3$) (%)	66	66	66	66	66	66	66	66	66	66	66	66
Channel coding overhead (turbo code at 2/3) (%)	33	33	33	33	33	33	33	33	33	33	33	33
Percentage of RE available for data (worst case)	20	20	20	20	20	20	20	20	20	20	20	20
Percentage of RE available for data (best case)	40	40	40	40	40	40	40	40	40	40	40	40
Minimum throughput (QPSK) with overhead (Mbit/s)	0.41	1.03	1.72	3.44	5.16	6.88	0.35	0.86	1.44	2.88	4.32	5.76
Maximum throughput (64QAM) with overhead (Mbit/s)	2.44	6.10	10.17	20.34	30.51	40.68	2.07	5.18	8.64	17.28	25.92	34.56

Table 14.27 LTE downlink throughput per cell considering overhead and inefficiencies

	Normal CP						Extended CP					
Channel bandwidth (MHz)	1.4	3	5	10	15	20	1.4	3	5	10	15	20
RB allocation inefficiency (%)	80	80	80	80	80	80	80	80	80	80	80	80
RB sub-utilization (%)	78	78	78	78	78	78	78	78	78	78	78	78
ARQ and H-ARQ (%)	88	88	88	88	88	88	88	88	88	88	88	88
Minimum throughput (QPSK) with overhead and inefficiency (Mbit/s)	0.23	0.57	0.94	1.89	2.83	3.78	0.19	0.47	0.79	1.58	2.37	3.16
Maximum throughput (64QAM) with overhead and inefficiency (Mbit/s)	1.34	3.35	5.58	11.17	16.75	22.34	1.14	2.85	4.74	9.49	14.23	18.98

Table 14.28 LTE downlink throughput per cell (sector) with MIMO

	Normal CP						Extended CP					
Channel bandwidth (MHz)	1.4	3	5	10	15	20	1.4	3	5	10	15	20
MIMO Spatial Multiplexing throughput gain for medium correlation (%)	25	25	25	25	25	25	25	25	25	25	25	25
Minimum Throughput (QPSK) with MIMO (Mbit/s)	0.28	0.71	1.18	2.36	3.54	4.72	0.24	0.59	0.99	1.98	2.97	3.95
Maximum Throughput (64QAM) with MIMO (Mbit/s)	1.68	4.19	6.98	13.96	20.94	27.92	1.42	3.56	5.93	11.86	17.79	23.72

14.24 Synchronization

LTE in TDD mode requires time synchronization between eNBs better than 3 μs, and between eNB and the time reference source better than 1.5 μs. The standard itself does not specify how this time synchronization can be achieved, but similarly to WiMAX, the easiest way is to use GPS (Global Positioning System).

14.25 Beyond 4G

The 3GPP specifications will continue to develop for quite a while. There still a long way to go to finalize the specifications for a real commercial launch and for testing interoperability between eNB and phone vendors, a task that took WIMAX several years after the first commercial deployment.

For the next generation, the quest for higher speeds will continue, but some other aspects may take a prominent role, such as Device to Device Communications (D2D). UEs that are in close range could exchange data directly without having to resort to the network, reducing network traffic.

Table 14.29 LTE uplink throughput per cell considering overhead

	Normal CP						Extended CP					
Channel bandwidth (MHz)	1.4	3	5	10	15	20	1.4	3	5	10	15	20
Number of sub-carrier groups	6	15	25	50	75	100	6	15	25	50	75	100
Total Resource Elements/frame (thousand)	10	25	42	84	126	168	9	22	36	72	108	144
Control channel RE/frame (thousand)	5.76	5.76	5.76	5.76	5.76	5.76	4.80	4.80	4.80	4.80	4.80	4.80
PRACH RE RE/frame (thousand)	1.01	1.01	1.01	1.01	1.01	1.01	0.86	0.86	0.86	0.86	0.86	0.86
RS RE/frame	0.50	1.26	2.1	4.2	6.3	8.4	0.43	1.08	1.80	3.60	5.40	7.20
PUSCH RE/frame (thousand)	2.81	17.17	33.13	73.03	112.9	152.8	2.54	14.86	28.54	62.74	96.94	131.1
Channel coding overhead (turbo code at $^1/_3$) (%)	66	66	66	66	66	66	66	66	66	66	66	66
Channel coding overhead (turbo code at 2/3) (%)	33	33	33	33	33	33	33	33	33	33	33	33
Percentage of RE available for data (worst case)	9	23	27	30	30	31	20	20	20	20	20	20
Percentage of RE available for data (best case)	19	46	53	58	60	61	40	40	40	40	40	40
Minimum throughput (QPSK) with overhead (Mbit/s)	0.19	1.17	2.25	4.97	7.68	10.39	0.35	0.86	1.44	2.88	4.32	5.76
Maximum throughput (64QAM) with overhead (Mbit/s)	1.13	6.90	13.32	29.36	45.40	61.44	2.07	5.18	8.64	17.28	25.92	34.56

Table 14.30 LTE uplink throughput per cell considering overhead and inefficiencies

	Normal CP						Extended CP					
Channel bandwidth (MHz)	1.4	3	5	10	15	20	1.4	3	5	10	15	20
RB allocation inefficiency (%)	75	75	75	75	75	75	75	75	75	75	75	75
RB sub-utilization (%)	70	70	70	70	70	70	70	70	70	70	70	70
ARQ and H-ARQ (%)	88	88	88	88	88	88	88	88	88	88	88	88
Minimum throughput (QPSK) with overhead and inefficiency (Mbit/s)	0.09	0.54	1.04	2.29	3.55	4.80	0.16	0.40	0.67	1.33	2.00	2.66
Maximum throughput (64QAM) with overhead and inefficiency (Mbit/s)	0.52	3.19	6.15	13.56	20.97	28.38	0.96	2.40	3.99	7.98	11.98	15.97

Table 14.31 LTE uplink throughput per cell (sector) with MIMO

	Normal CP						Extended CP					
Channel bandwidth (MHz)	1.4	3	5	10	15	20	1.4	3	5	10	15	20
MIMO Spatial Multiplexing throughput gain for medium correlation (%)	25	25	25	25	25	25	25	25	25	25	25	25
Minimum throughput (QPSK) with MIMO (Mbit/s)	0.11	0.67	1.30	2.87	4.43	6.00	0.20	0.50	0.83	1.66	2.49	3.33
Maximum throughput (64QAM) with MIMO (Mbit/s)	0.65	3.99	7.69	16.95	26.22	35.48	1.20	2.99	4.99	9.98	14.97	19.96

15

Broadband Standards Comparison

15.1 Introduction

There are two different groups of Broadband Standards proposed: the ones based on 3G technologies, like HSPA and EVDO, and the ones based on OFDM technology like WLAN, WiMAX, and LTE. Although throughput claims are similar for both groups, OFDM-based equipment present a more adequate solution for data transmission.

OFDM technologies have similar characteristics and have undergone several rounds of improvements to extend their throughputs.

This chapter presents the OFDM standards side by side. Similar parameters have different names across technologies but, in the tables presented here, they are shown as equivalents and the most representative name is chosen to represent the parameter. This allows the reader to compare parameters and better understand the differences between technologies. These tables can also be used as a quick reference guide for each technology.

15.2 Performance Tables

These tables establish the main technology parameters, such as FFT size, sub-carrier separation, and, consequently, symbol size and bandwidth.

The first set of tables presented here (Tables 15.1, 15.4, 15.7, 15.14 and 15.15) gives WLAN characteristics for 802.11-2007, which is the most successful technology today, being used worldwide, and that gave initial wireless access inside homes; it is even being used outdoors today. The tables also cover 802.11n-2009, the new version of this technology, conceived to provide higher throughput. This new technology, however, still has to prove itself and is struggling with backward compatibility issues.

The second set of tables (Tables 15.2, 15.5, 15.8, 15.10, 15.12 and 15.16) gives WiMAX 802.16d characteristics, including the OFDMA part that, although not implemented by vendors, was the basis for the next standard. Today, it is mainly used for point-to-point applications, as its cost advantage has been eroded over time by the most recent WiMAX 802.16e.

The third set of tables (Tables 15.3, 15.6, 15.9, 15.11, 15.13 and 15.17) gives the characteristics of WiMAX 802.16e-2005, the first scalable OFDM. These tables also include LTE, the European response to the WiMAX technology, which is being supported by the GSM-based community. The MBSFN (Mobile Broadcast Single Frequency Network) version of LTE is also included.

LTE, WiMAX and WLAN Network Design, Optimization and Performance Analysis, First Edition.
Leonhard Korowajczuk.

WiMAX is being adopted by small carriers and service providers, as it only requires cheaper and more readily available data processing equipment. LTE is being adopted by large, existing operators that plan to capitalize on their existing GSM infrastructure.

15.2.1 General Characteristics

Tables 15.1 to 15.3 define bandwidths and associated parameters for each technology.

15.2.2 Cyclic Prefix

Tables 15.4 to 15.6 cover the options given by the different technologies to avoid multipath. All solutions are based on the use of a Cyclic Prefix.

15.2.3 Modulation Schemes

Tables 15.7 to 15.9 present the modulation schemes used by each technology.

15.2.4 Framing

Table 15.10 and 15.11 cover framing solutions used by each technology.

15.2.5 Resource Blocks

Tables 15.12 and 15.13 cover how frames are divided into resource blocks.

15.2.6 Throughput

The last set of tables (Tables 15.14, 15.15, 15.16 and 15.17) analyzes the throughput provided by each technology. Technology marketing has claimed incredible throughputs that usually apply to a very small percentage of users, if any. These tables present a more realistic view; even so, the spectrum efficiency is shown for the best case to allow comparison to marketing claims. A more realistic figure applicable to the average user would be half of the spectrum efficiency used in these tables, and, even then, only for the best designed systems.

WLAN is the hardest technology to determine throughput, because of its frameless contention-based access. The final throughput depends on the average length of the packets and of the number of users. For this reason, throughput curves were included to allow an analysis of the effect of these parameters in the final throughput. Figures 15.1 to 15.6 present the maximum WLAN throughput for different package sizes. Figures 15.7 and 15.8 show throughput variations for different number of users.

For WiMAX and LTE, some assumptions, must be made to calculate framing overhead; Tables 15.16 and 15.17 list these assumptions (e.g. pilot to data and the control to data ratio). Tables 15.18 and 15.19 present ratios of data and control symbols to the total number of symbols.

Table 15.1 WLAN general characteristics

General			WLAN 802.11–2007 OFDM 11a/g			802.11n-2009 HT mix 11n	HT GF
Channel Bandwidth (MHz)	BW	input	5.0	10.0	20.0	20.0	40.0
Transmission Bandwidth (MHz)	Bt	15	4.14	8.28	16.56	17.81	35.94
Subcarrier frequency spacing (kHz)	Δf	Fs/NFFT	78.1	156.3	312.5	312.5	312.5
Symbol duration (μs)	Tb	1/Δf	12.8	6.4	3.2	3.2	3.2
Total number of subcarriers	Tsc	BW/Δf		64		64	128
Number of used sub-carriers	Nusc	Tsc-Nansc		53		57	115
Number of FFT null sub-carriers	Nfnsc	NFFT-Tsc		11		7	13
Number of actual null sub-carriers	Nansc	BW.Ngb		11		7	13
Null guard band ratio	Ngb	input	0.86	1.72	3.44	2.19	4.06
Null guard band ratio	Ngb	input	17.2%	17.2%	17.2%	10.9%	10.2%
Central sub-carrier	Csc	input		1		1	1
Number of reference signals (pilots)	Nrs	input		4		4	6
Number of data sub-carriers	Dsc	Nusc-Csc-Nrs		48		52	108
Maximum number of sub-channels/OFDM sub-carrier					does not apply		
FFT size	NFFT	input		64		64	128
Sampling factor (fraction)	n	input	1	1	1	1	1
Sampling factor (factor)	n	input	1	1	1	1	1
Sampling Frequency (MHz)	Fs	floor(n.BW/8000) 8000	5.0	10.0	20.0	20.0	40.0
Sampling Period (ns)	Ts	1/Fs	200	100	50	50	25
Standard time unit (ns)	Tu	min(Ts)		50		50	25
Symbol duration (time units)	Tb	Tb/Tu	256	128	64	64	128

Table 15.2 WiMAX general characteristics

General	WiMAX 802.16d-2004 constant OFDM													
	OFDM					OFDMA								
	WM	WM	WM	WM	WM	WM	WM	WM	WM	WH	WM	WM	WH	WM
Channel Bandwidth (MHz)	1.8	3.0	3.5	5.5	7.0	1.3	3.5	7.0	8.8	10.0	14.0	17.5	20.0	28.0
Transmission Bandwidth (MHz)	1.57	2.70	3.14	4.96	6.28	1.17	3.28	6.57	8.21	9.38	13.13	16.42	18.76	26.27
Subcarrier frequency spacing (kHz)	7.8	13.4	15.6	24.7	31.3	0.7	2.0	3.9	4.9	5.6	7.8	9.8	11.2	15.6
Symbol duration (μs)	128.0	74.4	64.0	40.5	32.0	1438.2	512.0	256.0	204.8	179.3	128.0	102.4	89.6	64.0
										PUSC				
Total number of subcarriers			224							1798				
Number of used sub-carriers			201							1681				
Number of FFT null sub-carriers			55							367				
Number of actual null sub-carriers			23							117				
Null guard band ratio	0.18	0.31	0.36	0.57	0.72	0.08	0.23	0.46	0.57	0.65	0.91	1.14	1.30	1.82
Null guard band ratio	10.3%	10.3%	10.3%	10.3%	10.3%	6.5%	6.5%	6.5%	6.5%	6.5%	6.5%	6.5%	6.5%	6.5%
Central sub-carrier			1							1				
Number of reference signals (pilots)			8							241				
Number of data sub-carriers			192							1439				
Maximum number of sub-channels/OFDM sub-carrier			19							103				
FFT size			256							2048				
Sampling factor (fraction)	8/7	86/75	8/7	316/275	8/7	8/7	8/7	8/7	8/7	8/7	8/7	8/7	8/7	8/7
Sampling factor (factor)	1.14	1.15	1.14	1.15	1.14	1.14	1.14	1.14	1.14	1.14	1.14	1.14	1.14	1.14
Sampling Frequency (MHz)	2.0	3.4	4.0	6.3	8.0	1.4	4.0	8.0	10.0	11.4	16.0	20.0	22.9	32.0
Sampling Period (ns)	500	291	250	158	125	702	250	125	100	88	63	50	44	31
Standard time unit (ns)			125							31				
Symbol duration (time units)	1024	595.3	512	324.051	256	46022	16384	8192	6554	5737	4096	3277	2867	2048

Table 15.3 WiMAX scalable and LTE general characteristics

General	WiMAX 802.16e-2005 scalable OFDM				LTE 3GPP TS 36.104						LTE 3GPP TS 36.104					
	OFDMA				OFDMA						OFDMA MBSFN					
Channel Bandwidth (MHz)	1.3	5.0	10.0	20.0	1.4	3.0	5.0	10.0	15.0	20.0	1.4	3.0	5.0	10.0	15.0	20.0
Transmission Bandwidth (MHz)	0.93	4.60	9.20	18.39	1.08	2.70	4.50	9.00	13.50	18.00	1.08	2.70	4.50	9.00	13.50	18.00
Subcarrier frequency spacing (kHz)		10.9375					15						7.5			
Symbol duration (μs)		91.4					66.7						133.3			
		PUSC, DL														
Total number of subcarriers	114	457	914	1829	93	200	333	667	1000	1333	187	400	667	1333	2000	2667
Number of used sub-carriers	85	421	841	1681	72	180	300	600	900	1200	144	360	600	1200	1800	2400
Number of FFT null sub-carriers	43	91	183	367	56	76	212	424	636	848	112	152	424	848	248	1696
Number of actual null sub-carriers	29	36	73	148	21	20	33	67	100	133	43	40	67	133	200	267
Null guard band ratio	0.32	0.40	0.80	1.61	0.32	0.30	0.50	1.00	1.50	2.00	0.32	0.30	0.50	1.00	1.50	2.00
Null guard band ratio	25.6%	7.9%	8.0%	8.1%	22.9%	10.0%	10.0%	10.0%	10.0%	10.0%	22.9%	10.0%	10.0%	10.0%	10.0%	10.0%
Central sub-carrier		1					1						1			
Number of reference signals (pilots)	12	60	120	240	4	10	17	33	50	67	8	20	33	67	100	133
Number of data sub-carriers	72	361	721	1441	67	169	282	566	849	1132	135	339	566	1132	1699	2266
Maximum number of sub-channels/OFDM sub-carrier	6	30	60	120	6	15	25	50	75	100	12	30	50	100	150	200
FFT size	128	512	1024	2048	128	256	512	1024	1536	2048	256	512	1024	2048	2048	4096
Sampling factor (fraction)		28/25			11/8	32/25	77/50	77/50	77/50	77/50	11/8	32/25	77/50	77/50	77/50	77/50
Sampling factor (factor)		1.12			1.38	1.28	1.54	1.54	1.54	1.54	1.38	1.28	1.54	1.54	1.54	1.54
Sampling Frequency (MHz)	1.4	5.6	11.2	22.4	1.92	3.84	7.68	15.36	23.04	30.72	1.92	3.84	7.68	15.36	23.04	30.72
Sampling Period (ns)	714	179	89	45	520.83	260.42	130.21	65.10	43.40	32.55	520.83	260.42	130.21	65.10	43.40	32.55
Standard time unit (ns)		45					32.55						32.55			
Symbol duration (time units)		2048					2048						4096			

Table 15.4 WLAN cyclic prefix

			WLAN 802.11–2007			802.11n-2009	
			OFDM			HT mix	HT GF
Cyclic prefix			11a/g			11n	
Channel bandwidth (MHz)	BW	input	5.0	10.0	20.0	20.0	40.0
			1 client, packet = 128 B, ST = 10 μs, DIFS = 30 μs, SIFS = 10 μs				
Reduced = nominal 1/32 (3.125%) of symbol duration	CP	input					
Duration (μs)	TCP	Tb/CP					
Duration (time units)	TCP	Tb/CP					
Duration (time units- first frame symbol- LTE only)	TCP	Tb/CP (adjusted)					
Multipath mitigation distance (km)	d_{mp}	c.CP					
Maximum Number of OFDM symbols per second (Msymbol/s)							
Normal = nominal 1/16 (6.25%) of symbol duration	CP	input					
Duration (μs)	TCP	Tb/CP					
Duration (time units)	TCP	Tb/CP					
Duration (time units- first frame symbol- LTE only)	TCP	Tb/CP (adjusted)					
Multipath mitigation distance (km)	d_{mp}	c.CP					
Maximum Number of OFDM symbols per second (Msymbol/s)							
Large = nominal 1/8 (12.5%) of symbol duration	CP	input				0.125	
Duration (μs)	TCP	Tb/CP				0.4	0.4
Duration (time units)	TCP	Tb/CP				15.625	7.8125
Multipath mitigation distance (km)	d_{mp}	c.CP				0.1	0.1
Maximum Number of OFDM symbols per second (Msymbol/s)						0.024	0.026
Extended = nominal 1/4 (25%) of symbol duration	CP	input			0.25		
Duration (μs)	TCP	Tb/CP	3.2	1.6	0.8	0.8	0.8
Duration (time units)	TCP	Tb/CP		16		16	32
Multipath mitigation distance (km)	d_{mp}	c.CP	1.0	0.5	0.2	0.2	0.2
Maximum Number of OFDM symbols per second (Msymbol/s)	MNs	1/(TCP+Tb)	0.063	0.125	0.250	0.250	0.250
Double = nominal 1/2 (50%) of symbol duration (us)	CP	input				0.5	0.5
Duration (μs)	TCP	Tb/CP				1.6	1.6
Duration (time units)	TCP	Tb/CP				32	64
Multipath mitigation distance (km)	d_{mp}	c.CP				0.5	0.5

Table 15.5 WiMAX cyclic prefix

	WiMAX 802.16d-2004 constant OFDM													
	OFDM					OFDMA								
Cyclic prefix	WM	WM	WM	WM	WM	WM	WM	WM	WM	WH	WM	WM	WH	WM
Channel bandwidth (MHz)	1.8	3.0	3.5	5.5	7.0	1.3	3.5	7.0	8.8	10.0	14.0	17.5	20.0	28.0
Reduced = nominal 1/32 (3.125%) of symbol duration			0.03125							0.03125				
Duration (μs)	4.0	2.3	2.0	1.3	1.0	44.9	16.0	8.0	6.4	5.6	4.0	3.2	2.8	2.0
Duration (time units)			8							64				
Duration (time units- first frame symbol- LTE only)														
Multipath mitigation distance (km)	1.2	0.7	0.6	0.4	0.3	13.5	4.8	2.4	1.9	1.7	1.2	1.0	0.8	0.6
Maximum Number of OFDM symbols per second (Msymbol/s)	0.008	0.013	0.015	0.024	0.030	0.001	0.002	0.004	0.005	0.005	0.008	0.009	0.011	0.015
Normal = nominal 1/16 (6.25%) of symbol duration			0.0625							0.0625				
Duration (μs)	8.0	4.7	4.0	2.5	2.0	89.9	32.0	16.0	12.8	11.2	8.0	6.4	5.6	4.0
Duration (time units)			16							128				
Duration (time units- first frame symbol- LTE only)														
Multipath mitigation distance (km)	2.4	1.4	1.2	0.8	0.6	27.0	9.6	4.8	3.8	3.4	2.4	1.9	1.7	1.2
Maximum Number of OFDM symbols per second (Msymbol/s)	0.01	0.01	0.01	0.02	0.03	0.001	0.002	0.004	0.005	0.005	0.007	0.009	0.011	0.015
Large = nominal 1/8 (12.5%) of symbol duration			0.125							0.125				
Duration (μs)	16.0	9.3	8.0	5.1	4.0	179.8	64.0	32.0	25.6	22.4	16.0	12.8	11.2	8.0
Duration (time units)			32							256				
Multipath mitigation distance (km)	4.8	2.8	2.4	1.5	1.2	53.9	19.2	9.6	7.7	6.7	4.8	3.8	3.4	2.4
Maximum Number of OFDM symbols per second (Msymbol/s)	0.007	0.012	0.014	0.022	0.028	0.001	0.002	0.003	0.004	0.005	0.007	0.009	0.010	0.014
Extended = nominal 1/4 (25%) of symbol duration			0.25							0.25				
Duration (μs)	32.0	18.6	16.0	10.1	8.0	359.6	128.0	64.0	51.2	44.8	32.0	25.6	22.4	16.0
Duration (time units)			64							512				
Multipath mitigation distance (km)	9.6	5.6	4.8	3.0	2.4	107.9	38.4	19.2	15.4	13.4	9.6	7.7	6.7	4.8
Maximum Number of OFDM symbols per second (Msymbol/s)	0.006	0.011	0.013	0.020	0.025	0.001	0.002	0.003	0.004	0.004	0.006	0.008	0.009	0.013

Table 15.6 WiMAX scalable and LTE cyclic prefix

	WiMAX 802.16e-2005 scalable OFDM	LTE 3GPP TS 36.104	LTE 3GPP TS 36.104
Cyclic prefix	OFDMA	OFDMA	OFDMA MBSFN
Channel bandwidth (MHz)	1.3 5.0 10.0 20.0	1.4 3.0 5.0 10.0 15.0 20.0	1.4 3.0 5.0 10.0 15.0 20.0
Cyclic Prefix			
Reduced = nominal 1/32 (3.125%) of symbol duration	0.03125		0.0352
Duration (μs)	2.9		4.7
Duration (time units)	4		144
Duration (time units- first frame symbol- LTE only)			160
Multipath mitigation distance (km)	0.9		1.4
Maximum Number of OFDM symbols per second (Msymbol/s)	0.011	0.015	0.007
Normal = nominal 1/16 (6.25%) of symbol duration	0.0625	0.0703	
Duration (μs)	5.7	4.7	
Duration (time units)	8	144	
Duration (time units- first frame symbol- LTE only)		160	
Multipath mitigation distance (km)	1.7	1.4	
Maximum Number of OFDM symbols per second (Msymbol/s)	0.010	0.014	
Large = nominal 1/8 (12.5%) of symbol duration	0.125		
Duration (μs)	11.4		
Duration (time units)	16		
Multipath mitigation distance (km)	3.4		
Maximum Number of OFDM symbols per second (Msymbol/s)	0.010		
Extended = nominal 1/4 (25%) of symbol duration	0.25	0.25	0.25
Duration (μs)	22.9	16.7	33.3
Duration (time units)	32	512	1024
Multipath mitigation distance (km)	6.9	5.0	10.0
Maximum Number of OFDM symbols per second (Msymbol/s)	0.009	0.012	0.006

Table 15.7 WLAN modulation schemes

	Wi-Fi 802.11–2007			802.11n-2009	
	OFDM			HT mix	HT GF
Modulation schemes	11a/g			11n	
Channel bandwidth (MHz)	5.0	10.0	20.0	20.0	40.0
Nominal Coding Scheme (Extended CP, DL)	Air Data Rate (Mbit/s) RSCC				
BPSK 1/2	1.5	3.0	6.0	6.0	13.5
BPSK 3/4	2.3	4.5	9.0	9.0	20.3
QPSK 1/2	3.0	6.0	12.0	12.0	27.0
QPSK 2/3					
QPSK 3/4	4.5	9.0	18.0	18.0	40.5
16QAM 1/2	6.0	12.0	24.0	24.0	54.0
16QAM 3/5					
16 QAM 3/4	9.0	18.0	36.0	36.0	81.0
16QAM 4/5					
64QAM 2/3	12.0	24.0	48.0	48.0	108.0
64QAM 3/4	13.5	27.0	54.0	54.0	121.5
64QAM 5/6				60.0	135.0

Table 15.8 WiMAX modulation schemes

	WiMAX 802.16d-2004 constant OFDM													
	OFDM					OFDMA								
Modulation schemes	WM	WM	WM	WM	WM	WM	WM	WM	WM	WH	WM	WM	WH	WM
Channel bandwidth (MHz)	1.8	3.0	3.5	5.5	7.0	1.3	3.5	7.0	8.8	10.0	14.0	17.5	20.0	28.0
Nominal Coding Scheme (Extended CP, DL)	Air Data Rate (Mbit/s) RSCC, BTC, CTC					Air Data Rate (Mbit/s) RSCC, BTC, CTC								
BPSK 1/2	0.6	1.0	1.2	1.9	2.4	0.4	1.1	2.2	2.8	3.2	4.5	5.6	6.4	9.0
BPSK 3/4														
QPSK 1/2	1.2	2.1	2.4	3.8	4.8	0.8	2.2	4.5	5.6	6.4	9.0	11.2	12.9	18.0
QPSK 2/3	1.6	2.8	3.2	5.1	6.4	1.1	3.0	6.0	7.5	8.6	12.0	15.0	17.1	24.0
QPSK 3/4	1.8	3.1	3.6	5.7	7.2	1.2	3.4	6.7	8.4	9.6	13.5	16.9	19.3	27.0
16QAM 1/2	2.4	4.1	4.8	7.6	9.6	1.6	4.5	9.0	11.2	12.8	18.0	22.5	25.7	36.0
16QAM 3/5	2.9	5.0	5.8	9.1	11.5	1.9	5.4	10.8	13.5	15.4	21.6	27.0	30.8	43.2
16 QAM 3/4	3.6	6.2	7.2	11.4	14.4	2.4	6.7	13.5	16.9	19.3	27.0	33.7	38.6	54.0
16QAM 4/5	3.8	6.6	7.7	12.1	15.4	2.6	7.2	14.4	18.0	20.6	28.8	36.0	41.1	57.6
64QAM 2/3	4.8	8.3	9.6	15.2	19.2	3.2	9.0	18.0	22.5	25.7	36.0	45.0	51.4	72.0
64QAM 3/4	5.4	9.3	10.8	17.1	21.6	3.6	10.1	20.2	25.3	28.9	40.5	50.6	57.8	81.0
64QAM 5/6	6.0	10.3	12.0	19.0	24.0	4.0	11.2	22.5	28.1	32.1	45.0	56.2	64.3	90.0

Table 15.9 WiMAX scalable and LTE modulation schemes

	WiMAX 802.16e-2005 scalable OFDM				LTE 3GPP TS 36.104						LTE 3GPP TS 36.104					
	OFDMA				OFDMA						OFDMA					
Modulation schemes											MBSFN					
Channel bandwidth (MHz)	1.3	5.0	10.0	20.0	1.4	3.0	5.0	10.0	15.0	20.0	1.4	3.0	5.0	10.0	15.0	20.0
Nominal Coding Scheme (Extended CP, DL)	Air Data Rate (Mbit/s) RSCC, BTC, CTC, LDPC				Air Data Rate (Mbit/s) CRC, RSCC, CTC						Air Data Rate (Mbit/s) CRC, RSCC, CTC					
BPSK 1/2	0.3	1.6	3.2	6.3	0.4	1.0	1.7	3.4	5.1	6.8	0.4	1.0	1.7	3.4	5.1	6.8
BPSK 3/4																
QPSK 1/2	0.6	3.2	6.3	12.6	0.8	2.0	3.4	6.8	10.2	13.6	0.8	2.0	3.4	6.8	10.2	13.6
QPSK 2/3	0.8	4.2	8.4	16.8	1.1	2.7	4.5	9.1	13.6	18.1	1.1	2.7	4.5	9.1	13.6	18.1
QPSK 3/4	0.9	4.7	9.5	18.9	1.2	3.0	5.1	10.2	15.3	20.4	1.2	3.1	5.1	10.2	15.3	20.4
16QAM 1/2	1.3	6.3	12.6	25.2	1.6	4.1	6.8	13.6	20.4	27.2	1.6	4.1	6.8	13.6	20.4	27.2
16QAM 3/5	1.5	7.6	15.1	30.3	1.9	4.9	8.1	16.3	24.5	32.6	1.9	4.9	8.1	16.3	24.5	32.6
16 QAM 3/4	1.9	9.5	18.9	37.8	2.4	6.1	10.2	20.4	30.6	40.8	2.4	6.1	10.2	20.4	30.6	40.8
16QAM 4/5	2.0	10.1	20.2	40.3	2.6	6.5	10.8	21.7	32.6	43.5	2.6	6.5	10.9	21.7	32.6	43.5
64QAM 2/3	2.5	12.6	25.2	50.4	3.2	8.1	13.6	27.2	40.8	54.4	3.2	8.1	13.6	27.2	40.8	54.4
64QAM 3/4	2.8	14.2	28.4	56.7	3.6	9.1	15.2	30.5	45.8	61.1	3.6	9.2	15.3	30.6	45.9	61.2
64QAM 5/6	3.1	15.8	31.5	63.0	4.0	10.1	16.9	33.9	50.9	67.9	4.1	10.2	17.0	34.0	51.0	68.0

Table 15.10 WIMAX framing

	WiMAX 802.16d-2004 constant OFDM													
	OFDM					OFDMA								
WiMAX framing	WM	WM	WM	WM	WM	WM	WM	WM	WM	WH	WM	WM	WH	WM
Channel bandwidth (MHz)	1.8	3.0	3.5	5.5	7.0	1.3	3.5	7.0	8.8	10.0	14.0	17.5	20.0	28.0
TDD														
Frame duration (ms)	2, 2.5, 4, 5, 8, 10, 12.5, 20					2, 2.5, 4, 5, 8, 10, 12.5, 20								
TDD sub-frame duration (ms)	flexible					flexible								
Sub-frame duration (ms)														
Slot duration (ms)														
Example	Tf = 10 ms, CP = 0.25, TTG/GP = 180 µs, RTG = 0 µs, DL/UL usage ratio = 0.5, PUSC					Tf = 10 ms, CP = 0.25, TTG/GP = 180 µs, RTG = 0 µs, DL/UL usage ratio = 0.5, PUSC								
Example														
DL TDD ratio DL/(DL+UP)														
OFDM Symbol total duration	160.0	93.0	80.0	50.6	40.0	1797.8	640.0	320.0	256.0	224.1	160.0	128.0	112.0	80.0
OFDM Symbols per frame-DL+UL	61	105	122	193	245	5	15	30	38	43	61	76	87	122
Total symbols per second (Msymbol/s)	1.2	2.1	2.5	3.9	4.9	0.8	2.5	5.0	6.4	7.2	10.3	12.8	14.6	20.5
Reference symbols (pilots) as % of total symbols-DL	4.0%					14.3%								
Average control symbols as % of total symbols-DL	4.6%					6.9%								
Reference symbols (pilots) as % of total symbols-UL	4.0%					33.3%								
Average control symbols as % of total symbols-UL	3.4%					4.6%								
TDD transition	1.8%					2.0%								
FEC overhead	20.0%					20.0%								
Wireless layer MAC overhead	5.0%					5.0%								
Total data Symbols per second-DL+UL (Msymbol/s)	0.9	1.5	1.8	2.8	3.6	0.5	1.4	2.8	3.5	4.0	5.7	7.1	8.1	11.3
Maximum data capacity-DL+ UL (Mbit/s)	5.3	9.2	10.6	16.8	21.4	2.8	8.4	16.7	21.2	23.9	34.0	42.3	48.5	67.9
Spectral Efficiency TDD (bits/Hz)														
Maximum DL	2.9	2.9	3.0	3.0	3.0	2.4	2.6	2.6	2.7	2.6	2.6	2.7	2.7	2.7
Average DL	1.0	1.0	1.0	1.0	1.0	0.8	0.9	0.9	0.9	0.9	0.9	0.9	0.9	0.9
Maximum UL	3.0	3.0	3.0	3.0	3.0	1.9	2.0	2.1	2.1	2.1	2.1	2.1	2.1	2.1
Average UL	1.0	1.0	1.0	1.0	1.0	0.6	0.7	0.7	0.7	0.7	0.7	0.7	0.7	0.7

Table 15.11 WiMAX scalable and LTE framing

Framing	WiMAX 802.16e-2005 scalable OFDM				LTE 3GPP TS 36.104						LTE 3GPP TS 36.104					
	OFDMA				OFDMA						OFDMA					
Channel Bandwidth (MHz)	1.3	5.0	10.0	20.0	1.4	3.0	5.0	10.0	15.0	20.0	1.4	3.0	5.0	10.0	15.0	20.0
Duplex	TDD				FDD						FDD MBSFN					
Frame duration (ms)	2, 2.5, 4, 5, 8, 10, 12.5, 20				10						10					
TDD sub-frame duration (ms)	flexible				5, 10						5, 10					
Sub-frame duration (ms)					1						1					
Slot duration (ms)					0.5						0.5					
Example	Tf = 10 ms, CP = 0.25, TTG/GP = 180 μs, RTG = 0 μs, DL/UL usage ratio = 0.51, PUSC				Tf = 10 ms, CP = 0.25, TTG/GP = 180 μs, RTG = 0 μs, DL/UL usage ratio = 0.5						Tf = 10 ms, CP = 0.25, TTG/GP = 180 μs, RTG = 0 μs, DL/UL usage ratio = 0.5					
DL TDD ratio DL/(DL+UP)					0.25, 0.38, 05, 075, 0.78, 0.89						0.25, 0.38, 05, 075, 0.78, 0.89					
OFDM Symbol total duration	114.3				83.3						166.7					
OFDM Symbols per frame-DL+UL	85				108						54					
Total symbols per second (Msymbol/s)	0.7	3.6	7.1	14.3	0.8	1.9	3.2	6.5	9.7	13.0	0.8	1.9	3.2	6.5	9.7	13.0
Reference symbols (pilots) as % of total symbols-DL	14.3%				5.6%						11.1%					
Average control symbols as % of total symbols-DL	6.9%				16.7%						16.7%					
Reference symbols (pilots) as % of total symbols-UL	33.3%				16.7%						33.3%					
Average control symbols as % of total symbols-UL	4.6%				8.3%						8.3%					
TDD transition	1.8%				5.0%						5.0%					
FEC overhead	20.0%				20.0%						20.0%					
Wireless layer MAC overhead	5.0%				5.0%						5.0%					
Total data Symbols per second-DL+UL (Msymbol/s)	0.4	2.0	4.0	7.9	0.5	1.1	1.9	3.8	5.6	7.5	0.4	1.0	1.6	3.2	4.8	6.4
Maximum data capacity-DL+ UL (Mbit/s)	2.4	11.9	23.7	47.4	2.7	6.8	11.3	22.6	33.9	45.1	2.3	5.8	9.6	19.3	28.9	38.6
Spectral Efficiency TDD (bits/Hz)																
Maximum DL	2.1	2.6	2.6	2.6	2.2	2.5	2.5	2.5	2.5	2.5	2.0	2.3	2.3	2.3	2.3	2.3
Average DL	0.7	0.9	0.9	0.9	0.7	0.8	0.8	0.8	0.8	0.8	0.7	0.8	0.8	0.8	0.8	0.8
Maximum UL	1.7	2.0	2.0	2.0	2.1	2.4	2.4	2.4	2.4	2.4	1.9	2.2	2.2	2.2	2.2	2.2
Average UL	0.6	0.7	0.7	0.7	0.7	0.8	0.8	0.8	0.8	0.8	0.6	0.7	0.7	0.7	0.7	0.7

Table 15.12 WiMAX resource blocks

	WiMAX 802.16d-2004 constant OFDM													
	OFDM					OFDMA								
Resource blocks	WM	WM	WM	WM	WM	WM	WM	WM	WM	WH	WM	WM	WH	WM
Channel Bandwidth (MHz)	1.8	3.0	3.5	5.5	7.0	1.3	3.5	7.0	8.8	10.0	14.0	17.5	20.0	28.0
Resource Blocks (sub-channel)										PUSC-DL				
Resource Block subcarriers	12	12	12	12	12	14	14	14	14	14	14	14	14	14
Resource Block Bandwidth (kHz)	93.8	161.3	187.5	296.3	375.0	9.7	27.3	54.7	68.4	78.1	109.4	136.7	156.2	218.8
Resource Block Symbols	4	4	4	4	4	4	4	4	4	4	4	4	4	4
Resource Block slots														
Number of Resource blocks in bandwidth	18	18	18	18	18	128	128	128	128	128	128	128	128	128
Number of Resource Blocks per frame	19	33	39	61	78	1	4	9	12	13	19	24	27	39
Total Resource Blocks	342	594	702	1098	1404	128	512	1152	1536	1664	2432	3072	3456	4992
Reference signal symbols per resource block	0	0	0	0	0	8	8	8	8	8	8	8	8	8
Data symbols per resource block	48	48	48	48	48	48	48	48	48	48	48	48	48	48

Table 15.13 WiMAX scalable and LTE resource blocks

	WiMAX 802.16e-2005 scalable OFDM				LTE 3GPP TS 36.104						LTE 3GPP TS 36.104					
	OFDMA				OFDMA						OFDMA					
Channel bandwidth (MHz)	1.3	5.0	10.0	20.0	1.4	3.0	5.0	10.0	15.0	20.0	1.4	3.0	5.0	10.0	15.0	20.0
														MBSFN		
Resource Blocks (sub-channel)		PUSC-DL														
Resource Block subcarriers	14	14	14	14	12	12	12	12	12	12	24	24	24	24	24	24
Resource Block Bandwidth (kHz)	153.1	153.1	153.1	153.1				180						180		
Resource Block Symbols	4	4	4	4	6	6	6	6	6	6	6	6	6	6	6	6
Resource Block slots					1	1	1	1	1	1	1	1	1	1	1	1
Number of Resource blocks in bandwidth	8	32	65	130	6	15	25	50	75	100	6	15	25	50	75	100
Number of Resource Blocks per frame	27	27	27	27	20	20	20	20	20	20	20	20	20	20	20	20
Total Resource Blocks	216	864	1755	3510	120	300	500	1000	1500	2000	120	300	500	1000	1500	2000
Reference signal symbols per resource block	8	8	8	8	4	4	4	4	4	4	8	8	8	8	8	8
Data symbols per resource block	48	48	48	48	68	68	68	68	68	68	136	136	136	136	136	136

Table 15.14 WLAN throughput

		Wi-Fi 802.11-2007			802.11n-2009	
		OFDM			HT MF	HT GF
Maximum throughput (Mbit/s)		11a/g			11n	
Bandwidth (MHz)	# users	5.0	10.0	20.0	20.0	40.0
Maximum rate considering access - Mbit/s	air	13.50	27.00	54.00	60.00	135.00
(1packet = 32 B, ST = 10 μs, DIFS = 30 μs, SIFS = 10 μs)	1	0.85	1.03	1.16	1.26	1.36
TDD ratio = 0.5 PER = 1% Control Frame overhead = 20%	5	0.51	0.69	0.76	0.82	0.89
Link length = 1000 m Uplink: RTS/CTS	10	0.45	0.51	0.54	0.59	0.63
Maximum rate considering access - Mbit/s	air	13.50	27.00	54.00	60.00	135.00
(1packet = 64 B, ST = 10 μs, DIFS = 30 μs, SIFS = 10 μs)	1	1.58	1.97	2.26	2.45	2.65
TDD ratio = 0.5 PER = 1% Control Frame overhead = 20%	5	0.97	1.33	1.49	1.61	1.75
Link length = 1000 m Uplink: RTS/CTS	10	0.85	0.99	1.07	1.16	1.26
Maximum rate considering access - Mbit/s	air	13.50	27.0	54.0	60.0	135.0
(1packet = 128 B, ST = 10 μs, DIFS = 30 μs, SIFS = 10 μs)	1	2.77	3.63	4.29	4.65	5.03
TDD ratio = 0.5 PER = 1% Control Frame overhead = 20%	5	1.77	2.13	2.86	3.10	3.36
Link length = 1000 m Uplink: RTS/CTS	10	1.27	1.87	2.08	2.25	2.44
Maximum rate considering access - Mbit/s	air	13.50	27.0	54.0	60.0	135.0
(1packet = 512 B, ST = 10 μs, DIFS = 30 μs, SIFS = 10 μs)	1	6.34	9.75	13.33	14.44	15.64
TDD ratio = 0.5 PER = 1% Control Frame overhead = 20%	5	3.91	6.44	8.04	8.71	9.44
Link length = 1000 m Uplink: RTS/CTS	10	3.65	4.73	5.55	6.01	6.51
Maximum rate considering access - Mbit/s	air	13.50	27.00	54.00	60.00	135.00
(1packet = 1024 B, ST = 10 μs, DIFS = 30 μs, SIFS = 10 μs)	1	8.08	13.56	20.52	22.23	24.08
TDD ratio = 0.5 PER = 1% Control Frame overhead = 20%	5	5.70	8.17	13.37	14.48	15.69
Link length = 1000 m Uplink: RTS/CTS	10	4.40	7.61	9.72	10.53	11.41
Maximum rate considering access - Mbit/s	air	13.50	27.0	54.0	60.0	135.0
(1packet = 2048 B, ST = 10 μs, DIFS = 30 μs, SIFS = 10 μs)	1	9.37	16.86	28.09	30.43	32.97
TDD ratio = 0.5 PER = 1% Control Frame overhead = 20%	5	6.42	11.76	16.72	18.11	19.62
Link length = 1000 m Uplink: RTS/CTS	10	6.18	9.01	15.57	16.87	18.27

Table 15.15 WLAN spectral efficiency

		Wi-Fi 802.11-2007			802.11n-2009	
		OFDM			HT MF	HT GF
Spectral efficiency (bits/Hz)	# users	11a/g			11n	
Bandwidth (MHz)		5.0	10.0	20.0	20.0	40.0
(1packet = 128 B, ST = 10 μs, DIFS = 30 μs, SIFS = 10 μs)	1	0.6	0.4	0.2	0.2	0.1
TDD ratio = 0.5 PER = 1% Control Frame overhead = 20%	5	0.4	0.2	0.1	0.2	0.1
Link length = 1000 m Uplink: RTS/CTS	10	0.3	0.2	0.1	0.1	0.1
(1packet = 2048 B, ST = 10 μs, DIFS = 30 μs, SIFS = 10 μs)	1	1.9	1.7	1.4	1.5	0.8242
TDD ratio = 0.5 PER = 1% Control Frame overhead = 20%	5	1.3	1.2	0.8	0.9	0.5
Link length = 1000 m Uplink: RTS/CTS	10	1.2	0.9	0.8	0.8	0.5

Table 15.16 WiMAX throughput and spectral efficiency

	WiMAX 802.16d-2004 constant OFDM													
	OFDM					OFDMA								
Throughput and spectral efficiency	WM	WM	WM	WM	WM	WM	WM	WM	WM	WH	WM	WM	WH	WM
Channel bandwidth (MHz)	1.8	3.0	3.5	5.5	7.0	1.3	3.5	7.0	8.8	10.0	14.0	17.5	20.0	28.0
			TDD							TDD				
Frame duration (ms)			2, 2.5, 4, 5, 8, 10, 12.5, 20							2, 2.5, 4, 5, 8, 10, 12.5, 20				
Sub-frame duration (ms)														
Slot duration (ms)														
Example			Tf = 10 ms, CP = 25%, 64QAM 5/6, PUSC							Tf = 10 ms, CP = 25%, 64QAM 5/6, PUSC				
OFDM Symbol total duration	160.0	93.0	80.0	50.6	40.0	1797.8	640.0	320.0	256.0	224.1	160.0	128.0	112.0	80.0
OFDM symbols per slot														
OFDM Symbols per frame	62	107	125	197	250	5	15	31	39	44	62	78	89	125
Total symbols per second (Msymbol/s)	1.2	2.2	2.5	4.0	5.0	0.8	2.5	5.2	6.6	7.4	10.4	13.1	15.0	21.0
Reference symbols (pilots) as % of total symbols-DL			4.0%							14.3%				
Average control symbols as % of total symbols-DL			4.6%							6.9%				
Reference symbols (pilots) as % of total symbols-UL			4.0%							33.3%				
Average control symbols as % of total symbols-UL			3.4%							4.6%				
FEC overhead			20.0%							20.0%				
Wireless layer MAC overhead			5.0%							5.0%				
Total data symbols per second-DL (Msymbol/s)	0.9	1.5	1.7	2.7	3.4	0.5	1.5	3.1	3.9	4.4	6.2	7.7	8.8	12.4
Maximum data capacity-DL (Mbit/s)	5.1	8.8	10.3	16.3	20.7	3.0	8.9	18.5	23.2	26.2	37.0	46.5	53.0	74.5
Total data symbols per second-UL (Msymbol/s)	0.9	1.5	1.7	2.7	3.5	0.4	1.2	2.4	3.1	3.4	4.9	6.1	7.0	9.8
Maximum data capacity-UL (Mbit/s)	5.2	9.0	10.5	16.5	20.9	2.3	7.0	14.6	18.3	20.7	29.1	36.6	41.8	58.7
Spectral Efficiency FDD (bits/Hz)														
Maximum DL	2.9	2.9	3.0	3.0	3.0	2.4	2.6	2.6	2.7	2.6	2.6	2.7	2.7	2.7
Average DL	1.0	1.0	1.0	1.0	1.0	0.8	0.9	0.9	0.9	0.9	0.9	0.9	0.9	0.9
Maximum UL	3.0	3.0	3.0	3.0	3.0	1.9	2.0	2.1	2.1	2.1	2.1	2.1	2.1	2.1
Average UL	1.0	1.0	1.0	1.0	1.0	0.6	0.7	0.7	0.7	0.7	0.7	0.7	0.7	0.7

Table 15.17 WiMAX scalable and LTE maximum throughput and spectral efficiency

Throughput and spectral efficiency	WiMAX 802.16e-2005 scalable OFDM				LTE 3GPP TS 36.104						LTE 3GPP TS 36.104					
	OFDMA				OFDMA						OFDMA					
Channel Bandwidth (MHz)	1.3	5.0	10.0	20.0	1.4	3.0	5.0	10.0	15.0	20.0	1.4	3.0	5.0	10.0	15.0	20.0
	TDD				FDD						MBSFN FDD					
Frame duration (ms)	2, 2.5, 4, 5, 8, 10, 12.5, 20				10						10					
Sub-frame duration (ms)					1						1					
Slot duration (ms)					0.5						0.5					
Example	Tf = 10 ms, CP = 25%, 64QAM 5/6, PUSC				Tf = 10 ms, CP = 25%, 64QAM 5/6						Tf = 10 ms, CP = 25%, 64QAM 5/6					
OFDM Symbol total duration	114.3				83.3						166.7					
OFDM symbols per slot					6						3					
OFDM Symbols per frame	87				120						60					
Total symbols per second (Msymbol/s)	0.7	3.7	7.3	14.6	0.9	2.2	3.6	7.2	10.8	14.4	0.9	2.2	3.6	7.2	10.8	14.4
Reference symbols (pilots) as % of total symbols-DL	14.3%				5.6%						11.1%					
Average control symbols as % of total symbols-DL	6.9%				16.7%						16.7%					
Reference symbols (pilots) as % of total symbols-UL	33.3%				16.7%						33.3%					
Average control symbols as % of total symbols-UL	4.6%				8.3%						8.3%					
FEC overhead	20.0%				20.0%						20.0%					
Wireless layer MAC overhead	5.0%				5.0%						5.0%					
Total data symbols per second-DL (Msymbol/s)	0.4	2.2	4.3	8.6	0.5	1.3	2.1	4.2	6.3	8.4	0.5	1.2	2.0	3.9	5.9	7.8
Maximum data capacity-DL (Mbit/s)	2.6	13.0	25.9	51.9	3.0	7.6	12.6	25.2	37.8	50.4	2.8	7.0	11.7	23.4	35.1	46.8
Total data symbols per second-UL (Msymbol/s)	0.3	1.7	3.4	6.8	0.5	1.2	2.0	4.1	6.1	8.1	0.4	1.1	1.8	3.6	5.5	7.3
Maximum data capacity-UL (Mbit/s)	2.1	10.2	20.4	40.8	2.9	7.3	12.2	24.3	36.5	48.6	2.6	6.6	10.9	21.8	32.8	43.7
Spectral Efficiency FDD (bits/Hz)																
Maximum DL	2.1	2.6	2.6	2.6	2.2	2.5	2.5	2.5	2.5	2.5	2.0	2.3	2.3	2.3	2.3	2.3
Average DL	0.7	0.9	0.9	0.9	0.7	0.8	0.8	0.8	0.8	0.8	0.7	0.8	0.8	0.8	0.8	0.8
Maximum UL	1.7	2.0	2.0	2.0	2.1	2.4	2.4	2.4	2.4	2.4	1.9	2.2	2.2	2.2	2.2	2.2
Average UL	0.6	0.7	0.7	0.7	0.7	0.8	0.8	0.8	0.8	0.8	0.6	0.7	0.7	0.7	0.7	0.7

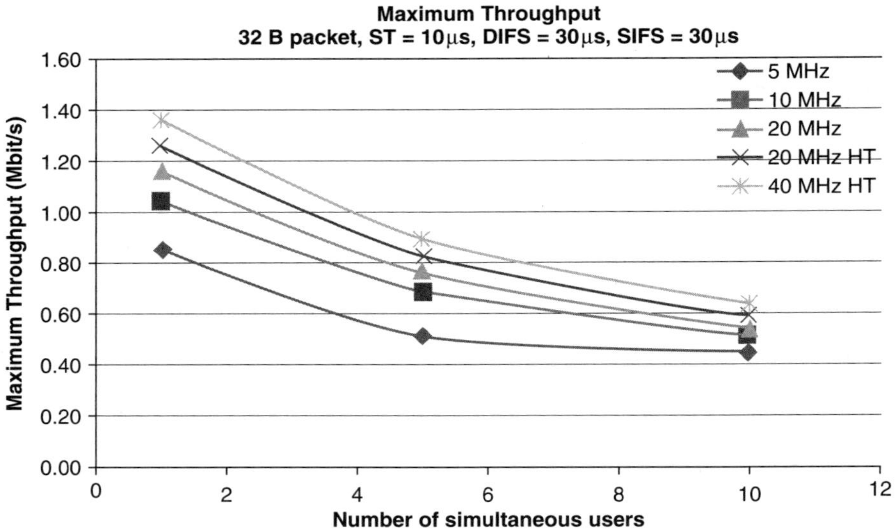

Figure 15.1 Maximum throughput for 32-byte packages.

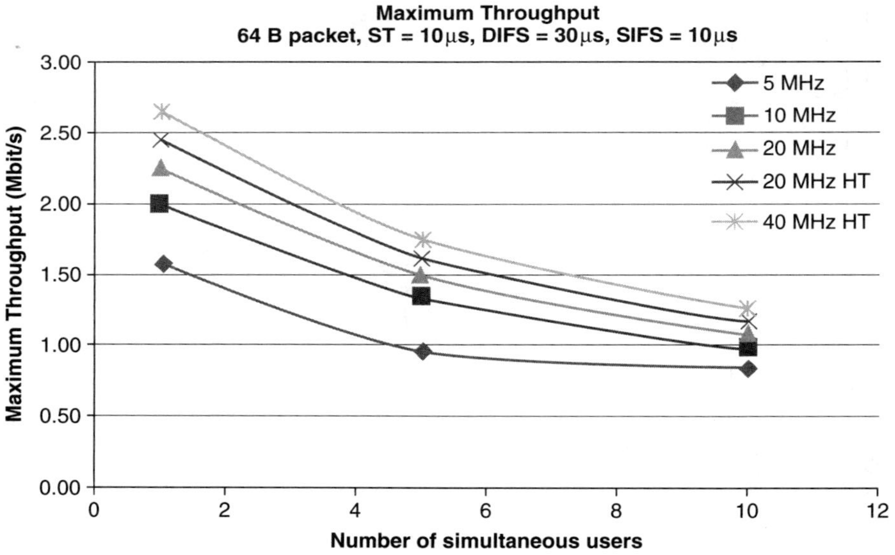

Figure 15.2 Maximum throughput for 64-byte packages.

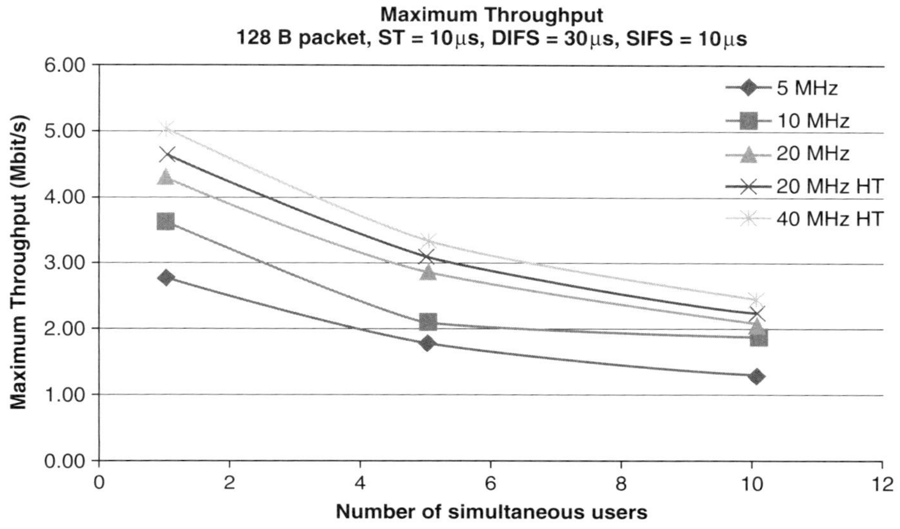

Figure 15.3 Maximum throughput for 128-byte packages.

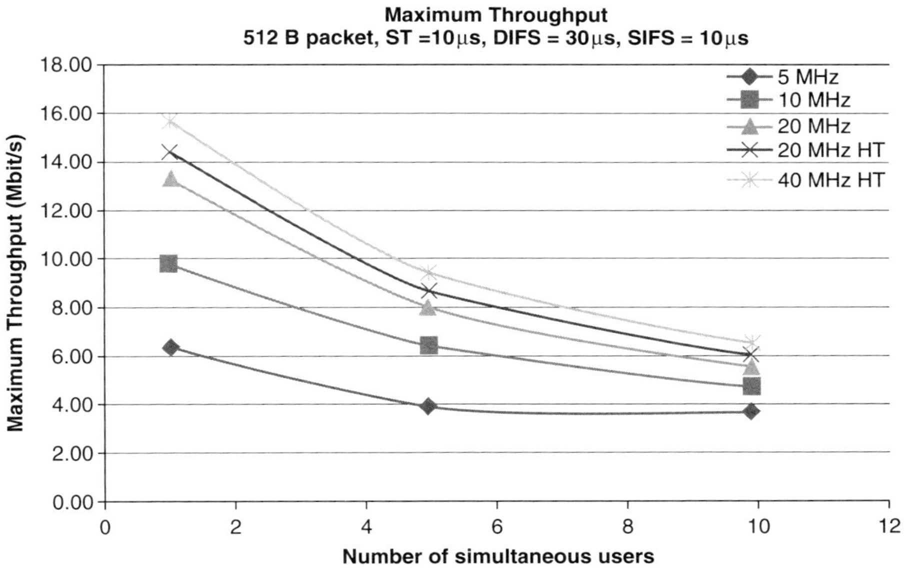

Figure 15.4 Maximum throughput for 512-byte packages.

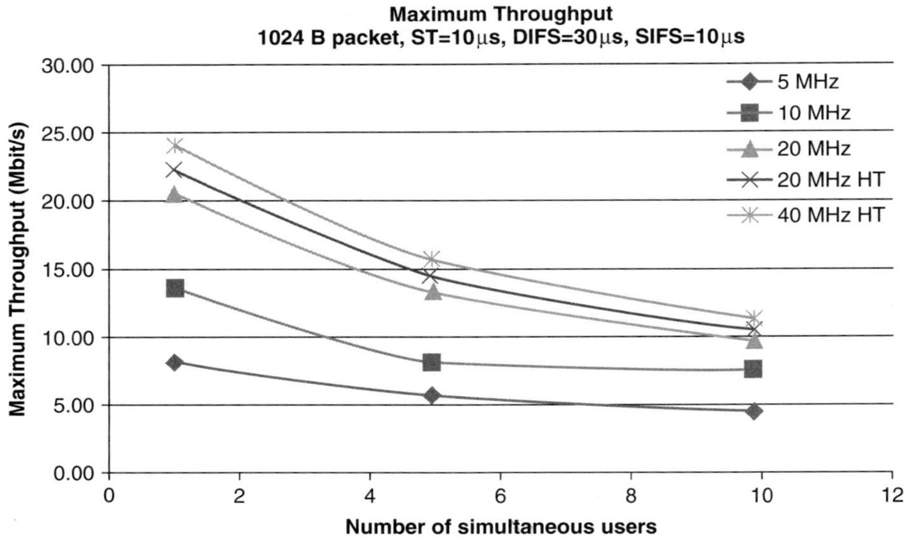

Figure 15.5 Maximum throughput for 1024-byte packages.

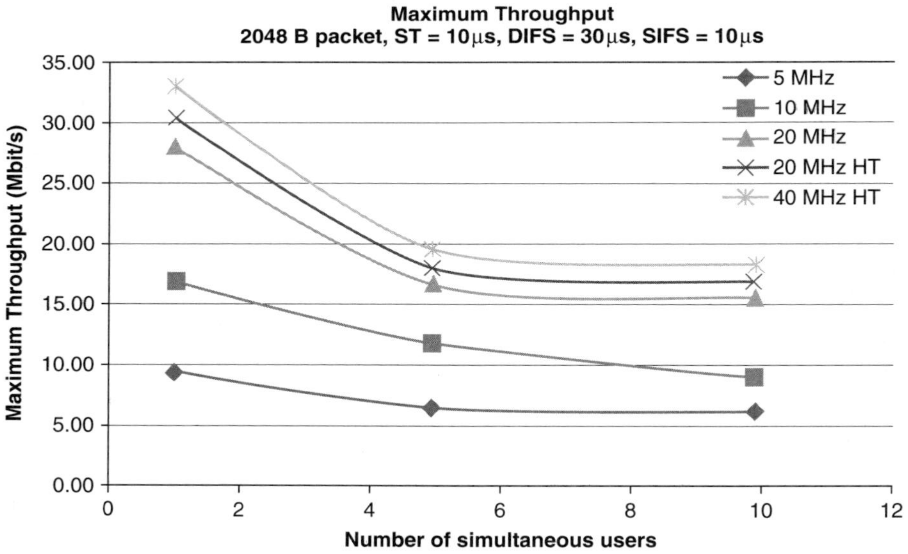

Figure 15.6 Maximum throughput for 2048-byte packages.

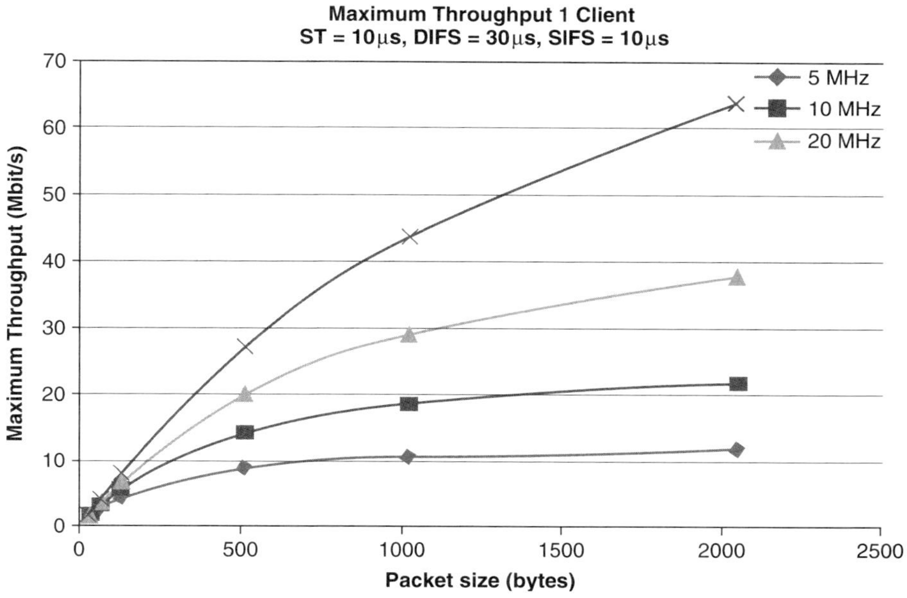

Figure 15.7 Maximum throughput for 1 client.

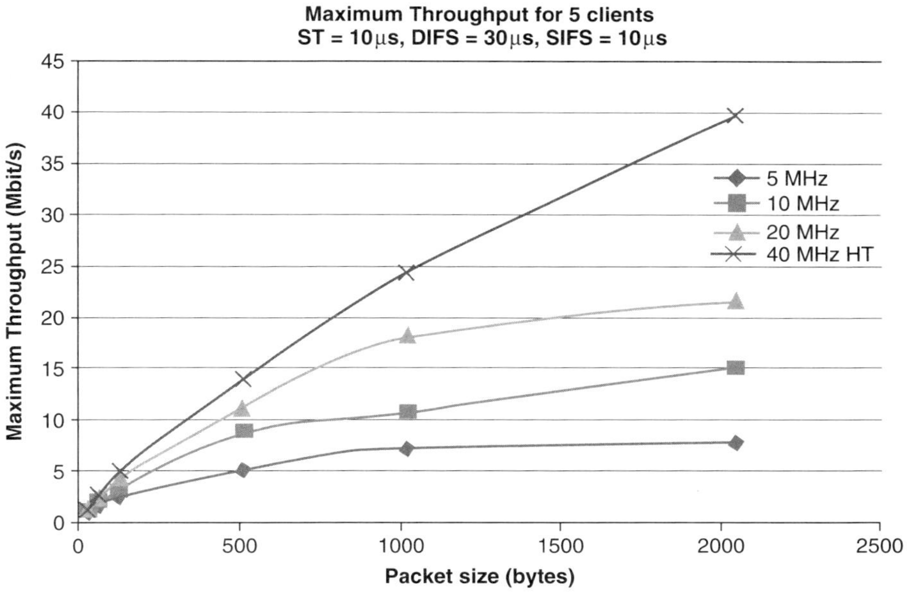

Figure 15.8 Maximum throughput for 5 clients.

Table 15.18 Pilot to data or symbol ratio

Pilot to data or symbol ratio	P/D	P/S	P/S
LTE DL (2/36)	0.0588235	0.0555556	5.6%
LTE UL (1/6)	0.2	0.1666667	16.7%
FUSC	0.108	0.0974729	9.7%
OFUSC/AMC DL	0.125	0.1111111	11.1%
PUSC	0.167	0.143102	14.3%
PUSC UL	0.5	0.3333333	33.3%
OPUSC UL	0.125	0.1111111	11.1%
AMC UL	0.125	0.1111111	11.1%
WLAN	0.0833333	0.0754717	7.5%

Table 15.19 Control to total symbols ratio

Control to total symbols ratio		
WiMAX OFDM DL	0.045977	4.6%
WiMAX OFDM UL	0.0344828	3.4%
WiMAX OFDMA DL	0.0689655	6.9%
WiMAX OFDMA UL	0.045977	4.6%
LTE DL	0.1666667	16.7%
LTE UL	0.0833333	8.3%

16

Wireless Network Design

16.1 Introduction

The design of a wireless network is a complex task that requires an in-depth knowledge of:

- RF propagation
- wireless technology
- equipments, software and rf components
- data protocols
- market modeling
- wireless design and optimization tools

It takes many years for an engineer to develop all these skills. The first chapters of this book gave the basic principles behind wireless technologies, described in detail wireless broadband networks, and explained how to model market and services.

A set of specialized tools is required to perform all these tasks. These tools should be able to consider all the intricacies of a wireless design.

The design tasks can be grouped in four main phases as shown in Figure 16.1. Each phase is described in detail in the following chapters.

These phases are applicable either to a greenfield or an operational network. They can be applied to network expansions and should be revised periodically to accommodate changes in traffic.

Real life statistics from operational systems can be applied to further enhance the network modeling. It is even possible to detect equipment hardware and software issues by comparing predicted and measured statistics. Figure 16.2 shows how this feedback is done over the life of the network.

16.2 Wireless Market Modeling

This phase provides the information according to which the design will be made. The input to this phase is the business plan, GIS and demographic data bases.

The market should be specified in terms of Service Classes to which the wireless service will be provided. Each Service Class should represent users that have similar behavior and operate in similar RF conditions.

LTE, WiMAX and WLAN Network Design, Optimization and Performance Analysis, First Edition.
Leonhard Korowajczuk.

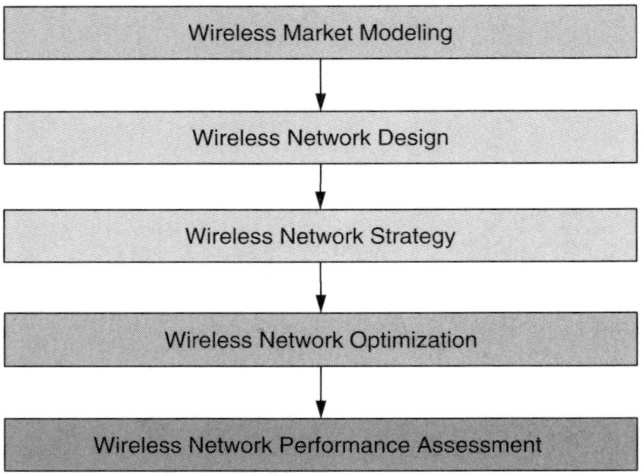

Figure 16.1 Design phases.

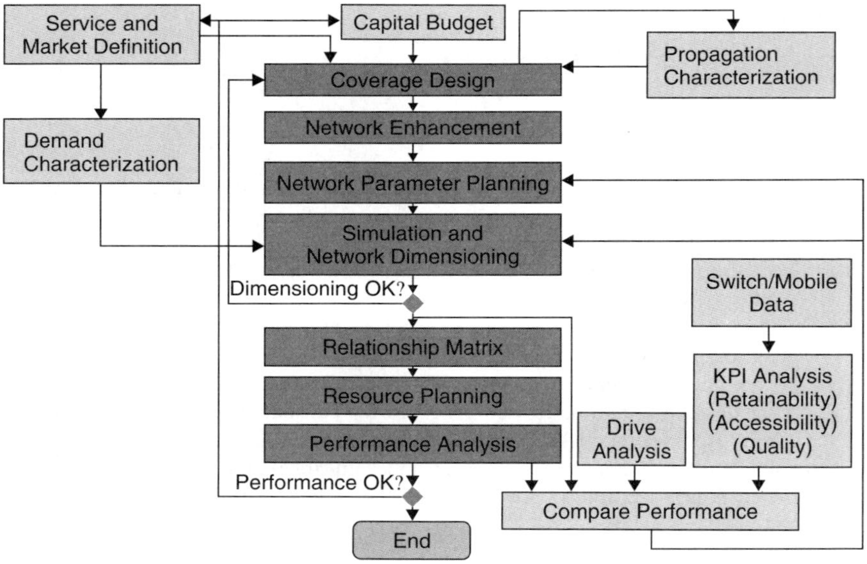

Figure 16.2 Prediction and operational data interaction.

The infrastructure should define the equipment to be used (hardware and software). This should be done even in cases in which the vendor is not yet defined, a generic parameter definition can be used but it needs to reflect the characteristics of the equipment that will be used (e.g. 802.16 d or e?; with MIMO or no MIMO?).

The tasks of this phase are listed below and will be described in the next chapters.

- Findings phase
- Area of interest definition
- Terrain databases
  - Topography
  - Morphology
  - Landmarks
  - Images
- Demographic databases
  - Households
  - Businesses
  - Commercial
  - Special events
- Service modeling
  - Applications
  - Real time services
  - Non-real time services
- Service Class modeling
  - Building heights
  - Environments
  - Terminal types
- Traffic distribution modeling
  - Traffic layers
  - Traffic variation per hour of the day
- Wireless infrastructure definition
  - Core equipment
  - Base station equipment
    - Radios
    - Antenna systems
    - Antenna
  - User terminals
    - Radios
    - Antenna systems
    - Antenna
- Backhaul definition
  - Radio
  - Antenna

16.3 Wireless Network Strategy

This phase establishes general design guidelines, which have also non-technical components. Spectrum usage may have significant limitations and the operator may have to comply with many regulations.

Base station deployment strategy depends on commercial agreements with property owners, whereas CEP deployment strategy depends on the installation philosophy.

Equipment vendor selection depends on many factors besides technical equipment quality, such as vendor soundness, past history, commercial and financial agreements, and so on. Backhaul interconnection depends on Internet Point-of-Presence availability and how much the operator wants to be independent from local carriers. All these factors affect the design and should be defined before this process can start. It is possible though that the design process will help consolidate some of the options.

The tasks of this phase are listed below:

- Spectrum usage strategy
- Deployment strategy
- Wireless infrastructure definition
  - Core equipment
  - Base Station equipment
    - Radios
    - Antenna systems
    - Antenna
  - Customer premises equipment
    - Radios
    - Antenna systems
    - Antenna
  - Backhaul equipment
    - Radio
    - Antenna
- Backhaul interconnection
- Landline access points
- Available site locations

16.4 Wireless Network Design

This phase establishes spectrum usage policies, deployment strategies, estimates the required number of sites, positions them and interconnects them.

The inputs to this phase re the business plan and market and infrastructure modeling. The main tasks of this phase are listed below:

- Field measurement campaign
  - CW measurements
    - Outdoor static
    - Indoor static
    - Indoor multi floor static
    - Outdoor dynamic
  - CPE measurements
    - No diversity
    - Diversity
- Measurement processing
  - Fast fading filtering
  - Distance filtering
  - Azimuth filtering
  - Error filtering

- Establish propagation models and parameters
  - Models calibration
- Locate sites
- Run initial predictions
- Static simulation
- Adjust design
- Design backhaul
- Perform predictions

16.5 Wireless Network Optimization

This phase adjusts and distributes network resources in an optimized way, targeting maximum throughput, but keeping in mind expandability and cost.

The input for this phase is the Wireless Network Design plan. The outputs of this phase are the setting and allocation of all network parameters. The main tasks of this phase are listed below:

- Optimize cell footprint
- Generate neighbor list
  - Topological
  - Interference
- Calculate handover thresholds
- Define paging groups
- Generate interference matrix
- Perform code planning
- Perform carrier planning
- Optimize cell footprint
- Generate backhaul interference matrix
- Perform backhaul planning

16.6 Wireless Network Performance Assessment

This phase predicts network performance as perceived by each group of users defined by the Service Classes.

The inputs to this phase are the Wireless Network Design plan, the Wireless Network Optimization plan and the business plan. The outputs are a performance report and action plan. Based on the results obtained in this phase, a redesign may be suggested, the deployment may get a green light, or a revision of the business plan may be required. The main tasks of this phase are listed below:

- Perform dynamic traffic simulation
- Generate KPIs
  - Area
  - Traffic
- Perform network performance plots
  - Thematic
  - Raster
- Calculate link performance
- Analyze SLA, CAPEX, OPEX and ROI compliance
- Prepare report

17

Wireless Market Modeling

Market modeling is an important and complex task that frequently is under-estimated in its importance and complexity. Chapter 3 explains the principles of market modeling, while this chapter gives a more operational approach based on the findings of Chapter 3. The following are the main tasks of this phase.

17.1 Findings Phase

In this phase, designers work in close proximity with the operator and gather as much information as possible about the operator plans, and the market itself. A business plan is not always available and, often, the operator has not evaluated all the intricacies of such a deployment.

The ideal scenario would be to start from a business plan or prepare one if it does not exist; alternatively a minimum amount of information should be gathered directly from the operator. A questionnaire, such as the one presented in Chapter 1, can be a good start.

The project budget is fundamental for a design. In the past we designed 100+ sites networks based on client's requirements, to find out that the available budget was for only ten sites.

17.2 Area of Interest (AoI) Modeling

The Area of Interest should be defined in the marketing plan. Unfortunately it is common that vague terms are used to define the AoI, such as: the whole city or the urban area.

The AoI should be defined geographically, through one or more polygons. This definition is very important as it limits the extent of the wireless design. The AoI should also specify for each area the type of service expected (e.g. outdoor, indoor). An AoI polygon can be discontinuous and have multiple phases. Figure 17.1 shows a hypothetical AOI.

17.3 Terrain Databases (GIS Geographic Information System)

A GIS database represents terrain characteristics, such as elevation, clutter type and height, landmarks, maps, and images. Each these items forms a database layer.

Terrain elevation is called topography, the clutter is called morphology; both are represented by raster files with a certain resolution. A raster file is composed of pixels that, together, define a specific region; in GIS databases, this region corresponds to an area on the globe, defined by a certain number lines (latitude) with a certain number of points in each line (longitude). The number of points defines the resolution of the database.

LTE, WiMAX and WLAN Network Design, Optimization and Performance Analysis, First Edition.
Leonhard Korowajczuk.

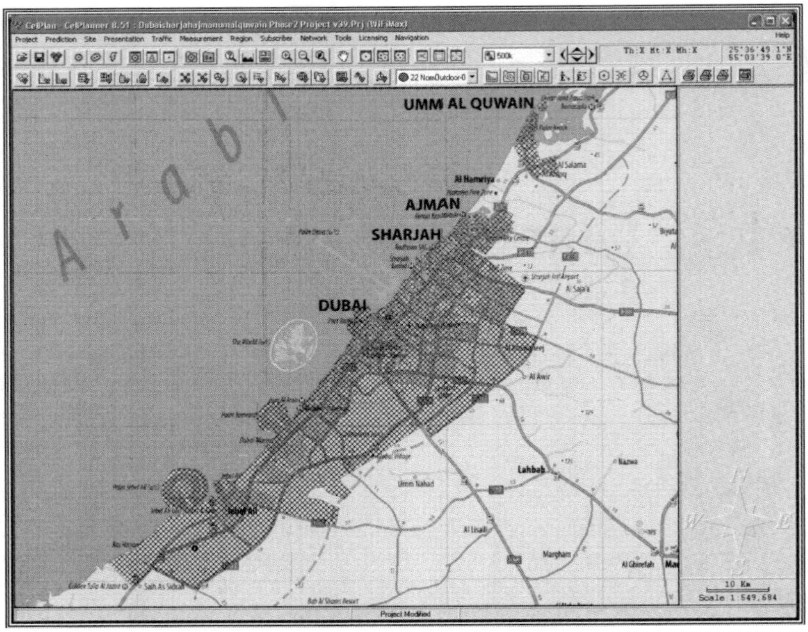

Figure 17.1 Area of Interest (AoI).

Landmarks (roads, streets, rivers, names) are usually given as vector files, in which a landmark, or part of it, is specified by specific coordinates and the paths that connect them.

Maps and images are also raster files, that is, geo-referenced pixels (points with image attributes, like color and brightness) are used to represent the desired image.

Because all this layers have to be geographically referenced, they are based on a certain projection and datum. The projection specifies how the round Earth is represented onto a flat surface and the datum specifies the point where the measurements originate. The more distant a point is from the datum, the more its location will be distorted; thus many countries choose to define their own datum.

The design software should be able to convert between projections and datum to a universal geographic representation. In this representation pixels are not square, and should follow the Earth's curvature, that is, pixels at the equator are approximately square but become trapezoidal in shape as they approach the poles.

Ideally, all GIS layers should be digitized from the same projection and datum, as conversions always introduce distortions. The layers should also be recently digitized, as terrain attributes, specially clutter and maps, change over time.

It is very common that operators decide not to invest in GIS databases, using cheap data with unknown origin or age. This compromises the design and may even make it useless.

The following sections briefly describe the steps to generate GIS databases.

17.3.1 Satellite/Aerial Photos for Area of Interest

It is important for the design to be done using up-to-date geographical data, mainly in areas with high recent growth. Satellite and Aerial Stereo Pairs should be used to obtain this information. One of the difficulties commonly found at this phase is to get Control Points to geo-reference the images, as this

Figure 17.2 Satellite image 2005.

may require a visit to establish coordinates of reference points. Figure 17.2 shows one component of a stereo pair and Figure 17.3 shows a satellite image used to digitize terrain databases.

Once the aerial photos have been geo-referenced, the relevant data can be digitized. The digitization is done by averaging values inside small rectangles, which are defined in meters or arc-seconds, and called pixels. The longitudinal dimension of the pixel defines the database resolution. The size of the database grows quadratically with the dimension of the pixel. It is essential that all layers of the GIS data use the same datum and coordinate system.

It is common for operators get parts of databases from different origins with unknown datum and coordinate systems, distorting the results and wasting all the work forward. A designer should verify these items before starting the design.

17.3.2 Topography

Topography represents the terrain at ground level, expressed as elevation above mean sea level (AMSL). The terrain elevation is represented in a raster file, whose pixels cover the database area. A good design tool should additionally interpolate between pixels in real time to increase the resolution of the database.

Figure 17.4 shows an example of topography with 30 meter horizontal resolution and 1 meter vertical resolution. The most common topography resolutions are shown in Table 17.1.

17.3.3 Digitize Landmarks

Main landmarks are roads, streets, rivers, railroads and borders. They are digitized as vectors and are used as reference to analyze geographically the predictions, mainly because vectors do not interfere with the colors representing the prediction. Figure 17.5 shows the digitization of streets (thin lines) and roads (thick lines).

Figure 17.3 Satellite image 2006.

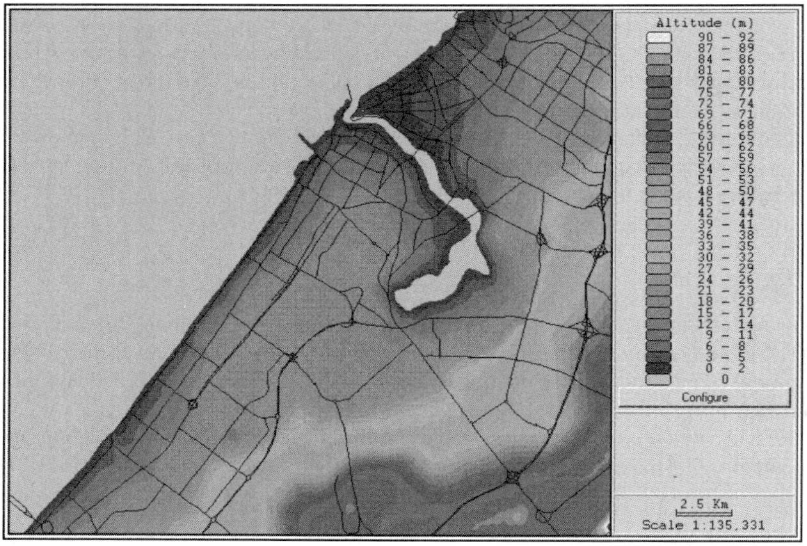

Figure 17.4 Topography.

Table 17.1 Topography database resolution requirements

Topography database resolution	Topography Horizontal (m)	Topography Vertical (m)	Relative cost per km^2
Low resolution	90	3	1
Medium resolution	30	2	2
High resolution	5	1	5
Very high resolution	1	1	20

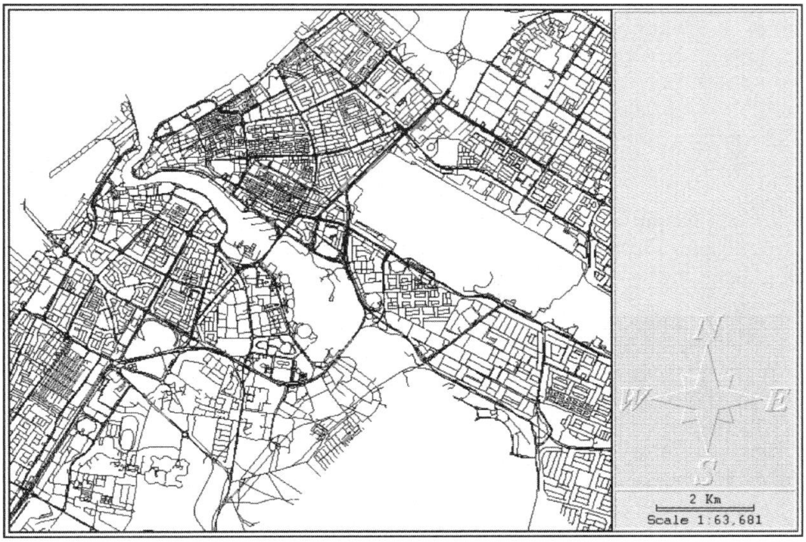

Figure 17.5 Landmark representation of streets and roads.

17.3.4 Morphology

Morphology represents all the clutter above ground, such as grass, trees and constructions and is defined by:

- *Number of classes*: Morphology classes group similar clutters. Generally commercially available morphologies use clutter classification that are not RF related and have to be processed to be used in wireless design application. Only morphology databases generated specifically for wireless design purposes have the right information required for RF predictions. Morphologies can be used also to distribute wireless users traffic, so it is possible to have a database for RF propagation and others for demographic analysis. Too many morphology classes may be prejudicial, as it will require the calibration of additional morphologies, with reduced number of samples per morphology. Morphologies represented with different resolutions, should be usually considered as different morphologies, as

Table 17.2 Morphology clutter types

0	Open, Water
1	Emergent Herbaceous Wetlands
2	Bare Rock/Sand/Clay, Quarries/Strip Mines/Gravel Pit, Transitional
3	Grasslands/Herbaceous, Pasture/Hay, Row Crops, Small Grains, Fallow
4	Shrub land, Orchards/Vineyards/Other
5	Deciduous Forest
6	Evergreen Forest
7	Mixed Forest
8	Urban/Recreational Grasses
9	Airports and parking lots
10	Roads
11	Streets
12	Low Intensity Residential
13	High Intensity Residential
14	Commercial/Industrial/Transport
15	Buildings

their propagation parameters will differ. Eight and a maximum of sixteen morphology classes are enough to represent different RF impacting clutter groups. A set of RF-related morphology classes is shown in Table 17.2.

- *Height information type*: In terms of clutter height information morphologies can be classified into:
  - *Flat morphology*: this database does not carry information about clutter heights, and is the morphology type most commonly used by the majority of design tools. Some tools include the morphology height into the topography height, distorting the information required for proper RF propagation analysis. The Okomura-Hata propagation model was initially developed for this type of database. It is impossible to do precise predictions with this morphology type.
  - *Canopy morphology*: this database is defined by the average of the morphology heights in a relatively large area, as if the area was covered by a cloth. Large height variations are smoothed, resulting in average heights per block. This morphology does not identify streets, so outdoor measurements or predictions are done considering that a street may exist at the measurement or prediction location. Legacy propagation models were conceived for flat or canopy morphologies. The big deficiency of this morphology type is that streets and build areas cannot be distinguished. Lee's propagation model was developed for this kind of database.
  - *Carved canopy morphology*: a methodology developed by CelPlan Technologies, this type of database adds carved roads, streets, and open areas into the above canopy database, significantly improving performance of propagation models that explore these features, such as the Korowajczuk 2D model. It is a good compromise between prediction quality and cost.
  - *Building height morphology*: this type of database represents buildings with its respective height, and is very expensive. It requires propagation models that can do 3D propagation analysis, such as Ray-tracing or the Korowajczuk 3D model.
  - *Horizontal and vertical resolution*: the resolution defines the size of the pixels used in the database representation. The horizontal resolution defines the size of length/width of the pixels, whereas the vertical resolution defines the height step. Note that the height step is not always directly connected with the actual height of the clutter, mainly in canopy representations where the height is an average.

Table 17.3 lists the most common morphology types and their respective resolutions.

Table 17.3 Morphology resolution requirements

Database Resolution	Morphology			Relative cost per km^2
	Horizontal (m)	Vertical (m)	Classes	
Low resolution	100	Flat	6	1
Medium resolution	30	Canopy	8	10
High resolution	5	Canopy with carved streets	12	100
Very high resolution	1	Building height	1	1000

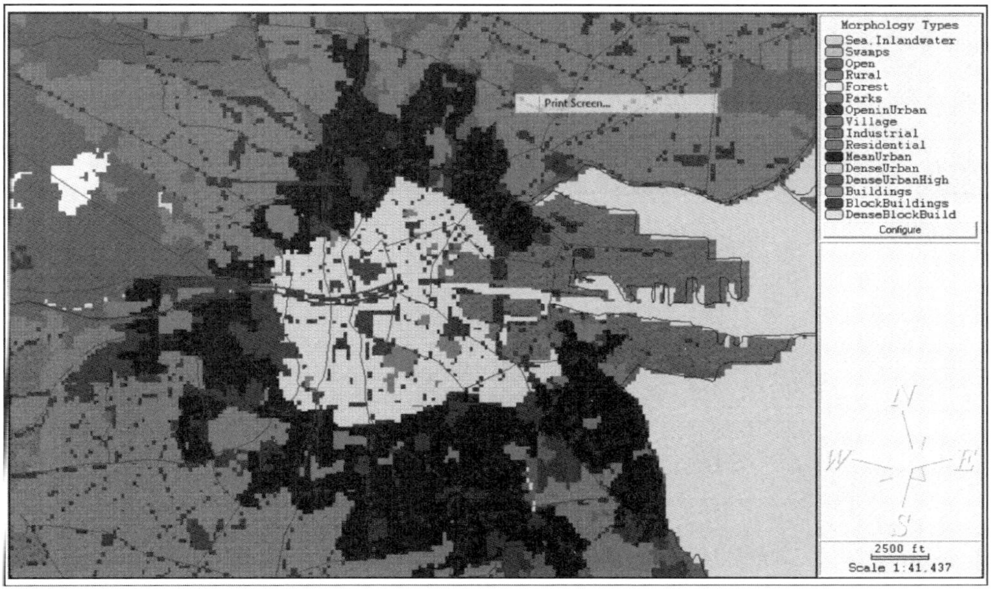

Figure 17.6 Example of canopy morphology.

Figure 17.6 shows an example of canopy morphology. Many design tools use a flat morphology and add height information through a morphology table, resulting in single heights per morphology. More advanced design tools record the height information per pixel, even when canopy databases are employed, this allows the same morphology type to have different heights throughout the canopy area.

The digitization of morphology can be very expensive when it comes to adding height information. Common canopy morphology is affordable but does not produce good results as RF canyons formed by roads and streets are not represented. CelPlan developed a process of oversampling canopy databases and then carving roads and streets with pre-defined widths.

Canopy databases are usually created with low resolution (15 to 30 m), which is not enough to represent narrower streets. To be able to properly carve these streets, the original database has first to be oversampled to 5 m or less so the streets can be carved using vector landmarks representation as a guideline. Figure 17.7 shows a bird's eye view of such a database and Figure 17.8 shows the profile along a street with gaps in clutter introduced by the carved streets.

Figure 17.7 Morphology with carved streets and roads.

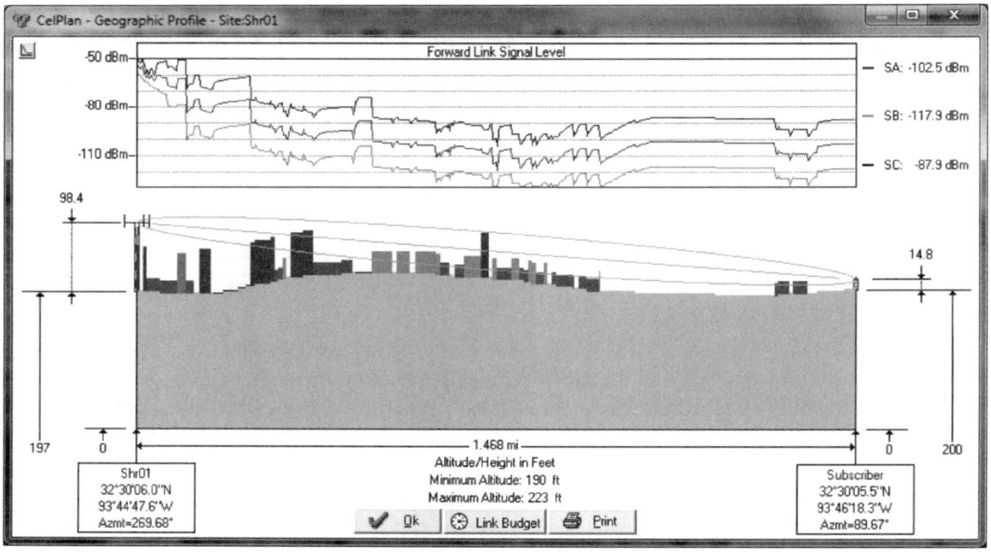

Figure 17.8 Profile along a street within canopy morphology with carved streets.

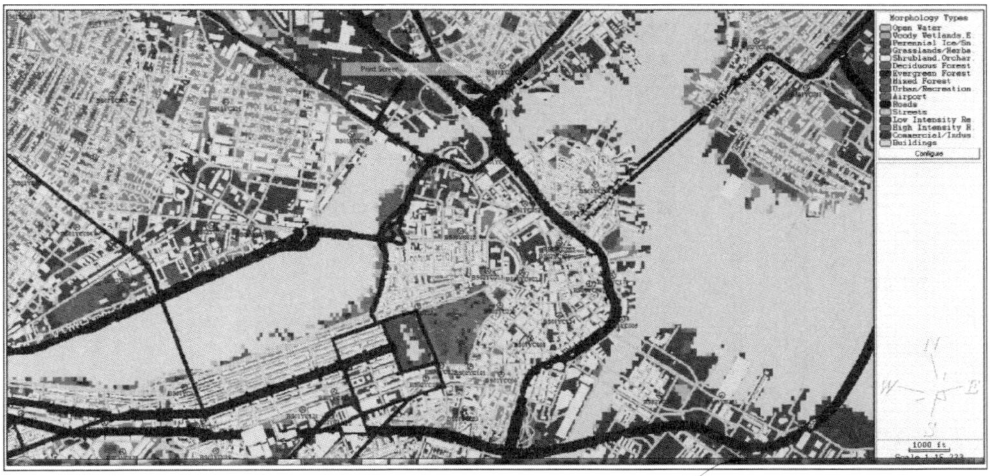

Figure 17.9 Example of building level morphology.

17.3.5 Buildings Morphology

Buildings require a very high resolution to be properly represented in a morphology database, as their shape should be well characterized. Usually, building level morphologies use a 3 m or better resolution. There are design tools that use ray tracing modeling to calculate predictions based only on the building information contained in such databases, ignoring other morphologies. Whereas this is the ideal approach to be taken for ray tracing and might work well in very dense urban areas, this usually leads to deficient results in most areas in which a blend of different morphology types is present. A propagation model should consider all morphology types present in the analyzed area.

Figure 17.9 shows a building-level morphology which includes other morphologies, besides buildings, such as, streets, roads, high and low intensity residential areas, forests and others.

17.3.6 Multiple Terrain Layers

Most networks cover an area that comprises different clutter characteristics, and in some areas may require better morphology databases than others, for example, downtown. To lower the cost of such a database, it is convenient to use different quality (resolution) databases for different areas, instead of purchasing very high resolution data for the whole market. However, the use of files with different resolutions implies that some or total overlap of covered areas will happen.

For this reason it is important that prediction tools be able to support multiple layers of topography and morphology data and automatically select the best resolution per pixel in real time. Figure 17.10 shows the configuration of multiple topography and morphology directories with different resolutions. If this cannot be done automatically by the tool, users must manipulate the data to eliminate these overlaps and smooth the transition between the databases, which requires specialized tools and is a very time-consuming process.

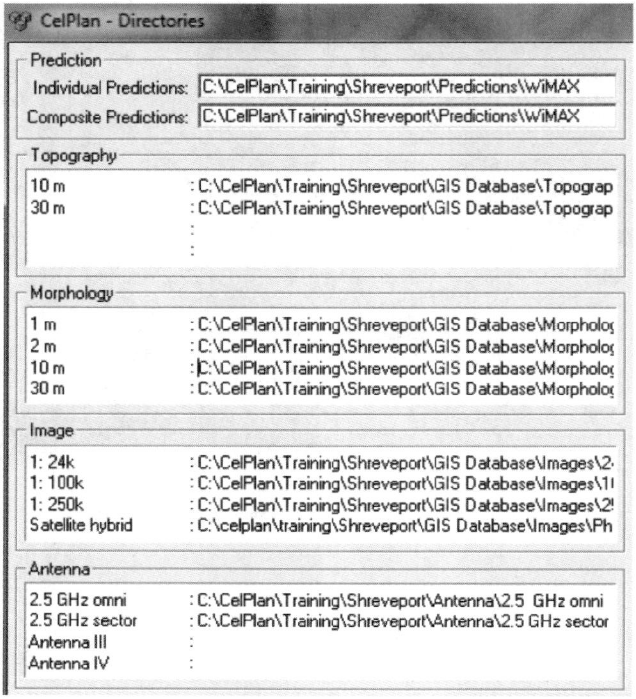

Figure 17.10 Multilayer topography and morphology definition.

17.3.7 Terrain Database Editing

On some occasions, there is a need to correct morphology data so a morphology editor is an important feature of the design tool. Figure 17.11 shows CelData, a database editor that is used to correct or improve morphology data. Typical editing required on morphology files is modification of morphology types and heights.

Another common task required of database editors is format conversion as frequently the terrain database purchased may not be in the desired format.

At CelPlan frequently we get third party databases without reference to date, projection or datum. Although educated guesses can be made and format conversion can be done, it is highly recommended not to use such databases in deployment designs.

17.3.8 Background Images

Background images are important for use as a reference to better understand predictions (e.g. geographical location of coverage area). The most common backgrounds are landmarks, scanned maps, and satellite photos in 2D and 3D.

Area maps and aerial photos are a good visual aid for network performance analysis as they provide geographical references to the designer; even though, its usage may slightly distort prediction colors. Figures 17.12 and 17.13 show a digitized map and a satellite photo, respectively.

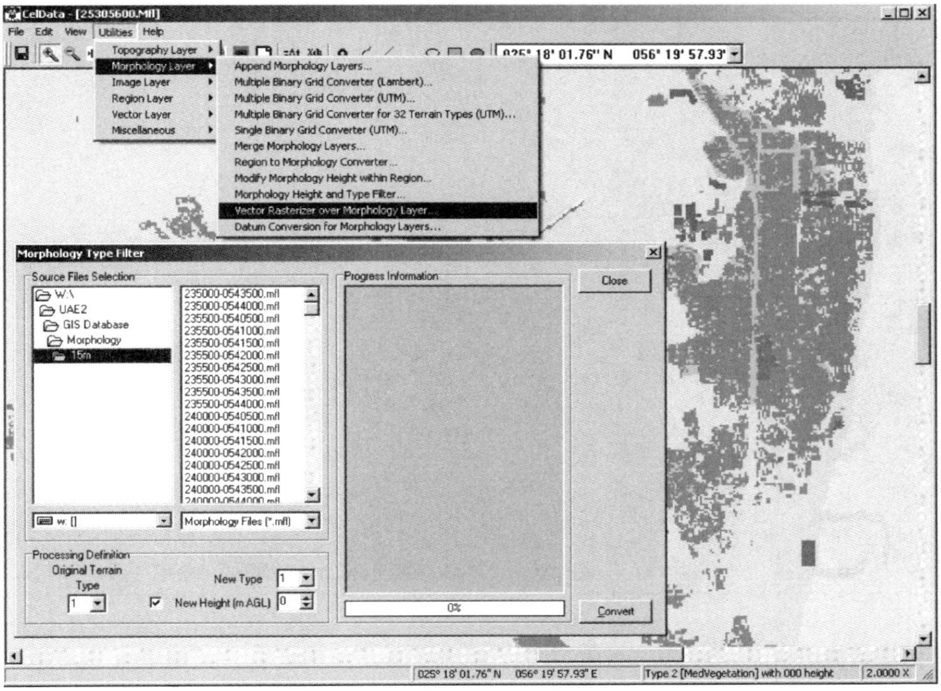

Figure 17.11 CelData morphology editor.

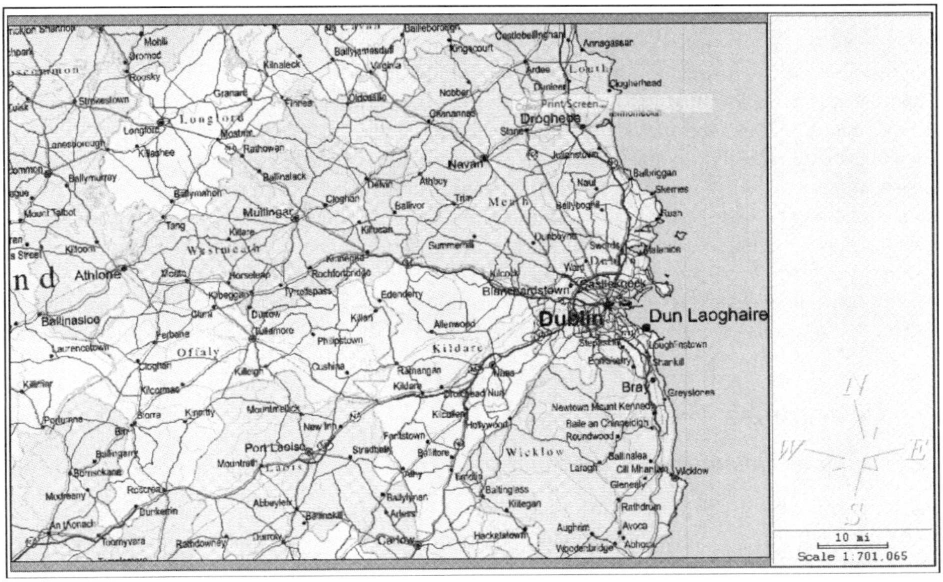

Figure 17.12 Example of a map used as background.

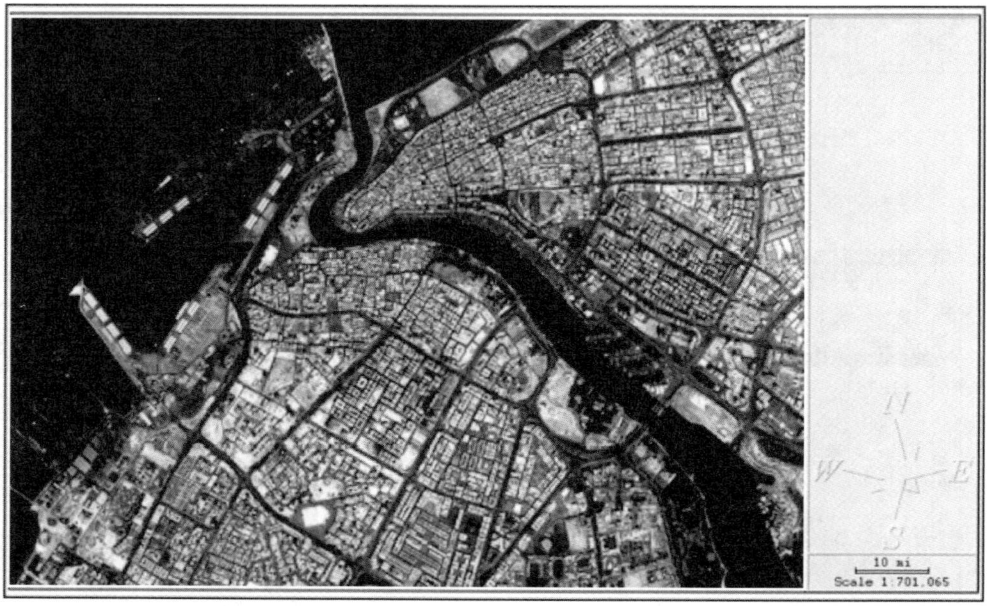

Figure 17.13 Example of satellite image used as background.

Map digitalization should be approached with care, as maps are usually printed based on certain projection and coordinate system. They should undergo the same process as other GIS layers before being digitized. Also, many maps are pictorial only and do not have any projection embedded, this type of imagery should not be used and it cannot be properly adjusted to the right projection. Sometimes, it is better to have no information than the wrong information.

The issue with backgrounds is to be able to see them through predictions, some kind of color blending should be done and this always distorts the prediction colors. The visualization can be done by blending or by representing the predictions with points, so the background can be seen in between points. Another possibility is to use multiple screens with synchronized pointers, so the prediction is loaded in one screen and the background in another, and the pointer correlates both information.

For this reason, landmark vectors are ideal for use as background as they do not change prediction colors and yet give nearly the same information. This is illustrated in Figure 17.14.

On the other hand, background images can be used as 2D layers or as 3D layers, which can be visualized from different angles as shown in Figure 17.15.

17.4 Demographic Databases

17.4.1 Obtain Demographic Information (Maps and Tables)

Local geographic, census and traffic institutes provide demographic data that can be used to calculate services demand. Typical demographic data is the distribution of population according to income levels, number of households, number of business, number of employees, business type, traffic on roads and streets, and so on. This data has to be analyzed and transformed into geographic regions that have traffic-related attributes.

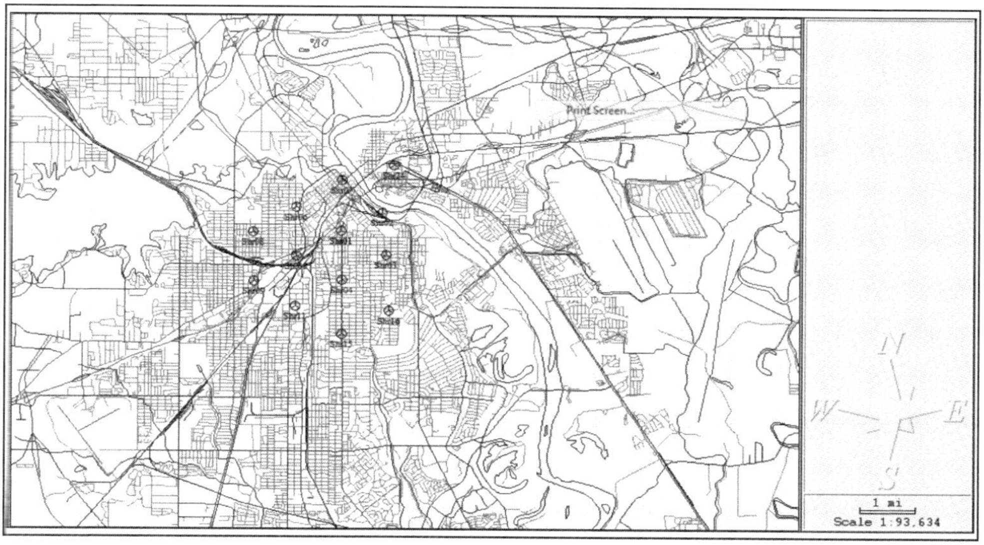

Figure 17.14 Example of landmarks used as background.

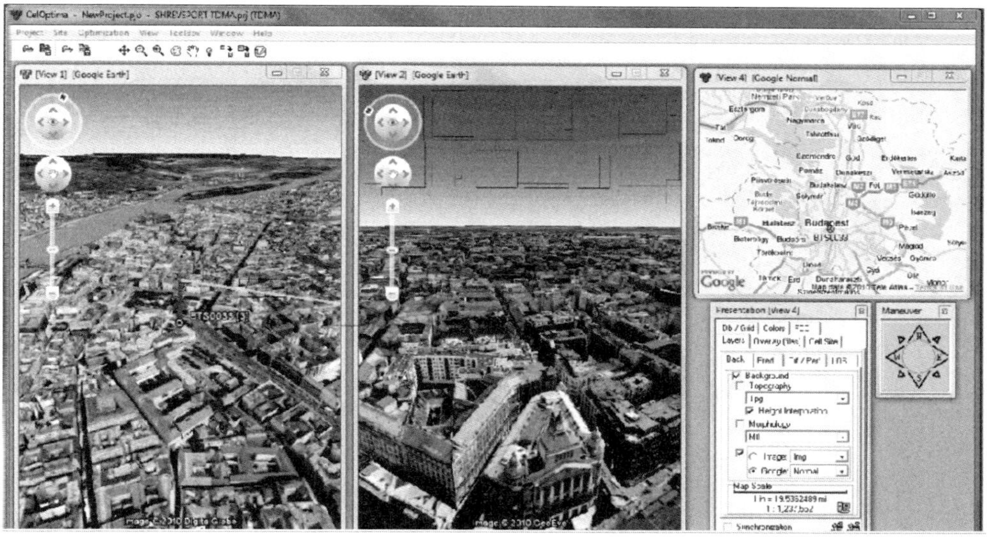

Figure 17.15 3D images from area with site location (left) and view from site in shown direction.

Several of these geographic layers have to be created to provide enough information for the generation of traffic grids. Typical layers are:

- Households distribution
- Businesses distribution with number of employees
- Commercial areas with number of clients
- Multi-story dwelling with population
- Vehicle density per road segment

17.4.2 Generate Demographic Regions

Based on the demographic data, it is possible to generate regions with specific attributes related to the services to be deployed. Regions are defined as closed polygons with specific attributes associated with the polygon. The design tool should be able to work with millions of polygons and support attributes that apply to a set of non-continuous polygons.

In principle, polygons should not overlap, but as not all data comes in this form, the design tool should be able to cope with the overlap, to avoid time-consuming manipulation of available data.

The regions that define users per service class are usually a result of the combination of several demographic layers that are processed by the design tool to obtain a raster distribution of users.

Figure 17.16 shows the number of households per region and Figure 17.17 shows how the USA divides the country into Tracts (T), where each tract comprises between 1500 and 8000 inhabitants (600 to 3000 housing units). A tract is divided further in Block Groups (BG), with an average of 7 BGs per T. Block Groups are subdivided in Blocks (B), with an average of 38 B per BG. Table 17.4 gives the number of 1990 US Census Regions, which include information such as number of households

Figure 17.16 Household demographic regions example.

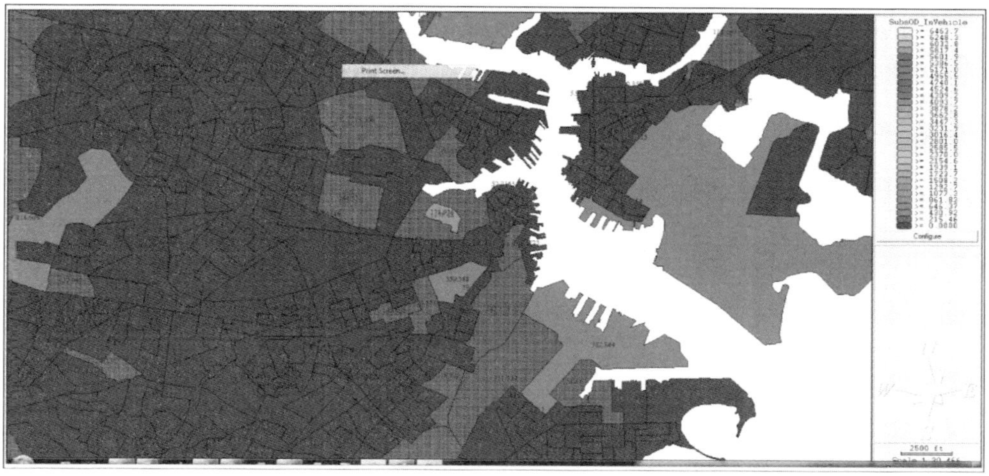

Figure 17.17 Business demographics region example.

Table 17.4 US Census regions

Regions US 1990	Number	Inhabitants		Households	
Track	30,000	1500	8000	600	3000
Block Group	211,267	213	1136	85	426
Block	8,200,000	5	29	2	11

and income level. It is possible to get this information for businesses as well, including data such as number of employees.

Traffic congestion maps can be used to calculate the number of vehicles per street per hour of the day as shown in Figure 17.18.

Designers should also create their own specific regions that might be important for the area being analyzed. Specific regions that might be needed include commercial areas, such as malls, or special events, like stadiums and arenas, as illustrated in Figure 17.19.

17.5 Service Modeling

This task establishes QoS parameters and traffic tonnage for the services to be offered to the client. Chapter 3 covers this procedure in terms of the analysis of available applications, their frequency of usage, and traffic requirements. Applications with similar QoS requirements can be grouped together and traffic requirements, totalized. In the example given here, two service classes were established. Tables 17.5 and 17.6 show the traffic demand for personal and business use for a mix of applications, as calculated in Chapter 3.

Figures 17.20 and 17.21 show an example of how to input into a planning tool the real time and non-real time mixed service configuration for business and personal use provided in Tables 17.5 and 17.6.

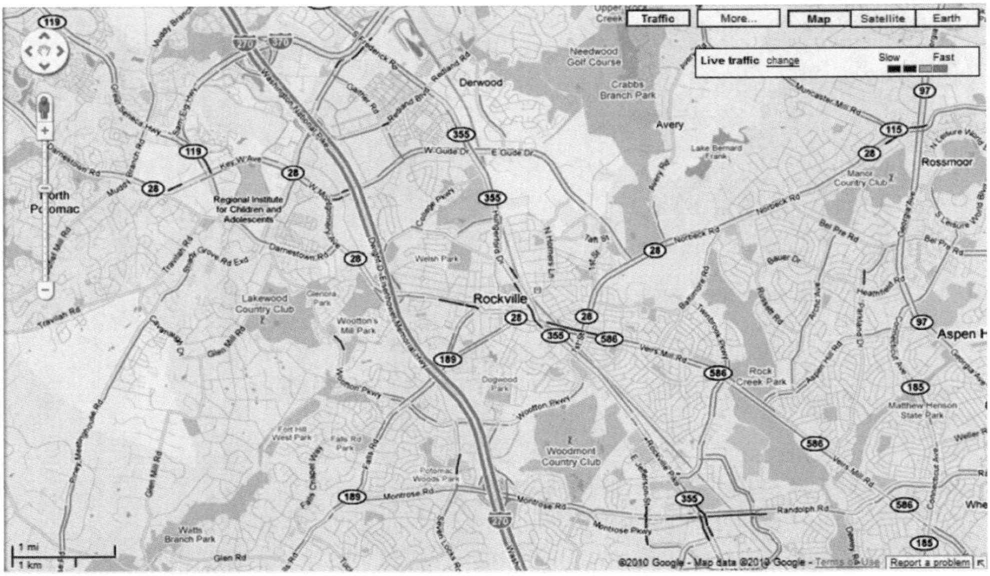

Figure 17.18 Vehicular traffic congestion map.

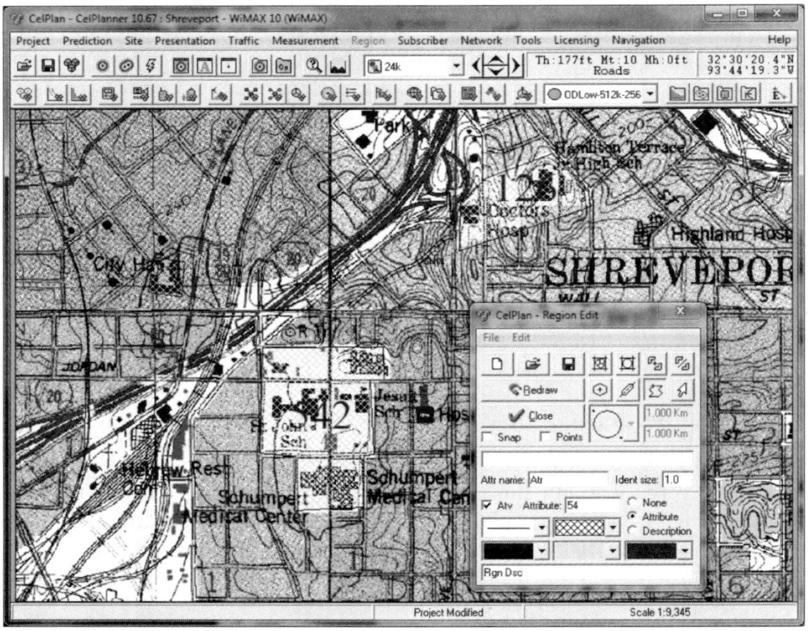

Figure 17.19 Commercial area region editing.

Table 17.5 Unconstrained personal services

Services	Unconstrained BH personal user incoming traffic		Unconstrained BH personal user outgoing traffic		QoS
	MiB	kbit/s	MiB	kbit/s	
Total NRT MiB	4.06	9.24	0.78	1.77	NRT
Total RT MiB	3.72	8.47	1.19	2.71	RT

Table 17.6 Unconstrained business services

Services	Unconstrained business BH user incoming traffic		Unconstrained business BH user outgoing traffic		QoS
	MiB	kbit/s	MiB	kbit/s	
Total NRT MiB	7.73	17.59	2.15	4.89	NRT
Total RT MiB	3.12	7.10	1.01	2.29	RT

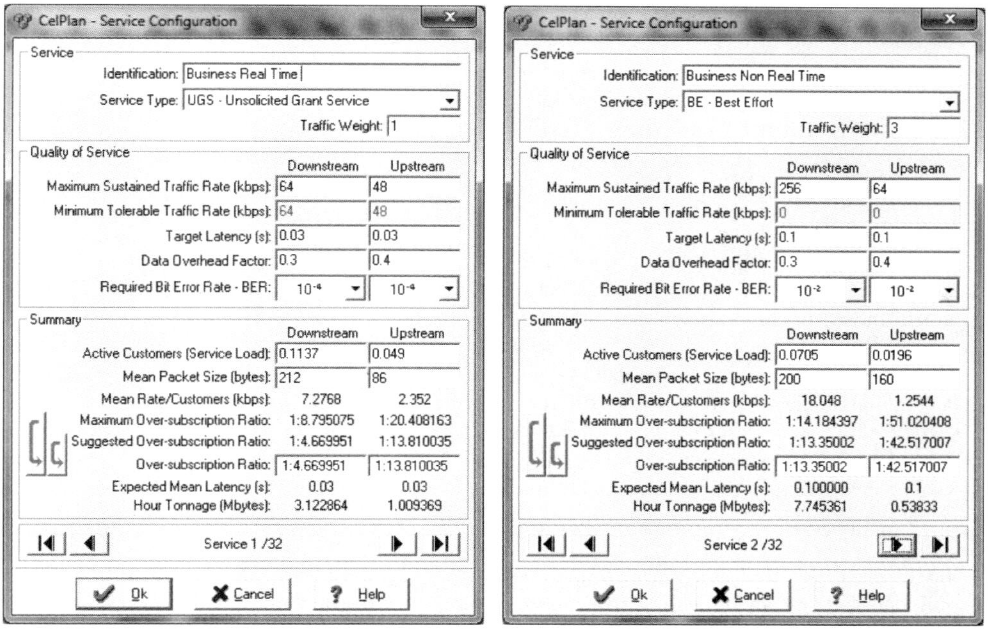

Figure 17.20 Mix service configuration for business users.

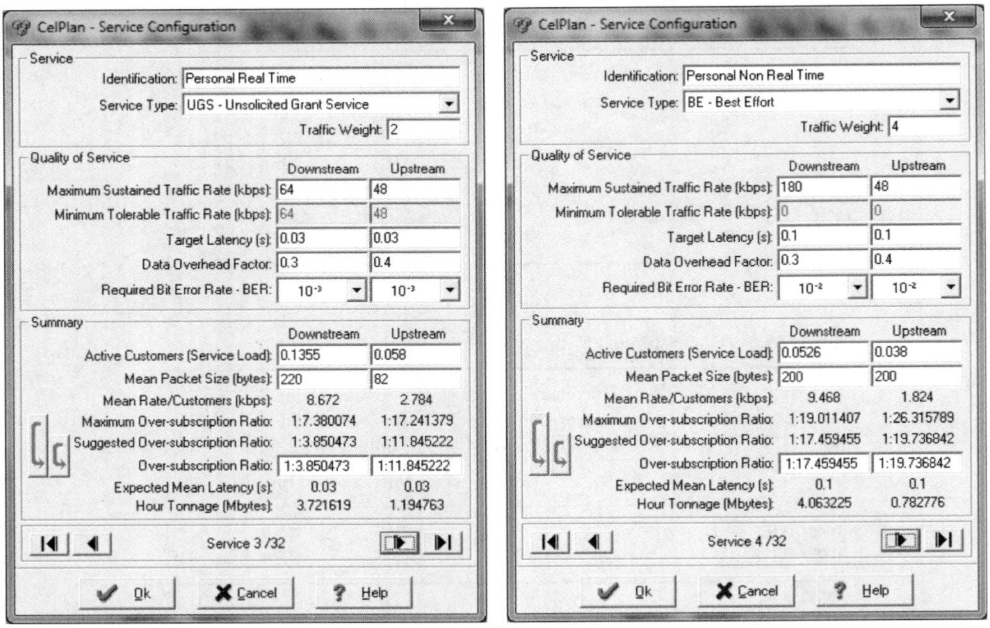

Figure 17.21 Mix service configuration for personal users.

17.6 Environment Modeling

In Chapter 3, when Service Classes (SC) were described, the example used had six different RF environments (Figure 3.23), three indoor for three different elevations; it is up to the designer to establish how many different heights should be analyzed.

Human body attenuation applies according to the terminal type, for example, it should not be considered for rooftop and desktop terminals but it applicable to portable terminals. Except for rooftop terminals, most other outdoor terminals are portable, hence the attenuation is applicable. Indoor terminals may be portable or desktop type, but in the example in Chapter 3 they are not being distinguished, so an average value between both should be applied.

Penetration attenuation applies only to indoor terminals and the SLA should specify the extent of this coverage. In some cases coverage is only promised for locations close to the window facing the site, in others, more encapsulated locations are considered.

Rain precipitation margin does not apply for frequencies below 10 GHz and should be automatically calculated by the tool for higher frequencies.

Shadow fading depends on the prediction resolution considered and expresses the signal variation inside the prediction bin.

Multipath fading should ideally be calculated using Ricean distribution as explained in Chapter 5.

Figure 17.22 shows an example of a dialog box for configuration of user terminal environment.

Table 17.7 gives suggested values for the different environmental parameters. Those values should be statistically added to the average prediction loss calculated by the prediction model. When the average prediction loss includes some of these factors, those should not be considered in the environment.

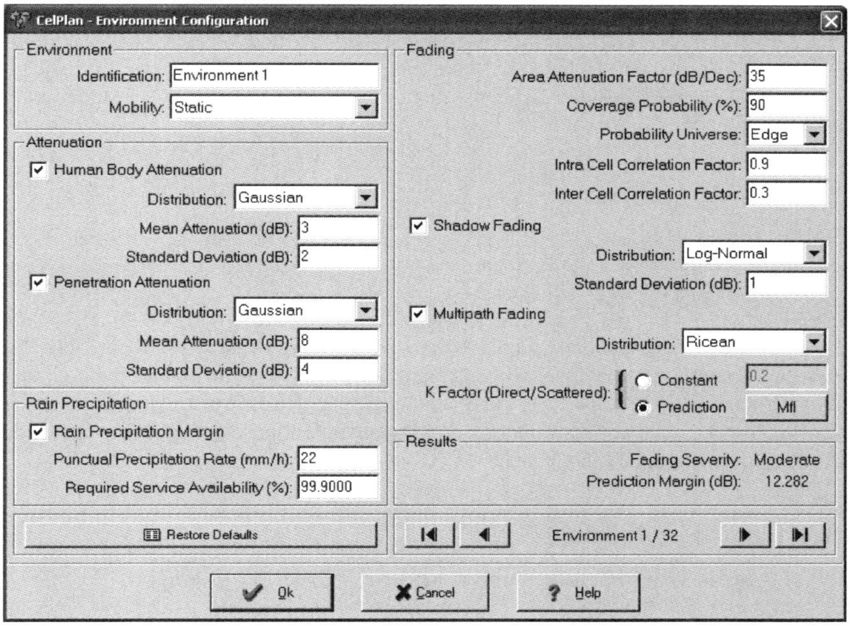

Figure 17.22 Environment configuration.

Table 17.7 Suggested environmental attenuations

Type	Environment	Distribution	Mean attenuation	Standard deviation
Human Body Attenuation	Portable Phone	Gaussian	3	2
Penetration Attenuation	Vehicle	Gaussian	4	3
Penetration Attenuation	Window	Gaussian	8	4
Penetration Attenuation	Indoor	Gaussian	16	7
Penetration Attenuation	Indoor concrete	Gaussian	22	12
Shadow Fading	resolution 3m	Log-normal	3	2
Multipath Fading		Ricean	f(k factor)	6

17.7 User Terminal Modeling

Chapter 3 also lists user terminal types. Seven types are given in the example used in that chapter, further divided into four groups, according to their location, mobility, and ease of use:

- Rooftop (R)
- Desktop (D)
- Laptop (L)
- Palmtop, Phone, Portable Multimedia Player (P)

User Terminals are composed of a Radio and an Antenna System (AS). Terminals should be specified for different AS heights, to represent different locations (building floors) where users could be located. The heights should be normalized to a few representative heights; in Chapter 3, three heights were chosen, that end up representing five different situations as some terminals are installed indoors and some on rooftop.

- Ground
- 3rd Floor
- 10th Floor
- Rooftop business (equivalent to 10th floor)
- Rooftop personal (equivalent to 3rd floor)

Examples of terminal configuration dialogue boxes are shown in Figures 17.23 to 17.26. Note that, in this example, the terminal dialogue defines height and radio type; the actual radio characteristics are given in separate dialogues: the Antenna System dialogue defines the antenna algorithms and the Performance dialogue defines user terminal receive sensitivity for different situations (more details on radio performance can be found in Chapter 11).

17.8 Service Class Modeling

Chapter 3 describes in detail the procedures to model a market. Many designers skip this step and proceed to configuration of the design tool directly, resulting in an incomplete and erroneous modeling.

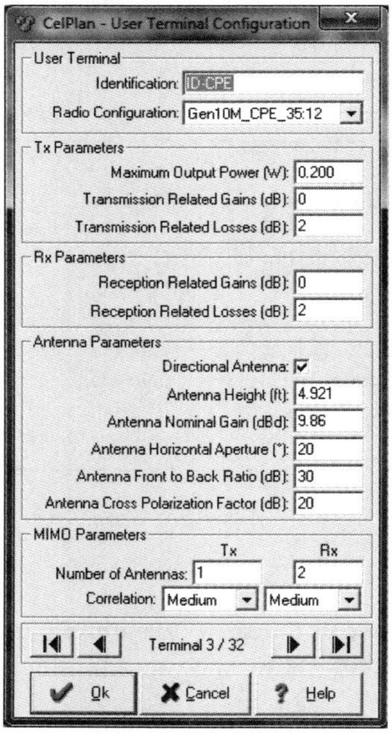

Figure 17.23 Terminal configuration.

The market modeling process is complex and requires an in-depth analysis, before the design tool can be configured, thus it is highly recommended to for the reader to at least browse through Chapter 3 before proceeding.

The configuration should start with the creation of Service Classes (SC). Only then should each SC component be configured, as, by then, the context to which they will be configured is known. The components to be configured are:

- Services
- Terminals
- Environments
- Traffic Grids

In the example used in Chapter 3, twenty-two SCs are identified as shown in Table 17.8 and Figure 17.27.

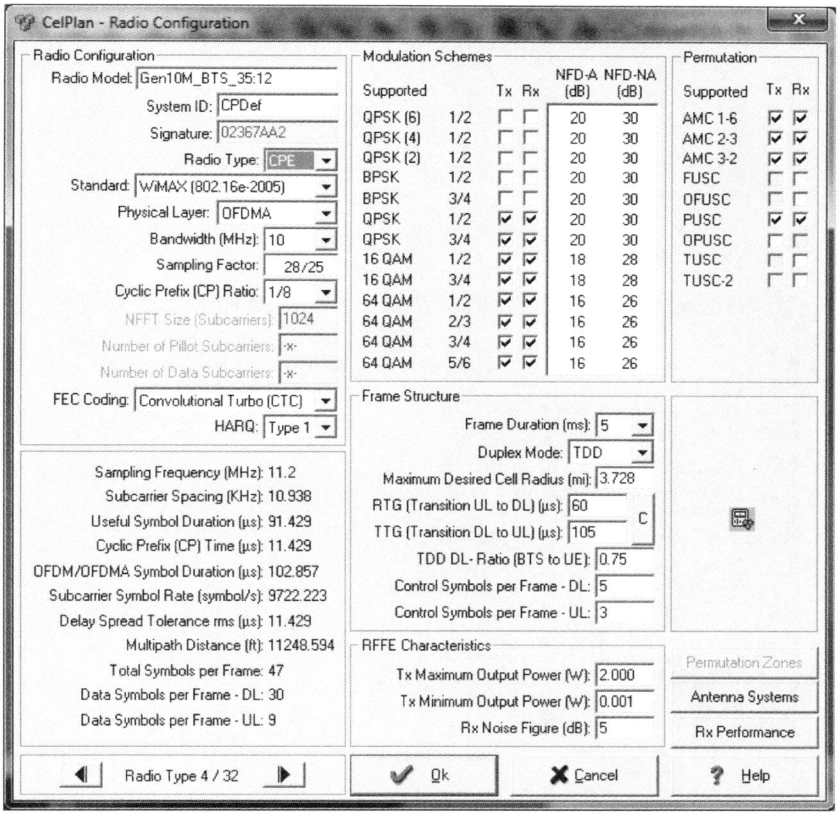

Figure 17.24 Radio characteristics 802.16e radio.

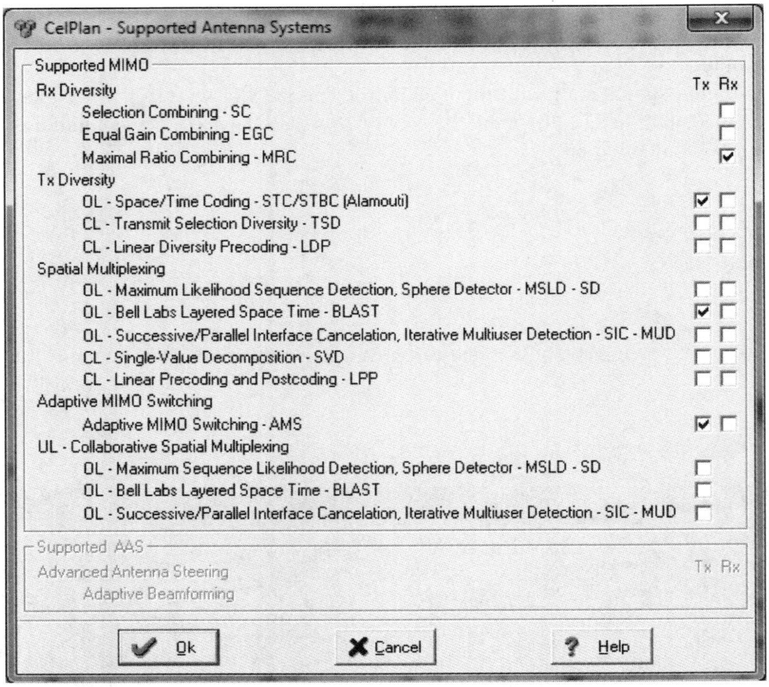

Figure 17.25 Supported antenna systems dialogue.

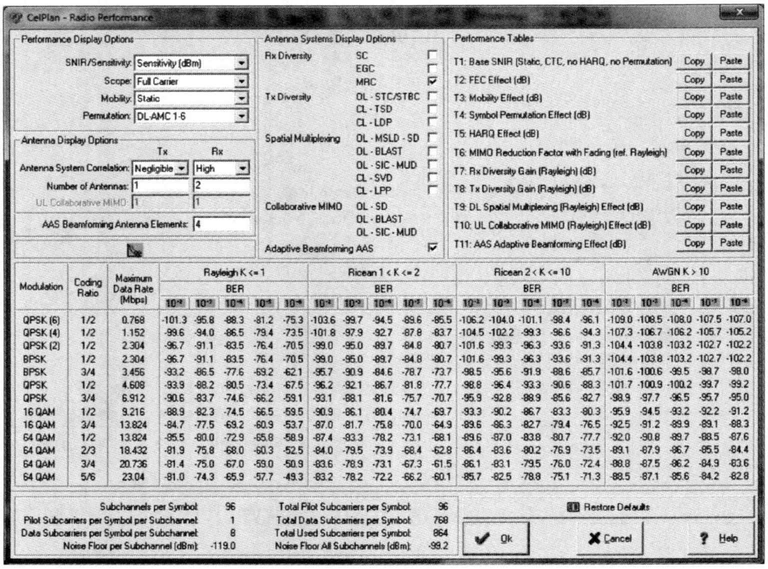

Figure 17.26 Radio performance dialogue.

Table 17.8 Example of prediction service classes

	Prediction Service Class	Service	Terminal	Environment	Height (m)	Traffic layer
1	Indoor ground P NRT	NRT	Desktop ground	Indoor ground	1.5	Indoor ground P
2	Indoor 3rd floor P NRT	NRT	Desktop 3rd Floor	Indoor third floor	10	Indoor 3rd floor P
3	Indoor 10th floor P NRT	NRT	Desktop 10th Floor	Indoor tenth floor	30	Indoor 10th floor P
4	Indoor ground B NRT	NRT	Desktop ground	Indoor ground	1.5	Indoor ground B
5	Indoor 3rd floor B NRT	NRT	Desktop 3rd Floor	Indoor third floor	10	Indoor 3rd floor B
6	Indoor 10th floor B NRT	NRT	Desktop 10th Floor	Indoor tenth floor	30	Indoor 10th floor B
7	Outdoor ground NRT	NRT	Portable	Outdoor	1.5	Outdoor ground
8	Outdoor rooftop P NRT	NRT	Rooftop	Outdoor rooftop P	10	Outdoor rooftop P
9	Outdoor rooftop B NRT	NRT	Rooftop	Outdoor rooftop B	30	Outdoor rooftop B
10	Vehicle NRT	NRT	Portable	Vehicle	1.5	Vehicle
11	Commercial NRT	NRT	Portable	Commercial	1.5	Commercial
12	Indoor ground P RT	RT	Desktop ground	Indoor ground	1.5	Indoor ground P
13	Indoor 3rd floor P RT	RT	Desktop 3rd Floor	Indoor third floor	10	Indoor 3rd floor P
14	Indoor 10th floor P RT	RT	Desktop 10th Floor	Indoor tenth floor	30	Indoor 10th floor P
15	Indoor ground B RT	RT	Desktop ground	Indoor ground	1.5	Indoor ground B
16	Indoor 3rd floor B RT	RT	Desktop 3rd Floor	Indoor third floor	10	Indoor 3rd floor B
17	Indoor 10th floor B RT	RT	Desktop 10th Floor	Indoor tenth floor	30	Indoor 10th floor B
18	Outdoor ground RT	RT	Portable	Outdoor	1.5	Outdoor ground
19	Outdoor rooftop P RT	RT	Rooftop	Outdoor rooftop P	10	Outdoor rooftop P
20	Outdoor rooftop B RT	RT	Rooftop	Outdoor rooftop B	30	Outdoor rooftop B
21	Vehicle RT	RT	Portable	Vehicle	1.5	Vehicle
22	Commercial RT	RT	Portable	Commercial	1.5	Commercial

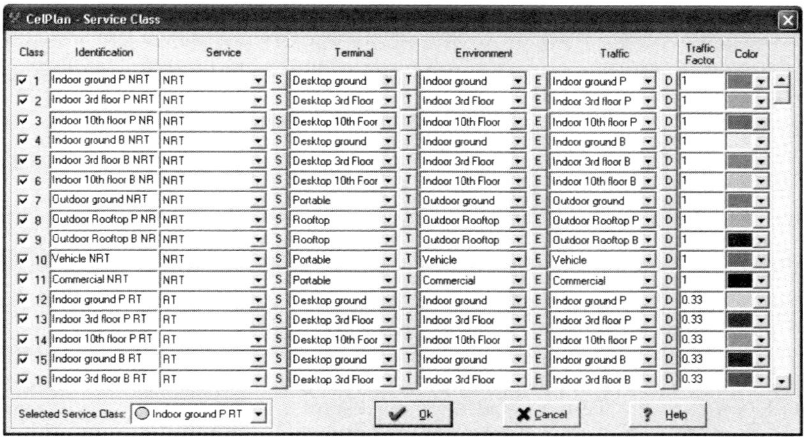

Figure 17.27 Service classes configuration dialogue box.

17.9 User Distribution Modeling

The best way to characterize the complex data traffic is to express it by type of user, where each type is associated with the actual traffic defined by a service specification.

Chapter 3 also makes the distinction between a subscriber (or subscription) and a user, in which a subscriber defines the existence of a terminal that can be used by one or more users. The number of subscriptions should be determined for existing networks and estimated for future networks, and, from it, the number of users can be estimated.

Figure 17.28 shows how data from a Census Block Group (CBG) can be divided into different user distribution layers.

17.9.1 User Distribution Layers

User distribution can be obtained from previously defined demographic regions. Different layers have to be generated for each user type defined as a service class.

Following the example of Chapter 3, the eleven user-type layers listed next should be created.

- Outdoor
- Vehicle
- Commercial
- Indoor Business Ground
- Indoor Business 3rd floor
- Indoor Business 10th floor
- Indoor Business Rooftop
- Indoor Personal Ground
- Indoor Personal 3rd floor
- Indoor Personal 10th floor
- Indoor Personal Rooftop

These layers should be prepared for the peak hour of each type and should contain the number of estimated users for the given type of service. One or more regions of each type might be defined within the Area of Interest.

To obtain the traffic layers expressed in users, the demographic regions have to be consolidated proportionally for each Service Class. This is done by creating raster traffic grids, defined in seconds. Three arc-sec (90 m) to one arc-sec (30 m) are typical resolutions for traffic grids.

The next step is to distribute demographic region attributes throughout the area of interest. This distribution can be done in a uniform manner, but, ideally, should be further enhanced by using morphology weights. The use of weights allows a more realistic representation of the real distribution of users, as, for example, it is expected that there will be more users in buildings then in a park or lake within the same region. Figure 17.29 shows an example of dialogues for configuration of weights per morphology for traffic distribution and the set-up of parameters for the creation of a traffic layer.

Figures 17.30 and 17.31 show examples of Business Outdoor and Vehicle Busy Hour Traffic. Even though both layers represent the busy hour of a service, that actual time of the day when that Busy Hour happens might be different for each service.

A second consideration when distributing traffic is that indoor traffic has to be distributed between the different building floors, thus, first, buildings must be classified according to their heights.

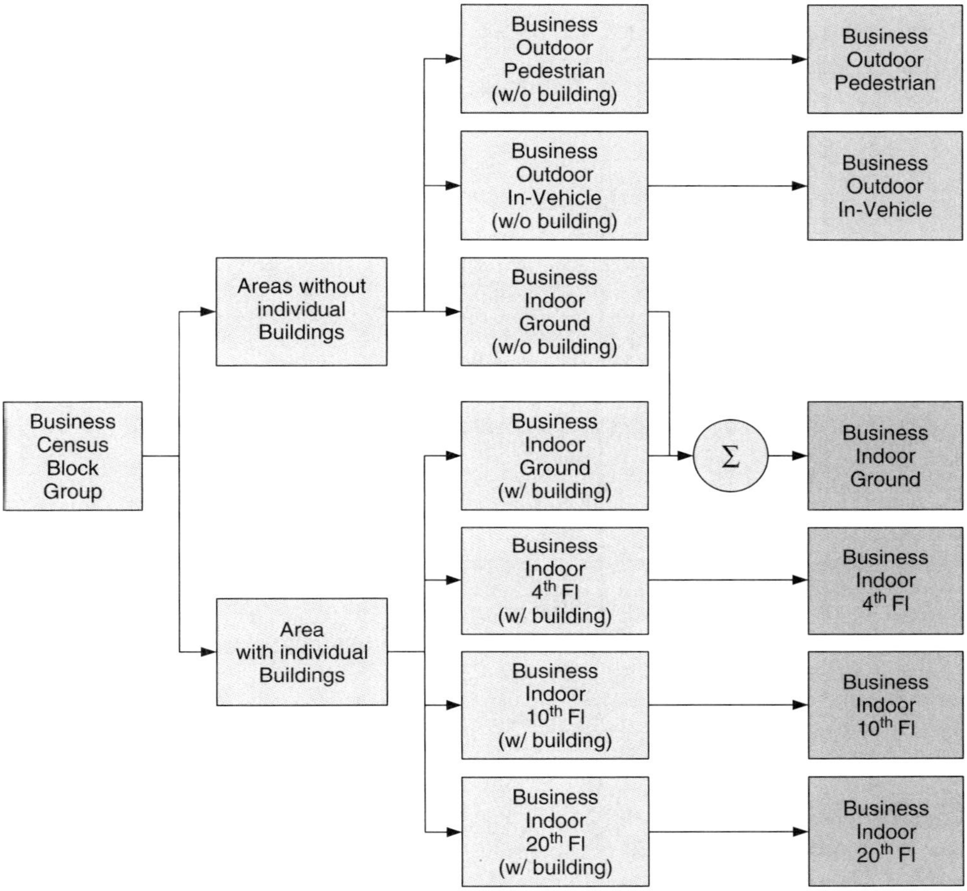

Figure 17.28 User distribution from a census block group.

Figure 17.32 shows building classification according to height and type of building (business or residential).

Areas represented in the morphology database as residential or commercial should be also considered when defining building areas, even if individual buildings are not represented in the database. In cases like this, the whole area is considered as a group of buildings, and the traffic can be distributed accordingly. Because buildings are not individually represented, the same area can also hold outdoor traffic (e.g. sidewalks and streets between the buildings, hence it should be considered in the outdoor traffic layer as well.

Figures 17.33 to 17.36 show indoor business busy hour traffic for different height ranges.

Figures 17.37 to 17.40 show indoor residential (personal) busy hour traffic for different height ranges.

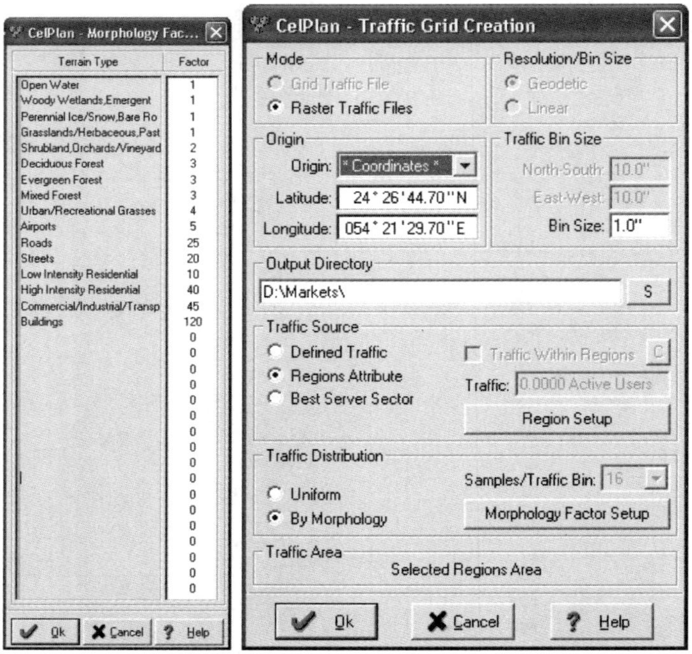

Figure 17.29 Traffic grid/raster generation.

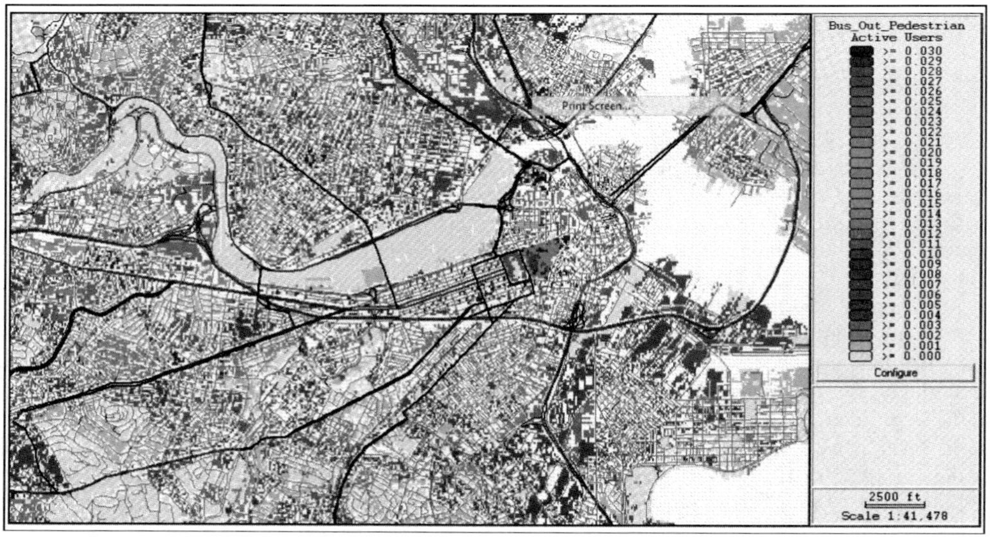

Figure 17.30 Business outdoor traffic.

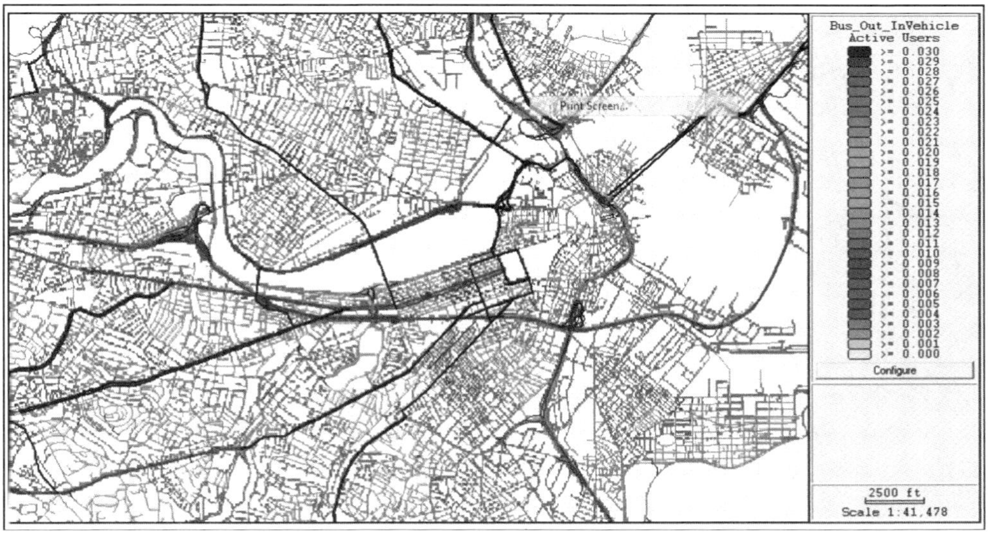

Figure 17.31 Business indoor vehicle traffic.

Figure 17.32 Buildings classified according to their building height and type (business).

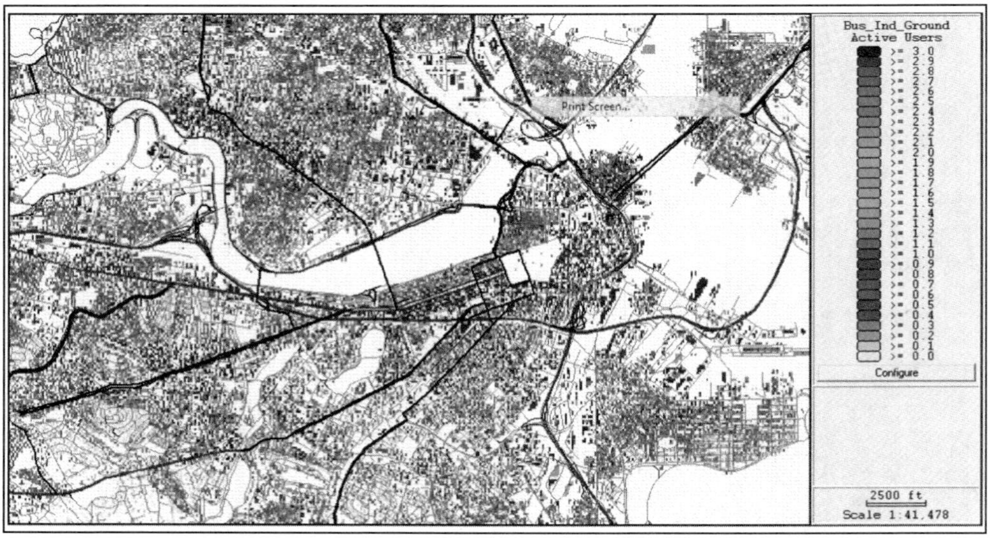

Figure 17.33 Business indoor ground up to 4th floor traffic.

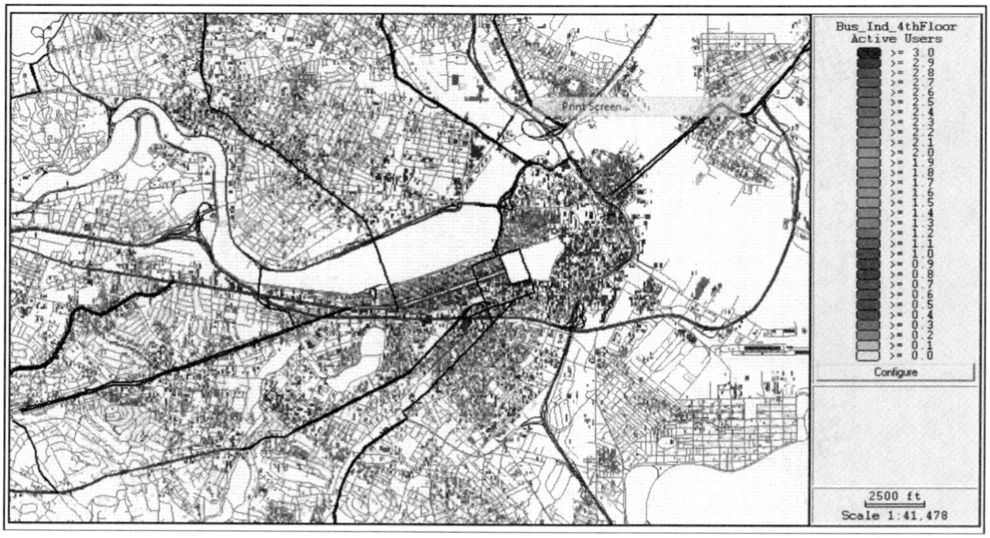

Figure 17.34 Business indoor up to 4th up to 9th floor traffic.

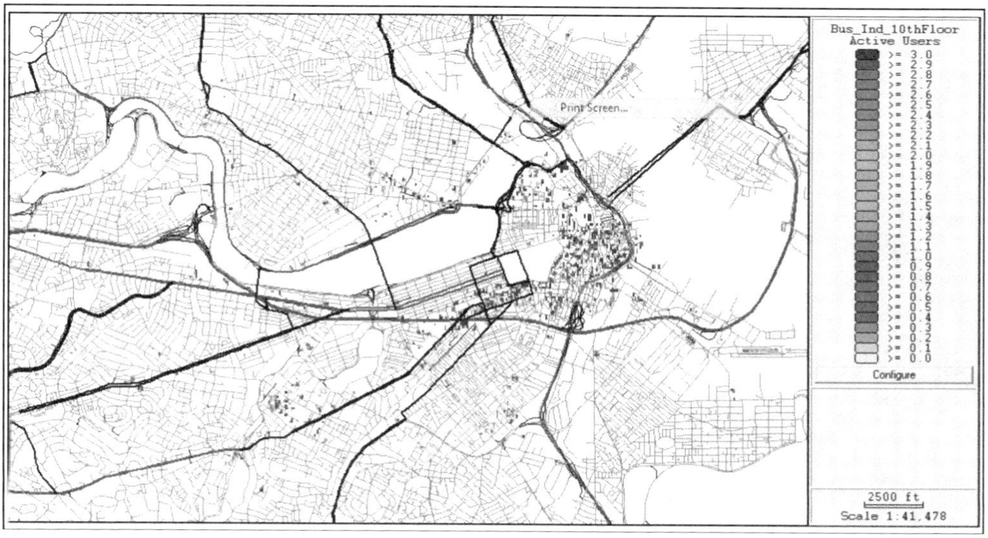

Figure 17.35 Business indoor 10th up to 19th floor traffic.

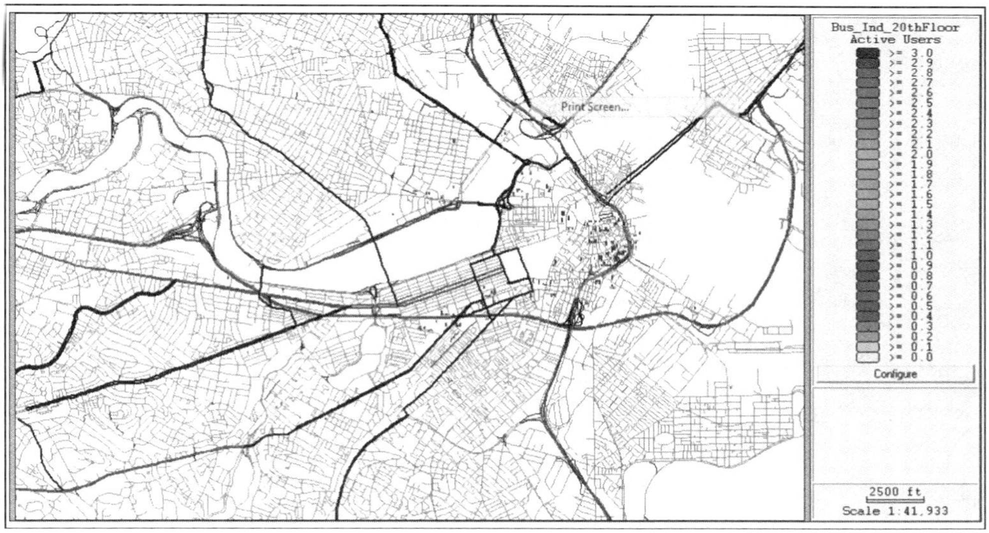

Figure 17.36 Business indoor above 20th floor traffic.

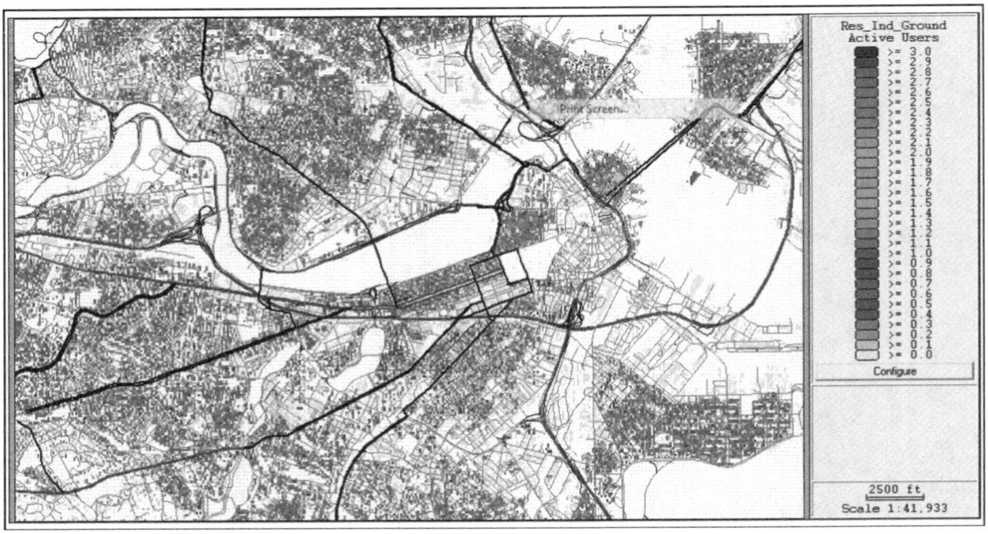

Figure 17.37 Residential indoor ground traffic.

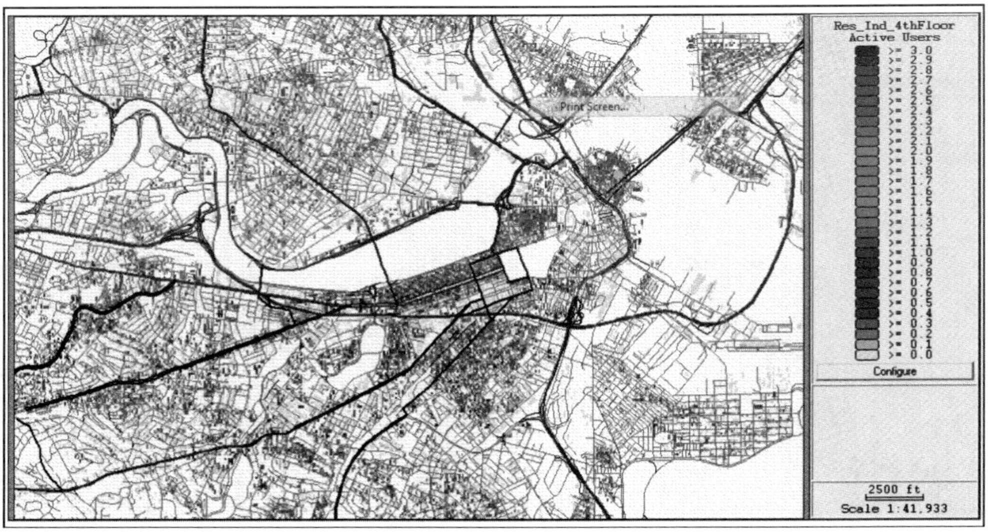

Figure 17.38 Residential indoor 4th floor traffic.

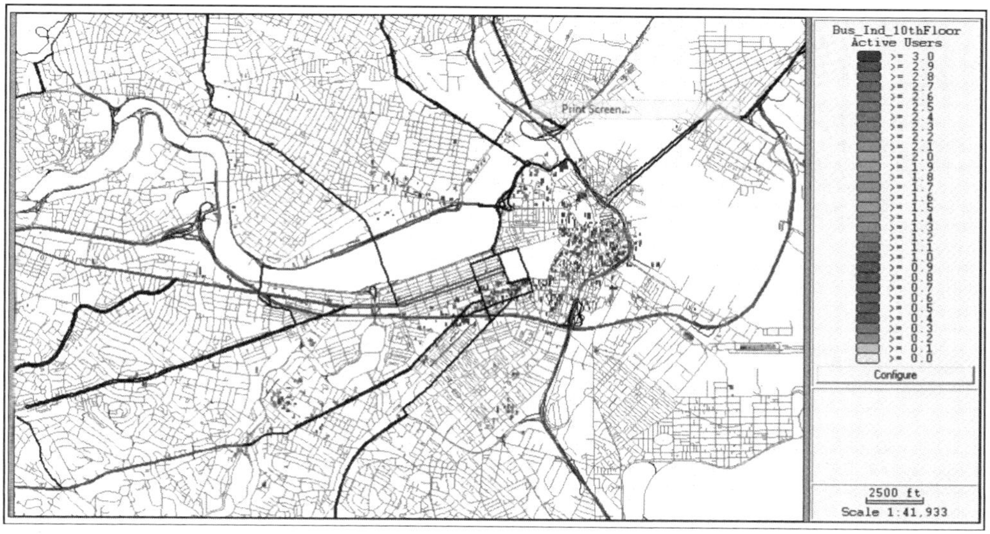

Figure 17.39 Residential indoor 10th floor traffic.

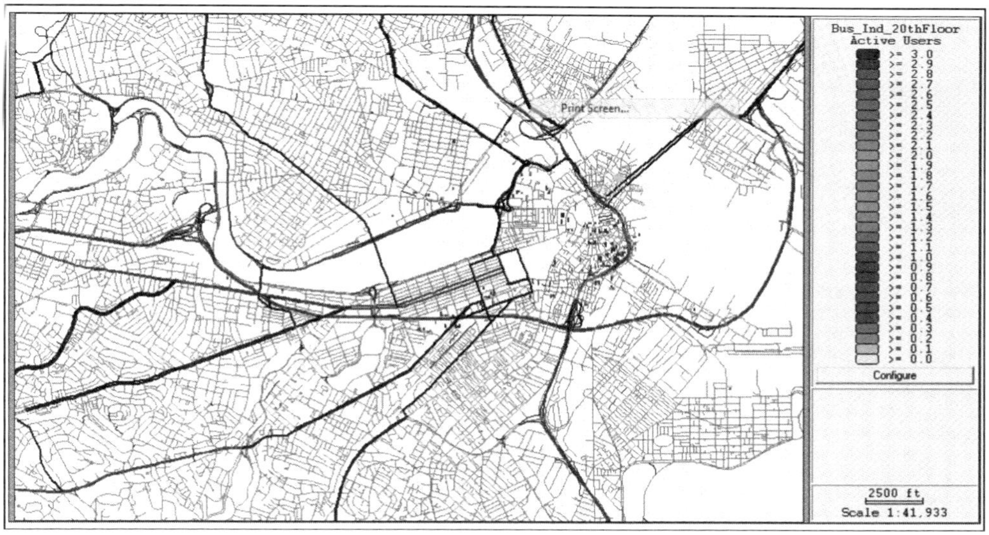

Figure 17.40 Residential indoor 20th floor traffic.

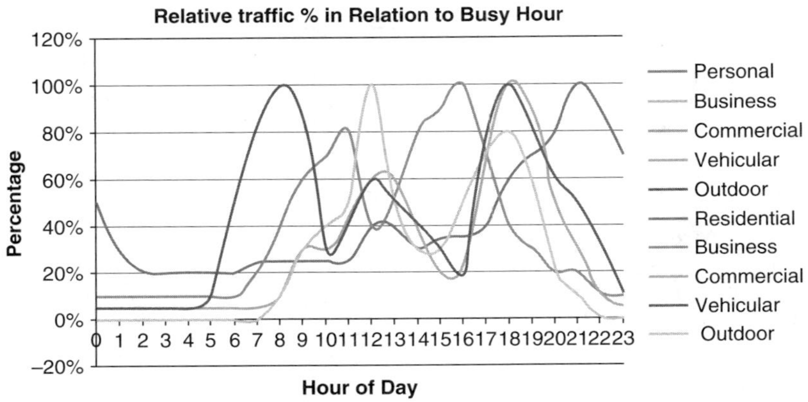

Figure 17.41 Hourly traffic variation.

Table 17.9 Combined traffic layers per service class

	SC traffic layer
1	Indoor ground P NRT
2	Indoor 3rd floor P NRT
3	Indoor 10th floor P NRT
4	Indoor ground B NRT
5	Indoor 3rd floor B NRT
6	Indoor 10th floor B NRT
7	Outdoor ground NRT
8	Outdoor rooftop P NRT
9	Outdoor rooftop B NRT
10	Vehicle NRT
11	Commercial NRT
12	Indoor ground P RT
13	Indoor 3rd floor P RT
14	Indoor 10th floor P RT
15	Indoor ground B RT
16	Indoor 3rd floor B RT
17	Indoor 10th floor B RT
18	Outdoor ground RT
19	Outdoor rooftop P RT
20	Outdoor rooftop B RT
21	Vehicle RT
22	Commercial

17.9.2 User Hourly Distribution

Figure 17.41 gives the busy hour traffic variation for different traffic layer types. Each of the traffic layers created in the previous sections consider the respective busy hour of each service. This chart helps designers to apply factors to those layers to represent the actual hour of the day that the designer wants to represent, that is, late at night, residential traffic is at its busy hour, but business traffic is at its lowest, thus running a simulation that considers the busy hour traffic for both layers would be erroneous, the use of factors allows designers to reduce business traffic to better represent reality.

17.10 Traffic Distribution Modeling

The different traffic layers should be combined according to the SC, as exemplified in Chapter 3. Table 17.9 lists the combined traffic grids generated according to the mix specified in Chapter 3.

18

Wireless Network Strategy

18.1 Define Spectrum Usage Strategy

Wireless broadband allows for the use of different channel bandwidths and the decision process a designer has to go through to choose one is fairly complex. Each market has different constraints and the designer may have to experiment with several possibilities before deciding on one. A broadband channel requires a lot of bandwidth, which, with the limited spectrum usually available, allows for only a few channels (it is common for networks to have only 1 to 3 channels).

A broadband wireless network has RF propagation characteristics and interference issues similar to those of wireless cellular networks. In the latter, to control interference, a minimum frequency reuse of 7 is required. In some cases, even higher reuse numbers may be required. This same principle is valid for wireless broadband networks and, when this reuse is not achievable, interference averaging and avoidance techniques should be used to reduce interference, as explained in Chapter 6.

Designers may have to experiment with different scenarios before settling for one. To make this task easier and yet productive, designers can construct simple scenarios without the use of GIS databases, with free space propagation, and uniformly spaced sites with a radius that approximates the smallest radius expected in the final deployment. Traffic should also be uniformly spread, considering the average density in the area.

By constructing a fictitious network such as this, it is possible to experiment on different scenarios and get outcomes for each. As an example, let's assume an operator has 20 +20 MHz of spectrum. The possible scenarios are:

- One FDD channel of 20 MHz.
- Two FDD channels of 10 MHz.
- Four FDD channels of 5 MHz.
- Two TDD channels of 20 MHz.
- Four TDD channels of 10 MHz.
- Eight TDD channels of 5 MHz.

Once the main configuration is selected (assuming that the choice for this first scenario analysis was of eight TDD channels of 5 MHz), a second level of scenarios can be built considering sectorization and zoning.

LTE, WiMAX and WLAN Network Design, Optimization and Performance Analysis, First Edition.
Leonhard Korowajczuk.

- Omni
- Three sectors, no zones
- Six sectors, no zones
- Three sectors, one zone
- Six sectors, one zone

Once again, assume the choice here was to use three sectored sites with one zone.

The designer can then choose channel frequencies, considering out-of-band attenuations and external interferences. For this, it is strongly recommend that the area be drive tested with a tool that can measure the spectrum at each location and collect samples to look for interferers, before deciding on which frequencies to use. This drive test should be repeated for different days of the week and different times of the day, preferably on expected busy hours.

In rare cases, more than one carrier can be used per sector. In fact, this is a strong argument favoring lower bandwidths, as spectrum is not wasted in sites with low traffic. At the end, the best scenario (configuration) is determined by the service classes established in the market modeling.

After this initial analysis of multiple scenarios, once the channel bandwidth is decided, and carriers and zones defined, cell IDs and other codes should be defined. Figure 18.1 shows a dialog with five carriers configured with different polarities (H and V) to be used for diversity.

18.1.1 Define Backhaul Spectrum Strategy

Backhaul is essential for the feasibility of the wireless network and broadband requires a large backhaul capacity. Backhaul can be done by connecting to a wireline backbone, through its access points, called PoP (Point of Presence). There are generally few of these points and backhaul information has to be

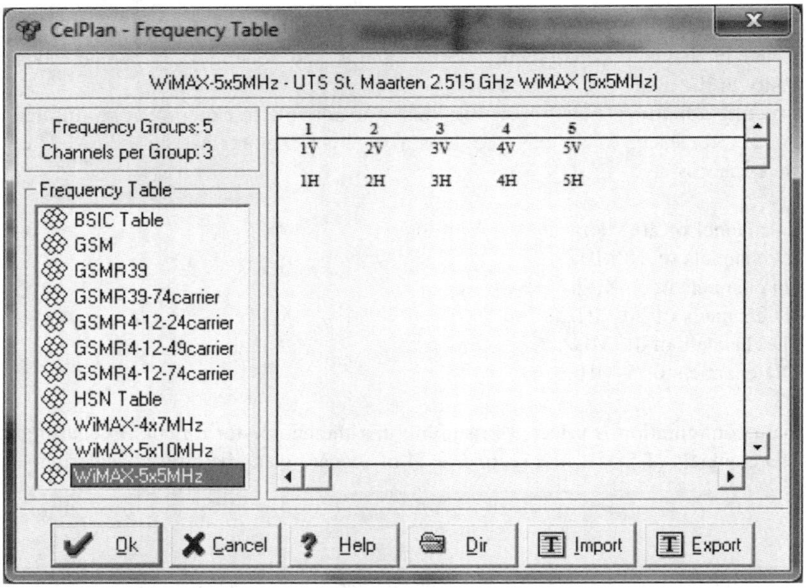

Figure 18.1 Carrier definition.

transported to these points by wireline (generally optical fiber) or by wireless connections. In small deployments the actual broadband spectrum can be used for backhaul, but it is more common to use higher frequencies to carry backhaul traffic.

Backhaul range and capacity are directly related to the frequency band used. Higher frequencies have shorter ranges but higher capacities. It is common to have two or three backhaul layers at different frequencies, so local interconnections can be done at higher frequencies, whereas long links will use lower frequencies. Typical backhaul frequencies are 7 GHz, 18 GHz, 23 GHz, and 39 GHz.

The reliability of the backhaul connection should be established also, by choosing the topology (radial or ring), redundancy (single or double) and the link availability.

Backhaul availability is traditionally specified by the time percentage the link is available over a certain period of time. Backhaul links above 10 GHz are affected by rain, snow, and fog and large margins should be considered to mitigate these effects. Typical availability percentages range from 99.9% (known as three nines or 87 minute a year) up to 99.999% (known as 5 nines or 5.25 seconds a year). These percentages do not consider link failures, which should be avoided by redundancy or topology.

Once backhaul frequencies are established, the precise channels should be chosen as resources.

18.2 Deployment Strategy

The deployment strategy defines how and where Cell Sites and CPEs will be installed. Cell site installations can be done in towers, on poles, or building rooftops. They may be done on existing or new infrastructure, or on a mix of both. CPEs can be self-installed or require professional installation. Backhaul can use leased lines or have its own, dedicated fiber lines installed.

All these definitions, although apparently simple, require in-depth analysis and are essential to start network design activities.

18.3 Core Equipment

Core equipment was described in Chapter 11. Its definition is important for the design process, as it may affect network architecture and interconnections. Many features to be considered during the design depend on core options.

18.4 Base Station Equipment

The main base station equipments considered in wireless designs are:

- radio modem
- RF head
- cable set
- antenna

Equipment offered by different vendors will be slightly different in characteristics. It is wise to design for specifications that are supported by more than one vendor, so more than one supplier can be used. Some operators request that the design be done for the best specification of each vendor, not realizing that no equipment will comply with the specifications in all places at all times and, consequently, the design will not be realistic.

Care should be taken not to use obsolete equipment or equipment with proprietary specifications, but at the same time claims of excessive performance should be scrutinized. This does not mean that these equipments should not be considered, as special conditions can make their deployment justified.

Vendor specifications should be compared against the theory and the conditions for which the specification are done should be well defined (e.g. a certain throughput is specified for a Gaussian or Rayleigh channel). Vendors are trying to sell their products so it is natural that they will specify them for the best possible situation, mainly in respect to performance.

Broadband equipments are recent developments and many vendors do not have all the data about their performance yet, mainly when they are still aggregating parts from other vendors. It is common for vendors not to be able to provide many of the required information about the equipment and designers will have to obtain this data from theory and literature. This is one of the reasons a network designer has to have a theoretical background of the technology; judgment calls will often be required and a solid theoretical knowledge will help on the decisions.

18.4.1 Base Station and Sector Controller

A Site is the location where a Base Station is installed. A Base Station has one or more Sectors (typically three). A Sector has one or more Radios, each supporting one carrier frequency or channel. A Radio is composed of a Sector Controller and an RF Head, which can be separated or integrated. Finally, the RF Head is connected to one or more antennas (which can also be integrated).

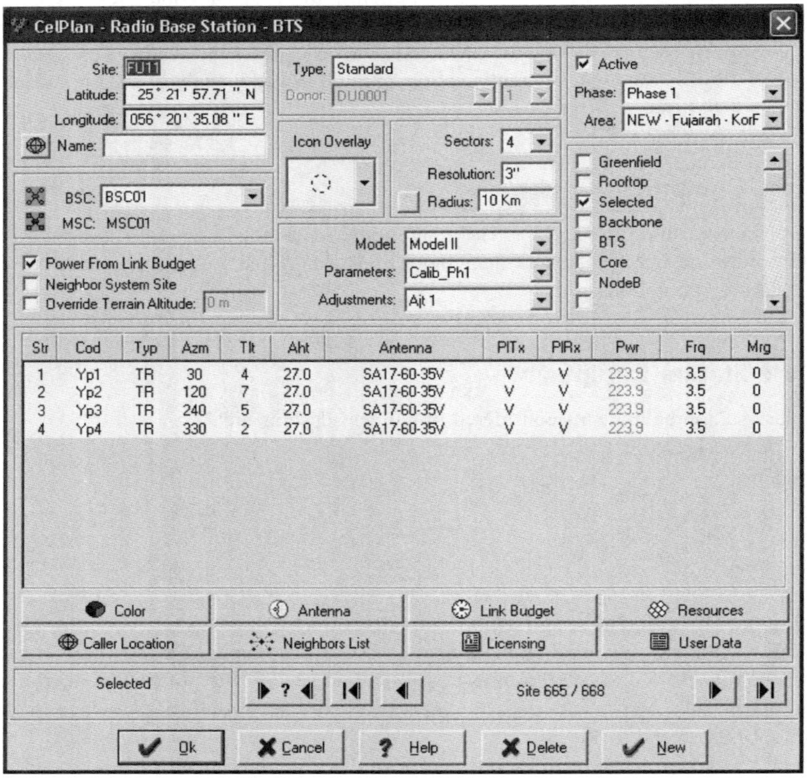

Figure 18.2 Base Station and Sector template.

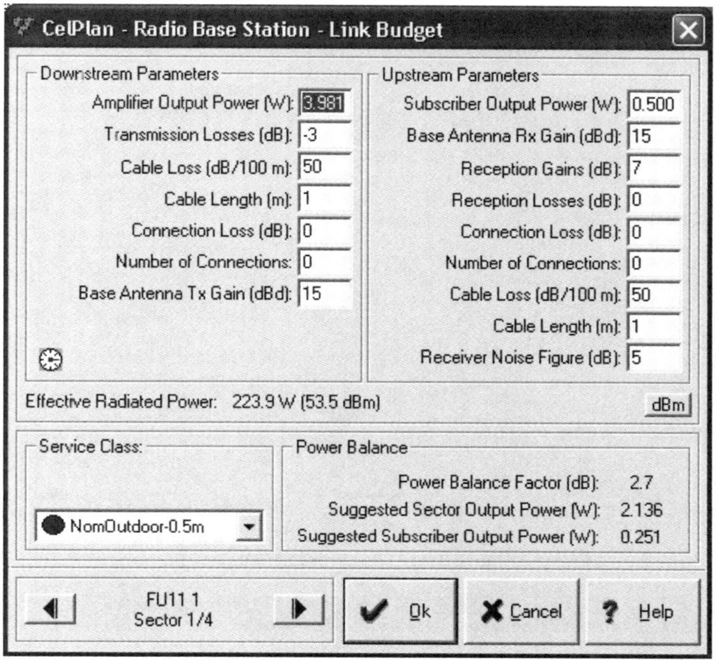

Figure 18.3 Link budget for 802.16e Sector Controller.

Typical Base Station and Sector Controller specifications are:

- Technology supported and release or profile complied
- Certification, if any
- FDD or TDD
- Number of Sectors supported
- Frequency Band supported
- Modulations and FEC codes supported
- Mean Output Power
- Receiver Sensitivity
- MIMO algorithms supported

Designers should configure templates for Base Stations, Link Budget data, Sector Resources, and Zones. This way, as new sites are created or added to the network, the template can be used, reducing the chance of introducing errors by configuring a site from scratch every time. Typical templates are shown in Figures 18.2 to 18.4.

18.4.2 Sector Radio and RF Head

The choice of the radio equipment depends on many factors that transcend network design and have to be done prior to the start of the design. Frequently the vendor is not defined, so the designer has to use estimated values for the radio parameters.

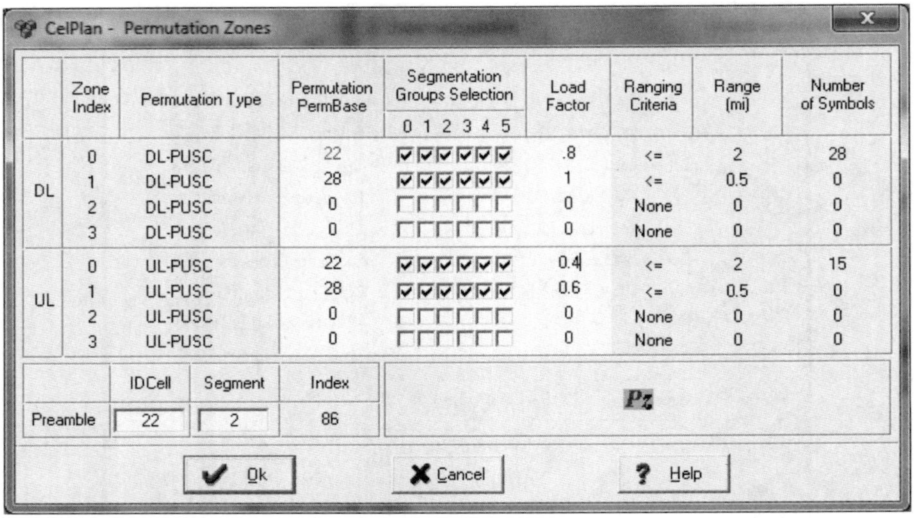

Figure 18.4 Zones configuration.

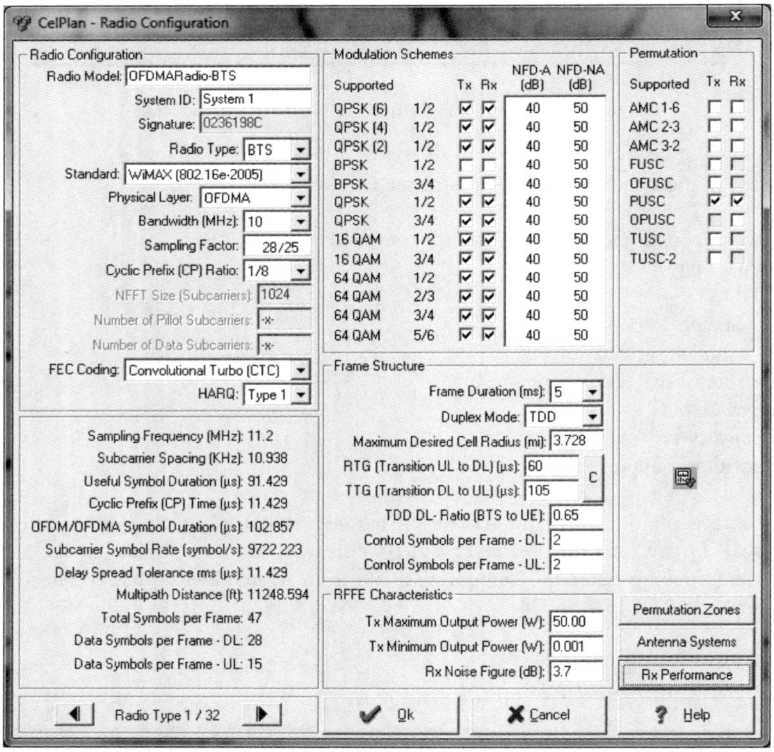

Figure 18.5 Base Station radio configuration.

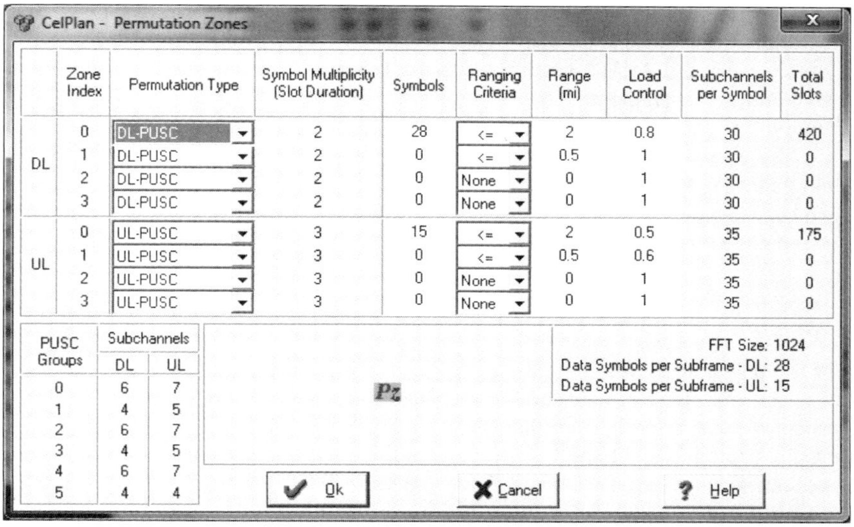

Figure 18.6 Base Station radio zone configuration.

The radio definition is very complex in WiMAX and LTE systems and the best way to explain it is by using actual configuration dialogs from a design tool. Due to this complexity it is important for the tool to have embedded default values, that serve as a starting point and that can be edited by the designer according to the characteristics of the equipment used. The relevant characteristics of a radio are exemplified in Figures 18.5 to 18.8.

The main characteristics of a radio that must be configured by designers are listed next; planning tools may add a few more specific fields for configuration but the ones listed below are essential to properly characterize the radio:

- *Technology*: for example, 802.16d, 802.16e, OFDM, OFDMA.
- *Modulation schemes*: as not all modulation schemes are mandatory, designers should specify which ones are supported by the radio and will be supported by the network.
- *Permutation*: also, not all permutations are mandatory, so designers must configure which ones are supported by each radio.
- *Frame structure*: designers must define the frame size (e.g. 5 ms), TDD ratio and transition time gaps.
- *RFFE (RF Front End) characteristics*: these include the Tx power and Rx noise figure.
- *Permutation zone*: designers must define if permutation zones will be used and their distribution in the frame.
- *Antenna system*: if antenna systems are supported by the radio, designer should specify which ones are supported on the DL and UL. The decision of which algorithm to apply, should be taken automatically by the tool.
- *RX (Receive) Performance*: designers must define as well as possible the receive sensitivity and throughput of the radio for different configurations. Often this information is not available in detail from equipment vendors and some assumptions have to be made. In the tool used as an example here (Figure 18.8), the receiver sensitivity is defined by 10 editable tables that provide sensitivities, gains, and losses for different configurations. The final RX sensitivity is calculated in real time during

simulation for the exact network configuration for each session. This approach allows designers to exercise various scenarios and see the "actual sensitivity" considered in each case (e.g. what's the sensitivity considered if MIMO Matrix A is used?); hence designers gain a better understanding of how the tool is calculating predictions and simulations and can make informed decisions before committing to the configuration. More details about performance tables are shown in Chapter 11.

The dialog shown in Figure 18.5 shows some user-editable fields, and some parameters that are automatically calculated based on the radio configuration. Figures 18.6 to 18.8 show a sample configuration of additional radio parameters such as permutation zones, antenna systems, and radio performance.

18.4.3 Antenna

In terms of equipment, designers must also define the type of antennas to be used and select a vendor. Many factors influence this decision, such as technical characteristics, price, availability, and support.

Once antennas have been defined, their patterns have to be digitized or obtained from vendors. Ideally antenna patterns should specify two patterns for each polarization (Azimuth and Elevation) in 1 degree increments. The prediction software must be able to generate a 3-D pattern from this information, so it can calculate the gain for any 3-D angle needed during predictions. Figures 18.9 and 18.10 show the representation of a typical antenna in a planning tool and its visualization in 3-D.

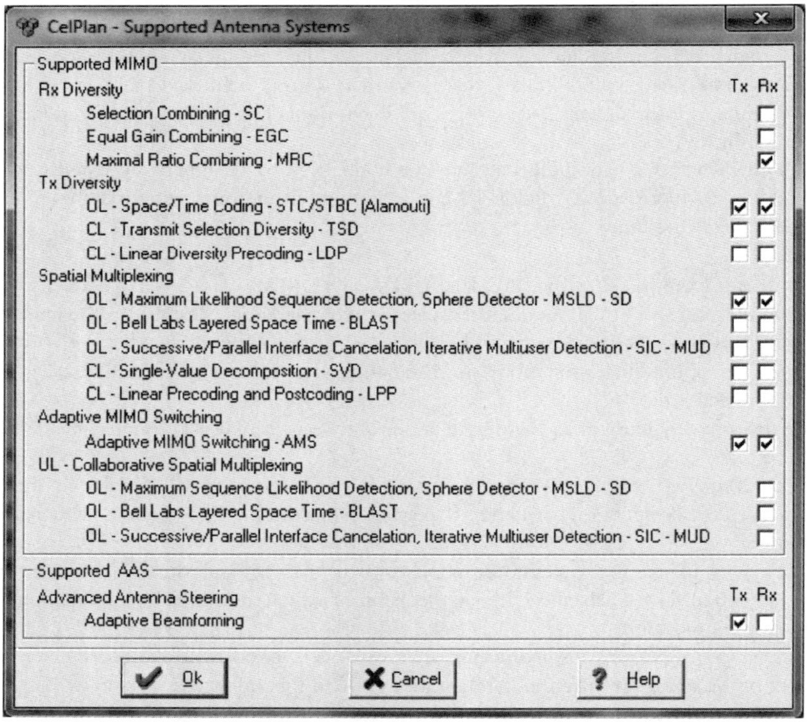

Figure 18.7 Base Station antenna system configuration.

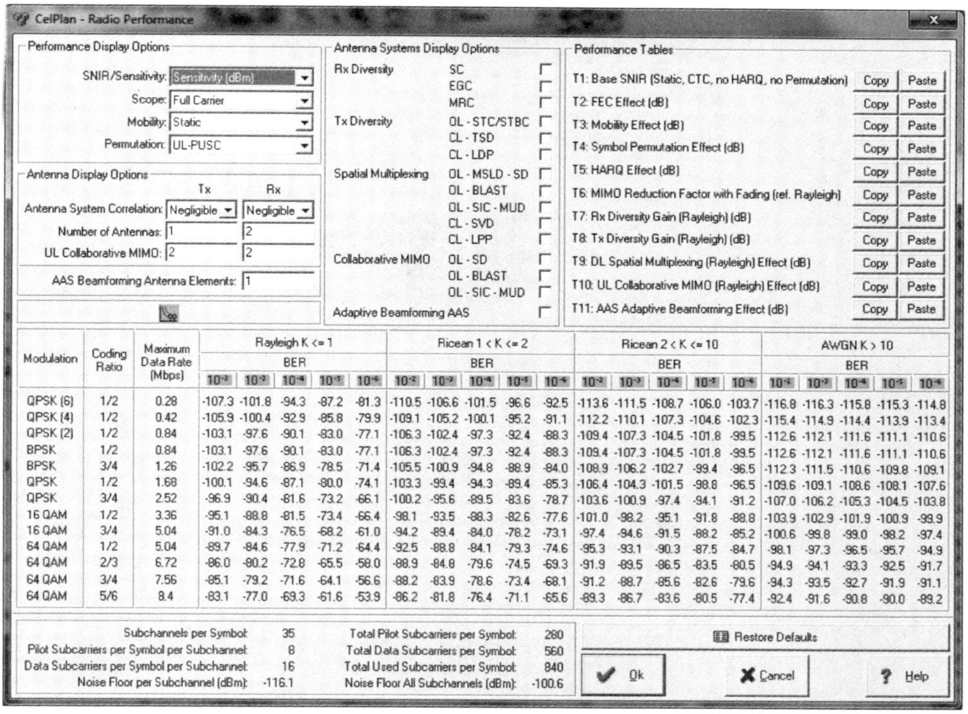

Figure 18.8 Base Station performance configuration.

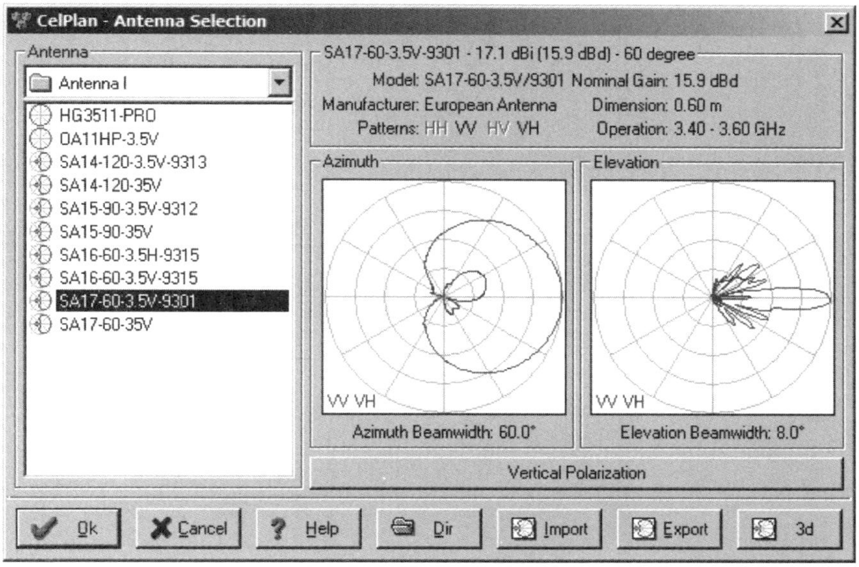

Figure 18.9 Antenna pattern.

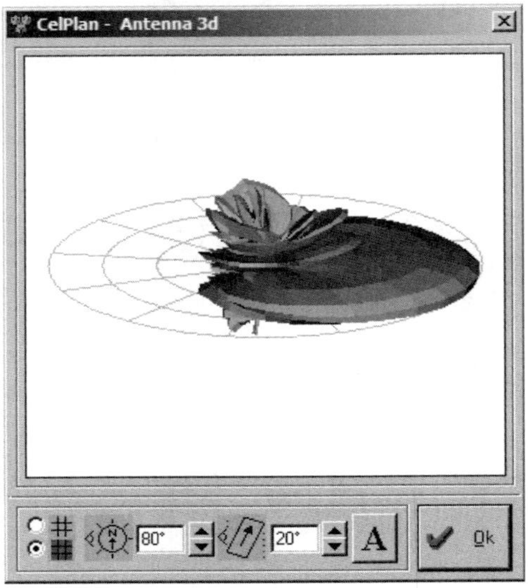

Figure 18.10 Antenna pattern 3-D view.

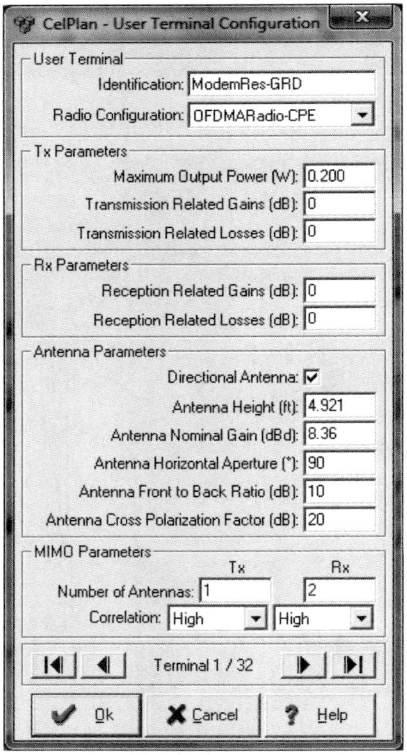

Figure 18.11 CPE terminal configuration.

18.5 Customer Premises Equipment (CPE)

Several options of CPEs are generally offered to customers. Many of them have very similar RF and performance characteristics and can be grouped in a single category when it comes to network design. As explained in Chapter 3, these categories can be mapped into Service Classes terminals.

CPE equipment must be fully compatible with the Base Station equipment for the network to operate properly, for example, "Do both CPE and BS support the same release of the technology?"

In the example in Chapter 3, four terminal types were identified:

- *Rooftop*: most of the time this type of terminal is an integrated unit with a directional antenna, RF head, and controller all within the same "box." It is placed on a rooftop and pointed towards the BTS.
- *Desktop*: desktop CPEs tend to also be an integrated unit that is placed on the customer's desk. It may have an omni antenna or a slightly directional antenna.
- *Laptop*: it can have an embedded radio or an external USB radio with an omni antenna.
- *Palmtop*: it has an embedded radio and antenna.

Figure 18.11 shows a typical CPE terminal configuration in a planning tool, with parameters to specify Tx power, antenna characteristics, and cable losses. Figures 18.12 and 18.13 show typical CPE radio

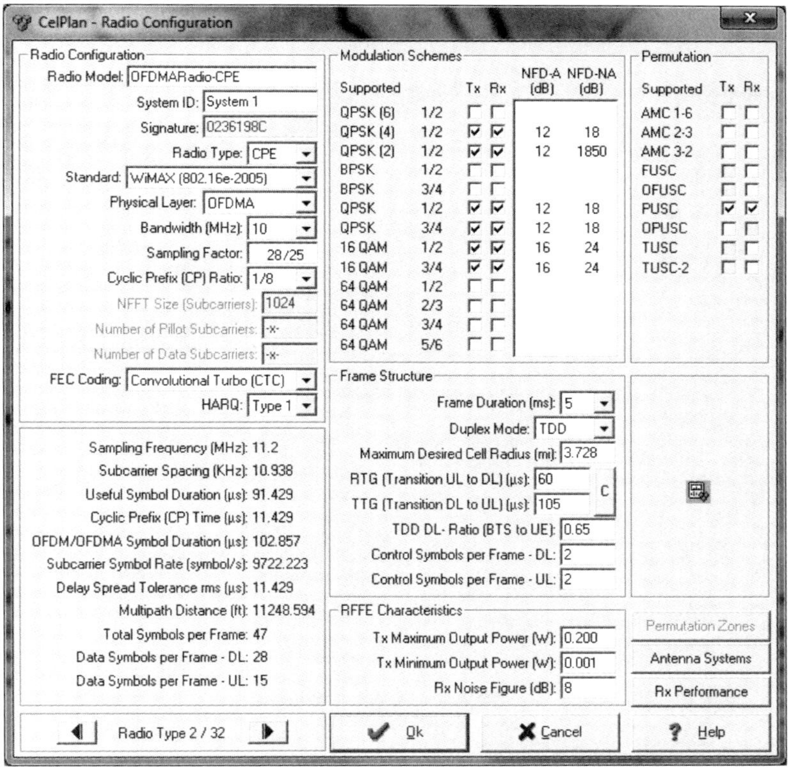

Figure 18.12 CPE radio configuration.

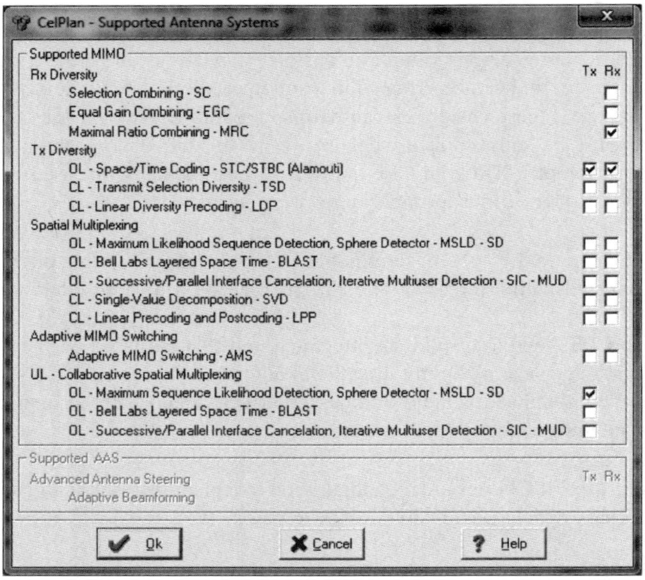

Figure 18.13 CPE antenna system configuration.

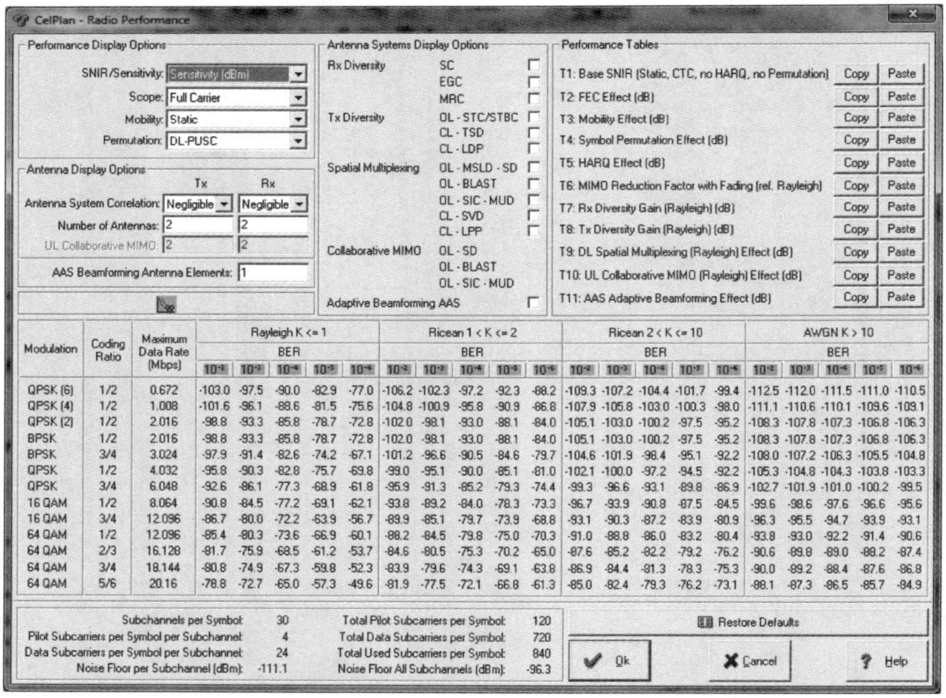

Modulation	Coding Ratio	Maximum Data Rate (Mbps)	Rayleigh K <= 1 BER					Ricean 1 < K <= 2 BER					Ricean 2 < K <= 10 BER					AWGN K > 10 BER				
			10^{-2}	10^{-3}	10^{-4}	10^{-5}	10^{-6}	10^{-2}	10^{-3}	10^{-4}	10^{-5}	10^{-6}	10^{-2}	10^{-3}	10^{-4}	10^{-5}	10^{-6}	10^{-2}	10^{-3}	10^{-4}	10^{-5}	10^{-6}
QPSK (6)	1/2	0.672	-103.0	-97.5	-90.0	-82.9	-77.0	-106.2	-102.3	-97.2	-92.3	-88.2	-109.3	-107.2	-104.4	-101.7	-99.4	-112.5	-112.0	-111.5	-111.0	-110.5
QPSK (4)	1/2	1.008	-101.6	-96.1	-88.6	-81.5	-75.6	-104.8	-100.9	-95.8	-90.9	-86.8	-107.9	-105.8	-103.0	-100.3	-98.0	-111.1	-110.6	-110.1	-109.6	-109.1
QPSK (2)	1/2	2.016	-98.8	-93.3	-85.8	-78.7	-72.8	-102.0	-98.1	-93.0	-88.1	-84.0	-105.1	-103.0	-100.2	-97.5	-95.2	-108.3	-107.8	-107.3	-106.8	-106.3
BPSK	1/2	2.016	-98.8	-93.3	-85.8	-78.7	-72.8	-102.0	-98.1	-93.0	-88.1	-84.0	-105.1	-103.0	-100.2	-97.5	-95.2	-108.3	-107.8	-107.3	-106.8	-106.3
BPSK	3/4	3.024	-97.9	-91.4	-82.6	-74.2	-67.1	-101.2	-96.6	-90.5	-84.6	-79.7	-104.6	-101.9	-98.4	-95.1	-92.2	-108.0	-107.2	-106.3	-105.5	-104.8
QPSK	1/2	4.032	-95.8	-90.3	-82.8	-75.7	-69.8	-99.0	-95.1	-90.0	-85.1	-81.0	-102.1	-100.0	-97.2	-94.5	-92.2	-105.3	-104.8	-104.3	-103.8	-103.3
QPSK	3/4	6.048	-92.6	-86.1	-77.3	-68.9	-61.8	-95.9	-91.3	-85.2	-79.3	-74.4	-99.3	-96.6	-93.1	-89.8	-86.9	-102.7	-101.9	-101.0	-100.2	-99.5
16 QAM	1/2	8.064	-90.8	-84.5	-77.2	-69.1	-62.1	-93.8	-89.2	-84.0	-78.3	-73.3	-96.7	-93.9	-90.8	-87.5	-84.5	-99.6	-98.6	-97.6	-96.6	-95.6
16 QAM	3/4	12.096	-86.7	-80.0	-72.2	-63.9	-56.7	-89.9	-85.1	-79.7	-73.9	-68.8	-93.1	-90.3	-87.2	-83.9	-80.9	-96.3	-95.5	-94.7	-93.9	-93.1
64 QAM	1/2	12.096	-85.4	-80.3	-73.6	-66.9	-60.1	-88.2	-84.5	-79.8	-75.0	-70.3	-91.0	-88.8	-86.0	-83.2	-80.4	-93.8	-93.0	-92.2	-91.4	-90.6
64 QAM	2/3	16.128	-81.7	-75.9	-68.5	-61.2	-53.7	-84.6	-80.5	-75.3	-70.2	-65.0	-87.6	-85.2	-82.2	-79.2	-76.2	-90.6	-89.8	-89.0	-88.2	-87.4
64 QAM	3/4	18.144	-80.8	-74.9	-67.3	-59.8	-52.3	-83.9	-79.6	-74.3	-69.1	-63.8	-86.9	-84.4	-81.3	-78.3	-75.3	-90.0	-89.2	-88.4	-87.6	-86.8
64 QAM	5/6	20.16	-78.8	-72.7	-65.0	-57.3	-49.6	-81.9	-77.5	-72.1	-66.8	-61.3	-85.0	-82.4	-79.3	-76.2	-73.1	-88.1	-87.3	-86.5	-85.7	-84.9

Figure 18.14 CPE radio performance configuration.

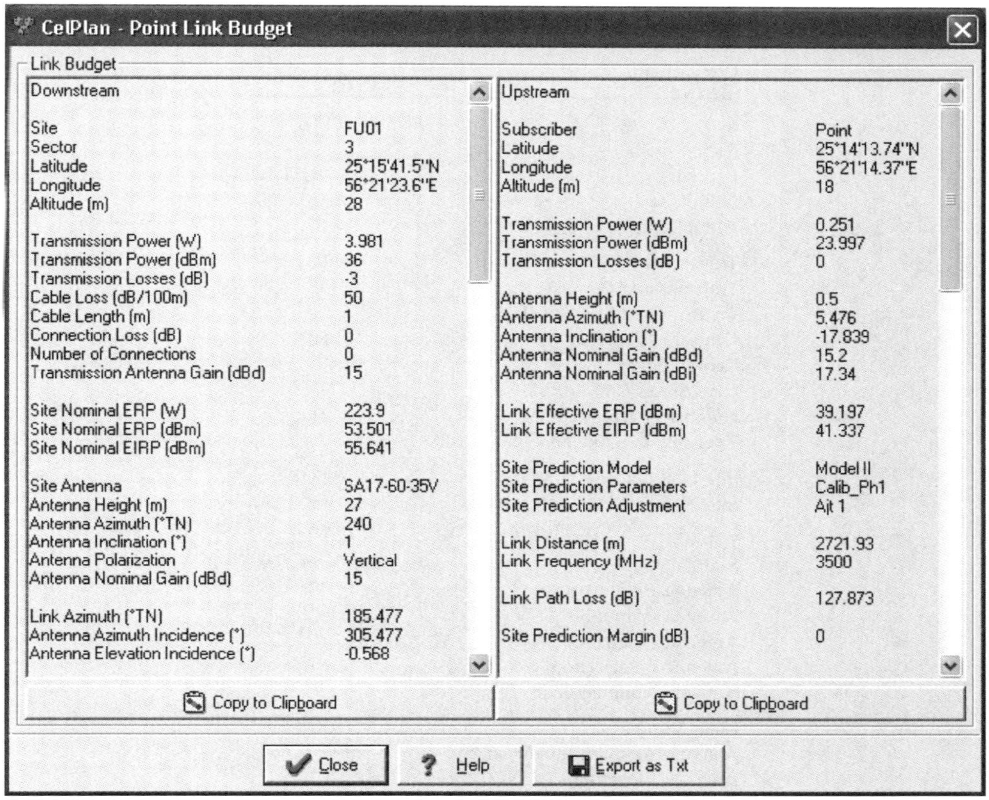

Figure 18.15 Example of a link budget between 802.16e sector controller and an arbitrary point.

and antenna system configurations. Figure 18.14 shows the expected CPE performance. More details about radio performance can be found in Chapter 11.

18.6 Link Budget

Among the parameters that affect the link budget, designers have the radio, user terminal, environment, and service configuration (each described in previous sections of this chapter). Figure 18.15 shows an example of a complete link budget dialog from a Sector Controller to a given point in the AOI. An understanding of the link budget is important for the designer to determine losses to be considered in the design.

The link budget example presented in Figure 18.15 is displayed in better detail in Tables 18.1 and 18.2.

18.7 Backhaul Equipment

Backhaul equipment is usually mounted at the Base Stations themselves to interconnect them to Land Line Access Points, also known as Point of Presence (PoP).

Table 18.1 Downstream link budget example

Downstream	
Site	FU01
Sector	3
Latitude	25°15′41.5″N
Longitude	56°21′23.6″E
Altitude (m)	28
Transmission Power (W)	3.981
Transmission Power (dBm)	36
Transmission Losses (dB)	−3
Cable Loss (dB/100m)	50
Cable Length (m)	1
Connection Loss (dB)	0
Number of Connections	0
Transmission Antenna Gain (dBd)	15
Site Nominal ERP (W)	223.9
Site Nominal ERP (dBm)	53.501
Site Nominal EIRP (dBm)	55.641
Site Antenna	SA17-60-35V
Antenna Height (m)	27
Antenna Azimuth (°TN)	240
Antenna Inclination (°)	1
Antenna Polarization	Vertical
Antenna Nominal Gain (dBd)	15
Link Azimuth (°TN)	185.477
Antenna Azimuth Incidence (°)	305.477
Antenna Elevation Incidence (°)	−0.568
Antenna Pattern Gain (dBd)	6.664
Antenna Effective Gain (dBd)	6.664
Antenna Effective Gain (dBi)	8.804
Link Effective ERP (dBm)	45.164
Link Effective EIRP (dBm)	47.304
Site Prediction Model	Model II
Site Prediction Parameters	Calib_Ph1
Site Prediction Adjustment	Ajt 1
Link Distance (m)	2721.93
Link Frequency (MHz)	3500
Link Path Loss (dB)	127.873
Site Prediction Margin (dB)	0
Environment Std Attenuation (dB/Dec)	30
Required Coverage Probability (%)	90
Coverage Probability Universe	Edge
Shadow Fading Model	Log-Normal
Standard Deviation (dB)	8.7
Multipath Fading Model	Rayleigh
K Factor (Direct/Scattered)	1.59
Resulting Fading Margin (dB)	14.414

Table 18.1 *(continued)*

Downstream	
Human Body Attenuation (dB)	12
Penetration Attenuation (dB)	15
Resulting Path Loss (dB)	154.287
Subscriber	Point
Latitude	25°14′13.74″N
Longitude	56°21′14.37″E
Altitude (m)	18
Antenna Height (m)	0.5
Antenna Azimuth (°TN)	5.476
Antenna Inclination (°)	−17.839
Antenna Nominal Gain (dBd)	15.2
Antenna Nominal Gain (dBi)	17.34
Reception Related Losses (dB)	12.3
Reception Related Gains (dB)	7
Downstream Signal Prediction (dBm)	−94.942
Radio Type	RedMAX-16e
System	WiMAX (802.16-OFDMA)
Bandwidth per Carrier (MHz)	5
Modulation Scheme	QPSK 3/4
Maximum Data Rate (Mbps)	4.652
Coding Gain (dB)	0.625
Reference Temperature (°K)	290
Receiver Noise Figure (dB)	5
Service Threshold (dBm)	−99.5
Link Carrier to Noise Ratio (dB)	7.903
Link Service Margin (dB)	4.558

18.7.1 Backhaul Radio Equipment

The selection of the radio equipment to be used for backhaul is both a technical and a commercial issue. Radios can be divided into a controller and an RF Head or be integrated into a single outdoor unit. They can provide dual streams for polarization redundancy and can be used as a single link or as a redundant link (1 + 1).

It is common for a backhaul design to employ multiple types of radio, depending on link distances and required throughput. The following is an example of radios that could be used in a possible deployment.

- *Radio A*: Used for Layer 1 short links, at 38 GHz with 16QAM modulation and 14 MHz bandwidth, providing 34 Mbps capacity, which is enough for three fully loaded 10 MHz WiMAX sectors.
- *Radio B*: Used for Layer 2 short and medium links, at 18 GHz with 32QAM modulation and 28 MHz bandwidth with 82 Mbps capacity, which should be enough for 12 three 10 MHZ WiMAX sectored sites.
- *Radio C*: Used for Layer 2 long links, at 7 GHz with 32QAM modulation and 28 MHz bandwidth with 82 Mbps capacity, which should be enough for 12 three 10 MHZ WiMAX sectored sites.

Table 18.2 Upstream link budget example

Upstream	
Subscriber	Point
Latitude	25°14′13.74″N
Longitude	56°21′14.37″E
Altitude (m)	18
Transmission Power (W)	0.251
Transmission Power (dBm)	23.997
Transmission Losses (dB)	0
Antenna Height (m)	0.5
Antenna Azimuth (°TN)	5.476
Antenna Inclination (°)	−17.839
Antenna Nominal Gain (dBd)	15.2
Antenna Nominal Gain (dBi)	17.34
Link Effective ERP (dBm)	39.197
Link Effective EIRP (dBm)	41.337
Site Prediction Model	Model II
Site Prediction Parameters	Calib_Ph1
Site Prediction Adjustment	Ajt 1
Link Distance (m)	2721.93
Link Frequency (MHz)	3500
Link Path Loss (dB)	127.873
Site Prediction Margin (dB)	0
Human Body Attenuation (dB)	12
Penetration Attenuation (dB)	15
Environment Std Attenuation (dB/Dec)	30
Required Coverage Probability (%)	90
Coverage Probability Universe	Edge
Shadow Fading Model	Log-Normal
Standard Deviation (dB)	8.7
Multipath Fading Model	Rayleigh
K Factor (Direct/Scattered)	1.59
Resulting Fading Margin (dB)	14.414
Resulting Path Loss (dB)	169.287
Site	FU01
Sector	3
Latitude	25°15′41.5″N
Longitude	56°21′23.6″E
Altitude (m)	28
Site Antenna	SA17-60-35V
Antenna Height (m)	27
Antenna Azimuth (°TN)	240
Antenna Inclination (°)	1
Antenna Polarization	Vertical
Antenna Nominal Gain (dBd)	15

Table 18.2 *(continued)*

Upstream	
Link Azimuth (°TN)	185.477
Antenna Azimuth Incidence (°)	305.477
Antenna Elevation Incidence (°)	−0.568
Antenna Pattern Gain (dBd)	6.664
Antenna Effective Gain (dBd)	6.664
Antenna Effective Gain (dBi)	8.804
Reception Antenna Gain (dBd)	15
Site Reception Gains (dB)	7
Site Reception Losses (dB)	0
Connection Loss (dB)	0
Number of Connections	0
Cable Loss (dB/100m)	50
Cable Length (m)	1
Upstream Signal Prediction (dBm)	−97.646
Radio Type	RedMAX-16e
System	WiMAX (802.16-OFDMA)
Bandwidth per Carrier (MHz)	5
Modulation Scheme	QPSK 1/2
Maximum Data Rate (Mbps)	3.102
Coding Gain (dB)	1.505
Reference Temperature (°K)	290
Receiver Noise Figure (dB)	5
Service Threshold (dBm)	−99.5
Link Carrier to Noise Ratio (dB)	5.2
Link Service Margin (dB)	1.854

Radio parameters are shown in the configuration dialogs of Figure 18.16.

18.7.2 Backhaul Antennas

Antennas for backhaul use have to be selected to provide enough gain for different link lengths.

Typical antennas used in microwave links are listed below.

- 38 GHz antennas: 0.3, 0.6 and 0.8 meter diameter.
- 18 GHz antennas: 0.3, 0.6, 0.8 and 1.2 meter diameter.
- 7 GHz antennas: 0.6, 0.8 and 1.2 meter diameter.

Figure 18.17 shows a typical antenna patterns for backhaul antennas.

18.7.3 Backhaul Network Layout Strategy

The preferred backhaul architecture should be established at this stage. Backhaul interconnections can be done as stars or rings.

Star networks use redundancy to improve reliability, whereas ring networks use alternative routes to divert traffic in case of failure.

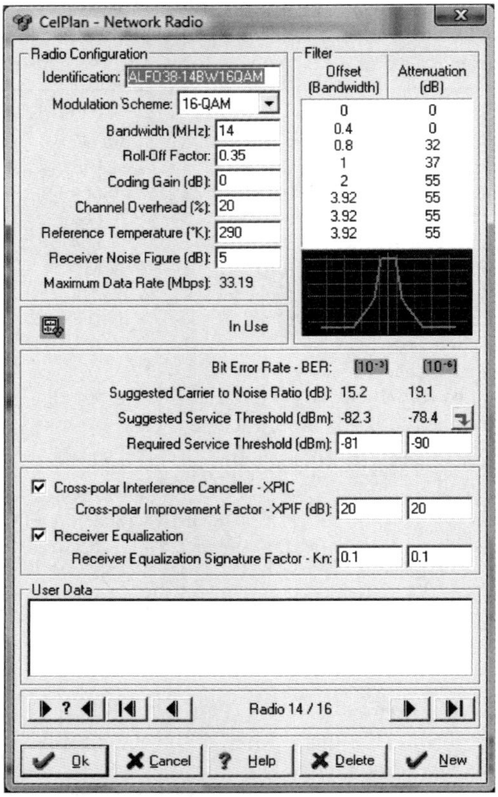

Figure 18.16 A 38 GHz microwave link radio configuration dialogue.

Figure 18.18 shows one example of a three-layered, star backhaul network. All links in this example are designed with redundant equipment, so if one radio fails, the other takes over.

18.8 Land Line Access Points of Presence (PoP)

Usually there are few PoPs per city and their location should be defined in the project, so the backhaul network can be properly designed.

18.9 List of Available Site Locations

A list of available, or negotiable, site locations should be prepared for use as site candidates during the design process. Design tools should allow multiple identifications of sites, so they can be activated and deactivated with ease.

Figure 18.19 depicts site classifications used in a typical project for easy identification of sites; for example, designers may need to see all sites that are part of the backbone network.

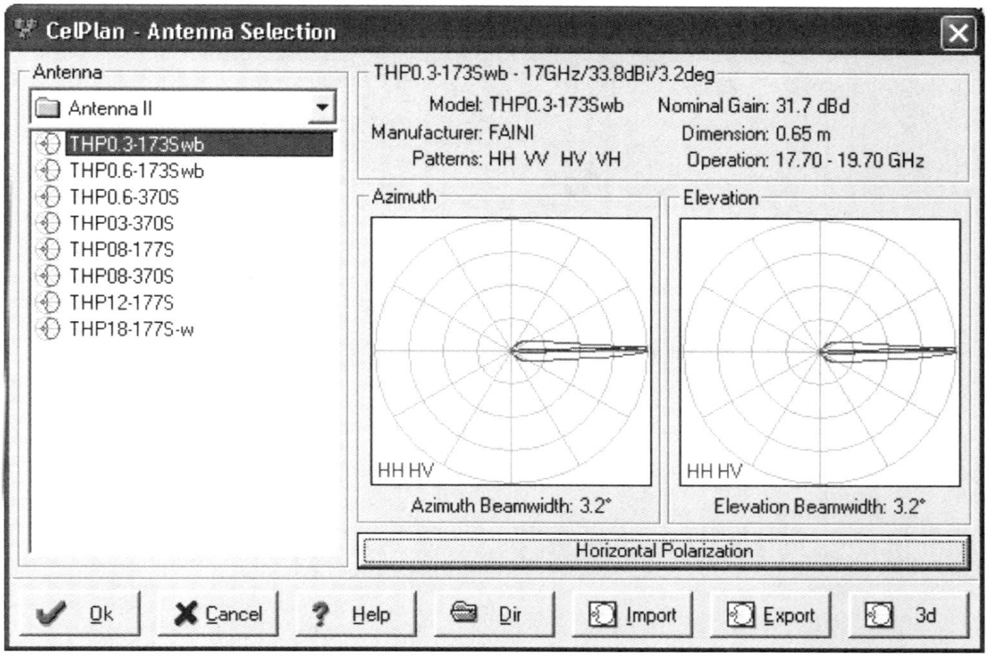

Figure 18.17 Backhaul antenna pattern.

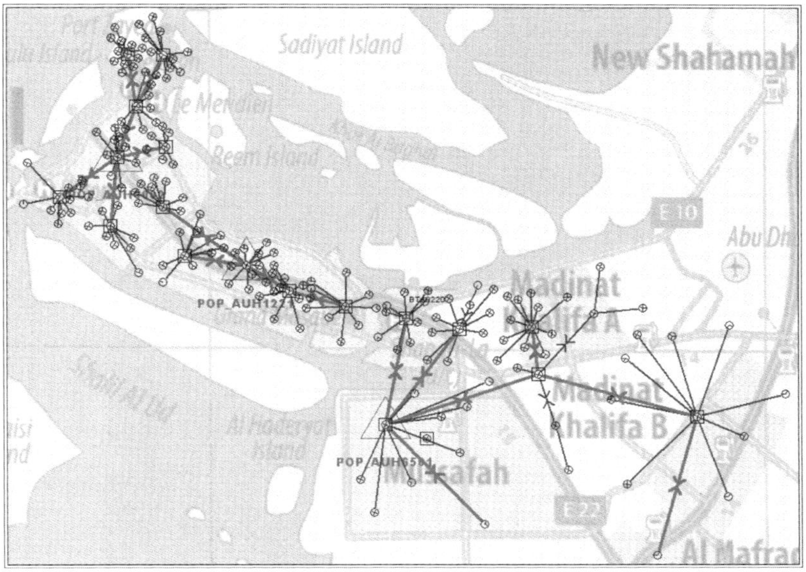

Figure 18.18 Backhaul radio links.

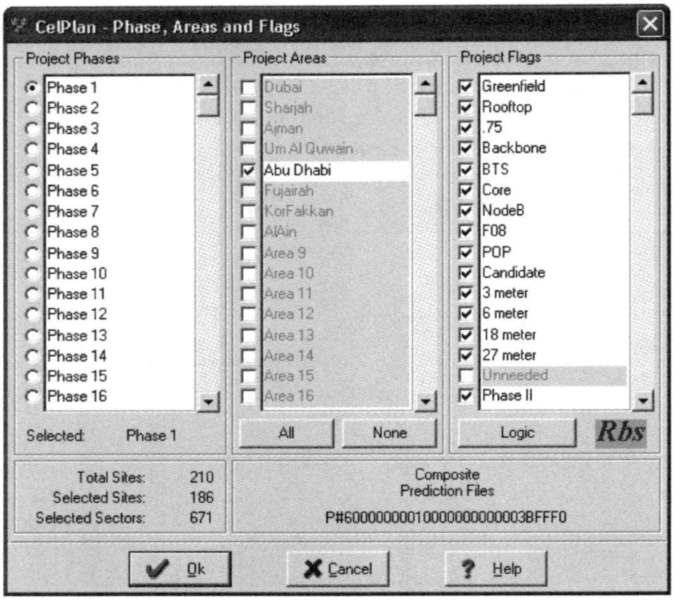

Figure 18.19 Project phases, areas and flags.

19

Wireless Network Design

This is the design phase in which sites locations are chosen; thus it is necessary to define the propagation model and its parameters to be used at each location. With this knowledge, the traffic requirements defined in Chapter 17, and the deployment strategy established in Chapter 18, it is possible to select the best site locations to provide the desired services.

19.1 Field Measurement Campaign

Chapter 6 describes the methodology to gather propagation measurements for propagation model calibration purposes. This chapter shows typical results obtained using collection and processing tools.

The first step in a drive test campaign is to select representative locations inside the AoI and in agreement with the deployment strategy. If using fractional morphology methodology, only a few locations need to be measured. In a large city, 4 to 8 different locations are usually sufficient. The exact number can be obtained by analyzing the results for initial sites and analyzing the dispersion between them.

The measurements themselves can usually be done over a short period of time; it is the post-processing that takes time to be done properly. This section uses the drive test tool CelSignal to exemplify CW measurement collection and the software CelTools as an example of a post-processing tool. The example used here is based on a campaign done at 1.8 GHz in a dense urban area.

The calibration of a propagation model requires a set of statistically representative path loss measurements, which should be spread over the network area and measure the signal at different distances and morphologies. A site only covers some of the morphologies represented in the GIS database and, for this reason more than one site should be measured. The antenna elevation in relation to the morphology height should also be considered when selecting representative sites, being the elevation below/above morphology used as a criterion.

To measure path loss at large distances, a narrowband signal, such as CW (Continuous Wave), should be used. It is very important to use narrowband filters at the receiver, so it is not overloaded by other signals transmitted in the vicinity.

Measurements should be done in all environments where the network will be used:

- Outdoor measurements.
- Indoor measurements at various heights and encapsulations.
- Rooftop measurements at various heights.

LTE, WiMAX and WLAN Network Design, Optimization and Performance Analysis, First Edition.
Leonhard Korowajczuk.

Measurements should be done over all morphologies, and be filtered for:

- Fast fading (so values are averaged over the coherence time, which varies with the frequency).
- Antenna angle (vertical and horizontal), so antenna nulls are avoided, as they can vary significantly from one antenna to the other and in relation to the specified pattern.
- Noise floor as it distorts the measured value and gives the same value for all measurements below it. Measurements should be filtered at 3 dB above noise floor.
- Special locations that are not well represented in the data base should be filtered (e.g. tunnels, elevated highways).
- Measurements should be equally distributed along the drive route, so one particular segment does not distort the statistics of the distribution. As an example, repetitive measurements while a vehicle is stationary should be eliminated (averaging).
- Measurements that give large errors during the first round of calibration generally represent deviations between the database and the actual environment.

Measurements should be also be geographically adjusted so they coincide with the morphology where they were measured. It is common that, due to GPS error, measurements collected on streets fall in constructed areas in the database, resulting in erroneous information. These measurements should be adjusted to fall within the morphologies where they were measured.

We used a CW (Continuous Wave: a single frequency) transmitter and a Spectrum analyzer as receiver. The spectrum analyzer was set to zero span with a bandwidth of 1 kHz, as in this configuration it becomes a time analyzer instead of a frequency analyzer. The spectrum analyzer and the collection SW were configured to collect samples every 2 ms. A GPS was used to collect location coordinates.

Figure 19.1 shows the CW receiving antenna location on the vehicle. A half-wave dipole antenna was used, so proper care was taken to provide a ground plane in the vicinity of the antenna base.

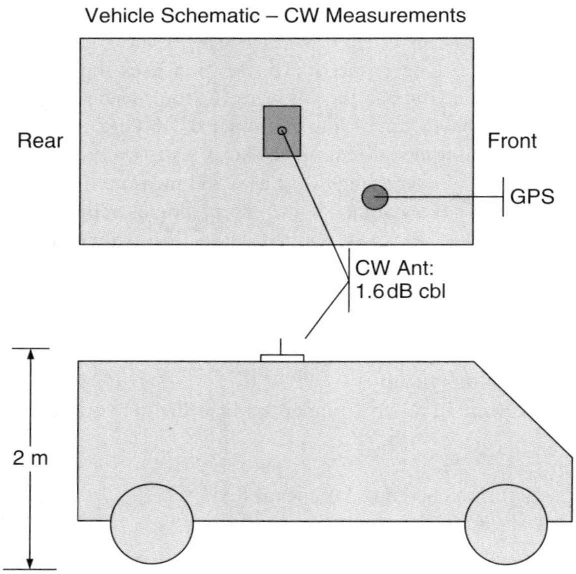

Figure 19.1 Measurement vehicle layout.

The same set-up can be used to do CPE measurements, so CW results can be compared to radio collected signal levels (RSSI, Receive Signal Strength Indicator), in single antenna and multiple antenna configurations. This data should be used to adjust radio performance tables. Additionally, indoor measurement should be done at different floors, so morphology penetration parameters can be established.

19.2 Measurement Processing

The first step in processing the collect data is to evaluate fast fading in several locations, which can be done by the collection of stationary measurements over a period of time. Figures 19.2 and 19.3 show collected samples over a period of 3 minutes in our example. The measurements were averaged for 100 ms, 200 ms, 300 ms, and 600 ms and the results are shown in the same figure.

In the plot of Figure 19.3, it is possible to identify several fading components: very fast variations below 100 ms, slower variations at 1 s, 3 s, and 10 s. Each of these variations is due to different effects, such as moving vehicles or foliage movement. This means that it is not feasible to remove

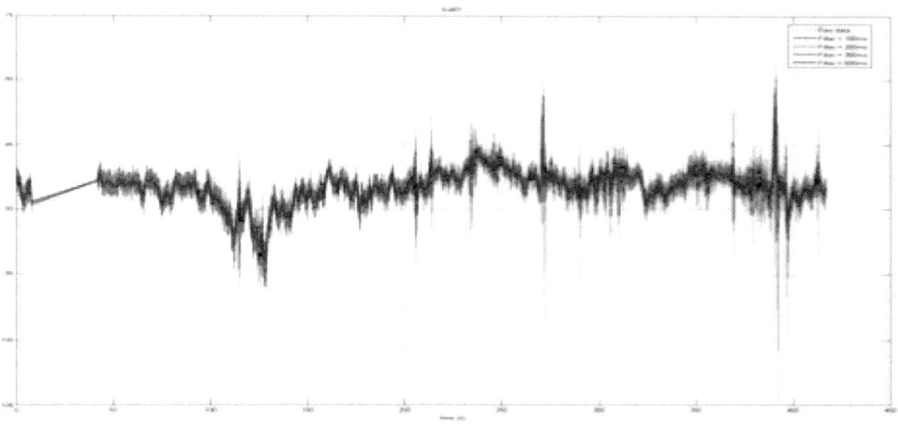

Figure 19.2 CW measurements every 2 ms and averaged values over 180 s.

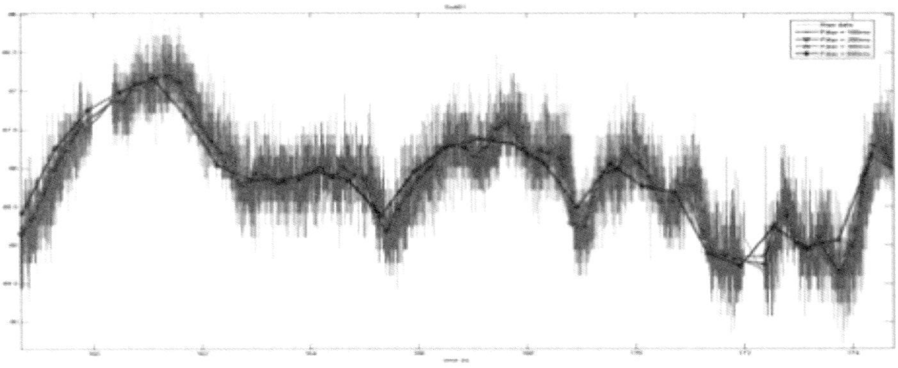

Figure 19.3 Detail of CW measurements over 16 s.

all fast fading from the measurement and that, at least part of it, should be treated as slow fading. In this exercise the decision was to use a filter of 100 ms, because this is approximately the size of a broadband symbol and yet the location span is small for a car in movement (1.67 m for 60 km/h).

Figure 19.4 shows static measurements collected every 2 ms and Figure 19.5 shows their distribution. Figure 19.6 shows the same measurements averaged over 100 ms periods and Figure 19.7 shows the distribution of these averaged measurements. Table 19.1 gives measurement characteristics when using different averaging periods.

Once the fast fading filter is defined, repetitive samples at the same location should be eliminated using a distance filter. In this example the filter used 2m distance as the criterion, that is, if two samples are less than 2m from each other, then the filter is applied.

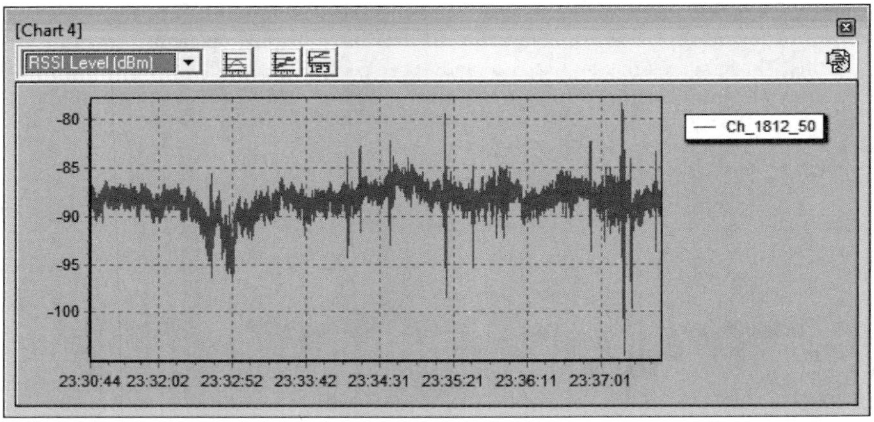

Figure 19.4 Static measurements at 2 ms.

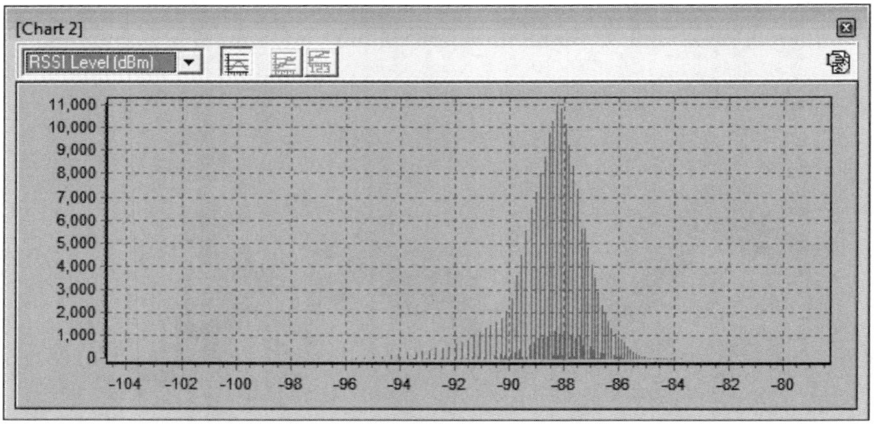

Figure 19.5 Static measurement at 2 ms distribution.

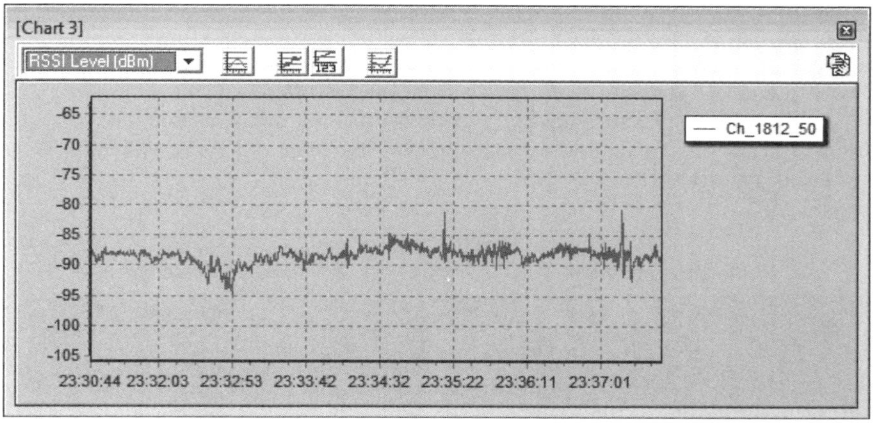

Figure 19.6 Static measurements at 2 ms averaged every 100 ms.

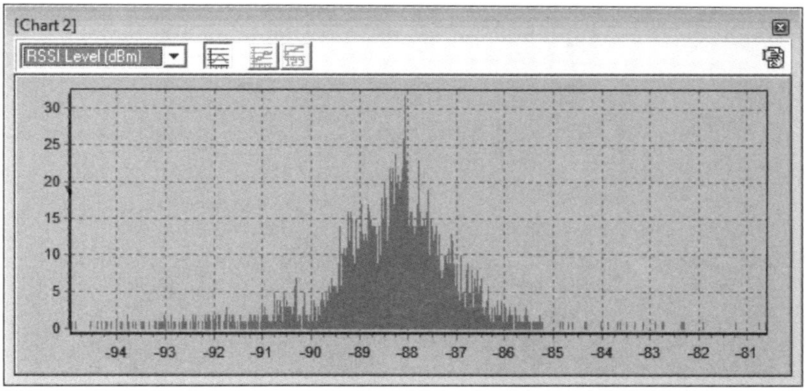

Figure 19.7 Static measurements at 2 ms averaged every 100 ms distribution.

Table 19.1 Static measurements' characteristics averaged for 100 ms, 600 ms, 1 s, 10 s and 60 s

File name	# Points	Min. Signal	Max. Signal	Average	Std. deviation
Sub01 – Raw	206,800	−105	−78	−88.12	1.44
Sub01 – 100	3724	−95	−81	−88.11	1.36
Sub-1 – 600	634	−94	−82	−88.1	1.31
Sub01 – 1000	384	−94	−83	−88.1	1.31
Sub01 – 10,000	39	−92	−86	−88.06	1.17
Sub01 – 60,000	7	−89	−87	−88.1	0.69

The next step is to analyze measurements in movement. The collected samples were plotted over the morphology and, as expected, GPS precision in an urban area is affected by buildings and foliage that block view to satellites and, consequently, introduce errors.

Figures 19.8 and 19.9 show location errors due to foliage and high rise buildings, before and after filtering with the post-processing software. Figure 19.10 shows location errors due to the inherent imprecision of the GPS, which can be corrected by, for example, moving the samples to vectors that represent streets and roads.

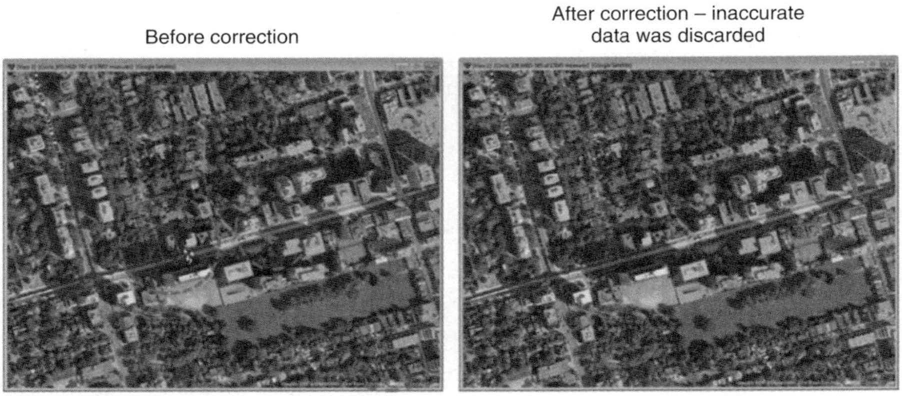

Figure 19.8 GPS errors caused by foliage, before and after filtering.

Figure 19.9 GPS errors due to high rise buildings, before and after filtering.

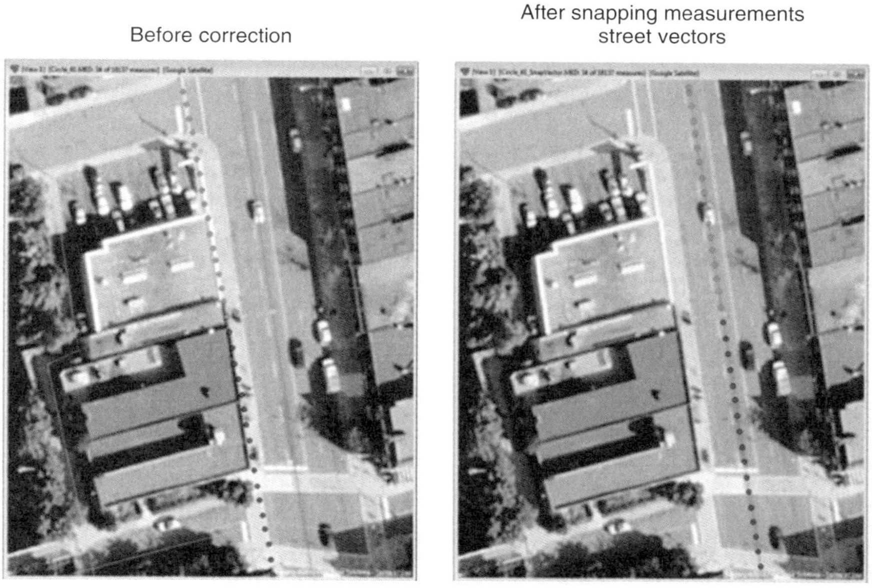

Figure 19.10 GPS errors due to imprecision, before and after correction.

Figure 19.11 shows collected measurement values, raw, averaged over time and averaged over time and distance.

After measurements are processed, they should be separated in two lots: a calibration lot and a control lot. The calibration lot is used for the actual model calibration, whereas the control lot is used to verify the reusability of the calibrated model in a similar environment. In this example, the measurements were split in 32 slices, with 25% of the slices used as control measurements and the rest for calibration.

Ideally the design tool should do this automatically, with the user specifying the percentage of samples used for which. The tool used in this example separates the groups with a slicing method, as this has the added advantage of keeping all types of paths in both lots. This is illustrated in Figure 19.12.

19.3 Propagation Models and Parameters

The model initially chosen for this exercise was the Korowajczuk model with constrained parameters. Figure 19.13 shows the propagation parameters obtained from calibration for a site that covers only few dense urban morphology types.

The propagation parameters obtained were then verified against the measurements used for calibration; the results are shown in Figure 19.14. In the example, the average deviation between measurements and prediction was 0.15 dB with a standard deviation of 6.343 dB. The measurement standard deviation over a grid of 5m was 1.1 dB. These are considered good results for a dense urban area with buildings.

The propagation parameters were then applied to the control lot (results are shown in Figure 19.15). The average deviation between measurements and prediction was also 0.15 dB with a standard deviation of 6.34 dB. The measurement standard deviation over a grid of 5 m was 0.99 dB. This shows that the obtained parameters have a very good reusability in the area.

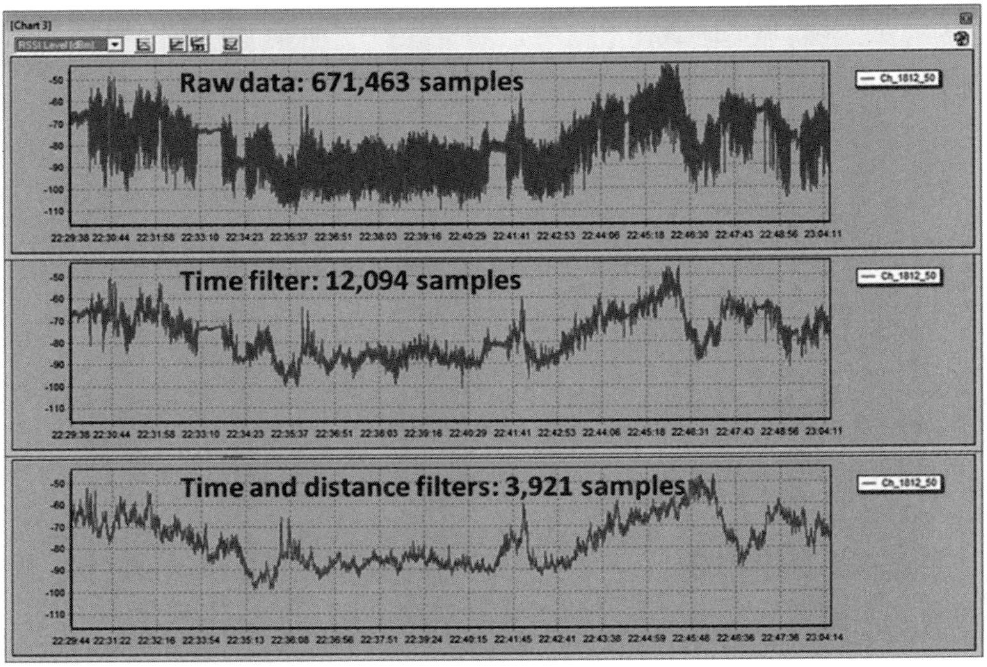

Figure 19.11 Drive test measurement collection with raw, time and distance averaging.

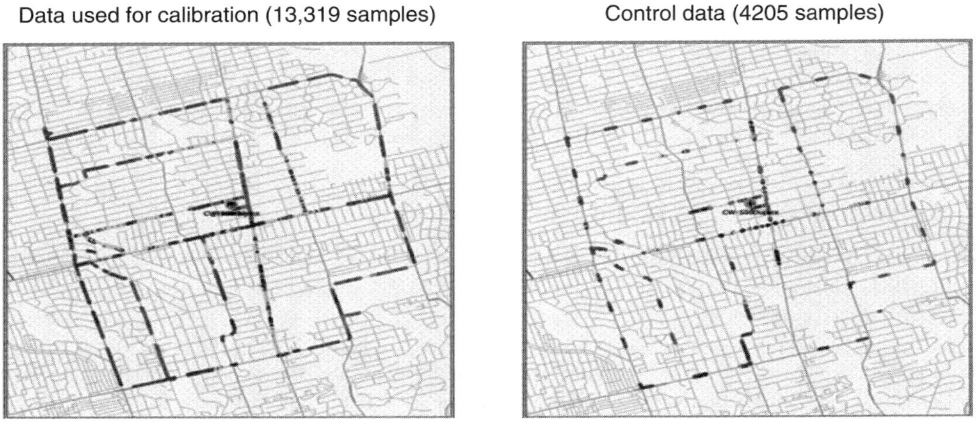

Figure 19.12 Measurement split into a calibration and a control lot.

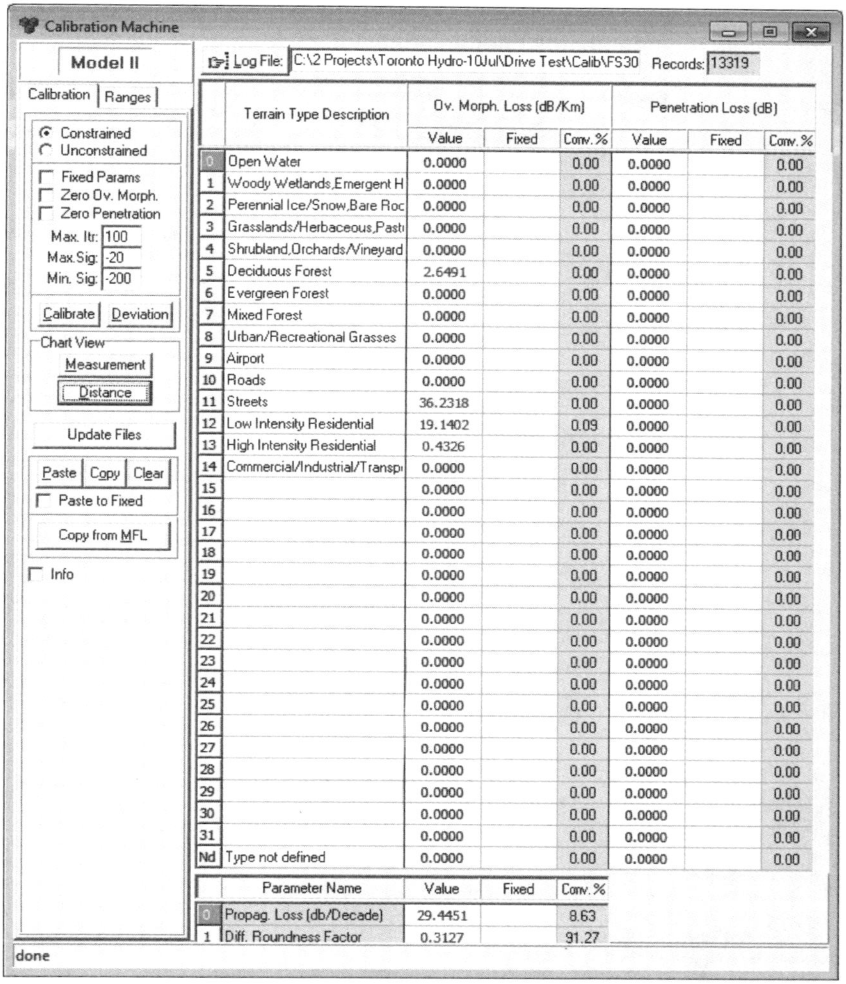

Figure 19.13 Calibration results using a constrained parameters approach.

19.3.1 Calibrate for Different Propagation Models

Database quality influences the result that propagation models provide. More advanced models provide best results but require better databases. When databases are not very accurate, less elaborate models may provide better results as they were conceived to work with less detailed terrain databases.

Ideally, the calibration should be done for several models and the one that produces the best results should be used. The models to be tested may vary from an area to another or even on a site-by-site basis.

19.3.2 Define Propagation Models and Parameters for Different Site Types

Also, depending on the selected model, some propagation parameters cannot be obtained from measurements and have to be estimated by the designer. At this stage, previous experience plays an important role in the design process.

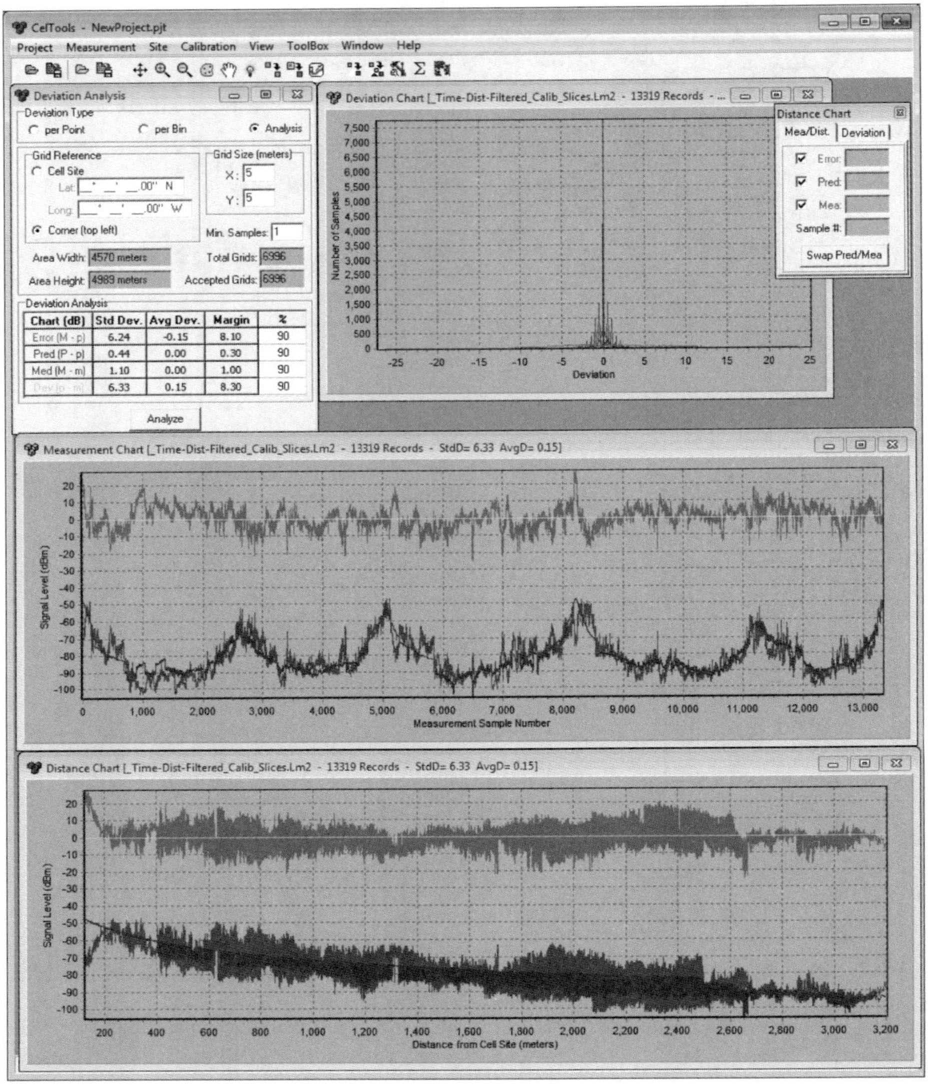

Figure 19.14 Propagation model calibration results.

19.4 Site Location

19.4.1 Simplified Site Distribution

Depending on the size of the AoI, greenfield designs may require distribution of 1000 sites or more. The sites must provide signal and traffic coverage, and usually follow an approximately regular distribution to minimize interference, use available and candidate sites as often as possible, and choose locations in pre-defined morphologies (for e.g., avoid sites in forests and water). For designs like this, planning

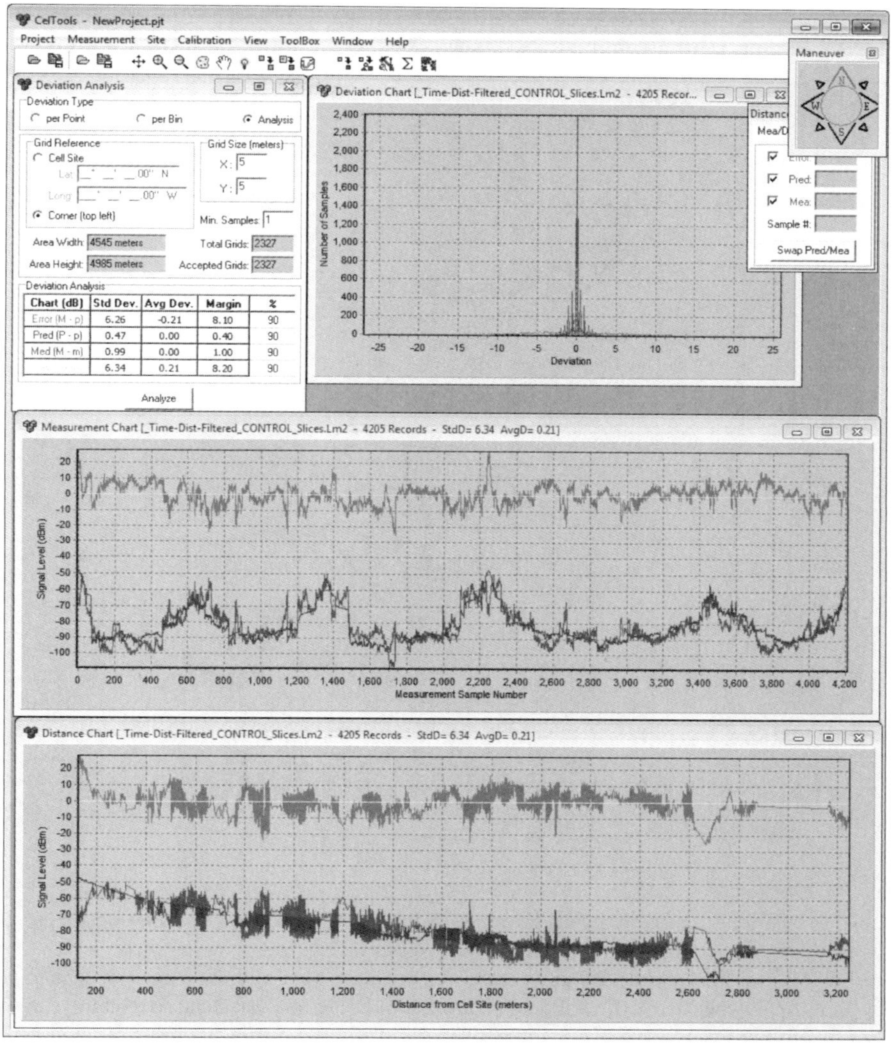

Figure 19.15 Control lot results applying the calibrated model.

tools sometimes provide a feature that allows designers to automatically populate an initial layout of sites. An example of such a feature can be seen in the dialog of Figure 19.16.

19.4.2 Advanced Cell Selection Procedure

In many cases, however, a more advanced procedure is required for site selection, in which, besides considering existing sites as candidates, it is also important to consider traffic distribution, backhaul feasibility, actual site coverage, costs, and available budget to select sites for the design. Basically,

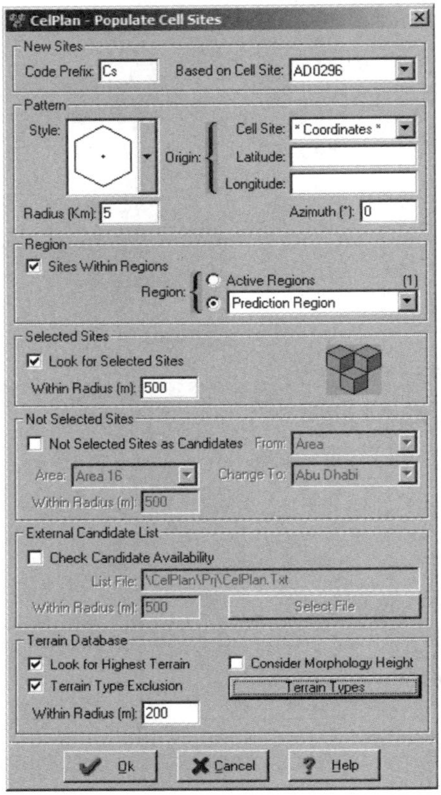

Figure 19.16 Populate cell sites dialogue.

the idea is for the tool to select the best coverage at the lowest cost. An example of this type of tool is given in Figure 19.17 that illustrates the software CelSelect.

For such tools, designers must initially obtain (or prepare) a list of sites with user-defined parameters such as deployment cost, fiber Point of Presence (PoP), and backhaul availability, as shown in Figure 19.18. Designers must also define coverage thresholds, create traffic layers, and configure deployment costs (Figure 19.19).

In this type of site selection process, site desirability is inversely proportional to:

- Cost of installation.
- Backhaul cost.
- Excess traffic above capacity.
- Coverage overlap with other cells above a desired target.

And site desirability is directly proportional to:

- Traffic coverage increment, limited to maximum site capacity.

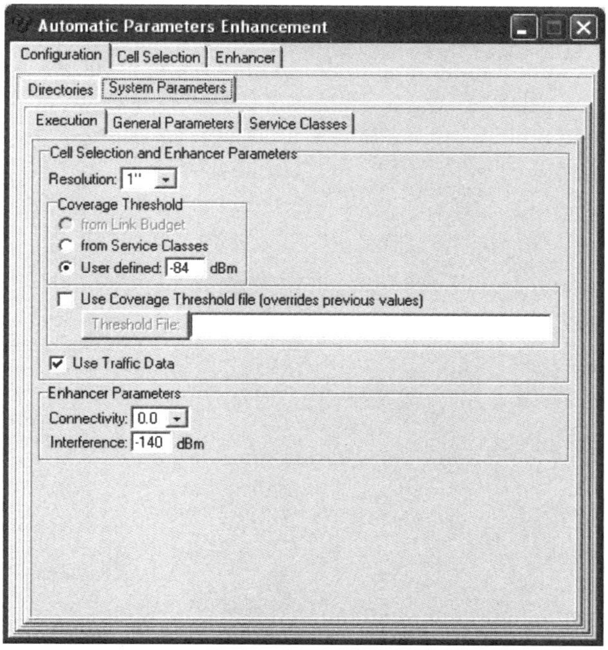

Figure 19.17 Parameters for automatic cell selection dialogue.

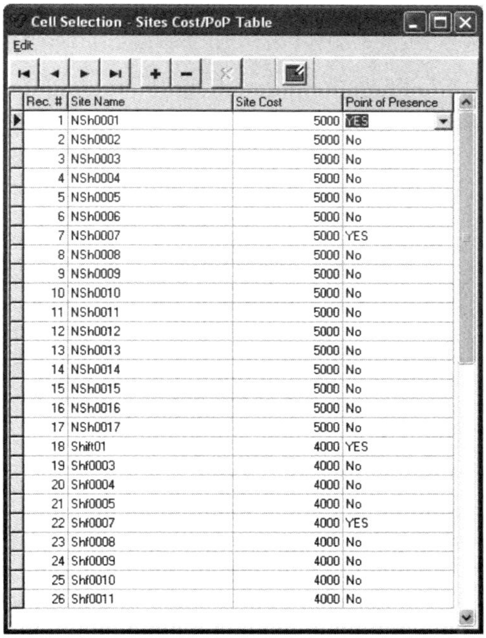

Cell Selection - Sites Cost/PoP Table

Edit

Rec. #	Site Name	Site Cost	Point of Presence
1	NSh0001	5000	YES
2	NSh0002	5000	No
3	NSh0003	5000	No
4	NSh0004	5000	No
5	NSh0005	5000	No
6	NSh0006	5000	No
7	NSh0007	5000	YES
8	NSh0008	5000	No
9	NSh0009	5000	No
10	NSh0010	5000	No
11	NSh0011	5000	No
12	NSh0012	5000	No
13	NSh0013	5000	No
14	NSh0014	5000	No
15	NSh0015	5000	No
16	NSh0016	5000	No
17	NSh0017	5000	No
18	Shift01	4000	YES
19	Shf0003	4000	No
20	Shf0004	4000	No
21	Shf0005	4000	No
22	Shf0007	4000	YES
23	Shf0008	4000	No
24	Shf0009	4000	No
25	Shf0010	4000	No
26	Shf0011	4000	No

Figure 19.18 Site cost table.

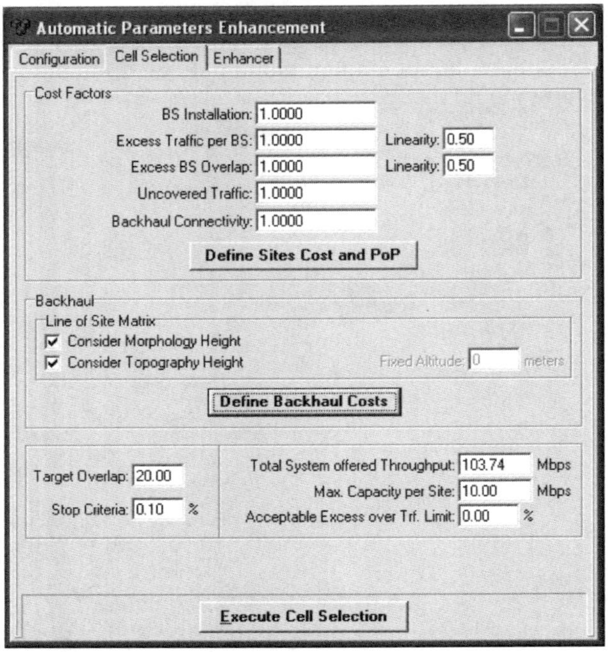

Figure 19.19 Cost parameters for automatic cell selection dialogue.

A linearity factor can be specified for excess traffic or excess overlap as illustrated in Figure 19.20. Backhaul availability is performed using Line of Sight (LOS) between all candidates and costs are associated to the backhaul based on PoP or number of links availability. A sample backhaul cost table is shown in Figure 19.21. The tool used in this example displays, as a result, the selected cells, in order of acceptance, with associated scores and absolute cost, as shown in Figure 19.22.

As an example of the application of such a tool, in Figure 19.23 we have 730 candidate sites in the Boston area, with 465 candidates inside the AoI. The offered throughput is 25 Gbps with maximum capacity per site of 10 Mbps. The example assumed the backhaul costs of Figure 19.21 and BS cost of $ 45,000.00. The LOS study is illustrated in Figure 19.24.

The stop criterion was 99% of maximum possible traffic using all candidates. Results are shown in Figure 19.25, where only 435 sites were selected from the original 730 sites.

Considering a monthly subscriber fee of $35.00, a monthly CAPEX return per subscriber of $8.7 and a time to recover the investment of 12 months, the results shows that, to be economically feasible, a BS has to cover a minimum of 429 users. After running this analysis, the study flagged 10 of the BSs as non-economical.

19.5 Run Initial Site Predictions

Once site locations have been selected, it is possible to calculate coverage predictions for each site and perform a traffic simulation to validate the simplified RF assumptions of the site selection tool.

Site traffic load is required for resource assignment (i.e. frequency planning), but, at the same time, interference affects the site traffic load. To solve this issue, a two step process should be undertaken:

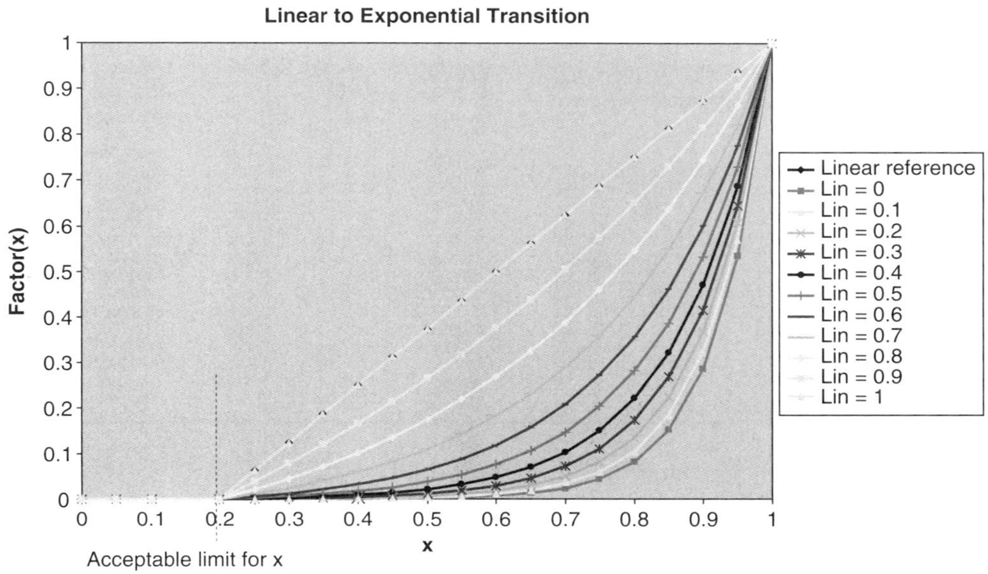

Figure 19.20 Site desirability curve.

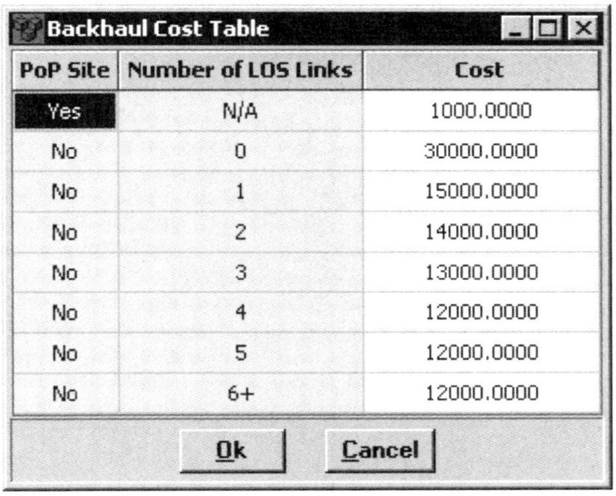

PoP Site	Number of LOS Links	Cost
Yes	N/A	1000.0000
No	0	30000.0000
No	1	15000.0000
No	2	14000.0000
No	3	13000.0000
No	4	12000.0000
No	5	12000.0000
No	6+	12000.0000

Figure 19.21 Backhaul cost table example.

(1) the traffic load should be calculated assuming an average interference value, then resources can be assigned to each site; and (2) once resources are assigned, an actual interference analysis can be done and the resulting traffic load can be verified.

This implies that for the initial traffic simulations, there is no need to calculate interference, noise rise can be used as an interference assumption. The use of this approach allows designers to run

Log Information

12:40:50 PM [719]:	Cell Selection results:						
12:40:50 PM [720]:	Site	IncrTraf	PercIncr	TotalATraf	TotalLimTraf	Score	AbsCost
12:40:50 PM [721]:	NSh0004	32.8723	0.0964	32.8723	32.8634	152.1462	5000.0000
12:40:50 PM [722]:	NSh0007	29.3086	0.0860	62.1810	62.1721	204.7179	6000.0000
12:40:50 PM [723]:	NSh0001	26.9194	0.0790	89.1003	89.0914	222.8878	6000.0000
12:40:50 PM [724]:	Shf0007	12.4246	0.0364	101.5249	101.5160	402.4283	5000.0000
12:40:50 PM [725]:	NSh0015	29.3490	0.0861	130.8739	130.8650	340.7274	10000.0000
12:40:50 PM [726]:	NSh0017	29.0633	0.0852	159.9372	159.9283	309.6689	9000.0000
12:40:50 PM [727]:	NSh0010	23.6924	0.0695	183.6296	183.6207	354.5434	8400.0000
12:40:50 PM [728]:	NSh0014	23.5450	0.0691	207.1747	207.1658	339.7744	8000.0000
12:40:50 PM [729]:	NSh0005	23.2007	0.0681	230.3754	230.3665	362.0578	8400.0000
12:40:50 PM [730]:	NSh0013	23.5380	0.0690	253.9134	253.9045	424.8442	10000.0000
12:40:50 PM [731]:	NSh0008	18.2909	0.0537	272.2043	272.1954	459.2441	8400.0000
12:40:50 PM [732]:	NSh0016	16.5058	0.0484	288.7102	288.7013	508.9108	8400.0000
12:40:50 PM [733]:	NSh0003	12.7183	0.0373	301.4285	301.4196	660.4646	8400.0000
12:40:50 PM [734]:	NSh0002	11.4320	0.0335	312.8605	312.8516	734.7820	8400.0000
12:40:50 PM [735]:	Shf0005	8.2900	0.0243	321.1505	321.1416	892.6372	7400.0000
12:40:50 PM [736]:	NSh0009	11.9791	0.0351	333.1296	333.1207	834.7867	10000.0000
12:40:50 PM [737]:	NSh0012	6.0939	0.0179	339.2235	339.2146	1312.7825	8000.0000
12:40:50 PM [738]:	NSh0006	0.8791	0.0026	340.1026	340.0937	9100.2986	8000.0000
12:40:50 PM [739]:	NSh0011	0.8226	0.0024	340.9252	340.9163	9725.2373	8000.0000

Figure 19.22 Site selection and ordering.

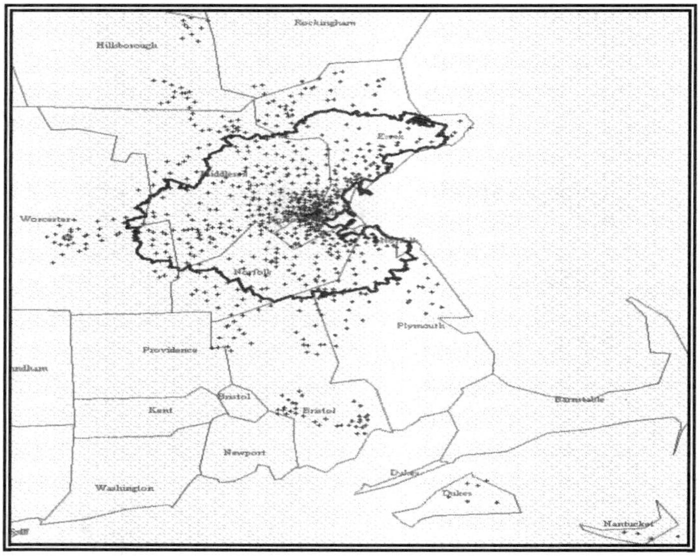

Figure 19.23 Sites and area of interest.

initial site predictions only for the estimated coverage area of the cell, so the prediction radius can be significantly reduced, thus expediting the whole design process.

Path loss predictions should be performed for the different user heights considered in the design. This means that path loss values should be stored for each sector and each prediction height specified in the Service Classes (SC).

Figures 19.26 to 19.31 display site predictions first just for one sector of a site and then as a composite of three sectors for different user heights and locations.

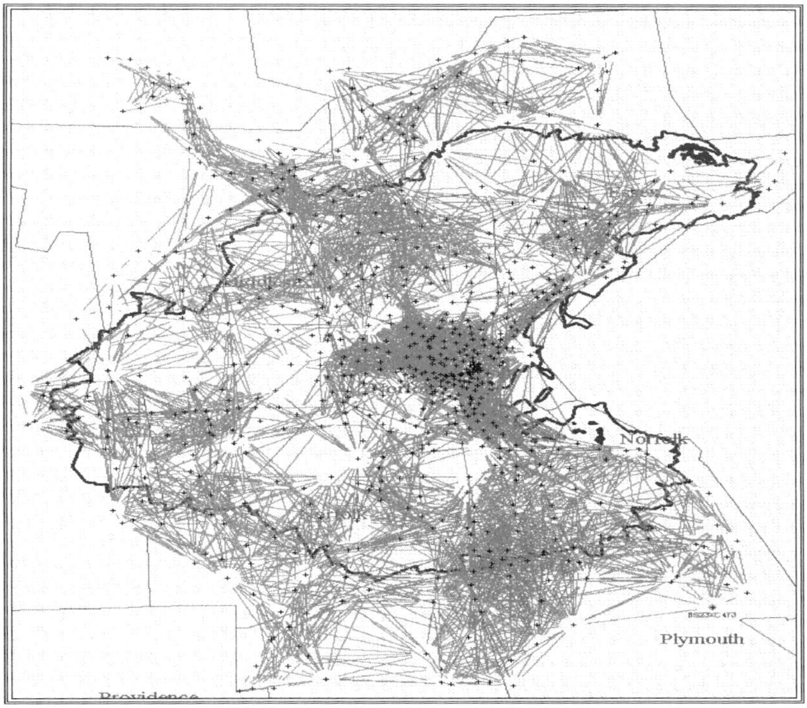

Figure 19.24 Line of sight study.

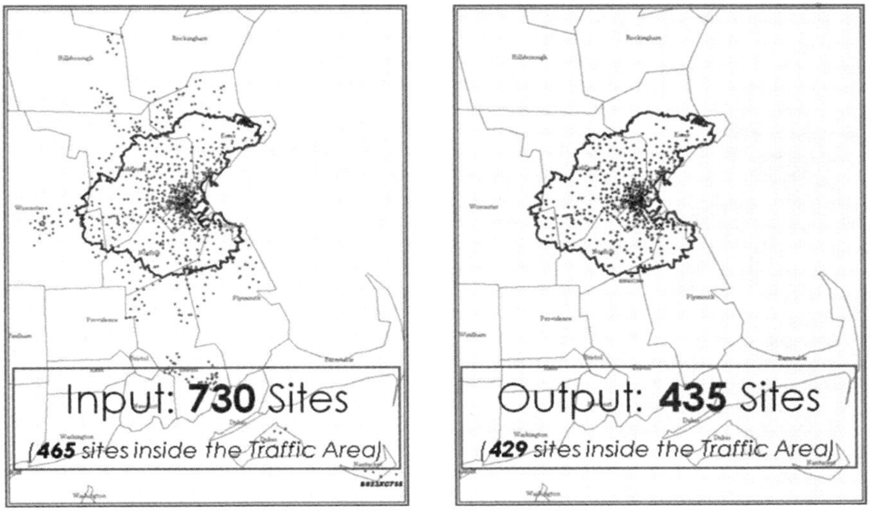

Figure 19.25 Original and selected sites.

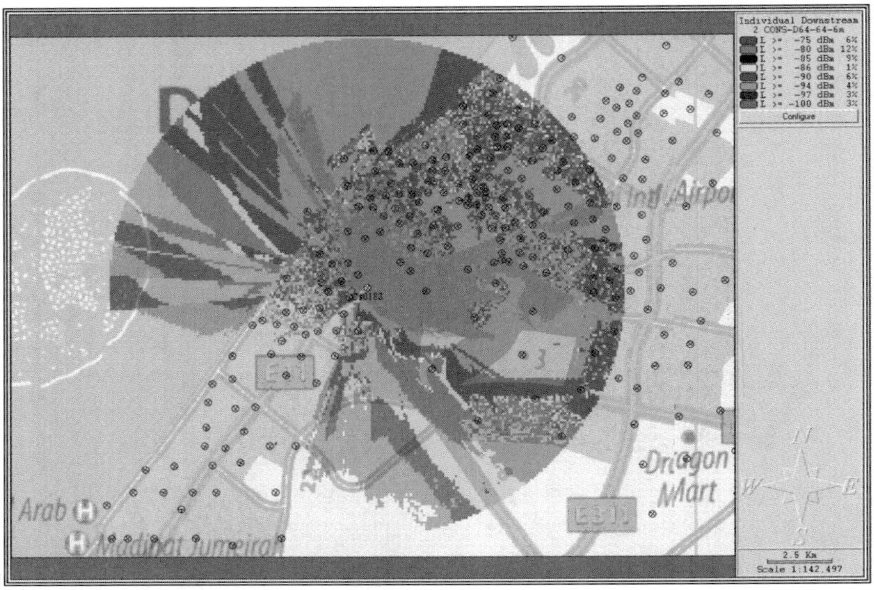

Figure 19.26 RSSI for a single sector at ground level outdoor.

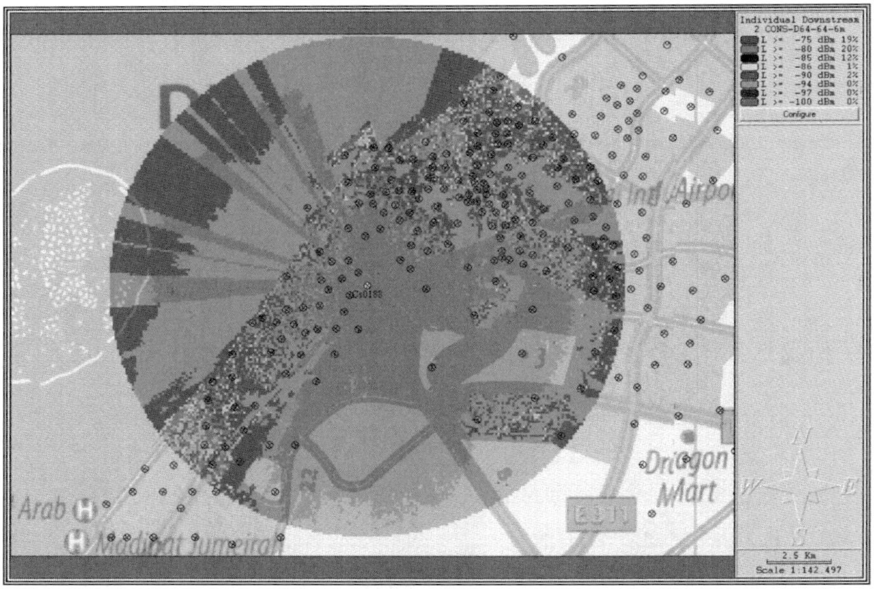

Figure 19.27 RSSI composite for all sectors at 6 m rooftop.

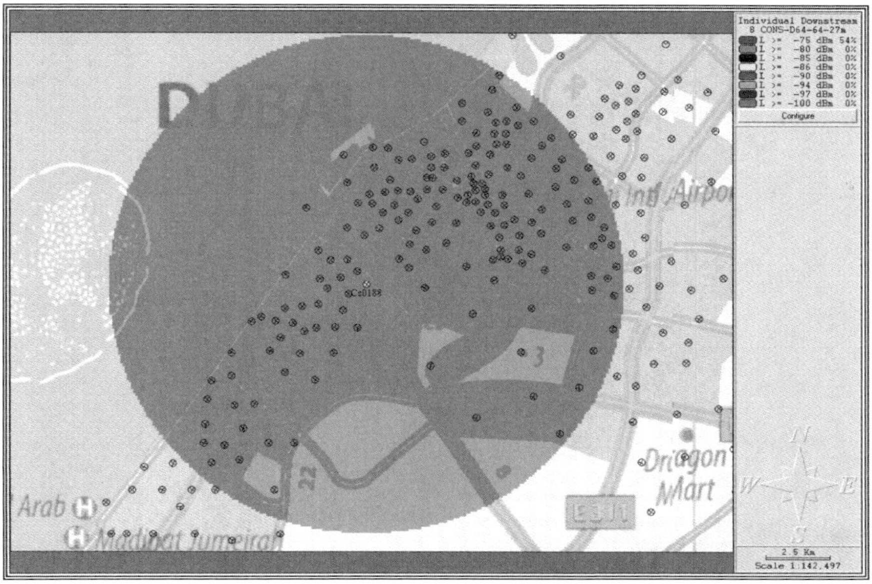

Figure 19.28 RSSI composite for all sectors at 27 m rooftop.

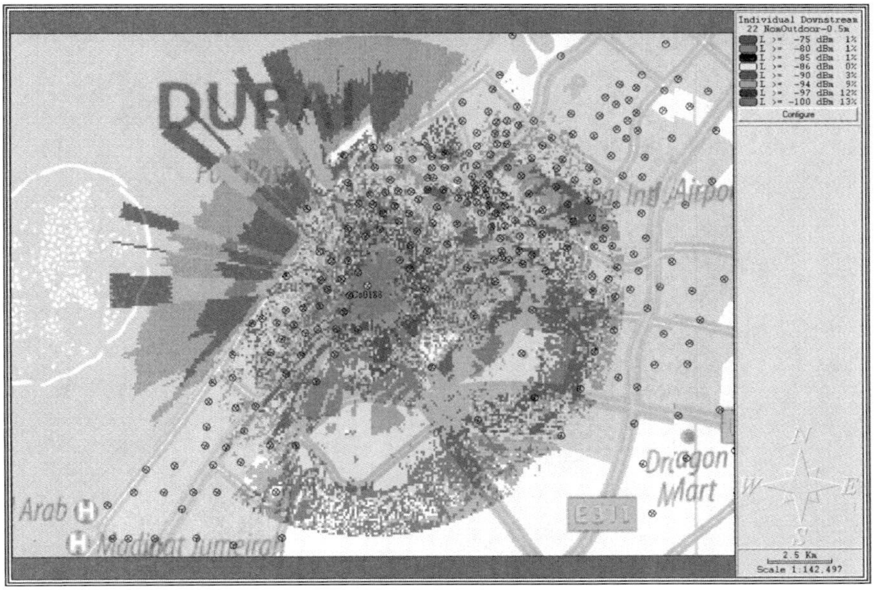

Figure 19.29 RSSI composite for all sectors at 0.5 m outdoor.

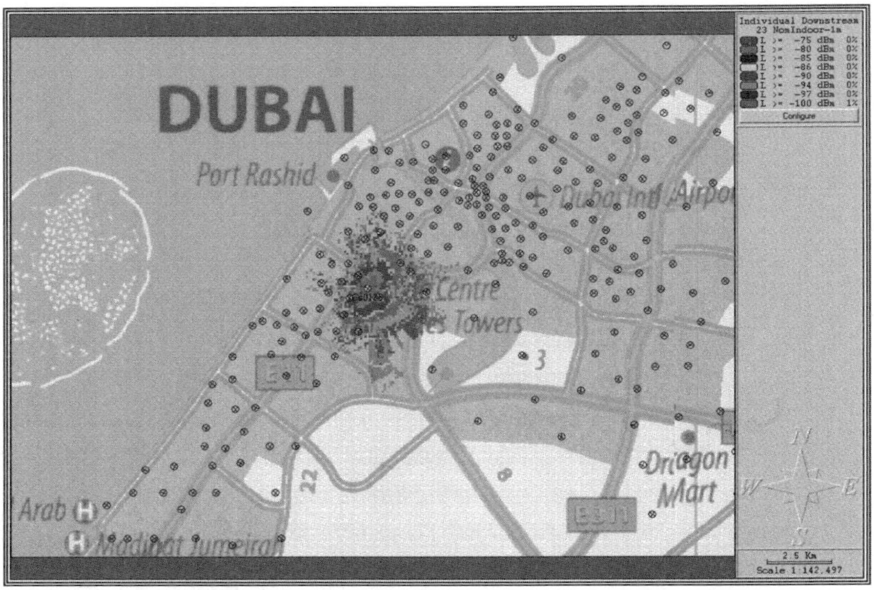

Figure 19.30 RSSI composite for all sectors at 1 m indoor.

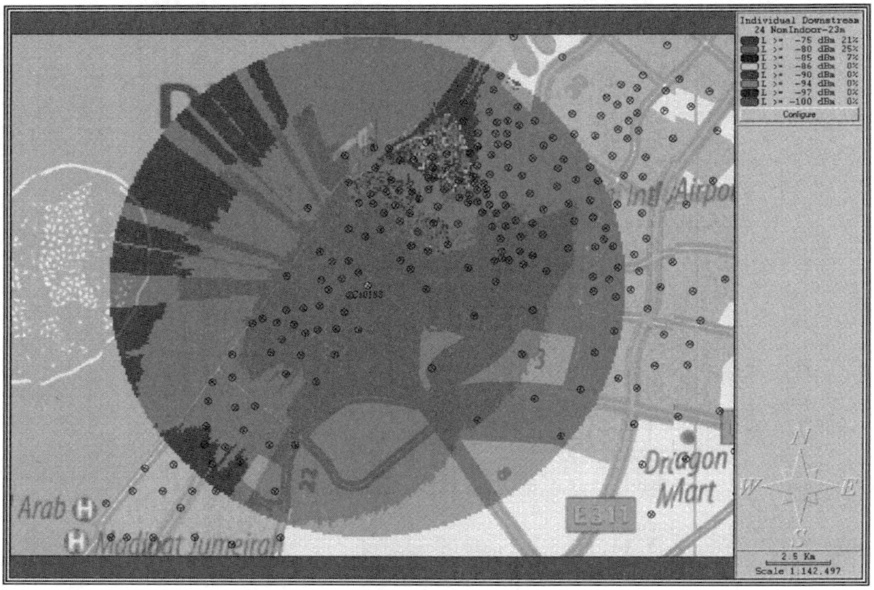

Figure 19.31 RSSI composite for all sectors at 23 m indoor.

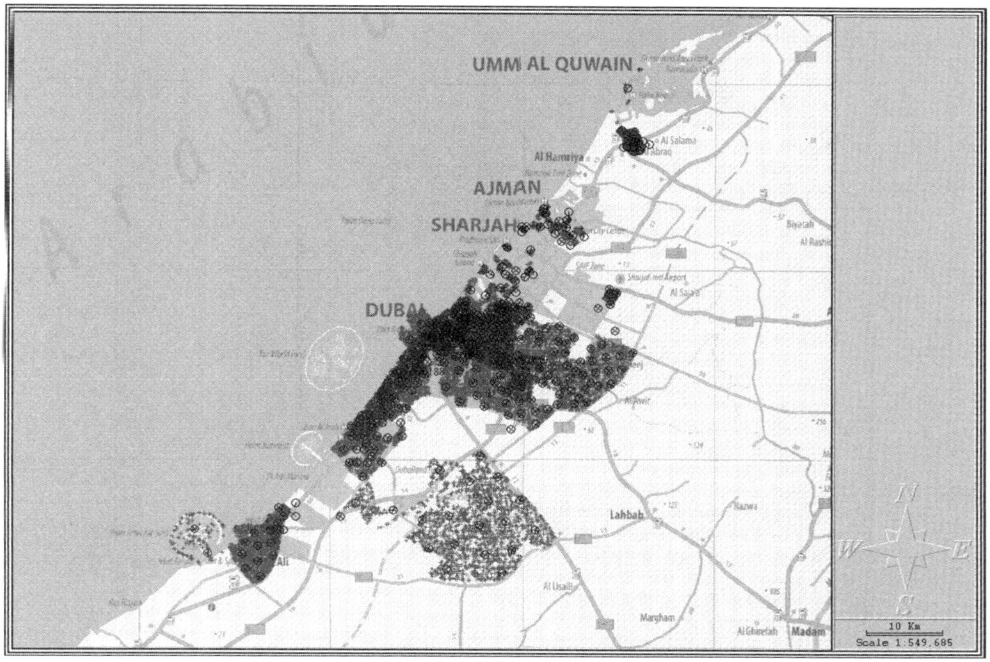

Figure 19.32 Traffic simulation (each session type is represented by the legend color).

19.6 Static Traffic Simulation

The type of simulation described here is static in terms of interference but is dynamic in terms of time, as results of one snapshot are carried on to the next.

19.6.1 Define Target Noise Rise Per Area

The traffic carried by a cell depends on the interference it suffers, while the interference depends on the traffic carried. As explained in Section 19.5, this chicken or egg problem can be solved by running the first traffic simulation with a constant Noise Rise figure that corresponds to the average interference expected at each site of the network. Ideally, Noise Rise values should be estimated by area.

The estimation of Noise Rise may require experimentation by the designer. The closer the designer estimates the actual Noise Rise value, fewer iterations will be required in the design.

19.6.2 Static Traffic Simulation

Once predictions are done, the static simulation can be performed. This simulation is called static because it does not consider the interference between the users in the network assuming just a flat increase in the noise floor (Noise Rise). Trafficwise the simulation is dynamic as it considers the traffic evolution from one snapshot to the next.

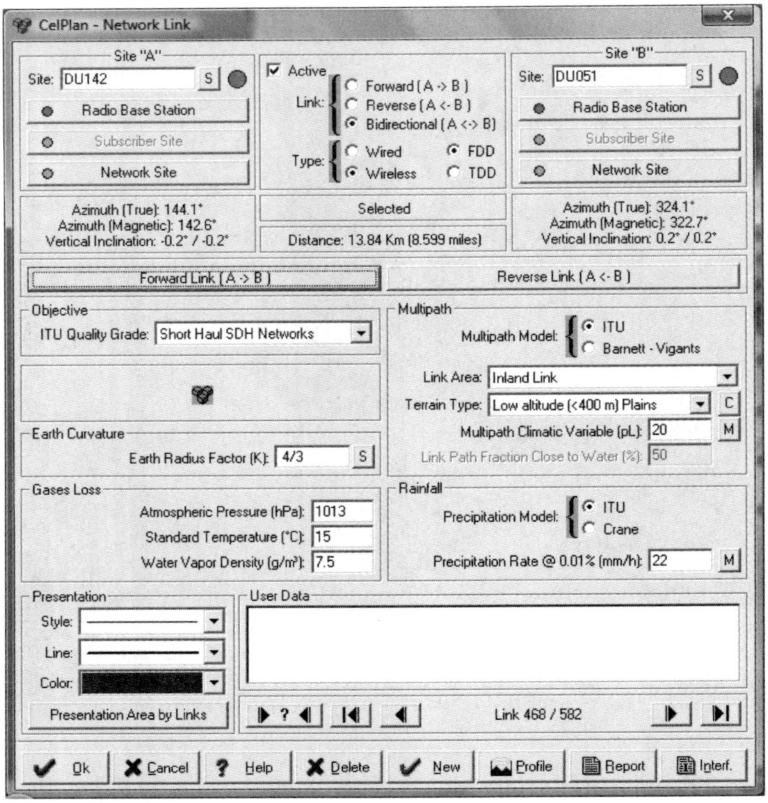

Figure 19.33 Microwave link configuration.

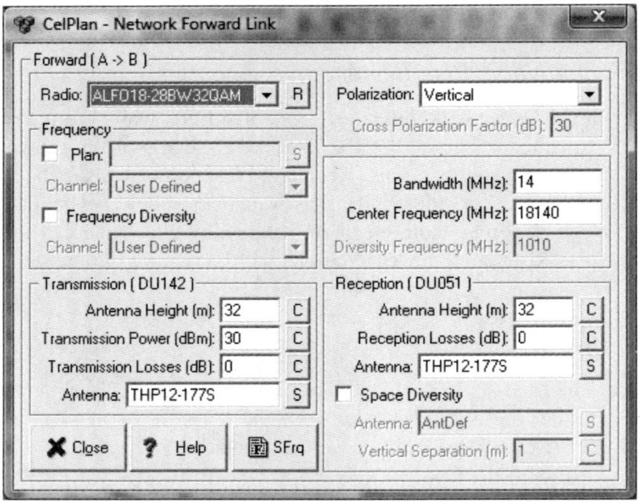

Figure 19.34 Forward link configuration.

The traffic simulation approach described here has the following steps:

- An instantaneous snapshot is made: Network users of each Service Class have their status statistically established based on the Service description and geographically located based on the traffic distribution of the Service Class.
- Users are allocated to cells and an inner loop is done, in which power is adjusted against interference. When convergence is reached, the snapshot statistics are stored.
- Another snapshot is done and the process is repeated (outer loop) until convergence of the snapshot statistics is reached.
- A report with the accumulated statistics is generated.

A more detailed description of the simulation process is given in the description of the Dynamic Simulation (Section 21.1).

Figure 19.32 shows an example of a static traffic simulation, with different SC session being established.

19.7 Adjust Design for Area and Traffic Coverage

The traffic report generated after a simulation should provide the traffic load and queued traffic for each cell. Sites can then be re-distributed to balance loads, removed when the load is too light, or added in congested areas.

19.8 Configure Backhaul Links and Perform Backhaul Predictions

Once the sites are defined, the backhaul network can be designed. Initially, links are established to connect sites and then configured according to their location and the requirements of the connection, as illustrated in Figures 19.33 to 19.35.

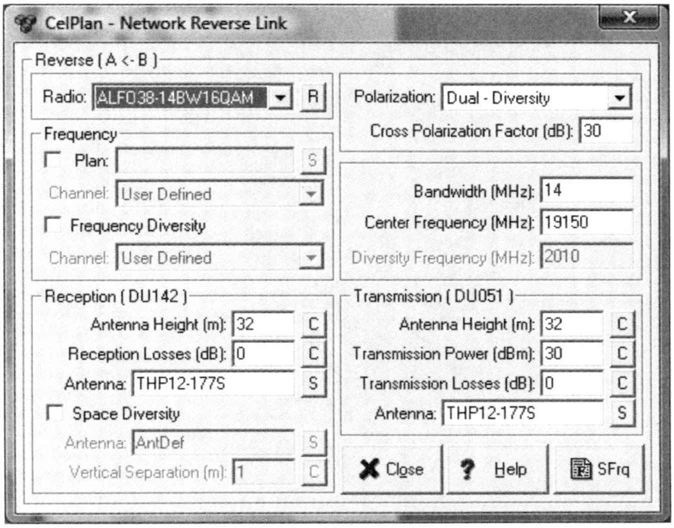

Figure 19.35 Reverse link configuration.

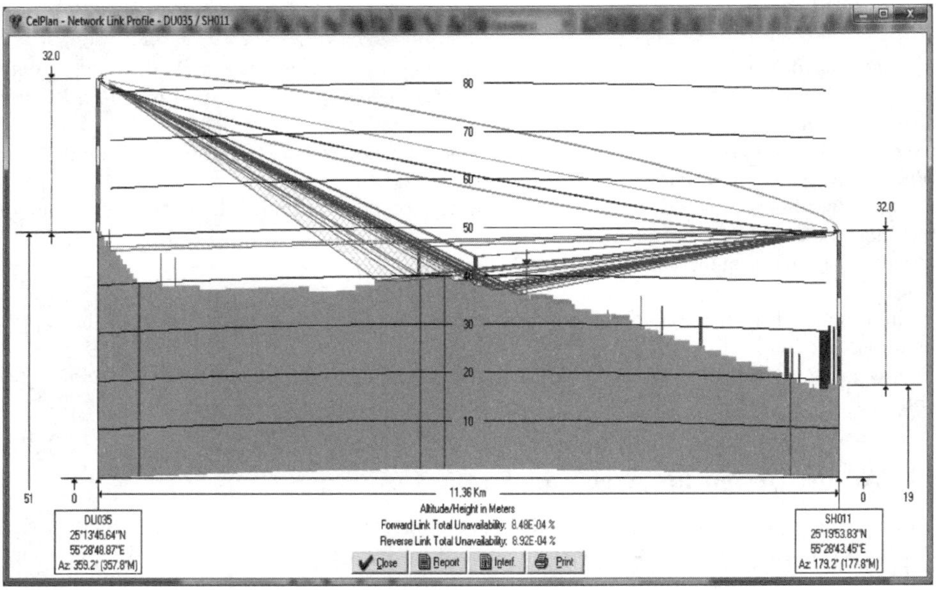

Figure 19.36 Link analysis profile.

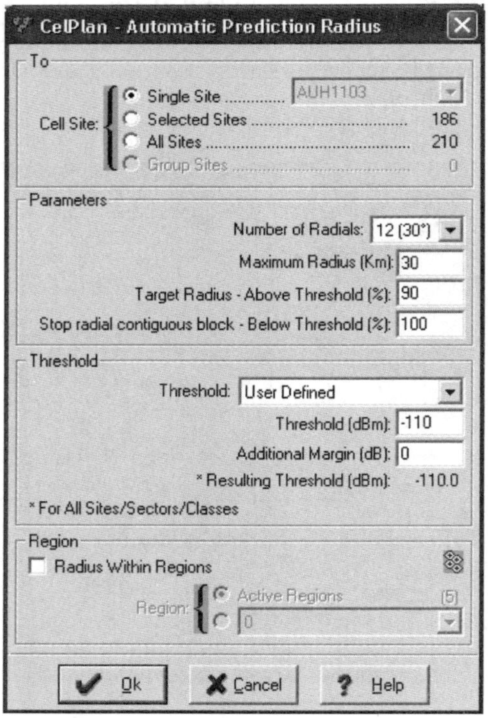

Figure 19.37 Automatic prediction radius calculation.

Microwave links are analyzed by calculating their RF performance. Each link profile is analyzed (Figure 19.36) for Fresnel zone obstructions and potential ground reflections.

19.9 Perform Signal Level Predictions with Extended Radius

After the redesign due to issues found during the first traffic simulation (e.g. sites eliminated/added), the path loss predictions should be rerun but, this time, with a larger radius to include areas where sites do not provide coverage but may act as interferers. The design tool should be able to suggest the best prediction radius for this analysis. Figure 19.37 illustrates an example of a prediction radius calculator.

20

Wireless Network Optimization

The first step in optimizing a network is to optimize the sites' footprint to minimize interference while assuring enough overlap for handovers. The next step is to generate a Neighbor list and calculate handover thresholds, if required. Then an Interference Matrix (IM) should be generated, based on signal outage. Finally, resources can be optimally assigned considering the relationships of the IM.

20.1 Cell Enhancement or Footprint Optimization

Wireless systems are implemented with overlap between cells to provide mobility from one cell to another and to avoid coverage "holes"; however, overlapping increases interference and, consequently, may diminish network capacity.

Achieving the proper amount of overlap while simultaneously minimizing interference and maintaining the coverage area is a very daunting task, but it may significantly increase network capacity. Footprint optimization tools, also known as automatic cell planners (ACP), automatically shape cell footprints to enhance network performance by minimizing interference (considering diversity effects) while maintaining coverage, and balancing traffic distribution. In this section, we use the tool CelEnhancer as an example.

This enhancement of the cell footprint is done by selecting the best antenna parameters within limits specified by the designer. The antenna parameters that can usually be adjusted are:

- antenna type
- antenna azimuth
- antenna downtilt
- antenna height
- transmit power

The network can be optimized for all these parameters or just for some of them as chosen by the designer.

Because the optimization process is not limited to the physical characteristics of cell sites, the tool should be designed to accommodate different optimization strategies. Enhancement should be performed at least twice in the network process: once before the resources are optimized and again when they are, that is, after the frequency plan has been created.

LTE, WiMAX and WLAN Network Design, Optimization and Performance Analysis, First Edition.
Leonhard Korowajczuk.

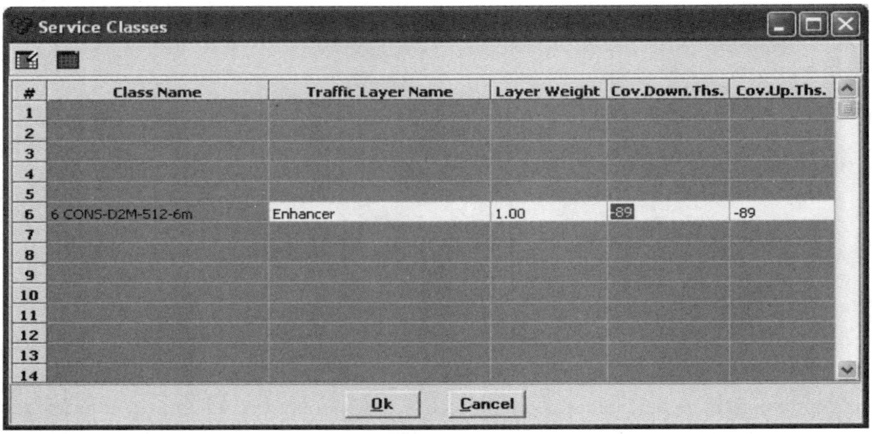

Figure 20.1 Service class and traffic configuration for enhancement purposes.

The first step in the enhancement process is to determine the target (Service Class) of the optimization, as shown in Figure 20.1.

For cell footprint enhancement, users should be able to define what the goal of the optimization is, for example, traffic x interference, as, most of the times, there is a trade-off between these parameters.

The tool used in this example, CelEnhancer, uses an Improvement and Deterioration Table to allow designers to weight the importance of coverage, traffic, and interference in the network. The use of weights allows the tool to calculate scores and grades for each of the possible changes and thus select the best one. Figure 20.2 shows a typical configuration.

According to the configuration presented in Figure 20.2, improvement (reduction) of interference is ten times more important than an improvement (increase) in the coverage area; whereas deterioration (increase) of interference is four times worse than deterioration (reduction) in the coverage area. The way in which priorities are given might vary from one tool to another but it is important that designers have the capability to decide what the goal of the enhancement is.

The table in the example also relates each parameter to itself; that is, avoiding deterioration in the coverage is five times more important than trying to improve the coverage. To give some flexibility for the tool, a small percentage of reduction of the total coverage area (1%) was allowed but no increase of interference was accepted.

Because changes in the cell footprint affect the traffic carried by each sector, designers should also be able to define what kind of traffic variation they are willing to accept for a sector. Limiting traffic variation avoids footprint changes that could drastically affect the current traffic pattern of the system, by establishing a maximum increase/decrease of the traffic currently carried by each sector.

The design of this example considers a heavily loaded network that has its capacity limited by interference and not coverage, in this case, a variation of 100% of best server traffic is acceptable, meaning that cell footprints may change the best service area of the site significantly if this benefits the overall interference level of the system.

The last thing to be configured by designers is the list of enhancement parameters, that is, which changes should be tested for which sectors. In the tool used in this example, these parameters are stored in a table where each line represents one change to be tested for a given sector; therefore the same sector may appear many times in the table to test different parameters.

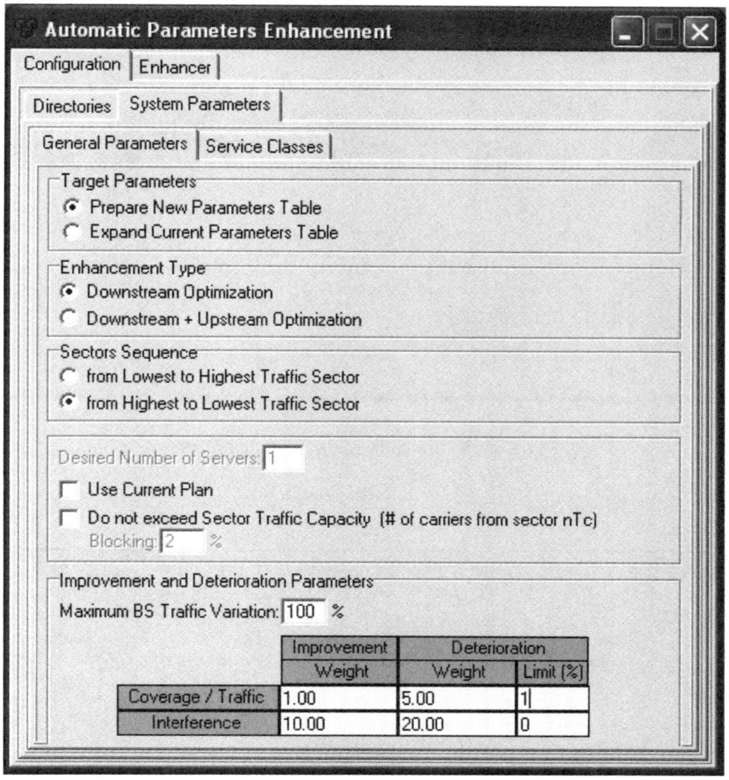

Figure 20.2 Enhancement parameters.

In the example of Figure 20.3 the parameters selected for optimization were:

- Downtilt – from 0 to 7 degrees (in increments of 1 degree).
- Azimuth – a variation of up to 30 degrees of the original sector azimuth (in increments of 5 degrees).

Figure 20.4 shows the log window of the tool used in the example. This window is displayed while the enhancement process is running and shows the scores and results obtained for each of the parameters tested.

The selection method for the modification to be made may vary from tool to tool but it will always involve some type of cost function or score system, which allows the tool to judge which changes will most benefit the system as a whole. In this figure, the tool shows the parameter being tested, the possible values, and the score obtained in each case.

Based on the score and traffic information, the tool calculates a grade for each possible value. The grade represents how much each specific change would affect the overall performance of the network; thus the current value always has grade 0. The suggested value is selected by choosing the highest grade.

CelEnhancer Parameters Table

Edit

Rec. #	Site Ident.	Sector #	Parameter	Enabled	Priority	Antenna Type List	Range From	Range to	Step	Current Value	Suggested Value
1	AJM2001	1	Tilt	Yes	1		0	7	1	1.00	1.00
2	BTS0911	4	Azimuth	Yes	1		290	320	5	335.00	335.00
3	AJM2001	3	Tilt	Yes	1		0	7	1	7.00	7.00
4	AJM2001	4	Tilt	Yes	1		0	7	1	0.00	0.00
5	AJM2007	1	Tilt	Yes	1		0	7	1	2.00	2.00
6	AJM2007	2	Tilt	Yes	1		0	7	1	0.00	0.00
7	AJM2007	3	Tilt	Yes	1		0	7	1	0.00	0.00
8	AJM2008	1	Tilt	Yes	1		0	7	1	7.00	7.00
9	AJM2008	2	Tilt	Yes	1		0	7	1	6.00	6.00
10	AJM2008	3	Tilt	Yes	1		0	7	1	7.00	7.00

Figure 20.3 Enhancement parameters table.

[Parameter 1 of 3048] [Site: BTS0886 Sector: 3 Priority= 1 Traffic= 1.00 Erl Class= 1 CONS-VOIP-6m] --> Tilt
Current Value: 4.00 Range: 0.00 1.00 2.00 3.00 5.00 6.00 7.00
Suggested Value= 4.00

Parameter	Score	SystemTrf(BS_Radius)	BS_Traffic	Grade
4.00	911632.57	1650.69	64.58	0
0.00	919666.56	1650.69	60.74	-0.176255
1.00	915163.65	1650.69	64.79	-0.077467
2.00	912840.13	1650.69	64.79	-0.026492
3.00	912333.23	1650.69	60.74	-0.015371
5.00	912311.29	1650.69	56.19	-0.01489
6.00	913986.33	1650.69	48.29	-0.051638
7.00	915961.31	1650.69	39.89	-0.094967

[Parameter 2 of 3048] [Site: BTS0886 Sector: 3 Priority= 1 Traffic= 1.00 Erl Class= 1 CONS-VOIP-6m] --> Azimuth
Current Value: 245.00 Range: 200.00 205.00 210.00 215.00 220.00 225.00 230.00
Suggested Value= 245.00

Parameter	Score	SystemTrf(BS_Radius)	BS_Traffic	Grade
245.00	911632.57	1650.69	64.58	0
200.00	934387.4	1650.69	50.04	-0.49921
205.00	931185.19	1650.69	50.04	-0.428958
210.00	928813.51	1650.69	50.74	-0.376927
215.00	926064.92	1650.69	54.59	-0.316626
220.00	923251.66	1650.69	55.29	-0.254907
225.00	920883.83	1650.69	55.29	-0.20296
230.00	918051.37	1650.69	59.84	-0.14082

Figure 20.4 Sample log window of enhancement process.

Some of the possible values are discarded because they violate the required enhancement objective (e.g. Interference Increment Exceeded).

After the enhancement is completed, the suggested values must be transferred to the project for generation of the frequency plan and calculation of the KPI (Key Performance Indicator).

20.2 Resource Optimization

20.2.1 Neighbor List

The neighbor list is very important for a proper network operation, as it will define handover options. The neighbor list should consider topological (or natural) neighbors, as indicated in Figure 20.5, as well as RF neighbors.

The calculation of RF neighbors is usually done by calculating a Neighborhood Matrix (NM). The design tool must be able to generate the NM and rate the neighbors by traffic. In the tool used in this example, neighbors are established by the amount of traffic being overlapped. This is shown in Figure 20.6. Finally, both lists should be mixed, and neighbors limited to the maximum number supported by the AP hardware. The final neighbors are declared on a sector basis as shown in Figure 20.7.

20.2.2 Handover Thresholds

For systems that support handover, thresholds must be calculated for each sector based on the signal level in the border between two cells using the NM. This is shown in Figure 20.8.

20.2.3 Paging Groups

Paging areas should be defined by the operator and the design tool should automatically allocate sites to these groups, with border sites belonging to more than one group.

20.2.4 Interference Matrix for Downstream and Upstream for All PSC

Once signal level predictions have been calculated for each enhanced site, it is possible to calculate the interaction between sites and, based on this, optimize the frequency plan. Chapter 6 described the concept of the outage matrix that stores this interaction.

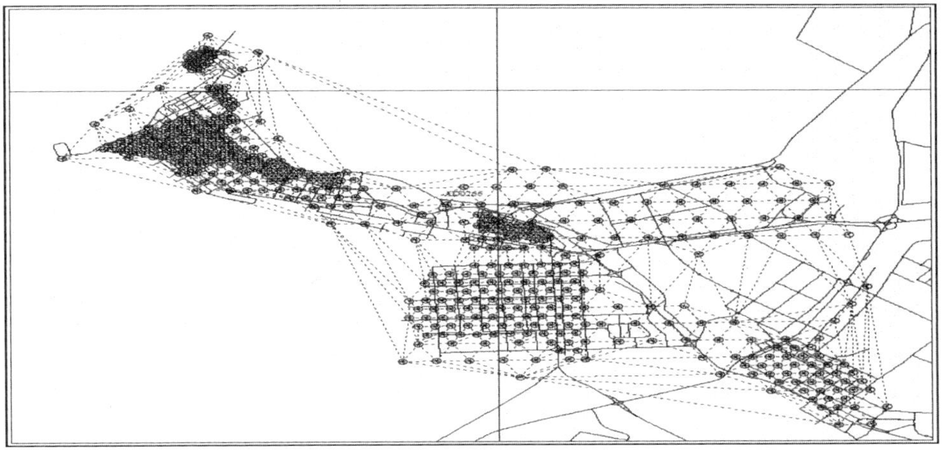

Figure 20.5 Natural neighbors.

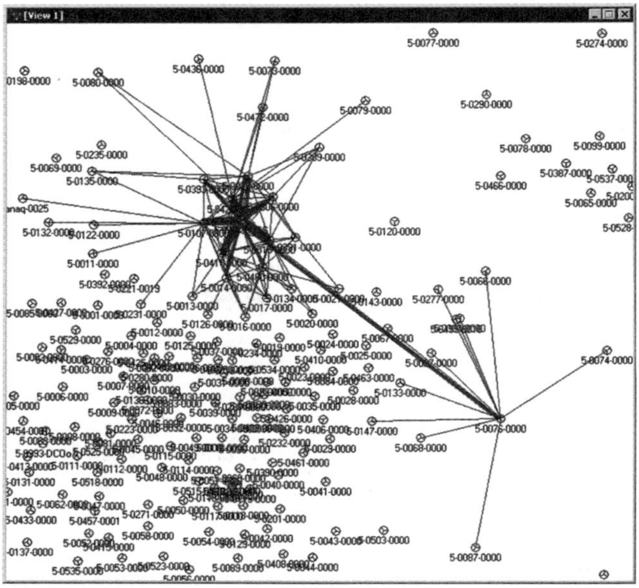

Red-Primary Neighbors

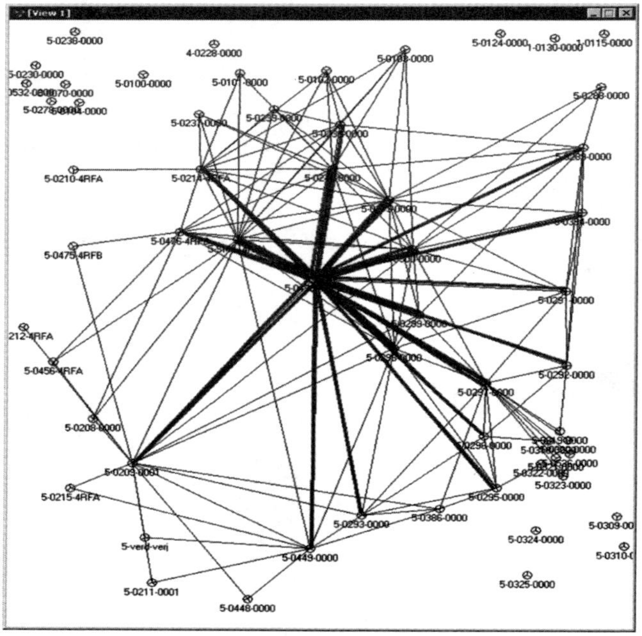

Blue-Secondary Neighbors

Figure 20.6 Interference neighbors from the interference matrix.

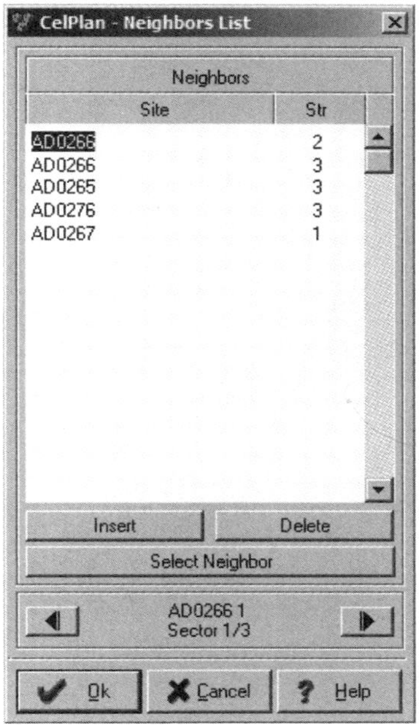

Figure 20.7 Neighbor list for a specific sector.

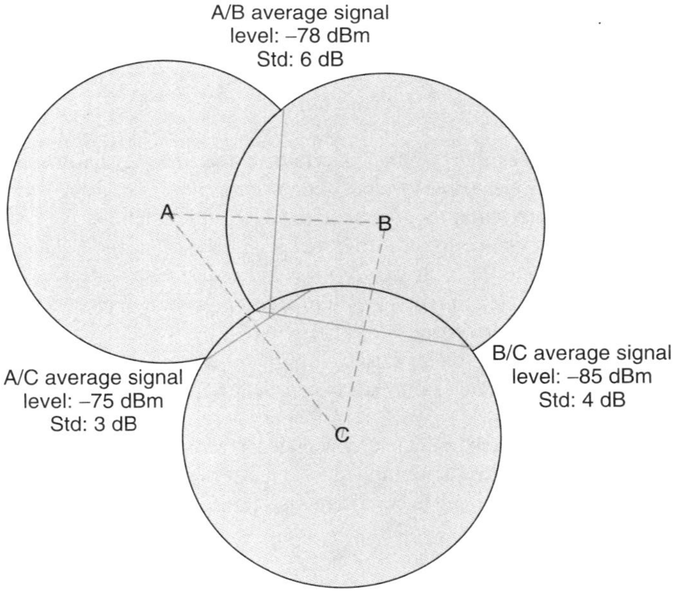

Figure 20.8 Handover threshold calculation algorithm per neighbor.

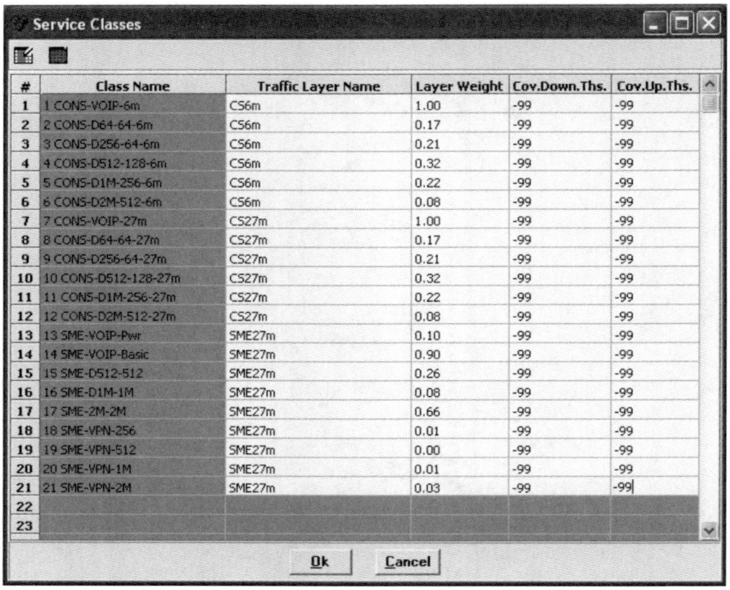

Service Classes

#	Class Name	Traffic Layer Name	Layer Weight	Cov.Down.Ths.	Cov.Up.Ths.
1	1 CONS-VOIP-6m	CS6m	1.00	-99	-99
2	2 CONS-D64-64-6m	CS6m	0.17	-99	-99
3	3 CONS-D256-64-6m	CS6m	0.21	-99	-99
4	4 CONS-D512-128-6m	CS6m	0.32	-99	-99
5	5 CONS-D1M-256-6m	CS6m	0.22	-99	-99
6	6 CONS-D2M-512-6m	CS6m	0.08	-99	-99
7	7 CONS-VOIP-27m	CS27m	1.00	-99	-99
8	8 CONS-D64-64-27m	CS27m	0.17	-99	-99
9	9 CONS-D256-64-27m	CS27m	0.21	-99	-99
10	10 CONS-D512-128-27m	CS27m	0.32	-99	-99
11	11 CONS-D1M-256-27m	CS27m	0.22	-99	-99
12	12 CONS-D2M-512-27m	CS27m	0.08	-99	-99
13	13 SME-VOIP-Pwr	SME27m	0.10	-99	-99
14	14 SME-VOIP-Basic	SME27m	0.90	-99	-99
15	15 SME-D512-512	SME27m	0.26	-99	-99
16	16 SME-D1M-1M	SME27m	0.08	-99	-99
17	17 SME-2M-2M	SME27m	0.66	-99	-99
18	18 SME-VPN-256	SME27m	0.01	-99	-99
19	19 SME-VPN-512	SME27m	0.00	-99	-99
20	20 SME-VPN-1M	SME27m	0.01	-99	-99
21	21 SME-VPN-2M	SME27m	0.03	-99	-99
22					
23					

Ok Cancel

Figure 20.9 Service class and traffic configuration for optimization purposes.

Figures 20.9 and 20.10 show examples of, respectively, how the minimum signal coverage threshold can be defined for each service class, and other general configuration parameters for the creation of the interference matrix.

20.2.5 Interference Matrix

The interference matrix expresses the potential interference that the signals from (downstream) and to (upstream) one sector controller have to other sector controllers. The best way to represent this interference is by an outage in relation to a pre-specified QoS, multiplied by the traffic affected. This operation should be repeated for all pixels in the area of interest and the values accumulated in a matrix. The calculations are done for each service class because of their different traffic patterns.

Other methods of weighting interference rely on arbitrary ratios, like cost figures, and fail to represent the real effect of the interference.

Once an interference matrix has been generated, it provides a good indication of potential interference between any pair of sectors. This information can then be used to optimize the distribution of resources for each sector.

In the example, the outage matrix calculates outages for all sector pairs, which are assumed to use the same resource, and the outage is multiplied by the affected traffic. Expressing interference by traffic outage allows us to add interference contributions.

20.2.5.1 Pixel Outage

As explained in Section 6.3.3.2, the tool used as an example classifies signals at each pixel in three categories: best server, server/interferer, and noise. The pixel traffic is divided between the best server

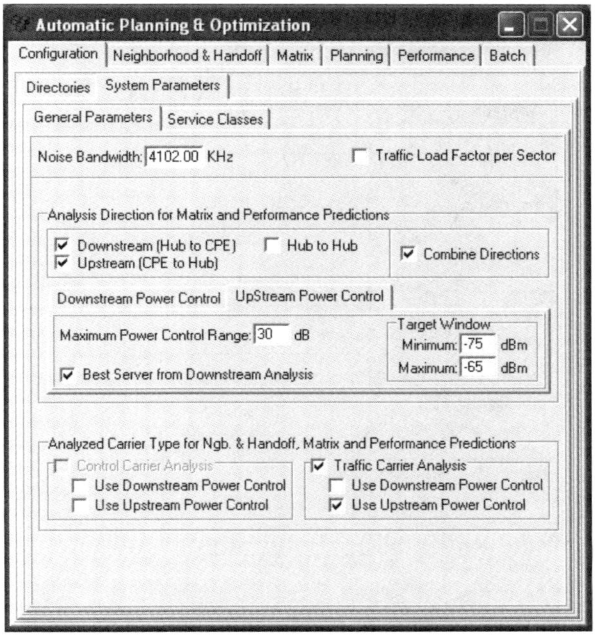

Figure 20.10 General parameter configuration for the optimization process.

and the other servers according to a mobility factor. The server traffic is divided between servers proportionally to their signal power. This is illustrated in Figure 20.11.

The traffic outage of each server interferer pair is calculated using the outage table (Figure 20.12) for co-channel, the adjacent channel, and the second adjacent channel. In this tool, each server is compared to the interference from the other servers and interferers and the resultant traffic outage is stored in the appropriate bin of the Interference Matrix (IM). The IM maps all sectors against all sectors as illustrated in Figure 20.13.

Several interference matrices can be created, three for each Service Class, one for the co-channel, one for the adjacent channel, and one for the second adjacent channel, as illustrated in Figure 20.14.

Interference matrices can be manipulated, multiplied, and scaled according to the hour of the day or other constraints. The combined matrixes can then be used to simulate different scenarios.

The Interference Matrix dialog is shown in Figure 20.15. The resulting interference matrix table is displayed in Figure 20.16 (traffic is expressed in mili-users). The interference relationship can be visualized also graphically, as shown in Figure 20.17, the colors in the matrix follow the coding scheme indicated in Figure 20.16.

20.2.6 Automatic Code Planning (Segmentation, CellID and PermBase)

The Interference Matrix should also be used as an input to the code planning process. WiMAX has the following codes, described in Chapter 13, that have to be planned:

- CellID: 32 codes for sector cells and 18 codes for omni cells.
- Segments: 3 segments for sector cells.
- PermBase: 32 codes in the downlink and 68 codes in the uplink.

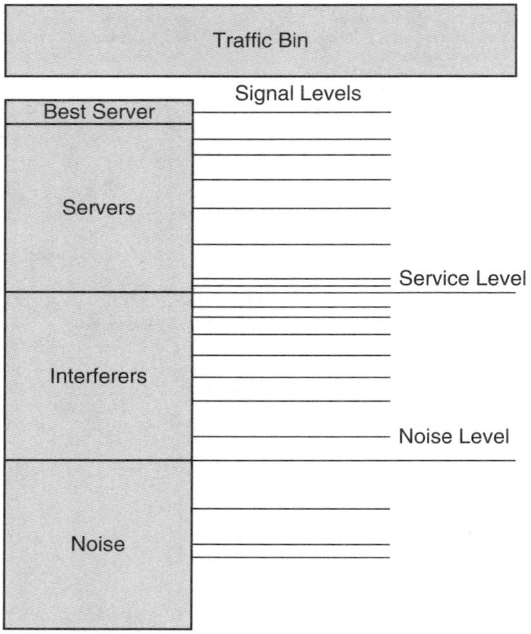

Figure 20.11 Pixel outage calculation.

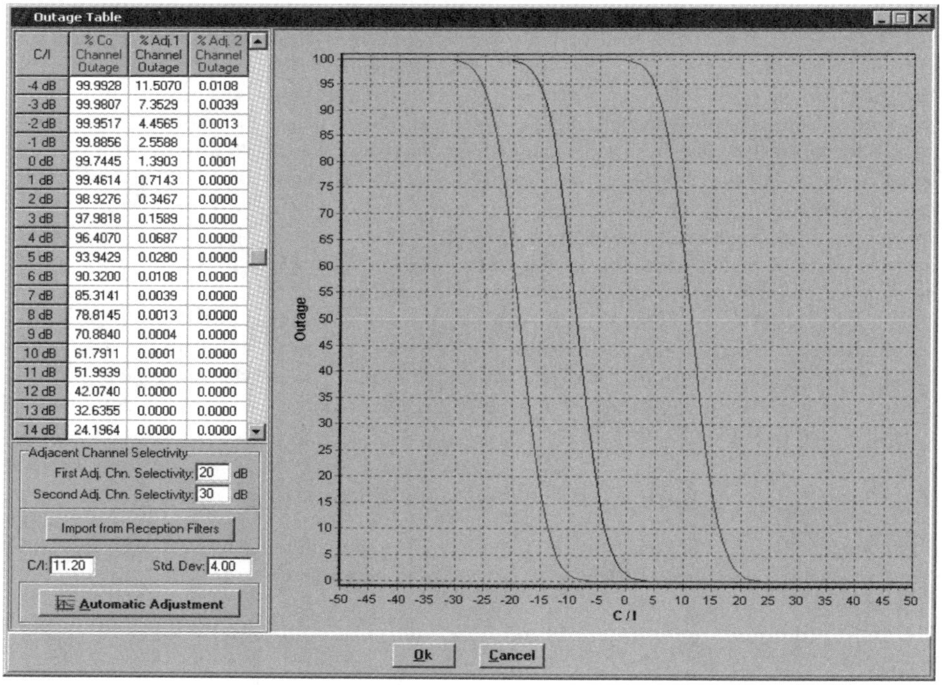

Figure 20.12 Outage table.

	S1	S2	S3	:	Sn
S1	$O_{1,1}$	$O_{1,2}$	$O_{1,3}$	:	$O_{1,n}$
S2	$O_{2,1}$	$O_{2,2}$	$O_{2,3}$	:	$O_{2,n}$
S3	$O_{3,1}$	$O_{3,2}$	$O_{3,3}$	:	$O_{3,n}$
:	:	:	:	:	:
Sn	$O_{n,1}$	$O_{n,2}$	$O_{n,3}$	:	$O_{n,n}$

Figure 20.13 Interference matrix.

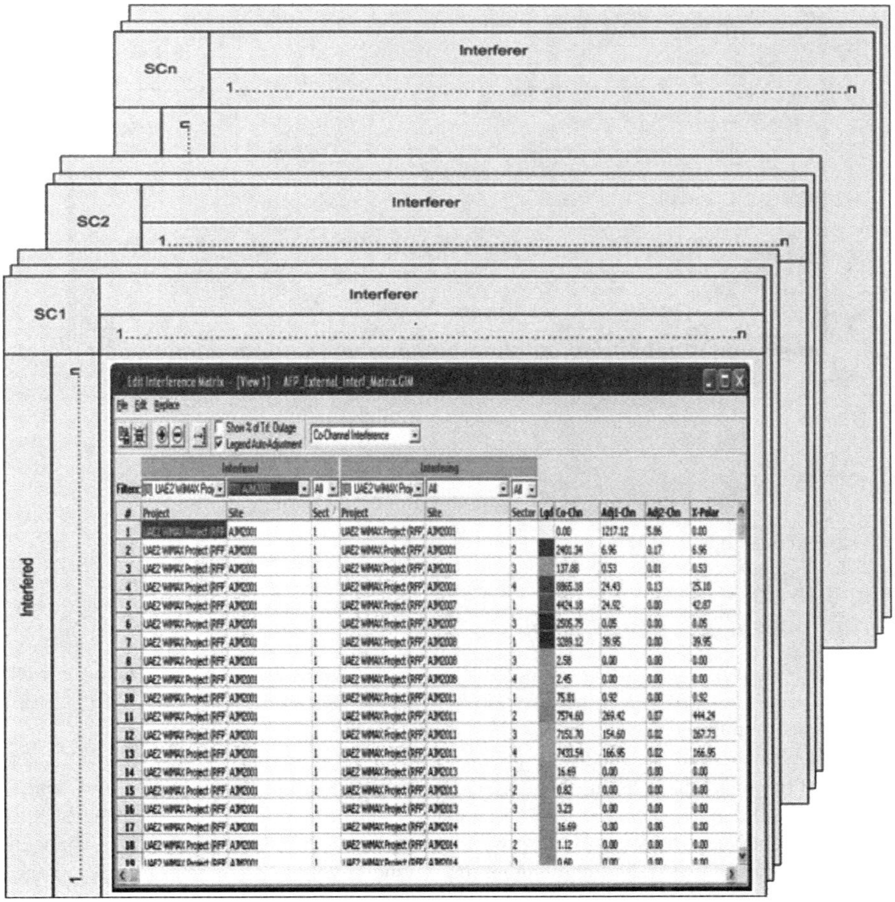

Figure 20.14 Set of interference matrixes.

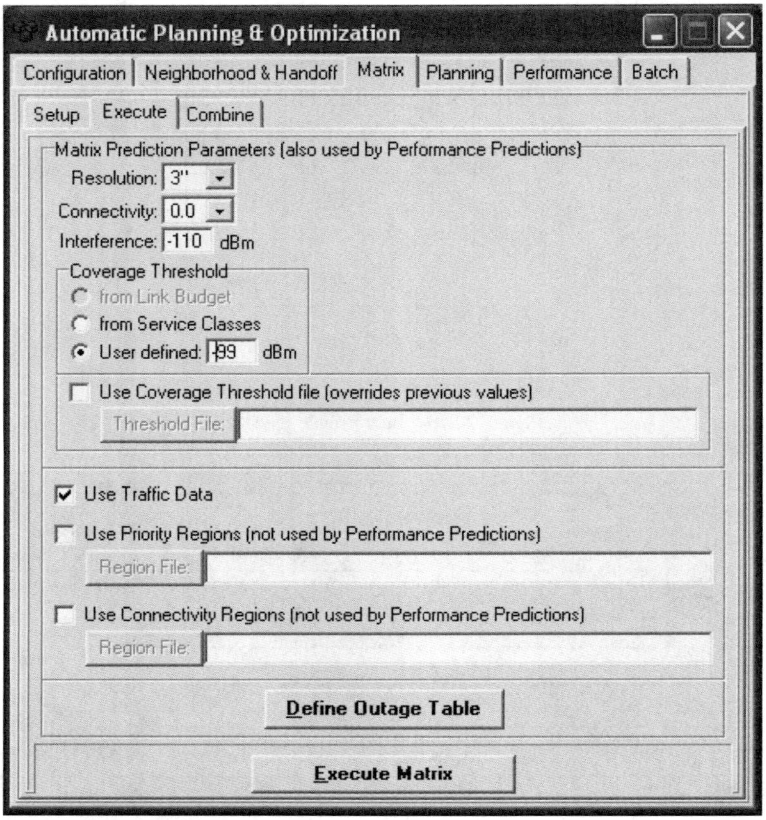

Figure 20.15 CelOptima matrix configuration screenshot.

LTE has the following codes:

- CellID: 168 codes of 3 CellIDs.

20.2.7 Automatic Carrier Planning

The first step in frequency planning is to establish the QoS criteria for the plan. In the tool used as an example, this criterion is given as a relationship between CNIR and traffic outage. This is done using the relationships established in Figure 20.12.

The next step is to define the channels available for planning. A sample channel table is illustrated in Figure 20.18. Prohibited channels (in shading) and adjacency relationships should be also represented; in the figure, prohibited channels are marked in shading. The figure also shows gaps in the spectrum as not available channels (NAV/NAH). If polarization is being planned as well as the channels (e.g. 802.16d fixed networks), the table might show the channels broken up for each polarization.

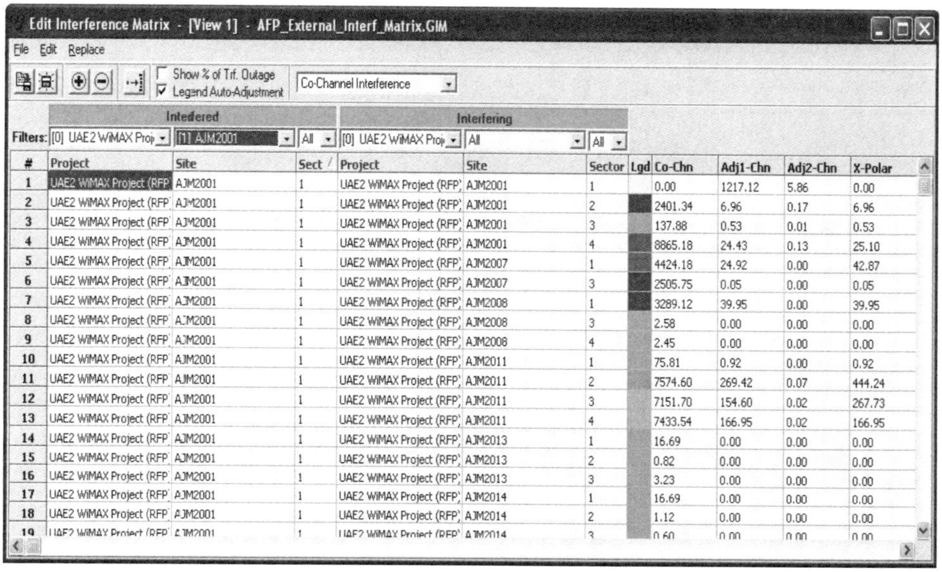

#	Project	Site	Sect	Project	Site	Sector	Lgd	Co-Chn	Adj1-Chn	Adj2-Chn	X-Polar
1	UAE2 WiMAX Project (RFP	AJM2001	1	UAE2 WiMAX Project (RFP)	AJM2001	1		0.00	1217.12	5.86	0.00
2	UAE2 WiMAX Project (RFP	AJM2001	1	UAE2 WiMAX Project (RFP)	AJM2001	2		2401.34	6.96	0.17	6.96
3	UAE2 WiMAX Project (RFP	AJM2001	1	UAE2 WiMAX Project (RFP)	AJM2001	3		137.88	0.53	0.01	0.53
4	UAE2 WiMAX Project (RFP	AJM2001	1	UAE2 WiMAX Project (RFP)	AJM2001	4		8865.18	24.43	0.13	25.10
5	UAE2 WiMAX Project (RFP	AJM2001	1	UAE2 WiMAX Project (RFP)	AJM2007	1		4424.18	24.92	0.00	42.87
6	UAE2 WiMAX Project (RFP	AJM2001	1	UAE2 WiMAX Project (RFP)	AJM2007	3		2505.75	0.05	0.00	0.05
7	UAE2 WiMAX Project (RFP	AJM2001	1	UAE2 WiMAX Project (RFP)	AJM2008	1		3289.12	39.95	0.00	39.95
8	UAE2 WiMAX Project (RFP	AJM2001	1	UAE2 WiMAX Project (RFP)	AJM2008	3		2.58	0.00	0.00	0.00
9	UAE2 WiMAX Project (RFP	AJM2001	1	UAE2 WiMAX Project (RFP)	AJM2008	4		2.45	0.00	0.00	0.00
10	UAE2 WiMAX Project (RFP	AJM2001	1	UAE2 WiMAX Project (RFP)	AJM2011	1		75.81	0.92	0.00	0.92
11	UAE2 WiMAX Project (RFP	AJM2001	1	UAE2 WiMAX Project (RFP)	AJM2011	2		7574.60	269.42	0.07	444.24
12	UAE2 WiMAX Project (RFP	AJM2001	1	UAE2 WiMAX Project (RFP)	AJM2011	3		7151.70	154.60	0.02	267.73
13	UAE2 WiMAX Project (RFP	AJM2001	1	UAE2 WiMAX Project (RFP)	AJM2011	4		7433.54	166.95	0.02	166.95
14	UAE2 WiMAX Project (RFP	AJM2001	1	UAE2 WiMAX Project (RFP)	AJM2013	1		16.69	0.00	0.00	0.00
15	UAE2 WiMAX Project (RFP	AJM2001	1	UAE2 WiMAX Project (RFP)	AJM2013	2		0.82	0.00	0.00	0.00
16	UAE2 WiMAX Project (RFP	AJM2001	1	UAE2 WiMAX Project (RFP)	AJM2013	3		3.23	0.00	0.00	0.00
17	UAE2 WiMAX Project (RFP	AJM2001	1	UAE2 WiMAX Project (RFP)	AJM2014	1		16.69	0.00	0.00	0.00
18	UAE2 WiMAX Project (RFP	AJM2001	1	UAE2 WiMAX Project (RFP)	AJM2014	2		1.12	0.00	0.00	0.00
19	UAE2 WiMAX Project (RFP	AJM2001	1	UAE2 WiMAX Project (RFP)	AJM2014	3		0.60	0.00	0.00	0.00

Figure 20.16 Interference matrix table.

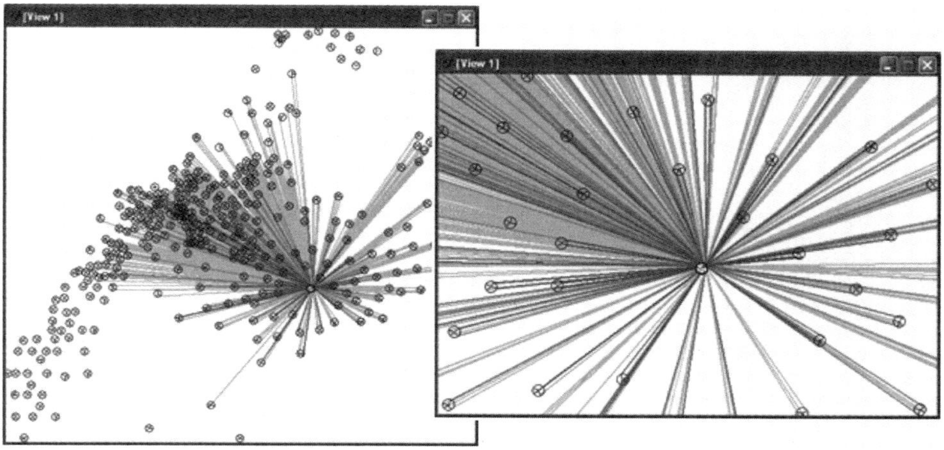

Figure 20.17 Interference matrix representation for a single site.

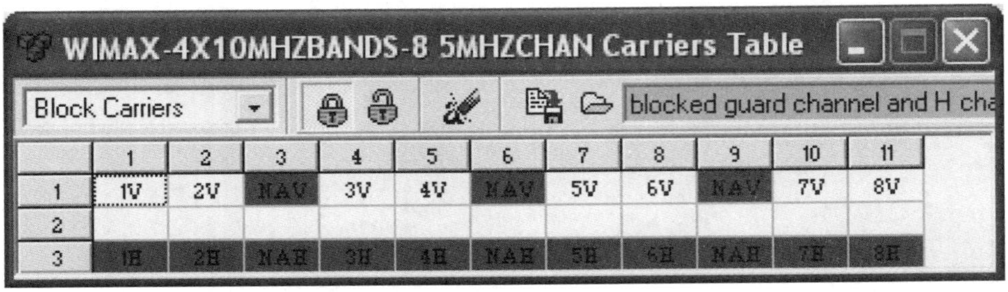

Figure 20.18 Channel table.

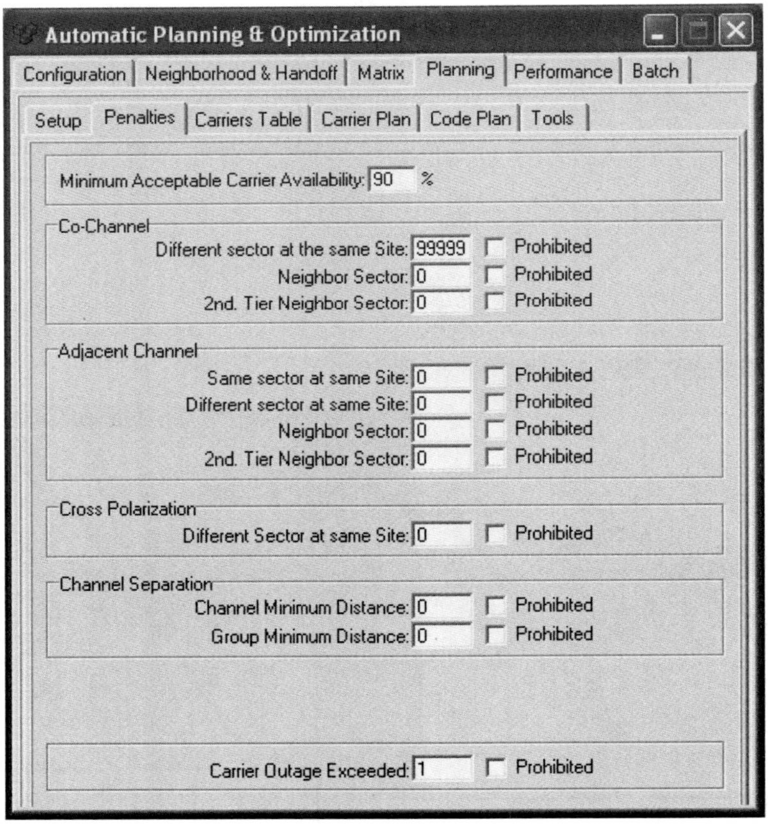

Figure 20.19 Penalties associated with the resource allocation.

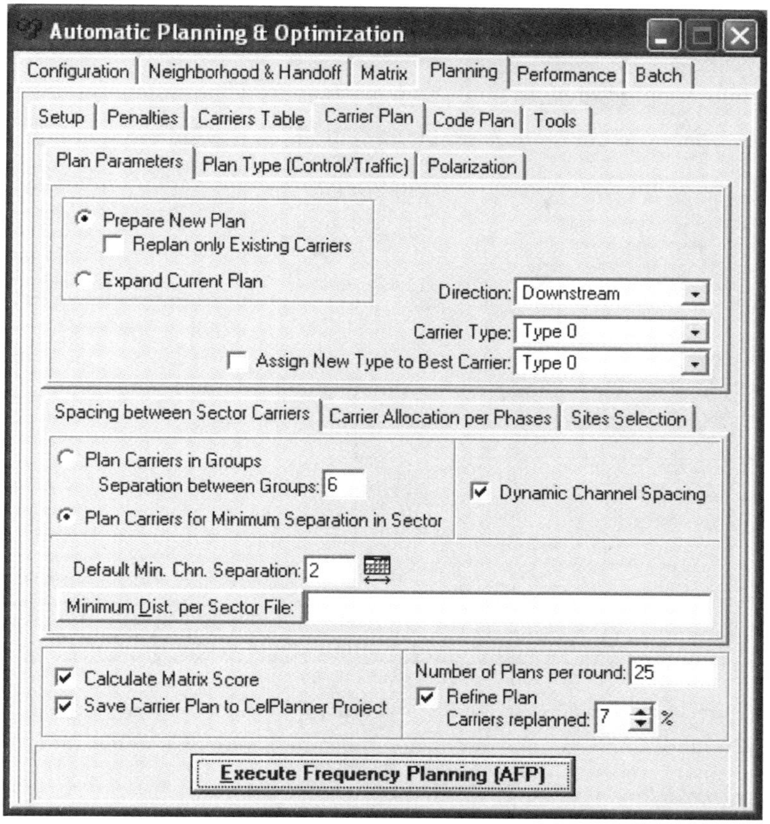

Figure 20.20 Frequency planning parameters.

The tool used as an example, CelOptima, implements advanced algorithms that assign channels automatically, minimizing network outage. A proprietary heuristic algorithm is used to perform this assignment. The algorithm has achieved highest marks in international performance tests.

Designers should be able to specify a set of rules, or guidelines, for the frequency plan. In this example, this is done through resource allocation penalties, as shown in Figure 20.19. These penalties are important as they instruct the tool to avoid certain situations that may lead to undesired handovers (handovers to wrong cells).

Finally, the parameters for frequency planning are defined in Figure 20.20.

Figure 20.21 depicts a snapshot of the actual frequency planning process in our sample tool, where each dot represents a one-iteration cycle and each plan quality is measured in terms of traffic outage.

20.2.8 Constrained Cell Enhancement

Once the code and carrier planning is finished, the cell enhancement process should be redone, this time considering constraints imposed by the resource assignment.

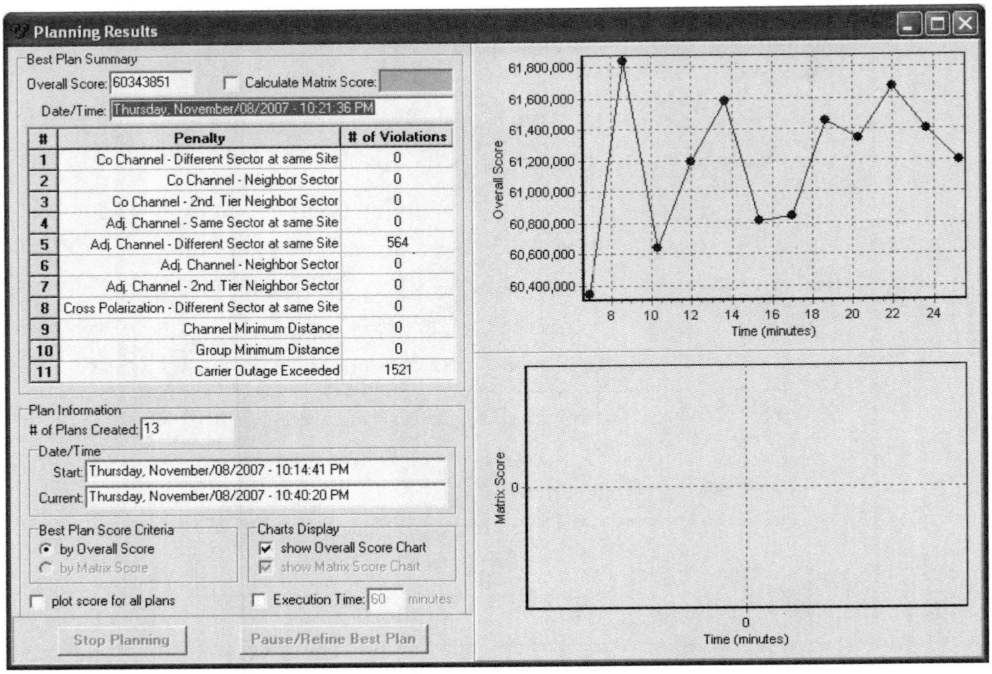

Figure 20.21 Optimization penalties and multiple iterations convergence display.

20.2.9 Backhaul Interference Matrix

Once all links have been designed, the Interference Matrix can be calculated. This matrix is similar to the one explained in Section 20.2.5 for the point-to-multipoint wireless network, but it considers the cell antenna as users.

20.2.10 Backhaul Automatic Channel Plan

After the matrix from Section 20.2.2 is created, a polarization and frequency plan can be done. This methodology for creating this plan is very similar to the one explained in Section 20.2.7 for the point-to-multipoint wireless network.

21

Wireless Network Performance Assessment

The outcome of a design and optimization should be verified by a performance assessment. This assessment may show deficiencies that need to be corrected in the design; show that it complies with the requirements or that it was over-designed and can be made more economical.

21.1 Perform Dynamic Traffic Simulation

The Traffic Simulation process should reproduce what happens in a live system and this is an extremely complex endeavor. There are different ways in which a traffic simulation can be performed; one of the most common is by use of Monte Carlo simulation to randomly place calls/sessions throughout the service area and then run the simulation by repeating this multiple times. Each of these iterations is called a snapshot. The tool used as an example, CelPlanner, does a very detailed simulation through a sequence of instantaneous traffic snapshots; this process is described in the following paragraphs. The process of creating a snapshot is described in Section 21.1.1.

Ideally, the design tool should automatically determine the throughput required throughout the market area based on the distribution of subscribers of each profile (Service Classes).

During traffic simulation, random draws must take place to distinguish users status as Connected (subscriber equipment is on, ongoing session) and Not Connected (not in a session). Next, connected subscribers are categorized as Active Sessions (in a burst, possibly holding traffic resources, for example, during a web access) or Dormant (in a session, but not in a burst, and possibly not using resources, for example, a reading interval between two web accesses). In the third status level, subscribers in Active Session are separated in two groups: Active Burst (transmitting and generating interference) and Idle (holding channels but not generating interference). Examples of Idle situation could include, for VoIP applications, the "silent" periods in the middle of a conversation. For the tool to be able to make these random draws, it needs the traffic pattern of the users being analyzed; in the tool used in the example, this pattern is configured as illustrated in Figures 3.5, 3.6, and 3.7.

The ratio between Active Burst and total Active Session subscribers should be user-defined (Activity Factor, in the example) and may vary per service. The terms connected/not-connected are used in this description in terms of a session being established; they apply both to connection oriented and connectionless models.

LTE, WiMAX and WLAN Network Design, Optimization and Performance Analysis, First Edition.
Leonhard Korowajczuk.

For the simulation, the planning tool should analyze compatibility between customer stations and base stations, and the SNIR available to define the modulation scheme selection for instantaneous data rate calculation. For situations in which not all the traffic demand can be served (contention), the tool should use some type of weighted-fairness algorithm for sharing resources.

In weighted fairness, the capacity is provided to different subscribers proportional to their Service Level agreement (SLA). In situations of contention (demand is higher than sector capacity), this is particularly relevant as it determines the priority with which users will be served or queued.

Because of the multiple schemes per radio and the multiple radios available per sector, link-adaptation mechanisms should also be simulated, that is, sectors may use different modulation-coding schemes for users depending on the available SNIR link quality.

The tool has to simulate resource management algorithms, as weighted fairness, that is, capacity is provided to different subscribers proportional to their Service Level agreement (SLA) weight. In situations of contention (demand is higher than sector capacity), this is particularly relevant as it determines the priority with which users will be served or queued. When there is contention in the traffic offered to a given sector, the tool should use the resource management algorithm (as weighted fairness) to decide how to share the resources.

Subscribers are initially assigned a share of the sector's usage time proportional to their service weight, or priority. Based on the user's data rate requirements (target data rate), some of the initially allocated sector's share may be left unused. Any unused sector's share should be re-allocated among subscribers that were not completely served in their target rate, following a weighted proportionality (or similar algorithm).

After the simulation is concluded, CelPlanner displays a dialogue box for visualization of results. Users can configure the display using two drop-down lists to filter specific carriers and service class.

Traffic reports should include information of how much traffic has been offered to the network (users demand), and how much traffic was fully served, partially served (queued), or rejected. It is important to have this information not only for each sector of the network but also separate for each service class as this allows designers to determine areas that need additional resources, but also identify "killer" services that might be consuming too many resources. Figure 21.1 illustrates the traffic simulation process.

The whole traffic simulation process is summarized next and illustrated in Figure 21.2:

- Market modeling is done through Service Classes.
  - Services
  - Terminals

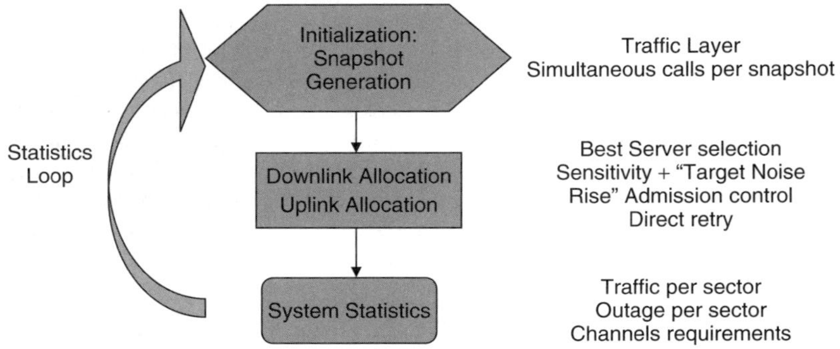

Figure 21.1 Traffic simulation overview.

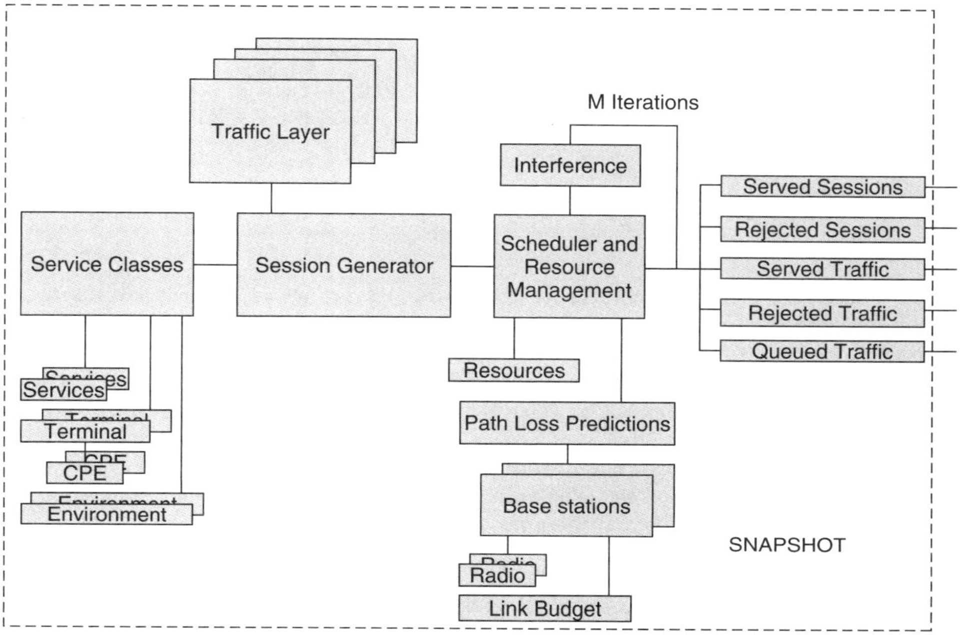

Figure 21.2 Dynamic traffic simulation process.

  - CPE radio
  - Environment
- Sessions are randomly generated in proportion to the traffic layers associated with them.
  - Snapshot
- Network is modeled by BTS and path loss predictions generated.
  - BTS
  - Radio
  - Link budget
- Scheduler and Resource Management algorithms assign calls, interference is replaced by an average noise rise.
- Statistics are recorded for snapshot.
  - Served Sessions
  - Rejected Sessions
  - Served Traffic
  - Rejected Traffic
  - Queued Traffic

21.1.1 Traffic Snapshot

A traffic snapshot represents a statistical instantaneous usage of the system. Most tools use the Monte Carlo method to draw sessions for the simulation. The tool used as an example applies Monte Carlo to draw service classes according to their traffic distribution and allocate sessions geographically.

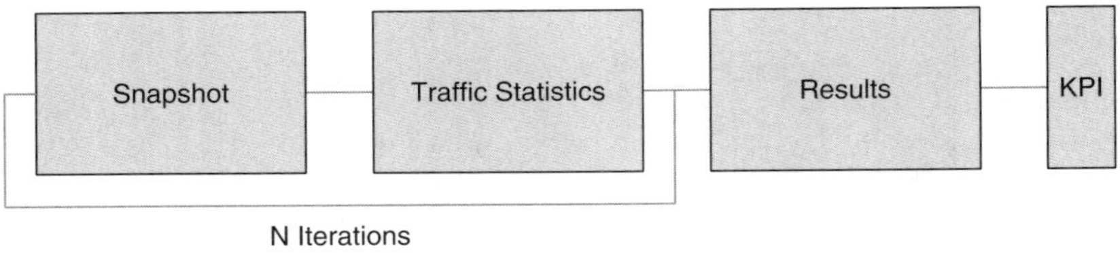

Figure 21.3 Illustration of traffic snapshot iterations.

Several snapshots should be done, after each one the statistics obtained should be gathered and accumulated. After the last snapshot the final results are obtained and can be expressed as KPIs (Key Parameter Indicators). This is illustrated in Figure 21.3.

Figure 21.4 shows sessions (data calls) placement during the simulation of one snapshot. The legend on the right identifies the Service Class to which each session belongs, each dot in the figure represents one active user.

During the simulation process, output power, data rate, load gain and sub-channelization gain should be automatically assigned and adjusted by the planning tool. The dynamicity in this type of simulation is provided by the iterative process and by the consideration of previous queues. The simulation time depends on the amount of traffic being simulated and the number of snapshots. The number of snapshots should be large enough, so all Service Classes are represented statistically.

The following statistics (average and standard deviation for each parameter) should be provided as results:

- Statistics are provided:
  - For All Classes
  - Per BTS
  - Total
- Per Service Class
  - Per BTS
  - Total
- All Results are expressed by:
  - Average
  - Standard Deviation
- Value are export to network Sectors

The following results are calculated:

- Offered Sessions
  - Total
  - Idle
  - Active-Down/Up
  - Active-Down and Up
- Served Sessions
  - Total
  - Idle
  - Active-Down/Up
  - Active-Down and Up

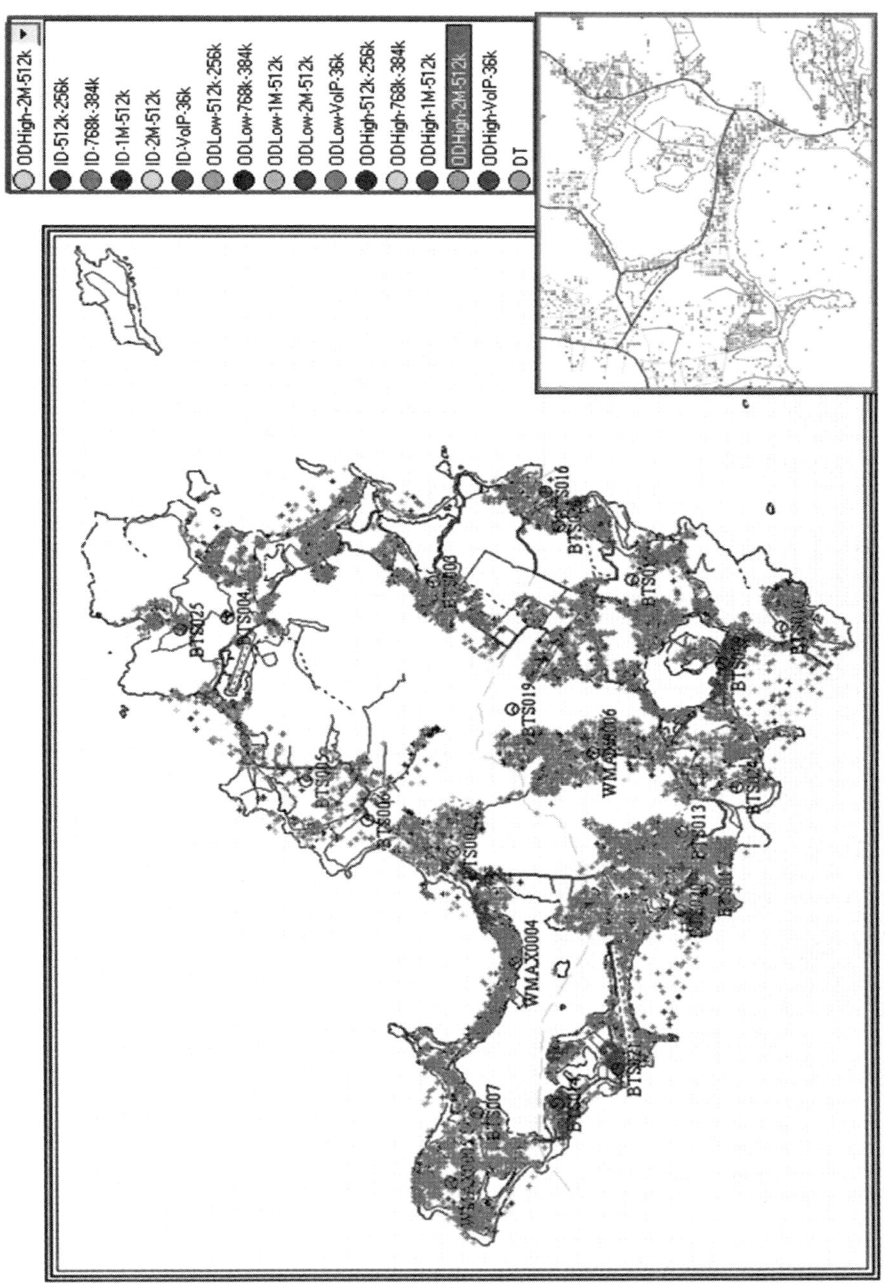

Figure 21.4 Traffic simulation (each session type is represented by the legend color).

- Rejected Sessions
  - Down
  - Up
  - Down/Up
- Offered Throughput Down/Up
- Served Throughput Down/Up
- Queued Throughput Down/Up
- Rejected Throughput Down/Up
- Served Load Factor

Figure 21.5 show a detail of the geographical distribution of sessions over several snapshots.

21.1.2 Traffic Report

An example of a traffic report is shown in Figure 21.6. This report shows the statistical information of offered traffic, served traffic, queued traffic and blocked traffic. The information is provided for all classes or for any individual class. The average value of each parameter is displayed as well as its standard deviation. The distribution is considered Gaussian.

21.2 Performance

Once snapshot statistics are obtained, the generation of KPIs can be done.

21.2.1 Generate Key Parameter Indicators (KPI)

The results of the traffic simulation should be analyzed and compared to the SLA.

21.2.1.1 Coverage Area KPI

The coverage area KPI specifies the minimum percentage of the area that should be covered for each different service.

Typical coverage KPIs are expressed as:

- 90% of area
- 90% of population

To run this type of analysis, designers must first establish the area of interest for each Service Class. Figure 21.7 and Figure 21.8 show, respectively, examples of selection of area of interest and coverage area KPIs in a planning tool.

21.2.1.2 Traffic Blockage KPI

This KPI expresses what percentage of sessions is accepted as blocked in what percentage of sectors. For example:

- 2% of session
- 10% of sectors

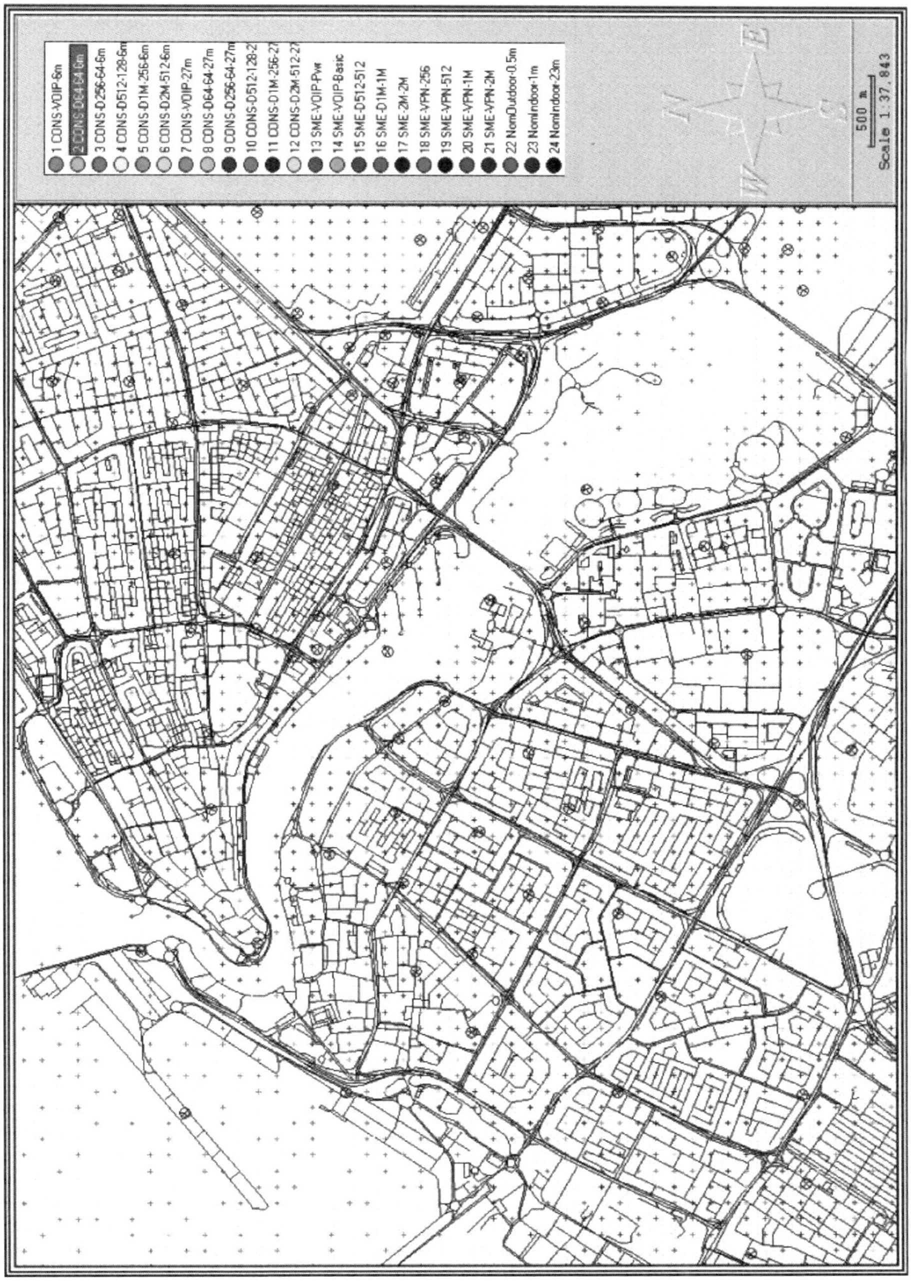

Figure 21.5 Traffic simulation sessions detail.

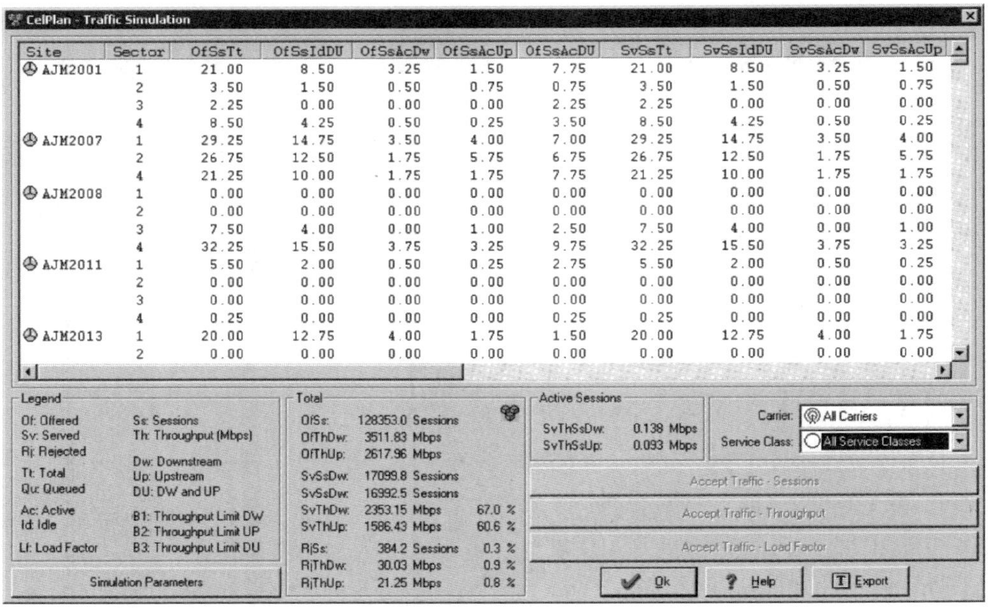

Figure 21.6 Traffic simulation results (part).

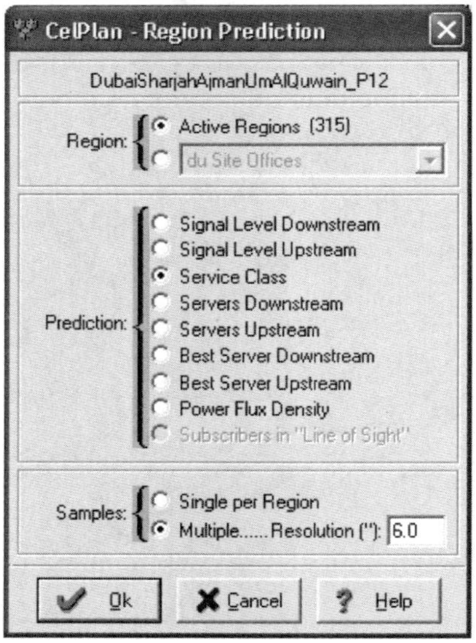

Figure 21.7 Coverage area calculation in CelPlanner.

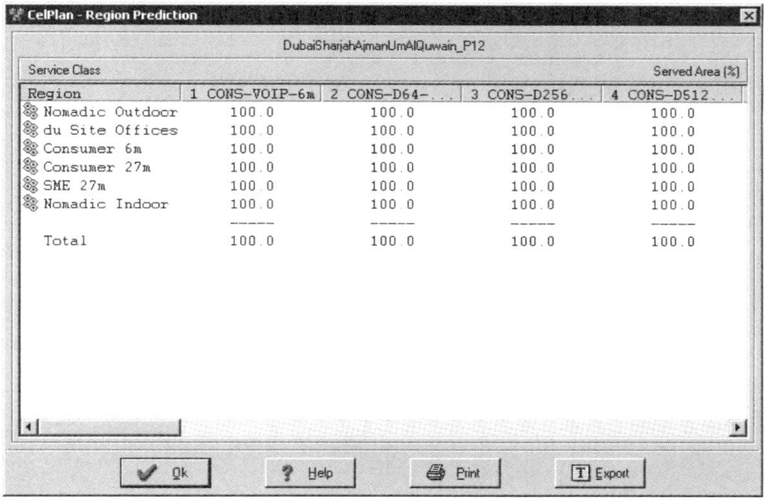

Figure 21.8 Coverage area results in CelPlanner (part 1).

In this example, the operator is stating that up to 10% of the sectors may present up to 2% blocking of sessions.

21.2.1.3 Traffic-Related Key Performance Indicators

This topic is described in more detail in Section 3.12 and it is presented here as a summary for completeness of KPIs analyses.

Traffic KPIs calculate how well the SLA proposed throughput (MSTR) is being fulfilled by the design. It is prohibitive to provide full MSTR access to all subscribers in the network as it demands a huge amount of infrastructure, and, economically, is not justifiable. For this reason SLAs should have a staggering clause that specifies the percentage of customers that will be served at a certain percentage of the MSTR (peak rate).

The Traffic Performance has to be calculated for at least two peak hours because SME traffic peaks at around 4 pm (day peak), whereas consumer traffic is around 10% of its peak at this hour. Consumer traffic peaks at around 8 pm (night peak), whereas the SME traffic is around 10% of its peak at this hour. This is illustrated in Figure 21.9.

One way to do this calculation for the two peak hours in planning tools is by applying a Traffic Factor to the demand layers when doing the traffic simulation.

21.2.1.4 Traffic KPI Calculation and Traffic KPI Tables

The traffic KPI calculation should be done based on the statistics obtained during the snapshots of the traffic simulation. The set of tables presented in this section illustrates a way of compiling the traffic simulation information to calculate different KPIs, which are described next. Note that each table presents a different percentage of MSTR and customers served; designers should adapt these values to match the requirements of the network being designed, the values given here are just for illustration purposes, although they can be said to be typical. This is illustrated in Figure 21.10.

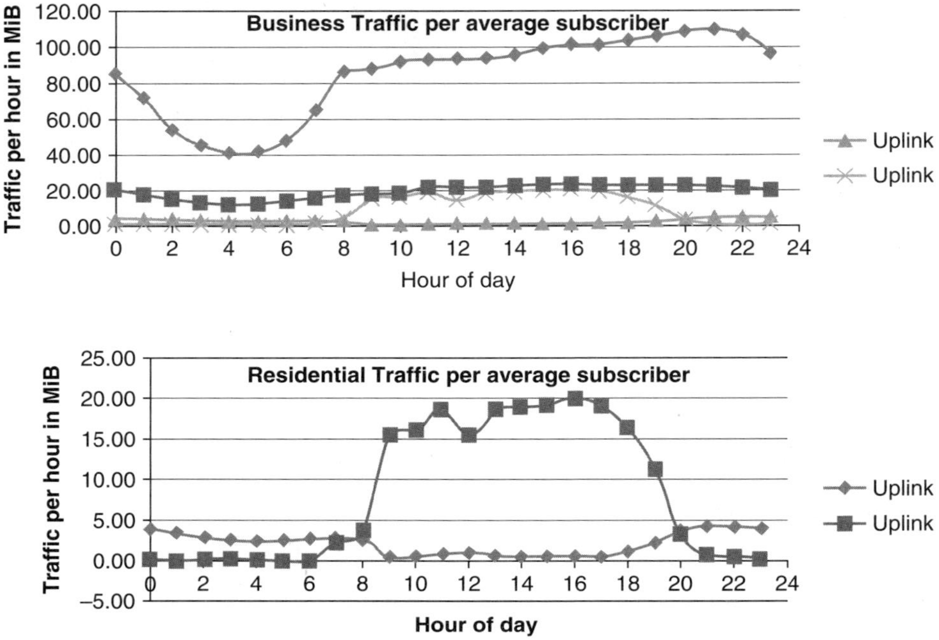

Figure 21.9 Relative traffic distribution at different hours of the day.

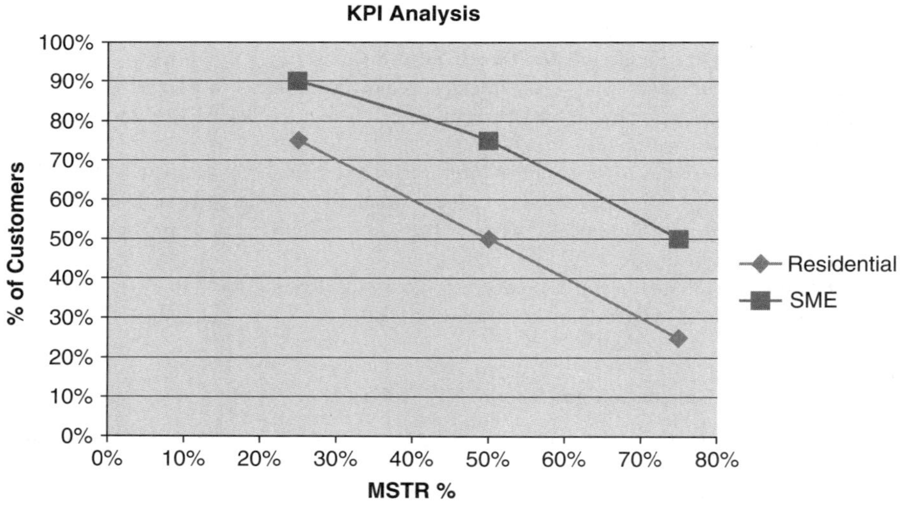

Figure 21.10 KPI specifications example.

21.2.1.5 Traffic Data Table

Table 21.1 summarizes the offered and served traffic per service class.

- Traffic Throughput KPI at 75% of MSTR. In Table 21.2, the target KPI is availability of 75% MSTR for at least 25% of residential customers and 50% SMEs. For this analysis, the table first presents the percentage of sessions served at 75% of the peak rate. Then it shows the percentage of users served with that rate; if this value is higher than the desired target (25% for residential and 50% for SME) the KPI is set to PASS, otherwise it FAILS.
- Traffic Throughput KPI at 50% of MSTR. In Table 21.3, the target KPI is availability of 50% MSTR for at least 50% of residential customers and 75% SMEs. For this analysis, the table first presents the percentage of sessions served at 50% of the peak rate. Then it shows the percentage of users served with that rate; if this value is higher than the desired target (50% for residential and 75% for SME) the KPI is set to PASS, otherwise it FAILS.
- Traffic Throughput KPI at 15% (consumer) and 25% (SME) of MSTR. In Table 21.4, the target KPI is availability of 15% MSTR for at least 75% of residential customers and 25% MSTR for at least 50% of SMEs. For this analysis, Table 21.4 first presents the percentage of sessions served at 15% or 25% of the peak rate, depending on the type of user (residential or SME). Then it shows the percentage of users served with that rate; if this value is higher than the desired target, the KPI is set to PASS, otherwise it FAILS.

Table 21.1 displays, as an example, the amount of traffic considered in the simulation for different service classes. Tables 21.2 to 21.4 are examples of KPI tables for different percentages of MSTR.

21.3 Perform Network Performance Predictions

This section presents examples of the most common types of predictions used to illustrate a wireless broadband network design. Several other types of predictions can also be generated for this type of network to cover specific aspects of the design. The ones presented here apply, however, to most types of design. It is common for designers to use paper sizes A1 or larger to print predictions such as these.

Designers should note that, depending on the type of service classes selected for the network, predictions may be quite different between them, for example, a prediction that shows the selected modulation scheme at each pixel looks completely different for an indoor user and a rooftop installation. Table 21.5 provides a list of common predictions/plots.

The following figures presented in this section show examples of these predictions for one service class. The predictions presented as examples were done using the CelPlanner™ design tool. It is not necessary for the designer to execute all predictions, as he should concentrate on the main ones and then drill deeper in eventual problem areas.

21.3.1 Topography

Figure 21.11 displays the composition of different resolution topography files.

21.3.2 Morphology

Figures 21.12 to 21.14 display the composition of different resolution morphology files.

Table 21.1 Traffic data per service class

Service Class Code	Class Name	Peak Throughput Downlink (kbps)	Peak Throughput Uplink (kbps)	Subscribers	Mean Active Users DL	Mean Active Users UL	Mean Offered Throughput (Mbps) DL	Mean Offered Throughput (Mbps) UL	Mean Served Throughput (Mbps) DL	Mean Served Throughput (Mbps) UL	Mean Offered Throughput per Sector (Mbps) DL	Mean Offered Throughput per Sector (Mbps) UL	Mean Served Throughput per Sector (Mbps) DL	Mean Served Throughput per Sector (Mbps) UL	Served/Offered Throughput per Sector Ratio DL	Served/Offered Throughput per Sector Ratio UL
C1	Consumer VoIP residential at 6 m height	36	36	11,384	524	525	24	27	24	27	0.04	0.04	0.04	0.04	1.00	1.00
C2	Consumer Data 64k/64k at 6 m height	64	64	1935	46	51	4	5	4	4	0.01	0.01	0.01	0.01	0.99	0.96
C3	Consumer Data 256k/64k at 6 m height	256	64	2391	62	66	20	6	18	6	0.03	0.01	0.03	0.01	0.90	0.96
C4	Consumer Data 512k/128k at 6 m height	512	128	3643	91	89	60	16	53	16	0.09	0.03	0.08	0.02	0.89	0.96
C5	Consumer Data 1M/256k at 6 m height	1024	256	2504	65	61	85	22	73	21	0.13	0.03	0.11	0.03	0.86	0.95
C6	Consumer Data 2M/512k at 6 m height	2048	512	911	22	23	58	17	46	16	0.09	0.03	0.07	0.02	0.80	0.95
C7	Consumer VoIP residential at 27 m height	36	36	15,550	709	712	33	37	33	37	0.05	0.06	0.05	0.06	1.00	1.00
C8	Consumer Data 64k/64k at 27 m height	64	64	2644	69	66	6	6	5	6	0.01	0.01	0.01	0.01	0.96	0.91
C9	Consumer Data 256k/64k at 27 m height	256	64	3266	79	77	26	7	21	6	0.04	0.01	0.03	0.01	0.81	0.92
C10	Consumer Data 512k/128k at 27 m height	512	128	4976	122	125	80	23	64	21	0.13	0.04	0.10	0.03	0.80	0.92
C11	Consumer Data 1M/256k at 27 m height	1024	256	3421	94	86	123	31	95	28	0.19	0.05	0.15	0.04	0.77	0.90
C12	Consumer Data 2M/512k at 27 m height	2048	512	1244	31	33	81	24	59	21	0.13	0.04	0.09	0.03	0.72	0.88
C13	SME Power VoIP at 27 m height	96	96	197	155	158	19	22	19	22	0.03	0.03	0.03	0.03	1.00	1.00
C14	SME Basic VoIP at 27 m height	48	48	1775	708	713	43	49	43	49	0.07	0.08	0.07	0.08	1.00	1.00
C15	SME Internet 512k/512K at 27 m height	512	512	513	25	26	17	19	16	17	0.03	0.03	0.02	0.03	0.96	0.91
C16	SME Internet 1M/1M at 27 m height	1024	1024	158	7	7	9	9	8	7	0.01	0.01	0.01	0.01	0.83	0.72
C17	SME Internet 2M/2M at 27 m height	2048	2048	1302	62	69	163	201	118	106	0.26	0.31	0.18	0.17	0.72	0.53
C18	SME VPN 256k/256k at 27 m height	256	256	15	15	15	5	5	4	4	0.01	0.01	0.01	0.01	0.84	0.72
C19	SME VPN 512k/512k at 27 m height	512	512	10	10	10	7	7	6	5	0.01	0.01	0.01	0.01	0.84	0.72
C20	SME VPN 1M/1M at 27 m height	1024	1024	15	15	15	20	22	16	15	0.03	0.03	0.02	0.02	0.80	0.70
C21	SME VPN 2M/2M at 27 m height	2048	2048	59	59	59	155	172	108	89	0.24	0.27	0.17	0.14	0.70	0.52
AC	All Classes			57,913	2970	2984	1037	727	833	523	1.62	1.14	1.30	0.82	0.80	0.72

Table 21.2 Traffic throughput KPI at 75% of peak rate

Service Class Code	Class Name	Peak Down (kbps)	Peak Up (kbps)	Target Percentage of Peak Rate	Target Rate (Mbps)		Target Sessions Percentage	Percentage of Sessions above Target Rate		Status		Minimum Throughput Achieved at Target Percentage of Sessions (kbps)		Average Throughput Achieved (kbps)	
					DL	UL		DL	UL	DL	UL	DL	UL	DL	UL
c2	Consumer Data 64k/64k at 6 m height	64	64	75	48	48	25	100	94%	PASS	PASS	64	64	63	61
c3	Consumer Data 256k/64k at 6 m height	256	64	75	192	48	25	75	94%	PASS	PASS	256	64	230	62
c4	Consumer Data 512k/128k at 6 m height	512	128	75	384	96	25	73%	93%	PASS	PASS	512	128	454	123
c5	Consumer Data 1M/256k at 6 m height	1024	256	75	768	192	25	66	91%	PASS	PASS	1024	256	876	243
c6	Consumer Data 2M/512k at 6 m height	2048	512	75	1536	384	25	57	93%	PASS	PASS	1995	512	1635	487
c8	Consumer Data64k/64k at 27 m height	64	64	75	48	48	25	93	80%	PASS	PASS	64	64	61	59
c9	Consumer Data256k/64k at 27 m height	256	64	75	192	48	25	57	83%	PASS	PASS	256	64	207	59
c10	Consumer Data512k/128k at 27 m height	512	128	75	384	96	25	57	81%	PASS	PASS	512	128	410	118
c11	Consumer Data 1M/256k at 27 m height	1024	256	75	768	192	25	53	77%	PASS	PASS	1007	256	793	231
c12	Consumer Data 2M/512k at 27 m height	2048	512	75	1536	384	25	46	73%	PASS	PASS	1883	512	1482	453
c15	SME Internet 512k/512K at 27 m height	512	512	75	384	384	50	99	81%	PASS	PASS	492	466	492	466
c16	SME Internet 1M/1M at 27 m height	1024	1024	75	768	768	50	64	47%	PASS	FAIL	852	740	852	740
c17	SME Internet 2M/2M at 27 m height	2048	2048	75	1536	1536	50	46	21%	FAIL	FAIL	1480	1079	1480	1079
c18	SME VPN 256k/256k at 27 m height	256	256	75	192	192	50	64	46%	PASS	FAIL	215	183	215	183
c19	SME VPN 512k/512k at 27 m height	512	512	75	384	384	50	67	45%	PASS	FAIL	433	367	433	367
c20	SME VPN 1M/1M at 27 m height	1024	1024	75	768	768	50	57	44%	PASS	FAIL	817	719	817	719
c21	SME VPN 2M/2M at 27 m height	2048	2048	75	1536	1536	50	43	20%	FAIL	FAIL	1432	1062	1432	1062

Table 21.3 Traffic throughput KPI at 50% of peak rate

Service Class Code	Class Name	Peak Down (kbps)	Peak Up (kbps)	Target Percentage of Peak Rate	Target Rate (Mbps) DL	Target Rate (Mbps) UL	Target Sessions Percentage	Percentage of Sessions above Target Rate DL	Percentage of Sessions above Target Rate UL	Status DL	Status UL	Minimum Throughput Achieved at Target Percentage of Sessions (kbps) DL	Minimum Throughput Achieved at Target Percentage of Sessions (kbps) UL	Average Throughput Achieved (kbps) DL	Average Throughput Achieved (kbps) UL
c2	Consumer Data 64k/64k at 6 m height	64	64	50	32	32	50	100	100%	PASS	PASS	63.0	61.5	63.0	61.5
c3	Consumer Data 256k/64k at 6 m height	256	64	50	128	32	50	96	100%	PASS	PASS	230.1	61.6	230.1	61.6
c4	Consumer Data 512k/128k at 6 m height	512	128	50	256	64	50	96	100%	PASS	PASS	454.3	122.6	454.3	122.6
c5	Consumer Data 1M/256k at 6 m height	1024	256	50	512	128	50	92	100%	PASS	PASS	876.4	243.5	876.4	243.5
c6	Consumer Data 2M/512k at 6 m height	2048	512	50	1024	256	50	87	100%	PASS	PASS	1635.4	486.7	1635.4	486.7
c8	Consumer Data64k/64k at 27 m height	64	64	50	32	32	50	100	98%	PASS	PASS	61.2	58.5	61.2	58.5
c9	Consumer Data256k/64k at 27 m height	256	64	50	128	32	50	84	99%	PASS	PASS	206.6	59.2	206.6	59.2
c10	Consumer Data512k/128k at 27 m height	512	128	50	256	64	50	84	99%	PASS	PASS	410.3	117.9	410.3	117.9
c11	Consumer Data 1M/256k at 27 m height	1024	256	50	512	128	50	81	97%	PASS	PASS	793.4	231.4	793.4	231.4
c12	Consumer Data 2M/512k at 27 m height	2048	512	50	1024	256	50	78	96%	PASS	PASS	1482.3	452.8	1482.3	452.8
c15	SME Internet 512k/512K at 27 m height	512	512	50	256	256	100	100	99%	PASS	FAIL	258.5	17.0	491.7	466.3
c16	SME Internet 1M/1M at 27 m height	1024	1024	50	512	512	100	93	75%	FAIL	FAIL	0.0	0.0	851.9	740.2
c17	SME Internet 2M/2M at 27 m height	2048	2048	50	1024	1024	100	78	54%	FAIL	FAIL	0.0	0.0	1480.3	1078.8
c18	SME VPN 256k/256k at 27 m height	256	256	50	128	128	100	91	75%	FAIL	FAIL	0.0	0.0	214.7	183.1
c19	SME VPN 512k/512k at 27 m height	512	512	50	256	256	100	94	77%	FAIL	FAIL	0.0	0.0	432.6	366.7
c20	SME VPN 1M/1M at 27 m height	1024	1024	50	512	512	100	87	74%	FAIL	FAIL	0.0	0.0	816.9	719.3
c21	SME VPN 2M/2M at 27 m height	2048	2048	50	1024	1024	100	75	53%	FAIL	FAIL	0.0	0.0	1432.1	1062.1

Table 21.4 Traffic throughput KPI at 15% (consumer) and 25% (SME) of peak rate

Service Class Code	Class Name	Peak Down (kbps)	Peak Up (kbps)	Target Percentage of Peak Rate	Target Rate (Mbps) DL	Target Rate (Mbps) UL	Target Sessions Percentage	Percentage of Sessions above Target Rate DL	Percentage of Sessions above Target Rate UL	Status DL	Status UL	Minimum Throughput Achieved at Target Percentage of Sessions (kbps) DL	Minimum Throughput Achieved at Target Percentage of Sessions (kbps) UL	Average Throughput Achieved (kbps) DL	Average Throughput Achieved (kbps) UL
c2	Consumer Data 64k/64k at 6 m height	64	64	15	9.6	9.6	90	100	100%	PASS	PASS	57.4	50.4	63.0	61.5
c3	Consumer Data 256k/64k at 6 m height	256	64	15	38.4	9.6	90	100	100%	PASS	PASS	156.6	50.5	230.1	61.6
c4	Consumer Data 512k/128k at 6 m height	512	128	15	76.8	19.2	90	100	100%	PASS	PASS	307.2	99.6	454.3	122.6
c5	Consumer Data 1M/256k at 6 m height	1024	256	15	153.6	38.4	90	100	100%	PASS	PASS	545.1	194.1	876.4	243.5
c6	Consumer Data 2M/512k at 6 m height	2048	512	15	307.2	76.8	90	99	100%	PASS	PASS	951.7	396.0	1635.4	486.7
c8	Consumer Data64k/64k at 27 m height	64	64	15	9.6	9.6	90	100	100%	PASS	PASS	49.9	42.8	61.2	58.5
c9	Consumer Data256k/64k at 27 m height	256	64	15	38.4	9.6	90	98	100%	PASS	PASS	105.6	43.9	206.6	59.2
c10	Consumer Data512k/128k at 27 m height	512	128	15	76.8	19.2	90	99	100%	PASS	PASS	213.6	86.5	410.3	117.9
c11	Consumer Data 1M/256k at 27 m height	1024	256	15	153.6	38.4	90	98	100%	PASS	PASS	388.4	163.7	793.4	231.4
c12	Consumer Data 2M/512k at 27 m height	2048	512	15	307.2	76.8	90	98	100%	PASS	PASS	721.8	310.1	1482.3	452.8
c15	SME Internet 512k/512K at 27 m height	512	512	25	128.0	128.0	90	100	100%	PASS	PASS	428.8	345.2	491.7	466.3
c16	SME Internet 1M/1M at 27 m height	1024	1024	25	256.0	256.0	90	99	93%	PASS	PASS	553.9	312.8	851.9	740.2
c17	SME Internet 2M/2M at 27 m height	2048	2048	25	512.0	512.0	90	95	84%	PASS	FAIL	728.9	362.9	1480.3	1078.8
c18	SME VPN 256k/256k at 27 m height	256	256	25	64.0	64.0	90	99	93%	PASS	PASS	131.4	77.8	214.7	183.1
c19	SME VPN 512k/512k at 27 m height	512	512	25	128.0	128.0	90	100	95%	PASS	PASS	287.0	177.6	432.6	366.7
c20	SME VPN 1M/1M at 27 m height	1024	1024	25	256.0	256.0	90	98	92%	PASS	PASS	469.8	301.7	816.9	719.3
c21	SME VPN 2M/2M at 27 m height	2048	2048	25	512.0	512.0	90	94	83%	PASS	FAIL	669.5	336.4	1432.1	1062.1

Table 21.5 Composite predictions plots

1	Topography
2	Morphology
3	Image
4	Landmarks
5	Region Consumer 6 m
6	Region Consumer 27 m
7	Region SME 27 m
8	Region nomadic outdoor 1 m
9	Region nomadic indoor
10	Region Density Consumer 6 m
11	Region Density Consumer 27 m
12	Region Density SME 27 m
13	Region Density nomadic outdoor 1 m
14	Region Density nomadic indoor
15	Traffic Consumer 6 m
16	Traffic Consumer 27 m
17	Traffic SME 27 m
18	Traffic nomadic outdoor 1 m
19	Traffic nomadic indoor
20	Traffic Simulation
21	Composite Signal Level dBm up
22	Composite Signal Level dBm down
23	Composite S/N down
24	Composite S/N up
25	Preamble Signal Level dBm
26	Preamble S/N
27	Preamble margin
28	MAP margin
29	MAP S/N
30	MIMO selection
31	Zone Selection
32	Best Server down
33	Best Server up
34	Number of Servers down
35	Number of Servers up
36	Modulation Scheme selection down
37	Modulation Scheme selection up
38	Maximum Data Rate down
39	Maximum Data Rate up
40	Maximum Sub-channel data rate down
41	Maximum Sub-channel data rate up
42	Interference down
43	Interference up
44	Noise Rise down
45	Noise Rise up
46	Service Down/Up
47	Service Margin
48	Service Classes
49	Frequency Plan
50	Microwave Links

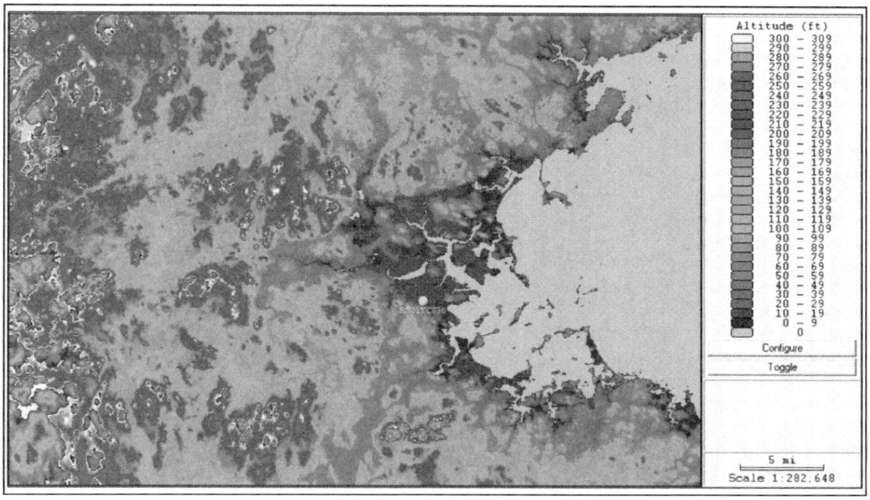

Figure 21.11 Topography plot sample.

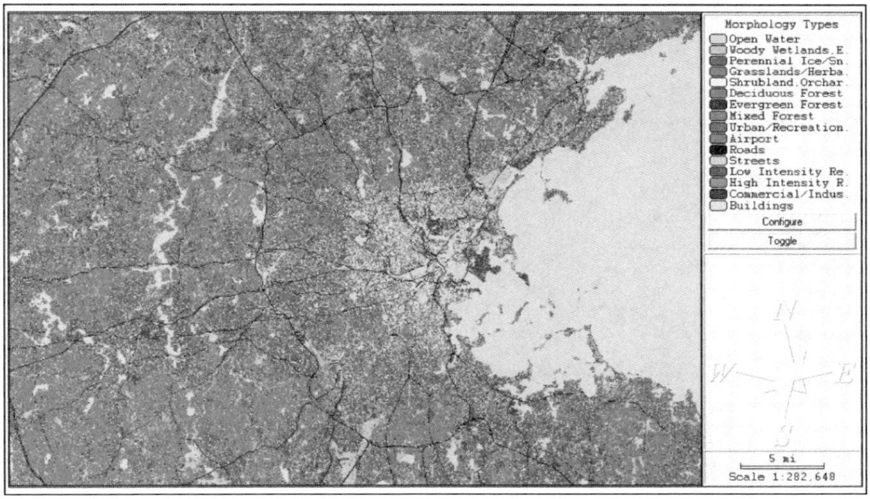

Figure 21.12 Morphology plot sample.

21.3.3 Image

Figure 21.15 displays a background image used as an underlay to predictions.

21.3.4 Landmarks

Figure 21.16 displays vectors representing roads and streets in the area of interest.

Figure 21.13 Morphology plot detail.

Figure 21.14 Morphology buildings with 1 m resolution.

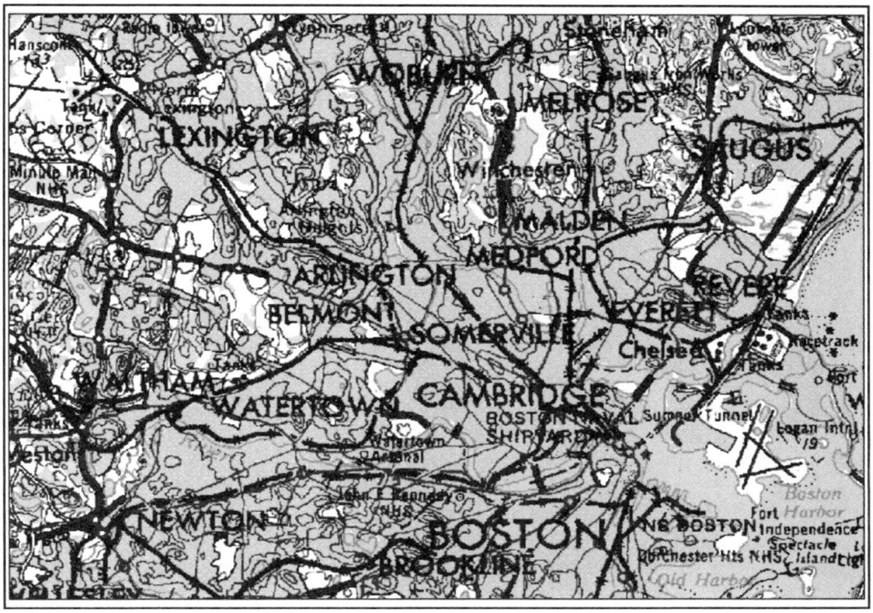

Figure 21.15 Image plot.

Figure 21.16 Landmarks in the AOI.

Figure 21.17 Census block with residential data.

Figure 21.18 Census block with business data.

21.3.5 Demographic Region

Demographic regions are used as input to calculate traffic. Figures 21.17 and 21.18 show details of residential and business census blocks.

21.3.6 Traffic Layers

Traffic layers generated from demographics are shown in Figures 21.19 and 21.20, respectively for residential and business areas.

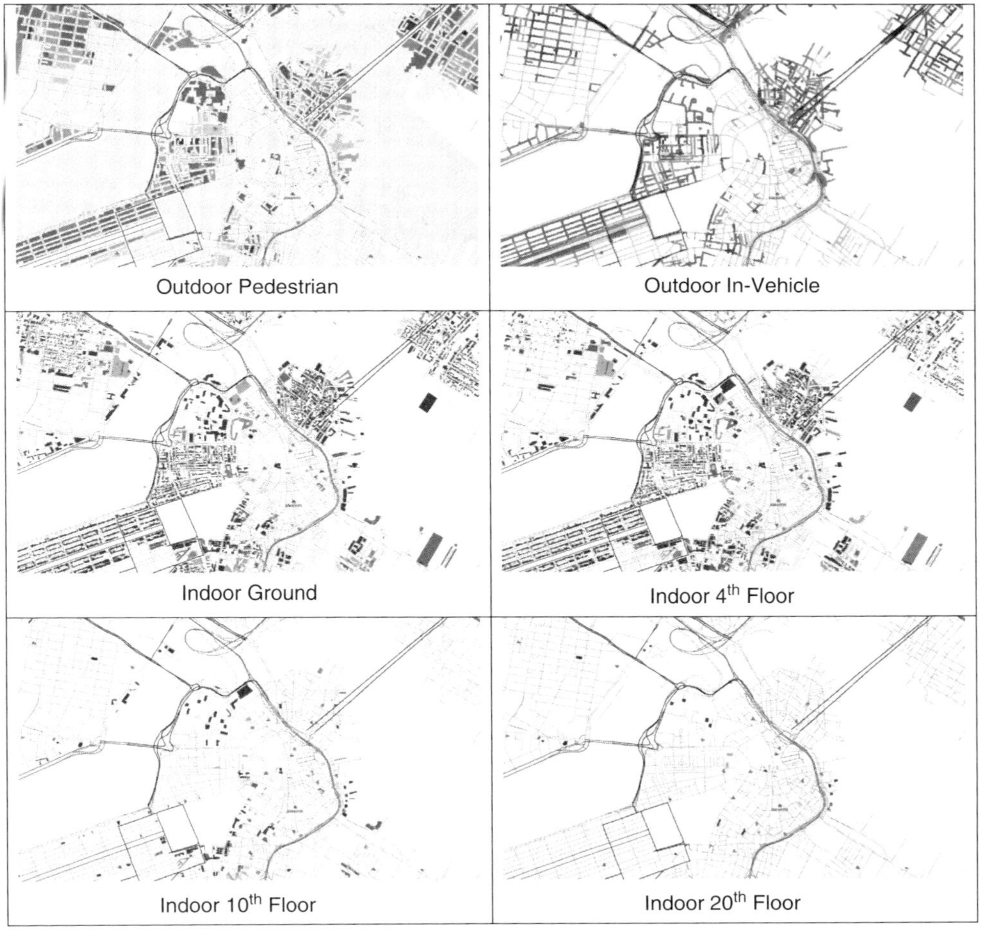

Figure 21.19 Residential traffic layers.

21.3.7 Traffic Simulation Result

This plot is used to indicate the areas in which data sessions were placed during traffic simulation. In this example (Figure 21.21), the tool uses different colors to indicate data sessions belonging to different service classes, black indicates no service due to lack of coverage, lack of resources or excessive interference.

21.3.8 Composite Signal Level

Similarly to single site signal level predictions, this prediction varies with the service class but it should be service-independent, that is, even if the site does not provide any service, the signal level should be calculated for the whole area. It is, in fact, a combination of all single site signal level

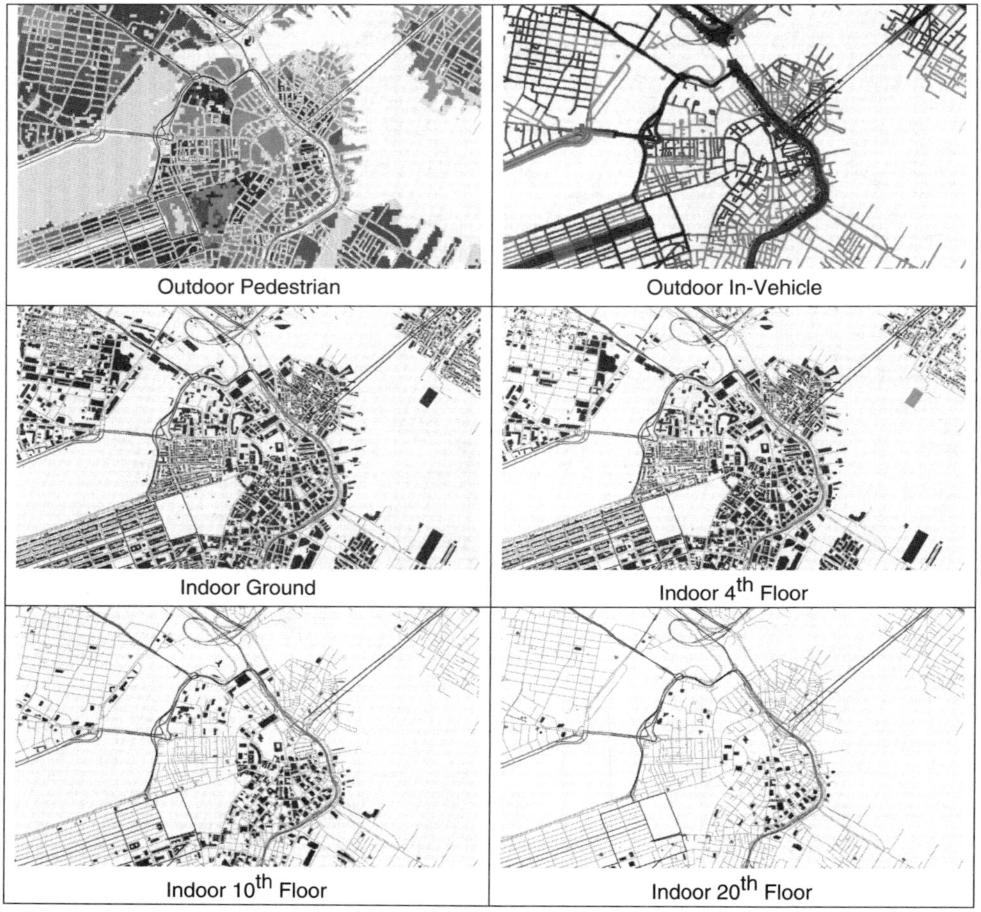

Figure 21.20 Business traffic layers.

prediction files, displaying the strongest signal level predicted for each pixel. Figures 21.22 and 21.23 show, respectively, for each predicted pixel, the signal level received by the subscriber radio in the downstream direction and the signal level received by the Sector radio in the upstream direction.

21.3.9 Composite S/N

This prediction shows the S/N of each pixel, where "S" is the signal from the best compatible server sector at the pixel. "N" is the sum of noise floor and interference for the sector's worst zone. This prediction is analyzed separately for downstream and upstream. Figures 21.24 and 21.25 show the expected CNIR (S/N) relative to each location in the area of interest.

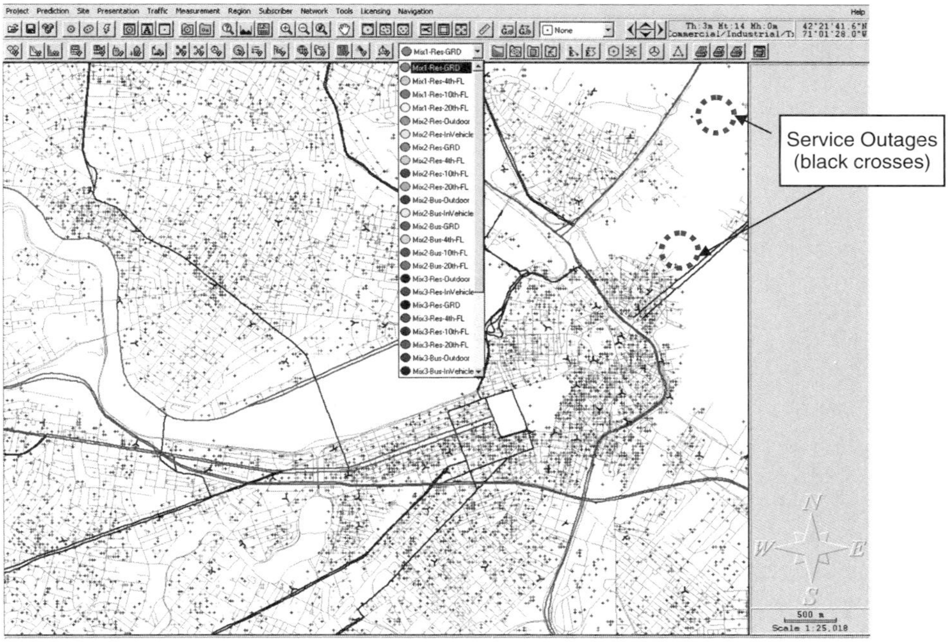

Figure 21.21 Traffic simulation depiction.

Figure 21.22 Composite signal level downstream at 4th floor.

Figure 21.23 Composite signal level upstream at 4th floor.

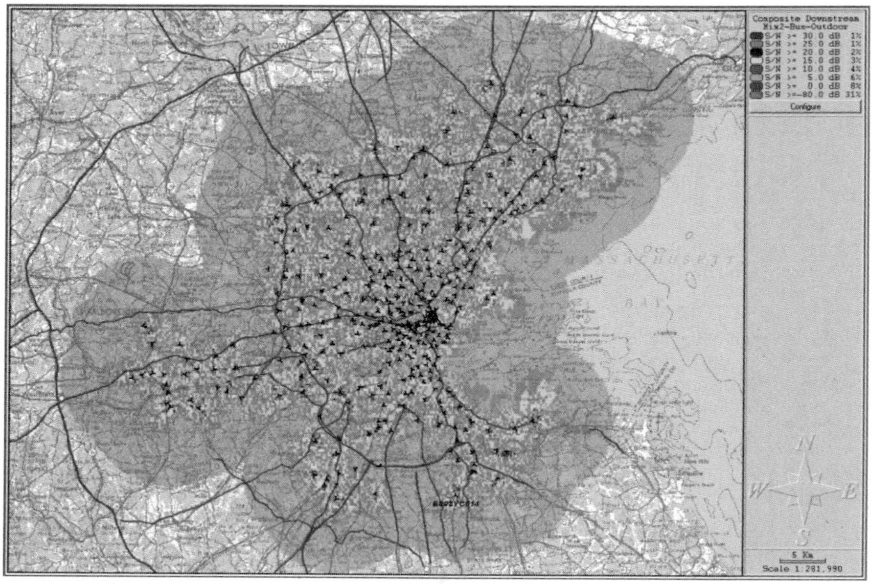

Figure 21.24 Composite S/N plot sample.

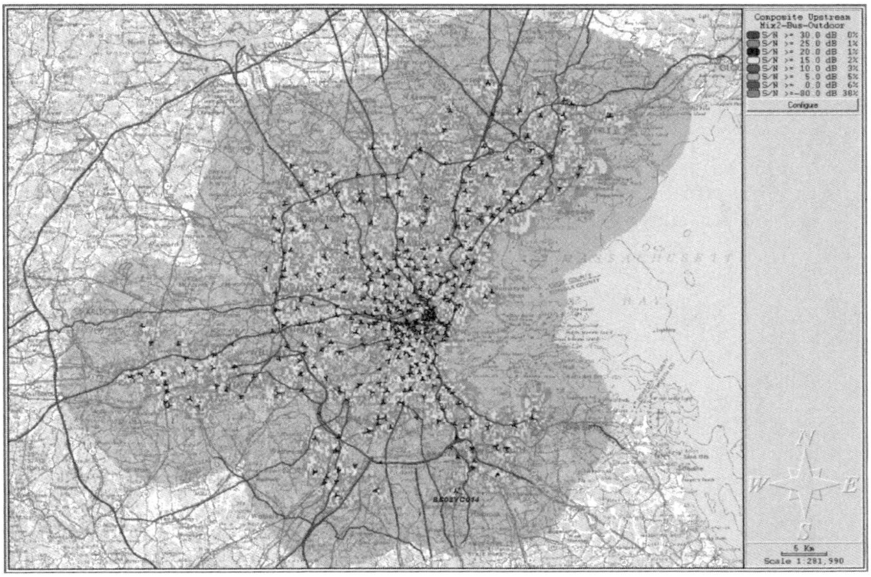

Figure 21.25 Composite S/N plot sample detail.

21.3.10 Preamble

A WiMAX frame starts with a preamble spread over the first symbol of the OFDM carrier. This preamble is a PN code that allows the identification of the FFT size, frame synchronization, and cell and segment identification. Preamble predictions are service-independent and are only created for the downlink. This prediction is not available for LTE and WLAN (due to lack of a preamble). Figure 21.26 shows this prediction.

21.3.11 Preamble SNIR

This prediction shows the SNIR of each pixel, assuming the preamble's boosted power, that is, "S" is the signal from the best compatible server sector at the pixel, boosted by 3 dB. The interference calculation does not account for the load factor (which could reduce interference). This prediction is not available for LTE or WLAN. Figure 21.27 shows this prediction.

21.3.12 Preamble Margin

This prediction shows, in each pixel, the maximum preamble margin among all radios of all sectors that are compatible with the selected class. For a given sector-radio-zone trio, the margin is defined as the difference between the achievable preamble SNIR and the required threshold set for BPSK $^1/_2$ at the selected radio. The margin can be negative if the required SNIR is greater than the obtained at the pixel. The required SNIR is determined through the Rx Performance table, depending on the service class configuration (required BER, environment). This prediction is not available for LTE or WLAN. Figure 21.28 shows this prediction.

Figure 21.26 Preamble prediction.

Figure 21.27 Preamble S/N.

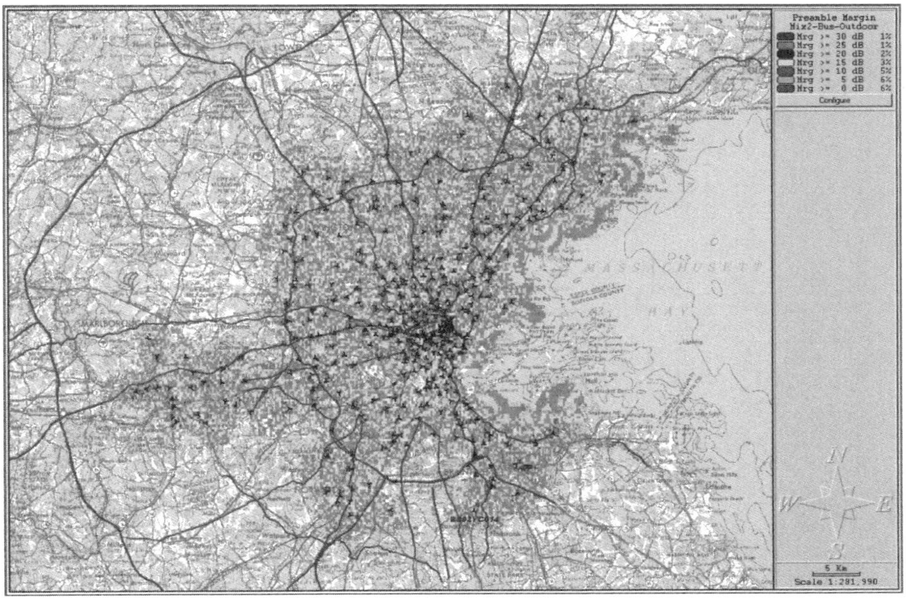

Figure 21.28 Preamble margin.

21.3.13 MAP (Medium Access Protocol) Margin

This prediction shows, in each pixel, the maximum MAP margin among all radios of all sectors that are compatible with the selected class. For a given sector-radio-zone trio, the margin is the difference between the achievable MAP SNIR and the required threshold set for QPSK $1/2$ at the selected radio. The margin can be negative if the required SNIR is greater than the obtained at the pixel. Figure 21.29 shows this prediction.

21.3.14 MAP S/N

Medium Access Protocol (MAP) predictions are service-independent and are only created for the downlink. It shows the SNIR of each pixel, where "S" is the signal from the best compatible server sector at the pixel. The interference calculation does not account for the load factor (which could weigh down the interference). Figure 21.30 shows this prediction.

21.3.15 Best Server

Figures 21.31 and 21.32 show the best server area of each site or sector. At each pixel, the tool determines which from all the possible serving sectors, is the sector that provides best service to the selected class. To fully understand this prediction, it is important for designers to know what is the planning tool criterion to determine "best service", some tools even allow users to choose what should be the main criterion (e.g. higher throughput, higher signal level).

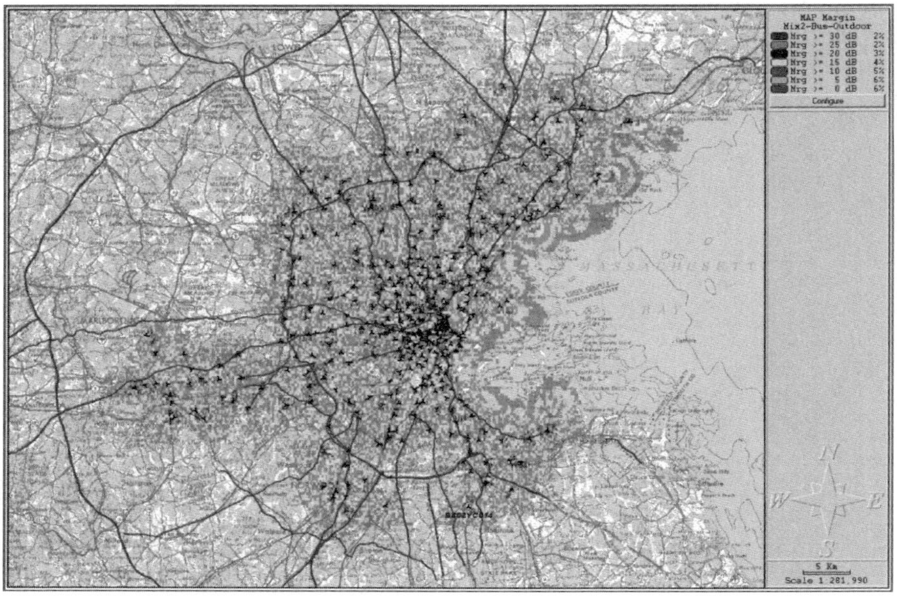

Figure 21.29 MAP margin.

Figure 21.30 MAP S/N.

Figure 21.31 Best server plot downstream.

Figure 21.32 Best server plot upstream.

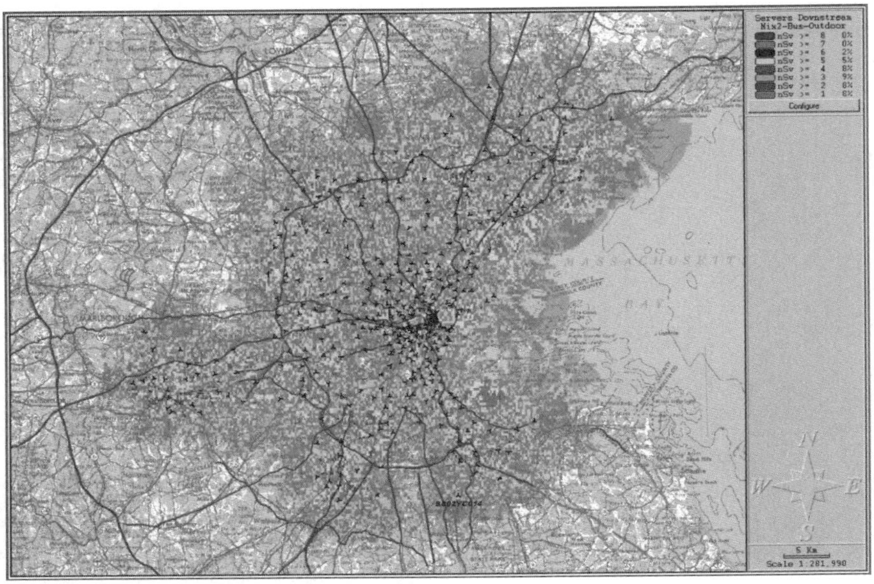

Figure 21.33 Number of servers downstream.

21.3.16 Number of Servers

This is also a service-independent prediction because it only considers radio compatibility and not service availability. This prediction counts the number of sectors that satisfy the minimum rate requirement (possible servers) for the selected class independently for each direction (up and downlink). Figures 21.33 and 21.34 display the prediction for downstream and upstream.

21.3.17 Radio Selection

The radio selection prediction should be calculated for the downlink only as the uplink radio is determined by the service class selected for presentation of the prediction. This prediction may change significantly from tool to tool depending on how multiple radios are identified in the tool. In the tool used as an example, the prediction indicates the carrier index (from 1 to 12) of the best radio selected at the best server sector. Figure 21.35 shows the carrier index for a sector. Figure 21.36 illustrates the radio selection.

21.3.18 Zone Selection

This prediction indicates the zone index (from 0 to 3) of the best zone, of the best radio selected for the best server sector. Zones are not declared in the LTE standard, but their implementation is left to the discretion of vendors. Figure 21.37 illustrates the zone selection.

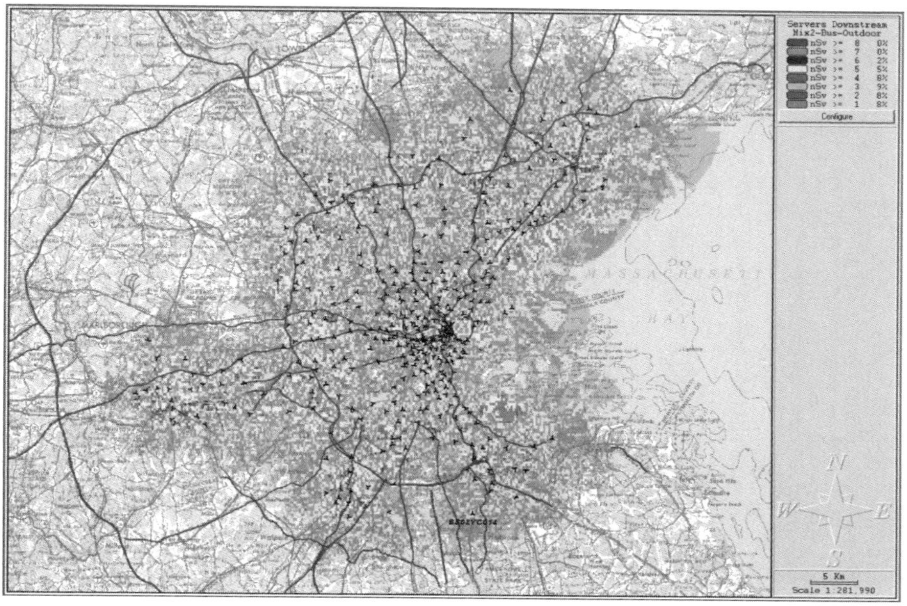

Figure 21.34 Number of servers upstream.

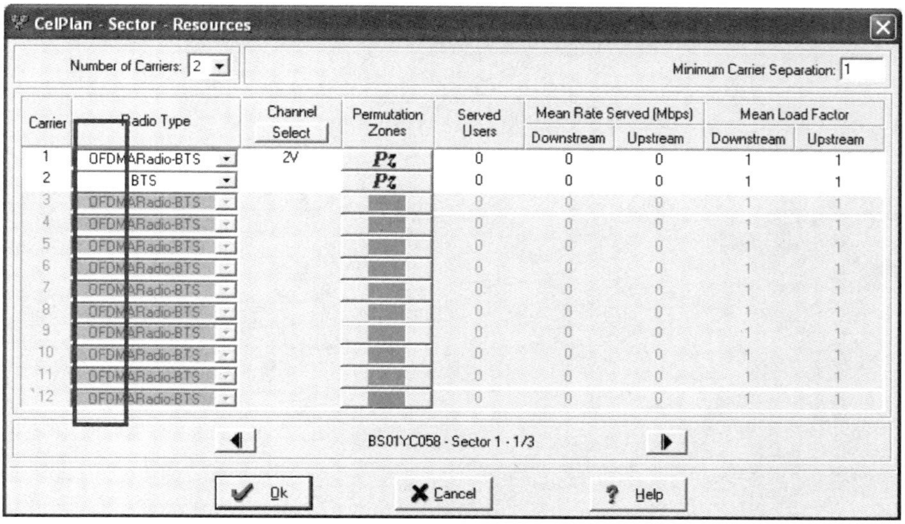

Figure 21.35 Multi-carrier radio index.

Figure 21.36 Radio selection.

Figure 21.37 Zone selection.

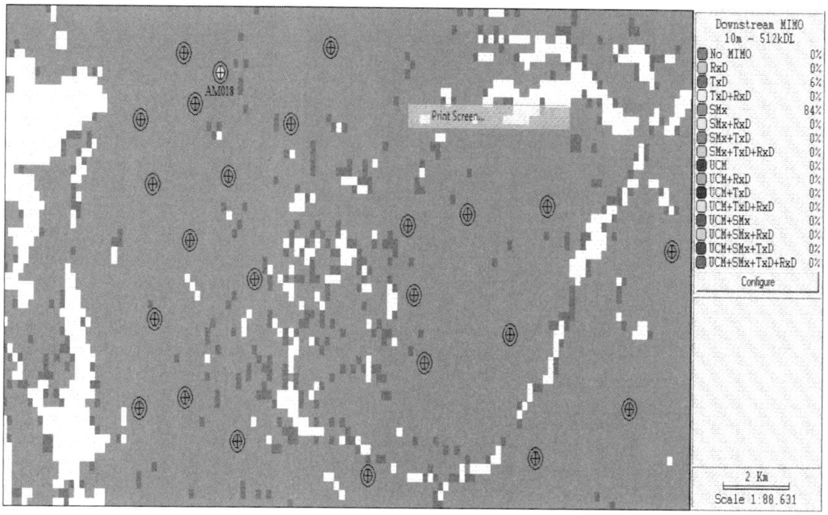

Figure 21.38 MIMO selection.

21.3.19 MIMO Selection

This prediction indicates the MIMO key corresponding to the best zone selected in the pixel for each direction. The following abbreviations are used in the legend:

- RxD = Rx Diversity
- TxD = Tx Diversity
- SMx = Spatial Multiplexing
- UCM = UL Collaborative MIMO

Figure 21.38 illustrates the MIMO selection.

21.3.20 Modulation Scheme Selection

Figures 21.39 and 21.40 show the coding-modulation scheme most likely selected at each location. This is a service-dependent prediction as it should only be calculated for points where two-way service is available, thus, designers not only have the information of the selected modulation scheme but they also know the service area, that is, if the prediction was service independent, designers could see a large area of QPSK "service" on the downstream that, in reality, does not exist as there is no service there for the upstream, hence, a data session would not be established. The selected scheme applies to the downlink best server sector of the location. Of the schemes that satisfy the sensitivity test, the one that provides maximum data rate is shown as the "selected scheme" for each pixel.

21.3.21 Payload Data Rate

Figures 21.41 and 21.42 show the maximum achievable data-rate available at each pixel, considering the class and direction (downlink or uplink) being analyzed. The tool should calculate the maximum

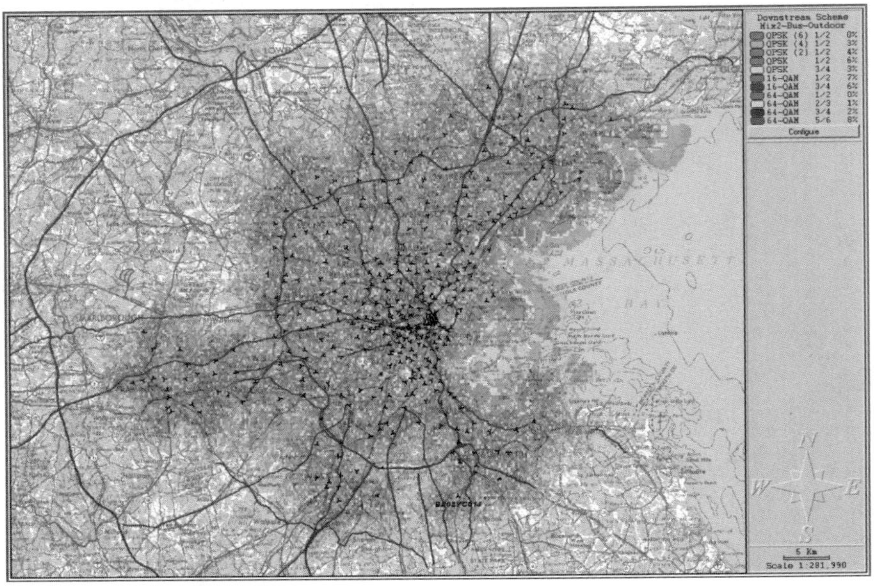

Figure 21.39 Modulation scheme plot downstream.

Figure 21.40 Modulation scheme plot upstream.

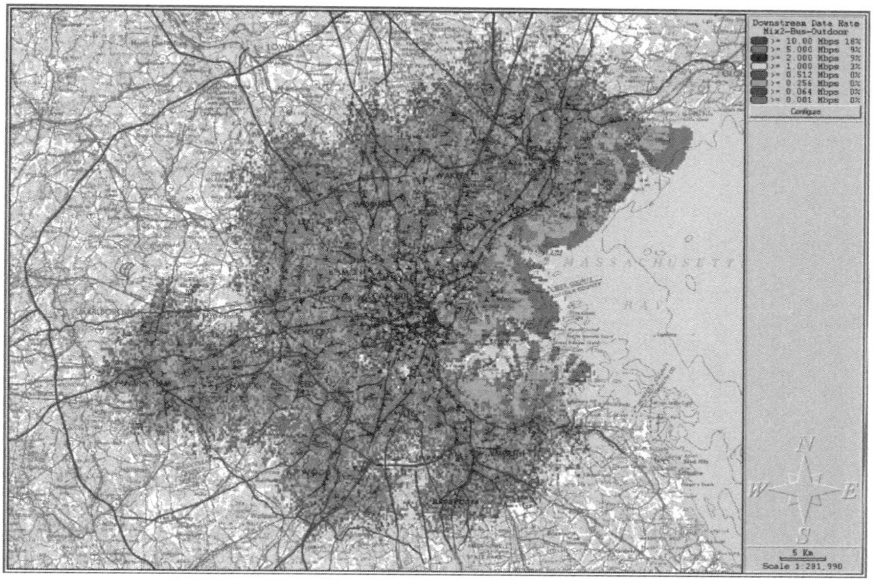

Figure 21.41 Payload data rate downstream plot sample.

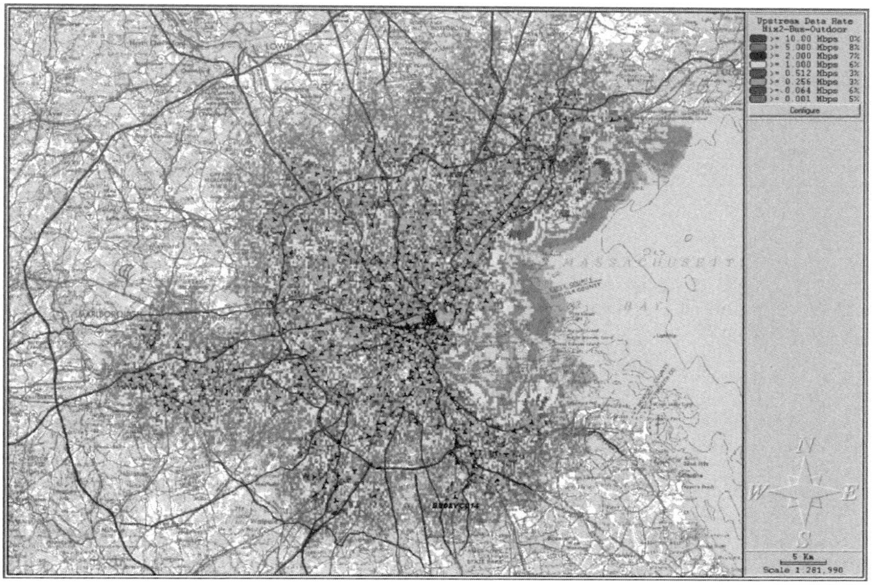

Figure 21.42 Payload data rate upstream plot sample.

Figure 21.43 Maximum data rate per sub-channel downlink.

achievable data rate considering the best server of the pixel in the downlink direction. From this, it is able to calculate the actual payload data rate, with the maximum scheme selection data rate minus the overhead factor, divided by the TDD ratio.

21.3.22 Maximum Data Rate Per Sub-Channel

This prediction indicates the data rate per sub-channel from the best zone selected in the pixel. In OFDM systems, this prediction is equivalent to the Max Data Rate/User prediction. Figures 21.43 and 21.44 illustrate this prediction.

21.3.23 Interference

Interference predictions are service-related and should be calculated separately for downstream and upstream as they might vary quite significantly for each direction. The interference is usually calculated in respect to the best server sector-radio unless otherwise specified.

21.3.23.1 Interference

Sometimes, planning tools offer different options for interference prediction calculation. The tool used for these examples, allows designers to choose one the following three options: I, C/I, I/N; and may also choose to consider Thermal Noise Floor. Figure 21.45 shows the different options selection mentioned in the example.

The selected interference analysis configuration affects how downstream and upstream Interference predictions are calculated. Table 21.6 describes the formulas for each of the possible configurations (without and with thermal noise considerations).

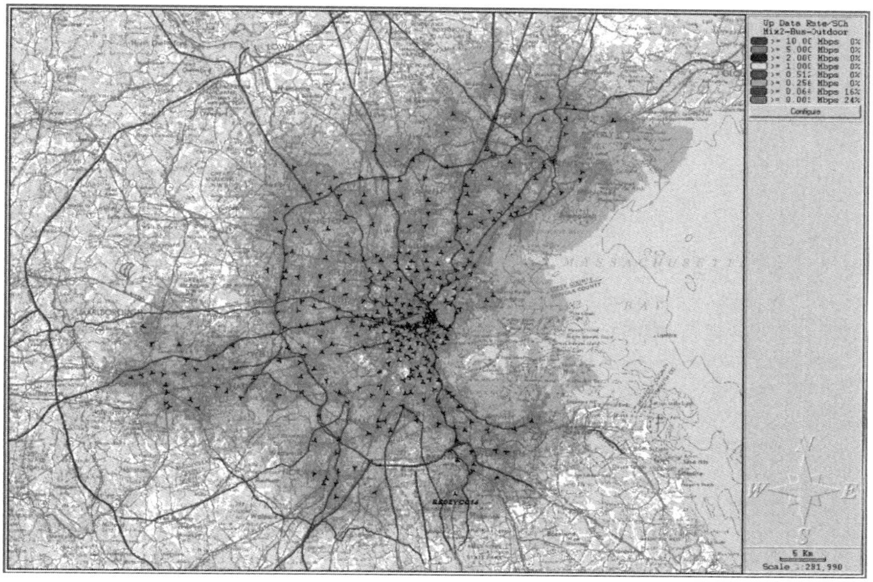

Figure 21.44 Maximum data rate per sub-channel uplink.

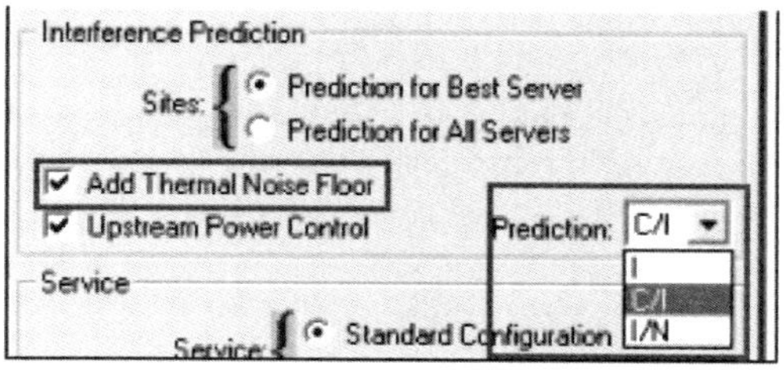

Figure 21.45 Interference configuration dialogue.

Table 21.6 Interference calculations

Selection	Without thermal noise	With thermal noise
I	$I_{(dBm)}$	$(I_{(W)} + N_{(W)})_{(dBm)}$
C/I	$BS_{(dBm)} - I_{(dBm)}$	$BS_{(dBm)} - (I_{(W)} + N_{(W)})_{(dBm)}$
I/N	$I_{(dBm)} - N_{(dBm)}$	$(I_{(W)} + N_{(W)})_{(dBm)} - N_{(dBm)}$*

Note: * the I/N prediction indicates how much the interference contributes to increase noise at each location When this prediction considers thermal noise, it is equivalent to the noise rise prediction.

Figure 21.46 Interference downstream.

Figures 21.46 and 21.47 illustrate these predictions.

21.3.24 Noise Rise

The Noise Rise prediction shows the ratio (I+N)/N at each pixel, considering the selected service class and analysis direction (forward/reverse link). The interference can be calculated considering the signal coming from the best server of the pixel or from all possible servers. It is illustrated in Figures 21.48 and 21.49.

21.3.25 Downstream/Upstream Service

This prediction shows, for each pixel, if there is two-way service from the same server-sector, service in only one direction (down or uplink), or no service. It is illustrated in Figure 21.50.

21.3.26 Service Margin

This prediction shows, in each pixel, the maximum link margin among all radios of all sectors that are compatible with the selected class. For a given sector-radio-zone trio, the margin is defined as the difference between the achievable SNIR and the required threshold for the selected radio. The margin can be negative if the required SNIR is greater than the obtained at the pixel. The required SNIR can be determined through the Rx Performance table, depending on the service class configuration (required BER, environment). The maximum margin among all zones of a radio is selected for each direction (up and downlink), and the minimum value between both directions should be selected as the resulting margin. Figure 21.51 illustrates this prediction.

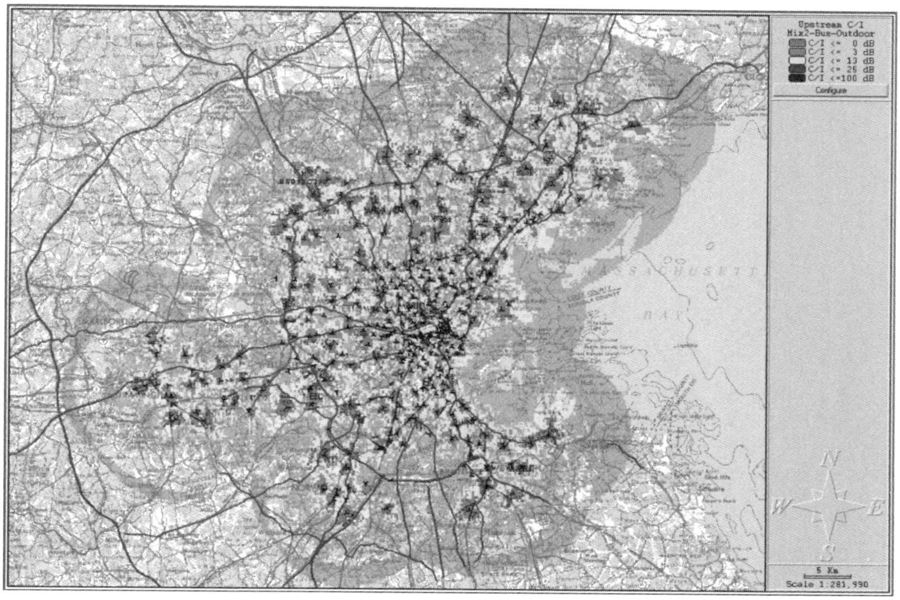

Figure 21.47 Interference upstream.

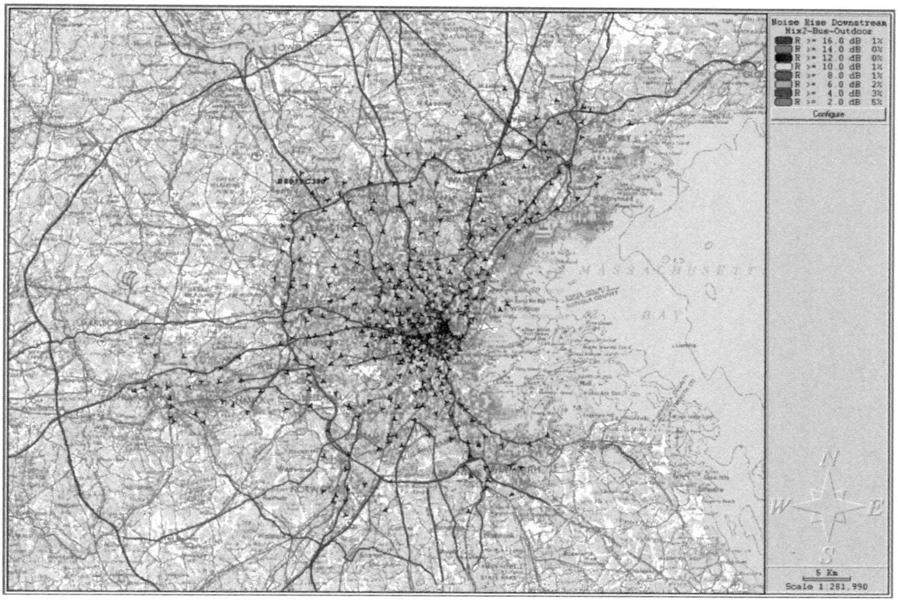

Figure 21.48 Noise Rise downstream.

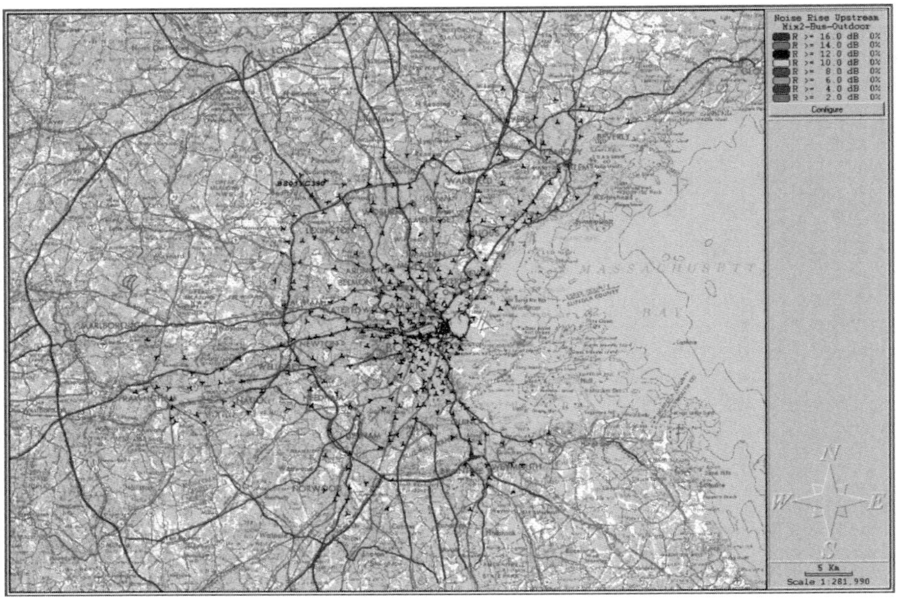

Figure 21.49 Noise Rise upstream.

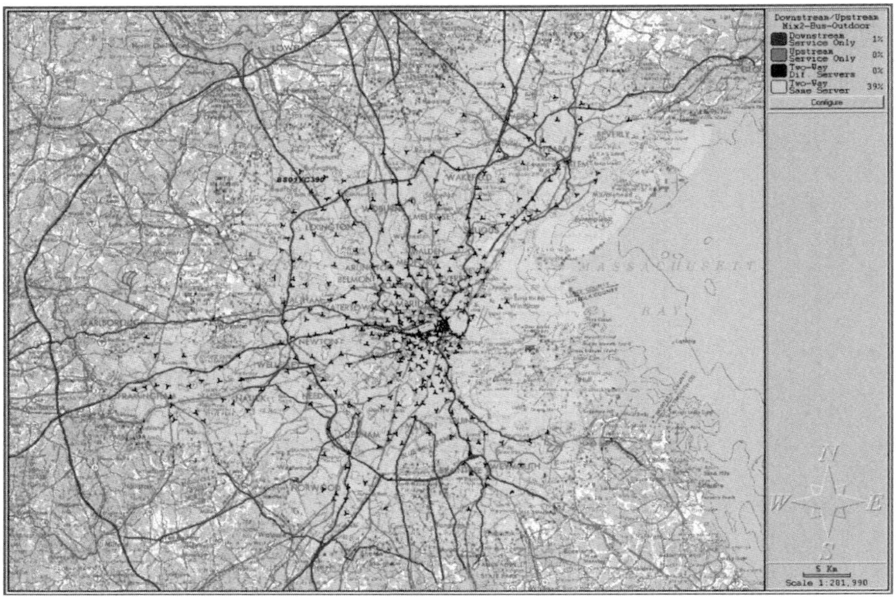

Figure 21.50 Downstream/upstream service.

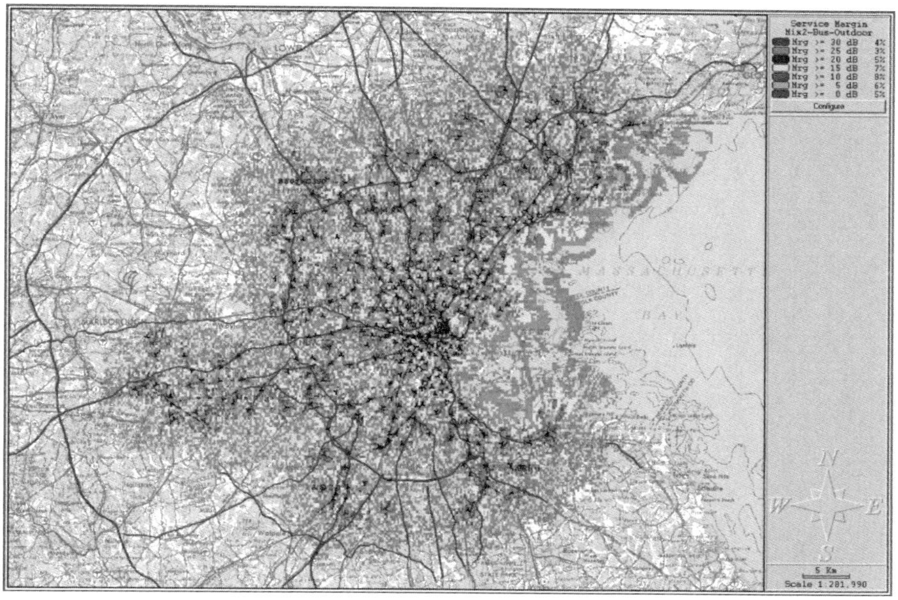

Figure 21.51 Service margin.

21.3.27 Service Classes

This prediction shows, for each pixel, the class with the highest downstream sensitivity with two-way service from the same server-sector. This prediction combines all classes in the same prediction. Figure 21.52 illustrates this prediction.

21.3.28 Channel (Frequency) Plan

It is important for designers to be able to graphically visualize the frequency plan as this gives a different perception of the plan versus looking at a spreadsheet that list channels per sector/site. A graphical representation allows designers to see, at the same time, which channel is being used where and where the sectors using each channel are physically located. There are numerous ways of displaying this information. In the example shown in Figure 21.53 channels are represented as a vector overlaid on top of a background image.

21.4 Backhaul Links Performance

Besides all the predictions for the wireless broadband network, for a complete analysis of the design, designers should also generate performance reports for the backhaul links to verify compliance with the specifications, as shown in Figure 21.54.

Figure 21.52 Service Class.

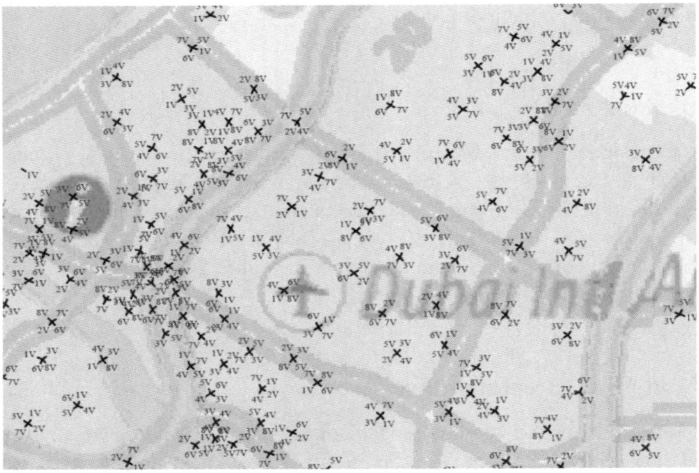

Figure 21.53 Channel Plan Plot sample – detail.

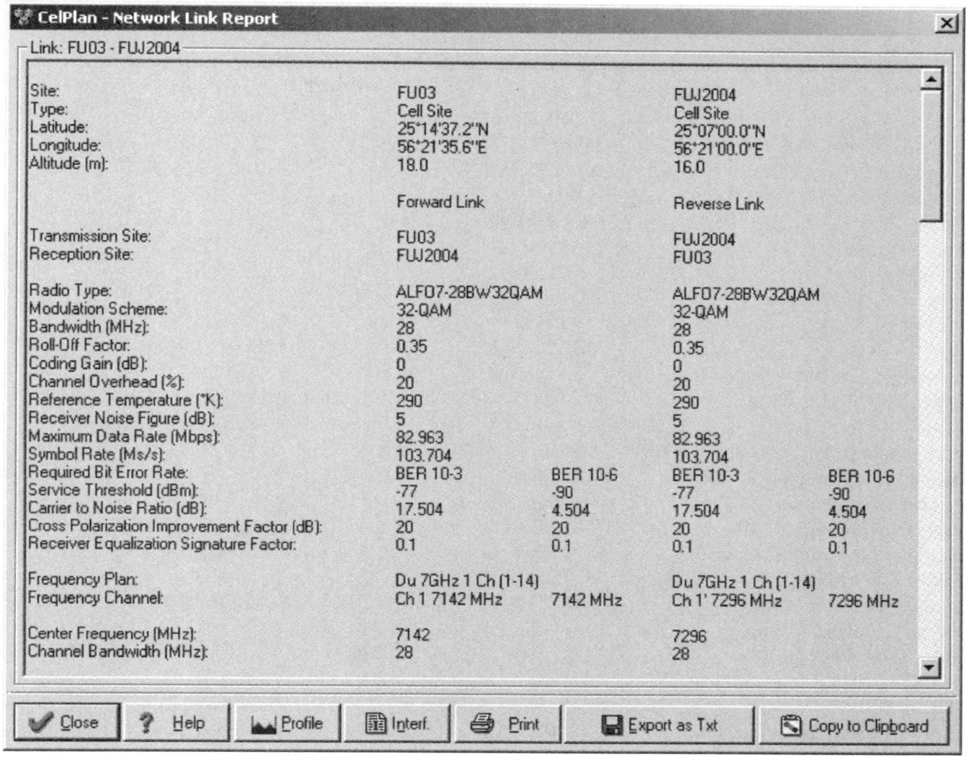

Figure 21.54 Link performance report.

The example shown in Figure 21.54 is displayed as a table in Table 21.7 for completeness.

Interference between links can also be analyzed and presented in the form of reports. Figure 21.55 shows a sample report containing this kind of analysis.

21.4.1 Backhaul Traffic Analysis

The last type of analysis for a network design should be backhaul traffic as the traffic aggregated by each link should be verified against its radio capacity. This traffic is not constant and a margin should be left for the variation, generally 25% above capacity for a single link.

For aggregated links, such a high margin is not required as traffic variations will most likely compensate for each other, in these cases a smaller margin of 10% may be sufficient. The actual margin should be given by the designer based on his understanding of possible traffic variations.

Table 21.7 Link performance table

Site:	FU03		FUJ2004	
Type:	Cell Site		Cell Site	
Latitude:	25°14′37.2″N		25°07′00.0″N	
Longitude:	56°21′35.6″E		56°21′00.0″E	
Altitude (m):	18		16	
	Forward Link		Reverse Link	
Transmission Site:	FU03		FUJ2004	
Reception Site:	FUJ2004		FU03	
Radio Type:	ALFO7-28BW32QAM		ALFO7-28BW32QAM	
Modulation Scheme:	32-QAM		32-QAM	
Bandwidth (MHz):	28		28	
Roll-Off Factor:	0.35		0.35	
Coding Gain (dB):	0		0	
Channel Overhead (%):	20		20	
Reference Temperature (°K):	290		290	
Receiver Noise Figure (dB):	5		5	
Maximum Data Rate (Mbps):	82.963		82.963	
Symbol Rate (Ms/s):	103.704		103.704	
Required Bit Error Rate:	BER 10-3	BER 10-6	BER 10-3	BER 10-6
Service Threshold (dBm):	−77	−90	−77	−90
Carrier to Noise Ratio (dB):	17.504	4.504	17.504	4.504
Cross Polarization Improvement Factor (dB):	20	20	20	20
Receiver Equalization Signature Factor:	0.1	0.1	0.1	0.1
Frequency Plan:	Du 7 GHz 1 Ch (1–14)		Du 7 GHz 1 Ch (1–14)	
Frequency Channel:	Ch 1 7142 MHz	7142 MHz	Ch 1′ 7296 MHz	7296 MHz
Center Frequency (MHz):	7142		7296	
Channel Bandwidth (MHz):	28		28	
Link Polarization:	Vertical		Vertical	
Transmission Power (dBm):	30		30	
Transmission Losses (dB):	1		1	
Transmission Antenna:	THP18-071S		THP18-071S	
Transmission Antenna Height (m):	30		30	
Transmission Antenna Gain (dBd):	34.86		34.86	
Transmission Antenna Gain (dBi):	37		37	
Transmission Power EIRP (dBm):	66		66	
Reception Losses (dB):	1		1	
Reception Antenna:	THP18-071S		THP18-071S	
Reception Antenna Height (m):	30		30	
Reception Antenna Gain (dBd):	34.86		34.86	
Reception Antenna Gain (dBi):	37		37	

Table 21.7 (*continued*)

Link Distance (m):	14143.659		14143.659	
Azimuth - True (°):	184.043		4.038	
Azimuth - Magnetic (°):	182.691		2.708	
Transmission Inclination (°):	0.008		−0.008	
Reception Inclination (°):	0.008		−0.008	
Free Space Distance (m):	14143.659		14143.659	
Center Frequency (MHz):	7142		7296	
Free Space Loss (dB):	132.528		132.713	
Earth Radius Factor:	4/3			
Effective Radius (m):	8502056			
Diffraction:	No Diffraction		No Diffraction	
Diffraction Loss (dB):	0		0	
Clearance Target (%):	60			
Minimum Clearance (m):	9.555		9.578	
Minimum Clearance Point (m):	311.845		311.845	
Terrain Reflection Dispersion (°):	0.5			
Reflection Area 1 (m):	593.511		593.511	
Reflection Area 2 (m):	653.868		653.868	
Reflection Area 3 (m):	694.1–714.2		694.1–714.2	
Reflection Area 4 (m):	774.582		774.582	
Reflection Area 5 (m):	2987.7–3068.1		2987.7–3068.1	
Reflection Area 6 (m):	3128.5–3249.2		3128.5–3249.2	
Reflection Area 7 (m):	3410.2–3450.4		3410.2–3450.4	
Reflection Area 8 (m):	3510.8–3571.1		3510.8–3571.1	
Reflection Area 9 (m):	3631.5–3852.8		3631.5–3852.8	
Reflection Area 10 (m):	3913.1–4033.9		3913.1–4033.9	
Reflection Area 11 (m):	4094.217		4094.217	
Reflection Area 12 (m):	4154.574		4154.574	
Reflection Area 13 (m):	4214.9–4235		4214.9–4235	
Reflection Area 14 (m):	4295.4–4436.2		4295.4–4436.2	
Reflection Area 15 (m):	4496.6–4717.9		4496.6–4717.9	
Reflection Area 16 (m):	4778.3–5120.3		4778.3–5120.3	
Atmospheric Pressure (hPa):	1013		1013	
Standard Temperature (°C):	15		15	
Water Vapor Density (g/m^3):	7.5		7.5	
Atmospheric Gases Loss (dB):	0.139		0.141	
Total Path Loss (dB):	132.666		132.854	
Reception Signal Level (dBm):	−30.666		−30.853	
	BER 10-3	BER 10-6	BER 10-3	BER 10-6
Service Threshold (dBm):	−77	−90	−77	−90
Link Gross Margin (dB):	46.334	59.334	46.146	59.146

(*continued overleaf*)

Table 21.7 *(continued)*

Objective ITU Quality Grade:	Short Haul SDH Networks			
Multipath Model:	ITU			
Multipath Link Area:	Inland Link			
Multipath Terrain Type:	Low altitude (< 400 m) Plains			
Multipath Climate Variable:	20			
Multipath Occurrence Factor:	5.92E+00		6.03E+00	
	BER 10-3	BER 10-6	BER 10-3	BER 10-6
Fading Outage (%):	1.38E-04	6.90E-06	1.46E-04	7.34E-06
Selective Fading Outage (%):	8.06E-06	8.06E-06	8.17E-06	8.17E-06
Composite Fading Outage (%):	1.46E-04	1.50E-05	1.55E-04	1.55E-05
ITU Error Performance Objective (%):	1.60E-04	1.28E-02	1.60E-04	1.28E-02
Fading Outage (s/Month):	3.616	0.181	3.848	0.193
Selective Fading Outage (s/Month):	0.212	0.212	0.215	0.215
Composite Fading Outage (s/Month):	3.828	0.393	4.063	0.408
ITU Error Performance Objective (s/Month):	4.205	336.384	4.205	336.384
Precipitation Model:	ITU			
Precipitation Rate @ 0.01% (mm/h):	22			
Rainfall Attenuation (dB):	1.469		1.564	
	BER 10-3	BER 10-6	BER 10-3	BER 10-6
Unavailability due to Rain (%):	8.40E-04	8.40E-04	8.40E-04	8.40E-04
Unavailability due to Rain (s/Year):	265.013	265.013	265.013	265.013
	BER 10-3	BER 10-6	BER 10-3	BER 10-6
Unavailability due to Fading (%):	1.46E-04	1.50E-05	1.55E-04	1.55E-05
Unavailability due to Rain (%):	8.40E-04	8.40E-04	8.40E-04	8.40E-04
Total Unavailability (%):	9.86E-04	8.55E-04	9.95E-04	8.56E-04
ITU Unavailability Objective (%):	1.65E-02	1.65E-02	1.65E-02	1.65E-02
Unavailability due to Fading (s/Year)	45.939	4.717	48.754	4.892
Unavailability due to Rain (s/Year):	265.013	265.013	265.013	265.013
Total Unavailability (s/Year):	310.952	269.73	313.767	269.905
ITU Unavailability Objective (s/Year):	5203.44	5203.44	5203.44	5203.44
	* PASS *	* PASS *	* PASS *	* PASS *

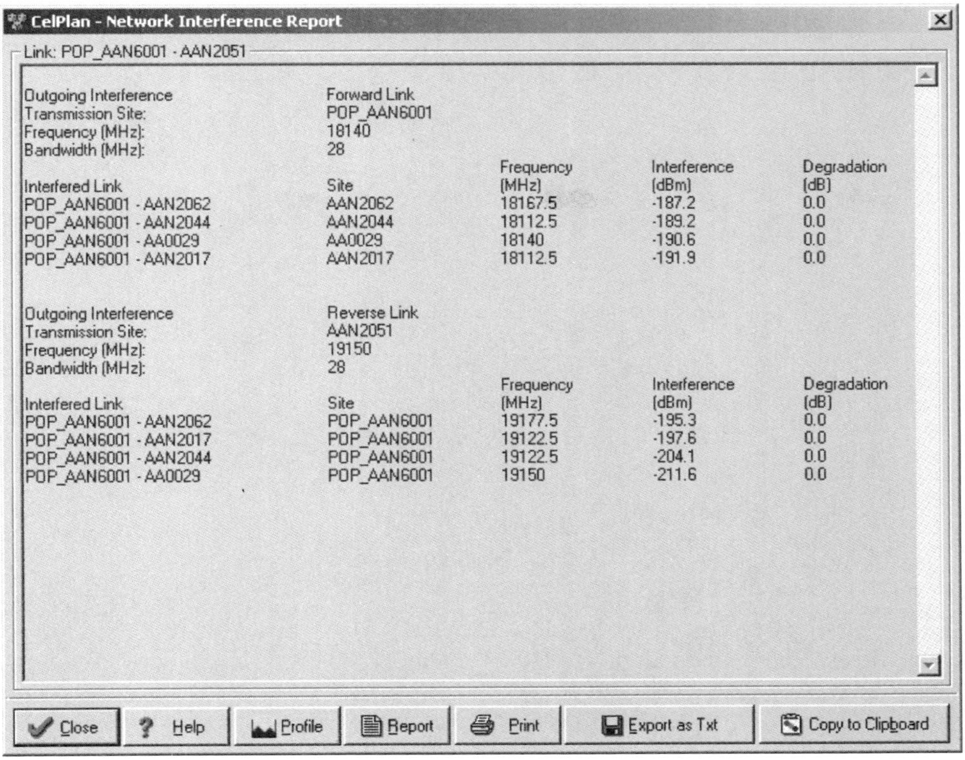

Figure 21.55 Network links interference report.

21.5 Analyze Performance Results, Analyze Impact on CAPEX, OPEX and ROI

The result of the first design iteration has to be analyzed to see if there is room to reduce or need to increase the number of sites. The actual noise rise of the network is now known with a good confidence, and this makes any redesign easier to do.

Once the CAPEX, OPEX and ROI figures are satisfactory, the design can go to the deployment phase.

In the deployment phase, the actual sites for deployment are sought based on Site Search Ring Areas and actual deployment heights are defined.

The design should be continuously updated as the build-up is done.

22

Basic Mathematical Concepts Used in Wireless Networks

There are few basic mathematical concepts that are essential to the understanding of the techniques used in wireless networks, this chapter covers most of them. It is very helpful to understand how these concepts evolved historically and what they tried to achieve. For this reason, we give a historical background of each concept, keeping the mathematics to a minimum, and, instead, emphasizing the principles involved and their outcome.

22.1 Circle Relationships

A circle is a very important geometric figure used to represent many relationships. It is used, for example, to represent locations on the Earth's globe (coordinates) and phase and amplitude relationships.

The calculation of the circle length, or circle linearization, was not an easy task and efforts have been made since antiquity to express it as a function of the circle's radius. It was soon established that the ratio between the length and the radius was a constant, but the exact value was unknown and it could only be approximated by use of fractions.

Another challenge was to calculate the area encompassed by the circle, or circle quadrature. Again it was suspected that there was a constant relationship between the area and the radius, similar to the one used in the circle linearization, but that constant could not be calculated either.

It was Archimedes (287–212BC) who demonstrated that the same constant applied to both relationships and he called it π, deriving it from the Greek word "περίμετρος" meaning perimeter. He calculated π to be between 3 10/71 (3.1429) and 3 1/7 (3.1408) by approximating it by the perimeter of a 96-side polygon inscribed and outscribed to a circle. The circle perimeter is given by Equation (22.1) and its area by Equation (22.2).

$$C = 2\pi R \qquad \text{(22.1) Circle perimeter}$$

$$A = \pi R^2 \qquad \text{(22.2) Circle area}$$

LTE, WiMAX and WLAN Network Design, Optimization and Performance Analysis, First Edition.
Leonhard Korowajczuk.

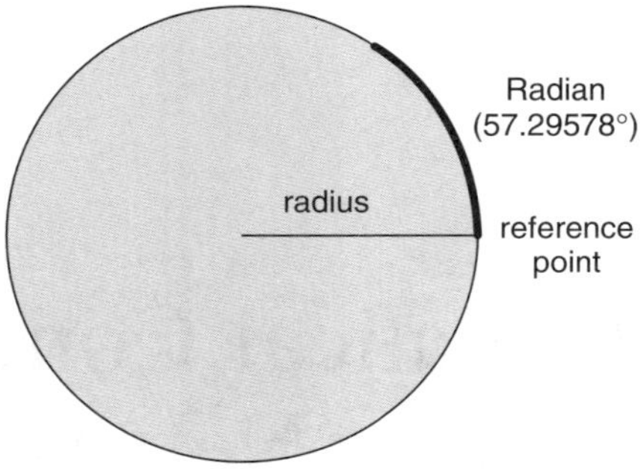

Figure 22.1 Circle representation.

Where:

$C =$ circle length.
$A =$ circle inscribed area.
$R =$ circle radius.
$\pi =$ constant pi $= 3.14159\ 26535\ 89793\ldots\ldots$

The circle circumference can be used to map distances from a reference point on the circle. Graphically this reference point is traditionally represented on the right side of the circle, as illustrated in Figure 22.1.

This distance can then be defined by the length of the circumference between the reference point and the destination point or by the angle formed from the center of the circle between the reference and the destination point.

Traditionally a circle is divided into 360 equal parts called degrees (represented by°); each degree is divided into 60 minutes (represented by ') and each minute into 60 seconds (represented by '').

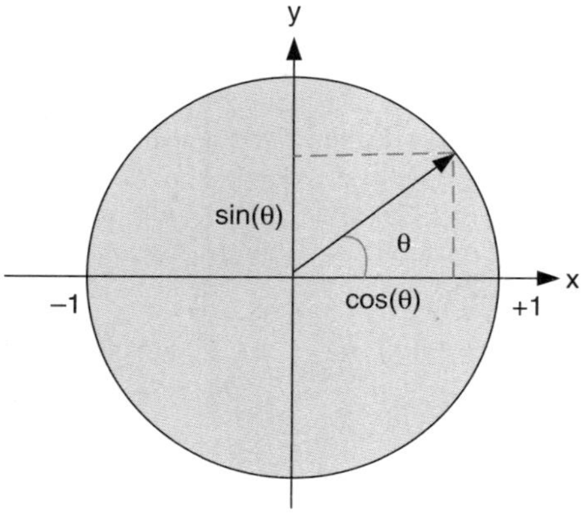

Figure 22.2 Circle location projections on orthogonal axis.

Seconds are presented with decimals In some applications, decimal degrees are used instead of minutes and seconds.

Over time (since 1700), it was perceived that it was advantageous to represent the angles in terms of distance from the reference point expressed in number of radius lengths. A distance of one radius over the circle length is called a radian, so a circle has a length of 2π radians. This allows us to express a location using radians, with 1 radian being approximately equal to 57° 17' 44.806'' (degrees, minutes, and seconds) or 57.29578° (decimal degrees). One degree (1°) is approximately equal to 0.0175 radians. A radian expresses also the length of the circle encompassed by the angle. A radian is illustrated in Figure 22.1.

A location on the circle can be projected on two orthogonal axes centered within the circle. One axis crosses the reference point and the angle to the location is measured from this point. The projection of the location to this axis is defined by a function cos (θ) and the projection on the orthogonal axis is defined by a function sin (θ). This is illustrated in Figure 22.2.

22.2 Numbers and Vectors

Numbers are abstract concepts which took a while to be fully developed. Some indigenous tribes counted only to two or three and after those numbers, they just used the concept of "many". Babylonians recorded objects as integers, but not very long afterwards the concept of fractions came about. Non-integer numbers were represented initially only by fractions. A fraction division either terminates after some digits or repeats the same sequence over and over.

22.2.1 Rational and Irrational Numbers

Numbers resulting from fractions were called rational in contrast to numbers that could not be expressed by a fraction, which were called irrational numbers.

Rational numbers can be represented by a ratio of two integer m/n and terminates after a finite number of digits or repeats the same sequence of numbers over and over. The set of rational numbers is represented by **Q** (quotient).

Irrational numbers cannot be represented by a ratio of two integer m/n and they never end. Examples of irrational numbers are: $\sqrt{2}$ (1.414213562), π (3.141599265...), e (2.71828182...) and ϕ (1.618033987...), the latter known as the Golden Ratio.

Numbers were used in algebraic equations and were represented geometrically on a line segment. This representation covered rational and irrational numbers, as illustrated in Figure 22.3.

Some equations, however, resulted in negative numbers and it took a while for these results to be accepted by all. They were finally accepted due to problems dealing with debt.

The geometrical representation of negative numbers required an infinite straight line, with zero as a reference point, as illustrated in Figure 22.4. This representation can be easily applied to express quantities and lengths and the basic arithmetic operations can be easily executed graphically.

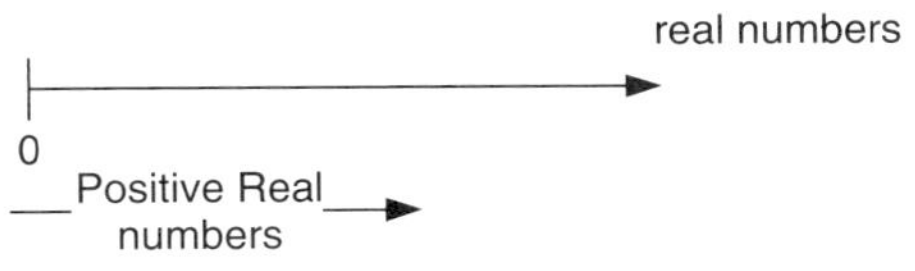

Figure 22.3 Initial representation of real numbers.

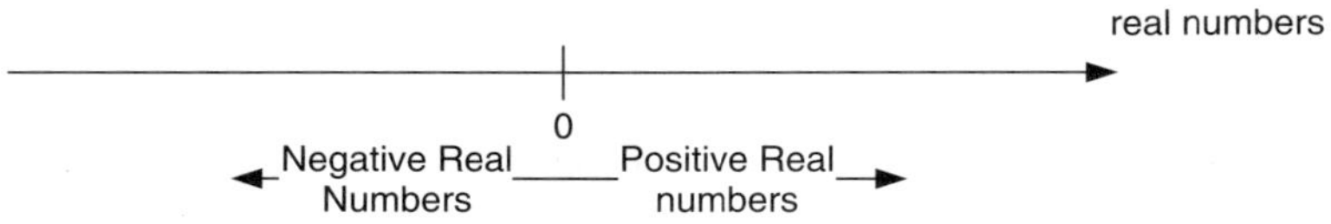

Figure 22.4 Real numbers representation.

22.2.2 *Imaginary Numbers* ($i = \sqrt{-1}$)

Algebraic equations express functions of a variable. These functions have result in a zero for certain values of its variable, which are called equation roots. The most common type of equation is the polynomial, characterized by expressions of finite length constructed from variables and constants using only, addition, subtraction, multiplication, and non negative, integer exponents. The degree of a polynomial equation is defined by the highest exponent of its variable. A solution for second degree polynomials of the type $ax^2 + bx + c = 0$ was soon found to be $x = \frac{-b \pm \sqrt{b^2 - 4ac}}{2a}$. A solution for third degree polynomials, such as $ax^3 + bx^2 + cx + d = 0$, known as cubic equation, was harder to find, so all efforts concentrated on the reduced cubic $x^3 + ax = b$.

In 1545, Girolamo Cardano proposed a formula to solve the reduced cubic equation, but it would give in some cases only one real solution and two solutions that contained $\sqrt{-1}$. These solutions were initially discarded as impossible, even though in some cases they should have provided real values as a result. At that time it was suspected that polynomial equations with a degree "n" should have n solutions. Rafael Bombelli, in 1572, was able to manipulate the results of Cardan's formula to calculate the expected real values. This led to the acceptance of $\sqrt{-1}$, but it was called an imaginary number. In 1674, Gottfried Leibniz proposed that $\sqrt{-1}$ be represented by the letter "i " (for "imaginary"), formalizing its acceptance in the calculations, even without a physical meaning for it.

The quest to provide geometric solutions for the cubic was open, but it failed in presenting solutions with imaginary numbers. René Descartes in 1637 stated that those solutions were impossible. It was John Wallis in 1685 who gave the first insight into the possibility that those solutions might exist, but outside the realm of real numbers. Real numbers were represented by a straight line, described previously, and they may lie in a complex plane that adds directionality to the real numbers.

At the same time, efforts were made to represent vectors geometrically. Vectors have a magnitude and a direction, so there was a need to add a second dimension to the real numbers representation, forming a plane.

Vectors were then drawn in this plane and were defined by their magnitude and an angle, as shown below. The vector $\vec{a}$ can then be represented by its magnitude "a" and its argument (angle) θ, as expressed by Equation (22.3) and represented in Figure 22.5.

$$\vec{a} = a\llcorner\theta \qquad \text{(22.3) Vector representation}$$

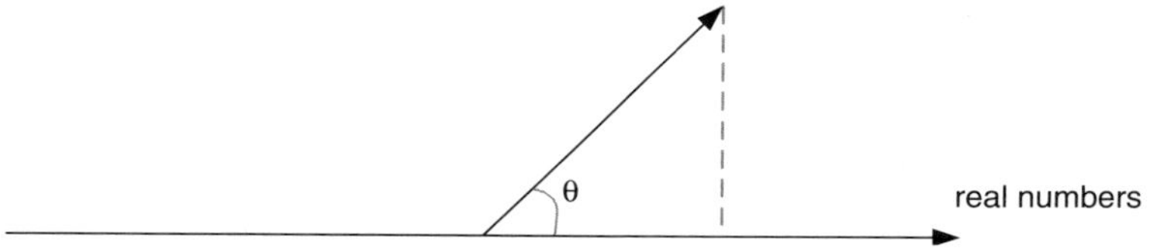

Figure 22.5 Vector representation over the real numbers axis.

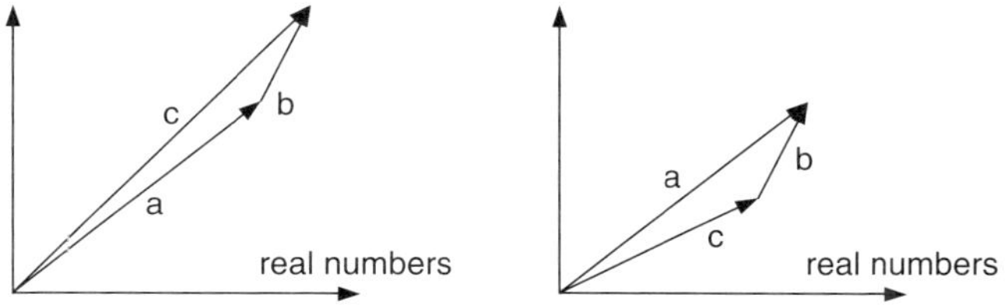

Figure 22.6 Vector addition (left) and subtraction (right).

The plane where vectors were represented was called "the complex plane" (as it was very difficult to conceive), and the additional axis was termed as the "imaginary axis" (as nobody knew precisely what it meant).

In 1797, it was Casper Wessel who gave a physical explanation for $\sqrt{-1}$ while working with vectors. He analyzed vector addition, subtraction and multiplication. Addition and subtraction can be intuitively done if we consider an additional dimension to the real numbers line, creating a 2D plane, so each real number will have an additional component that would define the vector direction, as illustrated below.

The sum of two vectors was done by positioning the starting point of the second vector at the terminal point of the first vector. The vector that connects the start of the first vector with the end of the second is the vector resultant from the sum. This operation can be done by adding the projections of each vector in the Cartesian coordinates, as illustrated in Figure 22.6.

Wessel relied on the DeMoivre formula, in 1722, to perform vector products. DeMoivre formula is shown in Equation (22.4).

$$(\cos\theta + i\sin\theta)^n = \cos(n\theta) + i\sin(n\theta) \qquad \text{(22.4) DeMoivre formula}$$

For two vectors this can be expressed by Equation (22.5).

$$(\cos\theta + i\sin\theta)^2 = \cos(2\theta) + i\sin(2\theta) \qquad \text{(22.5) DeMoivre formula for } n = 2$$

This means that if we want to multiply two vectors, we need to multiply their magnitudes and add their arguments (angles). This is illustrated in Figure 22.7.

Wessel's breakthrough that allowed him to implement this graphically was to observe that the product of two numbers had the same ratio to each of the numbers as the other number has to unity. He argued that the length of the product of two vectors should be the product of the numbers, and the direction of the resultant vector should differ in direction from one vector by the same amount the other vector differs from a unity vector used as a reference. This meant that the resultant vector direction will be the sum of the angles of the product vectors to a reference unit vector.

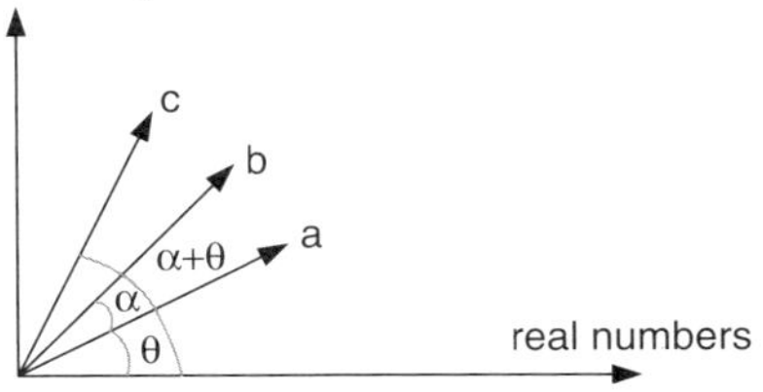

Figure 22.7 Unitary vector M.

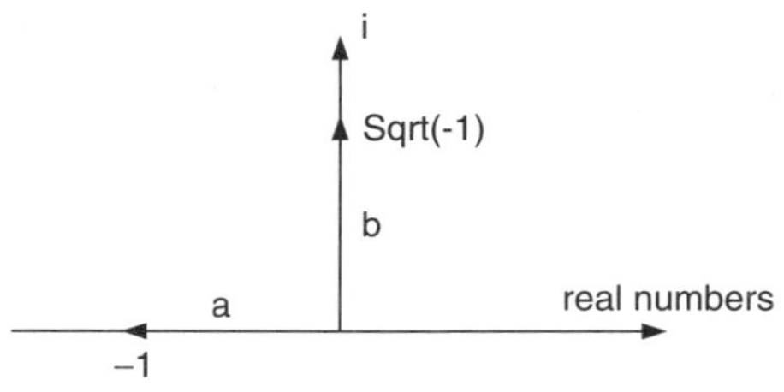

Figure 22.8 Physical interpretation of an imaginary number.

This said, he stated that the vector in Equation (22.6)

$$\vec{a} = 1\llcorner 180^\circ = -1 \qquad \text{(22.6) 180° vector}$$

was the product of two vectors described in Equation (22.7).

$$\vec{\imath} = 1\llcorner 90^\circ \qquad \text{(22.7) 90° vector}$$

and that this vector value would be $\vec{\imath} = \sqrt{-1}$, representing the long sought unit of this axis. This is illustrated in Figure 22.8.

The vector representation coincided with the algebraic calculations of a cubic equation and was able to express what Wallis has suggested, by placing the roots of the cubic in the complex plane. This imaginary axis was represented by "i" and replaced by "j" when dealing with electrical signals, so it is not to be confused with the current represented traditionally by "i".

Multiplying a number (real or complex) by $\sqrt{-1}$ corresponds to a rotation in the complex plane of 90° CCW (counter-clockwise). In reality, the axis "i" corresponded to a shift of 90° in relation to the real numbers, and represented a degree of orthogonality.

22.3 Functions Decomposition

The decomposition of complex functions into series of more elementary functions allows us to calculate their values more easily.

When analyzing the response of a system to a complex signal it is much easier to predict the system response by analyzing its response to the more simple components of the complex signal. The decomposition of signals in polynomials is presented next.

22.3.1 Polynomial Decomposition

In 1712, Brook Taylor demonstrated that a differentiable function around a given point can be approximated by a polynomial whose coeficients are the derivatives of the function at that point. The Taylor series is then a representation of a function as an infinite sum of terms calculated from its derivatives at a single point. If this series is centered at zero, it is called a MacLaurin series.

A function $f(x)$, differentiable at a point a, can be approximated by the polynomial in Equation (22.8).

$$f(x) = f(a) + \frac{f'(a)}{1!}(x-a) + \frac{f''(a)}{2!}(x-a)^2 + \frac{f^{(3)}(a)}{3!}(x-a)^3 + \cdots\cdots$$ (22.8) Approximation of differentiable function *f(x)*

When $a = 0$ we get the MacLaurin series (1725), given by Equation (22.9).

$$f(x) = f(0) + f'(0)x + \frac{f''(0)}{2!}x^2 + \frac{f^{(3)}(0)}{3!}x^3 + \cdots\cdots$$ (22.9) MacLaurin series

We can decompose $\sin(x)$ and $\cos(x)$ using the MacLaurin series, as both functions are differentiable over and over. The derivatives of sin and cos are shown in Equations (22.10) and (22.11).

$$\sin'(x) = \cos(x)$$ (22.10) Differentiation of sin (x)

$$\cos'(x) = -\sin(x)$$ (22.11) Differentiation of cos (x)

Sine of 0° is given by Equation (22.12):

$$\sin(0) = 0$$ (22.12) Sin of 0

Cosine of 0° is given by Equation (22.13):

$$\cos(0) = 1$$ (22.13) Cos of 0

Cosine and sine can then be decomposed as shown in Equations (22.14) and (22.15).

$$\cos(\theta) = 1 - \frac{\theta^2}{2!} + \frac{\theta^4}{4!} - \frac{\theta^6}{6!}$$ (22.14) Cosine decomposition

$$\sin(\theta) = \theta - \frac{\theta^3}{3!} + \frac{\theta^5}{5!} - \frac{\theta^7}{7!}$$ (22.15) Sine decomposition

22.3.2 *Exponential Number* (e)

The power function $f(x) = a^x$ is very useful to represent variations that are experienced in real life and its representation by polynomials has been sought.

When we apply Taylor's expansion to the exponential function we have to calculate its derivative at a point, which is given by Equation (22.16).

$$f'(x_0) = \lim[(f(x_0 + \delta) - f(x_0))/\delta]$$ (22.16) Function derivative at a point when $\delta \to 0$

The derivative of the power function is given by Equation (22.17):

$$f'(a^{x_o}) = \lim[(a^\delta - 1)/\delta]$$ (22.17) Derivative of a^{x_o} when $\delta \to 0$

Calculating the above limit when $\delta \to 0$, we see that it is less than 1 for $a = 2$ and greater than 1 for $a = 3$. It is possible then to find a value of a, for which this limit is equal to 1 and the derivative of the function will then be itself. This value is irrational and is approximated by 2.718281 ....

This mathematical constant is a unique real number with its derivative at a point $x = 0$ exactly equal to itself. This number is called the exponential number and is represented by "*e*".

The MacLaurin expansion of $e^{i\theta}$ is given by Equation (22.18).

$$e^{i\theta} = 1 + i\theta - \frac{\theta^2}{2!} - \frac{i\theta^3}{3!} + \frac{\theta^4}{4!} + \frac{i\theta^5}{5!} - \frac{\theta^6}{6!} - \frac{i\theta^7}{7!} + \cdots$$ (22.18) MacLaurin expansion of $e^{i\theta}$

This expansion can be divided into a real part and an imaginary part, as shown in Equations (22.19) and (22.20):

$$\mathrm{Re}\{e^{i\theta}\} = 1 - \frac{\theta^2}{2!} + \frac{\theta^4}{4!} - \frac{\theta^6}{6!} + \cdots$$ (22.19) Real part of $\{e^{i\theta}\}$

$$\mathrm{Im}\{e^{i\theta}\} = \theta - \frac{\theta^3}{3!} + \frac{\theta^5}{5!} - \frac{\theta^7}{7!} + \cdots$$ (22.20) Imaginary part of $\{e^{i\theta}\}$

The real part is equal to the expansion of the cosine and the imaginary part to the expansion of the sine, so we can write Equation (22.21):

$$e^{i\theta} = \cos(\theta) + i\sin(\theta)$$ (22.21) Euler's formula $\{e^{i\theta}\}$ as a function f sin and cos

This amazing relationship is called Euler's formula and from it Euler's identity can be derived. This identity is shown in Equations (22.22) and (22.23).

$$e^{i\pi} = \cos(\pi) + i\sin(\pi) = -1$$ (22.22) Euler's identity derivation

$$e^{i\pi} - 1 = 0$$ (22.23) Euler's identity

The modulus of $e^{i\theta}$ is given by Equation (22.24):

$$\sqrt{\cos^2(\theta) + \sin^2(\theta)}$$ (22.24) Modulus of $e^{i\theta}$

The argument is given by Equation (22.25).

$$\tan^{-1}\left(\frac{\sin(\theta)}{\cos(\theta)}\right)$$ (22.25) Argument of $e^{i\theta}$

Varying θ we find that the values $e^{i\theta}$ are represented by a unitary circle in the complex plane, shown in Figure 22.9.

22.4 Sinusoids

Sinusoids are very important functions as they are used to represent RF waveforms and used to decompose complex waveforms. A sinusoid is defined by a uniformly rotating vector which takes a period T to complete one rotation, resulting in a rotation frequency of $f = 1/T$, as shown in Figure 22.10.

The sinusoid can be defined in Cartesian coordinates and the instantaneous position of the rotating vector can be projected on the x (real) and y (imaginary) axis. These projections vary over time and, the function that captures the variation of the projection on the x axis is called "cosine", whereas the one that captures the variation of the projection on the y axis is called "sine", as shown in Figures 22.11 and 22.12. The cosine waveform is given by Equation (22.26).

$$f(t) = \cos(2\pi f t) = \cos(\omega t)$$ (22.26) Cosine waveform

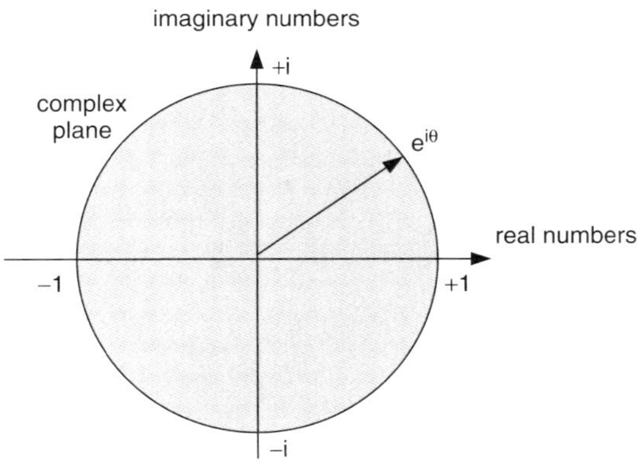

Figure 22.9 Representation of $e^{i\theta}$.

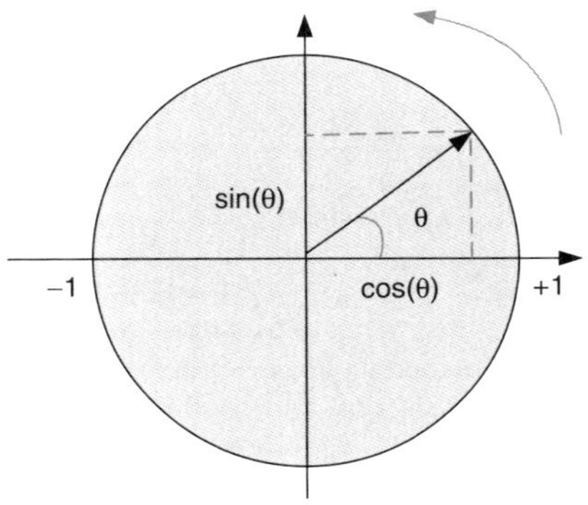

Figure 22.10 Rotating vector generating sinusoids.

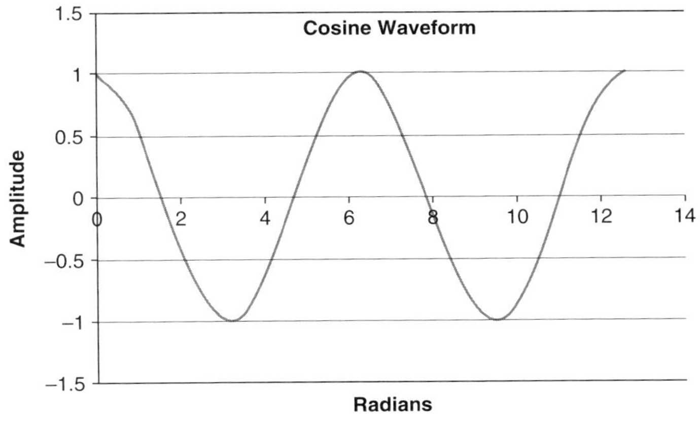

Figure 22.11 Cosine waveform.

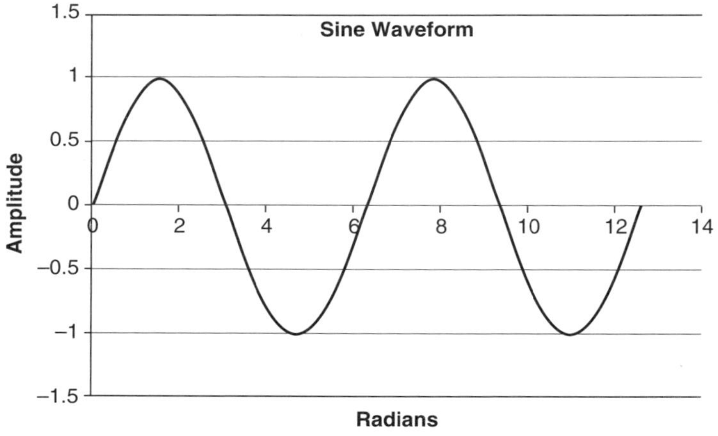

Figure 22.12 Sine waveform.

The sine waveform is given by Equation (22.27):

$$f(t) = \sin(2\pi f t) = \sin(\omega t) \qquad \text{(22.27) Sine waveform}$$

where f is the rotation frequency, 2π is the circle perimeter expressed in radians and consequently $2\pi ft$ is the total traveled distance by the vector in duration of a time "t". We nickname $2\pi f$ as ω, and it represents the velocity of the rotation expressed in radians (angular velocity).

22.4.1 Positive and Negative Frequencies ($+\omega$, $-\omega$)

Figure 22.13 shows the vector traveling in time. It starts from the origin and travels counter-clockwise (CCW) on the reference circle. This movement generates helicoids over time. The counter-clockwise rotation is assumed positive, resulting in a positive value of ω.

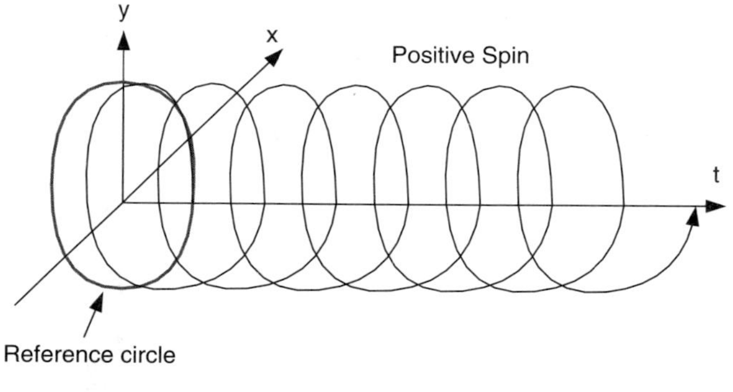

Figure 22.13 Sinusoid generated by a counter-clockwise rotation resulting in a positive ω.

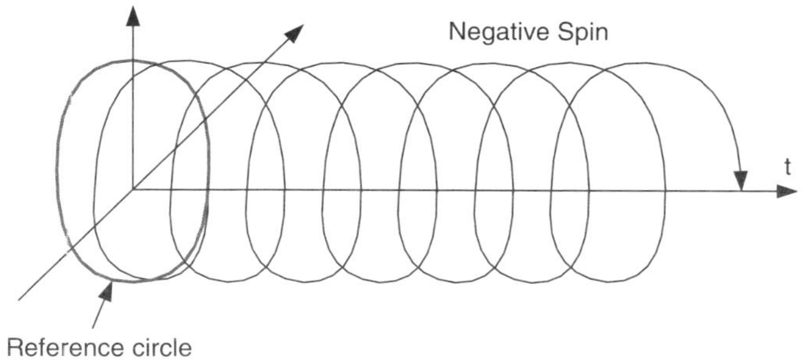

Figure 22.14 Sinusoid generated by a clockwise rotation resulting in a negative ω.

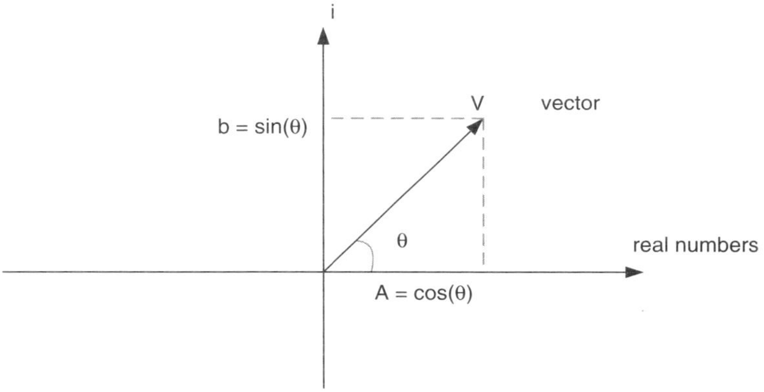

Figure 22.15 Complex plane used to represent vectors.

The clock wise (CW) rotation is negative and results in a negative value of ω, as shown in Figure 22.14.

Both sinusoids have the same frequency, but as 2π is positive, the frequency "f" has to be negative. This means that positive and negative frequencies represent the same sinusoid, with opposite phases. Positive sinusoids and negative sinusoids are orthogonal to each other.

As seen in Section 22.2, the complex plane is used to represent vectors. Vectors are represented by a real number in the real numbers axis and another number in the imaginary axis, as shown in Figure 22.15.

The vector $\vec{z} = a + ib$ shown in Figure 22.15 can be represented un Cartesian form: (a,b) where Equations (22.28) and (22.29) give the parameters for a Cartesian representation.

$$a = V\cos(\theta) \quad \text{(22.28) Vector in Cartesian form value of } a$$

$$b = V\sin(\theta) \quad \text{(22.29) Vector in Cartesian form value of } b$$

In polar form $m \llcorner a$, where "m" is the magnitude and "a" is the argument (angle) defined by Equations (22.30) and (22.31).

$$m = \sqrt{a^2 + b^2} \quad \text{(22.30) Vector magnitude in polar form}$$

$$a = \tan^{-1}\left(\frac{b}{a}\right) \quad \text{(22.31) Vector argument in polar form}$$

The circular motion described above is defined by Equations (22.32) and (22.33).

$$x(t) = \mathrm{e}^{i\omega t} \quad \text{(22.32) CCW (counter-clock wise) motion}$$

$$x(t) = \mathrm{e}^{-i\omega t} \quad \text{(22.33) CW (clockwise) motion}$$

Real sine and cosine require positive and negative frequencies to be represented. Cosine and sine functions can then be represented by Equations (22.34) and (22.35).

$$\cos(\omega t) = \frac{(\mathrm{e}^{i\omega t} + \mathrm{e}^{-i\omega t})}{2} \quad \text{(22.34) Real sine}$$

$$\sin(\omega t) = \frac{(\mathrm{e}^{i\omega t} - \mathrm{e}^{-i\omega t})}{2} \quad \text{(22.35) Real cosine}$$

22.5 Fourier Analysis

The decomposition of periodic functions in a sum of sinusoidal functions (sine, cosine or complex exponentials) is very convenient and is the basis of the processing of digital signals. In 1807, Fourier tried to solve heat equations by modeling a heat source by the superposition of sine and cosine, developing the Fourier series. It was later found that the same technique could be applied to many other applications.

The Fourier series is defined by Equations (22.36), (22.37) and (22.38):

$$f(x) = \frac{a_0}{2} + \sum_{n=1}^{\infty}[a_n \cos(nx) + b_n \sin(nx)] \quad \text{(22.36) Fourier series}$$

Where

$$a_n = \frac{1}{\pi}\int_{-\pi}^{+\pi} f(x)\cos(nx)\,dx \quad \text{(22.37) Fourier series coefficient for } n \geq 0$$

$$b_n = \frac{1}{\pi}\int_{-\pi}^{+\pi} f(x)\sin(nx)\,dx \quad \text{(22.38) Fourier series coefficient for } n \geq 1$$

It can also be represented in exponential form by Equations (22.39) and (22.40):

$$f(x) = \sum_{n=-\infty}^{\infty} c_n \mathrm{e}^{inx} \quad \text{(22.39) Fourier series in exponential form}$$

Where

$$c_n = \frac{1}{2\pi}\int_{-\pi}^{+\pi} f(x)\mathrm{e}^{-inx}\,dx \quad \text{(22.40) Fourier series in exponential form coefficient}$$

The Fourier series for a square wave, defined in Equation (22.41), is given by Equation (22.42):

$$f(x) = \begin{cases} 0 - \pi < x < 0 \\ 10 < x < \pi \end{cases} \qquad \text{(22.41) Square waveform}$$

$$f(x) = \frac{1}{2} + \frac{2}{\pi}\left(\sin(x)\frac{1}{3}\sin(3x) + \frac{1}{5}\sin(5x) + \cdots..\right)$$

(22.42) Fourier series of a square waveform

Observing the Fourier series, we can note that functions are represented by a DC component and a set of harmonic frequencies multiplied by coefficients.

Harmonic frequencies are integer multiples of the fundamental frequency f, that is, if $f = 1$ Hz, the harmonics are 2, 3, 4, 5 ... Hz. These frequencies and their coefficients represent the spectrum of the signal being decomposed. The magnitude of each spectrum component is given by Equation (22.43).

$$\text{magnitude}(f_n) = \sqrt{a_n^2 + b_n^2} \qquad \text{(22.43) Spectrum magnitude}$$

The following properties of sinusoidal waves explain the functionality of the Fourier series:

- The area under one entire period of a cosine (or a sine) wave is zero.
- The area under one period of a wave that is a product of a cosine (or sine) by a sine (or cosine) of any frequency (same or different) is equal to zero. Note that the period must be a multiple of both frequencies.
- As a consequence, the area under one period of a wave that is a product of a cosine (or a sine) and any of its harmonics is zero. Note that the period is the one of the fundamental frequency.
- The area under one period of a wave, which is a product of two cosine or sine waves of the same frequency is not zero.

This means that cosine and sine can be used as filters, to obtain the coefficients of a waveform. This property is fundamental in the digital processing of signals.

22.5.1 Fourier Transform

The Fourier Transform (FT) is an operation that transforms a function of one real variable into a function of another real variable. In signal processing, it transforms a function of time (time domain) into a function of frequency (frequency domain). We can say that it takes a signal in time and provides its spectrum. This is achieved by decomposing the signal into its sinusoidal (oscillatory) functions, or finding the components of the Fourier series that would represent the function.

The Fourier Transform of a function $f(t)$ is given by Equation (22.44):

$$F(f) = \int_{-\infty}^{\infty} f(t)\mathrm{e}^{-iwt}\, dt \qquad \text{(22.44) Fourier Transform}$$

The above equation states that if we multiply the time function by harmonic frequencies one at a time and then integrate the results, we will get the spectrum of the signal, because the multiplication acts as a filter.

A Discrete Fourier Transform (DFT) applies to periodic functions with a defined period duration, represented by its fundamental frequency. This allows the calculation to be done over a reduced number of samples and applies to digitized signals.

The following are the main parameters of a Discrete Fourier Transform:

- Total number of samples: N. The total number of samples collected over a period of the input function.
- Sampling Frequency: f_s. The rate with which the samples of the signal amplitude are collected for each harmonic are:

$$\text{Sample Time: } \tau = \frac{1}{f_s}$$

$$\text{Fundamental frequency: } f_0 = \frac{f_s}{N}$$

$$\text{Sequence time length: } T = N\tau$$

Sample index: "k" refers to the kth sample

Harmonic index: "n"

Reflects the number of frequencies that are integer multiples of the fundamental frequency. The sampling theorem states that a signal has to be sampled at least twice its highest frequency, as expressed in Equations (22.45) and (22.46).

$$fsn \leq fs2 \qquad \text{(22.45) Sampling theorem}$$

$$n \leq \frac{N}{2} \qquad \text{(22.46) Harmonic index}$$

This equation states that we can only detect half the harmonics of the total number of samples. However, index "n" itself goes from $-(N - 1)$ to $+(N - 1)$ and spans both sides of the spectrum reflecting positive and negative frequency components.

The Discrete Fourier Transform (DFT) can then be represented by its "n" components, each one expressed by Equation (22.47):

$$F(f_n) = \sum_{k=0}^{N-1} f(k\tau)\mathrm{e}^{-i2\pi \frac{n}{N} k} \qquad \text{(22.47) Discrete Fourier Transform}$$

N has to vary from 0 to $N-1$ to provide the complete spectrum.

A Discrete Fourier Transform (DFT) requires $N^2 + N$ multiplications, which makes the algorithm slow for large N. In 1948, Cooley and Tukey simplified the computational algorithm creating what is called the Fast Fourier Transform (FFT), as many calculations were repeated over and over for the harmonic frequencies. This algorithm reduced the number of computations to $2N$.

The Inverse Fourier Transform (IFT) reverses the direction of the transformation. It is easier to implement digitally, by generating samples of the harmonic frequencies involved and adding them together.

22.6 Statistical Probability Distributions

Wireless communications use several variables that can only be defined statistically, as their values vary with time and location. Statistical distributions have to be used to define these variables and the outcome of operations done with them. Choosing the distribution that fits a physical event implies an understanding of the causes of the event and what type of event each distribution represents.

Probability distributions identify the probability of obtaining each possible value of a random variable and are represented by a probability density function (pdf).

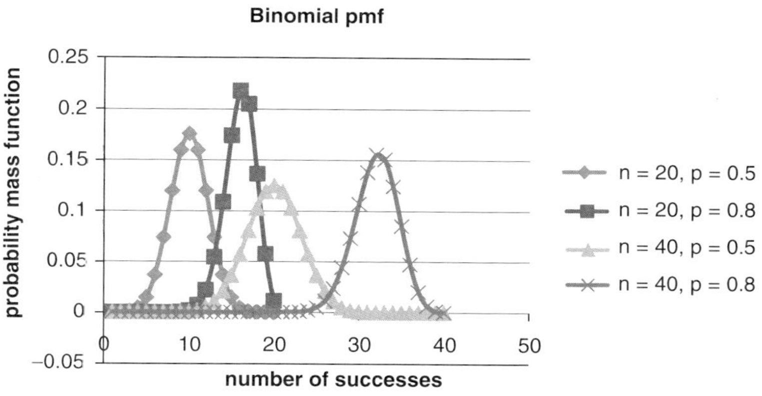

Figure 22.16 Binomial pmf.

The cumulative distribution function gives the probability that a random variable is smaller than a certain variable value.

The main distributions used in telecommunications are now presented.

22.6.1 Binomial Distribution

This is discrete probability distribution of the number of successes (k) in a sequence of (n) independent yes/no experiments (e.g. coin flip) with a probability (p). An example is the probability of rolling six on a dice three times in a row. It only applies to integer numbers.

The probability mass function (pmf) is given by the Equation (22.48) and is shown in Figure 22.16.

$$f(k; n, p) = \frac{n!}{k!(n-k)!} p^k (1-p)^{n-k} \qquad \text{(22.48) Binomial distribution pmf}$$

where:

k = number of successes.
n = number of experiments.
p = probability of successful outcome.

The binomial distribution for $n = 1$ is known as the Bernoulli distribution.

The cumulative distribution function is shown in Figure 22.17.

The binomial distribution approximates the normal distribution for large "n" and the Poisson distribution for large "n" and small "p".

22.6.2 Poisson Distribution (Law of Large Numbers)

This is a discrete probability distribution that expresses the probability of a number of events (n) occurring in a fixed interval of time, supposing that these events occur with a known average rate (λ) and independently of the interval since the last event. The interval can be time, area, distance and any other parameter.

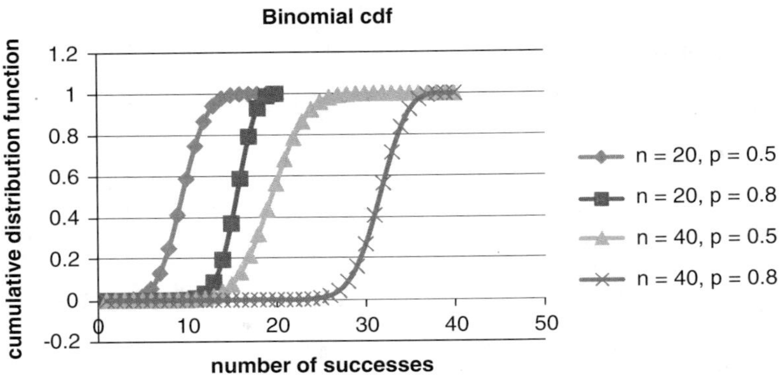

Figure 22.17 Binomial cdf.

The distribution is defined by:

$n =$ number of occurrences in the interval.
$\lambda =$ expected number of occurrences in the interval.

The probability of n occurrences is defined by the probability mass function defined in Equation (22.49) and shown in Figure 22.18.

$$f(n;\lambda) = \frac{\lambda^n e^{-\lambda}}{n!} \qquad \text{(22.49) Poisson pmf distribution}$$

For large "n" it can be approximated by a binomial distribution.

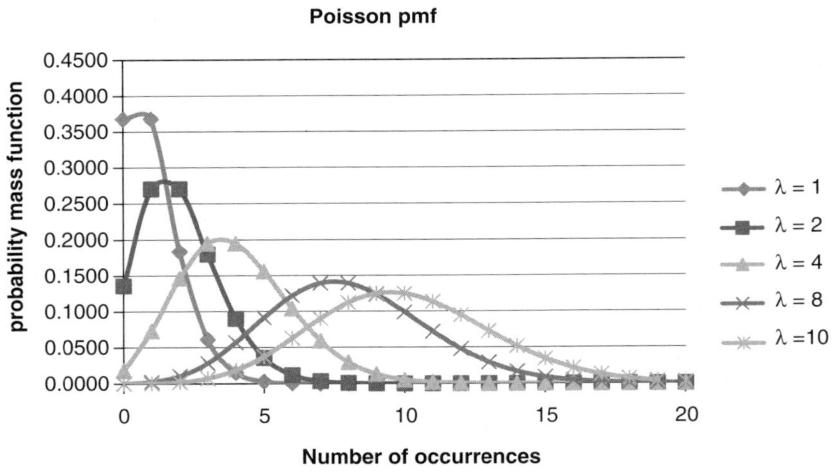

Figure 22.18 Poisson pmf.

Poisson's process (Simeón-Denis Poisson, 1781–1840) is a stochastic (random) process in which events occur continuously and independently of each other. Examples are phone calls arrival, web page requests, etc.

Application example: If a server receives an average of 100 web page requests per hour, what is the probability of receiving 80 requests in one hour?

22.6.3 Exponential Distribution

This is a continuous probability distribution that describes the times between events in a Poisson process. A Poisson process has events occurring continuously and independently at a constant average rate. The parameter λ is the rate parameter. The probability density function is shown in Equation (22.50) and in Figure 22.19.

$$f(x;\lambda) = \lambda e^{-\lambda x} \ for\ x \geq 0\ and\ 0\ for\ x < 0 \qquad \text{(22.50) Exponential pdf distribution}$$

The cumulative distribution function is shown in Equation (22.51) and Figure 22.20.

$$f(x;\lambda) = 1 - e^{-\lambda x} \ for\ x \geq 0\ and\ 0\ for\ x < 0 \qquad \text{(22.51) Exponential cdf distribution}$$

The mean and the standard deviation are given by: $1/\lambda$

An example of an exponential distribution is the time duration of voice calls.

22.6.4 Normal or Gaussian Distribution

This is a continuous probability function that describes data that clusters around a mean value (μ). It was first mathematically defined by Carl Friedrich Gauss (1777–1855). It usually applies to outcomes that have a large number of components that can influence it. A normal distribution will represent well a variable that:

- Has a strong tendency for the variable to have a central value.
- Has equal probability of having positive and negative deviations.
- Has a frequency of deviations that falls off rapidly as the deviations increase.

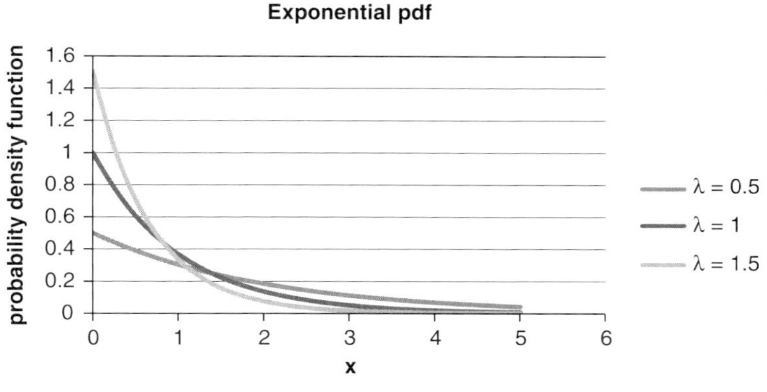

Figure 22.19 Exponential pdf.

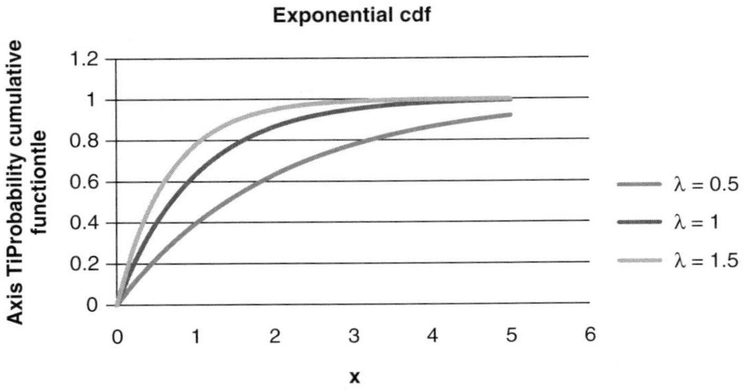

Figure 22.20 Exponential cdf.

The central limit theorem states that the sum of random variables with finite means and variances approaches a normal distribution for a large number of variables.

The variance (var) of a random variable is a measurement of the variation of a variable, and is defined as the mean of the squared deviation from the variable mean value. The standard deviation (σ) is the square root of the variance. Both are used to represent the variability or dispersion of a variable.

The probability density function is given by Equation (22.52) and Figure 22.21.

$$\aleph(\mu, \sigma^2) = \frac{1}{\sqrt{2\pi\sigma^2}} e^{\frac{-(x-\mu)^2}{2\sigma^2}} \qquad \text{(22.52) Gaussian pdf distribution}$$

The area under the curve adds up to 1. The curve is bell-shaped, although not all bell-shaped curves fit a normal distribution. The inflection point of the bell curve happens for $\sigma = 1$.

The standard normal distribution is the one that has a $\mu = 0$ and $\sigma^2 = 1$.

The cumulative Distribution Function is given by Equation (22.53) and Figure 22.22.

$$F(x; \mu, \sigma^2) = \frac{1}{2}\left[1 + \text{erf}\left(\frac{x-\mu}{\sigma\sqrt{2}}\right)\right] \qquad \text{(22.53) Gaussian cdf distribution}$$

Where:

$\mu =$ mean value of the variable.
$\sigma =$ standard deviation of the variable.
$\text{erf}(x) =$ error function.

22.6.4.1 Mean

Mean is the expected value of a random variable. A mean value can be of the following types:

- Arithmetic mean is used for sets of numbers that are interpreted according to their sum. Example: people or income. It is expressed by Equation (22.54).

$$\overline{x} = \frac{1}{n}\sum_{i=1}^{n} x_i \qquad \text{(22.54) Arithmetic mean}$$

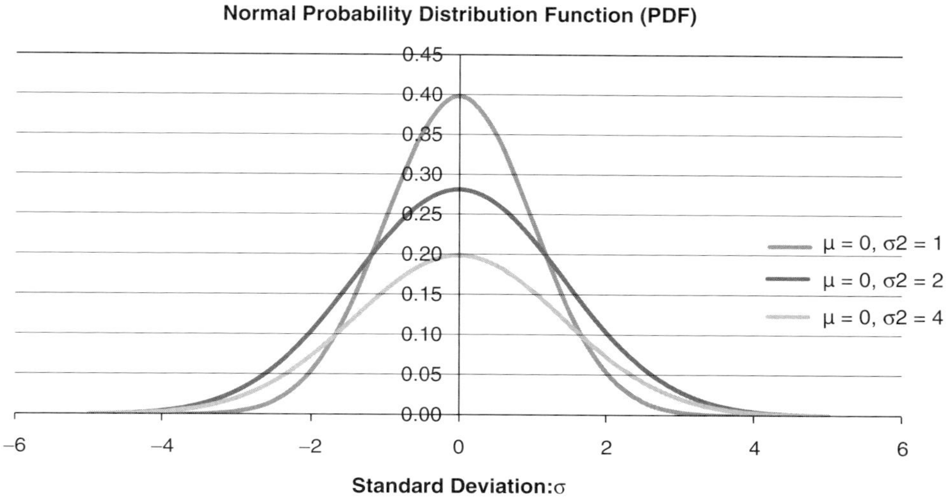

Figure 22.21 Normal pdf.

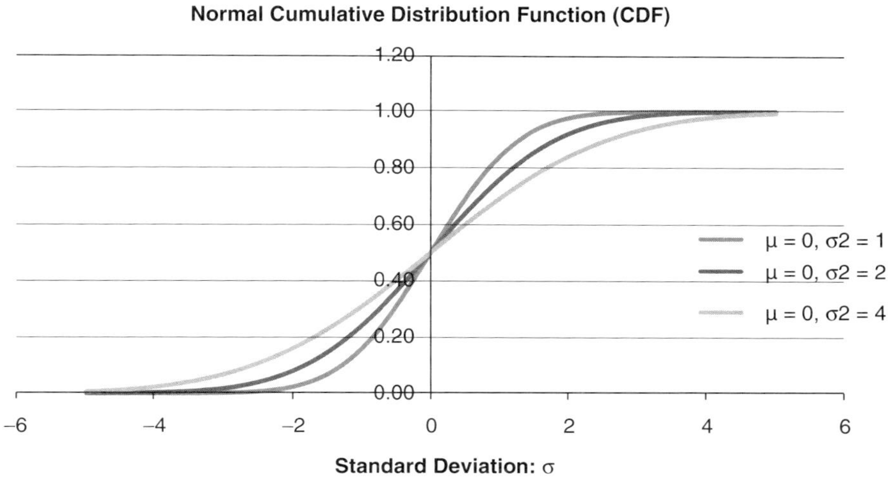

Figure 22.22 Normal cdf.

- Geometric mean is used for sets of numbers that are interpreted according to their products. Example: rates of growth. It is expressed by Equation (22.55).

$$\overline{x} = \left(\prod_{i=1}^{n} x_i\right)^{\frac{1}{n}} \qquad (22.55) \text{ Geometric mean}$$

- Harmonic mean is used for sets of numbers that are defined by a unit. Example; m/s, s... It is expressed by Equation (22.56).

$$\overline{x} = \left(\sum_{i=1}^{n} \frac{1}{x_i}\right)^{-1} \qquad (22.56) \text{ Harmonic mean}$$

22.6.4.2 Median

Median is the value that divides the sample into two equal groups.

22.6.4.3 Mode

Mode is the value that occurs more frequently in a data set.

The standard normal curve has the following probabilities for each step of standard deviations, as shown in Figure 22.23.

Table 22.1 gives the probability between numbers of standard deviations around the mean value.

Table 22.2 gives the required number of standard deviations for different density probabilities.

Typical examples of a normal distribution are environmental attenuations, such as the human body or indoor penetration. A received signal is a combination of several multipaths and is best represented by a normal distribution. SNR predictions result from the combination of signals from many sources and, according to the central limit theorem, have a normal distribution also.

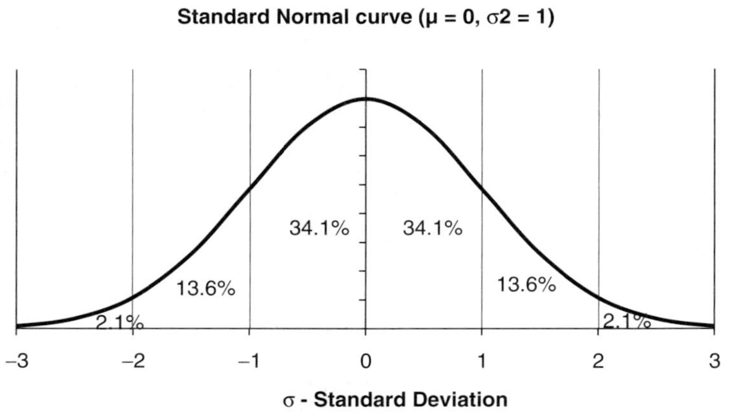

Figure 22.23 Standard normal curve.

Table 22.1 Probability density for different standard deviations

Standard deviations from mean value	Probability density *erf*(n/sqrt(2))
1	0.682689491
2	0.95449973
3	0.997300204
4	0.999936658
5	0.999999427
6	0.999999998

Table 22.2 Standard deviations for different probability densities

Probability density *erf*(n/sqrt(2))	Standard deviations from mean value
0.5	0.644897
0.6	0.841621
0.7	1.036433
0.8	1.281551
0.9	1.644853
0.95	1.959963
0.98	2.326347
0.99	2.575829
0.995	2.807033
0.999	3.090232
0.9999	3.890509

A log-normal distribution is a random variable distribution whose logarithm is normally distributed. In this distribution the value of "x" in the normal distribution should be replaced by $\log x$. The log-distance path loss model has a log-normal distribution.

A typical example of a log-normal distribution is shadow fading.

22.6.5 *Rayleigh Distribution*

This is a continuous probability distribution used to represent variables that are a combination of independent normally distributed variables. It is named after John William Strutt, 3rd Baron Rayleigh (1842–1919).

The probability density function is given by Equation (22.57) and in Figure 22.24.

$$P(x) = x\frac{e^{-x^2/2\sigma^2}}{\sigma^2} \qquad \text{(22.57) Rayleigh pdf distribution}$$

The cumulative distribution function is given by Equation (22.58) and Figure 22.25.

$$C(x) = 1 - e^{-x^2/2\sigma^2} \qquad \text{(22.58) Rayleigh cdf distribution}$$

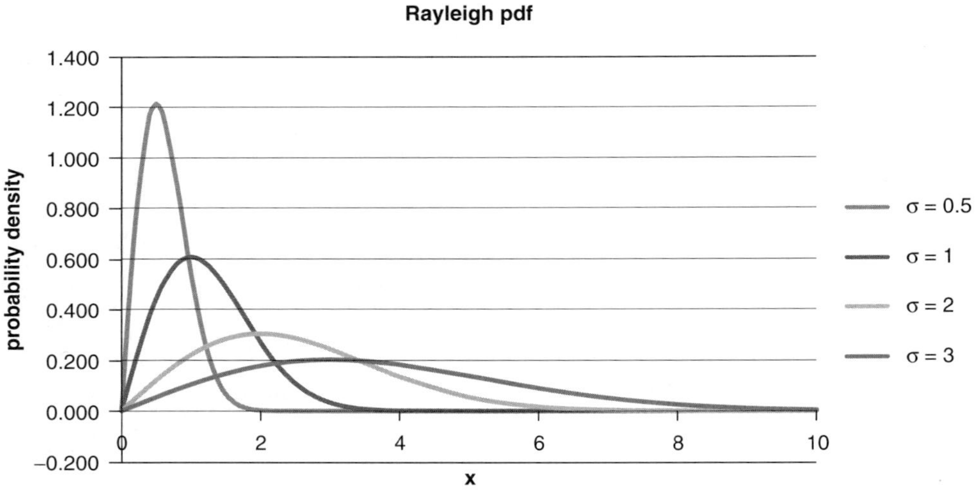

Figure 22.24 Rayleigh pdf.

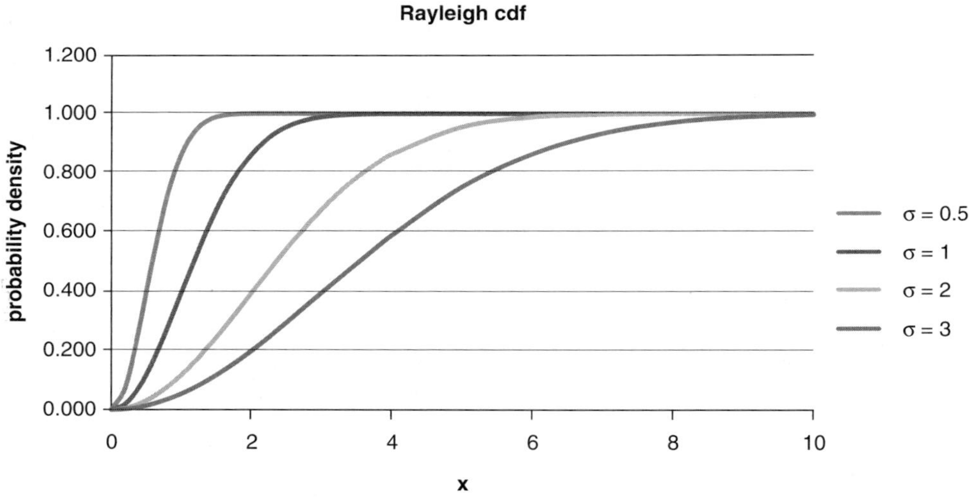

Figure 22.25 Rayleigh cdf.

The mean is given by Equation (22.59):

$$\sigma\sqrt{\frac{\pi}{2}} = 1.253\sigma \qquad \text{(22.59) Rayleigh distribution mean}$$

The median is given by Equation (22.60):

$$\sigma\sqrt{\ln 4} = 1.744\sigma \qquad \text{(22.60) Rayleigh distribution median}$$

The mode is given by Equation (22.61):

$$\sigma \qquad \text{(22.61) Rayleigh distribution mode}$$

The variance is given by Equation (22.62):

$$(4-\pi)\frac{\sigma^2}{2} = 0.429\sigma^2 \qquad \text{(22.62) Rayleigh distribution variance}$$

A typical example of a Rayleigh distribution is the fast fading for non-line of sight links.

22.6.6 Rice Distribution

This is a continuous probability function that describes a sum of independent random variables in which one of them may take a predominant role. It was developed by Stephen Rice (1907–1986).

The probability distribution function is given by Equation (22.63) and shown in Figure 22.26.

$$f(x|A,\sigma) = \frac{x}{\sigma^2} e^{\frac{-(x^2+A^2)}{2\sigma^2}} I_0\left(\frac{xA}{\sigma^2}\right) \qquad \text{(22.63) Rice pdf distribution}$$

A: denotes the peak amplitude of the dominant signal.
$I_0(.)$: is the modified Bessel function of first kind and zero order.
σ: is the standard deviation.

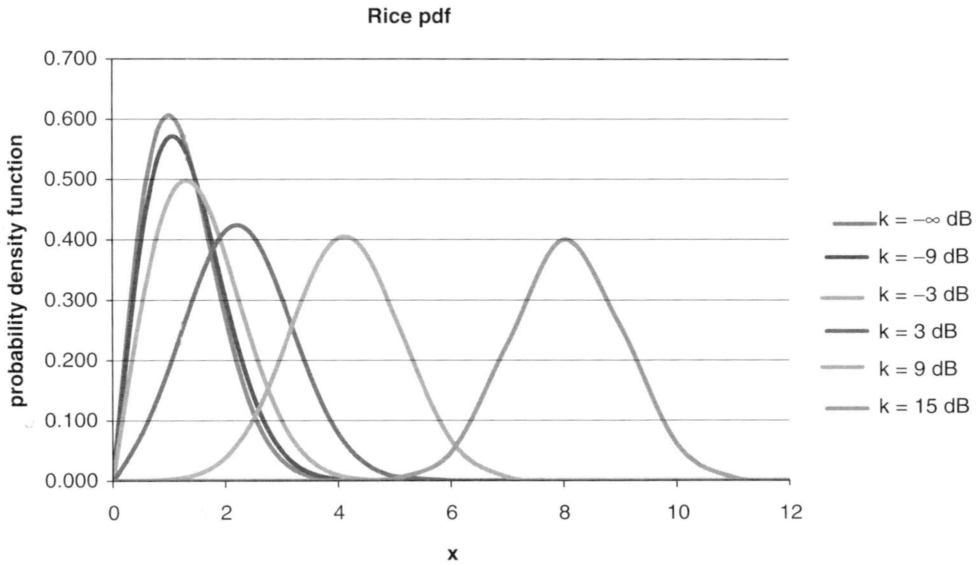

Figure 22.26 Rice pdf.

The Ricean distribution can be described by a parameters k which defines the ration between the predominant signal and the multipath variance and is defined by Equation (22.64).

$$k = \frac{A^2}{2\sigma^2} \qquad \text{(22.64) Rice } k \text{ factor}$$

The k factor or amplitude factor is usually expressed in dB. For $k <= -3$ dB the Ricean distribution is a Rayleigh distribution and for $k > 6$ dB it becomes a normal distribution around the mean.

The Ricean distribution is used to represent fading over a wide area, as the specific distribution at each point can be adjusted by the k factor.

22.6.7 Nakagami Distribution

This is a two-parameter, continuous, probability distribution. It is based on a spread parameter (ω) and a shape parameter (μ). The shape parameter represents the sum of μ independent exponentially distributed random variables with a mean of ω/μ. It was developed in 1960 by M. Nakagami. For $\mu = 1$ the Rayleigh distribution is obtained, whereas, for higher values of (μ), it moves towards a Gaussian distribution.

The probability density function is shown in Equation (22.65) and Figure 22.27.

$$f(x; \mu, \omega) = \frac{2\mu^{\mu}}{\Gamma(\mu)\omega^{\mu}} x^{2\mu-1} e^{-\mu x^2/\omega} \qquad \text{(22.65) Nakagami pdf distribution}$$

The Nakagami distribution can be used to represent wide area fading, in the same way as a Rice distribution is used, but its two parameters configurations makes its usage less practical.

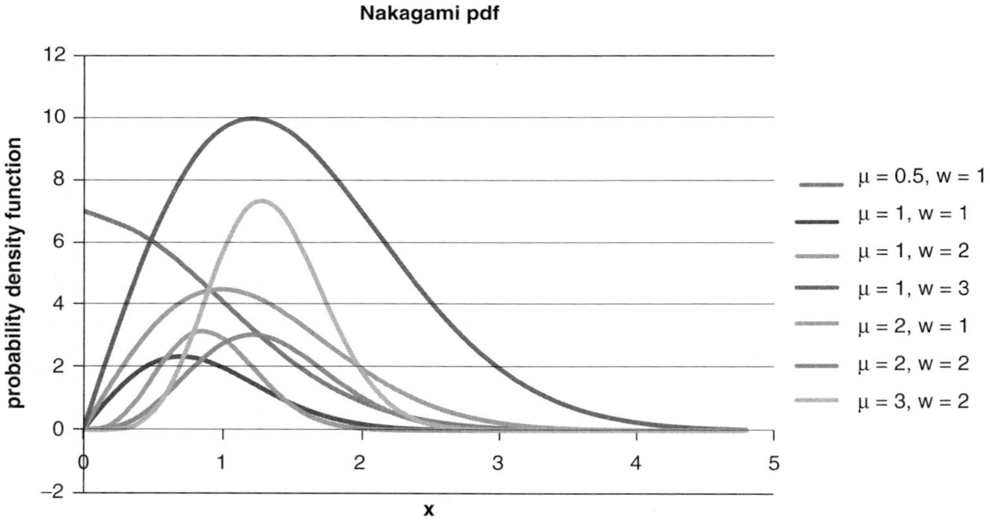

Figure 22.27 Nakagami pdf.

22.6.8 *Pareto Distribution*

The Pareto principle states that a market with a high freedom of choice will create a certain degree of inequality by favoring the upper 20% of the items against the other 80%, resulting in a long tail distribution.

The probability density function is given by Equation (22.66) and Figure 22.28.

$$f(x; \alpha, k) = \alpha k^{\alpha}/x^{\alpha+1} \qquad \text{(22.66) Pareto pdf distribution}$$

Where :

k = minimum value of x.
α = shape parameter or tail index.

The cumulative distribution is given by Equation (22.67) and Figure 22.29.

$$F(x; \alpha, k) = 1 - \left(\frac{k}{x}\right)^{\alpha} \qquad \text{(22.67) Pareto cdf distribution}$$

The mean value of the distribution is given by Table 22.3.

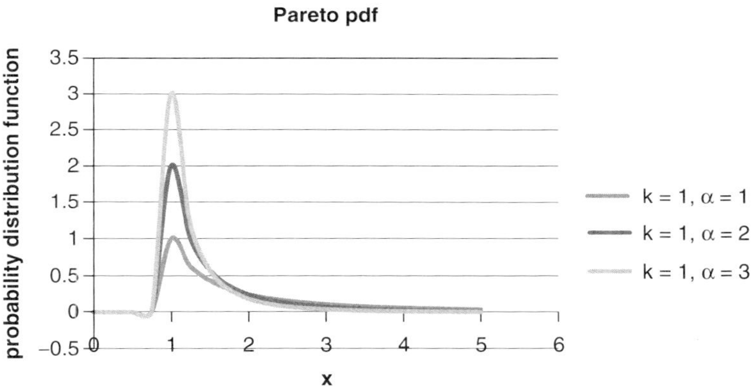

Figure 22.28 Pareto pdf.

Table 22.3 Pareto distribution mean value

k	α	mean
1	1.2	6.0
1	1.4	3.5
1	1.6	2.7
1	1.8	2.3
1	2	2.0

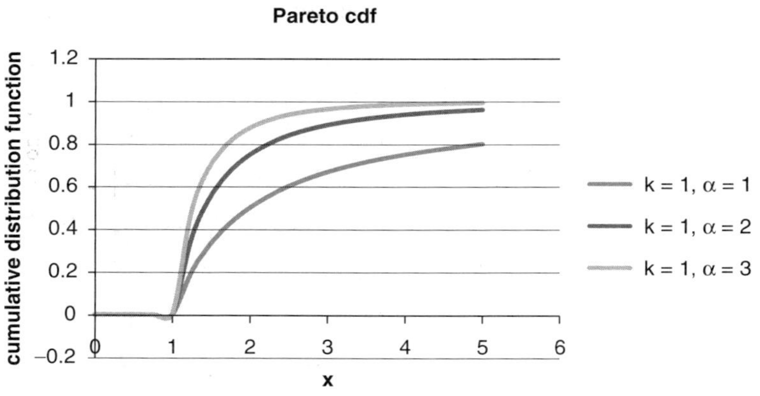

Figure 22.29 Pareto cdf.

It can be seen that when α decreases, a large part of the distribution is in the tail. Many retailers found that their sales followed a long-tailed distribution in terms of items offered. When offering only the most popular items, they were serving only a fraction of the demand and, to cover a larger percentage, they had to significantly increase the number of products.

The Pareto distribution is one of the simplest distributions representing long tail events. The distribution of book sizes has a long tail distribution and the same applies to ftp and www files. The www reading time follows the same distribution of www file sizes and, consequently, the time between reads has also a long tail distribution.

The Pareto distribution is used to represent traffic arrival and duration for www and FTP type data transfers.

22.6.8.1 Self-Similarity

A self-similar object is approximately similar to parts of itself. A self-similar time series has the property that, when aggregated, leading to a shorter time series, it results in a new series that has the same auto-correlation of the original one. The traditional distribution used to estimate inter-arrival time (Poisson) did not provide good approximations. It was found that long-tailed distributions, like Pareto, approximate well those events.

Appendix

List of Equations

LTE, WiMAX and WLAN Network Design, Optimization and Performance Analysis, First Edition.
Leonhard Korowajczuk.

Further Reading

1 The Business Plan

Edward F. McQuarrie (2006) *The Market Research Toolbox*, London: Sage Publications.

Michael R. Hyman and Jeremy J. Sierra (2010) *Marketing Research Kit*, Wiley Publishing, Inc.

2 Data Transmission

Charles M. Kozierok (2005) *TCP/IP Guide*, No Starch Press.

Eric A. Hall (2000) *Internet Core Protocols*, O'Reilly.

Gilbert Held (2003) *The ABCs of TCP/IP*, Auerbach Publications.

Gunnar Heine (2008) *QoS: Issues and their Resolution in Evolved 3G and Next Generation Networks*, Inacon.

Mark A. Miller (2004) *Internet Technology Handbook*, John Wiley & Sons, Ltd.

Richard Huzenlaub (2010) *IP for Telecom Professionals: Advanced Issues*, Inacon.

Robert G. Cole and Ravi Ramaswamy (2000) *Wide-Area Data Network Performance Engineering*, Artech House.

Tom Shaughnessy (2000) *Cisco: A Beginner's Guide*, Osborne/McGraw-Hill.

William Stallings (2001) *High-Speed Networks and Internets*, Prentice Hall.

William Stallings (2004) *Computer Networking with Internet Protocols and Technology*, Prentice Hall.

3 Market Modeling

Leonhard Korowajczuk et al. (2004) *Designing cdma2000 Systems*, John Wiley & Sons, Ltd.

Syed Ahson and Mahammad Ilyas (2008) *WiMAX Applications*, CRC Press.

4 Signal Processing Fundamentals

Bernard Sklar (2000) *Digital Communications: Fundamentals and Applications*, Prentice Hall.

Leonhard Korowajczuk et al. (2004) *Designing cdma2000 Systems*, John Wiley & Sons, Ltd.

Savo G. Glisic (2004) *Advanced Wireless Communications: 4G Technologies*, John Wiley & Sons, Ltd.

Savo G. Glisic (2007) *Advanced Wireless Communications: 4G Cognitive and Cooperative Broadband Technology*, John Wiley & Sons, Ltd.

LTE, WiMAX and WLAN Network Design, Optimization and Performance Analysis, First Edition.
Leonhard Korowajczuk.

5 RF Channel Analysis

David Parsons (1992) *The Mobile Radio Propagation Channel*, Pentech Press.

F. Pérez Fontán and P. Mariño Espiñeira (2008) *Modeling the Wireless Propagation Channel: A Simulation Approach with MATLAB*, John Wiley & Sons, Ltd.

J.D. Parsons (2000) *The Mobile Radio Propagation Channel*, John Wiley & Sons, Ltd.

Leonhard Korowajczuk et al. (2004) *Designing cdma2000 Systems*, John Wiley & Sons, Ltd.

Les Barclay (ed.) (2003) *Propagation of Radiowaves*, The Institution of Electrical Engineers.

Mohinder Jankiraman (2004) *Space-time Codes and MIMO Systems*, Artech House.

Rodney Vaughan and Jorgen Bach Andersen (2003) *Channel Propagation and Antenna for Mobile Communications*, IEE.

6 RF Channel Performance Prediction

Leonhard Korowajczuk et al. (2004) *Designing cdma2000 Systems*, John Wiley & Sons, Ltd.

Molten Tolstrup (2008) *Indoor Radio Planning*, John Wiley & Sons, Ltd.

Yan Zhang (ed.) (1977) *WiMAX Network Planning and Optimization*, CRC Press.

7 OFDM

Hyung G. Myung and David J. Goodman (2008) *Single Carrier FDMA*, John Wiley & Sons, Ltd.

8 OFDM Implementation

John Bard and Vincent J. Kovarik, Jr. (2007) *Software Defined Radio*, John Wiley & Sons, Ltd.

Savo G. Glisic (2004) *Advanced Wireless Communications: 4G Technologies*, John Wiley & Sons, Ltd.

Tzi-Dar Chiueh and Pei-Yun Tsai (2007) *OFDM Baseband Receiver Design for Wireless Communications*, John Wiley & Sons, Ltd.

9 Wireless Communications Network (WCN)

Leonhard Korowajczuk et al. (2004) *Designing cdma2000 Systems*, John Wiley & Sons, Ltd.

10 Antenna and Advanced Antenna Systems

ARRL (2000) *UHF/Microwave: Experimenter's Manual*, The ARRL, Inc.

ARRL (2007) *Antenna Book*, The ARRL, Inc.

David Tse and Pramod Viswanath (2005) *Fundamentals of Wireless Communications*, Cambridge University Press.

Ezio Biglieri et al. (2007) *MIMO Wireless Communications*, Cambridge.

George Tsoulos (ed.) (2006) *MIMO System Technology for Wireless Communications*, CRC Press.

John D. Kraus (1988) *Antennas*, McGraw-Hill.

Savo G. Glisic (2004) *Advanced Wireless Communications: 4G Technologies*, John Wiley & Sons, Ltd.

11 Radio Performance

Tzi-Dar Chiueh and Pei-Yun Tsai (2007) *OFDM Baseband Receiver Design for Wireless Communications*, John Wiley & Sons, Ltd.

12 Wireless LAN

Anurag Kumar, D. Manjunath, and Joy Kuri (2008) *Wireless Networking*, Morgan Kaufmann.
Bruce Fette et al. (2008) *RF & Wireless Technologies*, Newpress.
Chandra, Dobkin, Bensky, Olexa, Lide, Dowla (2009) *Wireless Networking: Know it All*, Elsevier.
Gunnar Heine (2010) *IEEE 802.11n Design Details and Protocol Analysis*, Inacon.
Jochen Schiller (2000) *Mobile Communications*, Pearson Education Ltd.
Juha Heiskala and John Terry (2002) *OFDM Wireless LANs: A Theoretical and Practical Guide*, SAMS.
Savo G. Glisic (2004) *Advanced Wireless Communications: 4G Technologies*, John Wiley & Sons, Ltd.
Walter R. Bruce III and Ron Gilster (2002) *Wireless LANs*, Hungry Minds, Inc.

13 WiMAX

Carl Eklund, Roger B. Marks, Subbu Ponnuswamy, Kenneth L. Stanwood, and Nico J.M. van Waes (2006) *WirelessMAN: Inside the 802.16 Standard for Wireless Metropolitan Networks*, IEEE Press.
Gunnar Heine (2007) *WiMAX from A to Z*, Inacon.
IEEE Communications Magazine (2008) Mobile WiMAX: A Technology Update, vol. 46, no. 10.
Mustafa Ergen (2009) *Mobile Broadband Including WiMAX and LTE*, Springer.

14 Universal Mobile Telecommunication System – Long Term Evolution (UMTS-LTE)

Farooq Khan (2009) *LTE for Mobile Broadband*, Cambridge University Press.
Gunnar Heine (2010a) *Long Term-Evolution: Signaling & Protocol Analysis*.
Gunnar Heine (2010b) *SAE from A to Z*, Inacon.
Gunnar Heine (2010c) *IMS Architecture Details and System Engineering*, Inacon.
IEEE Communications Magazine (2009a) LTE Part 1, Core Network, vol. 47, no. 2.
IEEE Communications Magazine (2009b), LTE Part II: Radio Access, vol. 47, no. 4.
Moray Rumney (ed.) (2008) *LTE and the Evolution 4G Wireless: Design and Measurement Challenges*, Agilent Technologies.
Mustafa Ergen (2009) *Mobile Broadband Including WiMAX and LTE*, Springer.
Stefan Bahrenburg (2009) *Long Term-Evolution A-Z*, Inacon.
Stefan Blomeier (2010) *Long Term-Evolution: Internetworking*, Inacon.
Stefania Sesia, Issam Toufik, and Matthew Baker (eds) (2009) *LTE: The UMTS Long Term Evolution: From Theory to Practice*, John Wiley & Sons, Ltd.

15 Broadband Standards Comparison

Mustafa Ergen (2009) *Mobile Broadband Including WiMAX and LTE*, Springer.
Stefania Sesia, Issam Toufik, and Matthew Baker (eds) (2009) *LTE: The UMTS Long Term Evolution: From Theory to Practice*, John Wiley & Sons, Ltd.

16 Wireless Network Design

Yan Zhang (ed.) (1977) *WiMAX Network Planning and Optimization*, CRC Press.

17 Wireless Market Modeling

Leonhard Korowajczuk et al. (2004) *Designing cdma2000 Systems*, John Wiley & Sons, Ltd.

18 Wireless Network Strategy

Leonhard Korowajczuk et al. (2004) *Designing cdma2000 Systems*, John Wiley & Sons, Ltd.

19 Wireless Network Design

Ajay R. Mishra (ed.) (2004) *Fundamentals of Cellular Network Planning and Optimization, 2G/2.5G/3G... Evolution to 4G*, John Wiley & Sons, Ltd.
Ajay R. Mishra (2007) *Advanced Cellular Network Planning and Optimization, 2G/2.5G/3G... Evolution to 4G*, John Wiley & Sons, Ltd.
Clint Smith and Curt Gervelis (2003) *Wireless Network Performance Handbook*, McGraw-Hill.
Leonhard Korowajczuk et al. (2004) *Designing cdma2000* Systems, John Wiley & Sons, Ltd.
Yan Zhang (ed.) (1977) *WiMAX Network Planning and Optimization*, CRC Press.
Zerihun Ababte (2009) *WiMAX RF System Engineering*, Artech House.

20 Wireless Network Optimization

Ajay R. Mishra (ed.) (2007) *Advanced Cellular Network Planning and Optimization, 2G/2.5G/3G... Evolution to 4G*, John Wiley & Sons, Ltd.
David W. Corne, Martin J. Oates, and George D. Smith (eds) (2000) *Telecommunications Optimization: Heuristic and Adaptive Techniques*, John Wiley & Sons, Ltd
Leonhard Korowajczuk et al. (2004) *Designing cdma2000 Systems*, John Wiley & Sons, Ltd.
Vladimir Mitlin (2006) *Performance Optimization of Digital Communication Systems*, Auerbach Publications.
Yan Zhang (ed.) (1977) *WiMAX Network Planning and Optimization*, CRC Press.
Zerihun Ababte (2009) *WiMAX RF System Engineering*, Artech House.

21 Wireless Network Performance Assessment

Leonhard Korowajczuk et al. (2004) *Designing cdma2000 Systems*, John Wiley & Sons, Ltd.
Syed Ahson, and Mahammad Ilyas (2008) *WiMAX: Technologies, Performance Analysis and QoS*, CRC Press.

22 Basic Mathematical Concepts Used in Wireless Networks

J.F. James (2002) *A Student's Guide to Fourier Transforms with Applications in Physics and Engineering*, Cambridge University Press.
Julius O. Smith III (2007) *Mathematics of the Discrete Fourier Transform*, W3K Publishing.
Lester L. Helms (1997) *Introduction to Probability Theory with Contemporary Applications*, Dover Publications.
Paul J. Nahin (1998) *An Imaginary Tale: The Story of* $\sqrt{-1}$, Princeton University Press.

Index

LTE, WiMAX and WLAN Network Design, Optimization and Performance Analysis, First Edition.
Leonhard Korowajczuk.